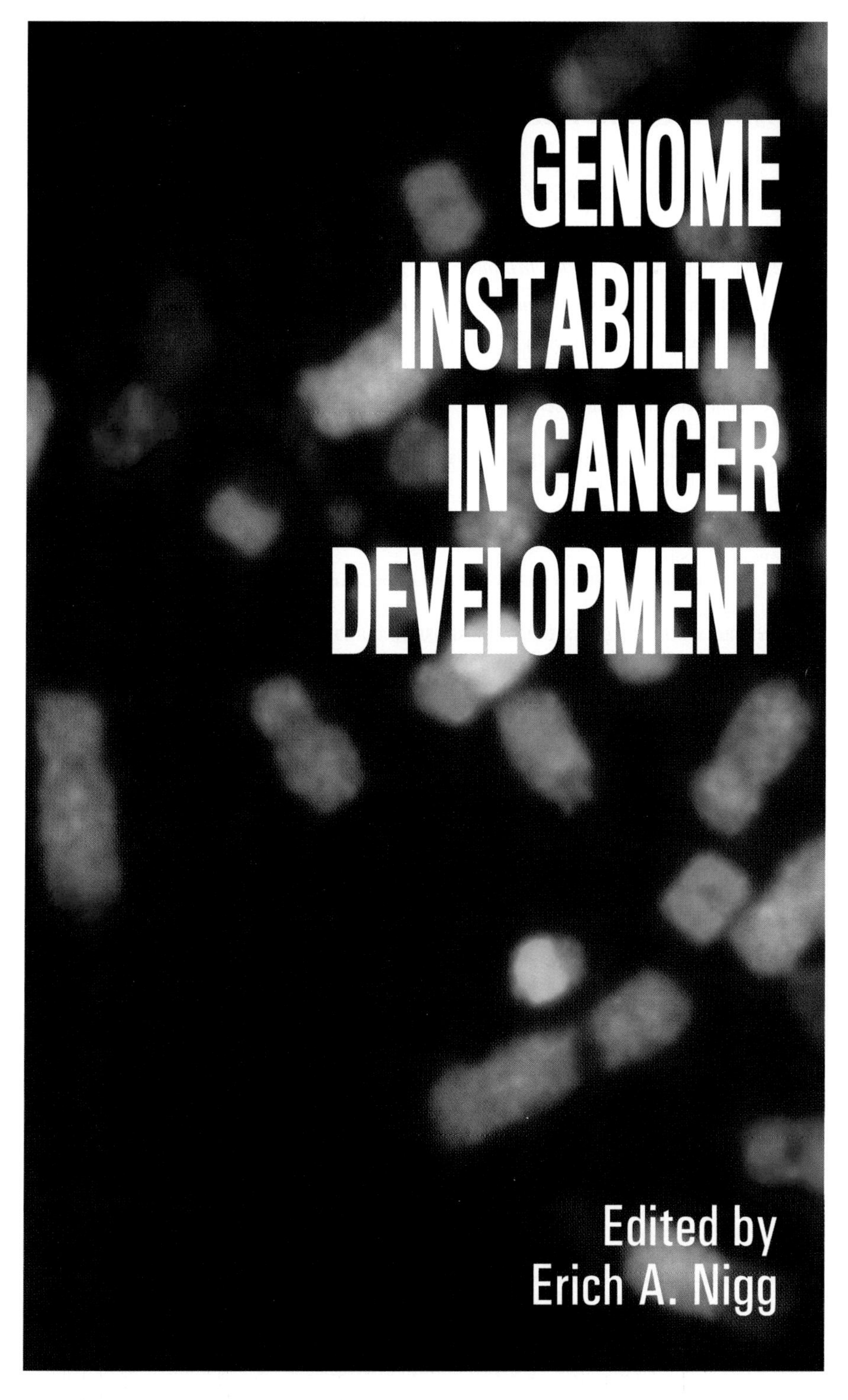

Image (kindly provided by Dr. M. Speicher): Metaphase spread of small cell lung cancer cell line H187 after M FISH hybridization. The metaphase spread is hyper diploid and has 53 chromosomes. Numerous structural aberrations, such as translocations, insertions and deletions are visible (see also Chapter 1.2).

ADVANCES IN EXPERIMENTAL MEDICINE AND BIOLOGY

Recent Volumes in this Series

Volume 491
THE MOLECULAR IMMUNOLOGY OF COMPLEX CARBOHYDRATES-2
Edited by Albert M. Wu

Volume 492
NUTRITION AND CANCER PREVENTION: New Insights into the Role of Phytochemicals
Edited under the auspices of the American Institute for Cancer Research

Volume 493
NEUROIMMUNE CIRCUITS, DRUGS OF ABUSE, AND INFECTIOUS DISEASES
Edited by Herman Friedman, Thomas W. Klein, and John J. Madden

Volume 494
THE NIDOVIRUSES (CORONAVIRUSES AND ARTERIVIRUSES)
Edited by Ehud Lavi, Susan R. Weiss, and Susan T. Hingley

Volume 495
PROGRESS IN BASIC AND CLINICAL IMMUNOLOGY
Edited by Andrzej Mackjewicz, Maciej Kurpisz, and Jan Zeromski

Volume 496
NONINVASIVE ASSESSMENT OF TRABECULAR BONE ARCHITECTURE AND THE COMPETENCE OF BONE
Edited by Sharmila Majumdar, Ph.D., and Brian K. Bay, Ph.D.

Volume 497
INTRACTABLE SEIZURES: Diagnosis, Treatment, and Prevention
Edited by W. McIntyre Burnham, Peter L. Carlen, and Paul A. Hwang

Volume 498
DIABETES AND CARDIOVASCULAR DISEASE: Etiology, Treatment, and Outcomes
Edited by Aubie Angel, Naranjan Dhalla, Grant Pierce, and Pawan Singal

Volume 499
FRONTIERS IN MODELING AND CONTROL OF BREATHING
Edited by Chi-Sang Poon and Homayoun Kazemi

Volume 500
BIOLOGICAL REACTIVE INTERMEDIATES VI: Chemical and Biological Mechanisms of Susceptibility to and Prevention of Environmental Diseases
Edited by Patrick M. Dansette, Robert Snyder, Marcel Delaforge, G. Gordon Gibson, Helmut Greim, David J. Jollow, Terrence J. Monks, and I. Glenn Sipes

---

A Continuation Order Plan is available for this series. A continuation order will bring delivery of each new volume immediately upon publication. Volumes are billed only upon actual shipment. For further information please contact the publisher.

GENOME INSTABILITY IN CANCER DEVELOPMENT

Edited by

Erich A. Nigg
Max-Planck Institute of Biochemistry,
Martinsried, Germany

ISBN-10 1-4020-3763-5 (HB)
ISBN-13 978-1-4020-3763-4 (HB)
ISBN-10 1-4020-3764-3 (e-book)
ISBN-13 978-1-4020-3764-1 (e-book)

Printed in the Netherlands.

9 8 7 6 5 4 3 2 1

springeronline.com

TABLE OF CONTENTS

Part 3. Cell Cycle Progression and Chromosome Aberration

Part 4. Genome Integrity Checkpoints

PREFACE

Research over the past decades has firmly established the genetic basis of cancer. In particular, studies on animal tumour viruses and chromosome rearrangements in human tumours have concurred to identify so-called 'proto-oncogenes' and 'tumour suppressor genes', whose deregulation promotes carcinogenesis. These important findings not only explain the occurrence of certain hereditary tumours, but they also set the stage for the development of anti-cancer drugs that specifically target activated oncogenes. However, in spite of tremendous progress towards the elucidation of key signalling pathways involved in carcinogenesis, most cancers continue to elude currently available therapies. This stands as a reminder that "cancer" is an extraordinarily complex disease: although some cancers of the haematopoietic system show only a limited number of characteristic chromosomal aberrations, most solid tumours display a myriad of genetic changes and considerable genetic heterogeneity. This is thought to reflect a trait commonly referred to as 'genome instability', so that no two cancers are ever likely to display the exact same genetic alterations.

Numerical and structural chromosome aberrations were recognised as a hallmark of human tumours for more than a century. Yet, the causes and consequences of these aberrations still remain to be fully understood. In particular, the question of how genome instability impacts on the development of human cancers continues to evoke intense debate. Is the observed instability merely a consequence of advanced tumour growth or does it constitute a prerequisite for the acquisition of an ever more aggressive cancer cell phenotype? At what time in the evolution of a tumour does genome instability arise and what are the implications of this trait for the design of therapeutic approaches ? To answer these important questions it will be indispensable to understand the mechanisms that give rise to genome instability. This information will then hopefully provide insight into the contribution of genome instability to cancer development and its relevance to therapy.

Recent years have seen a surge of renewed interest in the role of genome instability in cancer. Remarkable progress has been made towards understanding genome instability at the nucleotide level. Specifically, several hereditary cancer susceptibility syndromes have been linked to genetic defects in DNA repair systems, notably nucleotide excision repair (NER) and mismatch repair (MMR). Moreover, genetic connections have also been established between aneuploidy (numerical chromosome imbalances) and carcinogenesis. Taken together, these findings provide strong support for the hypothesis that genome instability is an important parameter in the aetiology and clinical behaviour of cancer. In expanding cell populations, genome instability is expected to increase the probability of acquiring critical mutations, notably the gain of activated oncogenes and the loss of tumour suppressor genes. Genome instability is also expected to favour the adaptation of incipient tumour cells to changing physiological conditions during tumour progression. And, last but not least, genome instability is likely to play an important role in the emergence of resistance to therapy.

This book explores the molecular origins of genome instability and discusses its impact on cancer development. It reviews both genetic and biochemical research on the mechanisms that allow cancer cells to accumulate critical mutations and thus evolve, through processes reminiscent of Darwinian selection, an ever increasingly aggressive behaviour. By bringing together authoritative reviews from experts in widely different but complementary fields, the book is meant to stimulate thought, discussion and experimentation. Hopefully, it will serve as a rich source of information for a wide audience, including advanced students, researchers and oncologists. My sincere thanks go to all authors for contributing excellent and comprehensive chapters, to Dr. M. Speicher for kindly providing the internal cover picture, to Ms Alison Dalfovo for expert secretarial assistance, and to Dr. Cristina Alves dos Santos and her colleagues at Springer Life Sciences for a very pleasant collaboration throughout the preparation of this book.

Martinsried, Spring 2005 Erich A. Nigg

LIST OF CONTRIBUTORS

Tarek Abbas
Department of Biochemistry and Molecular Genetics
University of Virginia School of Medicine
Charlottesville, VA 22908
USA

Jaan-Olle Andressoo
MGC Department of Cell Biology and Genetics
Center for Biomedical Genetics
Erasmus Medical Center
3000 DR Rotterdam
The Netherlands

Geoff Birrell
The Queensland Institute of Medical Research
300 Herston Rd
Herston Qld 4029
Australia

Nicholas E. Burgis
Division of Biological Engineering
Massachusetts Institute of Technology
Room 56-235
77 Mass Avenue
Boston, MA 02139
USA

Junjie Chen
Guggenheim Building, Room 1306
Mayo Clinic
200 First Street, SW
Rochester, MN 55905
USA

Phillip Chen
The Queensland Institute of Medical Research
300 Herston Rd
Herston Qld 4029
Australia

Michelle Debatisse
Institut Curie, FRE 2584,
26 rue d'Ulm,
75248 Paris,
France

Anindya Dutta
Byrd Professor of Biochemistry & Molecular Genetics
Professor of Pathology
Jordan Hall 1240, Box 800733
University of Virginia Health Sciences Center
Charlottesville, VA 22908
USA

Melanie Ehrlich
Human Genetics Program and Department of Biochemistry,
Tulane Cancer Center, SL31
Tulane Medical School
1430 Tulane Ave
New Orleans, LA 70112,
USA

Catherine M. Green
Genome Damage and Stability Centre
University of Sussex
Falmer, Brighton, BN1 9RQ
UK

Roger A. Greenberg
Department of Cancer Biology
Dana Farber Cancer Institute,
Boston, MA 02115
USA

Nuri Gueven
The Queensland Institute of Medical Research
300 Herston Rd
Herston Qld 4029
Australia

Ana Gutierrez del Arroyo
Molecular Oncology Laboratory
London Research Institute
Lincoln's Inn Fields Laboratories
Room 118
44 Lincoln's Inn Fields
London WC2A 3PX
UK

Jan H.J. Hoeijmakers
MGC Department of Cell Biology and Genetics
Center for Biomedical Genetics
Erasmus Medical Center
3000 DR Rotterdam
The Netherlands

Penny Jeggo
Genome Damage and Stability Centre,
University of Sussex,
Falmer, Brighton BN1 9RQ,
UK

Joseph Jiricny
Institute of Molecular Cancer Research
University of Zurich
August Forel Strasse 7
8008 Zurich
Switzerland

Gary D. Kao
Department of Radiation Oncology
University of Pennsylvania
Philadelphia 19104
USA

Sergei Kozlov
The Queensland Institute of Medical Research
300 Herston Rd
Herston Qld 4029
Australia

Martin F. Lavin
Radiation Biology and Oncology
The Queensland Institute of Medical Research
300 Herston Rd
Herston Qld 4029
Australia

Alan R. Lehmann
Genome Damage and Stability Centre
University of Sussex
Falmer, Brighton, BN1 9RQ
UK

Wilma Lingle
Tumor Biology Program
Mayo Clinic College of Medicine
Mayo Clinic Foundation
Rochester, MN 55905
USA

Lawrence A. Loeb
Department of Pathology
Joseph Gottstein Memorial Cancer Research Laboratory
University of Washington School of Medicine
Seattle WA 98195
USA

Zhenkun Lou
Department of Oncology
Mayo Clinic and Foundation
200 First Street, SW
Rochester, MN 55905
USA

Kara Lukasiewicz
Tumor Biology Program
Mayo Clinic College of Medicine
Mayo Clinic Foundation
Rochester, MN 55905
USA

Bernard Malfoy
Institut Curie
Section de Recherche
26 Rue d'ULM
Paris Cedex 5
France

Giancarlo Marra
Institute of Molecular Cancer Research
University of Zurich
August Forel Strasse 7
8008 Zurich
Switzerland

Tomohiro Matsumoto
Department of Gene Mechanisms
Graduate School of Biostudies
Kyoto University
Kitashirakawa-Oiwakecho
Sakyo-ku, Kyoto 606-8502
Japan

Lisiane B. Meira
Division of Biological Engineering
Massachusetts Institute of Technology
Room 56-235
77 Mass Avenue
Boston, MA 02139
USA

Cheng Peng
The Queensland Institute of Medical Research
300 Herston Rd
Herston Qld 4029
Australia

Gordon Peters
Molecular Oncology Laboratory
London Research Institute
Lincoln's Inn Fields Laboratories
Room 118
44 Lincoln's Inn Fields
London WC2A 3PX
UK

K. Lenard Rudolph
Department of Gastroenterology, Hepatology and Endocrinology
Medical School Hannover
Carl-Neuberg-Str. 1
D-30625 Hannover
Germany

Jeffrey L. Salisbury
Tumor Biology Program
Mayo Clinic College of Medicine
Mayo Clinic Foundation
Rochester, MN 55905
USA

Leona D. Samson
Division of Biological Engineering
Massachusetts Institute of Technology
Room 56-235
77 Mass Avenue
Boston, MA 02139
USA

Shaun Scott
The Queensland Institute of Medical Research
300 Herston Rd
Herston Qld 4029
Australia

Michael Speicher
Institute for Human Genetics
Technical University
Trogerstr. 32
81675 Munich
Germany

Ranga N. Venkatesan
Department of Pathology
Joseph Gottstein Memorial Cancer Research Laboratory
University of Washington School of Medicine
Seattle WA 98195
USA

Harm de Waard
MGC Department of Cell Biology and Genetics
Center for Biomedical Genetics
Erasmus Medical Center
3000 DR Rotterdam
The Netherlands

Mitsuhiro Yanagida
Department of Gene Mechanisms
Graduate School of Biostudies
Kyoto University
Kitashirakawa-Oiwakecho
Sakyo-ku, Kyoto 606-8502
Japan

Tim J. Yen
Fox Chase Cancer Center
7701 Burholme Ave
Philadelphia, PA 19111
USA

Wenge Zhu
Department of Biochemistry and Molecular Genetics
University of Virginia School of Medicine
Charlottesville, VA 22908
USA

Part 1

The Problem of Genome Instability

Chapter 1.1

THE MULTIPLICITY OF MUTATIONS IN HUMAN CANCERS

Ranga N. Venkatesan and Lawrence A. Loeb
Joseph Gottstein Memorial Cancer Research Laboratory, Department of Pathology, University of Washington, Seattle, USA

1. INTRODUCTION

Human cancer cells contain large numbers of mutations. These can be observed as alterations in chromosomal numbers (gains or losses) and structural integrity, by an analysis of the lengths of microsatellite sequences and mutations in oncogenes and tumour-suppressor genes. The question is how and when these mutations originate, what the consequences of these mutations are, and most importantly, whether they drive tumour progression. In order to account for the disparity between the infrequency of spontaneous mutations in normal somatic human cells and the large number of mutations in human cancers, we formulated the hypothesis that cancer cells express a mutator phenotype. The hypothesis states that an increase in mutation rate is an early step during tumorigenesis. As a result, random mutations are generated throughout the genome. Some of these mutations occur in genes that normally function to guarantee the accurate transfer of genetic information during each cell division. Among the many mutations produced, some are ones that impart a growth advantage and result in invasion and metastasis, the hallmarks of cancer. In this chapter we will focus on the multiple mutations in human tumours, postulated sources for these mutations, and the arguments, for and against, the mutator phenotype hypothesis.

2. CHROMOSOME NUMBERS AND CANCER

Changes in chromosome number, aneuploidy, may be a gross manifestation of genetic instability in tumours. During the early part of last

E.A. Nigg (ed.), Genome Instability in Cancer Development, 3-17.

century, using a light microscope and embryos of echinoderms (ascaris and sea urchins), Theodor Boveri made many remarkable observations on the numbers and structures of chromosomes (Boveri, 1902). Boveri postulated that (1) chromosomes are highly organised structures probably involved in heredity, (2) the egg and the sperm contribute equal number of chromosomes to the embryo, and (3) tumour growth may result from aberrant chromosome number or aneuploidy. Technical advances in human clinical cytogenetics led to the verification of Boveri's proposal that cancer cells possess abnormal chromosomal numbers and this has become one marker for grading human tumours. However, even though most solid and some hematopoietic cancers are aneuploid, the fundamental question that has remained unanswered is whether aneuploidy initiates tumorigenesis or is passively acquired during evolution of malignant cells, or simply stated, is aneuploidy the cause or an effect of cancer cell evolution? Duesberg and co-workers have argued that aneuploidy is the somatic event that initiates carcinogenesis (Duesberg et al., 1998). They present evidence that aneuploidy can be induced by treatment of cells with chemical carcinogens and that the induction of aneuploidy precedes the appearance of a transformed phenotype. Recently, Rahman and co-workers provided further evidence supporting the aneuploidy-cancer hypothesis: they reported that biallelic mutations in the spindle checkpoint gene *BUB1B* are associated with aneuploidy in a human disease, MVA (mosaic variegated aneupoidy). MVA is a rare recessive disease characterised by early onset of cancer. Thus in a rare inherited disease, a mutation in a gene that effects chromosome segregation is associated with human cancers (Hanks et al., 2004).

Considering the hundreds of genes encoded in each chromosome, it is difficult to understand how any cell with a different number of chromosomes can possibly maintain viability and how aneuploidy is compatible with live human births such as those seen in Down and Klinefelter syndromes. It seems likely that haploinsufficiency, over-expression, squelching and dominant negative interactions would render such cells less fit. Apparently, organisms have evolved buffering systems to tolerate changes in chromosome numbers that we have not even contemplated. Despite the observations that most cancer cells contain aneuploid karyotypes, the timing of acquisition of such events is unknown and hence their direct contribution toward development of a malignant phenotype remains obscure. The generalisation that aneuploidy initiates cancer is difficult to substantiate considering that most tumours are monoclonal and yet not all tumour cells within a tumour mass are aneuploid (Mitelman, 1994). Moreover, premalignant conditions such as Barrett's esophagus (Barratt et al., 1999) and ulcerative colitis exhibit mutations in multiple oncogenes but are not aneuploid (Rabinovitch et al., 1999). Thus, aneuploidy could be one of the manifestations of genetic instability and not causative in initiating carcinogenesis.

2.1 Chromosome Instability and Cancer

Chromosomal instability (CIN) results in gains, losses, deletions, insertions, translocations, amplifications, and rearrangements, and is frequently used to grade tumours with respect to prognosis (Lengauer et al., 1998). CIN is characterised by an increased frequency of chromosomal alterations often resulting in loss of heterozygosity (Rajagopalan et al., 2003). The majority of human tumours display the CIN phenotype. However, these tumours may contain larger numbers of other types of mutations that are more difficult to detect. The genes responsible for the maintenance of chromosomal stability in normal cells are beginning to be identified and their function is being delineated. Tumours with the CIN phenotype usually harbour mutations in oncogenes and/or tumour-suppressor genes, many of which are involved in the regulation of transcription. These tumours may display genetic instability as a result of altered global gene expression patterns and global changes in chromatin structure.

Large chromosomal rearrangements, a hallmark of the CIN phenotype, can be visualised by cytogenetic techniques and enhanced visualisation has been provided by spectral karyotyping (SKY) (Bayani et al., 2002). Using gene-specific probes labelled with different coloured fluorochromes, one can map specific segments of chromosomes and demonstrate multiple rearrangements within and between individual chromosomes. There are two widely used molecular techniques that examine populations of molecules at higher resolution. Comparative genomic hybridisation measures differences in hybridisation between fragments of DNA from different sources, each tagged with different fluorescent molecules. Localisation of signal can be achieved by using metaphase chromosomes as a scaffold. Using this technique, a large number of tumours have been shown to exhibit multiple changes in DNA copy number (Iwabuchi et al., 1995; Kallioniemi et al., 1994). The fact that benign tumours also exhibit extensive changes in DNA copy number (El-Rifai et al., 1998) suggests that changes in DNA copy number occur early during tumorigenesis. Loss of heterozygosity (LOH) in tumours permits one to scan small segments of the entire genome using a library of microsatellite markers. The finding that many cancers exhibit multiple changes suggests that in many DNA segments of the tumour genome, there is a modification (gain or loss) of segments of one of the parental alleles. Only a small proportion of the genome is interrogated by this technique since the PCR-amplified segments are about 1000 nucleotides in length. If one assumes that the sampling is representative, then the entire tumour genome may contain thousands of DNA segments that exhibit loss of heterozygosity.

Both comparative genomic hybridisation and measurements of loss of heterozygosity examine populations of DNA molecules and do not score for chromosomal alterations in individual tumour cells. However, both of these techniques have been applied to single metastatic cells in bone marrow and multiple alterations have been documented (Klein et al., 1999). Moreover, different single cells from the same tumour display alterations in different segments of the same chromosomes.

The majority of human cancers display a CIN phenotype and timing of its expression is presently debated even in the same tumour model, for example, colon cancer. Huang and coworkers reported that mutations in mismatch repair genes occur prior to mutations in the *APC* gene, a frequently mutated gene in colon cancer that is associated with the CIN phenotype. In addition, studies utilising microdissection that trace tumour evolution indicated that microsatellite instability was extensive in early adenomas, and additional instability was observed as adenomas progressed to adenocarcinomas (Shibata et al., 1996). In contrast, others have reported that there is no difference in the frequency and spectrum of mutations in the APC gene in colon tumours that exhibit extensive microsatellite instability versus others that do not (Homfray et al., 1998). These results have suggested that mutations in *APC* initiate carcinogenesis and may occur prior to microsatellite instability (Tomlinson and Bodmer, 1999).

2.2 Microsatellite Instability and Cancer

Studies on alterations in microsatellite sequences provided the first and strongest glimpse into the extensiveness of mutations in human cancers. Perucho and associates used oligonucleotides with random sequences as arbitrary primers in PCR-reactions and observed products of different lengths using DNA from human colon tumours compared to those obtained using DNA from adjacent normal tissues (Perucho, 1996). The PCR products contained microsatellite sequences with different numbers of repeats. Subsequent studies established that extensive microsatellite instability (MIN) was associated with hereditary nonpolyposis coli (HNPCC) (Fishel, 2001; Thibodeau et al., 1993), a disease caused by mutations in genes required for the repair of mismatches generated by erroneous DNA synthesis (Kolodner and Marsischky, 1999; Modrich, 1995). Unlike CIN, the MIN phenotype is observed in a small variety of tumours and frequently occurs early during tumorigenesis. The nature of these mutations and their consequences is considered in other chapters in this book. However it is clear that at least in some inherited human tumours (Cleaver and Kraemer, 1989), like HNPCC, MAP (Myh-associated polyposis), XP (Xeroderma pigmentosum) and Bloom's syndrome, the cells are predisposed to genetic instability at the nucleotide level and at least in these tumours, genetic instability clearly causes tumorigenesis. Furthermore, the cancers with a MIN phenotype display tissue specificity (Markowitz, 2000). However, with respect to genomic instability, it should be emphasised that this protocol only analyzes a small percentage of the microsatellite sequences in the genome. If one extrapolates these results and those obtained from studies with other tumours (Stoler et al., 1999) to the whole genome, it can be concluded that some tumours contain as many as

10,000 alterations in the number of repeats within microsatellites. It is usually assumed that microsatellite instability is generated by slippage of DNA polymerases during copying of repeats and thus represents a hot spot for mutagenesis. The extensiveness of microsatellite instability in tumours lacking mutations in mismatch repair genes [listed in (Jackson and Loeb, 2001)] provides an important indicator of the extensiveness of genomic instability in tumours.

2.3 Point Mutations and Cancer

Many agents that damage DNA produce single nucleotide substitutions (Singer, 1996), similar to misincorporations by DNA polymerases (Kunkel and Loeb, 1981). In addition, errors in DNA synthesis by trans-lesion DNA polymerases that copy past bulky adducts in DNA are predominantly single-base substitutions (Kunkel and Bebenek, 2000). The relationship of single base substitutions to gene amplification, large deletions and rearrangements has not been explored. Conceivably, single-base substitutions can initiate many of these events. Single base changes, (point mutations) are likely to be distributed randomly throughout the genome. The random distribution of these mutations renders current methods inadequate for their detection. Conventional DNA sequencing requires multiple copies of each DNA template and the sequences that are obtained score only for the predominant nucleotide at each position; micro-heterogeneity within a population of DNA templates would not be detected (Loeb et al., 2003). In order to detect heterogeneity by DNA sequencing, it is necessary to sequence multiple single DNA molecules obtained from the same tissue sample. As a result, the quantitation of random mutations in tumours and the evaluation of their contribution to a mutator phenotype in cancers are issues that have not been resolved. It should be noted that with each round of clonal selection, the non-selected random mutations are clonally fixed in all progeny derived from that cell (Figure 1). With absolute selection for mutations in an oncogene or a tumour-suppressor gene, the random mutations would also be present among all the progeny cells. In a branched structure for tumour evolution the random mutations would be distributed in groups of different cells throughout the tumour.

2.4 Mutational Requirements for Tumorigenesis

The number of specific mutations required to produce a tumour has been controversial. Based on age-associated increases in cancer incidence, it can be inferred that two mutations are required for tumour induction in retinoblastomas (Knudson, 1971; Knudson, 1985) and as many as 10 mutations in the case of prostate carcinomas (Ware, 1994). In retinoblastoma, the first event can be inherited and the second is a somatic mutation. Based on the transformed phenotype, it has been proposed that at

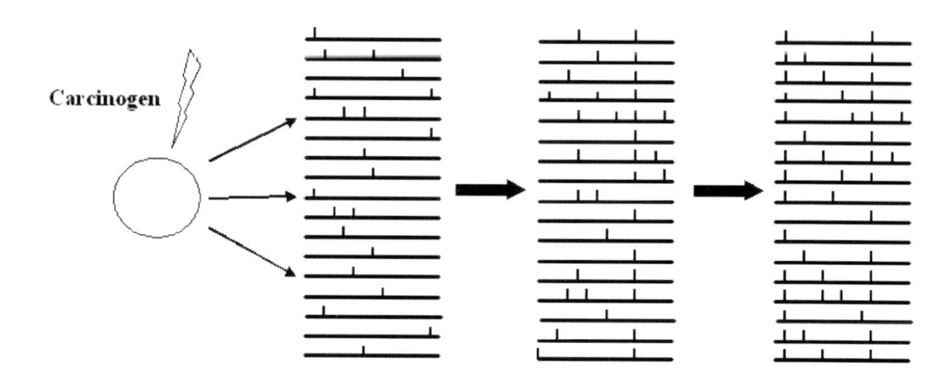

Figure 1. Random mutations are fixed as clonal after selection. Carcinogen or spontaneously-induced expression of a mutator phenotype is an early event generating genome-wide random mutations. The cells harbouring mutation(s) that provide a growth advantage proliferate and expand further, resulting in clonal fixation of that mutation and further generation of new mutations. Selection and clonal proliferation of only one clone is shown for simplicity. The circle represents human tissue, the horizontal lines are individual cellular genomes and vertical lines are random mutations.

least six clonal events are required for tumour induction (Hanahan and Weinberg, 2000). In culture, human cells require more mutations for neoplastic transformation than mouse cells (Rangarajan et al., 2004). How many random mutations are required in order to generate a small number of cancer-specific mutations that either inactivate a tumour-suppressor gene or activate an oncogene and hence confer a dominant phenotype for malignant growth is unknown. Thus, the large number of mutations generated by a mutator phenotype is not at variance with the small number of clonal mutations found in tumours.

3. MUTATOR PHENOTYPE AND CLONAL SELECTION

In order to account for the large numbers of mutations found in human tumours, we have previously advanced the hypothesis that precancer cells must exhibit a mutator phenotype (Loeb et al., 1974). In normal cells, DNA replication is an exceptionally accurate process and spontaneous mutations are very infrequent. As a result, normal mutation rates are insufficient to account for the multiple mutations observed in cancer cells (Jackson and Loeb, 1998). We have hypothesised that cancer cells express a mutator phenotype and that this occurs early during tumorigenesis. An early step in the evolution of a tumour is the introduction of mutations in genes that normally confer genetic stability. These mutant genes induce additional mutations throughout the genome, some of which result in increased fitness and clonal proliferation. Our initial hypothesis focused on mutations in DNA polymerase and DNA repair genes (Loeb et al., 1974). This hypothesis has been extended as it became apparent that mutations in a wide variety of genes including those involved in checkpoints, chromosome segregation, apoptosis and nucleotide metabolism would also produce a mutator phenotype (Loeb et al., 2003). In this model for tumour evolution, enhanced mutagenesis would be a driving force in promoting mutation accumulation for growth advantage, or simply stated, a mutator phenotype drives selection.

Nowell (Nowell, 1976; Nowell, 1993) proposed that a mutator phenotype could result from repetitive rounds of clonal selection, in which mutations in a single cell impart a proliferative advantage allowing progeny of that cell to repopulate the tumour. Successive rounds of clonal selection would drive tumour progression. With each round of clonal selection, all silent mutations would become clonal (Figure 1). Vogelstein and colleagues, in their description of tumorigenesis of colon cancer, have proposed an ordered succession of mutations in going from a small benign polyp to an adenocarcinoma (Vogelstein et al., 1988). If clonal selection were absolute, then all random mutations would be converted to clonal mutations. In an ordered succession model, each tumour cell would contain all of the mutations, but this has never been demonstrated.

It seems reasonable that both enhanced mutagenesis and repetitive rounds of clonal selection are operative in tumour progression. Miller and co-workers (Miller, 1996) provided evidence that these two processes are linked. They exposed bacteria to a mutagen and then carried out sequential rounds of selection for mutations that rendered the bacteria resistant to different agents. This protocol mimics the requirements for tumour proliferation under different conditions, *i.e.* the need to grow under reduced oxygen, reduced nutrition, ability to preferentially proliferate, etc. They observed that after three successive rounds of selection 100% of the bacteria exhibited a mutator phenotype (Mao et al., 1997). Each round of selection not only selected for mutants that were resistant to the selective agent but also for mutants that increase mutations in genes that render the cells resistant. Thus with successive rounds of selection there is a "piggy-backing" of mutant genetic instability genes.

4. EMERGING MECHANISMS FOR CAUSES OF GENETIC INSTABILITY

There are many pathways that lead to induction of genetic instability and some of these may be tumour-specific. Many inherited diseases associated with a high incidence of cancer, harbour recessive mutations in genes involved in maintenance of genomic integrity in normal cells. Figure 2 lists several genes which when mutated induce genetic instability. The mutations have been detected because they are clonal. Genetic instability can also be induced by over-expression or inappropriate expression of non-mutated enzymes involved in DNA metabolism.

Mammalian B-cell lymphocytes have the capacity to produce a diverse library of different antibodies. AID (activation-induced cytidine deaminase) has been identified as the key enzyme required for generation of high affinity antibodies. AID is a member of the RNA editing APOBEC-1 family, which is expressed specifically in the germinal centre-B cells (Muramatsu et al., 1999). Ectopic expression of AID and APOBEC-1 family members in *Escherichia coli,* a mouse pre-B cell line, a human B-cell and a non-B cell line resulted in elevation of the mutation frequency of DNA targets, suggesting that deamination by AID may not be restricted to antibody genes (Martin et al., 2002; Petersen-Mahrt et al., 2002). Liver-specific expression of APOBEC-1 in rabbits and mice has been demonstrated to cause dysplasia and hepatocellular carcinoma (Yamanaka et al., 1995). All of the above data suggests that deregulated expression of AID and APOBEC-1 family members may transduce tumorigenesis by induction of genome-wide random somatic mutagenesis.

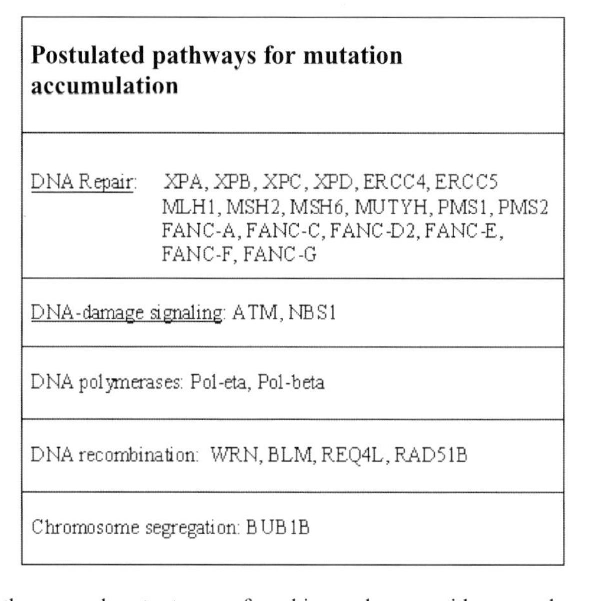

Postulated pathways for mutation accumulation
DNA Repair: XPA, XPB, XPC, XPD, ERCC4, ERCC5 MLH1, MSH2, MSH6, MUTYH, PMS1, PMS2 FANC-A, FANC-C, FANC-D2, FANC-E, FANC-F, FANC-G
DNA-damage signaling: ATM, NBS1
DNA polymerases: Pol-eta, Pol-beta
DNA recombination: WRN, BLM, REQ4L, RAD51B
Chromosome segregation: BUB1B

Figure 2. Pathways and mutant genes found in syndromes with a prevalence of cancers.

The classic replicative DNA polymerases (α, δ, and ε) stall upon encountering altered bases (DNA lesions) and it was previously unknown as to how the DNA synthesis could resume after stalling of replication forks at the site of DNA lesions. But, as discussed in detail elsewhere in this book, the recent discovery of a new family of DNA polymerases (Y-family: η, ι, κ, and ζ) that are able to use damaged DNA as a template (bypass) has provided new clues toward understanding the mechanism of bypass of stalled replication forks (Livneh, 2001; Masutani et al., 2000) (Friedberg et al., 2002; Goodman and Tippin, 2000). Based on accumulated evidence, a model has been proposed that suggests that a Y-family DNA polymerase is recruited to bypass specific DNA lesions, followed by resumption of processive DNA synthesis by replicative DNA polymerases (Friedberg et al., 2002; Haracska et al., 2001). The Y-family DNA polymerases synthesise DNA with very low fidelity on both damaged and undamaged DNA. Their fidelity has been estimated to be ~1000 to 4000-fold lower than the replicative DNA polymerases (Matsuda et al., 2000). Thus the benefits of ensuring continuous DNA synthesis even through the damaged DNA template may also result in generation of spontaneous mutations. The trans-lesion polymerases may be major constitutive sources for the generation of random mutations. It will be important to determine if the expression of some of these enzymes are elevated in specific tumours.

5. ARGUMENTS AGAINST A MUTATOR PHENOTYPE

It is instructive to consider the arguments that have been advanced against a mutator phenotype during tumorigenesis or more specifically, a mutator phenotype at the level of point mutations generated by mutations in DNA polymerases or DNA repair genes.

First is the concept of negative clonal selection, that is, most mutations result in reduced fitness and as a result, proteins cannot tolerate multiple amino acid substitutions. Recent studies involving random substitutions in a variety of proteins, including DNA polymerases, demonstrate that even these very highly conserved proteins are able to tolerate large numbers of substitutions without impairments in catalytic activity (Patel and Loeb, 2000). Studies from our laboratory have quantitated the probability of enzyme inactivation by single amino acid substitutions (Guo et al., 2004). The average probability of inactivation of human 3-methyladenine glycosylase by any single random amino acid replacement at any position in the protein is 0.34, *i.e* the protein tolerates substitutions in nearly all positions. In addition to mutation tolerance by individual proteins, cells have evolved redundant pathways for preserving vital activities.

Second, mathematical modelling of mutation accumulation in colonic stem cells suggests that one can account for 150,000 mutations per cell in adenocarcinoma of the colon based on normal mutation rates (Tomlinson et al., 1996). The cells that line the crypts of the colon undergo rapid cell divisions starting with the stem cells located near the base of the crypt that divide asymmetrically and give rise to well-differentiated cells that desquamate into the lumen of the intestine. The entire process can occur over thirty-six hours and, as a result, colonic epithelial cells can undergo 5000 divisions during adulthood. Based primarily on this continuous regenerative and replicative process, it has been calculated that normal cells could accumulate large numbers of mutations and thus the expression of a mutator phenotype in tumours may not be required for tumour progression. These calculations support the concept that colon cancers contain large numbers of random mutations (~10^{13}) and argued that since colonic stem cells undergo large number of cell generations, normal somatic mutations rates are sufficient to generate mutations that drive tumorigenesis. They suggest that premalignant cells need not express the mutator phenotype because increased mutation rates do not confer any advantages for tumour growth but rather natural selection provides the growth advantage (Tomlinson and Bodmer, 1999; Tomlinson et al., 2002). The two major difficulties with this model are that the mutation rate in stem cells is 100-fold lower than that exhibited by somatic cells (Cervantes et al., 2003) and that most tissues that do not shed progeny cells are therefore unlikely to undergo the enormous number of mutations that occur in colonic epithelium. In contrast to this formulation, other mathematical models that compare sporadic and hereditary forms of colorectal cancer indicate that the initial mutations involve mutations in genes associated with genetic instability (Komarova and Wodarz, 2003). It should also be noted that in most tissues other than colon and skin, early tumour cells compete with normal cells within a confined space.

Thirdly, a sequencing study of 3.2 megabases of exonic DNA from 12 tumour cell lines (Wang et al., 2002) revealed only 3 tumour-specific coding mutations. Assuming that this mutation frequency prevails throughout the genome, there would be 3000 mutations in the tumour genome. The authors concluded that, because cells lining the intestine rapidly proliferate, the small number of substitutions observed could result from normal mutation rates. Instead, they hypothesise that mutations in genes regulating chromosomal stability are more likely to initiate tumorigenesis. Of great interest would be the frequency of mutations that occur in introns and in non-expressed genes, the mutator phenotype hypothesis would predict that these would accumulate with successive rounds of replication. The limitation of this DNA sequencing protocol is the inability to detect random mutations that have occurred after the last round of clonal selection. Since a patient tumour sample is usually comprised of heterogeneous cellular genomes, only the most frequent mutation at any position would be detected.

Fourth is the lack of preponderance of clonal mutations in tumours that involve genes that regulate genetic stability. Futreal and coworkers have compiled a comprehensive catalogue of genes mutated in human cancer from published literature (Futreal et al., 2004). They reported that only 1% (291) of human genes harbour clonal mutations, predominantly within gene families including protein kinases, transcriptional regulators, and DNA binding proteins. Genes involved directly in maintenance of DNA sequence integrity, like mismatch repair, base excision repair and nucleotide excision repair, only accounted for a very minor fraction. Interestingly, the genes that signal DNA damage, and are involved in DNA repair-transactions also are present in small numbers. Two points emerge from the above study that argues against the mutator phenotype hypothesis: (1) Only a small number of genes are involved in tumorigenesis and (2) mutations in genes responsible for maintenance of DNA sequence integrity are rarely found in human tumours. One can argue, however, that most tumour cells contain multiple mutations in genetic stability genes and only those that were present in all of the cells in that tumour would be detected.

6. IMPLICATIONS OF A MUTATOR PHENOTYPE

The presence of large numbers of silent clonal mutations, and random mutations within any tumour can account for the rapid emergence of tumour resistance to chemotherapies. Clonal mutations could result in immediate resistance to chemotherapeutic agents. In contrast, resistance due to the presence of rare mutations within a tumour is likely to take more time to become manifested. Thus, the mutator phenotype hypothesis would predict that amongst the 5×10^8 cells that comprise a clinically detectable tumour, there are cancer cells, which harbour mutant genes rendering them resistant to any chemotherapeutic agent directed against the tumour. While chemotherapy might kill most of the cells within the tumour, the cells harbouring the resistant mutations would be able to proliferate and repopulate the tumour. The simultaneous utilisation of multiple chemotherapeutic agents might offer an advantage in obliterating cells with random mutations, since it would be infrequent for one cell to harbour mutations that render it resistant to two agents.

The overall number of mutations in a tumour may permit a new system of stratification in which the stage of tumour progression and/or the response of tumours to chemotherapy is based on the number of mutations within the tumour. Tumours with large numbers of mutations might be further along in tumour progression, and likely be more drug resistant. The question is whether additional mutagenesis would be likely to result in a more malignant phenotype or would be detrimental by inducing an error catastrophe needs to be explored. It is possible that many common cancer therapies are effective in part because they enhance mutagenesis.

The types of mutations that accumulate within a tumour should provide clues to both clonal lineage and to mechanisms for mutation accumulation. While sequential biopsies of human tumours is not feasible, tracing the lineage of animal tumours by mapping mutation frequencies in different genes has been proposed (Shibata et al., 1996). The types and spectra of mutations should provide a footprint of mechanisms that cause mutations in genetic stability genes. The accumulation of single-base substitutions provides a measure of errors during DNA replication that exceed mismatch and base excision repair capacities. Mutations that accumulate preferentially in mitochondrial DNA are likely to result from damage by oxygen reactive species (Fliss et al., 2000). Chromosome rearrangements imply deficits in double-strand break and recombination repair. The accumulation of large chromosomal gains, losses and duplications provides evidence pointing toward mutations in genes responsible for chromosome segregation.

7. CONSEQUENCES OF GENETIC INSTABILITY

Human cancers harbour at least two types of genetic instability, CIN and MIN, and can be classified into either category. The concept of genetic instability causing cancer is being debated and proponents of the theory emphasise that premalignant cells must express genetic instability early during tumorigenesis to accumulate advantageous mutations necessary for clonal selection. But what causes genetic instability is presently unknown (Marx, 2002). Tomlinson and Bodmer have argued that premalignant cells need not display genetic instability as an early event to acquire a malignant phenotype. They assert that a combination of normal somatic mutation rates and Darwinian selection is sufficient for normal cells to veer toward a tumorigenic pathway. They propose that since the majority of sporadic human tumours arise from a normal diploid cell with intact DNA repair, cell cycle checkpoints and apoptosis machinery, early expression of genetic instability or hypermutagensis may be detrimental to a cell's viability and not tolerated. Instead the premalignant cells offset the need for early expression of genetic instability by undergoing a large number of cell generations to select for malignant cells. They emphasise that genetic instability may assist tumorigenesis, as in certain inherited cancers such as HNPCC and XP, but may not be a universal prerequisite in all human tumours (Sieber et al., 2003). In contrast to their proposal, we argue that the existence of inherited diseases that arise from mutations in DNA repair genes and display high proclivity toward early onset, as well as the presence of large numbers of DNA alterations observed in tumours that do not arise from actively dividing tissues, provides strong evidence that genetic instability is a primary event in tumorigenesis. In summary, the fundamental question, whether genetic instability is required for tumorigenesis, still needs to be resolved.

ACKNOWLEDGEMENTS

Study in the author's laboratory was supported by grants from the U.S. National Cancer Institute CA78885 and National Institute of health CA102029.

REFERENCES

Barratt, M.T., C.A. Sanchez, L.J. Prevo, D.J. Wong, P.C. Galipeau, T.G. Paulson, P.S. Rabinovitch, and B.J. Reid. 1999. Evolution of neoplastic cell lineages in Barrett oesophagus. *Nat. Genet.* 22:106-109.

Bayani, J., J.D. Brenton, P.F. Macgregor, B. Beheshti, A.M. Nallainathan, J. Karaskova, B. Rosen, J. Murphy, S. Laframboise, B. Zanke, and J.A. Squire. 2002. Parallel analysis of sporadic primery ovarian carcinomas by spectral karyotyoping, comparative genomic hybridization and expression microarrays. *Cancer Res.* 62:3466-3476.

Boveri, T. 1902. Uber mehrpolige Mitosen als Mittel zur Analyse des Zellkerns. *Veh. Dtsch. Zool. Ges.* Wurzburg.

Cleaver, J.E., and K.H. Kraemer. 1989. Xeroderma pigmentosum. *In* Metabolic Basis of Inherited Disease. C.R. Scriver, A.L. Beudet, W.S. Sktm, and D. Valle, editors. McGraw-Hill, New York, NY. 2949-2971.

Duesberg, P., C. Rausch, D. Rasnick, and R. Hehlmann. 1998. Genetic instability of cancer cells is proportional to their degree of aneuploidy. *Proc. Natl. Acad. Sci. USA*. 95:13692-13697.

El-Rifai, W., M. Sarlomo-Rikala, S. Knuutila, and M. Miettinen. 1998. DNA copy number changes in development and progression in leiomyosarcomas of soft tissues. *Am. J. Pathol.* 153:985-990.

Fishel, R. 2001. The selection for mismatch repair defects in hereditary nonpolyposis colorectal cancer. *Cancer Res.* 61:7369-7374.

Fliss, M.S., H. Usadel, O.L. Caballero, L. Wu, M.R. Buta, S.M. Eleff, J. Jen, and D. Sidransky. 2000. Facile detection of mitochondrial DNA mutations in tumours and bodily fluids. *Science*. 287:2017-2019.

Friedberg, E.C., R. Wagner, and M. Radman. 2002. Specialized DNA polymerases, cellular survival, and the genesis of mutations. *Science*. 296:1627-1630.

Futreal, P.A., L. Coin, M. Marshall, T. Down, T. Hubbard, R. Wooster, N. Rahman, and M.R. Stratton. 2004. A census of human cancer genes. *Nat. Rev. Cancer*. 4:117-183.

Goodman, M.F., and B. Tippin. 2000. The expanding polymerase universe. *Nat. Rev. Mol. Cell. Biol.* 1:101-109.

Guo, H.H., J. Choe, and L.A. Loeb. 2004. Protein tolerance to random amino acid change. *Proc. Natl. Acad. Sci. USA*. 101:9205-9210.

Hanahan, D., and R.A. Weinberg. 2000. The hallmarks of cancer. *Cell*. 100:57-70.

Hanks, S., K. Coleman, S. Reid, A. Plaja, H. Firth, D. Fitzpatrick, A. Kidd, K. Mehes, R. Nash, N. Robin, N. Shannon, J. Tolmie, J. Swansbury, A. Irrthum, J. Douglas, and N. Rahman. 2004. Constitutional aneuploidy and cancer predisposition caused by biallelic mutations in BUB1B. *Nat. Genet.* 36:1159-1161.

Haracska, L., I. Unk, R.E. Johnson, E. Johansson, P.M. Burgers, S. Prakash, and L. Prakash. 2001. Roles of yeast DNA polymerases delta and zeta and of Rev1 in the bypass of abasic sites. *Genes Dev.* 15:945-954.

Homfray, T.F.R., S.E. Cottrell, M. Ilyas, A. Rowan, W.F. Bodmer, and I.P.M. Tomlinson. 1998. Defects in mismatch repair occur after *APC* mutations in the pathogenesis of sporadic colorectal tumours. *Hum. Mut.* 11:114-120.

Iwabuchi, H., M. Sakamoto, H. Sakunaga, and Y.Y. Ma. 1995. Genetic analysis of benign, Low-grade, and high-grade ovarian tumours. *Cancer Res.* 55:6172-6180.

Jackson, A.L., and L.A. Loeb. 1998. On the origin of multiple mutations in human cancers. *Cancer Biol.* 8:421-429.

Jackson, A.L., and L.A. Loeb. 2001. The contribution of endogenous sources of DNA damage to the multiple mutations in cancer. *Mutat. Res.* 477:187-198.

Kallioniemi, A., O.-P. Kallioniemi, J. Piper, M. Tanner, T. Stokke, L. Chen, H.S. Smith, D. Pinkel, J.W. Gray, and F.M. Waldman. 1994. Detecton and mapping of amplified DNA sequences in breast cancer by comparative gnomic hybridization. *Proc. Natl. Acad. Sci. USA*. 91:2156-2160.

Klein, C.A., O. Schmidt-Kittler, J.A. Schardt, K. Pantel, M.R. Speicher, and G. Riethmuller. 1999. Comparative gnomic hybridization, loss of heterozygosity, and DNA sequence analysis of single cells. *Proc. Natl. Acad. Sci. USA*. 96:4494-4499.

Knudson, A.G., Jr. 1971. Mutation and cancer: statistical study of retinoblastoma. *Proc. Natl. Acad. Sci. USA*. 68:820-823.

Knudson, A.G., Jr. 1985. Hereditary cancer, oncogenes and antioncogenes. *Cancer Res.* 45:1437-1443.

Kolodner, R.D., and G.T. Marsischky. 1999. Eukaryotic DNA mismatch repair. *Curr. Opin. Genet. Dev.* 9:89-96.

Komarova, N.L., and D. Wodarz. 2003. Evolutionary dynamics of mutator phenotypes in cancer: implications for chemotherapy. *Cancer Res.* 63:6335-6342.

Kunkel, T.A., and K. Bebenek. 2000. DNA replication fidelity. *Annu. Rev. Biochem.* 69:497-529.

Kunkel, T.A., and L.A. Loeb. 1981. Fidelity of mammalian DNA polymerases. *Science*. 213:765-767.

Lengauer, C., K.W. Kinzler, and B. Vogelstein. 1998. Genetic instabilities in human cancers. *Nature*. 396:643-649.

Livneh, Z. 2001. DNA damage control by novel DNA polymerases: translesion replication and mutagnesis. *J. Biol. Chem.* 276:25639-25642.

Loeb, L.A., K.R. Loeb, and J.P. Anderson. 2003. Multiple mutations and cancer. *Proc. Natl. Acad. Sci. USA*. 100:776-781.

Loeb, L.A., C.F. Springgate, and N. Battula. 1974. Errors in DNA replication as a basis of malignant change. *Cancer Res.* 34:2311-2321.

Mao, E.F., L. Lane, J. Lee, and J.H. Miller. 1997. Proliferation of mutators in a cell population. *J. Bacteriol.* 179:417-422.

Martin, A., P.D. Bardwell, C.J. Woo, M. Fan, M.J. Shulman, and M.D. Scharff. 2002. Activation-induced cytidine deaminase turns on somatic hypermutation in hybridomas. *Nature*. 415:802-806.

Marx, J. 2002. Debate surges over the origins of genomic defects in cancer. *Science*. 297:544-546.

Markowitz, S. 2000. DNA repair defects inactive tumour suppressor genes and induce hereditary and sporadic colon cancer. *J Clin Oncol.* 18:75S-80S.

Masutani, C., R. Kusumoto, S. Iwai, and F. Hanaoka. 2000. Mechanisms of accurate translesion synthesis by human DNA polymerase eta. *EMBO J.* 19:3100-3109.

Matsuda, T., K. Bebenek, C. Masutani, F. Hanaoka, and T.A. Kunkel. 2000. Low fidelity DNA synthesis by human DNA polymerase-eta. *Nature*. 404:1011-1013.

Miller, J.H. 1996. The relevance of bacterial mutators to understanding human cancer. *In* Cancer Surveys. Vol. 28. T. Lindahl, editor. Cold Spring Harbour Laboratory Press, Fairview, New York. 141-153.

Mitelman, F. 1994. Catalog of Chromosome Aberrations in Cancer. Wiley-Liss, New York, NY.

Modrich, P. 1995. Mismatch repair, genetic stability, and tumour avoidance. *Philosoph. Transact. Royal Soc.* 347:89-95.

Muramatsu, M., V.S. Sankaranand, S. Anant, M. Sugai, K. Kinoshita, N.O. Davidson, and T. Honjo. 1999. Specific expression of activation-induced cytidine deaminase (AID), a novel member of the RNA-editing deaminase family in germinal centre B cells. *J. Biol. Chem.* 274:18470-18476.

Nowell, P.C. 1976. The clonal evolution of tumour cell populations. *Science*. 194:23-28.

Nowell, P.C. 1993. Chromosomes and cancer: The evolution of an idea. *Adv. Cancer Res.* 62:1-17.

Patel, P.H., and L.A. Loeb. 2000. DNA polymerase active site is highly mutable: evolutionary consequences. *Proc. Natl. Acad. Sci. USA*. 97:5095-5100.

Perucho, M. 1996. Cancer of the microsatellite mutator phenotype. *Biol. Chem.* 377:675-684.

Petersen-Mahrt, S.K., R.S. Harris, and M.S. Neuberger. 2002. AID mutates *E. coli* suggesting a DNA deamination mechanism for antibody diversification. *Nature*. 418:99-103.

Rabinovitch, P.S., S. Dziadon, T.A. Brentnall, M.J. Emond, D.A. Crispin, R.C. Haggitt, and M.P. Bronner. 1999. Pancolonic chromosomal instability precedes dysplasia and cancer in ulcerative colitis. *Cancer Res.* 59:5148-5153.

Rajagopalan, H., M.A. Nowak, B. Vogelstein, and C. Lengauer. 2003. The significance of unstable chromosomes in colorectal cancer. *Nat. Rev. Cancer*:695-701.

Rangarajan, A., S.J. Hong, A. Gifford, and R.A. Weinberg. 2004. Species- and cell type-specific requirements for cellular transformation. *Cancer Cell.* 6:171-183.

Shibata, D., W. Navidi, R. Salovaara, Z.-H. Li, and L.A. Aaltonen. 1996. Somatic microsatellite mutations as molecular tumour clocks. *Nature Med.* 2:676-681.

Sieber, O.M., K. Heinimann, and I.P. Tomlinson. 2003. Genomic instability--the engine of tumorigenesis? *Nat. Rev. Cancer.* 3:701-708.

Singer, B. 1996. DNA damage: chemistry, repair, and mutagenic potential. *Regul. Toxicol. Pharmacol.* 23:2-13.

Stoler, D.L., N. Chen, M. Basik, M.S. Kahlenberg, M.A. Rodriguez-Bigas, N.J. Petrelli, and G.R. Anderson. 1999. The onset and extent of genomic instability in sporadic colorectal tumour progression. *Proc. Natl. Acad. Sci. USA.* 96:15121-15126.

Thibodeau, S.N., G. Bren, and D. Schaid. 1993. Microsatellite instability in cancer of the proximal colon. *Science.* 260:816-819.

Tomlinson, I., and W. Bodmer. 1999. Selection, the mutation rate and cancer: ensuring that the tail does not wag the dog. *Nature Med.* 5:11-12.

Tomlinson, I.P., P. Sasieni, and W. Bodmer. 2002. How many mutations in cancer? *Am. J. Pathol.* 100:755-758.

Tomlinson, I.P.M., M.R. Novelli, and W.F. Bodmer. 1996. The mutation rate and cancer. *Proc. Natl. Acad. Sci. USA.* 93:14800-14803.

Vogelstein, B., E.R. Fearon, S.E. Kern, S.R. Hamilton, A.C. Preisinger, M. Leppert, Y. Nakamura, R. White, A.M.M. Smits, and J.L. Bos. 1988. Genetic alterations during colorectal-tumour development. *N. Engl. J. Med.* 319:525-532.

Wang, T.L., C. Rago, N. Silliman, J. Ptak, S. Markowitz, J.K.V. Willson, G. Parmigiani, K.W. Kinzler, B. Vogelstein, and V.E. Velculescu. 2002. Prevalence of somatic alterations in the colorectal cancer cell genome. *Proc. Natl. Acad. Sci. USA.* 99:3076-3080.

Ware, J.L. 1994. Prostate cancer progression. *Amer. J. Pathol.* 145:983-993.

Yamanaka, S., M.E. Balestra, L.D. Ferrell, J. Fan, K.S. Arnold, S. Taylor, J.M. Taylor, and T.L. Innerarity. 1995. Apolipoprotein B mRNA-editing protein induces hepatocellular carcinoma and dysplasia in transgenic animals. *Proc. Natl. Acad. Sci. USA.* 92:8483-8487.

Chapter 1.2

MONITORING CHROMOSOME REARRANGEMENTS

Michael R. Speicher[1,2]

[1]Institute for Human Genetics, Technical University, Munich, Germany, [2]Institute für Human Genetics and GSF Research Centre for Environment and Health, Neuherberg, Germany

1. INTRODUCTION

It is well established that cells of solid tumours must accumulate a number of mutations in cancer genes to achieve malignant status. In fact, the last century witnessed the confirmation of Theodor Boveri's somatic hypothesis of cancer (Boveri, 1914). Boveri's hypothesis states that acquired mutations in somatic cells are the driving force behind the development of neoplastic tumours. For the majority of epithelial tumours it is believed that these mutations occur over a long time span of several decades. Each mutation has the potential to provoke a clonal expansion resulting in a larger number of cells that may form a substrate for subsequent mutations (Nowell, 1976). Whether normal rates of mutations together with clonal expansions are sufficient to account for the prevalence of cancer or whether some form of genetic instability is required to achieve the needed number of mutations is a matter of debate (Rajagopalan et al., 2003; Sieber et al., 2003). As there is growing evidence that genetic instability can contribute to cancer much research has recently focused on mechanisms leading to such instability. Of particular interest are neoplasias associated with chromosomal instability, which require sophisticated technologies to monitor chromosome rearrangements.

E.A. Nigg (ed.), Genome Instability in Cancer Development, 19-41.

2. THE NEED TO MONITOR CHROMOSOME REARRANGEMENTS: CHROMOSOMAL INSTABILITY AND ALTERATIONS IN TUMOUR CELLS

In colorectal cancer two forms of genetic instability have been identified (Lengauer et al. 1997, 1998). One form, observed in about 15% of colorectal cancers, is characterised by genetic instability at the nucleotide sequence level and has been dubbed microsatellite instability (MIN). MIN is caused by mutations in mismatch repair genes such as *MSH2*, *MLH1*, *MSH6*, or *PMS2* (reviewed by Vogelstein and Kinzler, 2004). However, the vast majority of colorectal cancers that is about 85%, show high rates of chromosome losses and gains, resulting in a dramatic karyotypic variability from cell to cell (Lengauer et al., 1997). This form of genetic instability has been termed chromosomal instability (CIN). CIN appears to be the most common instability in other solid tumours as well (Jallepalli and Lengauer, 2001) and the end result of CIN, aneuploidy, is observed in nearly all solid tumours (Duesberg and Li, 2003).

It was shown that CIN might occur at an early stage in tumorigenesis (Shih et al., 2001). Therefore it has been speculated that aneuploidy may be causative for malignant disease. In fact, the question was raised whether CIN may even be a prerequisite for carcinogenesis. Consequently, an "aneuploidy hypothesis" has evolved, which predicts that each chromosomal copy number change results in a massive alteration in gene dosage and may subsequently represent the primary cause for the genomic instability in neoplastic cells (Duesberg and Li, 2003). In addition, it has been suggested that genetic instability of cancer cells is proportional to their degree of aneuploidy (Duesberg et al., 1998; Li et al., 2000), which would make aneuploidy a driving force in tumorigenesis.

Further evidence that chromosomal changes may be causatively involved in the occurrence of neoplasias stems from the observation that telomere erosion associated with aging has a tremendous impact on chromosomal stability (Artandi et al., 2000; Artandi and DePinho, 2000). A dysfunctional telomere-induced genomic instability model of carcinogenesis was proposed. According to this model, telomere shortening coupled with somatic mutations inactivating retinoblastoma/INK4a/p53 checkpoints, may bypass the Hayflick limit. Continuous proliferation may cause dysfunctional telomeres and subsequent fusion-bridge-breakage cycles which may result in aneuploidy and complex non-reciprocal translocations and thus to cancer (Artandi et al., 2000; Artandi and DePinho, 2000).

At present, both the aneuploidy and the dysfunctional telomere-induced genomic instability models of carcinogenesis raise the possibility that chromosomal changes may be causatively involved in tumorigenesis and not a mere consequence of mutations within a tumour genome. It is interesting that David Hansemann already described mitotic disturbances more than a century ago (Hansemann, 1891). However, the two hypotheses mentioned above have in the past few years resulted in a considerable revitalisation of chromosome analysis in tumour samples, which may have the potential to elucidate the initiating steps in tumorigenesis. In addition, the analysis of tumours at different stages has the potential to unravel changes, which are specifically associated with disease progression. As a consequence, there is an increased demand to efficiently monitor chromosome rearrangements.

Tumour cytogenetics has revealed that karyotypic changes of tumour cells are unevenly distributed throughout the genome. However, different chromosomes, regions, and bands seem to be preferentially involved in the different neoplasia. Thus, a steadily increasing number of abnormalities are found to be associated with particular diseases or disease subtypes. Therefore, detailed description of tumour-genomes by cytogenetic methods may reveal important locations of genes, such as tumour-suppressor genes or oncogenes, which are critically involved in tumour initiation or progression.

However, although solid tumours comprise approximately 95% of all malignancies, they account for only a little over 25% of cases in published cytogenetic studies (Mitelman Database of Chromosome Aberrations in Cancer, 2004). The situation has improved over the last decade, partly as a result of the increased clinical expectations arising from an awareness of the value of cytogenetic studies in leukemias, and partly because of the successful application of fluorescence in situ hybridization (FISH) and molecular techniques. These have permitted studies even of sections from paraffin embedded blocks. As a consequence, the proportion of genetic and cytogenetic publications relating to solid tumours is increasing rapidly.

Despite advances in technologies, tumour genetics and tumour cytogenetics continue to represent demanding challenges. This is because of the dynamic nature of tumours. Cancers generally take decades to develop and during this period the tumour genome is always in a transient state, due to the underlying instability, and thus develops new changes at a frequent rate. Here, technologies will be reviewed which are at present widely used to monitor chromosome rearrangements in tumour cells.

3. MONITORING CHROMOSOME REARRANGEMENTS AND CHROMOSOMAL INSTABILITY OR THE DIFFERENCE BETWEEN "STATE" AND "RATE"

Currently there is a plethora of technologies available for chromosome analyses. However, it is important to keep in mind that there are fundamental differences whether the presence of chromosomal rearrangements or whether chromosomal instability is to be investigated. The existence of genetic alterations in a tumour, such as chromosomal rearrangements, even when frequent, does not mean that the tumour is genetically unstable. Instability is, by definition, a matter of rate, and the existence of a mutation (state) provides no information about the rate of its occurrence (Lengauer et al., 1998). As a consequence, chromosomal instability can only be assessed by single cell approaches, as the variability between different cells has to be analysed on a cell by cell basis. Therefore, technologies, which have the ability to unravel genetic alterations, may differ from approaches more suitable for the analysis of instability. Some technologies are summarised in Figure 1 and will be discussed in more detail in the following.

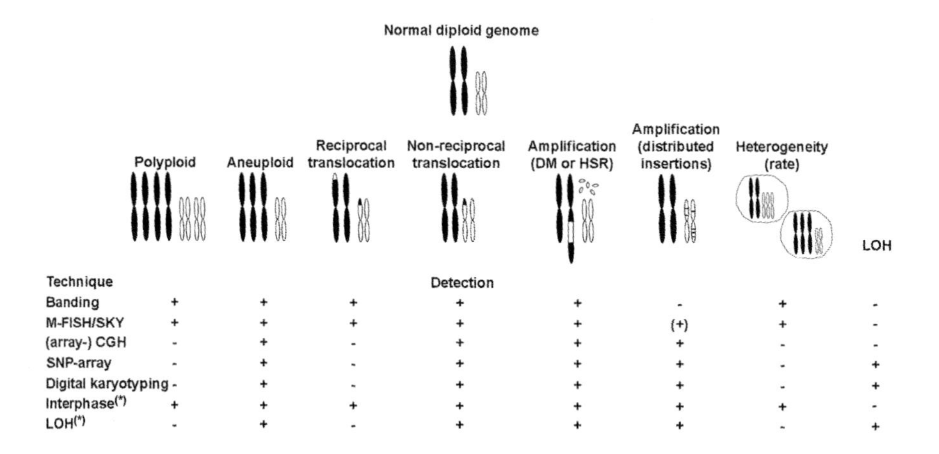

Technique	Polyploid	Aneuploid	Reciprocal translocation	Non-reciprocal translocation	Amplification (DM or HSR)	Amplification (distributed insertions)	Heterogeneity (rate)	LOH
				Detection				
Banding	+	+	+	+	+	-	+	-
M-FISH/SKY	+	+	+	+	+	(+)	+	-
(array-) CGH	-	+	-	+	+	+	-	-
SNP-array	-	+	-	+	+	+	-	+
Digital karyotyping	-	+	-	+	+	+	-	+
Interphase(*)	+	+	+	+	+	+	+	-
LOH(*)	-	+	-	+	+	+	-	+

Figure 1. Illustration of possible chromosomal aberrations in tumours and the respective technologies to detect the aberrations. (*) The detection rates of both interphase cytogenetics and LOH studies depend on probe selection. (This figure was modified from Albertson DG, et al. [2003].)

4. ANALYSIS OF CHROMOSOMAL REARRANGEMENTS IN METAPHASE SPREADS (KARYOTYPING)

4.1 Banding Analysis

Despite their technical difficulties, conventional cytogenetic studies of solid tumours still provide information that cannot be obtained by other means. For several decades chromosomes were analysed by traditional karyotyping, which depends on the analysis of characteristic banding patterns along the length of each chromosome. The major disadvantage of conventional cytogenetic banding methods is the limited resolution. Particularly problematic are the analysis of extensively rearranged chromosomes and marker chromosome identification. Furthermore, in tumour cells karyotyping is impeded by a lack of mitotic tumour cells, the low number of high-quality metaphase spreads and the complex nature of chromosomal changes. Therefore, additional technologies, which are mostly based on fluorescence in situ hybridization (FISH), have been introduced to complement traditional banding analyses.

4.2 FISH with One or Several Different Probes to Metaphase Spreads

In recent years the number of region-specific probes, which allow the specific staining of centromeres, subtelomeres, entire chromosomes or other regions within the genome has been steadily increased. This development of new probe sets was greatly facilitated by the human genome project from which well-characterised probes for any region within the genome have emerged. Furthermore, the evolution of different multicolour fluorescence *in situ* hybridization (FISH) technologies now allows the cohybridization of multiple DNA-probes of different colours.

Initially complex DNA probes proved difficult to use for FISH as they usually contain repeat sequences, such as Alu and LINES, which are present throughout the genome. Thus, the direct use of these probes resulted in a high background level of hybridization due to the repeated sequences in the specimen. As a result, the probe-specific signal was difficult or impossible to see. This problem was solved by a pre-incubation of the denatured complex probe with unlabelled genomic DNA or Cot-1 DNA. This pre-incubation, or pre-annealing, preferentially saturates the repetitive sequences and makes them unavailable for hybridization to the target (Landegent et al., 1987). This procedure has made complex genomic clones available for use with FISH and allowed the first specific staining of individual chromosomes in metaphase spreads and interphase nuclei (Pinkel et al., 1988; Cremer et al., 1988; Lichter et al., 1988). This specific FISH-staining of an individual chromosome was also dubbed "chromosome painting" by Pinkel et al. (1988).

These developments have paved the way for FISH-based automated karyotyping or the simultaneous analysis of multiple defined regions within the genome. Using appropriate instrumentation and image processing, the analysis can be performed two-dimensionally on metaphase spreads or three-dimensionally in intact interphase nuclei. FISH was rapidly introduced for a variety of applications and has been employed in both biology and medicine for karyotype analysis, gene mapping, DNA replication and recombination, clinical diagnosis and disease monitoring, radiation dosimetry, gene transcription and the study of chromatin organisation.

Chromosome paints allow the identification of even subtle chromosome rearrangements without the need for the interpretation of high-resolution chromosome banding by trained personnel. Today most chromosome paints are generated by flow sorting and subsequent PCR amplification of small numbers of flow-sorted chromosomes (reviewed by Langer et al., 2004). This development became possible due to the invention of new, universal

PCR-approaches, which allow unbiased amplification of large DNA-regions. The degenerate oligonucleotide-primed PCR (DOP-PCR) has evolved to a standard amplification procedure for this purpose (Telenius et al., 1992). These PCR-technologies have also paved the way for an alternative strategy for the generation of painting probes, i.e. chromosome microdissection. In this case a micromanipulator is used to scratch a small number of chromosomes from a glass slide. After transfer to an eppendorf-tube and PCR-amplification, microdissected probes have been shown to be highly effective as chromosome paints (Guan et al., 1994). Newer protocols allow the generation of painting probes even from single chromosomes (Gribble et al., 2004a; Thalhammer et al., 2004).

Due to the human genome project there is a growing number of BAC-clones available which were characterised both cytogenetically and on a molecular genetic level (Cheung et al., 2001). Easy to use internet-resources such as Ensembl Cytoview from the EBI and Wellcome Trust Sanger Institute (http://www.ensembl.org/), or the MapViewer (http://www.ncbi.nlm.nih.gov/mapview/) facilitate the selection of BAC-probes. Painting probes, arm-specific probes, chromosome centromere-specific probes, or other probes can be obtained from the University of Bari in Italy (http://www.biologia.uniba.it/rmc/index.html). With these resources available research groups are able to tailor probe sets for their specific needs to resolve complex rearrangements with an unprecedented resolution (Kraus et al., 2003a, 2003b).

4.3 Twenty-four Colour Karyotyping

Initially, painting probes were often used to verify a chromosomal rearrangement which was obvious or suspected in banding analysis. However, the deciphering of complex rearrangements with involvement of unidentifiable chromosomal material is inefficient if only one or two chromosome paints per hybridization are used. Therefore, technologies were developed which allow the simultaneous hybridization of multiple chromosome painting probes in different colours. This development was propelled by the introduction of new, spectrally resolvable fluorochromes, which cover the entire spectrum from the UV to the far infrared range, and highly sensitive imagers such as cooled charge coupled device (CCD) cameras.

The discrimination of many more targets than the number of spectrally resolvable fluorochromes can be achieved using either combinatorial or ratio-labelling strategies. The various labelling strategies and some other multicolour-karyotyping technologies have been recently reviewed (Fauth and Speicher, 2001; Maierhofer et al., 2002). Multicolour technologies have been particularly applied to chromosome painting probes. For a FISH-based karyotyping, the initial requirement was for 24 different colours, as this allows the simultaneous visualisation of all human chromosomes in a single experiment. The fluorochrome-specific optical filter-based multiplex (M-)

FISH (Speicher et al., 1996) and the interferometer-based spectral karyotyping (SKY) (Schröck et al., 1996) both use the combinatorial labelling strategy for probe discrimination. In contrast, combined binary ratio labelling (COBRA; Tanke et al., 1999) employs a mixture of both the combinatorial and the ratio labelling strategy.

The twenty-four colour karyotyping technologies have proven to be powerful tools in clinical cytogenetics and research applications, e.g. in the deciphering of complexly rearranged tumour karyotypes (see published examples by Veldman et al., 1997; Uhrig et al., 1999; Speicher et al., 2000).

As an example, Figure 2 depicts two metaphase spreads from the same non-small cell tumour to illustrate the complexity involved in the analysis of tumour metaphase spreads.

However, these technologies have also some limitations, for example the determination of the exact band origin of a marker or the breakpoints of intrachromosomal rearrangements. Furthermore, the detection sensitivity for small (< 3 Mb) intrachromosomal rearrangements is poor. This is mainly due to an inherent hybridization artifact where the blending of colours by fluorescence flaring at the interface of the translocation segments generates incorrect classification. This problem has already been addressed several times and solutions to diminish it by increasing the number of fluorochromes for probe labelling have been proposed (Azofeifa et al., 2000; Fauth and Speicher, 2001; Jentsch et al., 2003).

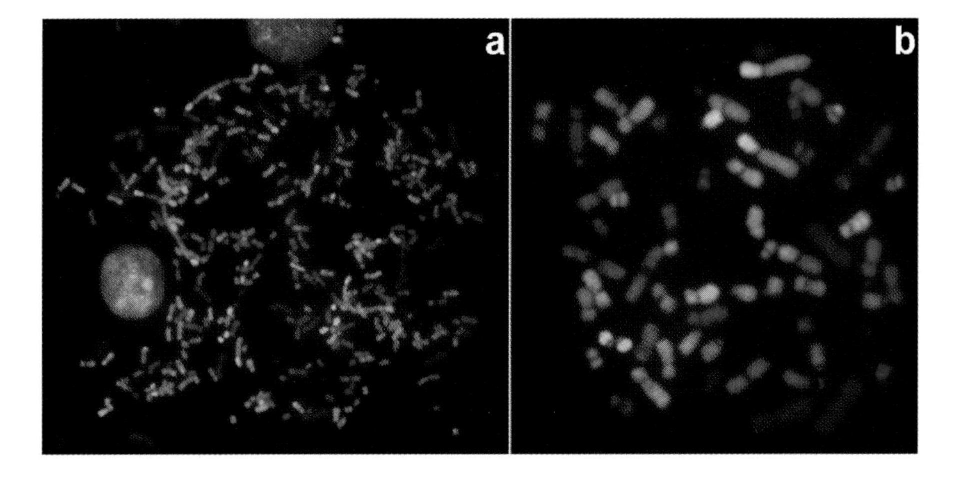

Figure 2. Two metaphase spreads from the same small cell lung cancer sample after M-FISH. The metaphase shown in (a) is in the pentaploid range, the metaphase in (b) in the diploid range.

The continuing development of multicolour technologies have paved the way for discussions concerning the maximum number of probes that can be simultaneously hybridized and how many of these in a single experiment would be useful. For selected applications it has already been demonstrated that more than 24 colours can be generated easily. For example, with both the COBRA (Wiegant et al., 2000) and the M-FISH (Karhu et al., 2001) technologies, 42 colour experiments were realised for the differential painting of all chromosome arms. The simultaneous multicolour-FISH analysis of all chromosomes in human-mouse hybrid cell lines, representing 45-colour experiments (24 colours for the human and 21 colours for the mouse chromosomes) has been reported (Langer et al., 2001b). Theoretically, both the combinatorial and the ratio labelling strategy allow the simultaneous discrimination of even higher numbers of different colours. Thus, the limiting factor will be the hybridization efficiency of multiplex probe sets rather than the combination of colours. In our opinion, the multitude of possible fluorochrome combinations would be better used for strategies to increase the resolution of existing probe sets (as discussed by Azofeifa et al., 2000) than to increase the number of simultaneously hybridized probes further as has been suggested (Müller et al., 2002).

4.4 Other Multicolour Technologies to Metaphase Spreads

Another solution to increase sensitivity is to replace painting probes by region-specific probes. This can again be done in a multiplex-fashion to simultaneously screen multiple regions in a single hybridization. A case in point were this was done quite extensively is the screening of subtelomeric regions. Multicolour FISH assays have been published to simultaneously check the telomere integrity of 8 (Granzow et al., 2000), 12 (Brown et al., 2001) or 24 chromosomes (Fauth et al., 2001; Henegariu et al., 2001). Thus, if regions of interest within a genome are known beforehand, it may be advisable not to use painting probes but rather a set of well-defined region specific probes.

An important limitation of painting probes is their poor sensitivity for the detection of intrachromosomal rearrangements. Deletions or duplications will be detected only if they result in significant size differences of the two homologous chromosomes. Pericentric inversions will be detected only if they result in a considerable displacement of the centromere, which changes the shape of the chromosome. Paracentric inversions are not detectable by painting probes at all.

Therefore, strategies for the generation of FISH based multicolour banding patterns, also dubbed “bar coding” (Lengauer et al., 1993), were developed. To this end, various multicolour banding strategies were developed for individual chromosomes (Chudoba et al., 1999) or for the entire genome (Müller et al., 1997, 1998).

Furthermore, painting probes can be co-hybridized with a variety of other region-specific probes. Depending on the application, painting probes can then, in addition with other region specific probes, yield a detailed description of a chromosome. Other examples include combinations of painting probes for acrocentric chromosomes with their respective centromeric probes and rDNA probes for the analysis of marker chromosomes (Langer et al., 2001a).

4.5 Reverse Painting

Reverse painting is a powerful tool for the analysis of rearranged chromosomes (Carter et al., 1992). The aberrant chromosome is either purified by flow sorting or by microdissection. In the same way as normal chromosome painting probes are generated, the collected aberrant chromosome is amplified using DOP-PCR, labelled, but hybridized onto normal metaphase spreads. The hybridization pattern of the reverse-painting probe elucidates the composition of the aberrant chromosome and furthermore reveals the localisation of breakpoints. The resolution can be increased drastically by hybridization of the reverse painting probes to arrays, a new technology referred to as "array painting" (Fiegler et al., 2003a; Gribble et al., 2004b).

5. ANALYSIS OF CHROMOSOMAL REARRANGEMENTS IN INTERPHASE CELLS (INTERPHASE CYTOGENETICS)

The stronghold of FISH, regardless whether it is performed with a painting probe or a region specific probe, is the option to analyse the genome or part of the genome at single-cell resolution. DNA analyses at single cell resolution should be instrumental for the understanding of cancer cell biology, cancer evolution, for chromosomal mosaic analysis and rare cell events (e.g. minimal residual disease). It is well established that during carcinogenesis cells in the tumour tissue become increasingly heterogeneous and disorganised in their structural properties. In order to understand the underlying molecular mechanisms it is often necessary to analyse the cells individually and within their natural tissue context. Advanced single cell tools should facilitate the identification of genetically different tissues. Rare cell events, e.g. early pathological lesions in neoplasia or minimal residual disease, should greatly benefit from such an analysis of individual cells within their natural tissue context, e.g. within a tissue section. Furthermore, such an analysis even allows a detailed phenotype-genotype correlation which cannot be achieved by another technology and should provide otherwise inaccessible information on essential biological processes. Interphase cytogenetics represents the most important technology to provide this information.

Advantages of interphase FISH over metaphase FISH include that a large number of cells can be scored thereby increasing the likelihood of detecting chromosomal aberrations present at a low-level or mosaic state. For example, interphase FISH has evolved to the method of choice for the analysis of CIN, i.e. for the analysis of the "rate" of chromosomal changes.

In contrast, as metaphase analysis is done rather on smaller cell samples, metaphase analysis is the preferred method to establish the state of chromosomal rearrangements. Therefore, interphase cytogenetics is considered to be the most efficient molecular cytogenetic tool to assess the instability rate. Furthermore, interphase FISH allows the detection of chromosomal aberrations preferentially present in uncultured cells. Painting probes applied to interphase nuclei have the disadvantage that they yield large signals, which often overlap. Therefore, the accurate scoring of signals may be a difficult and tedious task. For this reason, region-specific probes, such as YAC- or BAC-clones are preferable for most applications in interphase cytogenetics.

Most interphase studies, when done with cell suspensions require a special preparation to preserve the three-dimensional (3D) structure of cell nuclei. Protocols for the preparation of 3D preserved cell nuclei include the avoidance of especially harsh treatments (e.g. hypotonic treatment), prevention of drying up of cells, special fixation steps (e.g. use of 4% paraformaldehyde/0.3 x PBS instead of unbuffered formalin), permeabilisation steps (e.g. treatment in 0.5% Triton X-100 in PBS, 20% glycerol in PBS, repeated freezing/thawing in liquid nitrogen, mild pepsin digestion) and a special storage of slides. A detailed description of 3D-preservation protocols was summarised by Solovei et al. (2002).

Current techniques for 3D microscopy generate 3D data by "optical sectioning" of the specimen. Up to date most 3D fluorescence microscopy is done using confocal microscopy, widely regarded as gold standard. If confocal microscopy is being used, nuclei are scanned with a defined axial distance (often in the range of 200 nm) between light optical sections. If a multi-channel laser scanning confocal microscope (e.g. a three-channel Zeiss LSM 410) is employed, optical section images are sequentially collected for each fluorochrome. This results in stacks of 8-bit grey scale 2D images. These confocal images are further processed with appropriate software and tools are available for 3D reconstructions of image stacks. Multiple examples of such 3D-images generated using confocal microscopy have been published (e.g. Cremer et al., 2003; Chin et al., 2004; Argilla et al., 2004; Bolzer et al., 2005). Confocal 3D-microscopy has been used to elucidate both the topology of chromosome territories in tumour cells and also to analyse the timing of telomere crisis during tumorigenesis and the significance of telomerase activity in developing tumours.

The study of chromosome topology appears to be especially promising as there is increasing evidence that the 3D organisation of the genome has an essential impact on nuclear functions such as replication, gene expression and gene silencing. This is mirrored by the fact that nuclei of a defined cell type are generally characterised by a distinct morphology and chromatin texture, suggesting a cell type specific three-dimensional higher order chromatin architecture, which can change during malignant transformation.

In the studies aimed at unravelling the topology of chromosome territories in tumour cells, it was established that a gene density-correlated radial arrangement of chromosome territories represents a common feature present in most cell types. Surprisingly, such a gene density-correlated radial arrangement was also found in tumour cells, irrespective of chromosomal rearrangements and imbalances. Although the gene density-correlated radial arrangement was attenuated in tumour cell nuclei as compared to normal cell nuclei, these findings suggest a basic stability of large-scale genome architecture in malignant cells (Cremer et al., 2003). However, at present it is an unanswered question whether a premalignant state or tumour evolution involve local or large-scale reorganisation of the interphase genome.

Evidence for a role of 3D topology in initiation stems from recent studies, which demonstrated that in human lymphocytes oncogene loci have different spatial proximities, which may explain the occurrence of translocations or other recurrent structural aberrations commonly seen in malignant diseases (Roix et al., 2004; reviewed by Misteli, 2004). Under the assumption of a spatially sensitive difference between a low and high probability for a translocation, one might expect that already small-scale changes in the intranuclear chromatin arrangement can lead to profound cytogenetic events, such as structural or numerical rearrangements. These chromosomal changes, in turn, may result in dramatic changes in the gene expression profile of the respective cell, which may trigger the malignant transformation.

Confocal 3D-microscopy has also been used for the analysis of telomerase activity in tumour tissues. The dysfunctional telomere-induced genomic instability model of carcinogenesis predicts that during malignant transformation cells have to transit through telomere crisis. Using histologically defined breast cancer samples and appropriate 3D-analysis tools, it was recently shown that transition through telomere crisis and immortalisation in breast cancer may occur during progression from usual ductal hyperplasia to ductal carcinoma in situ (Chin et al., 2004). 3D-studies also helped to unravel that, when viral oncoproteins commandeer multiple cellular functions, including the Rb and p53 tumour suppressors, tumorigenesis may not be so critically dependent on telomerase activity (Argilla et al., 2004). These examples demonstrate that fundamental processes in tumorigenesis can now be directly analysed within the natural tissue context on tumour samples.

Deconvolution microscopy represents an alternative to confocal microscopy. Deconvolution refers to a wide-field image restoration by computational methods used to reduce out-of-focus fluorescence in 3D microscopy images. Deconvolution microscopy facilitates the simultaneous use of multiple different colour channels for the analysis of multiple probes. This approach employs a normal epifluorescence microscope with a motorised stage and represents therefore, compared to confocal microscopy, a low-cost alternative. A DNA counterstain is used for volume labelling of the nuclei offering the opportunity for a simultaneous segmentation of nuclei. Deconvolution microscopy has already been used for multicolour

FISH applications, which were applied for imaging deep into specimens, such as thick (30 μm) paraffin-embedded tissue sections. This allowed the detection of portions within the same tumour sample with different chromosomal patterns and various degrees of chromosomal instability (Maierhofer et al., 2003). Only an interphase-based technology may contribute to the identification of areas with different chromosomal and instability patterns and opens avenues for new strategies for an improved characterisation of tumour tissue section. This approach is in particular most promising for the analysis of CIN, e.g. the rate of instability, directly in a tissue section. Figure 3 exemplifies the use of 3D deconvolution microscopy on neuroblastoma sections. Different cell populations, with or without *N-myc* amplification, can clearly be distinguished.

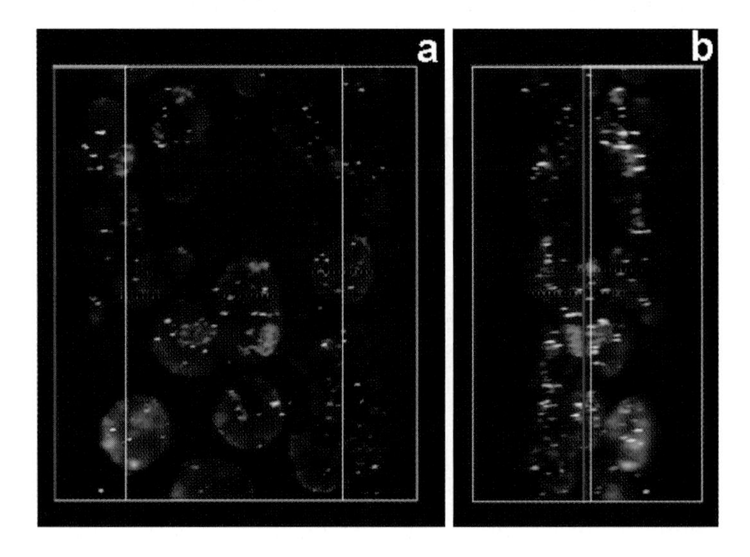

Figure 3. Application of 3D-deconvolution microscopy to a neuroblastoma tissue section. Four differently labelled probes for chromosomal regions 1p36 (green), 2p24 (red), 11q23 (pink), and 17q25 (yellow) were hybridized. The 2p24 probe covers the *N-myc* gene, which is frequently amplified in neuroblastoma, often as double minute (DM) chromosomes or as homogenously staining regions (HSRs). In some nuclei, the *N-myc* amplification is clearly visible as a large red region, which has a similar appearance as a chromosome territory. This hybridization pattern indicates that the *N-myc* amplification is present in form of a homogenously staining region. At the same time it is obvious that the tissue is heterogeneous as some cells do not show a *N-myc* amplification. The X-Y view is shown in (a), the Y-Z view (i.e. tilting of the section by 90°) is shown in (b).

The use of painting probes has allowed invaluable insights into the three-dimensional organisation of the interphase genome. Painting probes have provided direct visual evidence that chromosomes are organised into territories, a concept, which had been proposed previously, based on evidence from microirradiated nuclei (Cremer et al., 1982). Combined with multicolour technologies, the simultaneous assessment of multiple chromosome territories and their relation to each other is now possible. This has resulted in new models about the higher-order genome organisation. The emerging view is that chromosomes are compartmentalised into discreet territories and that the position of a chromosome may influence gene regulation. Current views about the three-dimensional organisation of the genome in interphase nuclei have been extensively reviewed recently (Cremer and Cremer, 2001).

Of particular importance, deconvolution microscopy has been used for the simultaneous 24-colour painting of all human chromosome types (22 autosomes, X and Y) in three-dimensionally (3D)-preserved prometaphase rosettes and postmitotic (G0) nuclei from male human fibroblasts (Bolzer et al., 2005). The application of 24-colour interphase FISH allowed to establish complete 3D maps of higher order chromosome arrangements in fibroblasts. The data demonstrated a probabilistic, yet highly non-random chromosome size correlated radial pattern: Small chromosomes show a high probability for their location towards the centre of the nucleus or rosette, while large chromosomes are generally positioned towards the nuclear rim. This pattern deviates strongly from the gene density correlated radial chromosome arrangements previously reported for lymphocytes and several other human cell types (Bolzer et al., 2005).

6. ANALYSIS OF CHROMOSOMAL REARRANGEMENTS BY COMPARATIVE GENOMIC HYBRIDIZATION (CGH)

The majority of human cancers as well as many developmental abnormalities harbour chromosomal imbalances, many of which result in the gain and/or loss of genomic material. Comparative genomic hybridization (CGH), a technique that detects and maps changes in copy number of DNA sequences, has been widely used for the analysis of tumour genomes and constitutional chromosomal aberrations since it was first reported (Kallioniemi et al., 1992; du Manoir et al., 1993). CGH provides a genome-wide screening for chromosomal imbalance in a single hybridization directly from DNA samples without requiring the sample material to be mitotically active. In CGH, DNA from a test (e.g. tumour) and a reference genome (genomic DNA from a normal individual) are differentially labelled and both have been traditionally hybridized to metaphase chromosome spreads. Hybridization of repetitive sequences is blocked by the addition of Cot-1 DNA. The fluorescence ratio of the test and reference hybridization signals is determined at different positions along each chromosome and provides information on the relative copy number of sequences in the test genome compared with a normal diploid genome. CGH describes the state but not the rate of chromosomal changes in a tumour genome.

CGH does not detect balanced chromosome rearrangements. Furthermore, CGH does not provide any information regarding the way in which chromosome segments involved in gains and losses are arranged in marker chromosomes of the test genome (Figure 4a-c). Chromosomal imbalances can only be detected if they are present in the majority of cells.

One recent exciting development is the CGH analysis of individual cells (Klein et al., 1999; Wells et al., 1999; Voullaire et al., 1999), which has revealed new insights into genomic copy number changes of individual disseminated cells (often referred to as micrometastasis) (Klein et al., 2002; Schmidt-Kittler et al., 2003; Gangnus et al., 2004). This approach is even feasible on cells on which previously interphase-FISH experiments have been performed (Langer et al., 2005).

Conventional CGH has been used extensively to map DNA copy number changes to chromosomal positions. Overviews over existing CGH-data can easily be retrieved from existing publicly accessible databases (e.g. www.progenetix.net, www.ncbi.nlm.nih.gov/sky/skyweb.cgi, amba.charite.de/cgh/). In the past few years, microarray-based formats for CGH (matrix or array CGH) have been reported and are beginning to be widely used in preference to chromosome-based CGH. As discussed below, arrays made from large genomic clones and cDNAs have been used most often for this purpose.

7. ANALYSIS OF CHROMOSOMAL REARRANGEMENTS BY ARRAY BASED TECHNOLOGIES

The introduction of microarray CGH provided a powerful tool to precisely detect and quantify genomic aberrations and map these directly onto the sequence of the human genome. In fact, some argue that the rapidly evolving matrix-/array-comparative genomic hybridization (CGH) technologies (Solinas-Toldo et al., 1997; Pinkel et al., 1998; Snjiders et al., 2001; Fiegler et al., 2003b; reviewed by Albertson and Pinkel, 2003) will replace chromosome analysis. The transfer of the CGH-technology from metaphase spreads to arrays provides a number of advantages over the use of chromosomes, including higher resolution, direct mapping of aberrations to the genome sequence and higher throughput. Furthermore, the evaluation of fluorescence intensities on arrays is amenable to automated analysis making array CGH-based *in vitro* diagnostic devices possible. However, similar to chromosome CGH the array technologies do not allow the identification of balanced structural changes and remain technologies for the assessment of the state of copy number changes in a tumour genome. A comparison between chromosome and array-CGH is depicted in Fig. 4d-e.

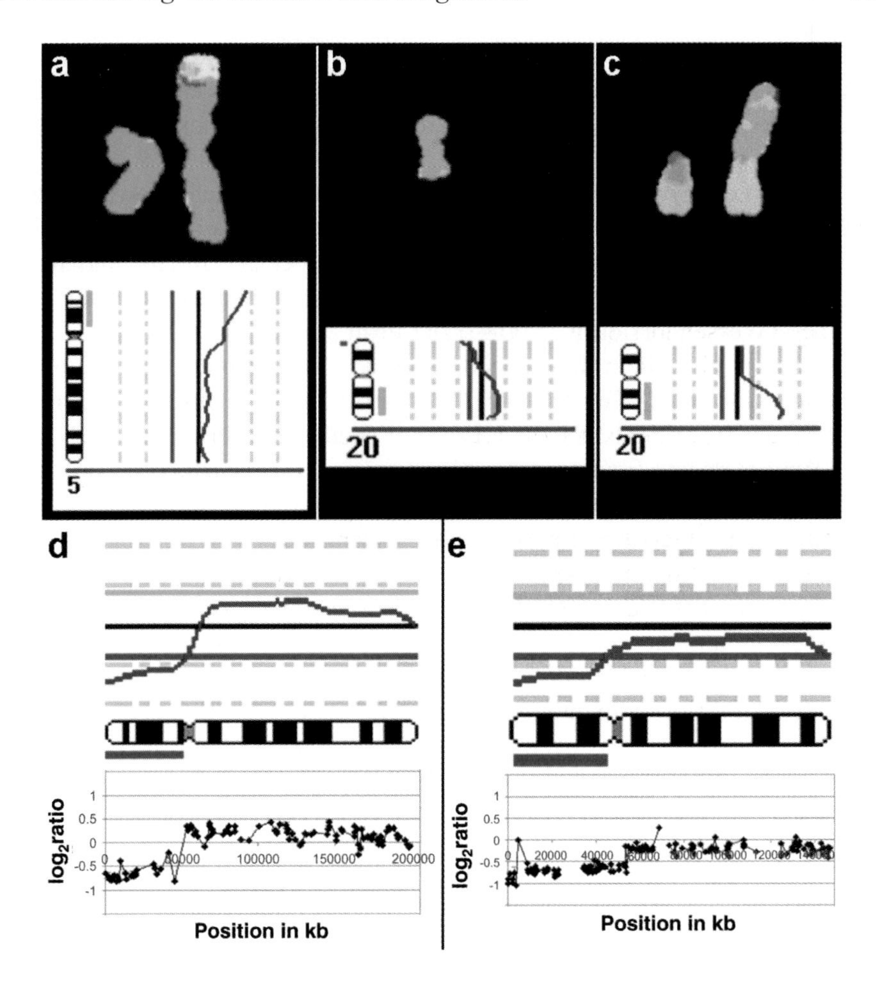

Figure 4. Comparison between M-FISH and chromosome CGH results (a-c) and between chromosome and array-CGH (d-e). (a-c) Three examples are shown to illustrate that CGH does not provide any information regarding the way in which chromosome segments involved in gains and losses are arranged in marker chromosomes of the test genome. The data were obtained with mammary fibroblasts after transfection with hTERT and SV40 (for details see Fauth et al., 2004) (a) CGH indicates a gain of 5p-material. This is often attributed to the presence of an i(5p). M-FISH revealed that this 5p overrepresentation was caused by a der(5)dup(5)t(5;11). (b-c) In both examples CGH shows an overrepresentation of 20q. In (b) this is the result of an i(20q), however in (c), the underlying cause is a complex rearrangement due to a der(14;20)dup(20). (d-e) Array CGH results are indicated in a co-ordinate system illustrating, on the abscissa, the position of the clones in kb according to the draft genomic sequence and on the ordinate, the mean log_2 ratios of the triplicate spots normalised to the median log_2 ratio for the genome (for details, see Snijders et al., 2001). Chromosome and array-CGH obtained with DNA from a prostate cancer cell line are shown for chromosomes 5 (d) and 8 (e) (for further details see Kraus et al., 2003c).

For applications involving the analysis of tumours, the technology must provide reliable detection of single copy gains and losses in mixed cell populations such as tumour and normal cells, accurate quantification of high level copy number gains, and confident interpretation of aberrations affecting only a single array element. Further, one would like to minimise the amount of specimen material required for an analysis. These requirements can be met if there are good signal-to-noise ratios in the hybridization.

In the past several years, a number of different approaches towards array-based CGH have been undertaken. Genome-wide array-CGH for human (Snijders et al., 2001; Cai et al., 2002; Fiegler et al., 2003b) and mouse (Hodgson et al., 2001) and arrays specifically designed for certain disease entities (Wessendorf et al., 2002; Fritz et al., 2002; Wilhelm et al., 2002; Veltman et al., 2002) have been reported. Furthermore, a number of platforms for array CGH have been described and have used large genomic clones such as cosmids, P1s and BACs or smaller clones such as cDNAs as the array elements (reviewed by Albertson and Pinkel, 2003). The use of DNA from large insert clones (e.g., BACs) provides substantially more intense signals than use of smaller clones such as cDNAs. Therefore, BAC-clones have –in theory- the potential for better performance to detect single copy gains and losses. However, the accurate assessment of copy number changes in tumour genomes using oligonucleotide arrays has also been reported. High-density single nucleotide polymorphism (SNP) arrays have been used to detect copy number alterations. By combining genotyping with SNP quantitation, it was even possible to distinguish loss of heterozygosity events caused by hemizygous deletion from those that occur by copy-neutral events (Zhao et al., 2004).

8. FISH FOR THE ANALYSIS OF TUMOR ANIMAL MODELS

The mouse has evolved to the primary mammalian genetic model organism. One of the most important applications includes the modelling of human cancer. Many of these models are associated with chromosomal aberrations and a detailed high-resolution chromosome analysis is an important cornerstone of the analyses.

Flow sorting has also been extensively used to generate painting probes for species other than man. This offers new opportunities for analyzing chromosomes of distant species, which often don't have chromosomal morphological landmarks, such as size differences or characteristic positions of the centromeres, which facilitate the analysis of human chromosomes. If applied in a multicolour-fashion, colour-coding can assist in karyotyping the complete chromosome set of other species. This has already been broadly demonstrated for the karyotype of the mouse (Liyanage et al., 1996; Jentsch et al., 2000, 2003). For example, mouse chromosome analysis assisted in understanding the role of p53 in Brca1-associated tumorigenesis (Xu et al., 1999), uncovered the interplay between p53 and telomere dysfunction in promoting genetic instability (Artandi et al., 2000), and elucidated the function of the DNA repair protein Ku80 as a caretaker gene (Difilippantonio et al., 2000).

9. MOLECULAR GENETIC STUDIES

Molecular genetic studies of isolated tumour DNA have been successful and have been used to detect common regions of allelic loss, mutation, or amplification. However, such molecular methods are highly focused; they target one specific gene or chromosome region at a time and leave the majority of the genome unexamined. Nevertheless, molecular genetic studies have the potential to confirm loss of DNA as seen in a CGH experiment, i.e. through the assessment of loss of heterozygosity (LOH). As LOH studies yield information with a better resolution, they have the potential of pinpointing directly putatively involved genes.

Digital PCR provides an approach for the identification of predefined mutations expected to be present in a minor fraction of a cell population (Vogelstein and Kinzler, 1999). At the molecular level, chromosomal instability is characterised by allelic imbalance (AI), representing losses or gains of defined chromosomal regions and, therefore digital PCR can be used to assess chromosomal stability. An impressive application of this approach was the demonstration that genetic instability may occur very early during colorectal neoplasia (Shih et al., 2001).

Another approach is digital karyotyping, that provides quantitative analysis of DNA copy number at high resolution (Wang et al., 2002). This approach involves the isolation and enumeration of short sequence tags from specific genomic loci. Analysis of human cancer cells by using this method identifies gross chromosomal changes as well as amplifications and deletions, including regions not previously known to be altered. Thus, similar to CGH, digital karyotyping provides a broadly applicable means for systematic detection of DNA copy number changes on a genomic scale. Currently it is not known which approach, array-/matrix-CGH or digital karyotyping, will have the better resolution.

10. FUTURE DEVELOPMENTS

In summary, chromosome analyses and FISH have boomed considerably in the last few years. In part, new efficient tools for multicolour-FISH caused this upsurge as it has broadened the range of applications considerably. A stronghold of FISH is its potential for single-cell analysis, which makes it an indispensable tool whenever aberrations have to be unravelled which occur only in a subset of cells. The molecular cytogenetic toolbox has recently been further complemented by new probe generations, such as peptide nucleic acid probes (Taneja et al., 2001), new detection systems, such as rolling circle amplification (Lizardi et al., 1998; Zhong et al., 2001), or new fluorescent probes such as semiconductor nanocrystals (Bruchez et al., 1998; Lacoste et al., 2000). Furthermore, cytogenetic data will be integrated with functional data, e.g. gene expression analyses obtained from microarrays (Geigl et al., 2004). Thus, it will be exciting to see how the expanding arsenal of molecular cytogenetic methods will contribute to the monitoring of chromosome rearrangements in the future.

ACKNOWLEDGEMENTS

Research in my laboratory is supported by the Bundesministerium für Bildung und Forschung (NGFN 2nd phase), the Deutsche Krebshilfe, the Wilhelm Sander Stiftung and the Deutsche Forschungsgemeinschaft. The images shown in Figure 3 were kindly provided by Dr. Isabell Jentsch.

REFERENCES

Albertson DG, Collins C, McCormick F, Gray JW (2003) Chromosome aberrations in solid tumours. *Nat. Genet.* 34:369-376.

Albertson DG, Pinkel D (2003) Genomic microarrays in human genetic disease and cancer. *Hum. Mol. Genet.* 12 Suppl 2:R145-52.

Argilla D, Chin K, Singh M, Hodgson JG, Bosenberg M, de Solorzano CO, Lockett S, DePinho RA, Gray J, Hanahan D (2004) Absence of telomerase and shortened telomeres have minimal effects on skin and pancreatic carcinogenesis elicited by viral oncogenes. *Cancer Cell* 6:373-385.

Artandi SE, Chang S, Lee SL, Alson S, Gottlieb GJ, Chin L, DePinho RA (2000) Telomere dysfunction promotes non-reciprocal translocations and epithelial cancers in mice. *Nature* 406:641-645.

Artandi SE, DePinho RA (2000) Mice without telomerase: what can they teach us about human cancer? *Nat. Med.* 6:852-855.

Azofeifa J, Fauth C, Kraus J, Maierhofer C, Langer S, Bolzer A, Reichman J, Schuffenhauer S, Speicher MR (2000) An optimized probe set for the detection of small interchromosomal aberrations by 24-color FISH. *Am. J. Hum. Genet.* 66:1684-1688.

Bolzer A, Kreth G, Solovei I, Koehler D, Saracoglu K, Fauth C, Müller S, Eils R, Cremer C, Speicher MR, Cremer T (2005) Non-random, probabilistic chromosome arrangements in human male fibroblast nuclei and prometaphase rosettes. *PLoS Biology*. In press.

Boveri T (1914) Zur Frage der Entstehung maligner Tumoren. Jena: *Verlag von Gustav Fischer*

Brown J, Saracoglu K, Uhrig S, Speicher MR, Eils R, Kearney L (2001) Subtelomeric chromosome rearrangements are detected using an innovative 12-colour FISH assay (M-TEL). *Nat. Med.* 7:497-501.

Bruchez M. Jr, Moronne M, Gin P, Weiss S, Alivisatos AP (1998) Semiconductor nanocrystals as fluorescent biological labels. *Science* 281:2013-2016.

Cai WW, Mao JH, Chow CW, Damani S, Balmain A, Bradley A (2002) Genome-wide detection of chromosomal imbalances in tumours using BAC microarrays. *Nat. Biotechnol.* 20:393-396.

Carter NP, Ferguson-Smith MA, Perryman MT, Telenius H, Pelmear AH, Leversha MA, Glancy MT, Wood SL, Cook K, Dyson HM, et al. (1992) Reverse chromosome painting: a method for the rapid analysis of aberrant chromosomes in clinical cytogenetics. *J. Med. Genet.* 29:299-307.

Cheung VG, Nowak N, Jang W, Kirsch IR, Zhao S, Chen XN, Furey TS, Kim UJ, Kuo WL, Olivier M, Conroy J, Kasprzyk A, Massa H, Yonescu R, Sait S, Thoreen C, Snijders A, Lemyre E, Bailey JA, Bruzel A, Burrill WD, Clegg SM, Collins S, Dhami P, Friedman C, Han CS, Herrick S, Lee J, Ligon AH, Lowry S, Morley M, Narasimhan S, Osoegawa K, Peng Z, Plazjer-Frick I, Quade BJ, Scott D, Sirotkin K, Thorpe AA, Gray JW, Hudson J, Pinkel D, Ried T, Rowen L, Shen-Ong GL, Strausberg RL, Birney E, Callen DF, Cheng JF, Cox DR, Doggett NA, Carter NP, Eichler EE, Haussler D, Korenberg JR, Morton CC, Albertson D, Schuler G, De Jong PJ, Trask BJ (2001) Integration of cytogenetic landmarks into the draft sequence of the human genome. *Nature* 409:953-958.

Chin K, de Solorzano CO, Knowles D, Jones A, Chou W, Rodriguez EG, Kuo WL, Ljung BM, Chew K, Myambo K, Miranda M, Krig S, Garbe J, Stampfer M, Yaswen P, Gray JW, Lockett SJ (2004) In situ analyses of genome instability in breast cancer. *Nat. Genet.* 36:984-988.

Chudoba I, Plesch A, Lörch T, Lemke J, Claussen U, Senger G (1999) High resolution multicolour-banding: a new technique for refined FISH analysis of human chromosomes. *Cytogenet. Cell Genet.* 84:156-160.

Cremer M, Küpper K, Wagler B, Wizelman L, v. Hase J, Weiland Y, Kreka L, Diebold J, Speicher MR, Cremer T (2003) Inheritence of gene-density related higher order chromatin arrangements in normal and tumour cell nuclei. *J. Cell Biol* 162:809-820.

Cremer T, Cremer C, Baumann H, Luedtke EK, Sperling K, Teuber V, Zorn C (1982) Rabl's model of the interphase chromosome arrangement tested in Chinese hamster cells by premature chromosome condensation and laser-UV-microbeam experiments. *Hum. Genet.* 60:46-56.

Cremer T, Lichter P, Borden J, Ward DC, Manuelidis L (1988) Detection of chromosome aberrations in metaphase and interphase tumor cells by in situ hybridization using chromosome-specific library probes. *Hum. Genet.* 80:235-246.

Cremer T, Cremer C (2001) Chromosome territories, nuclear architecture and gene regulation in mammalian cells. *Nat. Rev. Genet.* 2:292-301.

Difilippantonio MJ, Zhu J, Chen HT, Meffre E, Nussenzweig MC, Max EE, Ried T, Nussenzweig A (2000) DNA repair protein Ku80 suppresses chromosomal aberrations and malignant transformation. *Nature* 404:510-514.

Duesberg P, Rausch C, Rasnick D, Hehlmann R (1998) Genetic instability of cancer cells is proportional to their degree of aneuploidy. *Proc. Natl. Acad. Sci. USA.* 95:13692-13697.

Duesberg P, Li R (2003) Multistep carcinogenesis: a chain reaction of aneuploidizations. *Cell Cycle* 2:202-210.

du Manoir S, Speicher MR, Joos S, Schröck E, Popp S, Döhner H, Kovacs G, Robert-Nicoud M, Lichter P, Cremer T (1993) Detection of complete and partial chromosome gains and losses by comparative genomic in situ hybridization. *Hum. Genet.* 90:590-610.

Fauth C, Speicher MR (2001) Classifying by colours: FISH-based genome analysis. *Cytogenet. Cell Genet.* 93:1-10.

Fauth C, Zhang H, Harabacz S, Brown J, Saracoglu K, Lederer G, Rittinger O, Rost I, Eils R, Kearney L, Speicher MR (2001) A new strategy for the detection of subtelomeric rearrangements. *Hum. Genet.* 109:576-583.

Fauth C, O'Hare MJ, Lederer G, Jat PS, Speicher MR (2004) The order of genetic events determines critically aberrations in chromosome count and structure. *Genes, Chromosomes Cancer* 40:298-306.

Fiegler H, Gribble SM, Burford DC, Carr P, Prigmore E, Porter KM, Clegg S, Crolla JA, Dennis NR, Jacobs P, Carter NP (2003a) Array painting: a method for the rapid analysis of aberrant chromosomes using DNA microarrays. *J. Med. Genet.* 40:664-670.

Fiegler H, Carr P, Douglas EJ, Burford DC, Hunt S, Scott CE, Smith J, Vetrie D, Gorman P, Tomlinson IP, Carter NP (2003b) DNA microarrays for comparative genomic hybridization based on DOP-PCR amplification of BAC and PAC clones. *Genes Chromosomes Cancer* 36:361-374.

Fritz B, Schubert F, Wrobel G, Schwaenen C, Wessendorf S, Nessling M, Korz C, Rieker RJ, Montgomery K, Kucherlapati R, Mechtersheimer G, Eils R, Joos S, Lichter P (2002)

Microarray-based Copy Number and Expression Profiling in Dedifferentiated and Pleomorphic Liposarcoma. *Cancer Res.* 62:2993-2998.

Gangnus R, Langer S, Breit S, Pantel K, Speicher MR (2004) Genomic profiling of viable and proliferative micrometastatic cells from early stage breast cancer patients. *Clin. Cancer Res.* 10:3457-3464.

Geigl JB, Langer S, Barwisch S, Pfleghaar K, Lederer G, Speicher MR (2004) Analysis of gene expression patterns and chromosomal changes associated with aging. *Cancer Res.* 64:8550-8557.

Granzow M, Popp S, Keller M, Holtgreve-Grez H, Brough M, Schoell B, Rauterberg-Ruland I, Hager HD, Tariverdian G, Jauch A (2000) Multiplex FISH telomere integrity assay identifies an unbalanced cryptic translocation der (5)t(3;5)(q27;p15.3) in a family with three mentally retarted individuals. *Hum Genet* 107:51-57.

Gribble S, Ng BL, Prigmore E, Burford DC, Carter NP (2004a) Chromosome paints from single copies of chromosomes. *Chromosome Res.* 12:143-151.

Gribble SM, Fiegler H, Burford DC, Prigmore E, Yang F, Carr P, Ng BL, Sun T, Kamberov ES, Makarov VL, Langmore JP, Carter NP (2004b) Applications of combined DNA microarray and chromosome sorting technologies. *Chromosome Res.* 12:35-43.

Guan XY, Meltzer PS, Trent JM (1994) Rapid generation of whole chromosome painting probes (WCPs) by chromosome microdissection. *Genomics* 22:101-107.

Hansemann D (1891) Über pathologische Mitosen. *Arch. Pathol. Anat. Phys. Klein. Med.* 119:299-326.

Henegariu O, Artan S, Greally JM, Chen XN, Korenberg JR, Vance GH, Stubbs L, Bray-Ward P, Ward DC (2001) Cryptic translocation identification in human and mouse using several telomeric multiplex FISH (TM-FISH) strategies. *Lab. Invest.* 81:483-491.

Hodgson G, Hager JH, Volik S, Hariono S, Wernick M, Moore D, Nowak N, Albertson DG, Pinkel D, Collins C, Hanahan D, Gray JW (2001) Genome scanning with array CGH delineates regional alterations in mouse islet carcinomas. *Nat. Genet.* 29:459-464.

Jallepalli PV, Lengauer C (2001) Chromosome segregation and cancer: cutting through the mystery. *Nat. Rev. Cancer* 1:109-117.

Jentsch I, Adler ID, Carter NP, Speicher MR (2001) Karyotyping mouse chromosomes by multiplex-FISH (M-FISH). *Chromosome Res.* 9:211-214.

Jentsch I, Geigl J, Klein CA, Speicher MR (2003) Seven fluorochrome mouse M-FISH for high resolution analysis of interchromosomal rearrangements. *Cytogenet. Genome Res.* 103:84-88.

Kallioniemi A, Kallioniemi OP, Sudar D, Rutovitz D, Gray JW, Waldman F, Pinkel D (1992) Comparative genomic hybridization for molecular cytogenetic analysis of solid tumors. *Science* 258:818-821.

Karhu R, Ahlstedt-Soini M, Bittner M, Meltzer P, Trent JM, Isola JJ (2001) Chromosome arm-specific multicolour-FISH. *Genes Chromosomes Cancer* 30:105-109.

Klein CA, Schmidt-Kittler O, Schardt JA, Pantel K, Speicher MR, Riethmüller G. (1999) Comparative genomic hybridization, loss of heterozygosity, and DNA sequence analysis of single cells. *Proc. Natl. Acad. Sci. USA.* 96:4494-4499.

Klein CA, Blankenstein TJ, Schmidt-Kittler O, Petronio M, Polzer B, Stoecklein NH, Riethmuller G (2002) Genetic heterogeneity of single disseminated tumour cells in minimal residual cancer. *Lancet* 360:683-689.

Kraus J, Lederer G, Keri C, Seidel H, Rost I, Wirtz A, Fauth C, Speicher MR (2003a) A familial unbalanced subtelomeric translocation resulting in a monosomy 6q27→qter. *J. Med. Genet.* 40:e48.

Kraus J, Cohen M, Speicher MR (2003b) Multicolour-FISH fine-mapping unravels an insertion as a complex chromosomal rearrangement involving 6 breakpoints and a 5.89 Mb large deletion. *J. Med. Genet.* 40:e60.

Kraus J, Pantel K, Pinkel D, Albertson DG, Speicher MR (2003c) High-resolution genomic profiling of occult micrometastatic tumour cells. *Genes Chromosomes Cancer* 36:159-166.

Lacoste TD, Michalet X, Pinaud F, Chemla DS, Alivisatos AP, Weiss S (2000) Ultrahigh-resolution multicolour colocalization of single fluorescent probes. *Proc. Natl. Acad. Sci. USA.* 97:9461-9466.

Landegent JE, Jansen in de Wal N, Dirks RW, Baas F, van der Ploeg M (1987) Use of whole cosmid cloned genomic sequences for chromosomal localization by non-radioactive in situ hybridization. *Hum. Genet.* 77:366-370.

Langer S, Fauth C, Rocchi M, Murken J, Speicher MR (2001a) AcroM-FISH analyses of marker chromosomes. *Hum. Genet.* 109:152-158.

Langer S, Jentsch I, Gangnus R, Yan H, Lengauer C, Speicher MR (2001b) Facilitating haplotype analysis by fully automated analysis of all chromosomes in human-mouse hybrid cell lines. *Cytogenet. Cell Genet.* 93:11-15.

Langer S, Kraus J, Jentsch I, Speicher MR (2004) Multicolour chromosome painting in diagnostic and research applications. *Chromosome Res.* 12:15-23.

Langer S, Geigl JB, Gangnus R, Speicher MR (2005) Sequential application of interphase-FISH and CGH to single cells. *Lab. Invest.* In press.

Lengauer C, Speicher MR, Popp S, Jauch A, Taniwaki M, Nagaraja R, Riethman HC, Donis-Keller H, d'Urso M, Schlessinger D, Cremer T (1993) Chromosomal bar codes constructed by fluorescence in situ hybridization with *Alu*-PCR products of multiple YAC clones. *Hum. Mol. Genet.* 2:505–512.

Lengauer C, Kinzler KW, Vogelstein B (1997) Genetic instability in colorectal cancers. *Nature* 386:623-627.

Lengauer C, Kinzler KW, Vogelstein B (1998) Genetic instabilities in human cancers. *Nature* 396:643-649.

Li R, Sonik A, Stindl R, Rasnick D, Duesberg P (2000) Aneuploidy vs. gene mutation hypothesis of cancer: recent study claims mutation but is found to support aneuploidy. *Proc. Natl. Acad. Sci. USA.* 97:3236-3241.

Lichter P, Cremer T, Borden J, Manuelidis L, Ward DC (1988) Delineation of individual human chromosomes in metaphase and interphase cells by in situ suppression hybridization using recombinant DNA libraries. *Hum. Genet.* 80:224-234.

Liyanage M, Coleman A, du Manoir S, Veldman T, McCormack S, Dickson RB, Barlow C, Wynshaw-Boris A, Janz S, Wienberg J, Ferguson-Smith MA, Schröck E, Ried T (1996) Multicolour spectral karyotyping of mouse chromosomes. *Nat. Genet.* 14:312-315.

Lizardi PM, Huang X, Zhu Z, Bray-Ward P, Thomas DC, Ward DC (1998) Mutation detection and single-molecule counting using isothermal rolling-circle amplification. *Nat. Genet.* 19:225-232.

Maierhofer C, Jentsch I, Lederer G, Fauth C, Speicher MR (2002) Multicolour-FISH in two and three dimensions for clastogenic analyses. *Mutagenesis* 17:523-527.

Maierhofer C, Gangnus R, Diebold J, Speicher MR (2003) Multicolour deconvolution microscopy of thick biological specimens. *Am. J. Path.* 162:373-379.

Misteli T (2004) Spatial positioning; a new dimension in genome function. *Cell* 119:153-156.

Mitelman Database of Chromosome Aberrations in Cancer (2004). Mitelman F, Johansson B and Mertens F (Eds.), http://cgap.nci.nih.gov/Chromosomes/Mitelman

Müller S, Neusser M, Wienberg J (2002) Towards unlimited colours for fluorescence in-situ hybridization (FISH). *Chromosome Res.* 10:223-232.

Müller S, Rocchi M, Ferguson-Smith MA, Wienberg J (1997) Toward a multicolour chromosome bar code for the entire human karyotype by fluorescence in situ hybridization. *Hum. Genet.* 100:271-278.

Müller S, O'Brien PC, Ferguson-Smith MA, Wienberg J (1998) Cross-species colour segmenting: a novel tool in human karyotype analysis. *Cytometry* 33:445-452.

Nowell PC (1976) The clonal evolution of tumour cell populations. *Science* 194:23-28.

Pinkel D, Landegent J, Collins C, Fuscoe J, Segraves R, Lucas J, Gray JW (1988) Fluorescence in situ hybridization with human chromosome-specific libraries: detection of trisomy 21 and translocations of chromosome 4. *Proc. Natl. Acad. Sci. USA.* 85:9138-9142.

Pinkel D, Segraves R, Sudar D, Clark S, Poole I, Kowbel D, Collins C, Kuo WL, Chen C, Zhai Y, Dairkee SH, Ljung BM, Gray JW, Albertson DG (1998) High resolution analysis of DNA copy number variations using comparative genomic hybridization to microarrays. *Nat. Genet.* 20:207-211.

Rajagopalan H, Nowak MA, Vogelstein B, Lengauer C (2003) The significance of unstable chromosomes in colorectal cancer. *Nat. Rev. Cancer* 3:695-701.

Roix JJ, McQueen PG, Munson PJ, Parada LA, Misteli T (2004) Spatial proximity of translocation-prone gene loci in human lymphomas. *Nat. Genet.* 34:287-291.

Schmidt-Kittler O, Ragg T, Daskalakis A, Granzow M, Ahr A, Blankenstein TJ, Kaufmann M, Diebold J, Arnholdt H, Müller P, Bischoff J, Harich D, Schlimok G, Riethmüller G, Eils R, Klein CA (2003) From latent disseminated cells to overt metastasis: genetic analysis of systemic breast cancer progression. *Proc. Natl. Acad. Sci. USA.* 100:7737-7742.

Schröck E, du Manoir S, Veldman T, Schoell B, Wienberg J, Ferguson-Smith MA, Ning Y, Ledbetter DH, Bar-Am I, Soenksen D, Garini Y, Ried T (1996) Multicolour spectral karyotyping of human chromosomes. *Science* 273:494–497.

Shih IM, Zhou W, Goodman SN, Lengauer C, Kinzler KW, Vogelstein B (2001) Evidence that genetic instability occurs at an early stage of colorectal tumorigenesis. *Cancer Res.* 61:818-822.

Sieber OM, Heinimann K, Tomlinson IP (2003) Genomic instability--the engine of tumorigenesis? *Nat. Rev. Cancer.* 3:701-708.

Snijders AM, Nowak N, Segraves R, Blackwood S, Brown N, Conroy J, Hamilton G, Hindle AK, Huey B, Kimura K, Law S, Myambo K, Palmer J, Ylstra B, Yue JP, Gray JW, Jain AN, Pinkel D, Albertson DG (2001) Assembly of microarrays for genome-wide measurement of DNA copy number. *Nat. Genet.* 29:263-264.

Solinas-Toldo S, Lampel S, Stilgenbauer S, Nickolenko J, Benner A, Döhner H, Cremer T, Lichter P (1997) Matrix-based comparative genomic hybridization: biochips to screen for genomic imbalances. *Genes Chromosomes Cancer* 20:399-407.

Solovei I, Walter J, Cremer M, Habermann F, Schermelleh L, Cremer T (2002) FISH on three-dimensionally preserved nuclei. *In* FISH: a practical approach. J. Squire, B. Beatty, and S. Mai, editors. Oxford University Press, Oxford. 119-157.

Speicher MR, Ballard SG, Ward DC (1996) Karyotyping human chromosomes by combinatorial multi-fluor FISH. *Nat. Genet.* 12:368–375.

Speicher MR, Petersen S, Uhrig S, Jentsch I, Fauth C, Eils R, Petersen I (2000) Analysis of chromosomal alterations in non-small cell lung cancer by multiplex-FISH, comparative genomic hybridization, and multicolour bar coding. *Lab. Invest.* 80:1031-1041.

Taneja KL, Chavez EA, Coull J, Lansdorp PM (2001) Multicolour fluorescence in situ hybridization with peptide nucleic acid probes for enumeration of specific chromosomes in human cells. *Genes Chromosomes Cancer* 30:57-63.

Tanke HJ, Wiegant J, van Gijlswijk RPM, Bezrookove V, Pattenier H, Heetebrij RJ, Talman EG, Raap AK, Vrolijk (1999) New strategy for multi-colour fluorescence in situ hybridization: COBRA: COmbined Binary RAtio labelling. *Eur. J. Hum. Genet.* 7:2-11.

Telenius H, Pelmear AH, Tunnacliffe A, Carter NP, Behmel A, Ferguson-Smith MA, Nordenskjöld M, Pfragner R, Ponder BAJ (1992) Cytogenetic analysis by chromosome painting using DOP-PCR amplified flow-sorted chromosomes. *Genes Chromosomes Cancer* 4:257–263.

Thalhammer S, Langer S, Speicher MR, Heckl WM, Geigl JB (2004) Generation of chromosome painting probes from single chromosomes by laser microdissection and linker-adaptor PCR. *Chromosome Res.* 12:337-343.

Uhrig S, Schuffenhauer S, Fauth C, Wirtz A, Daumer-Haas C, Apacik C, Cohen M, Müller-Navia J, Cremer T, Murken J, Speicher MR (1999) Multiplex-FISH (M-FISH) for pre- and postnatal diagnostic applications. *Am. J. Hum. Genet.* 65:448-462.

Veldman T, Vignon C, Schröck E, Rowley JD, Ried T (1997) Hidden chromosome abnormalities in haematological malignancies detected by multicolour spectral karyotyping. *Nat. Genet.* 15:406-410.

Veltman JA, Schoenmakers EF, Eussen BH, Janssen I, Merkx G, van Cleef B, van Ravenswaaij CM, Brunner HG, Smeets D, van Kessel AG (2002) High-throughput analysis of subtelomeric chromosome rearrangements by use of array-based comparative genomic hybridization. *Am. J. Hum. Genet.* 70:1269-1276.

Vogelstein B, Kinzler KW (1999) Digital PCR. *Proc. Natl. Acad. Sci. USA.* 96:9236-9241.

Vogelstein B, Kinzler KW (2004) Cancer genes and the pathways they control. *Nat. Med.* 10:789-799.

Voullaire L, Wilton L, Slater H, Williamson R (1999) Detection of aneuploidy in single cells using comparative genomic hybridization. *Prenat. Diagn.* 19:846-851.

Wang TL, Maierhofer C, Speicher MR, Lengauer C, Vogelstein B, Kinzler KW, Velculescu VE (2002) Digital karyotyping. *Proc. Natl. Acad. Sci. USA.* 99:16156-16161.

Wells D, Sherlock JK, Handyside AH, Delhanty JD (1999) Detailed chromosomal and molecular genetic analysis of single cells by whole genome amplification and comparative genomic hybridization. *Nucleic Acids Res.* 27:1214-1218.

Wessendorf S, Fritz B, Wrobel G, Nessling M, Lampel S, Goettel D, Kuepper M, Joos S, Hopman T, Kokocinski F, Dohner H, Bentz M, Schwaenen C, Lichter P (2002) Automated screening for genomic imbalances using matrix-based comparative genomic hybridization. *Lab. Invest.* 82:47-60.

Wiegant J, Bezrookove V, Rosenberg C, Tanke HJ, Raap AK, Zhang H, Bittner M, Trent JM, Meltzer P (2000) Differentially painting human chromosome arms with combined binary ratio-labelling fluorescence in situ hybridization. *Genome Res.* 10:861-865.

Wilhelm M, Veltman JA, Olshen AB, Jain AN, Moore DH, Presti JC Jr, Kovacs G, Waldman FM (2002) Array-based comparative genomic hybridization for the differential diagnosis of renal cell cancer. *Cancer Res.* 62:957-960.

Xu X, Wagner KW, Larson D, Weaver Z, Li C, Ried T, Hennighausen L, Wynshaw-Boris A, Deng CX (1999) Conditional mutation of Brca1 in mammary epithelial cells results in blunted ductal morphogenesis and tumour formation. *Nat. Genet.* 22:37-43.

Zhao X, Li C, Paez JG, Chin K, Janne PA, Chen TH, Girard L, Minna J, Christiani D, Leo C, Gray JW, Sellers WR, Meyerson M (2004) An integrated view of copy number and allelic alterations in the cancer genome using single nucleotide polymorphism arrays. *Cancer Res.* 64:3060-3071.

Zhong XB, Lizardi PM, Huang XH, Bray-Ward PL, Ward DC (2001) Visualization of oligonucleotide probes and point mutations in interphase nuclei and DNA fibers using rolling circle DNA amplification. *Proc. Natl. Acad. Sci. USA.* 98:3940-3945.

Part 2

DNA Repair and Mutagenesis

Chapter 2.1

NUCLEOTIDE EXCISION REPAIR AND ITS CONNECTION WITH CANCER AND AGEING

Jaan-Olle Andressoo, Jan H.J. Hoeijmakers and Harm de Waard
MGC Department of Cell Biology and Genetics, Center for Biomedical Genetics, Erasmus Medical Center, Rotterdam, The Netherlands.

1. INTRODUCTION

Based on the instructions for life encrypted in the nucleotide sequence of the DNA all cellular components such as RNA, proteins, and indirectly through them all organelles, lipids, membranes and metabolites are continuously turned over, folded, remodelled and otherwise modified. However, DNA itself is irreplaceable, and therefore damage to DNA, if not repaired, can have serious consequences. It may interfere with the regular DNA metabolism leading to permanent changes in the genetic code and inborn defects in subsequent generations when occurring in the germ line or to cancer when affecting oncogenes or tumour suppressor genes in the soma. Alternatively, injury to our genes can hamper vital cellular processes such as transcription and replication and thereby -as argued in this chapter- strongly contribute to the onset of ageing-related pathology. DNA is under continuous attack by a plethora of damaging agents from exogenous and even more importantly also from endogenous origin. Well known genotoxic agents are UV, the short wave component of sunlight, X-rays and numerous chemicals as well as highly reactive compounds derived from cellular oxidative metabolism such as reactive oxygen species (ROS). To avoid the deleterious consequences of genetic damage the DNA of all organisms also encodes an intricate network of specialised and partially interwoven DNA repair pathways that is tightly integrated in a highly sophisticated genome care-taking apparatus. Many of these genome guardians are addressed in this book. In this chapter we primarily focus on the nucleotide excision repair (NER) pathway, which is primarily specialised in removing lesions that distort the helical structure, such as most bulky adducts and intrastrand crosslinks. In addition it may assist in the elimination of the highly

E.A. Nigg (ed.), Genome Instability in Cancer Development, 45-83.

cytotoxic interstrand crosslinks. The versatility and importance of NER is reflected by the astounding diversity of associated inborn pathologies which range from more than ~1000-times elevated cancer risk to dramatically accelerated segmental ageing, but paradoxically at the same time protection from cancer. In the first part of the chapter, we discuss the core reaction mechanism of NER and its implications for other repair pathways and cellular processes. Next, we comparatively analyze NER-associated disorders and provide clues to the causative etiology. In the final section we will attempt to integrate etiology of the different syndromes and the underlying molecular mechanisms in terms of the associated cancer susceptibility and the various accelerated ageing phenotypes in light of the most recent spectacular progress and development of novel concepts in the field.

1.1 Consequences of DNA damage

The most well known sources of DNA damage are exogenous physical and chemical agents, such as UV, ionizing radiation and toxious chemicals in food, or e.g. as pollution in the air, including cigarette smoke. However, an other continuous stress to our genome is exerted by our own respiration: the mitochondrial respiratory chain generates ATP using electron transport starting from NADH but at the same time produces various reactive oxygen species (ROS) in every cell of our body. In addition, NADPH-oxidases, cyclooxygenases and lipid peroxidation generate ROS (Balaban et al., 2005), whereas also other reactive chemical species, including nitric oxides, and alkylating species, are produced by the cellular metabolism. The magnitude of the endogenous source of damage is generally undervalued, but conservative estimates indicate that each day up to 50.000 lesions are spontaneously inflicted upon our genome in every nucleus (Lindahl, 1993). The vital importance of basic DNA repair mechanisms, including NER, to maintain the genomic integrity is underscored by their extreme conservation from unicellular bacteria and yeast to man (Hoeijmakers, 2001). As highlighted in other chapters of this book, defects in DNA repair systems are either embryonic lethal or result in severe pathology later in life. In the few exceptions to this rule there is generally redundancy stressing the vital importance of DNA repair mechanisms particularly in mammals in view of their massive genome. The effects of genetic injury on the cellular level range from accumulation of mutations, being the prime cause of cancer, to premature cellular senescence and cell death, which can contribute to pathological ageing (Campisi, 2005; Friedberg et al., 1995; Hasty et al., 2003; Hoeijmakers, 2001). By way of overview Figure 1A depicts the major classes of DNA damage and the corresponding repair pathways involved in removal of these lesions. The severe physiological consequences of defects in these processes are depicted in Figure 1B.

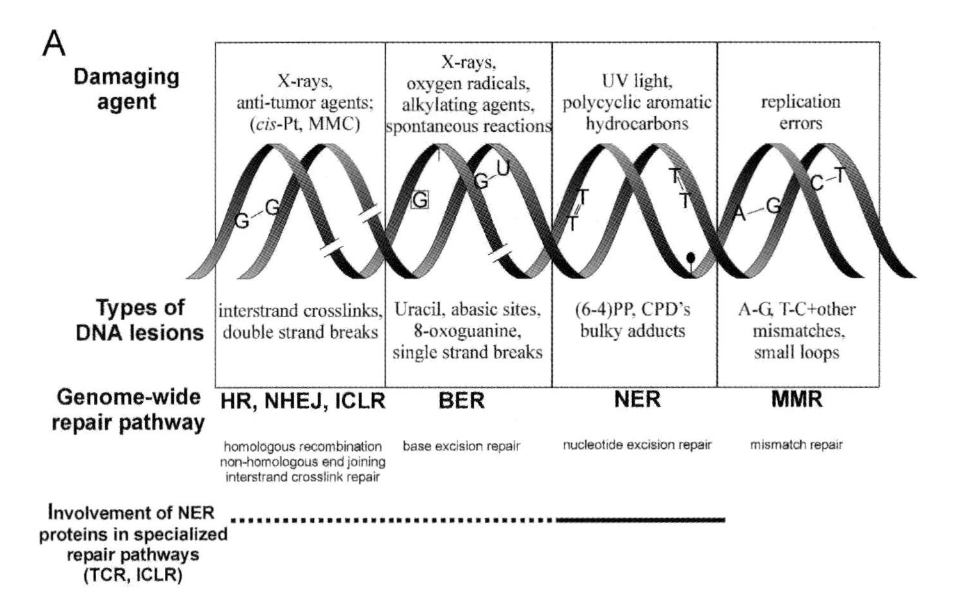

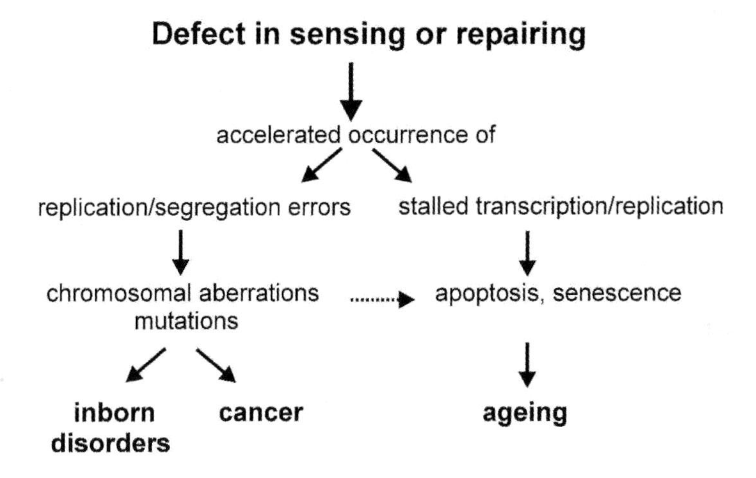

Figure 1. **A.** The major sources of DNA damage. TCR-general transcription-coupled repair, ICLR-interstrand crosslink repair, HR-homologous recombination, NHEJ-nonhomologous endjoining, BER-base excision repair, NER-nucleotide excision repair, MMR-mismatch repair, CisPt-cisplatin, MMC-mitomycin C, CPD-cyclobutane pyrimidine dimer, (6-4)PP-6-4 pyrimidine-pyrimidone photoproduct. Dotted line indicates the mechanistically yet poorly understood involvement of some NER proteins in other or more general repair trails. **B.** The major consequences of DNA damage and defective repair at the molecular and biomedical level. Dotted arrow: significance of connection uncertain.

Research over the last decade has revealed that different DNA repair pathways share common factors and/or act in a highly concerted fashion.

One such a repair pathway intimately interwoven with other repair systems is transcription coupled repair (TCR), which is one of the main topics of this chapter. In contrast to specialised repair mechanisms that continuously scan the entire genome for specific lesions, TCR deals with the unique problem caused by a lesion in the transcribed strand that physically blocks an elongating RNA-polymerase. Obviously, it is of urgent importance to the cell to resolve this problem, as there is an apparent need to get the encoded gene product synthesised. This stalled transcription machinery has to be either removed or has to backtrack from the lesion in order to make the damaged base(s) accessible for repair, after which transcription can proceed. For example, within the context of TCR certain NER proteins, such as CSB and XPG, can facilitate removal of DNA lesions normally repaired by base excision repair (BER) (Cooper et al., 1997; de Waard et al., 2003). As discussed below the TCR pathway is of key importance for preventing early death from endogenously generated DNA damage. Additional examples of functional links between NER and other cellular processes such as other repair systems and their clinical impact will be discussed.

2. MECHANISM OF NER AND TCR

The mechanism of nucleotide excision repair can be divided in 2 subpathways: the global genome NER (GG-NER), that operates genome wide and is able to identify helix-distorting lesions anywhere in the genome and transcription-coupled NER (TC-NER), which focuses on NER-type damage that has escaped the GG-NER system and that instead has been detected by an elongating RNA polymerase and physically blocks the vital process of transcription. As argued above there is reason to believe that the latter repair process also deals with other non-NER type damage that interferes with progression of transcription elongation.

2.1 Global genome NER

The first identified step in global genome NER (GG-NER) is damage recognition by the heterodimer XPC/hHR23B (Riedl et al., 2003; Sugasawa et al., 1998; Volker et al., 2001), which operates genome-wide and binds with higher affinity to helix-distorting lesions or partially unwound normal DNA structures than to intact double stranded DNA (dsDNA) (Figure 2)(Winkler et al., 2001; Wood, 1999). Since damage recognition is highly dependent on the degree of DNA helix distortion, DNA lesions that only mildly disturb the helical structure are poorly recognised by XPC/hHR23B and as a consequence are inefficiently repaired by GG-NER. One important example of such type of damage is the most abundant UV-induced lesion, the cyclobutane pyrimidine dimer (CPD). However, it is likely that other examples of this category of DNA damage exist that hitherto have escaped detection (Jaspers et al., 2002). The persistence of this type of damage may have major physiological consequences as again exemplified by CPD lesions. Because of their slow repair these lesions turn out to be responsible for the vast majority of deleterious effects exerted by the UV component of sunlight, including mutagenesis and the consequent long term carcinogenesis as well as acute sunburn and immunosuppression exerted by UV (Jans et al., 2005). For this reason the -albeit slow- GG-NER of this photolesion in humans is enabled by the activity of a specific protein complex: the damaged DNA binding complex (DDB) (Hwang et al., 1999; Tan and Chu, 2002; Tang et al., 2000), composed of the DDB1 (damaged DNA binding protein 1, p125) and DDB2 proteins (damaged DNA binding protein 2, p48, XPE). This XPE-DDB-complex binds to CPDs (Hwang and Chu, 1993; Payne and Chu, 1994; Reardon et al., 1993) and induces bending of the DNA (Fujiwara et al., 1999), presumably highlighting this lesion for GG-NER. It has been shown that this complex acts before XPC and that its presence is necessary for XPC to bind to these lesions (Fitch et al., 2003; Wakasugi et al., 2002). In contrast, the other major, although less abundant, UV product, 6-4-photoproduct (6-4PP) causes significant helix distortion (Gunz et al., 1996) and is thereby efficiently recognised by XPC-hHR23B, without the need of the XPE-DDB-complex (Hwang et al., 1998). Interestingly, in rodents, in contrast to primates, expression of the p48 subunit of XPE-DDB is not upregulated upon UV (Hwang et al., 1999; Tan and Chu, 2002), probably due to lack of a p53 responsive element in the *p48-XPE* promoter (Tan and Chu, 2002) and therefore, rodents poorly repair CPDs by GG-NER. Similarly, XPC is upregulated by UV in a p53-dependent manner, as determined both on the RNA and the protein level (Adimoolam and Ford, 2002; Amundson et al., 2002; Ng et al., 2003). Thus, the damage recognition step is likely the rate-limiting factor in the GG-NER reaction. In relation to this it should be noted that the above proteins appear under tight control and that the window of up-regulating the critical DNA damage recognition factors XPC and XPE (and thereby the entire pathway) may be limited due to adverse effects of having too much of these strong DNA binding proteins around. For instance, stable overexpression of the XPC or its yeast equivalent RAD4 even in heterologous systems such as *E.coli* appears to be lethal ((Ng et al., 2003) and references therein). This limits the prospects of improving the GG-NER machinery without interfering with other vital DNA transactions.

Subsequent to damage recognition, the 10-subunit transcription factor TFIIH (Giglia-Mari et al., 2004) and presumably also at this stage the structure-specific endonuclease XPG are recruited to the lesion (de Laat et al., 1999; Riedl et al., 2003; Volker et al., 2001; Yokoi et al., 2000). TFIIH contains the XPB and XPD helicases, that locally open the DNA helix around the lesion (Drapkin et al., 1994; Evans et al., 1997b; Schaeffer et al., 1994; Schaeffer et al., 1993b; Winkler et al., 2000). This reaction is different from the multiple actions of TFIIH in transcription initiation (including promoter opening and phosphorylation of RNA polymerase II C-terminal domain and a number of transcription activators) (Drane et al.,

2004; Egly, 2001), since specific dephosphorylation of XPB is needed (Coin et al., 2004). The role of XPG in this step is likely to be structural, stabilising the open DNA helix (Evans et al., 1997b; Mu et al., 1996). After the XPC complex has left the scene (Hwang et al., 1998; Riedl et al., 2003; You et al., 2003), the damage is verified and located by XPA (Rademakers et al., 2003; Volker et al., 2001), who likely properly organises the incision machinery around a damage by positioning the single-strand binding protein, replication protein A (RPA) to bind the non-damaged strand. At the same time this prevents reannealing of the two single strands and may help stabilise the open NER reaction intermediate (de Laat et al., 1998; Evans et al., 1997a; Li et al., 1995; Stigger et al., 1998; You et al., 2003). Finally, the ERCC1/XPF endonuclease assembles and together with XPG, cleave 5' and 3' of the lesion respectively in the damaged strand only, thereby excising a 24-32 nt single stranded DNA (ssDNA) fragment containing the DNA damage (Mu et al., 1996; O'Donovan et al., 1994; Sijbers et al., 1996). Using the undamaged strand as a template, filling of the ssDNA gap is performed by repair replication presumably using the regular DNA replication machinery, consisting of RPA, proliferating cell nuclear antigen (PCNA), a sliding clamp, replication factor C (RFC), the clamp loader, and likely DNA polymerase δ and ε (Aboussekhra et al., 1995; Budd and Campbell, 1995; Shivji et al., 1995). These factors replace the recognition/incision NER machinery that has excised the damaged piece of DNA. Finally, the resulting nick is sealed by DNA ligase I (Araujo et al., 2000; Barnes et al., 1992) (Figure 2). The entire GG-NER reaction takes *in vivo* ~ 4-5 minutes as determined by photobleaching methods of NER factors tagged in living cells with green-fluorescent proteins (Hoogstraten et al., 2002).

2.2 Transcription-coupled repair

Damage in the transcribed strand of active genes can arrest the transcription machinery, and in effect inactivate an entire gene copy. Stalled RNA polymerases appear to constitute a potent signal for p53-dependent and -independent apoptosis and thus may be highly cytotoxic triggering cell death (Conforti et al., 2000; Ljungman and Lane, 2004; Ljungman and Zhang, 1996; van den Boom et al., 2002; Yamaizumi and Sugano, 1994). To avoid such a consequence and allow resumption of transcription, the cell is equipped with the transcription-coupled NER (TC-NER) system (Bohr et al., 1985; Mellon et al., 1986; Mellon et al., 1987). Since damage recognition in TC-NER does not depend on helix distortion, but instead on blockage of RNA polymerase II (Figure 2), the spectrum of lesions recognised by TC-NER and GG-NER is thought to differ. For example, whereas CPDs are poorly (and in rodents almost not at all) repaired by GG-NER, these persisting lesions efficiently block transcription and accordingly are selectively eliminated from the transcribed strand of active genes by TC-NER, when detected by an elongating RNA polymerase (Bohr et al., 1985).

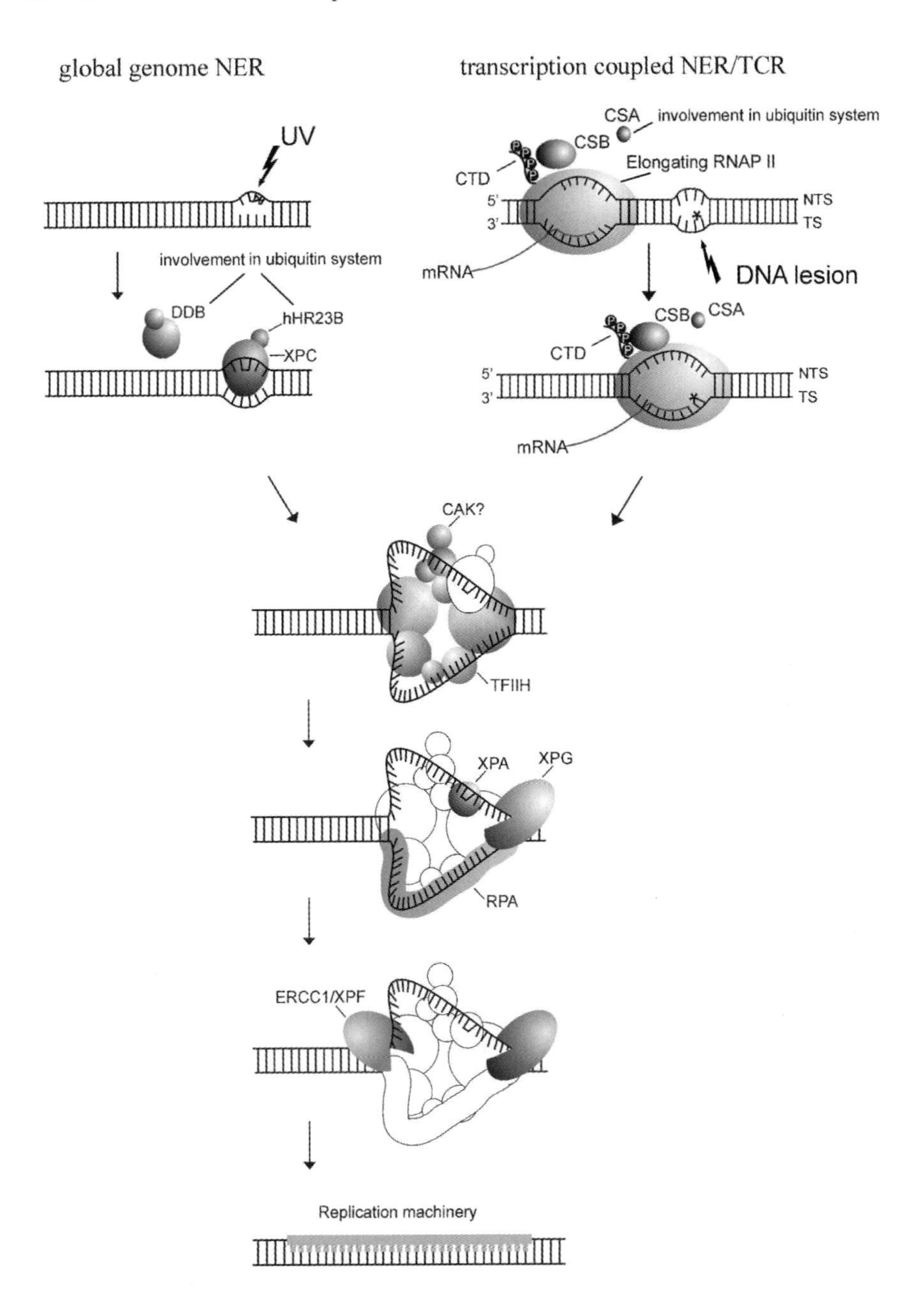

Figure 2. NER reaction. After damage recognition step by hHR23B/XPC (for some lesions facilitated by the XPE-UV-DDB complex) or elongating RNA-PolII respectively, the GG-NER and TC-NER pathways utilise the common core NER reaction, which involves recruitment of TFIIH and XPA followed by melting of the DNA around the lesion and excision by structure-specific endonucleases XPG at 3' and XPF/ERCC1 at 5' side of the lesion respectively; and gap-filling and ligation by the replication machinery. The role of proteins in NER indicated with a question mark is uncertain. For more detailed explanation see text.

The actual repair of the lesion that is responsible for the arrest of transcription is initiated by the CSA and CSB proteins, which facilitate the assembly of the core NER reaction (Figure 2) involving TFIIH, XPG, XPA, XPF/ERCC1 and the subsequent repair replication factors. Consequently, patients carrying completely inactivating mutations in the *CSA, CSB* or *XPA* genes are devoid of TC-NER, whereas mutations in TFIIH and *XPG* or *XPF* are frequently not completely inactivating and therefore cause partial defects. As a consequence, cells from those patients fail to recover RNA synthesis after induction of UV damage, consistent with the idea that the transcription block remains. In view of the notion that the clinical features of XPA patients (which are completely deficient in both GG- as well as TC-NER) are much milder compared to e.g. CSA, CSB mutations, (which cause defective TC-NER but allow still normal GG-NER), it has been put forward that mutations in the latter proteins result in defective transcription-coupled repair of all kinds of transcription-arresting DNA lesions (denoted here as TCR). This much broader spectrum of DNA damage addressed by TCR would include not only helix-distorting lesions classically repaired by NER but likely also other transcription-interfering lesions, such as a number of the ROS-induced DNA injuries. It has been suggested that CSA, CSB, as well as TFIIH and XPG participate in either the removal or back-tracking of the stalled RNA-Pol from the damaged site in order to make the lesion accessible to other repair factors. Whether the reaction is then completed by the remaining core NER machinery, including XPA, RPA and ERCC1/XPF, independent of the lesion type, or whether the injury is open to repair by the corresponding specific GG-repair machinery, i.e. GG-NER or GG-BER is still unknown. This will be more extensively discussed in the last section of this chapter, when we have presented the severe clinical consequences of defects in the various NER pathways.

Several recent lines of evidence point to a major involvement of the ubiquitin system in both the GG-NER and the TCR reaction (Groisman et al., 2003; Ng et al., 2003). Finally, it should be noted that the GG-NER system is relatively well understood due to the establishment of an *in vitro* reaction (Sibghat-Ullah et al., 1989; Wood et al., 1988) and reconstitution using recombinant factors (Aboussekhra et al., 1995). However, all efforts to set up an equivalent cell-free assay for the presumably more complex TC-NER mechanism have met with little success. Consequently, the initial steps in this reaction are not known.

2.3 Multifunctionality of NER/TCR proteins

A theme that becomes more and more a rule rather than the exception is the notion that NER proteins tend to have additional roles outside of the context of the NER reaction itself. The 10-subunit TFIIH complex was originally purified and characterised as a basal transcription initiation factor required for promoter opening and phosphorylation of RNA polymerase II, enabling transcription of all structural genes (Egly, 2001). Since the discovery of its distinct roles in this process as well as in GG-NER and TC (-NE)R its multifunctionality has been complemented with documented involvements in activated RNA polymerase II transcription as a protein kinase for a series of transcription activators (Drane et al., 2004; Keriel et al., 2002) as well as in basal transcription initiation of rRNA genes by RNA polymerase I (Hoogstraten et al., 2002; Iben et al., 2002). In view of these findings it may be anticipated that the complex is implicated in RNA polymerase III transcription as well. The identification of the CAK subcomplex of TFIIH carrying the cdk7, cyclin H and MAT1 proteins opens the additional possibility that the complex has a still unknown role in cell cycle regulation (Egly, 2001; Harper and Elledge, 1998). Another example of a NER factor with multiple engagements is the ERCC1/XPF complex. Cell biological and molecular studies advocate a role of this structure-specific endonuclease in removal of interstrand cross links via homologous recombination and perhaps also in double strand break repair ((Niedernhofer et al., 2001) and references therein). Obviously, also the single strand binding protein RPA has multiple essential functions at least in DNA replication, homologous recombination and other repair processes such as long patch BER and likely in interstrand crosslink repair (ICLR). Evidence for multiple repair functions has also been reported for the XPG endonuclease such as in stimulating repair of some BER lesions (Dianov et al., 2000). Thus, most NER factors are also employed for other purposes, which complicates the clinical picture when inborn defects occur in the corresponding genes. This will become apparent in the next sections.

3. NER-ASSOCIATED DISORDERS

Because of the overriding importance of NER for the removal of sunlight (UV)-induced DNA lesions, inborn NER disorders are recognised because of a striking photosensitivity, making this the hallmark feature of all NER syndromes. In view of the enhanced cytotoxic effects of UV in these patients frequently (depending on the degree of solar exposure) accelerated photoageing of the skin is noted. A third feature frequently -although not uniformly- noted among NER-deficient patients is accelerated neurodegeneration, pointing to an important role of this pathway in long-term survival of neurons. On the other hand, inborn defects in NER cause in addition an exceptional variety of distinct pathologies, ranging from >1000 elevated cancer risk (notably sun-induced skin carcinogenesis) as observed in xeroderma pigmentosum to dramatic accelerated ageing (with an apparent reduced risk for cancer) as manifested by disorders like Cockayne syndrome and combinations of these. To understand the mechanisms of NER-associated disease, one has to be aware of the distinct disease etiology associated with each disorder. In this section, we will describe and compare the pathology of the different NER-associated disorders. In the last section we will combine the clinical knowledge with recent integrated insights from cell and molecular biology, biochemistry and several recently generated knock in and knock out mouse models to put the pathological complexity into a logical, coherent mechanistic perspective.

Mutations in the NER pathway can lead to six rare, autosomal recessive syndromes: the already mentioned prototype repair disorder xeroderma pigmentosum (XP), as well as XP with Desanctis-Cacchione syndrome (XP-DSC), the premature ageing conditions Cockayne syndrome (CS), trichothiodystrophy (TTD), cerebro-oculo-facio-skeletal syndrome (COFS), combinations like XPCS and XPTTD and the yet poorly characterised UV-sensitive syndrome UV^s (Bootsma et al., 2002; Broughton et al., 2001; Cleaver et al., 1999; Graham et al., 2001; Hoeijmakers, 2001; Horibata et al., 2004). Although the NER reaction involves at least 30 proteins, mutations in 11 NER proteins have so far been found to be associated with human pathology. It is to be expected that inherited defects in many of the remaining factors will sooner or later be found to cause additional human syndromes. Intriguingly, mutations in a single NER gene can give rise to multiple distinct disorders. The most extreme example of this kind is XPD, in which specific mutant alleles up to now have been linked with 6 NER conditions. An overview of NER-associated diseases and causative genes is presented in Table 1.

Table 1. NER-associated human diseases and causative genes. XP can be caused by the mutations in *XPA, -C, -D, -E, -F, -G* and *-V* genes, CS syndrome can be caused by mutations in *CSB,* and *-A* genes, TTD can be caused by mutations in *XPD, XPB* and *TTDA* genes, COFS can be triggered by mutations in *XPD, XPG,* and *CSB* genes, XP combined with TTD (XP/TTD) has sofar been found to be associated exclusively with *XPD,* XPCS is triggered by mutations in either *XPD, XPB* and *XPG* genes and UV^s can result from a *CSB* defect (Broughton et al., 2001; Cleaver et al., 1999; Giglia-Mari et al., 2004; Graham et al., 2001; Hoeijmakers, 2001; Horibata et al., 2004; Meira et al., 2000). Note that distinct mutations in the *XPD* gene are causative for 6 clinically different conditions. XP-V stands for XP-variant, which is actually not a NER-deficiency syndrome but is caused by a mutation in a translesion DNA polymerase, Pol η, which is able to bypass UV-induced CPD lesions in the DNA template in a relatively error-free fashion. Inactivation of this damage tolerance gene causes XP cutaneous features including skin cancer predisposition.
[a] Two patients have been reported with features pointing to XP-DSC (Colella et al., 2000).

Causative gene:	***XPA***	*XPB*	***XPC***	*XPD*	*XPE*	*XPF*	*XPG*	*XPV*	*CSA*	*CSB*	*TTDA*
CS									X	X	
COFS				X			X			X	
TTD		X		X							X
XP	X		X	X	X	X	X	X			
XP-DSC	X			X						[a]	
XPCS		X		X			X				
XPTTD				X							
UV^S										X	

3.1 Cockayne syndrome (CS)

One of the most devastating NER syndromes is CS (and the even more dramatic COFS form). CS is a very rare autosomal, recessive disorder (estimated frequency less than 1 per 10^5) characterised by progressive postnatal growth failure, neurological dysfunction and symptoms reminiscent of segmental accelerated ageing. An average life span of 12.5 years has been calculated for the reported patients, which may represent a bias towards the more severe cases (Nakura et al., 2000; Nance and Berry, 1992). In general, older CS patients have a very characteristic appearance, including overall "aged" look: large ears, protruding nose and sunken eyes (generally denoted as "bird-like facies"), thin hairs, unsteady, wide-based gait, progressive hearing impairment and cachectic appearance due to loss of subcutaneous fat tissue (Nance and Berry, 1992). Normal *in utero* development of a CS patient is followed by profound growth failure and overall developmental delay, which generally begins within the first year of life. Soon after birth, the brain of CS patients fails to grow and remains extraordinarily small throughout life, but remarkably, it is not grossly malformed. The above findings and the almost exclusive postnatal timing of growth impairment in the central nervous system (CNS) make it unlikely that the cause of extreme microcephaly (small head and brain) of CS results from pre-natal premature curtailment of neurogenesis, disordered neuronal migration, or grossly aberrant connectivity. Postnatal interference with the proliferation, branching, and deployment of neuronal processes seems more plausible (Rapin et al., 2000). In the light of an inborn deficit in NER particularly TC(-NE)R, we suggest that the postnatal increase in oxidative DNA damage load (Randerath et al., 1997a; Randerath et al., 1997b; Randerath et al., 2001) may, at least in part, explain the almost exclusive postnatal onset of CS. The earliest common neurological symptom in CS is delayed psychomotor development. The progressive gait disorder is a manifestation of the combination of spasticity of the legs, (mainly cerebellar) ataxia, tremors, contractors of the hips, knees and ankles often accompanied by kyphosis of the vertebral column (Nance and Berry, 1992).

All CS patients are mentally retarded, yet, this feature varies from mild to severe retardation (Nance and Berry, 1992). It is important to note that the early onset of cataracts, neurological dysfunction and microcephaly is associated with poor prognosis and survival. To date, among ~200 CS cases (Rapin et al., 2000) no patients have been reported with normal neuronal functioning but severe other CS symptoms, arguing that progressive neuronal failure may be among the primary causes of the systemic pathological outcome. This notion is supported by post-mortem pathological findings, which in general reveal lack of overall chronic tissue degeneration or cell-death (necrosis or apoptosis) in any organ system except for the CNS and peripheral nervous system (PNS)(Nance and Berry, 1992). Due to progressive neuronal decline including retinal degeneration and hearing impairment and cachexia, patients gradually lose ability to move and make contact with the outside world, become passive and fail to feed actively. Progressive failure to thrive is often followed by increased susceptibility to infectious diseases such as pneumonia/respiratory infections, which are often reported as an ultimate cause of death. Secondary to cachexia, renal or hepatic failure have also been noted as a cause of death in several cases (Nance and Berry, 1992).

Most frequent radiological findings include intracranial calcifications and vertebral anomalies including kyphosis. Osteoporosis was noted in a few patients pointing to premature ageing of the skeleton. It should be noted though that data in the clinical reports are often from different stages of the disease. That might explain some at first glance conflicting observations. For example, the notion that bone age in 10 CS patients was found to be advanced, in 5 normal and in 6 delayed may well reflect late, medium and early stages of CS respectively, where delayed bone age may mirror an overall developmental delay prior to the onset of premature ageing in CS. Visual or auditory evoked potentials in CS were always found abnormal. This is reminiscent of profound cataracts, retinal degeneration and deafness observed in CS. Nerve conduction velocity analysis (EMG/NCV) is impaired in most patients. Muscle biopsies have shown variable changes, none thought to be primary.

Analysis of CNS by MRI or CT scans has revealed increased ventricular size and/or cerebellar and cerebral atrophy and/or calcifications in basal ganglia and elsewhere (Nance and Berry, 1992). Calcifications as well as appearance of neurofibrillary tangles reported in some cases of CS (Takada and Becker, 1986), are features of normal ageing, but appear early in CS. Most of the brain anomalies in CS are associated with white matter, or so-called glial compartment. The glial cells, more specifically oligodendrocytes are the cells, which isolate axons of the neurons (grey matter) by wrapping them into a myelin sheet. Proper myeliniation is required for high velocity conduction as well as neuronal survival. It has been proposed that demyelination is the primary neuronal defect in CS (Brooks, 2002). Nevertheless, recent post-mortem examinations of CS patients have revealed neuronal loss within several neuronal populations, such as those in the Meynert nucleus, putamen/caudate, thalamus, globus pallidus, dentate nucleus, granule cells and Purkinje cells (Itoh et al., 1999). Except in the cerebellum, these changes may be secondary because they were found adjacent to the demyelinated lesions. The other changes are likely to be primary since demyelination was not reported in those areas. Many CS patients display hypogonadism, such as undescended testis. It is tempting to speculate that underdeveloped gonadal axis may contribute to neuronal loss as gonadal steroid hormones are implicated in survival of several neuronal populations, such as hippocampal neurons, also implicated in CS (Azcoitia et al., 2003; Hayashi, 1999). Astrocytes, the second of the three glial populations in the CNS, are also affected in CS. They are found pleomorphic, a few are multinucleated, and many are bizarre and irregularly shaped with swollen, lobulated, hyperchromatic nuclei (Rapin et al., 2000).

Interestingly, similar bizarre astrocytes and Purkinje cell loss is found in ataxia-telangiectasia (AT) patients (Lindenbaum et al., 2001). AT-mutated (ATM) protein is a key regulator of signaling downstream of DNA damage, mainly double strand DNA breaks. Thus, the cellular signaling as a response to defective DNA repair in CS, or defective signaling on its own in AT can lead to similar pathology.

Laboratory tests of hematological and immunological parameters as well as thyroid, adrenal, and hepatic function do not show gross abnormalities in CS. Glucose tolerance tests, basal or stimulated growth hormone levels, and responses to insulin, arginine, and glucagon have not revealed the causes of the dramatic dwarfing and cachexia (Rapin et al., 2000), nor did growth hormone therapy result in significant progress in growth (Nance and Berry, 1992).

Finally, 2 genetic complementation groups have been identified in the classical form of the disease: CSB comprising ~80% and CSA accounting for the remainder of all UV-sensitive patients (Table 1). In addition there are also non-UV sensitive patients (which, except for the sun-UV sensitivity, show many of the other CS features) as well as very rare patients with CS and XP (see below).

3.2 Cerebro-oculo-facio-skeletal syndrome COFS

COFS can be regarded as a severe form of CS. Its incidence is even lower than CS. The even more devastating features of COFS in comparison to CS include COFS-syndrome eye defects (i.e., microcornea with optic atrophy) which are more severe than those usually associated with CS (i.e., pigmentary retinopathy (Graham et al., 2001)). Additional symptoms include reduced birth weight pointing to a prenatal onset of the disease, early microcephaly with subsequent brain atrophy, reduced white matter, patchy grey matter, hypotonia, deep-set eyes and cataracts. Movement is markedly decreased, joint contractures common. Failure to thrive in COFS is more pronounced than in CS, generally leading to death within first years after birth. Like in case of CS patients (and XP patients with DeSanctis-Cacchione syndrome, see below) a frequent ultimate cause of death is pneumonia/respiratory infections. As shown in Table 1 COFS patients fall into 3 NER complementation groups: XP-D, XP-G and CS-B.

3.3 Trichothiodystrophy (TTD)

The clinical manifestations of TTD patients, including post natal developmental delay, cachexia, neurodemyelination, CNS intracranial calcifications, cerebellar ataxia, mental retardation, microcephaly, sensorineural deafness and cause of death, are largely overlapping with those of CS (Itin et al., 2001). Distinguishing hallmarks of TTD from CS are scaling skin, and brittle hair and nails. The latter is caused by greatly reduced content of cysteine-rich matrix proteins in the hair-shafts, leaving the hair fragile and vulnerable to physical breakdown (Lehmann, 2001). Pathological changes in the epidermis include hyperkeratosis (thickened keratin layer responsible for the scaling skin) and acanthosis (thickening of the epidermal layer) (Itin et al., 2001). Below we will discuss the molecular basis for this characteristic feature of TTD. Genetically, NER-deficient TTD patients have been assigned into 3 complementation groups all encoding subunits of the TFIIH transcription-repair complex: XP-D which contains the vast majority of the patients, XP-B (1 family) and TTD-A (3 independent cases). A significant proportion of TTD patients fail to display photo (UV) sensitivity, and they have no overt NER defect. The responsible gene(s) still await discovery.

3.4 Xeroderma Pigmentosum (XP) and XP with DeSanctis-Cacchione syndrome (XP-DSC)

Unlike in the case of CS, TTD and COFS, XP is always associated with clinical and cellular sensitivity to ultraviolet radiation and defective repair of UV-induced DNA lesions. The incidence of this autosomal, recessive condition is estimated to be 1: 10^5, although in some regions the incidence is higher due to a higher frequency of consanguinity. First symptoms of sun sensitivity in XP become evident at average age of 2 years, when intense freckling and/or sun-burn is first noted. XP patients display a more than 1000-fold elevated risk to develop sun-induced malignant skin neoplasms such as squamous cell carcinomas (SCC) and basal cell carcinomas (BCC) (Figure 3). Yet, the frequency of metastasis appears to be quite low (5 out of 112 XP patients with SSC). Interestingly, only 5% of XP patients are

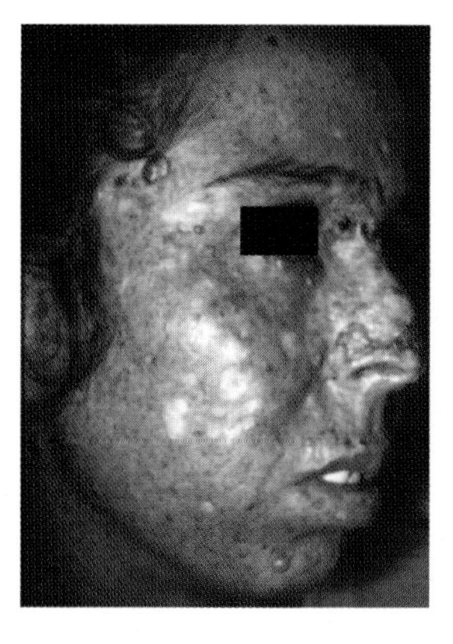
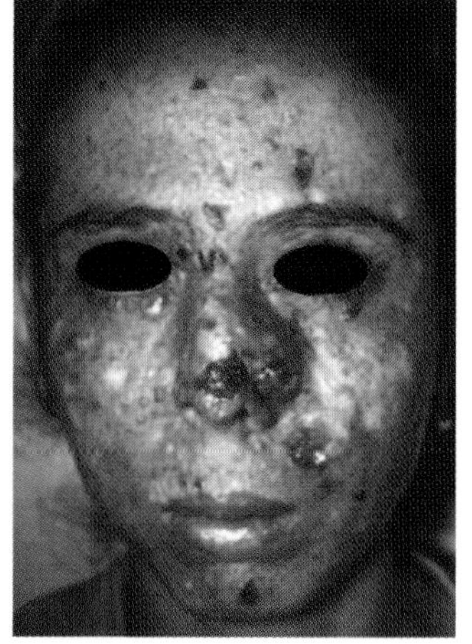

Figure 3. Typical skin abnormalities in an adolescent XP patient. Note sun-induced pigmentary changes including freckling, dryness and atrophy and (pre)malignant lesions and tumours. (courtesy of Pr Mohamed Denguezli, Sousse, Tunisia, www.atlas-dermato.org).

reported to develop melanomas. While 97% of SCC and BCC appear on sun-exposed areas such as face, head or neck, only 65% of melanomas were associated with this area, indicating that induction of a melanoma involves more complex and probably systemic factors. Generally accelerated photoageing of the skin is noted. Among ocular tissues, the eyelids, conjunctiva and cornea receive substantial amounts of UV radiation and subsequently are strongly affected in XP patients. Anomalies of the eyelid include sunburn, atrophy of the skin, loss of lashes or even the whole eyelid (Kraemer et al., 1987). Corneal abnormalities include corneal clouding and/or vascularisation. Neoplasms of the eye are exclusively associated with conjunctiva, eyelid and/or cornea whereas SCC is the most frequently occurring neoplasm.

Neurological abnormalities are reported in about 20% of XP patients (XP with DeSanctis-Cacchione syndrome (XP-DSC)). Although extraneurologic features such as number and aggressiveness of skin tumours between XP and XP-DSC patients appear similar, the average onset of sun sensitivity for XP-DSC is 6 months versus 2 years for classical XP (Kraemer et al., 1987). 80% of XP-DSC patients are mentally retarded, whereas less than a quarter of the patients display concomitant microcephaly, growth retardation, gait anomalies such as spasticity and ataxia; and sensorineural deafness, all of which have a progressive character (Brooks, 2002; Itoh et al., 1999; Kraemer et al., 1987). As in case of CS, COFS and TTD, the earlier the onset of neuronal features in XP-DSC, the more pronounced retardation of growth and sexual development (Itoh et al., 1999; Kraemer et al., 1987; Rapin et al., 2000) again strongly suggesting a link between endogenous DNA damage, repair, neuronal deployment and survival, and somatic development and maintenance. Interestingly, the above notion is supported by studies in model organisms, such as the fruit fly *(Drosophila melanogaster)*, showing that lowering the oxidative damage load by over-expressing the ROS scavenger enzyme superoxide dismutase 1 (SOD1) in only the motorneurons results in a ~40% longer lifespan of the fly, compared to wt (Parkes et al., 1998).

What is the difference between CS and the non-XP features of XP-DSC? XP-DSC patients do not develop CS-specific symptoms such as demyelination of CNS and PNS, retinal degeneration and calcifications of the basal ganglia and other brain areas. Instead, XP-DSC patients exhibit degeneration of specific populations of neurons on top of sun-induced skin freckling and/or skin cancer. In general, CS is associated with more severe symptoms, including microcephaly and cachexia. The most important difference is however, the primary cell-type affected in the CNS. Except for the neuronal loss in the cerebellum, demyelination in other areas of the CNS leaves the neurons in CS patients relatively intact. In XP-DSC myelin is not affected, yet besides neuronal death in the cerebellum (resulting in CS-like ataxia) several other neuronal populations die in other areas of the CNS, such as in the cortex and substantia nigra, resulting in progressive intellectual deterioration, dementia and gait anomalies (Brooks, 2002; Itoh et al., 1999; Rapin et al., 2000). Why the CS defect primarily affects the myelinating cells (oligodendrocytes) and XP DSC defect the neuronal cells, and how this results in often overlapping phenotypes remains to be elucidated. Since neuronal conductivity is a function of proper myelination (Brooks, 2002) and neurons and not oligodendrocytes establish the cellular connections both within CNS and with the soma, it is tempting to speculate that at least a subset of overlapping features of CS and severe XP-DSC are caused by a defect in neuronal functioning.

3.5 XP combined with CS (XPCS)

In rare cases (to date 9 patients described in the literature) a combined XPCS pathology has been reported (Lindenbaum et al., 2001). There is a remarkable degree of clinical variation in XPCS. The three patients with XPCS carrying a defect in the *XPB* gene (see Table 1) showed a remarkably milder CS phenotype with survival between the third and the fifth decade of life compared to those in groups XP-D and XP-G. Two patients in XP-D group and the remaining four in XP-G group all displayed very severe disease and died well before puberty (Lindenbaum et al., 2001). Unfortunately, an overall chronological pathology record for most of the XPCS patients is missing. Pathology of patient XP20BE (XPG-XPCS, see Figure 4) has been documented the best and will be described briefly.

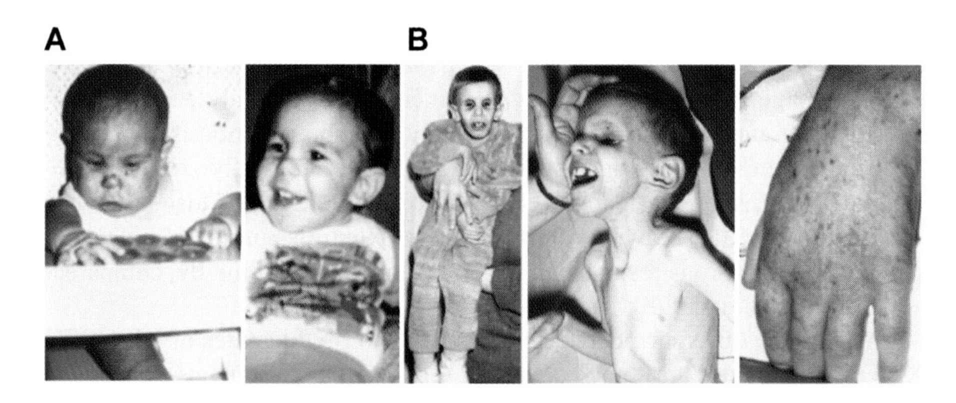

Figure 4. XPCS patient XP20BE (XPG-XPCS). **A.** Progressive pathology of Cockayne syndrome features. Note the normal fullness of the face at 4 months and 1.5 years of age and the typical CS appearance with deep-set eyes, prominent ears and profound cachexia at the age of 6 years. B. Age 6 years. XP pigmentary changes and CS-specific wrinkling of the skin of the hand showing signs of premature ageing. Death occurred at the age of 6.2 years. Reprinted from (Lindenbaum et al., 2001) with permission from the European Paediatric Neurology Society.

Electromyogram (EMG) analysis indicated primary neuropathic but notably also primary myopathic features, suggesting that muscle cell degeneration can also occur independently of axonal loss in the PNS of XPCS. Patient XP20BE died at the age of 6.2 years because of profound cachexia and pneumonia. His brain weight was 350g, while the expected brain weight of a child at that age is 1200g. Most of the pathological findings in the brain were typical for CS. However, in the midbrain the substantia nigra had focal neuronal loss, a feature characteristic for XP-DSC. Neuronal loss was noted also in hippocampus and certain brainstem nuclei. The cerebellum displayed typical CS features, including neuronal loss in Purkinje and internal granular layers. Taken together, loss of myelinated fibers and neurons was profound with resultant dementia, ataxia and notably dysmetria (Lindenbaum et al., 2001). Is neuronal cell death observed in XP20BE patient primary or secondary to demyelination? Most of the demyelinating lesions are found outside of the cerebellum. Purkinje cells are innervated mostly by granular cells within the cerebellum and not by neurons from other brain areas. Thus, the loss of Purkinje cells (and granular cells) is likely the primary neuronal defect and not secondary to an oligodendrocyte defect. Taken together, the extreme CNS pathology seen in XPCS and CS likely results from the primary DNA repair defect in oligodendrocytes, some neuronal populations and likely to some extent, a combination of these cell types. In conclusion, XP and CS features are both fully represented in XPCS, indicating that those quite different pathological traits are not affecting each other and can co-exist independently.

3.6 XP combined with TTD (XPTTD)

Recently, two patients have been identified with a combined form of XPTTD (Broughton et al., 2001). Both these patients are still alive and disease etiology of this condition is still largely unexplored. Since it usually takes time and exposition to sunlight (UV) before the typical XP cutaneous features develop a gradient will be expected between classical TTD patients without XP and those exhibiting XP cutaneous symptoms. Because of the short life span of many TTD patients, their frequent hospitalisation as well as their abnormal skin architecture due to the ichthyosis, protecting against UV irradiation, the incidence of XPTTD cases may in fact be underestimated.

3.7 UV-sensitive syndrome (UV^s)

UV-sensitive syndrome (UV^s) is a rare autosomal recessive disorder characterised by photosensitivity and mild freckling but without neurological abnormalities or skin tumours. UV^s cells show UV hypersensitivity and defective transcription-coupled DNA repair of UV damage. In view of the mild features of this disorder UV^s may well be an under diagnosed condition. Thus far only in one case the causative gene (*CSB*) has been identified (Horibata et al., 2004).

A summary of pathological features of all above NER disorders is presented in Table 2.

Table 2. Clinical symptoms of NER disorders.

Clinical symptoms	XP	XP DSC	UV[s]	CS	TTD	XPCS	XP/TTD	COFS
UV sensitivity	++	++	+	++(*)	++(*)	++	++	?
Photoageing of skin	++	++	?	-	-	?	?	?
Increased freckling	++	++	+	-	-	++	++	?
Skin cancer	++	++	-	-	-	++	+	?
Cachectic dwarfism	-	+	-	++	++	++	+	+++
Microcephaly	-	+	-	++	++	++	?	+++
Progressive cognitive impairment	-	+	-	++	++	++	++	+++
Sensorineural deafness	-	+	-	++	++	++	-	+++
Eye abnormalities	-	+	-	++	++	++	?	+++
Skeletal abnormalities	-	?	-	+	+	+	?	++
Spasticity	-	+	-	++	++	++	?	+++
Ataxia	-	+	-	++	++	++	-	+++
Axonal neuropathy	-	++	-	+/-	?	+	?	?
Demyelinating neuropathy	-	-	-	++	++	++	?	?
Myopathy	-	-	-	-	-	+/-	-	?
Brain calcification	-	-	-	++	++	++	?	+++
Hypogonadism	-	+	-	++	++	++	?	?
Brittle hair and nails	-	-	-	-	++	-	+	?
Hyperkeratosis	-	-	-	-	++	-	+	?
Progeria	-	+/-	-	++	++	++	?	?

(*) ~50% of patients display this feature. Assembled from (Broughton et al., 2001; Horibata et al., 2004; Itin et al., 2001; Kraemer et al., 1987; Nance and Berry, 1992; Rapin et al., 2000).

4. MECHANISMS OF NER-ASSOCIATED DISEASE ETIOLOGY

Although NER is a ubiquitous repair mechanism likely auditing the genome in each cell of our body and removing numerous different types of lesions, NER-associated disorders display a perplexing diversity of pathologies, ranging from >1000 times elevated cancer risk (XP) to dramatic accelerated segmental ageing (CS, COFS, TTD) and combinations of these. Mutations in the single XPD locus present the most extreme case: some alleles bias towards a striking cancer predisposition, other towards very fast segmental aging, yet others both. In this section we will focus on molecular mechanisms of specific NER-associated pathologies and attempt to put them in a coherent context.

4.1 Cancer predisposition in NER disorders

A hallmark of specific types of NER-deficiencies is damage-induced genomic instability that translates at the level of the patient into increased carcinogenesis, highlighting the key importance of DNA damage. The overruling role of UV as a source of DNA damage makes this agent the dominant causative factor for this aspect of inborn defects in NER. However, also other DNA damaging agents should not be ignored, including spontaneously occurring DNA injury. Here we will discuss both aspects of the protective role of NER. It appears that of all syndromes linked with NER impairment only XP and GG-NER defects are linked with a strong cancer susceptibility.

4.1.1 Carcinogenesis induced by UV

Mutations in different genes involved in NER can cause XP (see Table 1), associated with severe photosensitivity, photoaging of the skin and a dramatic, (more than ~1000 fold) increased UV-induced skin cancer risk. Patients, with mutations in the *XPC* or *XPA* genes are the most cancer-prone, a phenotype faithfully mimicked by the corresponding mouse mutants (Berg et al., 1997; de Vries et al., 1995; Friedberg et al., 2000; Sands et al., 1995). Strikingly, while *XPA* mutations abolish the complete NER reaction, *XPC* mutations only cause a defect in GG-NER. Since XPC patients in general appear even more cancer-prone than XPA patients this finding shows that damage in the global genome is the prime cause of mutations. The somewhat lower cancer incidence in other XP groups can directly be explained by the residual GG-NER activity of the mutated proteins and additionally by the protective effect of the simultaneous impairment of the TCR pathway in some groups, that triggers apoptosis or likely cellular senescence upon induction of DNA damage and thus protects from cancer. The fact that many mutations found in skin cancer of XP patients are CC to TT tandem transitions (Bodak et al., 1999; Daya-Grosjean et al., 1993; D'Errico et al., 2000; Giglia et al., 1998; Spatz et al., 2001), (the molecular hallmark of UV mutations) indeed indicates that unrepaired UV-induced lesions are the main cause of increased skin cancer in these patients.

How do DNA lesions turn into mutations? During replication, helix-distorting NER-type DNA damage will generally pose a block to the high fidelity replication machinery. To prevent cell death as the consequence of a permanent arrest of replication at least two main pathways are present. *First*, a battery of specialised translesion polymerases are present, which allow bypass of the troubled region. Although each of these polymerases has specialised in relatively error-free replication over a specific set of lesions, this solution usually goes nevertheless at the expense of increased mutagenesis (covered in detail by C.M. Green and A.R. Lehmann in this book). As apparent from the cancer phenotype of GG-NER mutants, when the critical damage load is reached translesion polymerases fail to provide full protection. Moreover, one of the genes causing XP, *XPV* (XP variant form), is not implicated in NER but in fact is related to a defect in translesion synthesis by polymerase η (Table 1). This shows that the cancer phenotype in XP is derived either from replication over persisting UV-induced damage, because of a GG-NER, defect or from improper bypass of those lesions. Recent work using transgenic mice carrying UV-lesion-specific photolyase genes, which either remove selectively CPD or 6-4PP lesions, has identified CPDs as the most biologically relevant type of UV damage. CPDs appear mainly responsible for both the acute effects of UV exposure as well as the immune-suppression and most importantly the carcinogenic consequences. Likely this is due to the fact that these lesions are poorly repaired in the genome, with the exception of the transcribed strand of active genes. *Second*, an alternative solution thought to allow bypass over an injury in the DNA template is strand-switching involving recombinational bypass. In this largely unexplored mechanism replication utilises temporarily the newly synthesised strand of the non-damaged complementary template until the lesion in the damaged strand is bypassed. Obviously, both of these solutions do not involve actual removal of the original lesion. Thus, when not repaired, the same lesion may cause mutations in the next rounds of replication in the daughter cells that inherit the damage. In conclusion, mutations accumulate either from deficient GG-NER, leading to accumulation of damage and consequently a higher chance of mutations during normal replication, or from failure to correctly replicate over UV-damaged nucleotides due to defective translesion synthesis as in XP-V mutants. This in turn explains the extreme skin cancer predisposition characteristic of XP.

4.1.2 The cancer-protective role of TC-NER defects

In contrast, CS patients with a defect in TC-NER only, are not cancer-prone. In these patients GG-NER is functional and can act at least in part as a backup mechanism for TC-NER. Due to their defective TC-NER, CS-cells are hypersensitive to UV-light induced DNA damage (Brash et al., 2001; Conforti et al., 2000; Ljungman and Zhang, 1996), which provides, as a beneficial side effect, protection against cancer via elimination of the most damaged cells, that otherwise are at risk of acquiring oncogenic mutations. This reduces the risk of spontaneous cancer. Nevertheless, CS mouse mutants are found to exhibit a modestly elevated skin cancer rate upon chronic exposure to UV or the chemical carcinogen DMBA (van der Horst et al., 2002; van der Horst et al., 1997a). This apparent difference with the human syndrome might be explained at least in part by the inefficient GG-NER of CPD photolesions in rodents (as compared to man) and/or by the fact that CS patients are generally protected from being exposed to significant sun light due to frequent hospitalisation. However, also in mice the cancer-protective effect of a TC-NER deficit has been revealed by the lowered spontaneous cancer rate in *Csb*|Ink4a/ARF double knock out mice compared to Ink4a/ARF single mutants (Lu et al., 2001). Although a striking correlation exists between GG-NER capacity and cancer predisposition, the latter might be additionally influenced by other factors. In the remaining part of this paragraph we will discuss several examples of unanticipated roles of NER factors and their relation to cancer.

4.1.3 Spontaneous cancer

XP patients have also been noted to have a higher (10-20-fold) incidence of developing internal malignancies: these include brain tumours, leukemias and lung tumours (Bootsma et al., 2002; Kraemer et al., 1987). With the exception of lung cancer in XP patients that have smoking history, these neoplasms are not obviously related to exposure. Importantly, it was shown that upon aging XPC mice have a higher spontaneous mutation load, showing not only that NER protects against skin cancer induced by UV but also may help to avoid endogenous spontaneous tumours (Wijnhoven et al., 2000). Again, elevated mutagenesis was lower in XPA and absent in CSB mice, indicating that relative to XPC, XPA and to a larger extent CSB mice are protected against carcinogenesis in general due to a higher apoptotic rate.

4.1.4 Emerging additional gate-keeper roles of GG-NER specific damage recognisers.

Damage also triggers a cascade of signaling pathways leading to transient or permanent cell cycle arrest (the latter also called senescence), as well as apoptosis or necrosis. Since for repair purposes damage has to be recognised anyway it would seem efficient to utilise the same machinery for triggering the other cellular responses linked to genomic injury. Increasing evidence emerges that nature has followed this logic. Recent findings support the notion that XPC has an unknown additional function beyond DNA repair *per se*. Indeed, XPC-deficient human cells were found to display attenuated p53, p21 and activated caspase 3 responses to cis-platinum-induced cell cycle regulation, apoptosis and DNA repair (Wang et al., 2004). Furthermore, on a global transcriptional level, XPC-deficient cells appear to be primarily defective in induction of cell cycle and cell proliferation-related genes in response to cis-platinum treatment (Wang et al., 2004). Moreover, XPC has been shown to be up-regulated upon UV (Adimoolam and Ford, 2003; Amundson et al., 2002; Ng et al., 2003) and surprisingly also after ionising radiation, an agent inducing mainly non-NER type of DNA lesions (Amundson et al., 2002). This data suggests that in addition to NER XPC may have a more general function as a gatekeeper. The latter function may be related to its binding partner Rad23B, a protein containing an ubiquitin-like and two ubiquitin-associated domains. Rad23B has been reported to be involved in cell cycle dependent regulation of the stability of a number of proteins, including p53 (Elder et al., 2002; Glockzin et al., 2003; Ng et al., 2003). Because XPC defects are associated with many types of cancers (reviewed in (Goode et al., 2002)), these findings suggest a more general role for XPC in initiating damage signaling in response to DNA damaging agents beyond UV alone.

Also the other initiator of global genome repair, XPE, might play a role outside of the GG-NER context. The phenotype of the DDB2-knockout mouse (KO) is consistent with a role of DDB2 in UV-mediated apoptosis.

Mice lacking DDB2 are prone to UV-induced skin cancers, predominantly squamous cell carcinomas. Interestingly, DDB2 KO embryonic fibroblasts display in comparison to wild type cells an increased resistance to killing by UV, correlating with a dampened p53-dependent apoptotic response. Since the caretaker role of DDB2 in GG-NER is anyway negligible in rodents (Tang et al., 2000), another function of DDB2 is suggested. The authors argue that a defective apoptotic response, rather than defective GG-NER, causes the higher cancer incidence observed in these mice. Moreover, increased relative UV resistance and decreased apoptosis have been observed in XP-E patient fibroblasts, which are also defective in DDB2 function (Itoh et al., 2003). Like XPC, XPE is induced by both UV and ionising radiation (Amundson et al., 2002; Hwang et al., 1999; Tan and Chu, 2002). Interestingly, XPE is part of a complex containing cullin 4A and Roc1 as well as the COP9 signalosome, a known regulator of cullin-based ubiquitin ligases, that can enhance ubiquitin-ligase activity upon UV irradiation (Groisman et al., 2003). This activity might in turn be used to modulate the activity of a large set of proteins that are involved in the cellular DNA damage response and, therefore, mutations in XPE will not only hamper DNA repair but also the DNA damage response, which might contribute to the development of cancer.

Since TFIIH is also involved in transcription (Schaeffer et al., 1993a), it might be that impaired transcription of specific genes contributes to the development of cancer in certain TFIIH mutants. It has been shown that specific mutations in XPD cause defective TFIIH-dependent transactivation of a subset of nuclear receptors (Drane et al., 2004; Keriel et al., 2002). If and how these defects contribute to carcinogenesis and/or other features associated with mutations in TFIIH subunits is currently unknown. Moreover, specific mutations in the *XPB* and *XPD* genes result in improper regulation of c-myc expression and thereby may modulate the development of malignancy (Liu et al., 2001).

Finally, also systemic factors might significantly impact the development of cancer. This is already indicated by the fact that unlike 97% of skin carcinomas, only 65% percent of melanomas are found in sun-exposed areas of the skin in XP. The progression of malignant cells into cancer can also be facilitated by the defective immune response upon UV, as observed in *Xpa*$^{-/-}$, but not in *Xpc*$^{-/-}$ mice (Boonstra et al., 2001; Miyauchi-Hashimoto et al., 1996). Taken together, future studies integrating functions of NER proteins both within and outside the DNA repair context will likely shed further light on the mechanism of cancer.

4.2 The accelerated ageing dimension of NER-associated pathology

Contrary to XP that is connected with cancer predisposition all CS forms of NER-associated syndromes, including the CS component in TTD, COFS and combinations with XP, are tightly linked with accelerated ageing. Here we will disentangle the various ageing features and the non-ageing symptoms and intend to provide a plausible scenario for their etiology.

4.2.1 The specific neuropathology in XP-DSC, CS and TTD

As discussed above, NER patients with developmental features suffer from primary degeneration of specific neuronal populations and/or oligodendrocytes (myelin-generating cells). Why do mutations in NER proteins specifically affect cells of neuronal origin? Several lines of evidence suggest neuronal type of cells to utilise NER in a specific manner. First, neurons and especially differentiated neurons are substantially more UV sensitive than e.g. fibroblasts or HeLa cells (James et al., 1982; Nouspikel and Hanawalt, 2002). It has been shown, that GG-NER activity declines dramatically during neuronal differentiation *in vitro* (Nouspikel and Hanawalt, 2000) and thus likely cannot support DNA repair when TC-NER or TCR activity is hampered by a mutation. The latter defect has been shown to render a number of cells hypersensitive to the cytotoxic effects of transcription-blocking lesions (D'Errico et al., 2005; Ljungman and Zhang, 1996; Yamaizumi and Sugano, 1994). Clinical studies have shown that in normal individuals demyelination (white matter loss) is a late response to CNS gamma irradiation (van der Maazen et al., 1993) and pathological comparison of the brains from the chemotherapeutically-treated or gamma-irradiated patients (and laboratory animals) with those from CS patients and normal ageing individuals revealed a remarkable degree of pathological similarity (D. Dickson personal communication)(Brooks, 2002). Specific vulnerability of neuronal tissue to endogenous damage is also supported by studies suggesting the involvement of ROS in the onset of Parkinson's disease, dementia and Alzheimer's disease (reviewed in (Betarbet et al., 2002; Butterfield et al., 2001)) as well as by studies with non-homologous end-joining (NHEJ)-defective (XRCC4 knock-out) mice, which display embryonic lethality likely due to massive neuronal apoptosis (Gao et al., 2000). Taken together these observations suggest that neurons and glia are hypersensitive to DNA damage of both exogenous (e.g. gamma rays) and endogenous (reactive metabolites, e.g. ROS) origin. Since ROS are most likely produced as a function of metabolic rate it is perhaps not surprising that in CS, TTD and XP-DSC patients degeneration of Purkinje neurons and the resultant cerebellar ataxia is an early event, as those cells are believed to be metabolically and transcriptionally among the most active cell-types in the brain (Brooks, 2002).

What causes the difference between neuronal loss in XP-DSC and oligodendrocyte deficit in CS, TTD and COFS? Clearly, there are qualitative differences between specific repair pathways in XP-DSC and CS, TTD and COFS. While 100% of XP-DSC patients are UV sensitive, enabling their assignment to a certain XP complementation group by cellular assays, about 50% of CS, TTD and COFS patients are not and thus the genes affected have remained unknown. A likely explanation here is that mutations in non-

UV sensitive CS, TTD and COFS patients specifically affect TCR of non-NER lesions or some other, non-NER function of the given protein. Because systematic efforts to sequence NER genes in those patients have not yet been performed and non-NER TCR capacity is to date difficult to measure, the latter explanation remains hypothetical. Since cell-type, differentiation status (de Waard et al., 2003; Nouspikel and Hanawalt, 2002) and likely metabolic type and rate determine both the lesion spectrum and repair activity in the given cell, it is tempting to speculate that neuronal loss in XP-DSC is caused by a TC-NER deficit of bulky adducts, combined with a GG-NER defect whereas oligodendrocyte loss in TTD, CS and perhaps COFS is related to inability to remove non-NER TCR targets from the transcribed genes. The overall concept underlying different pathology in NER and likely other DNA repair disorders will be largely determined by the following parameters:

Different cell types/organs/tissues will have a different spectrum of lesions due to intrinsic variation in metabolism and exposure. In addition, each has its own specific repair specificity and DNA damage response mechanism. In combination with a distinct mutation in a repair gene affecting the (multiple) function(s) of the encoded protein in different manners these factors will determine the specific phenotype of the NER syndrome in a given patient. This may not only explain the differences in neuronal phenotype but may also provide a plausible explanation for other premature aging features found to differ between these syndromes.

Moreover, recently it was found by the group of Hanawalt that in terminally differentiated cells such as neurons, TC-NER has a different mode, so-called differentiation associated repair or DAR. DAR preferentially repairs both transcribed and non-transcribed strands of genes (Nouspikel and Hanawalt, 2000; Nouspikel and Hanawalt, 2002). Therefore future studies addressing specific effects of XP, XP-DSC, CS, XPCS and TTD type of mutations on this repair trail are of great interest. Nevertheless, the basic mechanism of NER and/or the lesion spectrum that it deals with, likely remains similar in most of the tissue types, otherwise the plethora of different mutations in NER genes should result in even more diverse pathological outcomes. The role of GG-NER in neuronal phenotypes seems relatively limited because (i) GG-NER appears down-regulated upon differentiation (Nouspikel and Hanawalt, 2000) (ii) patients lacking XPC, the principal damage sensor in GG-NER, do no develop neuropathies, in contrast to XP and CS complementation groups in which the TC-NER pathway is significantly affected.

Finally, an interesting observation was recently made concerning the relation between the severity of the CS symptoms and the molecular defect. In CSB an inverse correlation was found between the severity of the defect at the molecular level: seemingly mild CSB mutations induce the extremely severe form of CS called COFS, whereas a virtually complete inactivation of the gene goes along with the mild UV-sensitive syndrome (Horibata et al., 2004).

4.2.2 Loss of subcutaneous fat tissue in NER disorders

A subset of XP-DSC, and all CS, COFS and TTD patients gradually loose subcutaneous fat tissue and become extremely cachectic. This is one of the clinically most important aspects of the disease as it largely determines the overall health status of patients. Since the endocrine axis in the above patients appears relatively normal (Nance and Berry, 1992) the causative deficit is likely cell-autonomous. Unlike most of neuronal cell types, adipocytes are constantly turned over and thus the defect may either lie in mature adipocytes, the adipocyte stem-cell compartment or both. Similar to many differentiating cells, GG-NER in adipocytes is down regulated as a function of differentiation status (Nouspikel and Hanawalt, 2002). Whether DAR occurs in adipocytes similar to neurons and how it compares to the DNA repair status in other differentiated tissues still needs to be determined. Recently, embryonic stem cells were found to be more vulnerable to genotoxic stress than e.g. fibroblasts or keratinocytes (de Waard et al., 2003), suggesting, that stem cells, in order to avoid damage accumulation and subsequent tissue malfunctioning and/or carcinogenesis have a lower apoptosis threshold than other cell types. In concordance with that notion, various CS mouse models display time-dependent loss of tubular germinal epithelium in the testis (J.O. Andressoo et al., submitted). Although hypogonadism is also a prominent feature in CS, TTD and COFS, to our knowledge histological examinations have not been performed and thus human-mouse comparisons of that tissue type cannot be made.

4.2.3 Other features of premature ageing in NER syndromes

Both CS patients and mice display developmental defects that can be interpreted as resulting from arrested development due to early onset of ageing. This holds for the reduced body weight, as well as the skeletal abnormalities that are related to early occurrence of osteoporosis. The latter feature is not clearly reflected in the CS mice but is apparent in the TTD mouse mutant and -as discussed below- in all double mutants. Also retinal degeneration is observed in CS patients as well as mice. Systematic analysis of defined cohorts of isogenic mouse strains with and without CS-mutations will provide a more complete picture of the ageing status of many organs and tissues. Such studies are ongoing and will reveal to which extent segmental ageing occurs in association with a CS phenotype.

5. MODELS FOR THE ONSET OF DEVELOPMENTAL AND ACCELERATED AGEING FEATURES IN NER DISORDERS

The relatively mild, non-cancer related disease etiology of XPA patients (carrying a complete defect in both GG-NER and TC-NER) and the very severe features associated with inactivation of e.g. CSB and CSA (with the defect in NER limited to TC NER) has led to the hypothesis that mutations in NER proteins resulting in symptoms different from XP, such as those observed in CS, TTD, XPCS, XPTTD and COFS, might be due to functions of those proteins outside the context of the classical NER pathway or outside of the NER spectrum of lesions.

During the past years multiple intriguing findings have triggered several mutually non-exclusive hypotheses. *The first* is the "transcription syndrome" hypothesis, based on TFIIH dual functionality in NER and transcription. Mutations in TFIIH subunits may, besides NER, also affect the basal and/or activated transcription initiation (Bergmann and Egly, 2001; Bootsma and Hoeijmakers, 1993; Vermeulen et al., 1994). Due to the exclusive association of mutations in 3 TFIIH subunits with TTD, this condition has been suggested to result at least in part from defects in basal transcription (Bergmann and Egly, 2001; de Boer et al., 1998; Giglia-Mari et al., 2004; Vermeulen et al., 1994). The latter hypothesis was indeed supported by early findings of reduced mRNA levels for proteins that are responsible for the cross-linking of keratin filaments in the skin of TTD mice. This explains why the skin of TTD is scaly and at the same time why the hair and nails are brittle as this is caused by the same lack of cross-linking between keratin filaments (de Boer et al., 1998). This feature appears to be independent of the extent of the repair defect in both TTD patients and in single TTD and XPA/TTD double mutant mice and therefore is distinct from the repair function of TFIIH or the NER status as such. The interpretation that TTD brittle hair is due to a basal transcription problem is also entirely consistent with the observed instability of TFIIH in TTD fibroblasts (Botta et al., 2002; Giglia-Mari et al., 2004; Vermeulen et al., 2001) and TTD MEF's (J.O. Andressoo et al., submitted). Another strong argument is the identification of several unusual TTD patients with more pronounced TTD features during episodes of high fever, which turned out to be due to a temperature-sensitive instability of TFIIH caused by the specific XPD^{TTD} mutation (Vermeulen et al., 2001). *Secondly*, the transcription-coupled repair (TCR) hypothesis has been put forward, based on the notion that CS, and XPCS cells from patients and mice are slightly but significantly more sensitive to several oxidative agents. Since most oxidative lesions are normally repaired by BER and not by NER, the existence of a general TCR pathway was suggested, in which proteins involved in CS are required for repair not only of transcription-stalling NER lesions but also BER, and perhaps other transcription-blocking injuries (Citterio et al., 2000b; Cooper et al., 1997; de Boer et al., 2002; de Waard et al., 2003). In this role the CSB protein may closely monitor elongating RNA polymerases for normal progression of the transcription process, explaining its close link with the elongation machinery (Bradsher et al., 2002; Citterio et al., 2000a; Hoogstraten et al., 2002). *Third*, the list of possible primary disease-causing mechanisms was further extended by reports indicating that CSB as well as TFIIH are components of RNA-PolI transcription (Bradsher et al., 2002;

Citterio et al., 2000a; Hoogstraten et al., 2002). *Fourth,* TFIIH complex has been shown to influence apoptosis (Wang et al., 1996) and it may, via its Cdk7/cyclinH containing CAK subcomplex also impinge upon cell-cycle regulation (Harper and Elledge, 1998). Which of those many processes primarily affects the outcome of a specific disease feature? Can e.g. neurodevelopmental defects in TTD, CS and XPCS be explained by the same basic mechanism or does each of them embody a distinct cause? Due to the inherent limitations of cellular and biochemical assays in predicting the systemic and time-dependent pathology the exact contribution of each of the above mechanisms to the specific disease feature has been difficult to disentangle. However, a number of patient-mimicking NER-defective mouse models have been generated and many of these, especially various genetic combinations, have turned out to be highly informative. First, we will discuss the relevant phenotypes of the single mutant mouse models.

6. LESSONS FROM REPAIR-DEFICIENT MOUSE MODELS WITH SEGMENTAL PREMATURE AGEING PHENOTYPES

Mice defective in the *Xpc* or *Xpa* genes (designated as *Xpc*$^{-/-}$ or *Xpa*$^{-/-}$ when homozygous mutants) completely lack GG-NER, or both the GG- and TC-NER respectively and, similar to the corresponding human patients display elevated UV-induced skin cancer predisposition (de Vries et al., 1995; Nakane et al., 1995; Sands et al., 1995). However, unlike a number of human XPA patients *Xpa*$^{-/-}$ mice do not show neuropathological features reminiscent to XP-DSC. Although the reason for this discrepancy is yet unclear, the development of neurodegenerative processes may require more time than the maximum murine life span of ~ 2-3 years enables. Similar reasoning may also explain the relatively mild phenotype of mice mimicking Cockayne syndrome who lack either the *Csa* or *Csb* genes (van der Horst et al., 2002; van der Horst et al., 1997b). Unlike CS patients, CS mice display only mildly accelerated ageing features and appear devoid of severe neuro-developmental traits, such as demyelination and profound failure to thrive (although they have reduced body weight). Mice mimicking a point mutation causative of TTD in the *Xpd* gene (XPD-R722W) show a more pronounced accelerated ageing phenotype, including early osteoporosis and cachexia, alongside the distinguishing hallmark of the disease, the brittle hair. Nevertheless, compared to human counterparts neurodevelopmental features in TTD mice are milder. Reminiscent of the CS mice, a recently generated mouse model for the combined form of XPCS (XPD - G602D) showed mild accelerated ageing features accompanied with XP-specific predisposition to UV-induced cancers (J.O. Andressoo et al., submitted). In conclusion, mouse models for NER disorders do phenocopy human pathology, albeit that the neurodevelopmental features, including accelerated segmental ageing, appear milder (Table 3, phenotypic group B).

Table 3. Phenotype of NER/TCR deficient mouse mutants.

	genotype	**Post natal cachexia**	**Cerebellar ataxia**	**Life span**	**Reference**
Phenotypic group A	*Xpf-/-*	severe	n.t.	~3 wks	(Tian et al., 2004)
	Xpg-/-	severe	yes	~3 wks	(Sun et al., 2001)
	Ercc1-/-	severe	yes	~3 wks	(Weeda et al., 1997) (Niederhofer et al., unpublished)
	Xpa-/-|*Csb-/-*	severe	yes	~3 wks	(de Boer et al., 2002; van der Horst et al., 1997b)
	Xpa-/-|*Csa-/-*	severe	n.t.	~3 wks	(van der Horst et al., 2002)
	Xpc-/-|*Csb-/-*	severe	n.t.	~3 wks	(I. van der Pluijm et al., submitted)
	Xpa-/-|TTD	severe	yes	~3 wks	(de Boer et al., 2002)
	Xpa-/-|XPCS	severe	yes	~3 wks	(J.O. Andressoo et al., submitted)
	Xpa-/- |Δex15*Xpg*	severe	n.t.	~3 wks	(Shiomi et al., 2005)
Phenotypic group B	*Csb-/-*	mild	n.t.	normal?	(van der Horst et al., 1997b)
	Csa-/-	mild	n.t.	normal?	(van der Horst et al., 2002)
	TTD	moderate	absent	reduced	(de Boer et al., 1998)
	XPCS	mild	absent	reduced	(J.O. Andressoo et al., submitted)
	Xpa-/-	absent	n.t.	normal?	(de Vries et al., 1995; Nakane et al., 1995)
	Xpc-/-	absent	n.t.	normal?	(Sands et al., 1995)

n.t.-not tested. Note: Although all the animals in group A have been reported to display features reminiscent to cerebellar ataxia this feature is only indicated as positive when confirmed by calbindin immunostaining of Purkinje neurons in the cerebellum. normal?- no gross differences in life span have been reported.

What is the mechanism causing acceleration of ageing in CS, XPCS and TTD? The premature segmental ageing cannot be merely explained by the NER defect, as completely NER-defective $Xpa^{-/-}$ mice fail to show these features, whereas the NER defect in CS, XPCS and TTD is generally partial. As apparent from the list above, multiple mechanisms may either alone or in combination stand for the outcome. Here, crossings of different NER-defective mice proved to be extremely informative. When $Xpa^{-/-}$ mice were crossed to either CS, XPCS or TTD mice the ageing features apparent in a mild form in the single mutants became dramatically exacerbated in all the double mutant mice. All the double mutants exhibited features reminiscent of human CS, XPCS and TTD, including severe progressive postnatal cachexia, loss of subcutaneous fat tissue, cerebellar ataxia, spasticity of movements and failure to thrive, and all shared a maximum life span of ~3 weeks (de Boer et al., 2002; van der Horst et al., 2002; van der Horst et al., 1997b)(J.O. Andressoo et al., submitted). The only anticipated additional effect of the XPA mutation at the molecular level is the conversion of a reduced level of GG-NER and/or TC-NER in CS, XPCS or TTD mice to a complete impairment in the double mutants. Moreover, mice double mutant for *Xpc* and *Csb* exhibited similar pathology and life span (van der Horst et al., 1997b)(I. van der Pluijm et al., submitted), confirming that the observed acceleration of ageing was due to endogenous DNA damage and not due to some unknown non-NER function of XPA. These findings show unequivocally that endogenous DNA damage is a major contributing factor to premature ageing in CS, XPCS and TTD alike and strongly suggest that a common defect in the same DNA repair trail is underlying the overlapping phenotypic outcome. Furthermore, mice deficient for the *Xpg* (the 3'-endonuclease of the NER reaction) display a strikingly similar phenotype and life span as the double mutants above (Harada et al., 1999; Shiomi et al., 2004; Weeda et al., 1997). Considering the notion that several XPG patients exhibit severe CS features, XPG protein must act within the same DNA repair trail similar to TFIIH, CSA and CSB. *Xpg* mice carrying a relatively mild mutation, deletion of exon15 (Δex15*Xpg*), develop only an XP-like, UV-sensitive phenotype, but fail to display notable CS features. Strikingly, similar to the other double mutants discussed above, Δex15*Xpg*/*Xpa-/-* mice display postnatal cachexia and a life span of ~3 weeks (Shiomi et al., 2005). Inactivation of the 5' NER endonuclease, *Ercc1/Xp,f* induces also a very severe premature ageing phenotype (Table 3, phenotypic group A) although the clinical features are in part different from the other NER mutants, notably this mutant shows dramatic premature aging of the liver and kidney. The latter is likely due to the additional role of the *Ercc1/Xpf* endonuclease in removal of the very toxic interstrand crosslinks (ICL). Milder mutations in this gene gave correspondingly longer lifespan ranging from ~4 months to ~14 months (A. Lalai, L. Niedernhofer et al., manuscript in preparation). Interestingly, principally the same clinical features emerge, albeit over a longer period of time compared to the KO mutant, demonstrating that these ageing symptoms can be compressed in time from over 1 year to ~3 weeks (A. Lalai, L. Niedernhofer, manuscript in preparation). These and other findings indicate that the severity of the mutation is at least in part determining the life span and extent of accelerating lesions. The notion that the phenotype of the Ercc1 mutant is in part distinct from that of the other NER mutants indicates that the type of lesion influences the clinical outcome. This may explain the segmental nature of premature aging syndromes in general: depending on the repair pathway affected and the corresponding lesion spectrum premature aging may differ between different organs and tissues. This is also consistent with the notion that organs and tissues will differ in terms of damage induction, exposure, response and protection systems, making them differentially dependent on the capacity and efficiency of the corresponding mechanisms. In line with this reasoning none of the BER, NHEJ, or HR mouse mutants display the particular phenotype of postnatal onset of progressive cachexia, neuropathies and a 3 week maximum life span exhibited by mutants in the NER or NER/ICLR pathways.

TCR is believed to be responsible not only for the removal of helix-distorting NER type of lesions but also for overcoming any transcriptional block by any kind of lesion. Consequently, due to the wider lesion spectrum of TCR, mutations in this trail can result in more severe pathology. In concordance with this notion it has been shown that some CS cells are hypersensitive to oxidative agents, indicating defective TCR of ROS-induced lesions normally repaired by a distinct repair trail:base excision

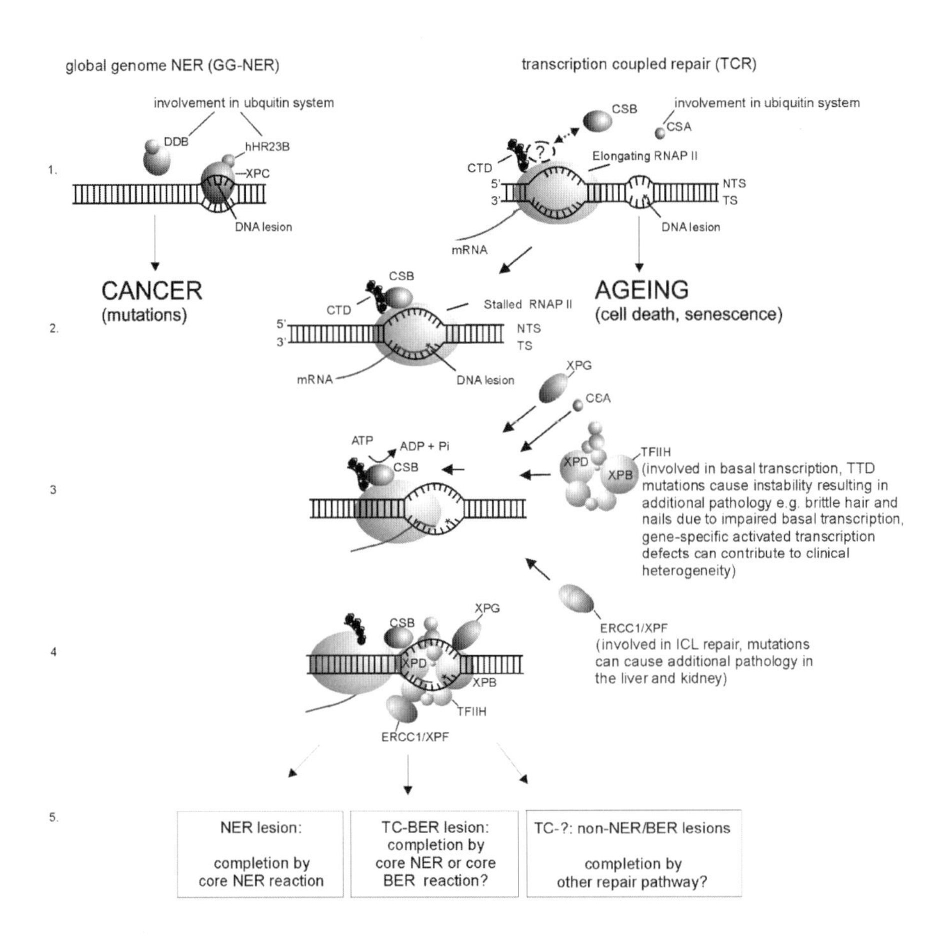

Figure 5. Model for GG-NER and TCR reaction and their physiological consequences. After the damage recognition step by hHR23B/XPC or elongating RNA-PolII respectively, GG-NER and TC-NER pathways utilise the common core NER reaction which involves recruitment of TFIIH and XPA followed by melting of the DNA around the lesion. In much less well understood TCR genetic evidence points to involvement of TFIIH, and XPG alongside with CS proteins in early steps of the reaction, such as removal of blocked RNA-PolII from the lesion. Whether in TCR all lesions are repaired by NER or are some delivered to specific repair trails is currently unknown.

repair (BER) (Cooper et al., 1997; de Boer et al., 2002; de Waard et al., 2003). Within the TCR context, TFIIH and CS proteins are thought to function in removal or back-tracking of the blocked RNA polymerase which subsequently enables repair proteins to access the lesion (Figure 5) (de Boer et al., 2002; Hoeijmakers, 2001; van den Boom et al., 2002). It has been suggested that in CS, XPCS and TTD the damage-stalled RNA polymerase complex may persist longer, causing gene inactivation and enhanced apoptosis or senescence and leading to functional decline and depletion of cell renewal capacity, which can ultimately lead to acceleration of ageing.

7. FINAL CONCLUSIONS

The intricacies of the GG-NER, TC-NER and presumably the broader TCR reaction combined with the occurrence of functions beyond repair of several of the NER factors and global genome repair defects translates into a complex and highly pleiotropic set of syndromes. However, by careful analysis of patients and mouse mutants we can come to the following simplifications. A defect in the global genome subpathway of NER, as observed in XPC patients, causes the cutaneous features of XP with the high incidence of skin cancer in the UV-exposed parts of the skin. When, in addition, the TC-NER pathway is affected, as in the case of XPA, the additional feature of accelerated neurodegeneration may be occurring as seen to an extreme extent in the DeSanctis-Cacchione form of XP. The CS features, including those apparent in TTD, may be the consequence of a defect in the broader TCR system, with TTD having on top of this the TFIIH instability leading to the typical brittle (unfinished) hair, nails and skin. This discloses some of the mechanistic intricacies that still need to be resolved. The problems caused by defective TCR can be further aggravated by a combination with a deficiency of the GG-NER subpathway, presumably because this increases the overall damage load. This may explain the very severe phenotype exhibited by some XPCS patients. Altogether this highlights the importance of the process that keeps our transcription going despite the continuous induction of DNA lesions: transcription-coupled repair.

ACKNOWLEDGEMENTS

We would like to specially thank N.G.J Jaspers, J. Mitchell, G. Garinis and S. Bergink for helpful discussions. This work was financially supported by the Netherlands Organisation for Scientific Research (NWO) through the foundation of the Research Institute Diseases of the Elderly, as well as grants from the NIH (1PO1 AG17242-02), NIEHS (1UO1 ES011044), EC (QRTL-1999-02002), and the Dutch Cancer Society (EUR 99-2004).

REFERENCES

Aboussekhra, A., M. Biggerstaff, M.K.K. Shivji, J.A. Vilpo, V. Moncollin, V.N. Podust, M. Protic, U. Hubscher, J.-M. Egly, and R.D. Wood. 1995. Mammalian DNA nucleotide excision repair reconstituted with purified components. *Cell.* 80:859-868.

Adimoolam, S., and J.M. Ford. 2002. p53 and DNA damage-inducible expression of the xeroderma pigmentosum group C gene. *Proc Natl Acad Sci U S A.* 99:12985-90.

Adimoolam, S., and J.M. Ford. 2003. p53 and regulation of DNA damage recognition during nucleotide excision repair. *DNA Repair (Amst).* 2:947-54.

Amundson, S.A., A. Patterson, K.T. Do, and A.J. Fornace, Jr. 2002. A nucleotide excision repair master-switch: p53 regulated coordinate induction of global genomic repair genes. *Cancer Biol Ther.* 1:145-9.

Araujo, S.J., F. Tirode, F. Coin, H. Pospiech, J.E. Syvaoja, M. Stucki, U. Hubscher, J.M. Egly, and R.D. Wood. 2000. Nucleotide excision repair of DNA with recombinant human proteins: definition of the minimal set of factors, active forms of TFIIH, and modulation by CAK. *Genes Dev.* 14:349-59.

Azcoitia, I., L.L. DonCarlos, and L.M. Garcia-Segura. 2003. Are gonadal steroid hormones involved in disorders of brain aging? *Aging Cell.* 2:31-7.

Balaban, R.S., S. Nemoto, and T. Finkel. 2005. Mitochondria, oxidants, and aging. *Cell.* 120:483-95.

Barnes, D.E., A.E. Tomkinson, A.R. Lehmann, A.D.B. Webster, and T. Lindahl. 1992. Mutations in the *DNA ligase I* gene of an individual with immunodeficienies and cellular hypersensitivity to DNA-damaging agents. *Cell.* 69:495-503.

Berg, R.J., A. de Vries, H. van Steeg, and F.R. de Gruijl. 1997. Relative susceptibilities of XPA knockout mice and their heterozygous and wild-type littermates to UVB-induced skin cancer. *Cancer Res.* 57:581-4.

Bergmann, E., and J.M. Egly. 2001. Trichothiodystrophy, a transcription syndrome. *Trends Genet.* 17:279-86.

Betarbet, R., T.B. Sherer, and J.T. Greenamyre. 2002. Animal models of Parkinson's disease. *Bioessays.* 24:308-18.

Bodak, N., S. Queille, M.F. Avril, B. Bouadjar, C. Drougard, A. Sarasin, and L. Daya-Grosjean. 1999. High levels of patched gene mutations in basal-cell carcinomas from patients with xeroderma pigmentosum. *Proc Natl Acad Sci U S A.* 96:5117-22.

Bohr, V.A., C.A. Smith, D.S. Okumoto, and P.C. Hanawalt. 1985. DNA repair in an active gene: removal of pyrimidine dimers from the *DHFR* gene of CHO cells is much more efficient than in the genome overall. *Cell.* 40:359-369.

Boonstra, A., A. van Oudenaren, M. Baert, H. van Steeg, P.J. Leenen, G.T. van der Horst, J.H. Hoeijmakers, H.F. Savelkoul, and J. Garssen. 2001. Differential ultraviolet-B-induced immunomodulation in XPA, XPC, and CSB DNA repair-deficient mice. *J Invest Dermatol.* 117:141-6.

Bootsma, D., and J.H.J. Hoeijmakers. 1993. Engagement with transcription. *Nature.* 363:114-115.

Bootsma, D., K.H. Kraemer, J.E. Cleaver, and J.H. Hoeijmakers. 2002. Nucleotide Excision Repair Syndromes:Xeroderma Pigmentosum, Cockayne Syndrome, and Trichothiodystrophy. In The Genetic Basis of Human Cancer. B. Vogelstein and K.W. Kinzler, editors. McGraw-Hill Medical Publishing Devision. 211-237.

Botta, E., T. Nardo, A.R. Lehmann, J.M. Egly, A.M. Pedrini, and M. Stefanini. 2002. Reduced level of the repair/transcription factor TFIIH in trichothiodystrophy. *Hum Mol Genet.* 11:2919-28.

Bradsher, J., J. Auriol, L. Proietti de Santis, S. Iben, J.L. Vonesch, I. Grummt, and J.M. Egly. 2002. CSB is a component of RNA pol I transcription. *Mol Cell.* 10:819-29.

Brash, D.E., N.M. Wikonkal, E. Remenyik, G.T. van der Horst, E.C. Friedberg, D.L. Cheo, H. van Steeg, A. Westerman, and H.J. van Kranen. 2001. The DNA damage signal for

Mdm2 regulation, Trp53 induction, and sunburn cell formation in vivo originates from actively transcribed genes. *J Invest Dermatol.* 117:1234-40.

Brooks, P.J. 2002. DNA repair in neural cells: basic science and clinical implications. *Mutat Res.* 509:93-108.

Broughton, B.C., M. Berneburg, H. Fawcett, E.M. Taylor, C.F. Arlett, T. Nardo, M. Stefanini, E. Menefee, V.H. Price, S. Queille, A. Sarasin, E. Bohnert, J. Krutmann, R. Davidson, K.H. Kraemer, and A.R. Lehmann. 2001. Two individuals with features of both xeroderma pigmentosum and trichothiodystrophy highlight the complexity of the clinical outcomes of mutations in the XPD gene. *Hum Mol Genet.* 10:2539-47.

Budd, M.E., and J.L. Campbell. 1995. DNA polymerases required for repair of UV-induced damage in *Saccharomyces cerevisiae. Molecular and Cellular Biology.* 15:2173-2179.

Butterfield, D.A., J. Drake, C. Pocernich, and A. Castegna. 2001. Evidence of oxidative damage in Alzheimer's disease brain: central role for amyloid beta-peptide. *Trends Mol Med.* 7:548-54.

Campisi, J. 2005. Senescent cells, tumor suppression, and organismal aging: good citizens, bad neighbors. *Cell.* 120:513-22.

Citterio, E., V. Van Den Boom, G. Schnitzler, R. Kanaar, E. Bonte, R.E. Kingston, J.H. Hoeijmakers, and W. Vermeulen. 2000a. ATP-dependent chromatin remodeling by the Cockayne syndrome B DNA repair-transcription-coupling factor. *Mol Cell Biol.* 20:7643-53.

Citterio, E., W. Vermeulen, and J.H. Hoeijmakers. 2000b. Transcriptional healing. *Cell.* 101:447-50.

Cleaver, J.E., L.H. Thompson, A.S. Richardson, and J.C. States. 1999. A summary of mutations in the UV-sensitive disorders: xeroderma pigmentosum, Cockayne syndrome, and trichothiodystrophy. *Hum Mutat.* 14:9-22.

Coin, F., J. Auriol, A. Tapias, P. Clivio, W. Vermeulen, and J.M. Egly. 2004. Phosphorylation of XPB helicase regulates TFIIH nucleotide excision repair activity. *EMBO J.* 23:4835-46.

Colella, S., T. Nardo, E. Botta, A.R. Lehmann, and M. Stefanini. 2000. Identical mutations in the CSB gene associated with either Cockayne syndrome or the DeSanctis-cacchione variant of xeroderma pigmentosum. *Hum Mol Genet.* 9:1171-5.

Conforti, G., T. Nardo, M. D'Incalci, and M. Stefanini. 2000. Proneness to UV-induced apoptosis in human fibroblasts defective in transcription coupled repair is associated with the lack of Mdm2 transactivation. *Oncogene.* 19:2714-20.

Cooper, P.K., T. Nouspikel, S.G. Clarkson, and S.A. Leadon. 1997. Defective transcription-coupled repair of oxidative base damage in Cockayne syndrome patients from XP group G. *Science.* 275:990-993.

Daya-Grosjean, L., C. Robert, C. Drougard, H. Suarez, and A. Sarasin. 1993. High mutation frequency in ras genes of skin tumors isolated from DNA repair deficient xeroderma pigmentosum patients. *Cancer Res.* 53:1625-9.

de Boer, J., J.O. Andressoo, J. de Wit, J. Huijmans, R.B. Beems, H. van Steeg, G. Weeda, G.T. van der Horst, W. van Leeuwen, A.P. Themmen, M. Meradji, and J.H. Hoeijmakers. 2002. Premature aging in mice deficient in DNA repair and transcription. *Science.* 296:1276-9.

de Boer, J., J. de Wit, H. van Steeg, R.J.W. Berg, M. Morreau, P. Visser, A.R. Lehmann, M. Duran, J.H.J. Hoeijmakers, and G. Weeda. 1998. A mouse model for the basal transcription/DNA repair syndrome trichothiodystrophy. *Mol. Cell.* 1:981-990.

de Laat, W.L., E. Appeldoorn, K. Sugasawa, E. Weterings, N.G. Jaspers, and J.H. Hoeijmakers. 1998. DNA-binding polarity of human replication protein A positions nucleases in nucleotide excision repair. *Genes Dev.* 12:2598-609.

de Laat, W.L., N.G.J. Jaspers, and J.H.J. Hoeijmakers. 1999. Molecular mechanism of nucleotide excision repair. *Genes Dev.* 13:768-785.

de Vries, A., C.T.M. van Oostrom, F.M.A. Hofhuis, P.M. Dortant, R.J.W. Berg, F.R. de Gruijl, P.W. Wester, C.F. van Kreijl, P.J.A. Capel, H. van Steeg, and S.J. Verbeek. 1995.

Increased susceptibility to ultraviolet-B and carcinogens of mice lacking the DNA excision repair gene *XPA*. *Nature*. 377:169-173.

de Waard, H., J. de Wit, T.G. Gorgels, G. van den Aardweg, J.O. Andressoo, M. Vermeij, H. van Steeg, J.H. Hoeijmakers, and G.T. van der Horst. 2003. Cell type-specific hypersensitivity to oxidative damage in CSB and XPA mice. *DNA Repair (Amst)*. 2:13-25.

D'Errico, M., A. Calcagnile, F. Canzona, B. Didona, P. Posteraro, R. Cavalieri, R. Corona, I. Vorechovsky, T. Nardo, M. Stefanini, and E. Dogliotti. 2000. UV mutation signature in tumor suppressor genes involved in skin carcinogenesis in xeroderma pigmentosum patients. *Oncogene*. 19:463-7.

D'Errico, M., M. Teson, A. Calcagnile, T. Nardo, N. De Luca, C. Lazzari, S. Soddu, G. Zambruno, M. Stefanini, and E. Dogliotti. 2005. Differential role of transcription-coupled repair in UVB-induced response of human fibroblasts and keratinocytes. *Cancer Res*. 65:432-8.

Dianov, G.L., T. Thybo, Dianova, II, L.J. Lipinski, and V.A. Bohr. 2000. Single nucleotide patch base excision repair is the major pathway for removal of thymine glycol from DNA in human cell extracts. *J Biol Chem*. 275:11809-13.

Drane, P., E. Compe, P. Catez, P. Chymkowitch, and J.M. Egly. 2004. Selective regulation of vitamin D receptor-responsive genes by TFIIH. *Mol Cell*. 16:187-97.

Drapkin, R., J.T. Reardon, A. Ansari, J.C. Huang, L. Zawel, K. Ahn, A. Sancar, and D. Reinberg. 1994. Dual role of TFIIH in DNA excision repair and in transcription by RNA polymerase II. *Nature*. 368:769-772.

Egly, J.M. 2001. The 14th Datta Lecture. TFIIH: from transcription to clinic. *FEBS Lett*. 498:124-8.

Elder, R.T., X.Q. Song, M. Chen, K.M. Hopkins, H.B. Lieberman, and Y. Zhao. 2002. Involvement of rhp23, a Schizosaccharomyces pombe homolog of the human HHR23A and Saccharomyces cerevisiae RAD23 nucleotide excision repair genes, in cell cycle control and protein ubiquitination. *Nucleic Acids Res*. 30:581-91.

Evans, E., J.G. Moggs, J.R. Hwang, J.M. Egly, and R.D. Wood. 1997a. Mechanism of open complex and dual incision formation by human nucleotide excision repair factors. *EMBO J*. 16:6559-73.

Evans, E., J.G. Moggs, J.R. Hwang, J.-M. Egly, and R.D. Wood. 1997b. Mechanism of open complex and dual incision formation by human nucleotide excision repair factors. *EMBO J*. 16:6559-6573.

Fitch, M.E., S. Nakajima, A. Yasui, and J.M. Ford. 2003. In vivo recruitment of XPC to UV-induced cyclobutane pyrimidine dimers by the DDB2 gene product. *J Biol Chem*. 278:46906-10.

Friedberg, E.C., J.P. Bond, D.K. Burns, D.L. Cheo, M.S. Greenblatt, L.B. Meira, D. Nahari, and A.M. Reis. 2000. Defective nucleotide excision repair in xpc mutant mice and its association with cancer predisposition. *Mutat Res*. 459:99-108.

Friedberg, E.C., G.C. Walker, and W. Siede. 1995. DNA repair and mutagenesis. ASM Press, Washington D.C.

Fujiwara, Y., C. Masutani, T. Mizukoshi, J. Kondo, F. Hanaoka, and S. Iwai. 1999. Characterization of DNA recognition by the human UV-damaged DNA-binding protein. *J Biol Chem*. 274:20027-33.

Gao, Y., D.O. Ferguson, W. Xie, J.P. Manis, J. Sekiguchi, K.M. Frank, J. Chaudhuri, J. Horner, R.A. DePinho, and F.W. Alt. 2000. Interplay of p53 and DNA-repair protein XRCC4 in tumorigenesis, genomic stability and development. *Nature*. 404:897-900.

Giglia, G., N. Dumaz, C. Drougard, M.F. Avril, L. Daya-Grosjean, and A. Sarasin. 1998. p53 mutations in skin and internal tumors of xeroderma pigmentosum patients belonging to the complementation group C. *Cancer Res*. 58:4402-9.

Giglia-Mari, G., F. Coin, J.A. Ranish, D. Hoogstraten, A. Theil, N. Wijgers, N.G. Jaspers, A. Raams, M. Argentini, P.J. Van Der Spek, E. Botta, M. Stefanini, J.M. Egly, R. Aebersold, J.H. Hoeijmakers, and W. Vermeulen. 2004. A new, tenth subunit of TFIIH is responsible for the DNA repair syndrome trichothiodystrophy group A. *Nat Genet*. 36:714-9.

Glockzin, S., F.X. Ogi, A. Hengstermann, M. Scheffner, and C. Blattner. 2003. Involvement of the DNA repair protein hHR23 in p53 degradation. *Mol Cell Biol.* 23:8960-9.

Goode, E.L., C.M. Ulrich, and J.D. Potter. 2002. Polymorphisms in DNA repair genes and associations with cancer risk. *Cancer Epidemiol Biomarkers Prev.* 11:1513-30.

Graham, J.M., Jr., K. Anyane-Yeboa, A. Raams, E. Appeldoorn, W.J. Kleijer, V.H. Garritsen, D. Busch, T.G. Edersheim, and N.G. Jaspers. 2001. Cerebro-oculo-facio-skeletal syndrome with a nucleotide excision-repair defect and a mutated XPD gene, with prenatal diagnosis in a triplet pregnancy. *Am J Hum Genet.* 69:291-300.

Groisman, R., J. Polanowska, I. Kuraoka, J. Sawada, M. Saijo, R. Drapkin, A.F. Kisselev, K. Tanaka, and Y. Nakatani. 2003. The ubiquitin ligase activity in the DDB2 and CSA complexes is differentially regulated by the COP9 signalosome in response to DNA damage. *Cell.* 113:357-67.

Gunz, D., M.T. Hess, and H. Naegeli. 1996. Recognition of DNA adducts by human nucleotide excision repair. Evidence for a thermodynamic probing mechanism. *J Biol Chem.* 271:25089-98.

Harada, Y.N., N. Shiomi, M. Koike, M. Ikawa, M. Okabe, S. Hirota, Y. Kitamura, M. Kitagawa, T. Matsunaga, O. Nikaido, and T. Shiomi. 1999. Postnatal growth failure, short life span, and early onset of cellular senescence and subsequent immortalization in mice lacking the xeroderma pigmentosum group G gene. *Mol Cell Biol.* 19:2366-72.

Harper, J.W., and S.J. Elledge. 1998. The role of Cdk7 in CAK function, a retro-retrospective. *Genes Dev.* 12:285-9.

Hasty, P., J. Campisi, J. Hoeijmakers, H. van Steeg, and J. Vijg. 2003. Aging and genome maintenance: lessons from the mouse? *Science.* 299:1355-9.

Hayashi, M. 1999. [Apoptotic cell death in child-onset neurodegenerative disorders]. *No To Hattatsu.* 31:146-52.

Hoeijmakers, J.H. 2001. Genome maintenance mechanisms for preventing cancer. *Nature.* 411:366-74.

Hoogstraten, D., A.L. Nigg, H. Heath, L.H. Mullenders, R. van Driel, J.H. Hoeijmakers, W. Vermeulen, and A.B. Houtsmuller. 2002. Rapid switching of TFIIH between RNA polymerase I and II transcription and DNA repair in vivo. *Mol Cell.* 10:1163-74.

Horibata, K., Y. Iwamoto, I. Kuraoka, N.G. Jaspers, A. Kurimasa, M. Oshimura, M. Ichihashi, and K. Tanaka. 2004. Complete absence of Cockayne syndrome group B gene product gives rise to UV-sensitive syndrome but not Cockayne syndrome. *Proc Natl Acad Sci U S A.* 101:15410-5.

Hwang, B.J., and G. Chu. 1993. Purification and characterization of a human protein that binds to damaged DNA. *Biochemistry.* 32:1657-66.

Hwang, B.J., J.M. Ford, P.C. Hanawalt, and G. Chu. 1999. Expression of the p48 xeroderma pigmentosum gene is p53-dependent and is involved in global genomic repair. *Proc Natl Acad Sci U S A.* 96:424-8.

Hwang, B.J., S. Toering, U. Francke, and G. Chu. 1998. p48 Activates a UV-damaged-DNA binding factor and is defective in xeroderma pigmentosum group E cells that lack binding activity. *Mol Cell Biol.* 18:4391-9.

Iben, S., H. Tschochner, M. Bier, D. Hoogstraten, P. Hozak, J.M. Egly, and I. Grummt. 2002. TFIIH plays an essential role in RNA polymerase I transcription. *Cell.* 109:297-306.

Itin, P.H., A. Sarasin, and M.R. Pittelkow. 2001. Trichothiodystrophy: update on the sulfur-deficient brittle hair syndromes. *J Am Acad Dermatol.* 44:891-920; quiz 921-4.

Itoh, M., M. Hayashi, K. Shioda, M. Minagawa, F. Isa, K. Tamagawa, Y. Morimatsu, and M. Oda. 1999. Neurodegeneration in hereditary nucleotide repair disorders. *Brain Dev.* 21:326-33.

Itoh, T., C. O'Shea, and S. Linn. 2003. Impaired regulation of tumor suppressor p53 caused by mutations in the xeroderma pigmentosum DDB2 gene: mutual regulatory interactions between p48(DDB2) and p53. *Mol Cell Biol.* 23:7540-53.

James, M., J. Mansbridge, and C. Kidson. 1982. Ultraviolet radiation sensitivity of proliferating and differentiated human neuroblastoma cells. *Int J Radiat Biol Relat Stud Phys Chem Med.* 41:547-56.

Jans, J., W. Schul, Y.G. Sert, Y. Rijksen, H. Rebel, A.P. Eker, S. Nakajima, H. van Steeg, F.R. de Gruijl, A. Yasui, J.H. Hoeijmakers, and G.T. van der Horst. 2005. Powerful skin cancer protection by a CPD-photolyase transgene. *Curr Biol*. 15:105-15.

Jaspers, N.G., A. Raams, M.J. Kelner, J.M. Ng, Y.M. Yamashita, S. Takeda, T.C. McMorris, and J.H. Hoeijmakers. 2002. Anti-tumour compounds illudin S and Irofulven induce DNA lesions ignored by global repair and exclusively processed by transcription- and replication-coupled repair pathways. *DNA Repair (Amst)*. 1:1027-38.

Keriel, A., A. Stary, A. Sarasin, C. Rochette-Egly, and J.M. Egly. 2002. XPD mutations prevent TFIIH-dependent transactivation by nuclear receptors and phosphorylation of RARalpha. *Cell*. 109:125-35.

Kraemer, K.H., M.M. Lee, and J. Scotto. 1987. Xeroderma pigmentosum. Cutaneous, ocular, and neurologic abnormalities in 830 published cases. *Arch Dermatol*. 123:241-50.

Lehmann, A.R. 2001. The xeroderma pigmentosum group D (XPD) gene: one gene, two functions, three diseases. *Genes Dev*. 15:15-23.

Li, L., X. Lu, C.A. Peterson, and R.J. Legerski. 1995. An interaction between the DNA repair factor XPA and replication protein A appears essential for nucleotide excision repair. *Mol Cell Biol*. 15:5396-402.

Lindahl, T. 1993. Instability and decay of the primary structure of DNA. *Nature*. 362:709-715.

Lindenbaum, Y., D. Dickson, P. Rosenbaum, K. Kraemer, I. Robbins, and I. Rapin. 2001. Xeroderma pigmentosum/cockayne syndrome complex: first neuropathological study and review of eight other cases. *Eur J Paediatr Neurol*. 5:225-42.

Liu, J., S. Akoulitchev, A. Weber, H. Ge, S. Chuikov, D. Libutti, X.W. Wang, J.W. Conaway, C.C. Harris, R.C. Conaway, D. Reinberg, and D. Levens. 2001. Defective interplay of activators and repressors with TFIH in xeroderma pigmentosum. *Cell*. 104:353-63.

Ljungman, M., and D.P. Lane. 2004. Transcription - guarding the genome by sensing DNA damage. *Nat Rev Cancer*. 4:727-37.

Ljungman, M., and F. Zhang. 1996. Blockage of RNA polymerase as a possible trigger for u.v. light-induced apoptosis. *Oncogene*. 13:823-31.

Lu, Y., H. Lian, P. Sharma, N. Schreiber-Agus, R.G. Russell, L. Chin, G.T. van der Horst, and D.B. Bregman. 2001. Disruption of the Cockayne syndrome B gene impairs spontaneous tumorigenesis in cancer-predisposed Ink4a/ARF knockout mice. *Mol Cell Biol*. 21:1810-8.

Meira, L.B., J.M. Graham, Jr., C.R. Greenberg, D.B. Busch, A.T. Doughty, D.W. Ziffer, D.M. Coleman, I. Savre-Train, and E.C. Friedberg. 2000. Manitoba aboriginal kindred with original cerebro-oculo- facio-skeletal syndrome has a mutation in the Cockayne syndrome group B (CSB) gene. *Am J Hum Genet*. 66:1221-8.

Mellon, I., V.A. Bohr, C.A. Smith, and P.C. Hanawalt. 1986. Preferential DNA repair of an active gene in human cells. *Proc Natl Acad Sci U S A*. 83:8878-8882.

Mellon, I., G. Spivak, and P.C. Hanawalt. 1987. Selective removal of transcription-blocking DNA damage from the transcribed strand of the mammalian *DHFR* gene. *Cell*. 51:241-249.

Miyauchi-Hashimoto, H., K. Tanaka, and T. Horio. 1996. Enhanced inflammation and immunosuppression by ultraviolet radiation in xeroderma pigmentosum group A (XPA) model mice. *J Invest Dermatol*. 107:343-8.

Mu, D., D.S. Hsu, and A. Sancar. 1996. Reaction mechanism of human DNA repair excision nuclease. *J. Biol. Chem.* 271:8285-8294.

Nakane, H., S. Takeuchi, S. Yuba, M. Saijo, Y. Nakatsu, T. Ishikawa, S. Hirota, Y. Kitamura, Y. Kato, Y. Tsunoda, H. Miyauchi, T. Horio, T. Tokunaga, T. Matsunaga, O. Nikaido, Y. Nishimune, Y. Okada, and K. Tanaka. 1995. High incedence of ultraviolet-B- or chemical-carcinogen-induced skin tomours in mice lacking the xeroderma pigmentosum group A gene. *Nature*. 377:165-168.

Nakura, J., L. Ye, A. Morishima, K. Kohara, and T. Miki. 2000. Helicases and aging. *Cell Mol Life Sci*. 57:716-30.

Nance, M.A., and S.A. Berry. 1992. Cockayne syndrome: Review of 140 cases. *Am. J. Med. Genet.* 42:68-84.

Ng, J.M., W. Vermeulen, G.T. van der Horst, S. Bergink, K. Sugasawa, H. Vrieling, and J.H. Hoeijmakers. 2003. A novel regulation mechanism of DNA repair by damage-induced and RAD23-dependent stabilization of xeroderma pigmentosum group C protein. *Genes Dev.* 17:1630-45.

Niedernhofer, L.J., J. Essers, G. Weeda, B. Beverloo, J. de Wit, M. Muijtjens, H. Odijk, J.H. Hoeijmakers, and R. Kanaar. 2001. The structure-specific endonuclease Ercc1-Xpf is required for targeted gene replacement in embryonic stem cells. *EMBO J.* 20:6540-9.

Nouspikel, T., and P.C. Hanawalt. 2000. Terminally differentiated human neurons repair transcribed genes but display attenuated global DNA repair and modulation of repair gene expression. *Mol Cell Biol.* 20:1562-70.

Nouspikel, T., and P.C. Hanawalt. 2002. DNA repair in terminally differentiated cells. *DNA Repair (Amst).* 1:59-75.

O'Donovan, A., A.A. Davies, J.G. Moggs, S.C. West, and R.D. Wood. 1994. XPG endonuclease makes the 3' incision in human DNA nucleotide excision repair. *Nature.* 371:432-435.

Parkes, T.L., A.J. Elia, D. Dickinson, A.J. Hilliker, J.P. Phillips, and G.L. Boulianne. 1998. Extension of Drosophila lifespan by overexpression of human SOD1 in motorneurons. *Nat Genet.* 19:171-4.

Payne, A., and G. Chu. 1994. Xeroderma pigmentosum group E binding factor recognizes a broad spectrum of DNA damage. *Mutat Res.* 310:89-102.

Rademakers, S., M. Volker, D. Hoogstraten, A.L. Nigg, M.J. Mone, A.A. Van Zeeland, J.H. Hoeijmakers, A.B. Houtsmuller, and W. Vermeulen. 2003. Xeroderma pigmentosum group A protein loads as a separate factor onto DNA lesions. *Mol Cell Biol.* 23:5755-67.

Randerath, E., G.D. Zhou, and K. Randerath. 1997a. Organ-specific oxidative DNA damage associated with normal birth in rats. *Carcinogenesis.* 18:859-66.

Randerath, K., G.D. Zhou, S.A. Monk, and E. Randerath. 1997b. Enhanced levels in neonatal rat liver of 7,8-dihydro-8-oxo-2'-deoxyguanosine (8-hydroxydeoxyguanosine), a major mutagenic oxidative DNA lesion. *Carcinogenesis.* 18:1419-21.

Randerath, K., G.D. Zhou, R.L. Somers, J.H. Robbins, and P.J. Brooks. 2001. A 32P-postlabeling assay for the oxidative DNA lesion 8,5'-cyclo-2'-deoxyadenosine in mammalian tissues: evidence that four type II I-compounds are dinucleotides containing the lesion in the 3' nucleotide. *J Biol Chem.* 276:36051-7.

Rapin, I., Y. Lindenbaum, D.W. Dickson, K.H. Kraemer, and J.H. Robbins. 2000. Cockayne syndrome and xeroderma pigmentosum. *Neurology.* 55:1442-9.

Reardon, J.T., A.F. Nichols, S. Keeney, C.A. Smith, J.S. Taylor, S. Linn, and A. Sancar. 1993. Comparative analysis of binding of human damaged DNA-binding protein (XPE) and *escherichia-coli* damage recognition protein (UvrA) to the major ultraviolet photoproducts - T[CS]TT[Ts]TT[6-4]T and T[Dewar]T. *J. Biol. Chem.* 268:21301-21308.

Riedl, T., F. Hanaoka, and J.M. Egly. 2003. The comings and goings of nucleotide excision repair factors on damaged DNA. *EMBO J.* 22:5293-303.

Sands, A.T., A. Abuin, A. Sanchez, C.J. Conti, and A. Bradley. 1995. High susceptibility to ultraviolet-induced carcinogenesis in mice lacking *XPC*. *Nature.* 377:162-165.

Schaeffer, L., V. Moncollin, R. Roy, A. Staub, M. Mezzina, A. Sarasin, G. Weeda, J.H.J. Hoeijmakers, and J.M. Egly. 1994. The ERCC2/DNA repair protein is associated with the class II BTF2/TFIIH transcription factor. *EMBO J.* 13:2388-2392.

Schaeffer, L., R. Roy, S. Humbert, V. Moncollin, W. Vermeulen, J.H. Hoeijmakers, P. Chambon, and J.M. Egly. 1993a. DNA repair helicase: a component of BTF2 (TFIIH) basic transcription factor. *Science.* 260:58-63.

Schaeffer, L., R. Roy, S. Humbert, V. Moncollin, W. Vermeulen, J.H.J. Hoeijmakers, P. Chambon, and J. Egly. 1993b. DNA repair helicase: a component of BTF2 (TFIIH) basic transcription factor. *Science.* 260:58-63.

Shiomi, N., S. Kito, M. Oyama, T. Matsunaga, Y.N. Harada, M. Ikawa, M. Okabe, and T. Shiomi. 2004. Identification of the XPG region that causes the onset of Cockayne syndrome by using Xpg mutant mice generated by the cDNA-mediated knock-in method. *Mol Cell Biol*. 24:3712-9.

Shiomi, N., M. Mori, S. Kito, Y.N. Harada, K. Tanaka, and T. Shiomi. 2005. Severe growth retardation and short life span of double-mutant mice lacking Xpa and exon 15 of Xpg. *DNA Repair (Amst)*. 4:351-7.

Shivji, M.K.K., V.N. Podust, U. Hubscher, and R.D. Wood. 1995. Nucleotide excision repair DNA synthesis by DNA polymerase epsilon in the presence of PCNA, RFC, and RPA. *Biochemistry*. 34:5011-5017.

Sibghat-Ullah, I. Husain, W. Carlton, and A. Sancar. 1989. Human nucleotide excision repair in vitro: repair of pyrimidine dimers, psoralen and cisplatin adducts by HeLa cell-free extract. *Nucleic Acids Res*. 17:4471-84.

Sijbers, A.M., W.L. de Laat, R.R. Ariza, M. Biggerstaff, Y.F. Wei, J.G. Moggs, K.C. Carter, B.K. Shell, E. Evans, M.C. de Jong, S. Rademakers, J. de Rooij, N.G. Jaspers, J.H. Hoeijmakers, and R.D. Wood. 1996. Xeroderma pigmentosum group F caused by a defect in a structure- specific DNA repair endonuclease. *Cell*. 86:811-822.

Spatz, A., G. Giglia-Mari, S. Benhamou, and A. Sarasin. 2001. Association between DNA repair-deficiency and high level of p53 mutations in melanoma of Xeroderma pigmentosum. *Cancer Res*. 61:2480-6.

Stigger, E., R. Drissi, and S.H. Lee. 1998. Functional analysis of human replication protein A in nucleotide excision repair. *J Biol Chem*. 273:9337-43.

Sugasawa, K., J.M. Ng, C. Masutani, S. Iwai, P.J. van der Spek, A.P. Eker, F. Hanaoka, D. Bootsma, and J.H. Hoeijmakers. 1998. Xeroderma pigmentosum group C protein complex is the initiator of global genome nucleotide excision repair. *Mol Cell*. 2:223-32.

Sun, X.Z., Y.N. Harada, S. Takahashi, N. Shiomi, and T. Shiomi. 2001. Purkinje cell degeneration in mice lacking the xeroderma pigmentosum group G gene. *J Neurosci Res*. 64:348-54.

Takada, K., and L.E. Becker. 1986. Cockayne's syndrome: report of two autopsy cases associated with neurofibrillary tangles. *Clin Neuropathol*. 5:64-8.

Tan, T., and G. Chu. 2002. p53 Binds and activates the xeroderma pigmentosum DDB2 gene in humans but not mice. *Mol Cell Biol*. 22:3247-54.

Tang, J.Y., B.J. Hwang, J.M. Ford, P.C. Hanawalt, and G. Chu. 2000. Xeroderma pigmentosum p48 gene enhances global genomic repair and suppresses UV-induced mutagenesis. *Mol Cell*. 5:737-44.

Tian, M., R. Shinkura, N. Shinkura, and F.W. Alt. 2004. Growth retardation, early death, and DNA repair defects in mice deficient for the nucleotide excision repair enzyme XPF. *Mol Cell Biol*. 24:1200-5.

van den Boom, V., N.G. Jaspers, and W. Vermeulen. 2002. When machines get stuck--obstructed RNA polymerase II: displacement, degradation or suicide. *Bioessays*. 24:780-4.

van der Horst, G.T., L. Meira, T.G. Gorgels, J. de Wit, S. Velasco-Miguel, J.A. Richardson, Y. Kamp, M.P. Vreeswijk, B. Smit, D. Bootsma, J.H. Hoeijmakers, and E.C. Friedberg. 2002. UVB radiation-induced cancer predisposition in Cockayne syndrome group A (Csa) mutant mice. *DNA Repair (Amst)*. 1:143-57.

van der Horst, G.T., H. van Steeg, R.J. Berg, A.J. van Gool, J. de Wit, G. Weeda, H. Morreau, R.B. Beems, C.F. van Kreijl, F.R. de Gruijl, D. Bootsma, and J.H. Hoeijmakers. 1997a. Defective transcription-coupled repair in Cockayne syndrome B mice is associated with skin cancer predisposition. *Cell*. 89:425-35.

van der Horst, G.T.J., H. van Steeg, R.J.W. Berg, A. van Gool, J. de Wit, G. Weeda, H. Morreau, R.B. Beems, C.F. van Kreijl, F.R. de Gruijl, D. Bootsma, and J.H.J. Hoeijmakers. 1997b. Defective transcription-coupled repair in Cockayne syndrome B mice is associated with skin cancer predisposition. *Cell*. 89:425-35.

van der Maazen, R.W., B.J. Kleiboer, I. Verhagen, and A.J. van der Kogel. 1993. Repair capacity of adult rat glial progenitor cells determined by an in vitro clonogenic assay after in vitro or in vivo fractionated irradiation. *Int J Radiat Biol*. 63:661-6.

Vermeulen, W., S. Rademakers, N.G. Jaspers, E. Appeldoorn, A. Raams, B. Klein, W.J. Kleijer, L.K. Hansen, and J.H. Hoeijmakers. 2001. A temperature-sensitive disorder in basal transcription and DNA repair in humans. *Nat Genet.* 27:299-303.

Vermeulen, W., A.J. van Vuuren, M. Chipoulet, L. Schaeffer, E. Appeldoorn, G. Weeda, N.G.J. Jaspers, A. Priestley, C.F. Arlett, A.R. Lehmann, M. Stefanini, M. Mezzina, A. Sarasin, D. Bootsma, J.-M. Egly, and J.H.J. Hoeijmakers. 1994. Three unusual repair deficiencies associated with transcription factor BTF2(TFIIH): Evidence for the existence of a transcription syndrome. *Cold Spring Harb. Symp. Quant. Biol.* 59:317-329.

Volker, M., M.J. Mone, P. Karmakar, A. van Hoffen, W. Schul, W. Vermeulen, J.H. Hoeijmakers, R. van Driel, A.A. van Zeeland, and L.H. Mullenders. 2001. Sequential assembly of the nucleotide excision repair factors in vivo. *Mol Cell.* 8:213-24.

Wakasugi, M., A. Kawashima, H. Morioka, S. Linn, A. Sancar, T. Mori, O. Nikaido, and T. Matsunaga. 2002. DDB accumulates at DNA damage sites immediately after UV irradiation and directly stimulates nucleotide excision repair. *J Biol Chem.* 277:1637-40.

Wang, G., L. Chuang, X. Zhang, S. Colton, A. Dombkowski, J. Reiners, A. Diakiw, and X.S. Xu. 2004. The initiative role of XPC protein in cisplatin DNA damaging treatment-mediated cell cycle regulation. *Nucleic Acids Res.* 32:2231-40.

Wang, X.W., W. Vermeulen, J.D. Coursen, M. Gibson, S.E. Lupold, K. Forrester, G. Xu, L. Elmore, H. Yeh, J.H. Hoeijmakers, and C.C. Harris. 1996. The XPB and XPD DNA helicases are components of the p53-mediated apoptosis pathway. *Genes Dev.* 10:1219-32.

Weeda, G., I. Donker, J. de Wit, H. Morreau, R. Janssens, C.J. Vissers, A. Nigg, H. van Steeg, D. Bootsma, and J.H.J. Hoeijmakers. 1997. Disruption of mouse ERCC1 results in a novel repair syndrome with growth failure, nuclear abnormalities and senescence. *Curr Biol.* 7:427-39.

Wijnhoven, S.W., H.J. Kool, L.H. Mullenders, A.A. van Zeeland, E.C. Friedberg, G.T. van der Horst, H. van Steeg, and H. Vrieling. 2000. Age-dependent spontaneous mutagenesis in Xpc mice defective in nucleotide excision repair. *Oncogene.* 19:5034-7.

Winkler, G.S., S.J. Araujo, U. Fiedler, W. Vermeulen, F. Coin, J.M. Egly, J.H. Hoeijmakers, R.D. Wood, H.T. Timmers, and G. Weeda. 2000. TFIIH with inactive XPD helicase functions in transcription initiation but is defective in DNA repair. *J Biol Chem.* 275:4258-66.

Winkler, G.S., K. Sugasawa, A.P. Eker, W.L. de Laat, and J.H. Hoeijmakers. 2001. Novel functional interactions between nucleotide excision DNA repair proteins influencing the enzymatic activities of TFIIH, XPG, and ERCC1-XPF. *Biochemistry.* 40:160-5.

Wood, R.D. 1999. DNA damage recognition during nucleotide excision repair in mammalian cells. *Biochimie.* 81:39-44.

Wood, R.D., P. Robins, and T. Lindahl. 1988. Complementation of the xeroderma pigmentosum DNA repair defect in cell-free extracts. *Cell.* 53:97-106.

Yamaizumi, M., and T. Sugano. 1994. U.v.-induced nuclear accumulation of p53 is evoked through DNA damage of actively transcribed genes independent of the cell cycle. *Oncogene.* 9:2775-84.

Yokoi, M., C. Masutani, T. Maekawa, K. Sugasawa, Y. Ohkuma, and F. Hanaoka. 2000. The xeroderma pigmentosum group C protein complex XPC-HR23B plays an important role in the recruitment of transcription factor IIH to damaged DNA. *J Biol Chem.* 275:9870-5.

You, J.S., M. Wang, and S.H. Lee. 2003. Biochemical analysis of the damage recognition process in nucleotide excision repair. *J Biol Chem.* 278:7476-85.

Chapter 2.2

DNA MISMATCH REPAIR AND COLON CANCER

Giancarlo Marra and Josef Jiricny
Institute of Molecular Cancer Research, University of Zurich, Zurich, Switzerland

1. INTRODUCTION

The Mismatch Repair (MMR) system is the major pathway responsible for repair of base-base mispairs and short insertion/deletion loops (IDLs) that arise during DNA replication and as intermediates of homologous recombination. Left unrepaired, these structures will give rise to base-substitution and frameshift mutations, respectively. With the exception of archaea, the MMR system is highly conserved from bacteria to man, and in all these organisms makes a critical contribution towards the maintenance of genomic stability. In the last decade, our understanding of the biochemical and structural aspects of MMR has made great advances. However, the precise biological functions of the key factors of the human MMR system, the MutS homologues (MSH) hMSH2, hMSH3 and hMSH6, and the MutL homologues (MLH) hMLH1 and hPMS2 (where PMS stands for Post Meiotic Segregation) have yet to be elucidated. The roles of other MSH and MLH/PMS homologues, such as hMSH4, hMSH5, hPMS1 and hMLH3, remain largely uncharacterised, in spite of the availability of viable mouse knock-out models. The functional relationship between MMR proteins and the DNA replication factors PCNA, RF-C, RPA and DNA polymerase δ, as well as exonucleolytic enzymes such as EXO1, also requires further study.

The recent advancements in the MMR field have been propelled by the discovery in 1993 of a causal link between inherited mutations in *MMR* genes and the common colon cancer predisposition syndrome Hereditary Non-polyposis Colon Cancer (HNPCC). In addition, many sporadic (*i.e.* non-familial) colon cancers have defective MMR that results from somatic transcriptional silencing of *hMLH1*. Finally, although MMR-deficient tumours appear to have a better prognosis than other colorectal cancers (CRCs), cells with a MMR defect were found to be resistant to certain drugs

E.A. Nigg (ed.), Genome Instability in Cancer Development, 85-123.

currently used in the treatment of cancer. There is thus an urgent need to identify tools that can be deployed in the facile diagnosis of this type of tumours, which will facilitate more effective clinical management.

In this chapter, we will focus on the biochemical aspects of the recognition and processing of DNA replication errors and on the human phenotype resulting from MMR deficiency.

2. MECHANISM OF HUMAN MMR

2.1 Overview

In this section, we shall discuss the mechanistic aspects of the human post-replicative MMR, however, comparisons with MMR in bacteria and yeast will be made in cases where information from the human system is unavailable. To distinguish between human and yeast MMR proteins, the names of the former are in capitals and carry the prefix "h" (*e.g.* hMSH2). Yeast proteins are denoted with only the first letter capitalised and the name followed by the letter "p" (*e.g.* Msh2p). Where the discussed properties are shared by proteins from both species, only capital letters are used (*e.g.* MSH2). Names of human and yeast genes are capitalised and appear in *italics* (*e.g. hMSH2, MSH2*). Readers interested in details of the MMR process in bacteria and yeast are referred to any of the large number of excellent reviews that appeared during the recent years (Bhagwat and Lieb, 2002; Harfe and Jinks-Robertson, 2000; Kolodner and Marsischky, 1999; Marti et al., 2002; Paques and Haber, 1999; Rasmussen et al., 1998).

The term "MMR" is often preceded by the adjective "postreplicative", which implicates the process in the correction of errors of DNA replication. The need for such an editing function derives from the fact that the fidelity of replicating DNA polymerases is insufficient to generate an error-free copy of genomic DNA. Single-base substitutions have been estimated to arise once in every 10^4-10^6 nucleotides incorporated. The intrinsic exonucleolytic proofreading activity of the replicative polymerases increase the fidelity of DNA synthesis by a further two orders of magnitude, i.e. to one error in 10^7 to 10^8 ((Loeb, 1991; Schaaper, 1993); for review, see (Kunkel and Bebenek, 2000)). MMR reduces the error rate to a range of 10^{-9} to 10^{-10}, which ensures that the human genome can be duplicated without mutations.

In addition to base-base mismatches, slippage of the primer strand with respect to the template can give rise to IDLs. These structures arise in particular in repetitive DNA sequences, such as mono-, di-, tri- or tetra-nucleotide repeats, also called microsatellites. Extrahelical nucleotides in the primer strand give rise to insertions in the progeny DNA, while their presence in the template strand leads to deletions. As in the case of base-base mismatches, frameshift error rate depends on the type of DNA polymerase and on its proofreading activity. However, the mutation rates due to IDLs are much more dependent on sequence context. Thus, while misaligned intermediates arise in short repeats or non-iterated sequences with a rate similar to that for base substitutions, their frequency rises substantially with increasing length of the repeat. This rise may be linked to an increased frequency of slippage in long microsatellites, but a more likely explanation is that the longer the repeat, the higher is the likelihood that the end of the primer will form a stable duplex from which the polymerase can extend. In addition, when the slipped strands reanneal such that the IDL is more than four nucleotides from primer terminus, the misalignment is less likely to be detected by the proofreading activity of the polymerase. The apparent error rate of the enzyme might thus increase by up to 100-fold. That microsatellite instability (MSI) is a phenotypic trait of organisms lacking MMR pays witness to the fact that IDLs are efficiently eliminated by this DNA repair pathway. Interestingly, deletions predominate over insertions in microsatellites of MMR-deficient tumours (see below), which suggests either that the formation of extra-helical loops in the template strand is more frequent than in the primer strand, and/or that the former lesions are less efficiently repaired.

Given that the primary role of MMR is to eliminate replication errors, it is likely that its function will be linked to the replication machinery. But because DNA synthesis proceeds asymmetrically, it is conceivable that the link to the leading and the lagging strands may not be identical. It is interesting to note in this context that error rates in the two strands are not identical (Kunkel and Bebenek, 2000). The major difference between leading and lagging strand synthesis is that the former is synthesised processively by a complex of DNA polymerase δ, PCNA and RF-C. In contrast, the lagging strand is produced as a series of Okazaki fragments, where RNA primers generated by DNA polymerase α/primase are extended until they are sufficiently long to be utilised by DNA polymerase δ holoenzyme, which completes their synthesis. The RNA primers are subsequently removed by the FEN1 endonuclease and the fragments are ligated together by DNA ligase I to produce the continuous lagging strand ((Hübscher et al., 2002; Hübscher and Seo, 2001) and refs. therein). DNA polymerase ε has also been found to be involved in DNA synthesis and in MMR (Kirchner et al., 2000; Pospiech and Syvaoja, 2003), but too little is known about its specific role in these processes, and it will therefore not be discussed further.

In order to learn how MMR and DNA replication may interact, we must first familiarise ourselves with the constituents of the two systems (Figure 1a below) and with their respective biological roles (Figure 1b below). The process of mismatch correction is initiated by the binding of one of two mismatch recognition complexes, hMSH2/hMSH6 (also termed hMutSα) or hMSH2/hMSH3 (hMutSβ), to base/base mismatches or strand misalignments that were generated by DNA polymerase δ but that escaped

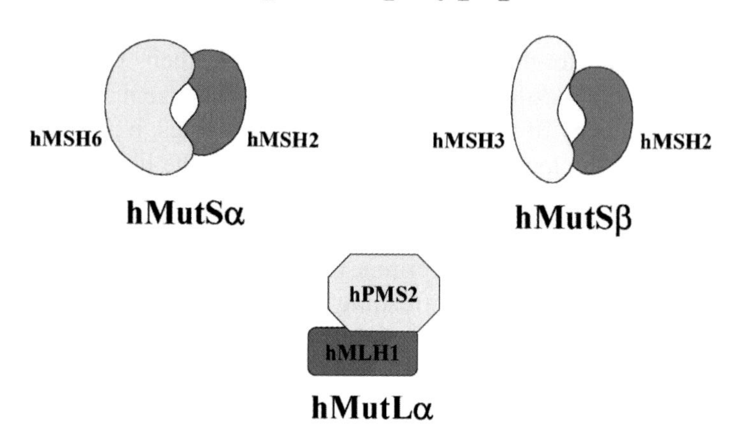

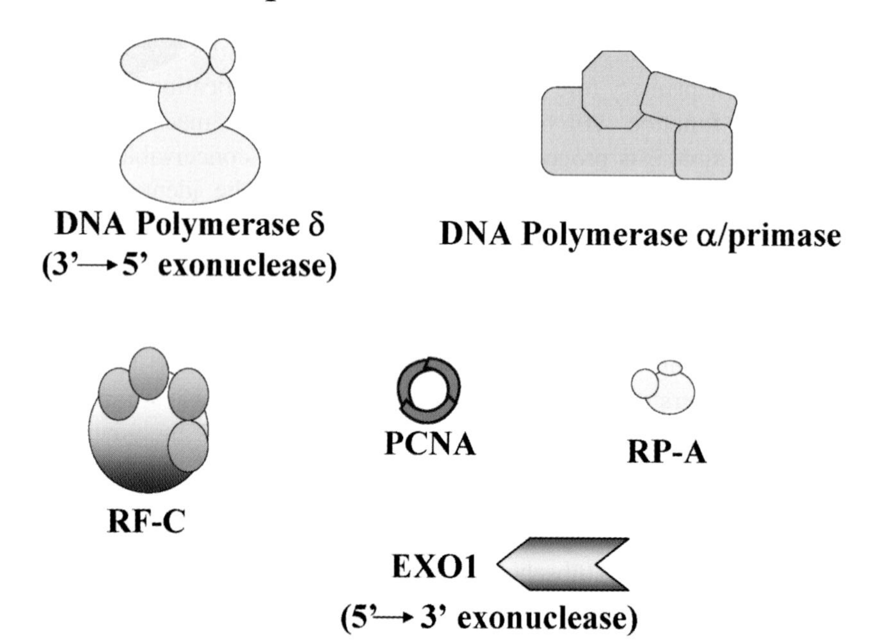

Figure 1a Factors involved in DNA mismatch repair (MMR) and replication and their respective biological roles.

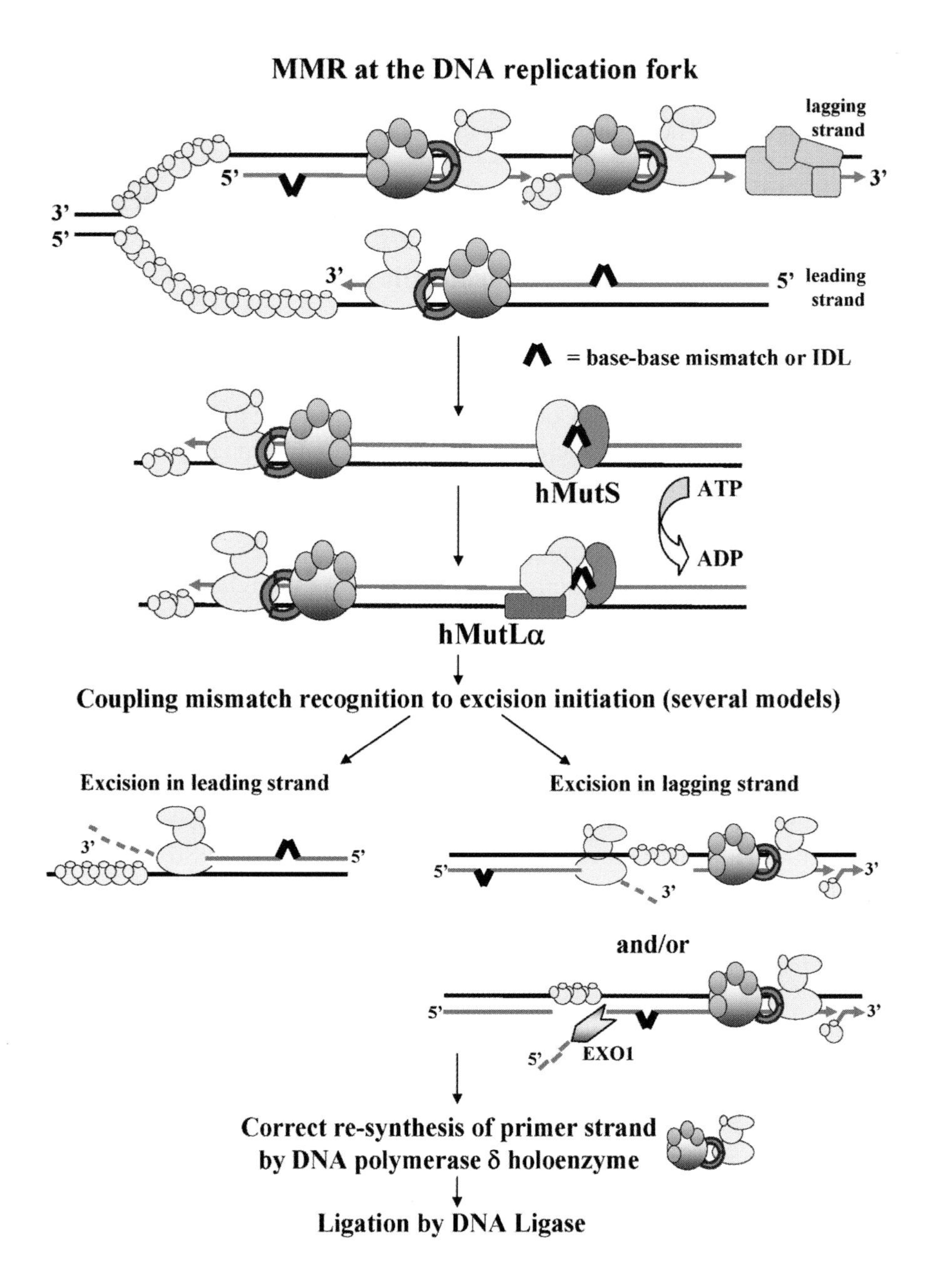

Figure 1b See text for a detailed description.

its proofreading activity. (Errors in the RNA/DNA primers of Okazaki fragments made by DNA polymerase α/primase, which lacks intrinsic or associated exonucleolytic activity, are presumably not relevant, because they are eliminated during the excision of these primers by FEN1 and associated factors (reviewed in (Hübscher and Seo, 2001)). The mismatch-bound hMutS heterodimer then undergoes an ATP-driven conformational change that is reputed to orchestrate its interaction with the hMLH1/hPMS2 heterodimer (hMutLα), another ATPase. Recruitment of other downstream factors leads to the exonucleolytic degradation of the primer strand and to its resynthesis by DNA polymerase δ. The remaining nick is most likely sealed by DNA ligase I. These functions are discussed below in greater detail.

2.2 The Mismatch and IDL Recognition Step: Specificity and Redundancy of hMSH2/hMSH6 and hMSH2/hMSH3 Heteroduplexes

Mismatch binding activity in human cell extracts was first detected in 1988 (Jiricny et al., 1988). It was ascribed to a polypeptide that could be covalently cross-linked to an oligonucleotide substrate containing a single G/T mismatch, whereby the protein/DNA complex migrated through denaturing polyacrylamide gels (SDS-PAGE) with an apparent molecular size of ~200 kDa. This mismatch binding activity was later purified to near homogeneity and shown to migrate through SDS-PAGE as two bands of apparent molecular size of ~160 and ~100 kDa. As the relative abundance of the larger polypeptide decreased during the lengthy purification procedure, it was assumed that it was proteolytically degraded to the ~100 kDa polypeptide. The protein was named GTBP (G/T binding protein) due to its high affinity for oligonucleotide substrates containing G/T mispairs (Hughes and Jiricny, 1992). In later studies it could be shown that the ~100 and ~160 kDA polypeptides were products of different genes, *hMSH2* and *hMSH6*, respectively (Drummond et al., 1995; Palombo et al., 1995). The hMSH2/hMSH6 heterodimer is frequently referred to as hMutSα (reviewed in (Jiricny, 1998) and (Marra and Schär, 1999)). Although gel-shift assays showed this factor to bind with appreciable affinities only to substrates containing the G/T or G/U mispairs, as well as IDLs of 1 or 2 extrahelical nucleotides, *in vitro* MMR repair assays showed that hMutSα supports the correction of G/T, A/C, G/G and A/A mispairs, and of IDLs of 1 to 4 extrahelical nucleotides with high efficiency. The remaining mismatches were repaired with intermediate efficiency, while C/C and larger IDLs were poor substrates for the MMR system (Fang and Modrich, 1993; Holmes et al., 1990; Thomas et al., 1991; Umar et al., 1994). However, hMutSα has been found to address also IDLs with 2-8 unpaired nucleotides, albeit much less efficiently than hMutSβ (Genschel et al., 1998). Differences in repair efficiency might be correlated with differences in the structural properties of the individual mismatches (reviewed in (Marra and Schär, 1999)), however, as reported in *E.coli* (Echols and Goodman, 1991; Kramer et al., 1984), it is also reasonable to believe that mismatches that are more likely to arise as DNA polymerase errors are better substrates for MMR than others that occur only rarely.

In vivo, hMSH6 competes for binding to hMSH2 with hMSH3. The hMSH2/hMSH3 heterodimer, termed hMutSβ, has been shown to address IDLs of 1-8 extrahelical nucleotides with high efficiency, but this role can be adopted also by hMutSα, albeit with lower efficiency – particularly where the larger IDLs are concerned (Acharya et al., 1996; Genschel et al., 1998; Palombo et al., 1996). It is unclear which factor will participate in the repair of these structures *in vivo*, given that hMutSα is around 10 times more abundant than hMutSβ in cultured human cells (Chang et al., 2000; Drummond et al., 1997; Genschel et al., 1998). The reverse is true in cells overexpressing hMSH3 as a result of amplification of the *DHFR/MSH3* locus through exposure to methotrexate. These cells have extremely low levels of hMutSα, because the overabundant hMSH3 sequesters all available hMSH2 and the partnerless hMSH6 is degraded (Drummond et al., 1997; Marra et al., 1998). Extracts of these cells are proficient in IDL repair as anticipated, but deficient in the repair of base-base mismatches, which demonstrates that hMutSβ is not involved in the repair of the latter substrates. These cells would thus be expected to display no MSI, in spite of having a mutator phenotype.

hMSH3 and hMSH6 compete for the same region of hMSH2 (Guerrette et al., 1998) and several studies showed that molecules of the former polypeptides that fail to form heterodimers with hMSH2 are destabilised. hMSH2 could be seen to remain stable in the absence of one of its partners, but our recent studies showed that this polypeptide is also degraded, provided however, that both hMSH6 and hMSH3 are absent. Although this finding is unlikely to be of biological significance, it implies that hMSH2 does not have other binding partners that might stabilise it.

As already mentioned above, the mismatch binding factor in human cell extracts could be covalently cross-linked to a polypeptide with a molecular size of ~160 kDa (Hughes and Jiricny, 1992; Jiricny et al., 1988) and subsequent experiments showed this polypeptide to be hMSH6 (Dufner et al., 2000; Iaccarino et al., 1998). These results invoked the asymmetric nature of the hMutSα heterodimer and implied that the two subunits play distinct roles during mismatch recognition. Indeed, substitution of a conserved phenylalanine in the putative mismatch recognition site located in the N-terminal part of hMSH6 abolished mismatch binding, whereas the same mutation in hMSH2 had no effect (Dufner et al., 2000). Substitution of the equivalent amino acid residues in the MutS protein of *E. coli* (Malkov et al., 1997) and in Msh6p of *S. cerevisiae* (Bowers et al., 1999) had a similar effect. These data were confirmed and extended when the crystal structures of *E.coli* MutS bound to a G/T mismatch (Lamers et al., 2000) and of *Thermus aquaticus* MutS bound to DNA containing an unpaired T (Obmolova et al., 2000) were determined. The MutS homodimer was found to encircle the mismatch-containing DNA as a pair of praying hands (Figure 2 below) with two openings, of approximately 30Å and 40Å, with the DNA passing through the latter (reviewed in (Jiricny, 2000)). The two MutS subunits interact at the base of the palms (carboxy-terminal domain V), which contain the composite ATP-binding domains. The mismatch-containing DNA is located between the touching fingertips and the top segments of the thumbs (amino-terminal domain I). The latter domain contains the highly conserved motif GXFY(E), the phenylalanine residue of which was mutated in the aforementioned studies. Remarkably, only one MutS subunit interacts with the mismatched nucleotide. Thus, the *E.coli* MutS homodimer is a functional heterodimer, which mirrors the asymmetry of the yeast (Bowers et al., 1999) and human (Dufner et al., 2000) MSH2/MSH6 heterodimers.

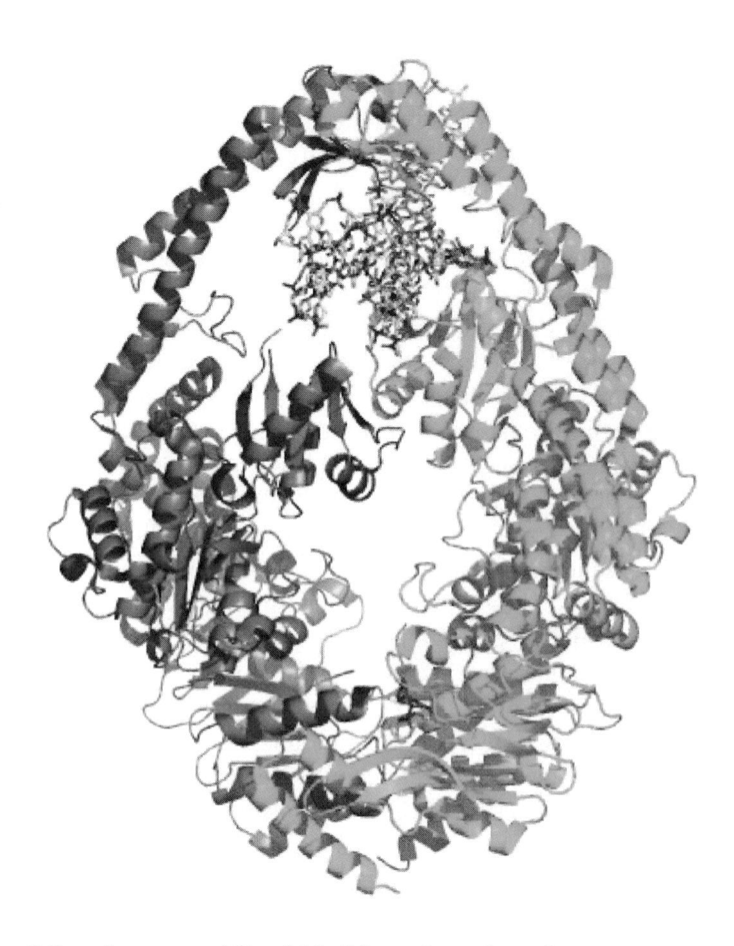

Figure 2 Crystal structure of *E. coli* MutS homodimer shown in cartoon representation, with the mismatch recognizing monomer in green and the other monomer in blue. DNA and ADP (magenta) in ball and stick representation. This figure was kindly provided by Dr. Titia K. Sixma, Netherlands Cancer Institute.

2.3 Adenosine Nucleotide Binding and ATPase Activity of MutS Homologues

The C-terminal domains of all MutS homologues characterised to date are highly conserved and resemble those of the ABC (ATP binding cassette) superfamily of ATPases. As revealed by the structural studies of the bacterial proteins (Lamers et al., 2000; Obmolova et al., 2000), the ATPase domains of the two subunits of the MutS dimers are intertwined. This helps explain how binding and hydrolysis of ATP in either subunit can bring about conformational changes that are transmitted to both DNA binding domains such that they change the affinity of the bound factor for mismatch-containing substrates (Iaccarino et al., 2000; Iaccarino et al., 1998). Thus, addition of ATP to DNA binding assays resulted in an apparent dissociation of hMutSα from the mismatch-containing oligonucleotide substrates (Gradia et al., 1997; Hughes and Jiricny, 1992; Iaccarino et al., 1998), presumably due to the release of the complex from the mispair and to its subsequent sliding off the end of the short duplexes. In the case of the *E. coli* MutS, this translocation was visualised in an electron microscopic study, where the factor was seen to be extruding a DNA loop from a circular plasmid, with the mismatch located within the loop (Allen et al., 1997). This translocation was reported to be dependent on ATP hydrolysis. A similar mechanism was also postulated for hMutSα, where ATP binding was reported to stimulate the dissociation from the mismatch, but translocation was proposed to require ATP hydrolysis (Blackwell et al., 1998).

In the above model, ATP hydrolysis is required for MutS translocation along the helix contour and for the formation of a mismatch-containing loop. An alternative model proposes that ATP binding by hMutSα represents a molecular switch, which enables the factor to leave the mismatch and diffuse along the DNA in an ATP-hydrolysis-independent fashion (Gradia et al., 1997). Should it fail to interact with the downstream MMR factors before the nucleotide is hydrolysed, the reverse switch triggered by ATP hydrolysis would release the factor from DNA and regenerate the ADP-bound form that can subsequently bind to other mismatched substrates with high affinity. Despite the differences between these models, they share the notion that the mismatch recognition factor dissociates from the mispair upon ATP binding. This hypothesis appears to be strengthened by the findings that mutations in the ATP-binding sites of different MutS homologues had little effect on mismatch binding, but brought about a considerable decrease in the ATP-driven dissociation of the proteins from the DNA (Alani et al., 1997; Haber and Walker, 1991; Iaccarino et al., 1998; Studamire et al., 1998). That the observed *in vitro* dissociation of MSH proteins from DNA substrates is of biological relevance was demonstrated by the finding that mutations in ATP-binding sites of either subunit of hMutSα, but particularly in hMSH6, result in MMR deficiency in an *in vitro* assay (Iaccarino et al., 1998), as well as *in vivo* (Iaccarino et al., 2000). Missense mutations in the ATP-binding site of *hMSH2* have also been detected in subjects affected by HNPCC (HNPCC mutation database at http://www.insight-group.org/).

It should be mentioned, however, that the translational mechanism has been questioned by the structural studies (Lamers et al., 2000; Obmolova et al., 2000), where the sharp kinking of mismatched DNA and the conformational changes of MutS upon DNA binding suggested that the protein might remain bound at the mismatch and thus help direct the subsequent repair events through interaction with the downstream MMR proteins. This hypothesis is supported by the evidence that bacterial MutS remains bound to mismatch-containing oligonucleotide substrates in the presence of MutL and ATP, and that this complex can activate the MutH endonuclease *in trans* (Junop et al., 2001). Furthermore, Wang & Hays demonstrated that MMR can be initiated despite the presence of a biotin-streptavidin blockade between the mismatch and the nick (Wang and Hays, 2004). Similar findings were described also for the yeast (Bowers et al., 2001) and human (Räschle et al., 2002) MMR factors.

2.4 MLH1/PMS2 (MutLα)

Human MLH1 can form heterodimers with hPMS2 (Li and Modrich, 1995), hPMS1 (Räschle et al., 1999) or hMLH3 ((Lipkin et al., 2000); our unpublished data). However, hMLH1/hPMS2, also called hMutLα, is the only complex with an essential role in MMR. Biochemical studies have demonstrated that the hMLH1/hPMS1 complex (hMutLβ) is not involved in MMR (Räschle et al., 1999). The activity of hMLH1/hMLH3 (hMutLγ) in *in vitro* MMR assays is very low, but still under investigation, whereas its yeast equivalent appears to act in conjunction with MutSβ to correct a small fraction of IDLs (Flores-Rozas and Kolodner, 1998; Harfe et al., 2000). Since germline alterations of both *hPMS1* and *hMLH3* have not been found associated with predisposition to cancer, hPMS1 and hMLH3 will not be extensively discussed in this review.

MutLα is essential for MMR, but its function remains enigmatic. Its weak ATPase activity suggests that it acts as a signalling rather than a catalytic molecule (Acharya et al., 2003; Räschle et al., 2002). Correspondingly, it has been given the epithet "molecular matchmaker", which implies that it facilitates interactions between the DNA-bound MutSα or MutSβ and the downstream factors of the MMR process. This concept originated in studies with the bacterial MutL, which was shown to interact with MutS and thus increase its DNA binding efficiency. Moreover, as mentioned above, the ATP-dependent interaction of MutS and MutL with heteroduplex DNA activated the latent endonucleolytic activity of MutH and the loading of UvrD (MutU) helicase at the site of the incision (reviewed in (Modrich and Lahue, 1996)); (Drotschmann et al., 1998; Yamaguchi et al., 1998). The crystal structure of the N-terminal 349 amino acid residues of bacterial MutL, which contains the nucleotide binding site, revealed that its binding to a non-hydrolyzable ATP analogue brings about a dramatic structural change and dimerisation (Ban et al., 1999; Ban and Yang, 1998). Assuming that the C-terminal region of MutL mediates constitutive dimerisation, the N-terminal part could function as an ATP-driven molecular gate that clamps MutL onto DNA (Ban et al., 1999; Mechanic et al., 2000; Spampinato and Modrich, 2000) and/or enables it to interact with different partners. Indeed, similar functions have been described for DNA gyrase and the chaperone HSP90, ATPases structurally related to MutL. This model is supported by a recent study, which describes the structure of the C-terminal dimerization domain of bacterial MutL (Guarne et al., 2004). In this study, a model of full-length MutL homodimer was proposed, in which a V-shaped C-terminal domain is situated opposite the saddle-shaped N-terminal domain. The two domains are connected by proline-rich linkers that generate a large central cavity of ~100 Å, capable of encircling up to 4 DNA duplexes simultaneously. This cavity might form around DNA upon ATP binding-dependent association of the N-terminal ATPase domains of the two MutL subunits dimerized at their C-termini. It has been proposed that MutL might utilise its ATPase and DNA-binding activities to activate the MutH endonuclease and the UvrD helicase (Guarne et al., 2004). The crystal structure of MutL can help us predict the functional importance of missense mutations in hMutLα. Indeed, missense mutations within or in the vicinity of the ATP binding domain of MutL homologues, that are similar to mutations identified in HNPCC families (HNPCC mutation database at http://www.insight-group.org/), have been found to impede MMR *in vitro* and *in vivo* (Aronshtam and Marinus, 1996; Ban et al., 1999; Räschle et al., 2002; Tran and Liskay, 2000).

2.5 Strand Discrimination and Excision

2.5.1 Requirement of Strand Discontinuities for Initiation of the Repair Process

The MMR system is often described by the adjective "long-patch", which implies that DNA replication errors are repaired through the resection of a DNA tract that is substantially longer than that excised during base excision repair (BER) and nucleotide excision repair (NER). This facet of the MMR system was thoroughly characterised by Modrich and colleagues, using the *E.coli* MMR system reconstituted from the individual purified components ((Modrich and Lahue, 1996) and refs. therein).

In all organisms, MMR has to be directed to the newly-synthesised strand. In *E.coli* and in some other gram-negative bacteria, this strand is transiently unmethylated on adenine residues within the sequence GATC, whereas the same sequence on the template strand is methylated. The enzyme that modifies this site, deoxyadenine methylase (Dam), lags behind the replicative polymerase by ~2 minutes. This "window of opportunity" is exploited by the MutH endonuclease, which cleaves the unmethylated strand at *dam*-sites upon activation by the MutS/MutL complex. Loading of the UvrD helicase and one of several exonucleases at the MutH-generated nick results in degradation of the newly-synthesised strand towards and a few nucleotides past the mismatch. In this system, the distance between the mispair and the strand-discrimination signal can be up to 1 kb. It should be noted that the repair efficiency was higher at shorter distances, and that in cases where the distance between the two signals was very short, the helicase function was dispensable (Mechanic et al., 2000).

The MMR system is highly conserved through evolution, but the MutH function is not present outside of gram-negative bacteria. In extracts of human cells, covalently-closed heteroduplex substrates were shown to be refractory to mismatch repair, which suggested that the mammalian MMR system lacked mismatch-activated endonucleases. However, a pre-existing nick within 1 kb of the mismatch was sufficient to activate the MMR process (Fang and Modrich, 1993; Holmes et al., 1990; Thomas et al., 1991). This gave rise to the hypothesis that in organisms lacking a MutH function, the exonucleolytic degradation of the heteroduplex might initiate at pre-existing strand interruptions such as the 3'-terminus of the leading strand and the 5'- and 3'-termini of Okazaki fragments on the lagging strand (Figure 1b above). Interestingly, gaps have been found to represent a more effective signal for MMR than nicks (Iams et al., 2002). Is it possible that more efficient loading of the MMR machinery in the gaps between Okazaki fragments is the reason why the lagging strand has lower error rates (Kunkel and Bebenek, 2000).

2.5.2 PCNA

If the MMR process is to be initiated as proposed by the above hypothesis, then it must be intimately linked to DNA replication. The first evidence for the existence of such a link came from *S. cerevisiae*, where Mlh1p and Msh2p were found to interact with POL30 (yeast PCNA) in a yeast two-hybrid screen (Umar et al., 1996) and where viable mutant alleles of the *pol30* gene were found to lead to MSI and to be epistatic with *MMR* gene mutations (Chen et al., 1999; Johnson et al., 1996; Kokoska et al., 1999; Umar et al., 1996). Moreover, $p21^{WAF1}$, a polypeptide known to bind PCNA and to inhibit PCNA-dependent DNA replication (Flores-Rozas et al., 1994; Li et al., 1994; Waga et al., 1994), inhibited MMR in an *in vitro* assay that did not require DNA synthesis (Umar et al., 1996), suggesting that PCNA is involved in MMR also at a step preceding the re-synthesis of the repair tract. What form this interplay takes is unclear at the moment, given that the yeast two-hybrid assays identified interaction between POL30 and Mlh1p and Msh2p (Umar et al., 1996), while human PCNA was found to interact with hMSH6 and hMSH3 *via* a highly-conserved motif at the N-termini of the two polypeptides (Clark et al., 2000; Flores-Rozas et al., 2000; Kleczkowska et al., 2001), the mutation or deletion of which reduced MMR efficiency in biochemical assays (Kleczkowska et al., 2001) and produced a mutator phenotype in yeast (Clark et al., 2000; Flores-Rozas et al., 2000).

More recently, PCNA was shown to be required for excision directed by a nick situated 3' from the mismatch, but was dispensable for 5'-nick-directed excision (Genschel and Modrich, 2003; Guo et al., 2004). These results could be explained by the fact that PCNA is known to bind to the 3'-termini of strand discontinuities with a given orientation (Figure 1b above). Thus, if an MSH-MLH complex assembled on a mismatch were to translocate in the same direction as the replicating DNA polymerase, it would encounter a PCNA homotrimer bound at the 3'-terminus. This interaction may be required to recruit or activate the 3'→5' exonuclease (see below). If the MSH-MLH complex travelled in the opposite direction, it would encounter a free 5'-terminus where a different exonuclease (5'→3') has to be loaded. But there are also other hypotheses. PCNA was proposed to interact with the mismatch binding complex prior to mismatch recognition and this interaction was suggested to help direct it to the mismatch and increase its binding affinity (Lau and Kolodner, 2003). This hypothesis is supported by the evidence that PCNA is loaded onto 3'-primer termini in replication intermediates and that it remains on the DNA after dissociation of the replication proteins (Stukenberg et al., 1994; Warbrick, 2000). These PCNA molecules might serve as docking sites for MSH2/MSH6 or MSH2/MSH3 on newly-replicated DNA, *i.e.* in regions likely to harbour mismatches or IDLs. The elucidation of the detailed mode of interaction of the MMR proteins with PCNA and with the rest of the replication machinery (Warbrick, 2000) must await the outcome of new experiments and the development of novel technologies that enable the visualisation of the multiprotein assemblies.

2.5.3 Degradation of the Newly-Synthesised Strand

At least two exonucleolytic functions are believed to be involved in MMR (Figure 1b above). One of these is the proofreading 3'→5' exonuclease of DNA polymerase δ (or DNA polymerase ε), the second is the 5'→3' exonuclease EXO1. DNA polymerase δ was found to be required for human MMR *in vitro* (Longley et al., 1997) and mutants of yeast DNA polymerase δ (POL3) lacking the 3'→5' exonuclease activity displayed an increased mutation rate, which implicated the proofreading subunit in MMR (Tran et al., 1999). The first evidence for the involvement of EXO1 in MMR was obtained in *S. pombe* (Rudolph et al., 1998; Szankasi and Smith, 1995), where *exo1* mutants were found to display a mutator phenotype. Genetic analyses in *S. cerevisiae* demonstrated an epistatic relationship of *Exo1* with *Msh2*, *Mlh1* and *Pms1* (yeast homologue of *hPMS2*) (Sokolsky and Alani, 2000; Tishkoff et al., 1997; Tran et al., 1999) and Exo1p was found to interact with Msh2p (Tishkoff et al., 1997) and Mlh1p (Tran et al., 2001), which strongly supported its involvement in MMR. Human EXO1 was identified as a homologue of *S. cerevisiae* Exo1p (Tishkoff et al., 1998; Wilson et al., 1998). Two forms of the human enzyme were characterised, hEXO1a and b, and shown to result from alternative splicing, but no functional differences between the two splice variants could be identified to date. Similarly to its yeast homologue, human EXO1 interacts with several MMR proteins, in particular hMSH2, hMSH3 and hMLH1 (Schmutte et al., 2001). Inactivation of EXO1 in mice resulted in MMR defects, MSI and increased cancer susceptibility (Wei et al., 2003), and genetic alterations in this gene have been found in humans with predisposition to colon cancer (see below).

Human EXO1 belongs to the RAD2 family of nucleases and exhibits a robust 5'→3' exonucleolytic activity. Similarly to other members of this family, it also possesses a 5' flap structure-specific endonuclease activity (Lee and Wilson, 1999). *In vitro* studies demonstrated that the processivity of hEXO1 in 5'→3' excision is activated by hMutSα, whereas termination of excision 60-230 nucleotides beyond the location of the mismatch is due to RPA-dependent displacement of the excision complex (*i.e.* EXO1 in a multiprotein complex containing a number of MMR proteins) from the helix (Genschel and Modrich, 2003). Stabilization of single-strand template tracts and modulation of MMR processivity by RPA explain the obvious and essential role of the latter protein during DNA replication and MMR (Lin et al., 1998) (Figure 1b above). Surprisingly, in an *in vitro* system, hEXO1 was also found to participate in mismatch-provoked excision directed by a strand break located 3' to the mismatch (Genschel et al., 2002). This cryptic 3'→5' exonuclease activity of EXO1 seems to be modulated by RF-C and PCNA loaded at the 3' termini (Dzantiev et al., 2004). It now remains to verify the importance of this activity *in vivo*.

In the above overview, we described the current knowledge of the biochemistry of the MMR process in human cells. The factors involved in this key metabolic process can be categorised as MMR-specific (MSH2/MSH6, MSH2/MSH3 and MLH1/PMS2) and DNA-replication specific (PCNA and its loading factor RF-C, RPA, DNA polymerase δ or ε and DNA Ligase I). The exonucleolytic functions associated with MMR can be assigned to both categories; the intrinsic proofreading activity of DNA polymerase δ can be thought of as a DNA-replication specific function that also participates in MMR, whereas EXO1 might be considered MMR-specific, even though its flap endonuclease activity may participate in the processing of Okazaki fragments during replication or in the resection of termini during double-strand break repair. It is possible, even likely, that other exonucleases may emerge as players in MMR. However, the two activities discussed above are sufficient to support the bi-directional model of MMR (Fang and Modrich, 1993; Iams et al., 2002). Indeed, the MMR- and DNA replication-specific factors listed above are sufficient to reconstitute both 5'→3' and 3'→5' MMR processes *in vitro* (Dzantiev et al., 2004). The situation *in vivo* could be more complex, inasmuch as the MMR process might be modulated by other, non-essential factors (Yuan et al., 2004).

The MMR-specific category of proteins can be further subdivided according to the phenotypic consequences associated with the loss of the respective functions. Thus, the loss of a MMR-specific protein such as MSH2 or MLH1 brings about a strong mutator phenotype associated with MSI and cancer predisposition. In cases of functional redundancy, such as with MSH6 and MSH3, the MMR defect is restricted to a specific subset of substrates, the MSI is more limited and cancer susceptibility, although present, becomes apparent at a more advanced age. The DNA replication-specific category cannot be similarly subdivided, as loss of these functions would result in cell death. In yeast, the presence of missense mutations in some of these factors, including Pol30 (PCNA), Pol32 (DNA polymerase δ) and Exo1p, was shown to bring about weak MMR defects (Amin et al., 2001). Although similar alleles may be present in the human population, their phenotypic expression in a diploid setting would be expected to be extremely rare.

3. MMR DEFICIENCY AND COLON CANCER

3.1 Microsatellite Instability (MSI)

When Perucho and colleagues carried out DNA fingerprinting analysis of normal and colorectal cancer (CRC) tissues of the same individuals, they found the expected quantitative changes due to allelic losses and gains in the tumour cell genome. In addition, they also identified qualitative alterations in a subset of CRCs, most of which took the form of additional, shorter PCR products that appeared in poly(A) stretches and in other repeated sequences, also called microsatellites (Ionov et al., 1993; Peinado et al., 1992). This phenomenon was observed in 12% of the CRCs investigated. Although additional bands of bigger size due to base insertion were also detected, the bias for deletions *versus* insertions was evident. Also, it was clearly documented that the mutation frequency was proportional to the length of the repeats. These peculiar somatic alterations were called USMs for "ubiquitous somatic mutations", since they were found everywhere in the cancer genome of a subset of CRCs (Ionov et al., 1993). They were not restricted to carcinomas, but also occurred in adenomas, the pre-cancerous lesions of the colon, suggesting that these events occurred before neoplastic transformation and were not the result of genetic instability of cancer cells during tumour progression (Shibata et al., 1994). In addition, these alterations were found to persist after transformation and to increase in number during tumour progression (Jacoby et al., 1995; Shibata et al., 1994).

As often happens in the case of major discoveries in the interactive world of science, this phenotype was observed by several groups at about the same time. Thibodeau and colleagues were using (CA)n microsatellites to study LOH in CRCs and found marked expansions or contractions of these repetitive units at several different loci, predominantly in tumours of the proximal (right side) colon (Thibodeau et al., 1993). Another group of collaborators, including Vogelstein' and de la Chapelle's laboratories, was studying microsatellites in an effort to locate the HNPCC gene(s) in familial clusters with early-onset CRCs. They too identified widespread alterations of short, repeated sequences in most of the familial cancers (Aaltonen et al., 1993) and succeeded in mapping the first HNPCC gene to chromosome 2p15-16 in two large families (Peltomäki et al., 1993).

The term microsatellite instability (MSI) is today generally accepted to designate the widespread genomic instability at these loci and has substituted the original terms USMs and RERs (replication errors in repeats). Since 1993, hundreds of studies about the frequency and characterization of MSI^{+} tumours in CRCs and in tumours of other organs have been published. It is clearly beyond the scope of this article to review these comprehensively. However, it appears that the frequency of MSI^{+} tumours of different organs is less than 20%, with CRCs and endometrial cancers predominating. Due to the use of different methodologies and varying sets of microsatellite markers, as well as to selection biases in the studied populations, the frequencies of MSI^{+} tumours described in the literature differ widely, (reviewed in (Maehara et al., 2001)). These circumstances have generated a great deal of confusion, but it is hoped that recent attempts at defining uniform criteria of MSI (type of markers to test, individuals to be tested, etc.) will resolve the situation (Umar et al., 2004).

But let us go back to the historic 1993. At the beginning of that year, Petes and colleagues reported that the fidelity of replication of repeated sequences was up to three orders of magnitude lower in yeast strains lacking MMR (Strand et al., 1993). This seminal work linked MSI to defective MMR, which triggered a race for the identification of human *MMR* genes and their genetic loci. Thanks to the sequence similarity between microbial and human *MMR* genes, *hMSH2* was cloned and characterised already at the end of that year (Fishel et al., 1993; Leach et al., 1993), within the locus on chromosome 2p, which was previously implicated in HNPCC by linkage analysis (Peltomäki et al., 1993). *hMLH1* was mapped to chromosome 3p (Lindblom et al., 1993) and cloned shortly thereafter (Bronner et al., 1994; Papadopoulos et al., 1994), and *hPMS1* and *hPMS2* were identified on chromosomes 2q and 7p, respectively (Nicolaides et al., 1994). A different pathway led to the discovery of hMSH6. This polypeptide turned out to be a component of the G/T binding protein (GTBP), a factor shown to bind with high affinity to oligonucleotide duplexes containing G/T mismatches. In DNA cross-linking experiments, the protein/DNA complex migrated with a molecular size of ~200 kDa (Jiricny et al., 1988). Partial purification of GTBP showed it to consist of two polypeptides of 160 and 100 kDa (Hughes and Jiricny, 1992), and protein sequencing showed that the 160 kDa polypeptide was a new member of the MutS family, named hMSH6, whereas the 100 kDa protein was shown to be hMSH2 (Drummond et al., 1995; Palombo et al., 1995). The *hMSH6* gene was localised to chromosome 2p and found mutated in human cells with mononucleotide repeat instability (Papadopoulos et al., 1995), which suggested that the hMSH2/hMSH6 complex was not essential for the repair of IDLs of 2 or more extrahelical bases. As discussed in Part 2, this role is fulfilled in part by the hMSH2/hMSH3 complex, the role of which in IDL repair was demonstrated in biochemical assays using recombinant proteins (Palombo et al., 1996) and human cell extracts (Drummond et al., 1997; Marra et al., 1998; Risinger et al., 1996). Interestingly, the *hMSH3* gene (initially called

REP3) was identified already in 1989 on chromosome 5q, as an open reading frame transcribed in the opposite direction from the *DHFR* gene, with which it shares the promoter region (Fujii and Shimada, 1989; Linton et al., 1989). *hMSH3* has often been found mutated in human cell lines, but not in the germline of HNPCC subjects (see below). Cells lacking this polypeptide do not display MSI, and it seems therefore unlikely that these mutations will segregate with HNPCC. The same applies for *hMLH3*, which was localised to chromosome 14q (Lipkin et al., 2000). As discussed in Part 2, the function of this polypeptide in human MMR appears to be only marginal and involvement in human cancer is thus unlikely to play a major role.

3.2 The Role of MSI in CRC Progression

Due to the widespread presence of microsatellites in our DNA, thousands of somatic mutations are expected to accumulate in the genome of tumour cells during their transformation. Some of these alterations might contribute to tumourigenesis by favouring clonal selection of cells with higher survival rate and proneness to invasion, whereas other mutations might be irrelevant to these processes. The relevance of a particular mutation in microsatellites to tumourigenesis can be arbitrary, however, if the mutation is located in a coding sequence such that it causes a frameshift mutation in the respective protein product of the gene, it is more likely to be biologically important than a similar mutation in an intergenic region. This is the case of deletions in poly(A) repeats of the coding regions of *TGFβ Receptor II (TGFβRII)* (Markowitz et al., 1995) and *BAX* (Rampino et al., 1997), which were among the first genes found to be affected by MSI in MMR-deficient CRCs. The relevance of these two alterations to cell transformation is unquestionable, because the products of these two genes are important players in the TGFβ signalling and apoptotic pathways, respectively; their absence will thus favour transformation rather than cell death. Microsatellites in other genes, the relevance of which to human cancer is not as clear-cut, are also found with high frequency in MMR-deficient CRCs (reviewed in (Mori et al., 2001) and (Woerner et al., 2003)). These findings, associated with a low incidence of MSI-independent mutations traditionally considered crucial in colon carcinogenesis, such as those in the *Adenomatous Polyposis Coli* (*APC*) and *p53* tumour suppressor genes, as well as in the *K-ras* oncogene (Ionov et al., 1993; Kim et al., 1994; Losi et al., 1997), point to the existence of MSI-specific cancer genes, the mutations in which are determinants of tumourigenesis in MSI^+ CRCs.

Positive selection pressure should lead to the frequent detection of mutations at microsatellite loci such as *TGFβRII* and *BAX*, as well as in *IGFRII* (Souza et al., 1999), *TCF4* (Duval et al., 1999), *axin* (Liu et al., 2000), *gastrin receptor* (Laghi et al., 2002) and *β_2-microglobulin* (Bicknell et al., 1996), which have been reported to be altered with lower frequency. For these mutations to exert a maximal effect, both alleles should be affected.

However, mutations in the coding sequences of above genes, including *TGFβRII* and *BAX*, were often found to be mono-allelic. This implies that inactivation of a single allele results in the production of insufficient amounts of a particular gene product (haploid-insufficiency), which is sufficient to affect the phenotype of the colonic epithelial cells.

Frameshift mutations associated with MSI have been detected also in mononucleotide repeats of the coding region of *hMSH3* and *hMSH6* (Malkhosyan et al., 1996; Percesepe et al., 1998) and of another DNA repair gene, *MBD4* (Methyl Binding Domain 4) (Riccio et al., 1999) in ~20% of MMR-deficient CRCs. These mutations, as well as those described in the paragraph above, are believed to be secondary, *i.e.* to arise during tumour progression in CRCs carrying primary alterations in *hMLH1* or *hMSH2*. What is the functional significance of a concomitant inactivation of two *MMR* genes? Based on the biochemical properties of the MSH homologues (part 2 of this chapter), in a tumour where hMSH2 is absent, there is no advantage in mutating the genes encoding its cognate partners, since hMSH6 and hMSH3 are degraded in the absence of hMSH2. However, this situation may be different in *hMLH1*-negative CRCs. Lack of hMLH1 results in a severe mutator phenotype (part 2 of this chapter), but a further increase in the mutation rate (2.5 times higher) and a change in the mutation spectrum (increase of G/C→A/T transitions and decrease of one base pair frameshifts) were observed by the additional inactivation of *hMSH6* in a cell line lacking hMLH1 (Baranovskaya et al., 2001). This finding implies that another, as yet unknown factor might partially substitute for hMLH1 in the repair of mismatches at G/C sites detected by hMutSα. Inactivation of *hMSH3* in a hMLH1-deficient tumours might play a role in tumour progression, as this alteration seems to be a predictor of metastatic disease in these CRCs (Plaschke et al., 2004), although this latter finding has still to be substantiated in a larger cohort of patients. The functional importance of a secondary mutation in *hMSH3* may be explained by the possible involvement of this protein in mitotic recombination, as was reported for Msh3p ((Nicholson et al., 2000) and refs. therein). Secondary mutations in *MBD4* could also augment the mutator phenotype of MLH1-deficient cells. MBD4 was identified as a member of a family of mammalian methyl-CpG binding proteins (Hendrich and Bird, 1998) and later shown to be a mismatch-specific thymine/uracil DNA glycosylase *in vitro* (Hendrich et al., 1999). The same protein (then called MED1) was identified in two hybrid assays as a partner of hMLH1 (Bellacosa et al., 1999). Although the functional significance of the MBD4/MLH1 interaction is unclear at present, its repair function may be important. G/T mismatches can originate through the spontaneous deamination of 5-methylcytosine to thymine, and MBD4 may function, together with TDG (Hardeland et al., 2003) to reduce mutagenesis at methylated CpGs. The concomitant inactivation of MBD4 and MMR may thus further decrease genomic stability and propel tumour progression. It is interesting to note in this regard that mutations in CpGs of *p53* in colon cancers are substantially more frequent than in other tumour types (Hollstein et al., 1991).

Changes in the length of microsatellites in non-coding sequences may also contribute towards the malignant phenotype, especially when they occur in gene regulatory regions. Shortening of a T_{11} repeat within the polypyrimidine stretch/accessory splicing signal of human *MRE11* leads to exon skipping and consequent premature stop codons in a high percentage of MMR-deficient CRCs (Giannini et al., 2004), which results in the functional impairment of the MRE11/NBS1/RAD50 complex that plays a pivotal role in the processing of DNA double strand breaks. An increased expression of the oncogene *c-myb* in CRCs might be the result of microsatellite shortening in a mononucleotide repeat present in its transcriptional attenuator region (Thompson et al., 1997). Mono- and di-nucleotide repeats are also frequently located within the 5'-UTRs and 3'-UTRs (Li et al., 2004), and deletions at these sites have been detected in MMR-deficient cancer cells ((Ruggiero et al., 2003; Suraweera et al., 2001) and our unpublished observations). Such sequence alterations might affect the rate of synthesis, stability and translational efficiency of mRNA. Interestingly, transcribed, non-coding mononucleotide repeats in the 5'UTR of *p21/WAF1* and in the 3'UTR of *BCL2*, two genes the products of which play a crucial role in cell cycle regulation and apoptosis, displayed a mutation frequency below the lower prediction limit, arguing for a negative selective pressure (Woerner et al., 2003).

In summary, counter-selection or maintenance of mono-allelic alterations might represent plausible mechanisms to retain the function of genes essential for cell fitness and survival. It cannot be excluded however, that in some of the alterations reported in this paragraph, selection may not be driven by functionality, but rather by genetic or genomic constrains, *i.e.* unknown features of the sequences or chromatin structure within or surrounding individual microsatellites that may contribute to their mutability (Zhang et al., 2001).

3.3 Chromosomal Instability and DNA Mismatch Repair

It is currently accepted that colon cancers can be divided into two main groups according to their type of genetic instability (reviewed in (Lengauer et al., 1998) and in (Grady, 2004), and references therein). In about 15% of CRCs, the instability is observed at the nucleotide level, *i.e.* base substitutions and MSI as consequences of MMR deficiency. Most of the remaining CRCs show instability at the chromosomal level, called CIN (Chromosomal Instability), resulting in losses and gains of whole chromosomes, or large portions thereof. This phenomenon was first described by Lengauer *et al.*, who reported that MMR-deficient colon cancer cell lines exhibited a normal rate of gross chromosomal alterations and a near-diploid chromosomal pattern, whereas MMR-proficient lines displayed increased rates of chromosomal changes and were aneuploid (Lengauer et al., 1997). The results of this study were consistent with previous karyotypic analysis of CRCs (Aaltonen et al., 1993; Bocker et al., 1996). A more recent study (Georgiades et al., 1999) has confirmed, by flow cytometric analysis, that MMR-deficient CRCs have a near diploid phenotype. However, the use of comparative genomic hybridization in this study revealed also several chromosome arm amplifications and deletions in MMR-deficient tumours, although the number of such events per genome was lower than in CIN^+ tumours. Thus, some chromosomal changes occur also in MSI^+ tumours, even though diploidy is maintained. MSI and CIN seem to be mutually exclusive, but exceptions have been reported in the colon cancer cell line KM12 (Camps et al., 2004) and in a small percentage (3.4%) of CRCs (Goel et al., 2003), where MSI and CIN coexist. Thus, either type of genetic instability may be sufficient for driving the transformation process, but MSI does not preclude CIN and *vice versa*. However, a portion of CRCs detected in cohorts of European (Georgiades et al., 1999, Chan et al., 2001; Abdel-Rahman et al., 2005a), Asian (Yao et al., 1999, Chan et al., 2001), Australian (Hawkins et al., 2001) and American (Goel et al., 2003) patients displayed neither MSI nor CIN, suggesting that other pathways may bring about cell transformation in the colon. This latter phenotype might be under-represented amongst the available colon cancer cell lines, possibly due to the failure of cells with this phenotype to adapt to growth *in vitro* (Georgiades et al., 1999), but this finding merits further evaluation, especially as non-MSI/non-CIN CRCs appear to be clinically more aggressive than those with MSI (Hawkins et al., 2001).

As mentioned above, the MSI phenotype is believed to be not only the consequence of the MMR defect, but also a driving force during transformation. It appears already in small, pre-cancerous adenomatous lesions of the colon (Jacoby et al., 1995; Shibata et al., 1994), presumably just after an initiation step consisting of activation of Wnt signalling, with the consequent dysregulation of the cell proliferation/apoptosis homeostasis (Jass et al., 2002; Polakis, 2000). The high mutation rate in these adenomas could accelerate their progression to carcinoma, since it has been estimated that most MMR-deficient adenomas can transform into carcinomas in less than 2 years, whereas MMR-proficient ones might take more than 10 years, and only 5-10% progress to carcinomas (Lynch et al., 1996; Vasen et al., 1995; Winawer et al., 1997). This is the reason why only few adenomas from unselected series display MSI, whereas the incidence of MSI^+ carcinomas approaches 15% of all CRCs. MSI persists after transformation and is associated with multiple, repeated changes in microsatellites throughout the tumour growth (Georgiades et al., 1999; Jacoby et al., 1995; Shibata et al., 1994). It has also been hypothesised that, in a later phase of transformation, the high frequency of mutations might make MMR-deficient cells more immunogenic, since a conspicuous intraepithelial infiltration of T lymphocytes and nodular aggregates of B cells are usually detected in

MMR-deficient CRCs (Jass, 2000). A robust immune reaction against MMR-deficient cells might decrease their potential for invasiveness and metastasis and this phenotype might explain why these CRCs have a better prognosis (Gryfe et al., 2000).

Thus, MSI^+ cells possess all the attributes necessary for malignant transformation, and can dispense with contributions from other mechanisms of genetic instability, such as CIN. A low frequency of chromosomal changes still occur (Georgiades et al., 1999), perhaps because of the increased recombination rate that might be expected in MMR-deficient cells (Ciotta et al., 1998; Harfe and Jinks-Robertson, 2000), but these changes are not required for selection due to the domination of the MSI phenotype. In the few cases where both MSI and CIN are present, CIN might result from the inactivation of genes involved in the maintenance of chromosomal stability. Indeed, in MSI^+ tumours, frameshifts have been reported also in repeat sequences of the DNA damage checkpoint genes *ATM* and *ATR* (Ejima et al., 2000; Menoyo et al., 2001; Vassileva et al., 2002) and loss of function of such genes might exacerbate the predisposition of MMR-deficient cells to perform illegitimate recombination events (Fang et al., 2004). In conclusion, there is little doubt that MSI is the principal driving force behind carcinogenesis in MMR-deficient cancers.

MSI represents the most frequent form of genetic instability with a pivotal role in cancer, as the incidence of other cancer syndromes etiologically related to failure of genomic integrity, such as Multiple Adenoma Syndrome of the colon associated with mutations of the base excision repair gene *MYH* (Al-Tassan et al., 2002; Sieber et al., 2003b), *Xeroderma pigmentosum*, *Ataxia telangectasia*, *Nijmegen breakage syndrome* and *Bloom's syndrome* (reviewed in (Hoeijmakers, 2001)), is substantially lower than that of HNPCC and *hMLH1*-deficient sporadic CRCs (see below).

In contrast, the role of CIN in CRCs with functional MMR is object of intensive debate (Rajagopalan et al., 2003; Sieber et al., 2003a). The discussion arose from divergent results regarding the incidence of chromosomal abnormalities in small pre-cancerous adenomatous lesions. Using different experimental approaches, these abnormalities were detected frequently in one study (Shih et al., 2001), but rarely in two others (Haigis et al., 2002; Sieber et al., 2002). There is a great deal of evidence supporting both hypotheses - that of CIN as an engine of tumourigenesis (Rajagopalan et al., 2003), or as one of the tumour features occurring during Darwinian natural selection for cells with increased reproductive fitness (Sieber et al., 2003a). However, the existence of a large portion of CRCs without CIN and MSI (Georgiades et al., 1999; Chan et al., 2001; Abdel-Rahman et al., 2005a; Goel et al., 2003; Hawkins et al., 2001; Yao et al., 1999) argues that tumourigenesis may progress also in the absence of genetic instability, unless a different form of it remains undetected by standard procedures in this group of tumours (Abdel-Rahman et al., 2001; Abdel-Rahman et al., 2005b). It is clear that much more work has to be done on this subject.

However, the current state of knowledge is already beneficial in cancer therapy.

3.4 Medical Genetics of Mismatch Repair Deficiency

Inheritance of a mutated allele of a *MMR* gene predisposes to cancers of the colon and endometrium, and, with a much lower frequency, to cancers of the stomach, bladder, urethra, renal pelvis, biliary tract, brain, sebaceous glands, small intestine and ovaries. Subjects carrying these mutations usually belong to families whose members are afflicted by a similar spectrum of cancers. This familial syndrome was named Hereditary Non-Polyposis Colon Cancer (HNPCC) (Lynch et al., 1993; Marra and Boland, 1995; Vasen et al., 1999), primarily to distinguish it from Familial Adenomatous Polyposis (FAP), in which the colons of affected individuals contain thousands of adenomatous polyps. The number of adenomatous polyps in HNPCC patients is generally less than five, but they arise about two decades earlier than in the general population, which suggests that the presence of a germline *MMR* mutation increases the transformation rate in normal colonic epithelium (de Jong et al., 2004a; Lindgren et al., 2002; Ponz de Leon et al., 1998). HNPCC is inherited in an autosomal dominant way, but MMR deficiency arises only when the wild-type *MMR* allele is mutated or lost. Thus, the disease is recessive at the somatic level. It has been estimated that about 1:2000 people carry germline *MMR* gene mutations and that ~5% of all CRCs are HNPCC. This would make HNPCC the most frequent cancer predisposition syndrome.

Two highly accurate tests are used for in the diagnosis of MMR-deficient tumours: MSI analysis and immunohistochemistry. In general, a tumour is considered MSI^+ when more than 30% of the microsatellite markers investigated show alleles absent in the control DNA extracted from normal tissues of the same patient. The deployment of recently-established guidelines for MSI analysis should increase the accuracy of the MSI analysis (Umar et al., 2004). Immunostaining for MMR proteins is also very useful, since these proteins are in most cases not expressed in MMR-deficient tumours due to bi-allelic alterations. In addition, this procedure helps identify the mutated *MMR* gene and thus facilitates the search for germline mutations (Figure 3 below). Using this approach, we identified aberrant patterns of MMR protein expression in 13.2% of 1048 consecutive, unselected CRCs (Truninger et al., 2005). Loss of expression of hMSH2, hMSH6, hMLH1 and hPMS2 was found in 1.4%, 0.5%, 9.8% and 1.5%, respectively (manuscript submitted).

About 40% of the MMR-deficient tumours were found in patients belonging to HNPCC families or with personal or family histories suggestive of disease inheritance. The remaining 60% appeared to be sporadic. As shown in previous immunohistochemical studies (de Jong et al., 2004b; Lindor et al., 2002; Plaschke et al., 2002; Wahlberg et al., 2002; Wright and Stewart, 2003; Young et al., 2001), most of the sporadic cases lacked hMLH1, the expression of which is silenced by somatic, bi-allelic methylation of the *hMLH1* promoter (Herman et al., 1998). Thus, sporadic, hMLH1-deficient tumours are the most frequent MMR-deficient

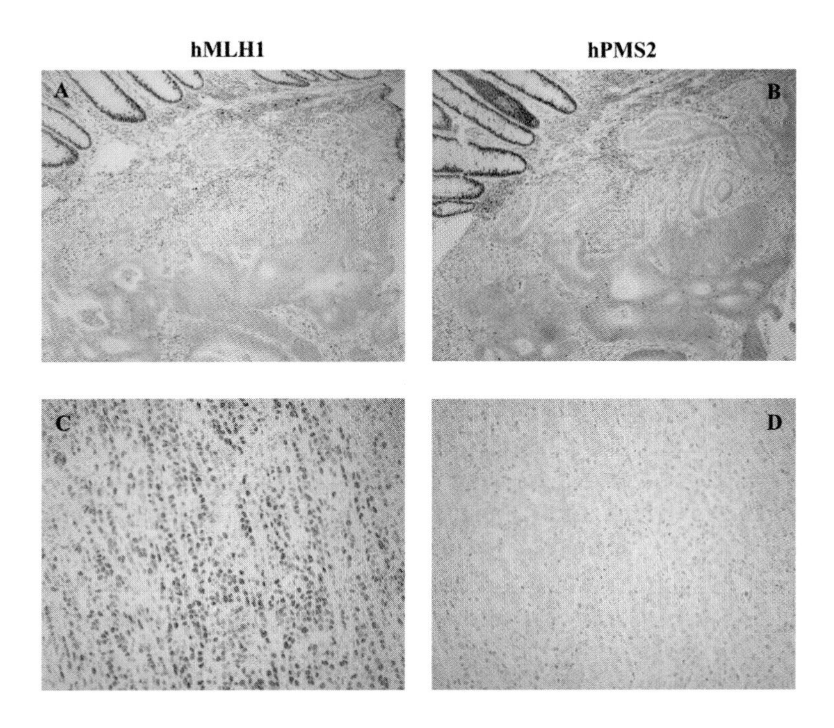

Figure 3. Immunohistochemical staining of colorectal tumours for MMR proteins.A) hMLH1 is absent from this colon cancer, but normal crypts (upper left corner of the picture) and proliferating stromal cells express this protein normally. B) The same tumour does not express hPMS2, because this protein is unstable in the absence of hMLH1. C) This colon cancer expresses hMLH1 normally, however, it is deficient in hPMS2 (D).

CRCs; they are almost invariably located in the proximal colon of old subjects, more frequently women. The incidence of this subset of sporadic CRCs is thus expected to increase worldwide, because of population ageing in developed and developing countries. In an attempt to identify adults likely to be affected by this late-onset form of CRC, we have examined normal colonic mucosa of persons with disease-free colon by quantitative, methylation-sensitive PCR. We found (Menigatti *et al.*, manuscript in preparation) that detectable levels of hypermethylation of the *hMLH1* promoter are present already in the normal mucosa of the proximal colon of some subjects, predominantly women, suggesting that epigenetic changes in this promoter begin several years before the peak of incidence of this type of CRCs.

As already mentioned, a considerable proportion of inherited forms of MMR-deficient CRCs fulfil the three principal criteria for diagnosis of

HNPCC, the so-called Amsterdam criteria (Vasen et al., 1999), which require that 1) there are at least 3 colon or endometrial cancers in a family; 2) one case should be a first-degree relative of the other two, and 3) at least one of them should be diagnosed before the age of 50. Most persons carrying germline mutations in *hMSH2* or *hMLH1* do belong to classical HNPCC families, but germline mutations in *MMR* genes have been found also in families with a lower burden of cancers and even in individuals without a family history of CRC. Germline mutations in *hMSH6* for example, result in an attenuated phenotype, the families often not fulfilling the above-mentioned criteria ((Hendriks et al., 2004) and refs. therein). The risk of CRCs is lower in women and, in both sexes, the mean age of diagnosis of CRC is around 12 years later than in those with *hMSH2* or *hMLH1* mutations. However, by the age of 70, men with *hMSH6* mutations have the same risk of CRC as those with *hMSH2* or *hMLH1* mutations. In addition, the risk of endometrial cancer, although delayed in onset, is significantly greater than in women with *hMSH2* or *hMLH1* mutations. These phenotypic differences can be explained by the biochemical role of hMSH6 in MMR, in particular its partial functional redundancy with hMSH3 (see Part 2 of this chapter). Correspondingly, the MSI phenotype in hMSH6-negative tumours is attenuated, as the hMSH2/hMSH3 heterodimer partially compensates for the absence of hMSH2/hMSH6 in IDL repair. The broader substrate spectrum of hMSH2/hMSH6, which includes both base mismatches and IDLs, might explain why tumours with a primary alteration in *hMSH3* have not been reported (HNPCC mutation database at http://www.insight-group.org/).

Since the discovery of the etiologic association between MMR deficiency and hereditary CRC, only a few cases with germline mutations in *hPMS2* have been described (De Rosa et al., 2000; De Vos et al., 2004; Hamilton et al., 1995; Miyaki et al., 1997; Nicolaides et al., 1994; Trimbath et al., 2001). The mutation carriers did not belong to classical HNPCC families, rather, they presented with CRCs and brain tumours in their first two decades of life, a condition also called Turcot's syndrome (reviewed in (Paraf et al., 1997) and (De Vos et al., 2004)). In most of these cases, both *hPMS2* alleles were mutated in the germline and MSI was detectable even in non-neoplastic tissues, suggesting a recessive way of inheritance. These findings resemble the childhood cancer syndromes characterised by gastrointestinal, haematological and brain cancers in subjects with compound heterozygous germline mutations in *hMLH1* ((Gallinger et al., 2004) and refs. therein). The severity of this syndrome is attributed to the fact that both alleles of a *MMR* gene are affected in the germline. These rare cases do not represent HNPCC. Rather, they may be compared to mouse knock-out models of MMR deficiency. Mice with bi-allelic germline inactivation of *MMR* genes are viable, but are mainly affected by lymphomas. Tumours of the small and large intestine were also detected with higher frequency in these mice, but brain tumours have not been reported (reviewed in (Jiricny and Marra, 2003)).

Besides the rare cases of a *hPMS2*-associated childhood cancer syndrome, the absence of involvement of *hPMS2* in HNPCC was contrary to expectations. Despite the existence of hPMS1 and hMLH3, both of which have been shown to interact with hMLH1, their participation in MMR has not been detected to date, and it is thus unlikely that they are functionally redundant with hPMS2 ((Räschle et al., 1999) and our unpublished observations). Correspondingly, cell lines lacking hPMS2 display MSI and their extracts are MMR-deficient to an extent similar to that observed in cells mutated in *hMLH1* or *hMSH2*. It might therefore be anticipated that, similarly to *hMSH2* and *hMLH1*, germline mutations in a single *hPMS2* allele would predispose to CRC. Indeed, loss of expression of hPMS2 has been recently identified in CRCs of adults from selected populations (de Jong et al., 2004b; Nakagawa et al., 2004; Plaschke et al., 2002; Rigau et al., 2003; Young et al., 2001). In our immunohistochemical study (Truninger et al., 2005), tumours lacking hPMS2 expression were identified with a frequency similar to that of hMSH2-defective CRCs. As the 1048 tumours were collected consecutively and without any selection bias, this frequency represents the true occurrence of hPMS2 deficiency. Individuals carrying PMS2-deficient colon cancers did not belong to HNPCC families as defined by Amsterdam Criteria, however, a closer examination of the clinical and pathological data showed several features of inheritance. Indeed, we identified heterozygous germline mutations in *hPMS2* in many of these subjects. The inheritance pattern of *hPMS2* mutations apparently differs from the autosomal dominant trait characteristic of *hMSH2* and *hMLH1* mutations in HNPCC kindred, and is most likely the main reason why hPMS2-negative tumours escaped detection to date. It is possible that the penetrance of the *hPMS2* mutations might be attenuated by the numerous *hPMS2* pseudogenes on the same chromosome. Recombination events among these sequences (Bailey et al., 2002), occurring in the germline or in somatic cells, might mitigate the severity of the *hPMS2* defects, mainly through mutation reversion.

A further example of non-mendelian inheritance has been recently described in several subjects with early onset CRC, both sporadic and with a family history of cancer not fulfilling the Amsterdam Criteria for HNPCC (Gazzoli et al., 2002; Miyakura et al., 2004; Suter et al., 2004). In these cases, an extensive and mono-allelic methylation of the *hMLH1* promoter was detected in the germline. *hMLH1* loss of heterozygosity was observed in the tumours, which indeed did not express hMLH1 and were MSI$^+$. This condition has been designated “germline epimutation”. The reversibility of epigenetic states might cause a mosaicism depending on the time of its occurrence during embryogenesis and on the cell types that are affected by it. Thus, the pattern of disease risk in these patients might vary from sporadic to complex traits.

Germline mutations that could be linked to an inherited predisposition to CRCs in other genes, the products of which are involved in MMR, have not been reported. The role of missense variants found in genes like *EXO1*

((Alam et al., 2003) and refs. therein) and *hMLH3* ((Hienonen et al., 2003) and refs. therein) is still under investigation. As suggested by studies in yeast (Amin et al., 2001; Schär, 2001), combination of mutations in the latter genes with heterozygous missense alterations in genes involved in DNA replication and DNA-replication checkpoints might give rise to a variable MMR defect that may in turn cause different degrees of cancer risk.

4. CONCLUSIONS AND UNANSWERED QUESTIONS

The past decade witnessed an unprecedented progress in our understanding of the eukaryotic MMR system. This progress has been fuelled to a great extent by the discovery of MSI in human tumours. Credit for the discoveries goes to a very large number of basic- and clinical research laboratories. Unfortunately, due to the enormous volume of literature on the subject, coupled with space restrictions with the focus of this chapter, the work of many of these laboratories has not been cited here.

In spite of this advancement, a number of problems remain. In the basic research field, we need to better understand the biochemistry of MMR. Of particular interest are the following aspects: 1) characterisation of the molecular transactions downstream from mismatch recognition, specifically the role of hMLH1/hPMS2; 2) the functional significance of the interactions between hMLH1 and other polypeptides, such as hPMS1 (Räschle et al., 1999), hMLH3 (Lipkin et al., 2000), MBD4 (Bellacosa et al., 1999), EXO1 (Schmutte et al., 2001), MRE11 (Her et al., 2002) and BLM (Langland et al., 2001; Pedrazzi et al., 2001); 3) mechanism of strand discrimination; 4) transition from replication to MMR, and 5) the recruitment of exonucleases. Needless to say, structural studies are likely to bring new insights into some of these molecular transactions, especially if they are complemented with biochemical approaches and *in vivo* evidence. The use of proteomics technologies may help identify factors that participate in MMR *in vivo*, yet may be dispensable in reconstitution assays *in vitro*.

In the genetic and clinical fields, the burning questions are: 1) why do heterozygous germline mutations in *MMR* genes predispose predominantly to cancers of the colon and uterus, rather than of other organs; 2) which factors trigger the somatic loss of the wild-type alleles of *MMR* genes in heterozygous cells, and which pathways are affected when the second hit occurs; 3) why do HNPCC tumours arise predominantly in the right colon, and why do sporadic, *hMLH1*-deficient tumours of older individuals occur more frequently in women; 4) which are the common mechanisms of cancer development between humans carrying compound germline mutations in *MMR* genes and knock-out mice; 5) what are the mechanisms of selection pressure and which genomic constrains target a subset of microsatellite sequences for frameshift mutagenesis? Analyses of the transcriptomes and proteomes of CRCs at different tumour stages and at the time of diagnosis may help answer some of the above questions, but it is also hoped that these techniques will open new doors to the discovery of novel, efficacious therapies of MMR-deficient cancers. These cancers seem to have a better prognosis, but they are still life threatening. They are tolerant to certain drugs, in particular alkylating agents and cisplatin (reviewed in (Stojic et al., 2004)), yet may be exquisitely sensitive to others. Only a profound understanding of the molecular transactions governing the genesis and survival of these tumours can help find the right cure.

ABBREVIATIONS

CIN (Chromosomal Instability); Colorectal Cancer (CRC); Exonuclease I (EXO I); Insertion/Deletion Loops (IDLs); Microsatellite Instability (MSI); Mismatch Repair (MMR); MutL Homolog (MLH); MutS Homolog (MSH); Post-Meiotic Segregation (PMS); Proliferating Cell Nuclear Antigen (PCNA); Replication Protein A (RPA); Replication Factor C (RF-C)

ACKNOWLEDGEMENTS

We thank Katja Bärenfaller, Petr Cejka, Elda Cannavo' and Ulrich Hübscher for critical comments.

REFERENCES

Aaltonen, L.A., P. Peltomäki, F.S. Leach, P. Sistonen, L. Pylkkanen, J.P. Mecklin, H. Jarvinen, S.M. Powell, J. Jen, S.R. Hamilton, and et al. 1993. Clues to the pathogenesis of familial colorectal cancer. *Science*. 260:812-6.

Abdel-Rahman, W.M., M. Ollikainen, R. Kariola, H.J. Järvinen, J-P. Mecklin, M. Nyström-Lahti, S. Knuutila, and P. Peltomäki. 2005a. Comprehensive characterization of HNPCC-related colorectal cancers reveals striking molecular features in families with no germline mismatch repair gene mutations. *Oncogene*. 24:1-10.

Abdel-Rahman, W.M., K. Katsura, W. Rens, P.A. Gorman, D. Sheer, D. Bicknell, W.F. Bodmer, M.J. Arends, A.H. Wyllie, and P.A.W. Edwards. 2001. Spectral karyotyping suggests additional subsets of colorectal cancers characterized by pattern of chromosome rearrangement. *Proc Natl Acad Sci USA*. 98:2538-2543.

Abdel-Rahman, W.M., H. Lohi, S. Knuutila, and P. Peltomäki. 2005b. Restoring mismatch repair does not stop the formation of reciprocal translocations in the colon cancer cell line HCA7 but further destabilizes chromosome number. *Oncogene*. 24:706-713.

Acharya, S., P.L. Foster, P. Brooks, and R. Fishel. 2003. The coordinated functions of the E. coli MutS and MutL proteins in mismatch repair. *Mol Cell*. 12:233-46.

Acharya, S., T. Wilson, S. Gradia, M.F. Kane, S. Guerrette, G.T. Marsischky, R. Kolodner, and R. Fishel. 1996. hMSH2 forms specific mispair-binding complexes with hMSH3 and hMSH6. *Proc Natl Acad Sci U S A*. 93:13629-34.

Al-Tassan, N., N.H. Chmiel, J. Maynard, N. Fleming, A.L. Livingston, G.T. Williams, A.K. Hodges, D.R. Davies, S.S. David, J.R. Sampson, and J.P. Cheadle. 2002. Inherited variants of MYH associated with somatic G:C-->T:A mutations in colorectal tumors. *Nat Genet*. 30:227-32.

Alam, N.A., P. Gorman, E.E. Jaeger, D. Kelsell, I.M. Leigh, R. Ratnavel, M.E. Murdoch, R.S. Houlston, L.A. Aaltonen, R.R. Roylance, and I.P. Tomlinson. 2003. Germline deletions of EXO1 do not cause colorectal tumors and lesions which are null for EXO1 do not have microsatellite instability. *Cancer Genet Cytogenet*. 147:121-7.

Alani, E., T. Sokolsky, B. Studamire, J.J. Miret, and R.S. Lahue. 1997. Genetic and biochemical analysis of Msh2p-Msh6p: role of ATP hydrolysis and Msh2p-Msh6p subunit interactions in mismatch base pair recognition. *Mol Cell Biol*. 17:2436-47.

Allen, D.J., A. Makhov, M. Grilley, J. Taylor, R. Thresher, P. Modrich, and J.D. Griffith. 1997. MutS mediates heteroduplex loop formation by a translocation mechanism. *EMBO J*. 16:4467-76.

Amin, N.S., M.N. Nguyen, S. Oh, and R.D. Kolodner. 2001. exo1-Dependent mutator mutations: model system for studying functional interactions in mismatch repair. *Mol Cell Biol*. 21:5142-55.

Aronshtam, A., and M.G. Marinus. 1996. Dominant negative mutator mutations in the mutL gene of Escherichia coli. *Nucleic Acids Res*. 24:2498-504.

Bailey, J.A., Z. Gu, R.A. Clark, K. Reinert, R.V. Samonte, S. Schwartz, M.D. Adams, E.W. Myers, P.W. Li, and E.E. Eichler. 2002. Recent segmental duplications in the human genome. *Science*. 297:1003-7.

Ban, C., M. Junop, and W. Yang. 1999. Transformation of MutL by ATP binding and hydrolysis: a switch in DNA mismatch repair. *Cell*. 97:85-97.

Ban, C., and W. Yang. 1998. Crystal structure and ATPase activity of MutL: implications for DNA repair and mutagenesis. *Cell*. 95:541-52.

Baranovskaya, S., J.L. Soto, M. Perucho, and S.R. Malkhosyan. 2001. Functional significance of concomitant inactivation of hMLH1 and hMSH6 in tumor cells of the microsatellite mutator phenotype. *Proc Natl Acad Sci U S A*. 98:15107-12.

Bellacosa, A., L. Cicchillitti, F. Schepis, A. Riccio, A.T. Yeung, Y. Matsumoto, E.A. Golemis, M. Genuardi, and G. Neri. 1999. MED1, a novel human methyl-CpG-binding endonuclease, interacts with DNA mismatch repair protein MLH1. *Proc Natl Acad Sci U S A*. 96:3969-74.

Bhagwat, A.S., and M. Lieb. 2002. Cooperation and competition in mismatch repair: very short-patch repair and methyl-directed mismatch repair in Escherichia coli. *Mol Microbiol*. 44:1421-8.

Bicknell, D.C., L. Kaklamanis, R. Hampson, W.F. Bodmer, and P. Karran. 1996. Selection for beta 2-microglobulin mutation in mismatch repair-defective colorectal carcinomas. *Curr Biol*. 6:1695-7.

Blackwell, L.J., D. Martik, K.P. Bjornson, E.S. Bjornson, and P. Modrich. 1998. Nucleotide-promoted release of hMutSalpha from heteroduplex DNA is consistent with an ATP-dependent translocation mechanism. *J Biol Chem*. 273:32055-62.

Bocker, T., J. Schlegel, F. Kullmann, G. Stumm, H. Zirngibl, J.T. Epplen, and J. Ruschoff. 1996. Genomic instability in colorectal carcinomas: comparison of different evaluation methods and their biological significance. *J Pathol*. 179:15-9.

Bowers, J., T. Sokolsky, T. Quach, and E. Alani. 1999. A mutation in the MSH6 subunit of the Saccharomyces cerevisiae MSH2-MSH6 complex disrupts mismatch recognition. *J Biol Chem*. 274:16115-25.

Bowers, J., P.T. Tran, A. Joshi, R.M. Liskay, and E. Alani. 2001. MSH-MLH complexes formed at a DNA mismatch are disrupted by the PCNA sliding clamp. *J Mol Biol*. 306:957-68.

Bronner, C.E., S.M. Baker, P.T. Morrison, G. Warren, L.G. Smith, M.K. Lescoe, M. Kane, C. Earabino, J. Lipford, A. Lindblom, and et al. 1994. Mutation in the DNA mismatch repair gene homologue hMLH1 is associated with hereditary non-polyposis colon cancer. *Nature*. 368:258-61.

Camps, J., C. Morales, E. Prat, M. Ribas, G. Capella, J. Egozcue, M.A. Peinado, and R. Miro. 2004. Genetic evolution in colon cancer KM12 cells and metastatic derivates. *Int J Cancer*. 110:869-74.

Chan, T.L., L.C. Curtis, S.Y. Leung, S.M. Farrington, J.W. Ho, A.S. Chan, P.W. Lam, C.W. Tse, M.G. Dunlop, A.H. Wyllie, and S.T. Yuen. 2001. Early-onset colorectal cancer with stable microsatellite DNA and near-diploid chromosomes. *Oncogene*. 20:4871-4876.

Chang, D.K., L. Ricciardiello, A. Goel, C.L. Chang, and C.R. Boland. 2000. Steady-state regulation of the human DNA mismatch repair system. *J Biol Chem*. 275:18424-31.

Chen, C., B.J. Merrill, P.J. Lau, C. Holm, and R.D. Kolodner. 1999. Saccharomyces cerevisiae pol30 (proliferating cell nuclear antigen) mutations impair replication fidelity and mismatch repair. *Mol Cell Biol*. 19:7801-15.

Ciotta, C., S. Ceccotti, G. Aquilina, O. Humbert, F. Palombo, J. Jiricny, and M. Bignami. 1998. Increased somatic recombination in methylation tolerant human cells with defective DNA mismatch repair. *J Mol Biol*. 276:705-19.

Clark, A.B., F. Valle, K. Drotschmann, R.K. Gary, and T.A. Kunkel. 2000. Functional interaction of proliferating cell nuclear antigen with MSH2-MSH6 and MSH2-MSH3 complexes. *J Biol Chem*. 275:36498-501.

de Jong, A.E., H. Morreau, M. Van Puijenbroek, P.H. Eilers, J. Wijnen, F.M. Nagengast, G. Griffioen, A. Cats, F.H. Menko, J.H. Kleibeuker, and H.F. Vasen. 2004a. The role of mismatch repair gene defects in the development of adenomas in patients with HNPCC. *Gastroenterology*. 126:42-8.

de Jong, A.E., M. van Puijenbroek, Y. Hendriks, C. Tops, J. Wijnen, M.G. Ausems, H. Meijers-Heijboer, A. Wagner, T.A. van Os, A.H. Brocker-Vriends, H.F. Vasen, and H. Morreau. 2004b. Microsatellite instability, immunohistochemistry, and additional PMS2 staining in suspected hereditary nonpolyposis colorectal cancer. *Clin Cancer Res*. 10:972-80.

De Rosa, M., C. Fasano, L. Panariello, M.I. Scarano, G. Belli, A. Iannelli, F. Ciciliano, and P. Izzo. 2000. Evidence for a recessive inheritance of Turcot's syndrome caused by compound heterozygous mutations within the PMS2 gene. *Oncogene*. 19:1719-23.

De Vos, M., B.E. Hayward, S. Picton, E. Sheridan, and D.T. Bonthron. 2004. Novel PMS2 pseudogenes can conceal recessive mutations causing a distinctive childhood cancer syndrome. *Am J Hum Genet*. 74:954-64.

Drotschmann, K., A. Aronshtam, H.J. Fritz, and M.G. Marinus. 1998. The Escherichia coli MutL protein stimulates binding of Vsr and MutS to heteroduplex DNA. *Nucleic Acids Res*. 26:948-53.

Drummond, J.T., J. Genschel, E. Wolf, and P. Modrich. 1997. DHFR/MSH3 amplification in methotrexate-resistant cells alters the hMutSalpha/hMutSbeta ratio and reduces the efficiency of base-base mismatch repair. *Proc Natl Acad Sci U S A*. 94:10144-9.

Drummond, J.T., G.M. Li, M.J. Longley, and P. Modrich. 1995. Isolation of an hMSH2-p160 heterodimer that restores DNA mismatch repair to tumor cells. *Science*. 268:1909-12.

Dufner, P., G. Marra, M. Räschle, and J. Jiricny. 2000. Mismatch recognition and DNA-dependent stimulation of the ATPase activity of hMutSalpha is abolished by a single mutation in the hMSH6 subunit. *J Biol Chem*. 275:36550-5.

Duval, A., J. Gayet, X.P. Zhou, B. Iacopetta, G. Thomas, and R. Hamelin. 1999. Frequent frameshift mutations of the TCF-4 gene in colorectal cancers with microsatellite instability. *Cancer Res*. 59:4213-5.

Dzantiev, L., N. Constantin, J. Genschel, R.R. Iyer, P.M. Burgers, and P. Modrich. 2004. A defined human system that supports bidirectional mismatch-provoked excision. *Mol Cell*. 15:31-41.

Echols, H., and M.F. Goodman. 1991. Fidelity mechanisms in DNA replication. *Annu Rev Biochem*. 60:477-511.

Ejima, Y., L. Yang, and M.S. Sasaki. 2000. Aberrant splicing of the ATM gene associated with shortening of the intronic mononucleotide tract in human colon tumor cell lines: a novel mutation target of microsatellite instability. *Int J Cancer*. 86:262-8.

Fang, W.H., and P. Modrich. 1993. Human strand-specific mismatch repair occurs by a bidirectional mechanism similar to that of the bacterial reaction. *J Biol Chem.* 268:11838-44.

Fang, Y., C.C. Tsao, B.K. Goodman, R. Furumai, C.A. Tirado, R.T. Abraham, and X.F. Wang. 2004. ATR functions as a gene dosage-dependent tumor suppressor on a mismatch repair-deficient background. *EMBO J.* 23:3164-74.

Fishel, R., M.K. Lescoe, M.R. Rao, N.G. Copeland, N.A. Jenkins, J. Garber, M. Kane, and R. Kolodner. 1993. The human mutator gene homolog MSH2 and its association with hereditary nonpolyposis colon cancer. *Cell.* 75:1027-38.

Flores-Rozas, H., D. Clark, and R.D. Kolodner. 2000. Proliferating cell nuclear antigen and Msh2p-Msh6p interact to form an active mispair recognition complex. *Nat Genet.* 26:375-8.

Flores-Rozas, H., Z. Kelman, F.B. Dean, Z.Q. Pan, J.W. Harper, S.J. Elledge, M. O'Donnell, and J. Hurwitz. 1994. Cdk-interacting protein 1 directly binds with proliferating cell nuclear antigen and inhibits DNA replication catalyzed by the DNA polymerase delta holoenzyme. *Proc Natl Acad Sci U S A.* 91:8655-9.

Flores-Rozas, H., and R.D. Kolodner. 1998. The Saccharomyces cerevisiae MLH3 gene functions in MSH3-dependent suppression of frameshift mutations. *Proc Natl Acad Sci U S A.* 95:12404-9.

Fujii, H., and T. Shimada. 1989. Isolation and characterization of cDNA clones derived from the divergently transcribed gene in the region upstream from the human dihydrofolate reductase gene. *J Biol Chem.* 264:10057-64.

Gallinger, S., M. Aronson, K. Shayan, E.M. Ratcliffe, J.T. Gerstle, P.C. Parkin, H. Rothenmund, M. Croitoru, E. Baumann, P.R. Durie, R. Weksberg, A. Pollett, R.H. Riddell, B.Y. Ngan, E. Cutz, A.E. Lagarde, and H.S. Chan. 2004. Gastrointestinal cancers and neurofibromatosis type 1 features in children with a germline homozygous MLH1 mutation. *Gastroenterology.* 126:576-85.

Gazzoli, I., M. Loda, J. Garber, S. Syngal, and R.D. Kolodner. 2002. A hereditary nonpolyposis colorectal carcinoma case associated with hypermethylation of the MLH1 gene in normal tissue and loss of heterozygosity of the unmethylated allele in the resulting microsatellite instability-high tumor. *Cancer Res.* 62:3925-8.

Genschel, J., L.R. Bazemore, and P. Modrich. 2002. Human exonuclease I is required for 5' and 3' mismatch repair. *J Biol Chem.* 277:13302-11.

Genschel, J., S.J. Littman, J.T. Drummond, and P. Modrich. 1998. Isolation of MutSbeta from human cells and comparison of the mismatch repair specificities of MutSbeta and MutSalpha. *J Biol Chem.* 273:19895-901.

Genschel, J., and P. Modrich. 2003. Mechanism of 5'-directed excision in human mismatch repair. *Mol Cell.* 12:1077-86.

Georgiades, I.B., L.J. Curtis, R.M. Morris, C.C. Bird, and A.H. Wyllie. 1999. Heterogeneity studies identify a subset of sporadic colorectal cancers without evidence for chromosomal or microsatellite instability. *Oncogene.* 18:7933-40.

Giannini, G., C. Rinaldi, E. Ristori, M.I. Ambrosini, F. Cerignoli, A. Viel, E. Bidoli, S. Berni, G. D'Amati, G. Scambia, L. Frati, I. Screpanti, and A. Gulino. 2004. Mutations of an intronic repeat induce impaired MRE11 expression in primary human cancer with microsatellite instability. *Oncogene.* 23:2640-7.

Goel, A., C.N. Arnold, D. Niedzwiecki, D.K. Chang, L. Ricciardiello, J.M. Carethers, J.M. Dowell, L. Wasserman, C. Compton, R.J. Mayer, M.M. Bertagnolli, and C.R. Boland. 2003. Characterization of sporadic colon cancer by patterns of genomic instability. *Cancer Res.* 63:1608-14.

Gradia, S., S. Acharya, and R. Fishel. 1997. The human mismatch recognition complex hMSH2-hMSH6 functions as a novel molecular switch. *Cell.* 91:995-1005.

Grady, W.M. 2004. Genomic instability and colon cancer. *Cancer Metastasis Rev.* 23:11-27.

Gryfe, R., H. Kim, E.T. Hsieh, M.D. Aronson, E.J. Holowaty, S.B. Bull, M. Redston, and S. Gallinger. 2000. Tumor microsatellite instability and clinical outcome in young patients with colorectal cancer. *N Engl J Med.* 342:69-77.

Guarne, A., S. Ramon-Maiques, E.M. Wolff, R. Ghirlando, X. Hu, J.H. Miller, and W. Yang. 2004. Structure of the MutL C-terminal domain: a model of intact MutL and its roles in mismatch repair. *EMBO J.* 23:4134-45.

Guerrette, S., T. Wilson, S. Gradia, and R. Fishel. 1998. Interactions of human hMSH2 with hMSH3 and hMSH2 with hMSH6: examination of mutations found in hereditary nonpolyposis colorectal cancer. *Mol Cell Biol.* 18:6616-23.

Guo, S., S.R. Presnell, F. Yuan, Y. Zhang, L. Gu, and G.M. Li. 2004. Differential requirement for proliferating cell nuclear antigen in 5' and 3' nick-directed excision in human mismatch repair. *J Biol Chem.* 279:16912-7.

Haber, L.T., and G.C. Walker. 1991. Altering the conserved nucleotide binding motif in the Salmonella typhimurium MutS mismatch repair protein affects both its ATPase and mismatch binding activities. *EMBO J.* 10:2707-15.

Haigis, K.M., J.G. Caya, M. Reichelderfer, and W.F. Dove. 2002. Intestinal adenomas can develop with a stable karyotype and stable microsatellites. *Proc Natl Acad Sci U S A.* 99:8927-31.

Hamilton, S.R., B. Liu, R.E. Parsons, N. Papadopoulos, J. Jen, S.M. Powell, A.J. Krush, T. Berk, Z. Cohen, B. Tetu, and et al. 1995. The molecular basis of Turcot's syndrome. *N Engl J Med.* 332:839-47.

Hardeland, U., M. Bentele, J. Jiricny, and P. Schär. 2003. The versatile thymine DNA-glycosylase: a comparative characterization of the human, Drosophila and fission yeast orthologs. *Nucleic Acids Res.* Vol. 31. 2261-71.

Harfe, B.D., and S. Jinks-Robertson. 2000. Mismatch repair proteins and mitotic genome stability. *Mutat Res.* 451:151-67.

Harfe, B.D., B.K. Minesinger, and S. Jinks-Robertson. 2000. Discrete in vivo roles for the MutL homologs Mlh2p and Mlh3p in the removal of frameshift intermediates in budding yeast. *Curr Biol.* 10:145-8.

Hawkins, N.J., I. Tomlinson, A. Meagher, and R.L. Ward. 2001. Microsatellite-stable diploid carcinoma: a biologically distinct and aggressive subset of sporadic colorectal cancer. *Br J Cancer.* 84:232-6.

Hendrich, B., and A. Bird. 1998. Identification and characterization of a family of mammalian methyl-CpG binding proteins. *Mol Cell Biol.* 18:6538-47.

Hendrich, B., U. Hardeland, H.H. Ng, J. Jiricny, and A. Bird. 1999. The thymine glycosylase MBD4 can bind to the product of deamination at methylated CpG sites. *Nature.* 401:301-4.

Hendriks, Y.M., A. Wagner, H. Morreau, F. Menko, A. Stormorken, F. Quehenberger, L. Sandkuijl, P. Moller, M. Genuardi, H. Van Houwelingen, C. Tops, M. Van Puijenbroek, P. Verkuijlen, G. Kenter, A. Van Mil, H. Meijers-Heijboer, G.B. Tan, M.H. Breuning, R. Fodde, J.T. Wijnen, A.H. Brocker-Vriends, and H. Vasen. 2004. Cancer risk in hereditary nonpolyposis colorectal cancer due to MSH6 mutations: impact on counseling and surveillance. *Gastroenterology.* 127:17-25.

Her, C., A.T. Vo, and X. Wu. 2002. Evidence for a direct association of hMRE11 with the human mismatch repair protein hMLH1. *DNA Repair (Amst).* 1:719-29.

Herman, J.G., A. Umar, K. Polyak, J.R. Graff, N. Ahuja, J.P. Issa, S. Markowitz, J.K. Willson, S.R. Hamilton, K.W. Kinzler, M.F. Kane, R.D. Kolodner, B. Vogelstein, T.A. Kunkel, and S.B. Baylin. 1998. Incidence and functional consequences of hMLH1 promoter hypermethylation in colorectal carcinoma. *Proc Natl Acad Sci U S A.* 95:6870-5.

Hienonen, T., P. Laiho, R. Salovaara, J.P. Mecklin, H. Jarvinen, P. Sistonen, P. Peltomäki, R. Lehtonen, N.N. Nupponen, V. Launonen, A. Karhu, and L.A. Aaltonen. 2003. Little evidence for involvement of MLH3 in colorectal cancer predisposition. *Int J Cancer.* 106:292-6.

Hoeijmakers, J.H. 2001. Genome maintenance mechanisms for preventing cancer. *Nature.* 411:366-74.

Hollstein, M., D. Sidransky, B. Vogelstein, and C.C. Harris. 1991. p53 mutations in human cancers. Science. Vol. 253. 49-53.

Holmes, J., Jr., S. Clark, and P. Modrich. 1990. Strand-specific mismatch correction in nuclear extracts of human and Drosophila melanogaster cell lines. *Proc Natl Acad Sci U S A*. 87:5837-41.

Hübscher, U., G. Maga, and S. Spadari. 2002. Eukaryotic DNA polymerases. *Annu Rev Biochem*. 71:133-63.

Hübscher, U., and Y.S. Seo. 2001. Replication of the lagging strand: a concert of at least 23 polypeptides. *Mol Cells*. 12:149-57.

Hughes, M.J., and J. Jiricny. 1992. The purification of a human mismatch-binding protein and identification of its associated ATPase and helicase activities. *J Biol Chem*. 267:23876-82.

Iaccarino, I., G. Marra, P. Dufner, and J. Jiricny. 2000. Mutation in the magnesium binding site of hMSH6 disables the hMutSalpha sliding clamp from translocating along DNA. *J Biol Chem*. 275:2080-6.

Iaccarino, I., G. Marra, F. Palombo, and J. Jiricny. 1998. hMSH2 and hMSH6 play distinct roles in mismatch binding and contribute differently to the ATPase activity of hMutSalpha. *EMBO J*. 17:2677-86.

Iams, K., E.D. Larson, and J.T. Drummond. 2002. DNA template requirements for human mismatch repair in vitro. *J Biol Chem*. 277:30805-14.

Ionov, Y., M.A. Peinado, S. Malkhosyan, D. Shibata, and M. Perucho. 1993. Ubiquitous somatic mutations in simple repeated sequences reveal a new mechanism for colonic carcinogenesis. *Nature*. 363:558-61.

Jacoby, R.F., D.J. Marshall, S. Kailas, S. Schlack, B. Harms, and R. Love. 1995. Genetic instability associated with adenoma to carcinoma progression in hereditary nonpolyposis colon cancer. *Gastroenterology*. 109:73-82.

Jass, J.R. 2000. Familial colorectal cancer: pathology and molecular characteristics. *Lancet Oncol*. 1:220-6.

Jass, J.R., V.L. Whitehall, J. Young, and B.A. Leggett. 2002. Emerging concepts in colorectal neoplasia. *Gastroenterology*. 123:862-76.

Jiricny, J. 1998. Replication errors: cha(lle)nging the genome. *EMBO J*. 17:6427-36.

Jiricny, J. 2000. Mismatch repair: the praying hands of fidelity. *Curr Biol*. 10:R788-90.

Jiricny, J., M. Hughes, N. Corman, and B.B. Rudkin. 1988. A human 200-kDa protein binds selectively to DNA fragments containing G.T mismatches. *Proc Natl Acad Sci U S A*. 85:8860-4.

Jiricny, J., and G. Marra. 2003. DNA repair defects in colon cancer. *Curr Opin Genet Dev*. 13:61-9.

Johnson, R.E., G.K. Kovvali, S.N. Guzder, N.S. Amin, C. Holm, Y. Habraken, P. Sung, L. Prakash, and S. Prakash. 1996. Evidence for involvement of yeast proliferating cell nuclear antigen in DNA mismatch repair. *J Biol Chem*. 271:27987-90.

Junop, M.S., G. Obmolova, K. Rausch, P. Hsieh, and W. Yang. 2001. Composite active site of an ABC ATPase: MutS uses ATP to verify mismatch recognition and authorize DNA repair. *Mol Cell*. 7:1-12.

Kim, H., J. Jen, B. Vogelstein, and S.R. Hamilton. 1994. Clinical and pathological characteristics of sporadic colorectal carcinomas with DNA replication errors in microsatellite sequences. *Am J Pathol*. 145:148-56.

Kirchner, J.M., H. Tran, and M.A. Resnick. 2000. A DNA polymerase epsilon mutant that specifically causes +1 frameshift mutations within homonucleotide runs in yeast. *Genetics*. 155:1623-32.

Kleczkowska, H.E., G. Marra, T. Lettieri, and J. Jiricny. 2001. hMSH3 and hMSH6 interact with PCNA and colocalize with it to replication foci. *Genes Dev*. 15:724-36.

Kokoska, R.J., L. Stefanovic, A.B. Buermeyer, R.M. Liskay, and T.D. Petes. 1999. A mutation of the yeast gene encoding PCNA destabilizes both microsatellite and minisatellite DNA sequences. *Genetics*. 151:511-9.

Kolodner, R.D., and G.T. Marsischky. 1999. Eukaryotic DNA mismatch repair. *Curr Opin Genet Dev*. 9:89-96.

Kramer, B., W. Kramer, and H.J. Fritz. 1984. Different base/base mismatches are corrected with different efficiencies by the methyl-directed DNA mismatch-repair system of E. coli. *Cell*. 38:879-87.

Kunkel, T.A., and K. Bebenek. 2000. DNA replication fidelity. *Annu Rev Biochem*. 69:497-529.

Laghi, L., G.N. Ranzani, P. Bianchi, A. Mori, K. Heinimann, O. Orbetegli, M.R. Spaudo, O. Luinetti, S. Francisconi, M. Roncalli, E. Solcia, and A. Malesci. 2002. Frameshift mutations of human gastrin receptor gene (hGARE) in gastrointestinal cancers with microsatellite instability. *Lab Invest*. 82:265-71.

Lamers, M.H., A. Perrakis, J.H. Enzlin, H.H. Winterwerp, N. de Wind, and T.K. Sixma. 2000. The crystal structure of DNA mismatch repair protein MutS binding to a G x T mismatch. *Nature*. 407:711-7.

Langland, G., J. Kordich, J. Creaney, K.H. Goss, K. Lillard-Wetherell, K. Bebenek, T.A. Kunkel, and J. Groden. 2001. The Bloom's syndrome protein (BLM) interacts with MLH1 but is not required for DNA mismatch repair. *J Biol Chem*. 276:30031-5.

Lau, P.J., and R.D. Kolodner. 2003. Transfer of the MSH2.MSH6 complex from proliferating cell nuclear antigen to mispaired bases in DNA. *J Biol Chem*. 278:14-7.

Leach, F.S., N.C. Nicolaides, N. Papadopoulos, B. Liu, J. Jen, R. Parsons, P. Peltomaki, P. Sistonen, L.A. Aaltonen, M. Nystrom-Lahti, and et al. 1993. Mutations of a mutS homolog in hereditary nonpolyposis colorectal cancer. *Cell*. 75:1215-25.

Lee, B.I., and D.M. Wilson, 3rd. 1999. The RAD2 domain of human exonuclease 1 exhibits 5' to 3' exonuclease and flap structure-specific endonuclease activities. *J Biol Chem*. 274:37763-9.

Lengauer, C., K.W. Kinzler, and B. Vogelstein. 1997. Genetic instability in colorectal cancers. *Nature*. 386:623-7.

Lengauer, C., K.W. Kinzler, and B. Vogelstein. 1998. Genetic instabilities in human cancers. *Nature*. 396:643-9.

Li, G.M., and P. Modrich. 1995. Restoration of mismatch repair to nuclear extracts of H6 colorectal tumor cells by a heterodimer of human MutL homologs. *Proc Natl Acad Sci U S A*. 92:1950-4.

Li, R., S. Waga, G.J. Hannon, D. Beach, and B. Stillman. 1994. Differential effects by the p21 CDK inhibitor on PCNA-dependent DNA replication and repair. *Nature*. 371:534-7.

Li, Y.C., A.B. Korol, T. Fahima, and E. Nevo. 2004. Microsatellites within genes: structure, function, and evolution. *Mol Biol Evol*. 21:991-1007.

Lin, Y.L., M.K. Shivji, C. Chen, R. Kolodner, R.D. Wood, and A. Dutta. 1998. The evolutionarily conserved zinc finger motif in the largest subunit of human replication protein A is required for DNA replication and mismatch repair but not for nucleotide excision repair. *J Biol Chem*. 273:1453-61.

Lindblom, A., P. Tannergard, B. Werelius, and M. Nordenskjold. 1993. Genetic mapping of a second locus predisposing to hereditary non-polyposis colon cancer. *Nat Genet*. 5:279-82.

Lindgren, G., A. Liljegren, E. Jaramillo, C. Rubio, and A. Lindblom. 2002. Adenoma prevalence and cancer risk in familial non-polyposis colorectal cancer. *Gut*. 50:228-34.

Lindor, N.M., L.J. Burgart, O. Leontovich, R.M. Goldberg, J.M. Cunningham, D.J. Sargent, C. Walsh-Vockley, G.M. Petersen, M.D. Walsh, B.A. Leggett, J.P. Young, M.A. Barker, J.R. Jass, J. Hopper, S. Gallinger, B. Bapat, M. Redston, and S.N. Thibodeau. 2002. Immunohistochemistry versus microsatellite instability testing in phenotyping colorectal tumors. *J Clin Oncol*. 20:1043-8.

Linton, J.P., J.Y. Yen, E. Selby, Z. Chen, J.M. Chinsky, K. Liu, R.E. Kellems, and G.F. Crouse. 1989. Dual bidirectional promoters at the mouse dhfr locus: cloning and characterization of two mRNA classes of the divergently transcribed Rep-1 gene. *Mol Cell Biol*. 9:3058-72.

Lipkin, S.M., V. Wang, R. Jacoby, S. Banerjee-Basu, A.D. Baxevanis, H.T. Lynch, R.M. Elliott, and F.S. Collins. 2000. MLH3: a DNA mismatch repair gene associated with mammalian microsatellite instability. *Nat Genet*. 24:27-35.

Liu, W., X. Dong, M. Mai, R.S. Seelan, K. Taniguchi, K.K. Krishnadath, K.C. Halling, J.M. Cunningham, L.A. Boardman, C. Qian, E. Christensen, S.S. Schmidt, P.C. Roche, D.I. Smith, and S.N. Thibodeau. 2000. Mutations in AXIN2 cause colorectal cancer with defective mismatch repair by activating beta-catenin/TCF signalling. *Nat Genet*. 26:146-7.

Loeb, L.A. 1991. Mutator phenotype may be required for multistage carcinogenesis. *Cancer Res*. 51:3075-9.

Longley, M.J., A.J. Pierce, and P. Modrich. 1997. DNA polymerase delta is required for human mismatch repair in vitro. *J Biol Chem*. 272:10917-21.

Losi, L., M. Ponz de Leon, J. Jiricny, C. Di Gregorio, P. Benatti, A. Percesepe, R. Fante, L. Roncucci, M. Pedroni, and J. Benhattar. 1997. K-ras and p53 mutations in hereditary non-polyposis colorectal cancers. *Int J Cancer*. Vol. 74. 94-6.

Lynch, H.T., T. Smyrk, and J.F. Lynch. 1996. Overview of natural history, pathology, molecular genetics and management of HNPCC (Lynch Syndrome). *Int J Cancer*. 69:38-43.

Lynch, H.T., T.C. Smyrk, P. Watson, S.J. Lanspa, J.F. Lynch, P.M. Lynch, R.J. Cavalieri, and C.R. Boland. 1993. Genetics, natural history, tumor spectrum, and pathology of hereditary nonpolyposis colorectal cancer: an updated review. *Gastroenterology*. 104:1535-49.

Maehara, Y., S. Oda, and K. Sugimachi. 2001. The instability within: problems in current analyses of microsatellite instability. *Mutat Res*. 461:249-63.

Malkhosyan, S., N. Rampino, H. Yamamoto, and M. Perucho. 1996. Frameshift mutator mutations. *Nature*. 382:499-500.

Malkov, V.A., I. Biswas, R.D. Camerini-Otero, and P. Hsieh. 1997. Photocross-linking of the NH2-terminal region of Taq MutS protein to the major groove of a heteroduplex DNA. *J Biol Chem*. 272:23811-7.

Markowitz, S., J. Wang, L. Myeroff, R. Parsons, L. Sun, J. Lutterbaugh, R.S. Fan, E. Zborowska, K.W. Kinzler, B. Vogelstein, and et al. 1995. Inactivation of the type II TGF-beta receptor in colon cancer cells with microsatellite instability. *Science*. 268:1336-8.

Marra, G., and C.R. Boland. 1995. Hereditary nonpolyposis colorectal cancer: the syndrome, the genes, and historical perspectives. *J Natl Cancer Inst*. 87:1114-25.

Marra, G., I. Iaccarino, T. Lettieri, G. Roscilli, P. Delmastro, and J. Jiricny. 1998. Mismatch repair deficiency associated with overexpression of the MSH3 gene. *Proc Natl Acad Sci U S A*. 95:8568-73.

Marra, G., and P. Schär. 1999. Recognition of DNA alterations by the mismatch repair system. *Biochem J*. 338 (Pt 1):1-13.

Marti, T.M., C. Kunz, and O. Fleck. 2002. DNA mismatch repair and mutation avoidance pathways. *J Cell Physiol*. 191:28-41.

Mechanic, L.E., B.A. Frankel, and S.W. Matson. 2000. Escherichia coli MutL loads DNA helicase II onto DNA. *J Biol Chem*. 275:38337-46.

Menoyo, A., H. Alazzouzi, E. Espin, M. Armengol, H. Yamamoto, and S. Schwartz, Jr. 2001. Somatic mutations in the DNA damage-response genes ATR and CHK1 in sporadic stomach tumors with microsatellite instability. *Cancer Res*. 61:7727-30.

Miyaki, M., J. Nishio, M. Konishi, R. Kikuchi-Yanoshita, K. Tanaka, M. Muraoka, M. Nagato, J.M. Chong, M. Koike, T. Terada, Y. Kawahara, A. Fukutome, J. Tomiyama, Y. Chuganji, M. Momoi, and J. Utsunomiya. 1997. Drastic genetic instability of tumors and normal tissues in Turcot syndrome. *Oncogene*. 15:2877-81.

Miyakura, Y., K. Sugano, T. Akasu, T. Yoshida, M. Maekawa, S. Saitoh, H. Sasaki, T. Nomizu, F. Konishi, S. Fujita, Y. Moriya, and H. Nagai. 2004. Extensive but hemiallelic methylation of the hMLH1 promoter region in early-onset sporadic colon cancers with microsatellite instability. *Clin Gastroenterol Hepatol*. 2:147-56.

Modrich, P., and R. Lahue. 1996. Mismatch repair in replication fidelity, genetic recombination, and cancer biology. *Annu Rev Biochem*. 65:101-33.

Mori, Y., J. Yin, A. Rashid, B.A. Leggett, J. Young, L. Simms, P.M. Kuehl, P. Langenberg, S.J. Meltzer, and O.C. Stine. 2001. Instabilotyping: comprehensive identification of frameshift mutations caused by coding region microsatellite instability. *Cancer Res*. 61:6046-9.

Nakagawa, H., J.C. Lockman, W.L. Frankel, H. Hampel, K. Steenblock, L.J. Burgart, S.N. Thibodeau, and A. de la Chapelle. 2004. Mismatch repair gene PMS2: disease-causing germline mutations are frequent in patients whose tumors stain negative for PMS2 protein, but paralogous genes obscure mutation detection and interpretation. *Cancer Res.* 64:4721-7.

Nicholson, A., M. Hendrix, S. Jinks-Robertson, and G.F. Crouse. 2000. Regulation of mitotic homeologous recombination in yeast. Functions of mismatch repair and nucleotide excision repair genes. *Genetics.* 154:133-46.

Nicolaides, N.C., N. Papadopoulos, B. Liu, Y.F. Wei, K.C. Carter, S.M. Ruben, C.A. Rosen, W.A. Haseltine, R.D. Fleischmann, C.M. Fraser, and et al. 1994. Mutations of two PMS homologues in hereditary nonpolyposis colon cancer. *Nature.* 371:75-80.

Obmolova, G., C. Ban, P. Hsieh, and W. Yang. 2000. Crystal structures of mismatch repair protein MutS and its complex with a substrate DNA. *Nature.* 407:703-10.

Palombo, F., P. Gallinari, I. Iaccarino, T. Lettieri, M. Hughes, A. D'Arrigo, O. Truong, J.J. Hsuan, and J. Jiricny. 1995. GTBP, a 160-kilodalton protein essential for mismatch-binding activity in human cells. *Science.* 268:1912-4.

Palombo, F., I. Iaccarino, E. Nakajima, M. Ikejima, T. Shimada, and J. Jiricny. 1996. hMutSbeta, a heterodimer of hMSH2 and hMSH3, binds to insertion/deletion loops in DNA. *Curr Biol.* 6:1181-4.

Papadopoulos, N., N.C. Nicolaides, B. Liu, R. Parsons, C. Lengauer, F. Palombo, A. D'Arrigo, S. Markowitz, J.K. Willson, K.W. Kinzler, and et al. 1995. Mutations of GTBP in genetically unstable cells. *Science.* 268:1915-7.

Papadopoulos, N., N.C. Nicolaides, Y.F. Wei, S.M. Ruben, K.C. Carter, C.A. Rosen, W.A. Haseltine, R.D. Fleischmann, C.M. Fraser, M.D. Adams, and et al. 1994. Mutation of a mutL homolog in hereditary colon cancer. *Science.* 263:1625-9.

Paques, F., and J.E. Haber. 1999. Multiple pathways of recombination induced by double-strand breaks in Saccharomyces cerevisiae. *Microbiol Mol Biol Rev.* 63:349-404.

Paraf, F., S. Jothy, and E.G. Van Meir. 1997. Brain tumor-polyposis syndrome: two genetic diseases? *J Clin Oncol.* 15:2744-58.

Pedrazzi, G., C. Perrera, H. Blaser, P. Kuster, G. Marra, S.L. Davies, G.H. Ryu, R. Freire, I.D. Hickson, J. Jiricny, and I. Stagljar. 2001. Direct association of Bloom's syndrome gene product with the human mismatch repair protein MLH1. *Nucleic Acids Res.* 29:4378-86.

Peinado, M.A., S. Malkhosyan, A. Velazquez, and M. Perucho. 1992. Isolation and characterization of allelic losses and gains in colorectal tumors by arbitrarily primed polymerase chain reaction. *Proc Natl Acad Sci U S A.* 89:10065-9.

Peltomäki, P., L.A. Aaltonen, P. Sistonen, L. Pylkkanen, J.P. Mecklin, H. Jarvinen, J.S. Green, J.R. Jass, J.L. Weber, F.S. Leach, and et al. 1993. Genetic mapping of a locus predisposing to human colorectal cancer. *Science.* 260:810-2.

Percesepe, A., P. Kristo, L.A. Aaltonen, M. Ponz de Leon, A. de la Chapelle, and P. Peltomäki. 1998. Mismatch repair genes and mononucleotide tracts as mutation targets in colorectal tumors with different degrees of microsatellite instability. *Oncogene.* 17:157-63.

Plaschke, J., S. Kruger, B. Jeske, F. Theissig, F.R. Kreuz, S. Pistorius, H.D. Saeger, I. Iaccarino, G. Marra, and H.K. Schackert. 2004. Loss of MSH3 protein expression is frequent in MLH1-deficient colorectal cancer and is associated with disease progression. *Cancer Res.* 64:864-70.

Plaschke, J., S. Kruger, S. Pistorius, F. Theissig, H.D. Saeger, and H.K. Schackert. 2002. Involvement of hMSH6 in the development of hereditary and sporadic colorectal cancer revealed by immunostaining is based on germline mutations, but rarely on somatic inactivation. *Int J Cancer.* 97:643-8.

Polakis, P. 2000. Wnt signaling and cancer. *Genes Dev.* 14:1837-51.

Ponz de Leon, M., G. Della Casa, P. Benatti, A. Percesepe, C. di Gregorio, R. Fante, and L. Roncucci. 1998. Frequency and type of colorectal tumors in asymptomatic high-risk individuals in families with hereditary nonpolyposis colorectal cancer. *Cancer Epidemiol Biomarkers Prev*. 7:639-41.

Pospiech, H., and J.E. Syvaoja. 2003. DNA polymerase epsilon - more than a polymerase. *ScientificWorldJournal*. 3:87-104.

Rajagopalan, H., M.A. Nowak, B. Vogelstein, and C. Lengauer. 2003. The significance of unstable chromosomes in colorectal cancer. *Nat Rev Cancer*. 3:695-701.

Rampino, N., H. Yamamoto, Y. Ionov, Y. Li, H. Sawai, J.C. Reed, and M. Perucho. 1997. Somatic frameshift mutations in the BAX gene in colon cancers of the microsatellite mutator phenotype. *Science*. 275:967-9.

Räschle, M., P. Dufner, G. Marra, and J. Jiricny. 2002. Mutations within the hMLH1 and hPMS2 subunits of the human MutLalpha mismatch repair factor affect its ATPase activity, but not its ability to interact with hMutSalpha. *J Biol Chem*. 277:21810-20.

Räschle, M., G. Marra, M. Nyström-Lahti, P. Schär, and J. Jiricny. 1999. Identification of hMutLbeta, a heterodimer of hMLH1 and hPMS1. *J Biol Chem*. 274:32368-75.

Rasmussen, L.J., L. Samson, and M.G. Marinus. 1998. Dam-directed DNA mismatch repair. Humana Press, Totowa, NJ. 205-228 pp.

Riccio, A., L.A. Aaltonen, A.K. Godwin, A. Loukola, A. Percesepe, R. Salovaara, V. Masciullo, M. Genuardi, M. Paravatou-Petsotas, D.E. Bassi, B.A. Ruggeri, A.J. Klein-Szanto, J.R. Testa, G. Neri, and A. Bellacosa. 1999. The DNA repair gene MBD4 (MED1) is mutated in human carcinomas with microsatellite instability. *Nat Genet*. 23:266-8.

Rigau, V., N. Sebbagh, S. Olschwang, F. Paraf, N. Mourra, Y. Parc, and J.F. Flejou. 2003. Microsatellite instability in colorectal carcinoma. The comparison of immunohistochemistry and molecular biology suggests a role for hMSH6 [correction of hMLH6] immunostaining. *Arch Pathol Lab Med*. 127:694-700.

Risinger, J.I., A. Umar, J. Boyd, A. Berchuck, T.A. Kunkel, and J.C. Barrett. 1996. Mutation of MSH3 in endometrial cancer and evidence for its functional role in heteroduplex repair. *Nat Genet*. 14:102-5.

Rudolph, C., O. Fleck, and J. Kohli. 1998. Schizosaccharomyces pombe exo1 is involved in the same mismatch repair pathway as msh2 and pms1. *Curr Genet*. 34:343-50.

Ruggiero, T., M. Olivero, A. Follenzi, L. Naldini, R. Calogero, and M.F. Di Renzo. 2003. Deletion in a (T)8 microsatellite abrogates expression regulation by 3'-UTR. *Nucleic Acids Res*. 31:6561-9.

Schaaper, R.M. 1993. Base selection, proofreading, and mismatch repair during DNA replication in Escherichia coli. *J Biol Chem*. 268:23762-5.

Schär, P. 2001. Spontaneous DNA damage, genome instability, and cancer--when DNA replication escapes control. *Cell*. 104:329-32.

Schmutte, C., M.M. Sadoff, K.S. Shim, S. Acharya, and R. Fishel. 2001. The interaction of DNA mismatch repair proteins with human exonuclease I. *J Biol Chem*. 276:33011-8.

Shibata, D., M.A. Peinado, Y. Ionov, S. Malkhosyan, and M. Perucho. 1994. Genomic instability in repeated sequences is an early somatic event in colorectal tumorigenesis that persists after transformation. *Nat Genet*. 6:273-81.

Shih, I.M., W. Zhou, S.N. Goodman, C. Lengauer, K.W. Kinzler, and B. Vogelstein. 2001. Evidence that genetic instability occurs at an early stage of colorectal tumorigenesis. *Cancer Res*. 61:818-22.

Sieber, O.M., K. Heinimann, P. Gorman, H. Lamlum, M. Crabtree, C.A. Simpson, D. Davies, K. Neale, S.V. Hodgson, R.R. Roylance, R.K. Phillips, W.F. Bodmer, and I.P. Tomlinson. 2002. Analysis of chromosomal instability in human colorectal adenomas with two mutational hits at APC. *Proc Natl Acad Sci U S A*. 99:16910-5.

Sieber, O.M., K. Heinimann, and I.P. Tomlinson. 2003a. Genomic instability--the engine of tumorigenesis? *Nat Rev Cancer*. 3:701-8.

Sieber, O.M., L. Lipton, M. Crabtree, K. Heinimann, P. Fidalgo, R.K. Phillips, M.L. Bisgaard, T.F. Orntoft, L.A. Aaltonen, S.V. Hodgson, H.J. Thomas, and I.P. Tomlinson. 2003b. Multiple colorectal adenomas, classic adenomatous polyposis, and germ-line mutations in MYH. *N Engl J Med*. 348:791-9.

Sokolsky, T., and E. Alani. 2000. EXO1 and MSH6 are high-copy suppressors of conditional mutations in the MSH2 mismatch repair gene of Saccharomyces cerevisiae. *Genetics*. 155:589-99.

Souza, R.F., S. Wang, M. Thakar, K.N. Smolinski, J. Yin, T.T. Zou, D. Kong, J.M. Abraham, J.A. Toretsky, and S.J. Meltzer. 1999. Expression of the wild-type insulin-like growth factor II receptor gene suppresses growth and causes death in colorectal carcinoma cells. *Oncogene*. 18:4063-8.

Spampinato, C., and P. Modrich. 2000. The MutL ATPase is required for mismatch repair. *J Biol Chem*. 275:9863-9.

Stojic, L., R. Brun, and J. Jiricny. 2004. Mismatch repair and DNA damage signalling. *DNA Repair (Amst)*. 3:1091-101.

Strand, M., T.A. Prolla, R.M. Liskay, and T.D. Petes. 1993. Destabilization of tracts of simple repetitive DNA in yeast by mutations affecting DNA mismatch repair. *Nature*. Vol. 365. 274-6.

Studamire, B., T. Quach, and E. Alani. 1998. Saccharomyces cerevisiae Msh2p and Msh6p ATPase activities are both required during mismatch repair. *Mol Cell Biol*. 18:7590-601.

Stukenberg, P.T., J. Turner, and M. O'Donnell. 1994. An explanation for lagging strand replication: polymerase hopping among DNA sliding clamps. *Cell*. 78:877-87.

Suraweera, N., B. Iacopetta, A. Duval, A. Compoint, E. Tubacher, and R. Hamelin. 2001. Conservation of mononucleotide repeats within 3' and 5' untranslated regions and their instability in MSI-H colorectal cancer. *Oncogene*. 20:7472-7.

Suter, C.M., D.I. Martin, and R.L. Ward. 2004. Germline epimutation of MLH1 in individuals with multiple cancers. *Nat Genet*. 36:497-501.

Szankasi, P., and G.R. Smith. 1995. A role for exonuclease I from S. pombe in mutation avoidance and mismatch correction. *Science*. 267:1166-9.

Thibodeau, S.N., G. Bren, and D. Schaid. 1993. Microsatellite instability in cancer of the proximal colon. *Science*. 260:816-9.

Thomas, D.C., J.D. Roberts, and T.A. Kunkel. 1991. Heteroduplex repair in extracts of human HeLa cells. *J Biol Chem*. 266:3744-51.

Thompson, M.A., R. Flegg, E.H. Westin, and R.G. Ramsay. 1997. Microsatellite deletions in the c-myb transcriptional attenuator region associated with over-expression in colon tumour cell lines. *Oncogene*. 14:1715-23.

Tishkoff, D.X., N.S. Amin, C.S. Viars, K.C. Arden, and R.D. Kolodner. 1998. Identification of a human gene encoding a homologue of Saccharomyces cerevisiae EXO1, an exonuclease implicated in mismatch repair and recombination. *Cancer Res*. 58:5027-31.

Tishkoff, D.X., A.L. Boerger, P. Bertrand, N. Filosi, G.M. Gaida, M.F. Kane, and R.D. Kolodner. 1997. Identification and characterization of Saccharomyces cerevisiae EXO1, a gene encoding an exonuclease that interacts with MSH2. *Proc Natl Acad Sci U S A*. 94:7487-92.

Tran, H.T., D.A. Gordenin, and M.A. Resnick. 1999. The 3'-->5' exonucleases of DNA polymerases delta and epsilon and the 5'-->3' exonuclease Exo1 have major roles in postreplication mutation avoidance in Saccharomyces cerevisiae. *Mol Cell Biol*. 19:2000-7.

Tran, P.T., and R.M. Liskay. 2000. Functional studies on the candidate ATPase domains of Saccharomyces cerevisiae MutLalpha. *Mol Cell Biol*. 20:6390-8.

Tran, P.T., J.A. Simon, and R.M. Liskay. 2001. Interactions of Exo1p with components of MutLalpha in Saccharomyces cerevisiae. *Proc Natl Acad Sci U S A*. 98:9760-5.

Trimbath, J.D., G.M. Petersen, S.H. Erdman, M. Ferre, M.C. Luce, and F.M. Giardiello. 2001. Cafe-au-lait spots and early onset colorectal neoplasia: a variant of HNPCC? *Fam Cancer*. 1:101-5.

Truninger K., M. Menigatti, J. Luz, A. Russell, R. Haider, J-O. Gebbers, F. Bannwart, H. Yurtsever, J. Neuweiler, H-M. Riehle, M.S. Cattaruzza, K. Heinimann, P. Schär, J. Jiricny, and G. Marra. 2005. Immunohistochemical analysis reveals high frequency of PMS2 defects in colorectal cancer. *Gastroenterology*. (in press)

Umar, A., C.R. Boland, J.P. Terdiman, S. Syngal, A. de la Chapelle, J. Ruschoff, R. Fishel, N.M. Lindor, L.J. Burgart, R. Hamelin, S.R. Hamilton, R.A. Hiatt, J. Jass, A. Lindblom, H.T. Lynch, P. Peltomaki, S.D. Ramsey, M.A. Rodriguez-Bigas, H.F. Vasen, E.T. Hawk, J.C. Barrett, A.N. Freedman, and S. Srivastava. 2004. Revised Bethesda Guidelines for hereditary nonpolyposis colorectal cancer (Lynch syndrome) and microsatellite instability. *J Natl Cancer Inst*. 96:261-8.

Umar, A., J.C. Boyer, and T.A. Kunkel. 1994. DNA loop repair by human cell extracts. *Science*. 266:814-6.

Umar, A., A.B. Buermeyer, J.A. Simon, D.C. Thomas, A.B. Clark, R.M. Liskay, and T.A. Kunkel. 1996. Requirement for PCNA in DNA mismatch repair at a step preceding DNA resynthesis. *Cell*. 87:65-73.

Vasen, H.F., B.G. Taal, F.M. Nagengast, G. Griffioen, F.H. Menko, J.H. Kleibeuker, G.J. Offerhaus, and P. Meera Khan. 1995. Hereditary nonpolyposis colorectal cancer: results of long-term surveillance in 50 families. *Eur J Cancer*. 31A:1145-8.

Vasen, H.F., P. Watson, J.P. Mecklin, and H.T. Lynch. 1999. New clinical criteria for hereditary nonpolyposis colorectal cancer (HNPCC, Lynch syndrome) proposed by the International Collaborative group on HNPCC. *Gastroenterology*. 116:1453-6.

Vassileva, V., A. Millar, L. Briollais, W. Chapman, and B. Bapat. 2002. Genes involved in DNA repair are mutational targets in endometrial cancers with microsatellite instability. *Cancer Res*. 62:4095-9.

Waga, S., G.J. Hannon, D. Beach, and B. Stillman. 1994. The p21 inhibitor of cyclin-dependent kinases controls DNA replication by interaction with PCNA. *Nature*. 369:574-8.

Wahlberg, S.S., J. Schmeits, G. Thomas, M. Loda, J. Garber, S. Syngal, R.D. Kolodner, and E. Fox. 2002. Evaluation of microsatellite instability and immunohistochemistry for the prediction of germ-line MSH2 and MLH1 mutations in hereditary nonpolyposis colon cancer families. *Cancer Res*. 62:3485-92.

Wang, H., and J.B. Hays. 2004. Signaling from DNA mispairs to mismatch-repair excision sites despite intervening blockades. *EMBO J*. 23:2126-33.

Warbrick, E. 2000. The puzzle of PCNA's many partners. *Bioessays*. 22:997-1006.

Wei, K., A.B. Clark, E. Wong, M.F. Kane, D.J. Mazur, T. Parris, N.K. Kolas, R. Russell, H. Hou, Jr., B. Kneitz, G. Yang, T.A. Kunkel, R.D. Kolodner, P.E. Cohen, and W. Edelmann. 2003. Inactivation of Exonuclease 1 in mice results in DNA mismatch repair defects, increased cancer susceptibility, and male and female sterility. *Genes Dev*. 17:603-14.

Wilson, D.M., 3rd, J.P. Carney, M.A. Coleman, A.W. Adamson, M. Christensen, and J.E. Lamerdin. 1998. Hex1: a new human Rad2 nuclease family member with homology to yeast exonuclease 1. *Nucleic Acids Res*. 26:3762-8.

Winawer, S.J., R.H. Fletcher, L. Miller, F. Godlee, M.H. Stolar, C.D. Mulrow, S.H. Woolf, S.N. Glick, T.G. Ganiats, J.H. Bond, L. Rosen, J.G. Zapka, S.J. Olsen, F.M. Giardiello, J.E. Sisk, R. Van Antwerp, C. Brown-Davis, D.A. Marciniak, and R.J. Mayer. 1997. Colorectal cancer screening: clinical guidelines and rationale. *Gastroenterology*. 112:594-642.

Woerner, S.M., A. Benner, C. Sutter, M. Schiller, Y.P. Yuan, G. Keller, P. Bork, M.K. Doeberitz, and J.F. Gebert. 2003. Pathogenesis of DNA repair-deficient cancers: a statistical meta-analysis of putative Real Common Target genes. *Oncogene*. 22:2226-35.

Wright, C.L., and I.D. Stewart. 2003. Histopathology and mismatch repair status of 458 consecutive colorectal carcinomas. *Am J Surg Pathol*. 27:1393-406.

Yamaguchi, M., V. Dao, and P. Modrich. 1998. MutS and MutL activate DNA helicase II in a mismatch-dependent manner. *J Biol Chem*. 273:9197-201.

Yao, J., K.W. Eu, F. Seow-Choen, V. Vijayan, and P.Y. Cheah. 1999. Microsatellite instability and aneuploidy rate in young colorectal-cancer patients do not differ significantly from those in older patients. *Int J Cancer*. 80:667-70.

Young, J., L.A. Simms, K.G. Biden, C. Wynter, V. Whitehall, R. Karamatic, J. George, J. Goldblatt, I. Walpole, S.A. Robin, M.M. Borten, R. Stitz, J. Searle, D. McKeone, L. Fraser, D.R. Purdie, K. Podger, R. Price, R. Buttenshaw, M.D. Walsh, M. Barker, B.A.

Leggett, and J.R. Jass. 2001. Features of colorectal cancers with high-level microsatellite instability occurring in familial and sporadic settings: parallel pathways of tumorigenesis. *Am J Pathol*. 159:2107-16.

Yuan, F., L. Gu, S. Guo, C. Wang, and G.M. Li. 2004. Evidence for involvement of HMGB1 protein in human DNA mismatch repair. *J Biol Chem*. 279:20935-40.

Zhang, L., J. Yu, J.K. Willson, S.D. Markowitz, K.W. Kinzler, and B. Vogelstein. 2001. Short mononucleotide repeat sequence variability in mismatch repair-deficient cancers. *Cancer Res*. 61:3801-5.

Chapter 2.3

BASE EXCISION REPAIR

Lisiane B. Meira, Nicholas E. Burgis and Leona D. Samson
Department of Biological Engineering and Center for Environmental Health Sciences, Massachusetts Institute of Technology, Cambridge, Massachusetts, USA

1. INTRODUCTION

From the simplest to the most complex organism, cells have to perform a myriad of tasks to ensure cellular maintenance, survival, reproduction and even cell death. The orchestration of these tasks is performed by the many cellular proteins coded for in our genetic material, or DNA. DNA has thus been called the blueprint of life and as such, was once expected to be a stable informational molecule. Paradoxically, DNA is quite dynamic and constantly subject to change. The processes of recombination and transposition can transfer large tracts of genetic material between distant locations within a genome and even between genomes. DNA is also subject to alterations in its sequence and chemical composition of its bases. Many of these alterations result from errors that occur during the processes of DNA replication or recombination or even DNA repair itself. The inherent dynamic nature of the DNA molecule can be largely beneficial in an evolutionary sense; it is indeed desirable that a fine balance between genetic change and genetic stability is achieved within any given population. However, excessive or uncontrolled genetic change is often detrimental and many pathological conditions have genetic instability as an underlying basis. Because maintaining genetic stability is extremely important, it is not surprising that cells evolved a multitude of mechanisms to repair DNA damage, along with back-up systems that control how cells respond to unrepaired DNA damage, such as cell cycle arrest and programmed cell death, or apoptosis.

DNA damage is classified into two major classes, namely "spontaneous" and "environmentally induced". Spontaneous DNA damage normally results from replication errors or from the inherent instability of the DNA molecule. It is believed that a very large number of lesions occur in any

E.A. Nigg (ed.), Genome Instability in Cancer Development, 125-173.

given genome every day from spontaneous decay, replication errors, and as a result of the cell's own metabolic processes; depurination alone leads to an estimated 2,000 to 10,000 apurinic/apyrimidinic (AP) sites in the genome of each human cell each day (Lindahl, 1993). Moreover, DNA reacts very readily with several chemical compounds and physical agents, many of which have been present in our environment since prebiotic times, and many that have been introduced into the highly industrialised and developed communities where most of us live. As mentioned above, one of the main cellular mechanisms of defence against DNA damage is DNA repair. DNA repair pathways are fine-tuned to ensure that different types of damage are efficiently repaired or tolerated by the cell. DNA repair proteins can act alone or most often are part of a multi-step pathway.

This chapter deals with the base excision repair pathway, or BER. The BER pathway is responsible for the repair of a large number of chemically altered bases, abasic sites and single-strand DNA breaks that occur via spontaneous and environmentally induced mechanisms. BER is initiated by DNA glycosylases that cleave the *N*-glycosyl bond between the base and the deoxyribose sugar, thus creating an abasic or AP site. At least 10 DNA glycosylases have been characterised and cloned in humans, and they each

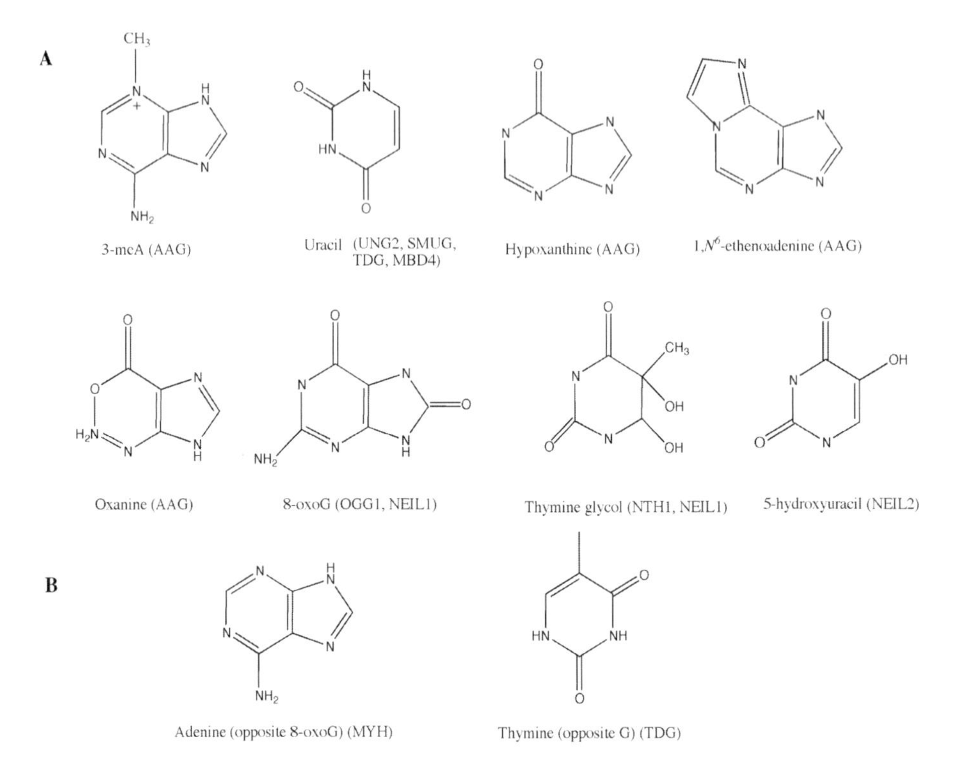

Figure 1. A) Examples of lesions repaired by the BER pathway; glycosylases that recognise and remove the lesions are mentioned in parentheses. B) The glycosylases MYH and TDG remove normal adenine opposite 8-oxo-G and normal thymine opposite guanine, respectively.

have a substrate specificity that can sometimes be broad and largely overlapping to excise a subset of deaminated, oxidised or alkylated bases (see Figure 1 above) (Hazra et al., 2002b; Takao et al., 2002a; Wood et al., 2001). The resulting abasic lesion is further processed by an AP endonuclease or by a bifunctional DNA glycosylase-associated AP lyase activity, before the actions of a DNA polymerase and a DNA ligase can complete restoration of the original sequence. Here we discuss recent advances in understanding BER pathways, the importance of these pathways to human health and what we have learned from the murine models available for the study of BER deficiencies.

2. TYPES OF DNA DAMAGE REPAIRED BY THE BER PATHWAY

The base excision repair pathway essentially removes and replaces nucleotides containing aberrant bases in DNA. Metabolically produced reactive nitrogen and oxygen species can modify the DNA bases due to oxidation, deamination and even alkylation at several positions in the base. Some of these lesions can lead to replication errors caused by anomalous base pairing and ultimately to mutation; other lesions can be cytotoxic if they block DNA polymerase extension. Additionally, exposure to exogenous mutagens can damage DNA in a way that is indistinguishable from spontaneous damage in that many chemicals can alkylate or oxidise DNA bases. A representative sampling of the types of base damage repaired by the BER pathway is discussed below and some types of BER substrates are shown in Figure 1.

2.1 Deaminated DNA Bases

Deamination can be defined as the loss of an exocyclic amino group, and three of the four DNA bases are subject to spontaneous deamination; deamination of adenine gives rise to hypoxanthine, deamination of cytosine yields uracil and deamination of guanine results in xanthine (Friedberg et al., 1995). Similarly, deamination of 5-methylcytosine, a naturally occurring derivative of cytosine, gives rise to thymine, producing T:G mispairs. Moreover, oxanine has been identified as an intracyclic guanine deamination product after treatment of DNA with nitric oxide or *N*-nitrosoindoles (Lucas et al., 1999); oxanine differs from guanine at the N^1 position with an oxygen replacing the nitrogen at that position (see Figure 1). All deaminated base lesions are mutagenic. Hypoxanthine pairs with cytosine during replication resulting in A:T to G:C transitions, (Hill-Perkins et al., 1986; Schouten and Weiss, 1999) while uracil pairs with adenine, causing C:G to T:A transitions (Coulondre et al., 1978; Duncan and Miller, 1980). Xanthine, when present in a DNA template pairs with C or T, depending on the polymerase used in the *in vitro* polymerase studies (Wuenschell et al., 2003). Recently, it was shown that dCTP, dATP, dTTP and to a lesser extent dGTP could be incorporated opposite oxanine in an oxanine-containing template (Hitchcock et al., 2004). Collectively, base deamination may be one of the major sources of mutagenic lesions in cells.

2.2 Oxidised DNA Bases

Oxidation of DNA by reactive oxygen species (ROS) represents a major source of spontaneous DNA damage. Major endogenous sources of ROS are metabolic processes, primarily oxidative metabolism in the mitochondria and pathological conditions such as inflammation (Bartsch et al., 2002; Ohshima and Bartsch, 1994). Exogenous sources of ROS include exposure to gamma-rays or radiomimetic chemicals. In addition, fatty acid radicals, aldehydes and other compounds that are formed during lipid peroxidation reactions can cause oxidative damage to DNA and result in the formation of etheno adducts of pyrimidines and purines (el Ghissassi et al., 1995). Etheno DNA adducts are highly cytotoxic and ethenoA and ethenoC were shown to reduce survival of a damaged phage by more than 65% (Basu et al., 1993). Etheno adducts are also highly mutagenic (Basu et al., 1993; Pandya and Moriya, 1996).

The most important ROS are the superoxide radical ($O_2\cdot-$), hydroxyl radical ($OH\cdot$) and hydrogen peroxide (H_2O_2). Major biologically relevant oxidative products include the highly mutagenic 8-oxo-7,8-dihydroguanine (8-oxoG) and ring-opened forms of purines (formamidopyrimidines) and thymine glycols, all of which are mutagenic and cytotoxic (Marnett, 2000; Rouet and Essigmann, 1985). Moreover, damage to DNA by oxidative stress not only damages bases but also damages the sugar-phosphate backbone causing single- and double-strand DNA breaks (SSBs or DSBs), that can result in lethality or recombinational events.

2.3 Alkylated DNA Bases

Alkylating agents are electrophilic compounds that can transfer alkyl groups to over a dozen nucleophilic sites in DNA. They are abundant in our environment (Calmels, 1987) and are also used in the clinic as chemotherapeutic agents. Some types of damage induced by alkylating agents are cytotoxic, because they block DNA polymerase (Beard et al., 1996; Doublie et al., 1998; Larson et al., 1985), while others are mutagenic, because the methylated base mispairs upon replication (Rebeck and Samson, 1991). The N7-position of guanine represents the major target of alkylation, gathering approximately 70% of all alkylation damage. Other base nitrogens and exocyclic oxygens can also be alkylated, largely depending on the type of mechanism by which the alkylating agents react. S_N1 alkylating agents attack both nitrogens and oxygens (eg. N-Methyl-N'-Nitro-N-Nitrosoguanidine; MNNG and 1-Methyl-1-Nitrosourea; MNU), while S_N2 agents attack mostly base nitrogens (eg. methyl methane sulfonate; MMS).

Alkylated bases can also be generated by nitrosamines formed as a consequence of reactive oxygen and nitrogen species (RONS) overproduction and etheno adducts can be generated endogenously by lipid peroxidation (Bartsch et al., 2002; el Ghissassi et al., 1995). Etheno adducts can also result from exposure to vinyl chloride, chloroacetaldehyde and certain chemotherapeutic alkylating agents (Park et al., 1993; Yang et al., 2000).

3. DNA GLYCOSYLASES

DNA glycosylases initiate BER by recognising and excising specific subsets of damaged bases. Many glycosylases possess a very broad substrate range and together they are responsible for removing a very impressive number of different damaged bases. Some known substrates for the mammalian glycosylases are summarised in Figure 1. Some DNA glycosylases, like AAG and UDG, are monofunctional and simply hydrolyze glycosylic bonds creating AP sites (Figure 2). However, some DNA glycosylases have an intrinsic AP lyase activity and cleave the AP site by one of two different mechanisms (Figure 2). In the first mechanism, the glycosylase/AP lyase uses an internal lysine as the active site nucleophile and cleaves the DNA strand by β-elimination, this cleavage generates a 3' phospho α,β-unsaturated aldehyde (3'PA) moiety that requires further processing by an AP-endonuclease (see Figure 3). The *Escherichia coli* endonuclease III (Nth) glycosylase is the prototype of this type of glycosylase; the mammalian NTH1 and OGG1, also belong to the Nth class (see Figure 2). The other type of mechanism, exemplified by the *E. coli* Fpg and Nei glycosylases, occurs by catalysis of a βδ-elimination at the AP site and removal of the deoxyribose terminus to produce a 3'phosphate at the break (Figure 2). The phosphate terminus must also be further processed and in *E.coli*, the exonuclease III (Xth) and the endonuclease IV (Nfo) AP-endonucleases possess not only a 3' phosphoesterase activity but also a phosphatase activity, so they can process the ends generated by both β- and βδ-elimination reactions. Until recently, it was not clear if this second mechanism was relevant for mammalian cells since the phosphatase activity of the mammalian AP-endonuclease, APE1, was shown to be very weak (Demple and Harrison, 1994; Xu et al., 2003) and no other dual function glycosylases apart from OGG1 and NTH1 had been described for mammalian cells. However, a new class of mammalian DNA glycosylases was recently identified and has been named NEIL (for Nei-like); the NEIL glycosylases remove a wide range of oxidised bases and catalyze βδ-elimination like the *E.coli* Fpg and Nei glycosylases (Hazra et al., 2002b; Takao et al., 2002a). Here, we discuss the implications of these new findings, as well as provide a summary of the properties of mammalian DNA glycosylases.

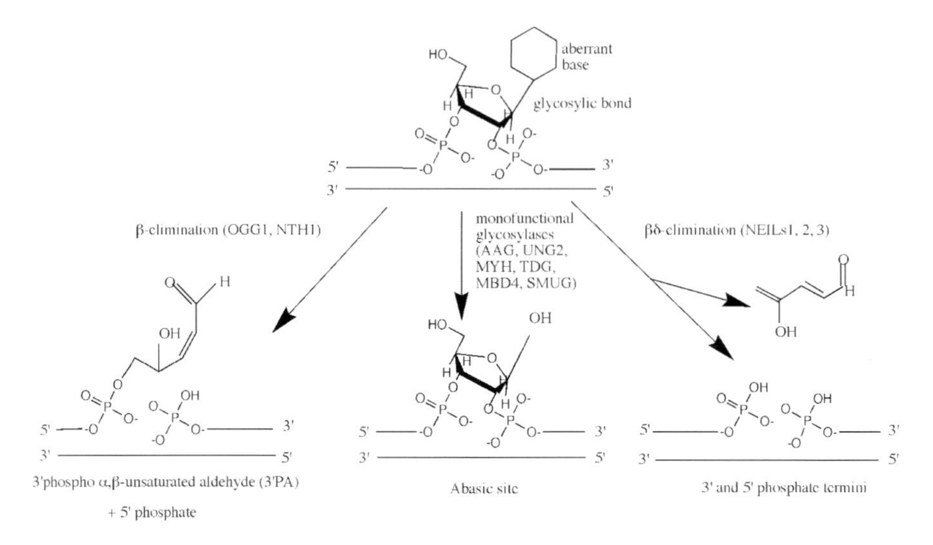

Figure 2. Different types of substrates generated by glycosylase action on aberrant bases. Monofunctional glycosylases hydrolyze the glycosidic bond creating abasic (AP) sites. Bi-functional DNA glycosylases have an intrinsic AP lyase activity and cleave the AP site by one of two different mechanisms: by β-elimination generating a 3' phospho α,β-unsaturated aldehyde (3'PA) moiety or by catalysis of a βδ-elimination at the AP site and removal of the deoxyribose terminus to produce a 3'phosphate at the break.

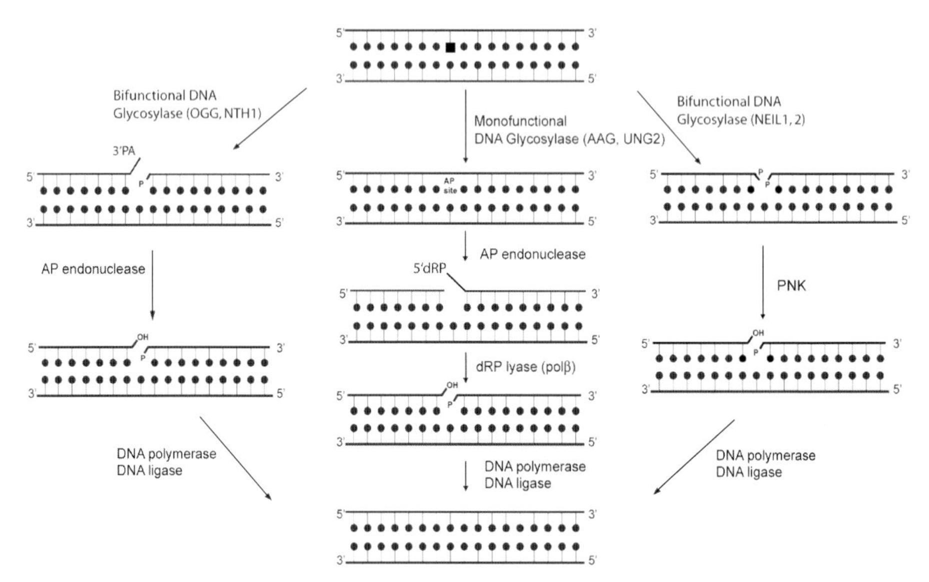

Figure 3. The short-patch BER pathway. The necessary steps for completion of repair are shown; whether damage removal is initiated by a mono-functional or bi-functional DNA glycosylase.

3.1 Monofunctional DNA glycosylases

3.1.1 Methyladenine DNA Glycosylase (AAG)

The human 3-methyladenine DNA glycosylase, known as alkyladenine DNA glycosylase (AAG), methylpurine DNA glycosylase (MPG) or alkylpurine DNA glycosylase (ANPG), is the only identified mammalian glycosylase that repairs simple alkylation DNA damage in mice and humans. Remarkably, this glycosylase excises a wide variety of DNA lesions, including the alkylated bases 3MeA, 7MeG, 3MeG; hypoxanthine, 1,N^6-ethenoadenine (εA) and 3,N^2-ethenoguanine (Dosanjh et al., 1994; O'Connor and Laval, 1990; Saparbaev et al., 2002; Saparbaev and Laval, 1994). Several of these substrates are mutagenic because of their mispairing properties. Some substrates, such as 3MeA, pose a serious problem to the cell because they block DNA synthesis (Boiteux et al., 1984; Larson et al., 1985). As already mentioned, one more Aag-specific substrate was recently described; Aag was shown to be the major glycosylase for the removal of oxanine, a second modified base resulting from the deamination of guanine (besides xanthine) (Hitchcock et al., 2004).

As for many glycosylases, the crystal structure of AAG complexed to DNA was solved (Lau et al., 1998; Lau et al., 2000) and, similar to uracil DNA glycosylase, it was found to excise the damaged bases by a nucleotide flipping mechanism where the nucleotide bearing the modified base to be excised was flipped into an active site pocket (Slupphaug et al., 1996). Tyrosine162 is essential for the flipping; this residue projects from a β-hairpin on the surface of the protein and inserts into the minor groove of DNA, thereby participating in flipping. Yeast cells expressing a mutant form of AAG, where Tyr162 is replaced by alanine are very sensitive to MMS and indeed the mutant protein was found to weakly bind to AAG-substrates in vitro (Lau et al., 2000). Important active site residues include Glu125, Arg182 and Val262, and cleavage of the *N*-glycosidic bond is initiated by Glu125. This residue is thought to deprotonate a water molecule, thus preparing a hydroxyl ion for nucleophilic attack of the electron deficient alkylated base. The water molecule is positioned in the active site by hydrogen bond interactions with the side chains of Glu125, Arg182 and the main chain carbonyl of Val262. This positions the water in close proximity to the damaged base, which is bound in the active site by stacking between Tyr159 and Tyr127 thus forming π–electron-stacking interactions with the aromatic side chains. Consistent with the reaction mechanism model, an E125Q AAG mutant protein has no detectable glycosylase activity *in vitro* or *in vivo* (Lau et al., 2000). The mechanism by which AAG discriminates between normal and damaged bases and how so many different substrates can be recognised and accommodated in the same active site pocket is still not completely clear. However, the structure of AAG bound to an εA damaged DNA revealed that the etheno adduct fits very snugly in the active site pocket by a combination of aromatic stacking interactions and hydrogen bonding between His-136 and the N^6 position of the εA adduct, suggesting that these interactions with the modified purine play a role in stabilising the damaged base in the active site. Recent in vitro studies comparing rates of excision, either catalyzed by AAG or spontaneously occurring, for damaged and undamaged DNA oligonucleotides, revealed that Hx is the preferred substrate of human AAG, and that AAG achieves a broad substrate specificity by a variety of mechanisms: use of acid catalysis (discriminates against pyrimidines), unfavourable interactions with the exocyclic amino groups of undamaged A and G and preferential base-flipping of non Watson-Crick base pairs (O'Brien and Ellenberger, 2004).

Mice deficient in the *Aag* glycosylase have been independently generated by two laboratories (Engelward et al., 1997; Hang et al., 1997). *Aag* mutant mice were born at the expected Mendelian ratio and showed no apparent phenotypic abnormalities after 2 years of age (Parsons and Elder, 2003), Meira and Samson, unpublished observations, also see Table 2). Cell extracts from *Aag* null tissues lack the activity to incise 3MeA, hypoxanthine, εA, and oxanine lesions (Engelward et al., 1997; Hitchcock et al., 2004) but so far, evidence for the *in vivo* relevance of this defect is lacking. Two independent reports have shown that urethane or vinyl carbamate (VC) treatment of *Aag* null animals leads to increased formation and persistence of εA lesions in both liver and lung DNA, and that Aag is the major repair mechanism for εA adducts *in vivo* (Barbin et al., 2003; Ham et al., 2004). However, long-term treatment of *Aag* null animals with VC did not lead to an increase in carcinogenesis compared to the wild type littermates (Barbin et al., 2003). One interesting twist is that absence of the glycosylase could actually have beneficial effects depending on the cell type or genetic context. One such example is found in the reported resistance of *Aag* null mouse bone marrow cells to alkylating agents (Roth and Samson, 2002). The absence of *Aag* also rescues MMS-induced lethality in β pol null mouse embryonic cells (Sobol et al., 2003) suggesting that initiation of BER can lead to cytotoxicity if toxic BER intermediates accumulate and cannot be repaired. These results obtained with mouse cells are in agreement with the hypothesis that an imbalance in base excision repair can have detrimental consequences and may result in a predisposition to cancer.

The paradigm of "Imbalanced Base Excision Repair" was initially based on the observation that overexpressing *MAG1*, the yeast counterpart of the *AAG* gene, in the yeast *Saccharomyces cerevisiae* results in increased mutation rates (Glassner et al., 1998). Evidence that a balance in expression levels of BER enzymes is important for human health has also been reported. By examining tissues from non-cancerous colons of patients suffering from ulcerative colitis (UC), a chronic inflammatory condition associated with a predisposition to colon cancer, Hofseth and coworkers have found an association between increased AAG and APE1 (apurinic/apyrimidinic endonuclease) and increased microsatellite instability

(MSI) (Hofseth et al., 2003). Increased MSI is frequently seen in association with mismatch repair defects and a high proportion of patients with a familial form of colon cancer called HNPCC (for Hereditary Non-Polyposis Colon Cancer), have increased MSI. In the case of UC, patients suffering from the disease had increased MSI but no associated mismatch repair defect. Increased MSI was also seen in cultured human erythroleukemia (K562) cells overexpressing either AAG or APE1 (Hofseth et al., 2003). Thus, imbalance in BER may contribute to the increase in carcinogenicity associated with chronic inflammatory conditions. Moreover, overexpression of *AAG* in breast cancer cells was also found to render these cells more sensitive to alkylating agents (Rinne et al., 2004), demonstrating biological relevance for the previous finding of the up-regulation of the *AAG* gene in breast cancer tissues (Cerda et al., 1998).

3.1.2 Uracil DNA Glycosylases

Uracil in DNA can result either from the deamination of cytosine, or from the misincorporation of dUMP during replication. Deamination of cytosine leads to mutagenic U:G mispairs that can result in C to T transitions. Misincorporation of dUMP instead of dTMP opposite adenine is potentially cytotoxic and mutagenic due to the formation of an AP site subsequent to the removal of uracil. The same can indeed be said for all glycosylases, so that removal of uracil or damaged bases, such as 3MeA will generate another lesion, the AP site.

Mammalian cells possess at least four uracil DNA glycosylases, and their substrate specificities can include other damaged bases and mismatched thymines. These enzymes are UNG, SMUG1, TDG and MBD4. The main uracil DNA glycosylase in mammalian cells is UNG, with UNG2 being the dominant enzyme for uracil removal in nuclear DNA and UNG1 acting on mitochondrial DNA (Kavli et al., 2002). UNG1 and UNG2 represent alternative splice variants from the *UNG* gene (see Figure 4A). UNG, SMUG and TDG all belong to the same protein superfamily (UDG) and share the same structure (Aravind and Koonin, 2000). MBD4 on the other hand, belongs to the HhH-GPD family, and possesses the hallmark helix-hairpin-helix (HhH) motif and Gly/Pro rich loop (GP) followed by a conserved aspartate (D) and will be discussed in more detail later. The structures for the Herpes simplex virus type 1 UDG and the human UDG were the first glycosylase structures solved (Mol et al., 1995; Savva et al., 1995). Structural and mutational analyses have revealed an exquisitely selective binding pocket and a nucleotide-flipping mechanism for DNA damage recognition. These studies suggested that UDG utilises a reaction mechanism similar to the mechanism mentioned previously for AAG.

The human and mouse UNG genes encode both mitochondrial (UNG1) and nuclear (UNG2) forms of the enzyme, both of which have a common catalytic domain but different N-termini (Nilsen et al., 1997). This is

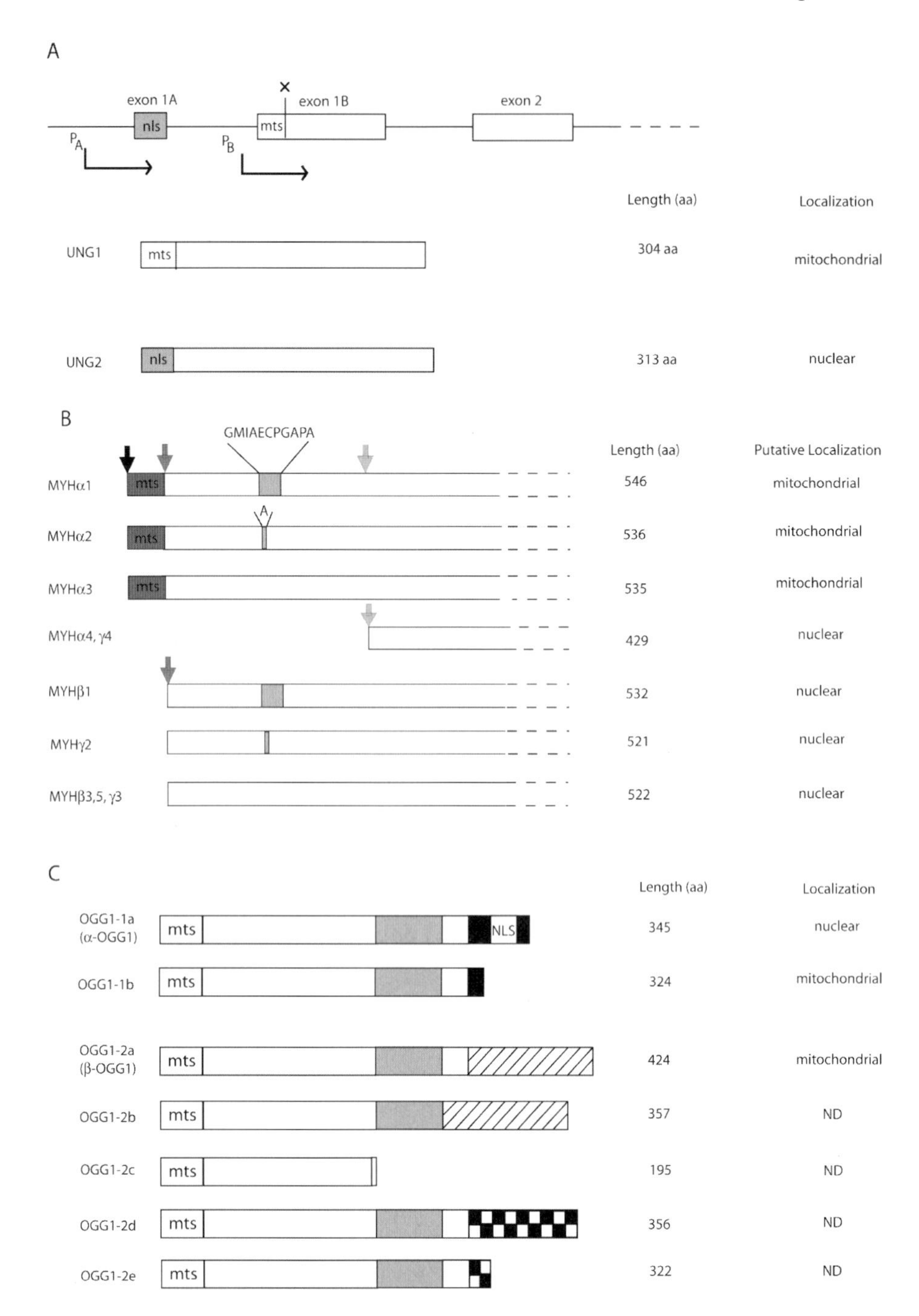

Figure 4. Schematic representation of the different protein forms for the glycosylases UNG, MYH and OGG1; where multiple forms are generated by differential promoter usage and/or alternative splicing.

A) The *UNG* gene. The mitochondrial form of the UNG gene, UNG1, is generated by transcription from promoter P_B from exon 1B, leading to a polypeptide with a mitochondrial target sequence (mts) in the N-terminus. UNG2 arises by usage of promoter P_A and splicing of exon1A into a consensus splice site (indicated by an "X") into exon 1B. UNG2 has a different N-terminus region (black box), where a nuclear localisation sequence (nls) can be found.

B) Predicted polypeptides encoded by the different MYH transcripts, which differ mainly in their N-terminus due to alternative splicing. The dark gray arrow represents a translation start site (tls) for the α forms, the medium gray arrow the tls for the β and γ forms and the light gray arrow represents tls used in forms α4 and γ4 due to an in-frame termination codon in the open reading frame from the first methionine normally used. The dark gray box represents a mts and the light gray boxes represent the amino acid insertions resulting from intronic sequences (intron 2) alternatively spliced into exon 3.
C) Predicted polypeptides encoded by the different OGG1 transcripts, which differ mainly in the C-terminus due to alternative splicing. All transcripts/polypeptides carry a weak mts, and only OGG1-1a carries a nls. The gray box represents a helix-hairpin-helix-PVD motif that seems to be essential to OGG1 glycosylase activity. The diagonal striped boxes represent a hydrophobic C-terminus present in types 2a and 2b. Type 2c is generated by skipping exons 4 to 6, which puts a stop codon in frame and generates a truncated protein. The checkered boxes in types 2d and 2e represent a different C-terminus arising from DNA sequence insertions between exons 6 and 8 for type 2d (100-base insertion) and 2e (53-base insertion).

achieved via alternative promoter usage and differential splicing (Nilsen et al., 1997) (Figure 4A). Using HeLa cell-free extracts and U:A mispair-containing oligonucleotides, UNG appears to be the dominant UDG activity, since antibodies against UNG inhibit more than 98% of the total UDG activity from HeLa cell extracts (Slupphaug et al., 1995). UNG removes uracil with a preference for ssDNA, but also removes uracil in U:G mispairs or in a U:A context. SMUG is very similar to UNG2 in that it also prefers ssDNA as a substrate but also removes U from dsDNA (reviewed in (Krokan et al., 2002)). Interestingly, it has been proposed that while UNG2 would repair misincorporated uracil post-replication, SMUG would be responsible for the removal of the bulk of the uracil resulting from deamination. This was largely based on the fact that UNG2 increases in S-phase and can be found in replication foci and also that SMUG was found to be the major UDG activity on U:G mispairs (Otterlei et al., 1999) (Nilsen et al., 2001). In a recent study, Kavli et al. characterised recombinant purified nuclear isoforms of SMUG1 and UNG2 and a picture emerged suggesting that UNG2 is the major enzyme to repair U resulting from both deamination of cytosine and postreplicative misincorporation, while SMUG1 would work much less efficiently but would display a broader substrate specificity (Kavli et al., 2002). The remaining 2 enzymes capable of repairing U (TDG and MBD4) seem to be limited to removing U only in the context of a mismatch and in dsDNA (Hardeland et al., 2001b; Hendrich et al., 1999). TDG and MBD4 also remove thymine glycols that are present opposite G but not opposite A (Krokan et al., 2002). These enzymes can therefore repair thymine glycols resulting from the oxidative deamination of 5meC.

UNG2 is one of two mammalian DNA glycosylases that have been shown to physically interact with proliferating cell nuclear antigen (PCNA), the loading clamp for the replicative polymerases (the other is MYH) (Otterlei et al., 1999) (see Table 1). UNG2 was also found to interact with RPA and to be largely localised in replication foci (Nagelhus et al., 1997; Otterlei et al., 1999). PCNA was found to stimulate a sub-pathway of BER called long-patch BER (Figure 5), where synthesis of 2-8 nt stretches begin

Table 1. BER interactions: interacting partners, interacting partner activity and function of interaction on BER.

BER enzyme	Interacting partner	Interacting partner activity	Interaction function for BER (known and putative)	References
(1) Glycosylases				
AAG	Estrogen receptor alpha	Transcription factor	Increased AAG acetylation, Stabilization of AAG substrate binding and enhanced catalysis, possible recruitment to actively transcribed DNA	(Likhite et al., 2004)
AAG	HR23B	Nucleotide excision repair (NER) factors, ubiquitin metabolic pathway associated	Stimulate DNA glycosylase activity, increased AAG binding to damaged DNA	(Miao et al., 2000)
UNG1	ND	NA	NA	
UNG2	Proliferating cell nuclear antigen (PCNA)	Loading clamp for the replicative polymerases	Recruitment of UNG2 to replication foci	(Otterlei et al., 1999)
UNG2	Replication protein A (RPA)	Required for DNA synthesis by Polδ or Polε	Recruitment of UNG2 to replication foci	(Otterlei et al., 1999)
SMUG	ND	NA	NA	
MBD4	ND	NA	NA	
TDG	APE1	Major AP endonuclease	Stimulate TDG turnover, coordinated completion of BER	(Waters et al., 1999)
TDG	XPC-HR23B	NER factors	Stimulate TDG turnover	(Shimizu et al., 2003)
TDG	SUMO proteins 1 and 2/3	Protein sumoylation	Stimulate TDG turnover	(Hardeland et al., 2002)
TDG	CBP/p300 acetylase	Transcription activator complex	Stimulate TDG turnover, possible recruitment to actively transcribed DNA	(Tini et al., 2002)
MYH	PCNA	Loading clamp for the replicative polymerases	Recruitment of MYH to replication foci	(Parker et al., 2001)
MYH	RPA	Required for DNA synthesis by Polδ or Polε	Recruitment of MYH to replication foci	(Parker et al., 2001)
OGG1	Protein kinase C (PKC)	Protein phosphorylation	Phosphorylation of OGG1, localization to nuclear matrix, recruitment to actively transcribed DNA	(Dantzer et al., 2002)
OGG1	APE1	Major AP endonuclease	Stimulate OGG1 turnover, coordinated completion of BER	(Vidal et al., 2001)

BER enzyme	Interacting partner	Interacting partner activity	Interaction function for BER (known and putative)	References
NTH1	XPG	Nucleotide excision repair	Enhanced substrate binding and transcription-coupled repair	(Klungland et al., 1999b)
NTH1	p53	Transcription factor, tumor suppressor	Enhanced NTH1 activity	(Oyama et al., 2004)
NTH1	Y box-binding protein 1 (YB-1)	Damage-inducible transcription factor	Enhanced NTH1 activity	(Marenstein et al., 2001a)
NTH1	APE1	Major AP endonuclease	Enhanced NTH1 activity	(Marenstein et al., 2003)
NTH1	PCNA	Loading clamp for the replicative polymerases	Recruitment to actively transcribed DNA	(Oyama et al., 2004)
NEIL1	PNK/XRCC1/Pol β/ DNA ligase IIIα complex	Post-base removal BER	Coordinated completion of BER	(Wiederhold et al., 2004)
NEIL2	p300	Protein acetylase	Modification of NEIL2 activity	(Bhakat et al., 2004)
NEIL3	ND	NA	NA	
(2) Steps post-base removal				
APE1	Pol β	DNA repair polymerase	Coordinated completion of BER, recruitment of Pol β	(Bennett et al., 1997)
Polynucleotide kinase (PNK)	DNA ligase IIIα/XRCC1	Ligation of DNA ends/scaffold protein	Coordinated completion of BER	(Wiederhold et al., 2004)
Flap endonuclease 1 (FEN1)	PCNA	Loading clamp for the replicative polymerases	Coordination of long-patch BER	(Cox, 1997)
			Stimulate FEN1 endonuclease activity	(Gary et al., 1999)
FEN1	APE1	Major AP endonuclease	Stimulate FEN1 endonuclease activity	(Dianova et al., 2001)
DNA Polymerase β (Pol β)	XRCC1	Scaffold protein	Coordinated completion of BER	(Caldecott et al., 1996; Kubota et al., 1996)
DNA Polymerase β (Pol β)	DNA ligase IIIα	Ligation of DNA ends	Coordinated completion of BER	(Caldecott et al., 1996)
DNA ligase I	Pol β	DNA repair polymerase	Coordinated completion of BER	(Dimitriadis et al., 1998; Prasad et al., 1996)
DNA ligase I	APE1	Major AP endonuclease	Stimulate ligase activity	(Ranalli et al., 2002)

NA, not available; ND, not determined/detected.

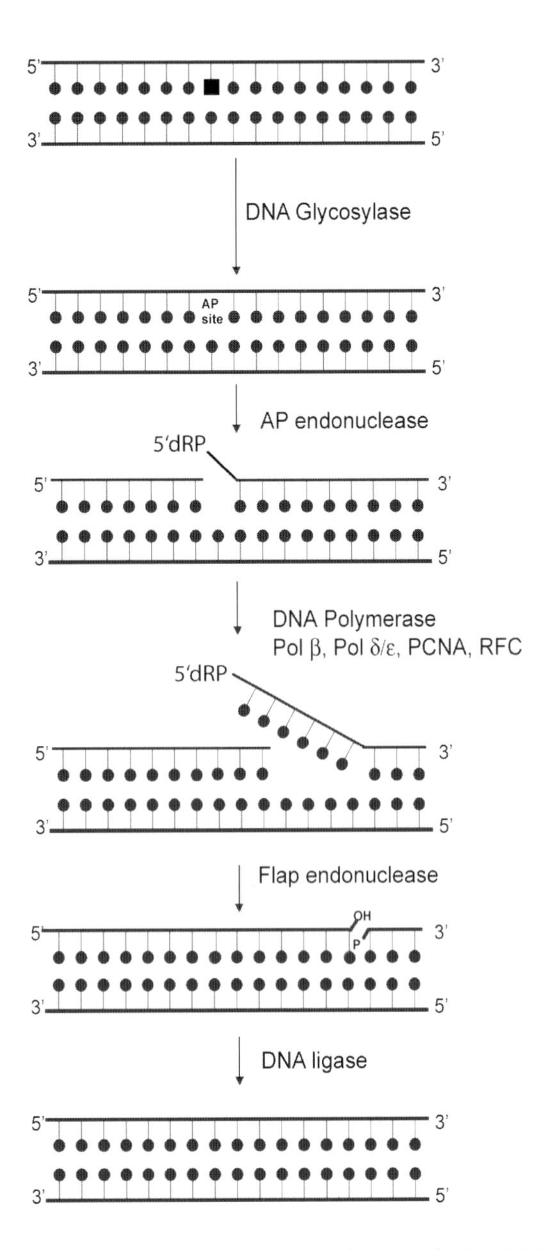

Figure 5. The long-patch BER pathway, where there is the resynthesis of 2-8 nucleotides at the site of damaged base removal. This pathway is PCNA and FEN1 dependent and uses either Pol β or the replicative Pols δ/ε.

at the damaged site in contrast to the so-called short-patch BER (Figure 3) where only 1 nt is inserted (reviewed in (Fortini et al., 2003)). RPA is required for DNA synthesis by Polδ or Polε, which are believed to be the main polymerases in the long-patch pathway (Fortini et al., 1998; Stucki et al., 1998). Thus, the long-patch BER may be replication associated and the interactions of UNG2 with PCNA may serve to recruit UNG2 to replication sites to facilitate the removal of misincorporated uracil.

Mouse models defective in uracil DNA glycosylases have revealed the importance of uracil repair in cancer and in immunoglobulin diversity and regulation (Table 2). Uracil DNA glycosylases are known to play a major role in immunity, but the exact nature of this role is still controversial. The *Ung* knock-out mouse was shown to be affected in class-switch recombination (CSR) and somatic hypermutation (SHM) (Nilsen et al., 2000; Nilsen et al., 2003; Rada et al., 2002), two mechanistically unrelated processes that depend on a protein named "activation-induced deaminase" (AID). Since AID is required for both CSR and SHM, two different hypotheses have been put forward on how AID could catalyze the two different reactions. AID is related to APOBEC-1, an mRNA editing cytosine deaminase that could edit different mRNAs producing proteins essential for initiating the CSR and SHM reactions (the mRNA editing hypothesis). However, the favoured hypothesis and the one strongly supported by biochemical and genetic evidence, is called the DNA deamination repair hypothesis and postulates that AID triggers antibody diversification by deaminating cytosine to uracil in DNA at specific regions within the immunoglobulin loci (Dickerson et al., 2003; Petersen-Mahrt et al., 2002). Thus, AID-induced deamination would create U:G mismatches in DNA that could be processed by alternative DNA repair pathways to produce either SHM or CSR (Petersen-Mahrt et al., 2002). Genetic evidence obtained with the UNG knock-out mouse lends support to the latter. The pattern of somatic hypermutation was altered in *Ung* null mice (Rada et al., 2002) and older animals (18 mo. old or older) have an increased incidence of B-cell lymphomas (Nilsen et al., 2003). The fact that they were B-cell lymphomas, and that AID is specifically expressed in B-cells, suggests that UNG normally modifies DNA that has been deaminated by AID in the immunoglobulin loci. Additional support for this model was given by findings in patients suffering from hyper-IgM syndrome (HIGM). In this syndrome, patients have increased susceptibility to bacterial infections, increased serum IgM, lymphoid hyperplasia, and severely depleted IgG and IgA serum concentrations. HIGM patients also have mutations in the UNG gene (Imai et al., 2003). However, in stark contrast to the DNA deamination repair hypothesis, the Honjo and Jaenisch groups reported that UNG repair activity was dispensable for immunoglobulin CSR (Begum et al., 2004). They found that catalytically inactive mutants of UNG can still rescue CSR in *UNG* deficient B cells, arguing against a repair role of UNG in CSR. Retroviral delivery of UNG mutants that lacked apparent uracil excision activity but retained the ability to bind DNA in *UNG*$^{-/-}$ B cells was sufficient to rescue CSR, suggesting that UNG would be necessary to CSR and antibody diversification by playing a structural role, rather than acting directly on the repair of deaminated cytosines at the immunoglobulin loci. One possible explanation for this apparent contradictory results can be found in recent evidence suggesting that the mismatch recognition factor MSH2 provides an alternate back-up pathway for binding of the U:G mismatches

Table 2. BER factors: available mouse models, phenotypes and potential involvement in human disease.

Gene	Activity	Knock-out phenotypes	Human disease related to deficiency/malfunction
(1) Glycosylases			
AAG	3-meA DNA glycosylase	Viable, no obvious phenotype, mutant bone marrow cells are resistant to alkylation	Ulcerative colitis (UC) associated with AAG and APE1 overexpression
UNG1	Uracil DNA glycosylase, mitochondrial form	NA	ND
UNG2	Uracil DNA glycosylase, nuclear form	Viable, B-cell lymphomas over 18 months, defective SHM and CSR	Hyper IgM syndrome (HIGM)
SMUG	Uracil DNA Glycosylase	NA	ND
MBD4	Uracil DNA glycosylase, repair of mismatched U/T:G residues at methylated CpG sites	Viable, increased C to T mutation frequency at CpG sites.	Mutations in colorectal, pancreatic and endometrial tumors.
TDG	Uracil DNA glycosylase, repair of mismatched U/T:G	NA	ND
MYH	Adenine DNA glycosylase	Viable; phenotypes of increased mutation frequency and tumor predisposition observed in the double Myh Ogg1 mutant animal.	Familial Adenomatous Polyposis (FAP)
OGG1	8-oxoG DNA glycosylase	Viable, mutant mice show accumulation of 8-oxoG and increase spontaneous mutation frequency in some tissues.	Polymorphisms associated with cancer risk.
NTH1	Thymine glycol DNA glycosylase	Viable, no obvious phenotype after 2 years.	ND
NEIL1, 2 and 3	Nei-like DNA glycosylases	NA	ND
(2) Steps post-base removal			
APE1	AP-endonuclease	Embryonic lethal, heterozygote cells and animals are more sensitive to increased oxidative stress.	Ulcerative colitis (UC) associated with AAG and APE1 overexpression
Pol β	DNA polymerase beta	Embryonic lethal	Variant forms associated with cancer and disease; overexpression found associated with cancer.
DNA ligase I	DNA ligase	Embryonic lethal	ND
DNA ligase III	DNA ligase	NA	ND
XRCC1	Ligase accessory factor	Embryonic lethal	Polymorphisms associated with increased cancer risk.
FEN1	Flap-endonuclease I, processing of the ends during long-patch BER	Embryonic lethal	ND

NA, not available; ND, not determined/detected.

that would follow AID-induced deamination (Rada et al., 2004). Thus, binding of the mismatch by a catalytically inactive UNG could actually enhance the alternate pathway for mismatch resolution, mediated by MSH2.

Mbd4 deficient mice (Millar et al., 2002) also display some interesting phenotypes (Table 2). The spleen and liver of animals between 3 and 6 months of age display a significantly higher frequency of C to T mutations when compared to the wild-type counterparts. When combined with the $Apc^{Min/+}$ background, the *Mbd4* deficiency accelerates intestinal neoplasia and decreases survival (Millar et al., 2002). Mbd4 physically interacts with Mlh1 (Bellacosa et al., 1999). However, no increase in mutations in the reporter *Dlb-1b* locus was observed in *Mbd4 Mlh1* double mutant animals, nor did deficiency in *Mbd4* alter survival or tumour predisposition of *Mlh1* null animals (Sansom et al., 2004) indicating that Mbd4 activity is not a modifier of MMR-dependent tumorigenesis and that mutations in *Mbd4* and *Mlh1* are epistatic.

The fourth DNA glycosylase that can remove uracil from DNA is TDG, a glycosylase that removes U and T from G:T and G:U mismatches, but that also has a wide range of substrates (reviewed in (Slupphaug et al., 2003)). TDG was found to interact with many proteins (see Table 1 for a list of interactions), and most of the reported interactions are thought to stimulate turnover of the protein, since the rate limiting step for TDG activity is the dissociation of the enzyme from the resulting AP site (Hardeland et al., 2000; Waters et al., 1999). TDG interacts with APE1 and this interaction was found to stimulate TDG's turnover (Waters et al., 1999). TDG also interacts with the global nucleotide excision repair factors XPC-HR23B (Shimizu et al., 2003) and with SUMO proteins 1 and 2/3, which are ubiquitin-like proteins that are increasingly found to be relevant for modulating a variety of cellular processes (Hardeland et al., 2002). TDG also interacts with the transcription activator complex (CBP/p300), possibly linking BER of TDG substrates to transcriptionally active regions in the genome (Tini et al., 2002). Direct evidence for a role of TDG in human cancer susceptibility is lacking, but defective repair of G:T mismatches could be an important underlying cause for tumour formation since G:C to A:T mutations at CpG sites in the p53 gene are very prominent in human cancers (Hardeland et al., 2001a).

3.1.3 MutY DNA Glycosylase

The mammalian homolog of the *E. coli MutY* gene *(MYH)* encodes an adenine DNA glycosylase that removes adenine incorporated opposite 8-oxoG in template DNA and harbours only weak, if any, lyase activity (Slupska et al., 1999; Williams and David, 1998; Williams and David, 1999). The MYH protein has also been shown to remove 2-hydroxyadenine (2-OH-A) in 2-OH-A:G mispairs (Ohtsubo et al., 2000), in contrast to the *E. coli MutY* which does not display activity with substrates containing 2-OH-A (Ohtsubo et al., 2000). *MYH* maps to the short arm of human chromosome 1 (Slupska et al., 1996) and in Jurkat cells at least three isoforms of MYH were identified, with molecular weights of 52, 53 and 57 kDa (Ohtsubo et al., 2000). The MYH isoforms are located in the nucleus (p52 and p53) and in the mitochondria (p57). Ohtsubo and coworkers demonstrated the existence of multiple *MYH* transcripts in HeLa cells using 5'RACE (5'-rapid amplification of cDNA ends) and RT-PCR (reverse transcription-polymerase chain reaction); the transcripts grouped into three major categories named hMYH α, β and γ, all with differing 5' exons (see Figure 4B) (Ohtsubo et al., 2000).

The *E. coli* MutY protein has two distinct domains, an N-terminal catalytic domain and a C-terminus structural domain (Manuel et al., 1996). Structural analysis of the catalytic domain of *E. coli* MutY revealed that MutY belongs to the HhH-GPD family of proteins containing the signature helix-hairpin-helix motif and a Gly/Pro rich loop followed by a catalytically essential aspartate (for review see (Fromme et al., 2004b)). However, in contrast to the other members of this superfamily, MutY possesses an additional C-terminal domain that is necessary for 8-oxoG recognition (Gogos et al., 1996; Noll et al., 1999). Overall, the catalytic domain of MutY is similar to *E. coli* Nth while the C-terminal domain is very similar to MutT, an enzyme that hydrolyses oxo-dGTP to oxo-dGMP and inorganic pyrophosphate (Volk et al., 2000). The crystal structure of the full-length *Bacillus stearothermophilus* MutY stalled at an A:oxoG base pair was recently reported (Fromme et al., 2004a). Fromme *et al.* show that MutY interacts at several sites within the A:oxoG pair; the catalytic domain interacts mostly with the strand containing the substrate adenine and the C-terminus domain interacts extensively with the 8-oxoG containing strand. Like other DNA glycosylases, MutY generates a sharp bend in the DNA at the site of the mismatch, flipping out the unmodified adenine and excising it by nucleophilic attack by water. Unlike the substrate adenine, the 8-oxoG residue is not flipped and remains buried in the DNA helix. An extensive array of hydrogen-bonding interactions are formed between the C-terminal recognition pocket of MutY and the 8-oxoG residue, resulting in great specificity for the modified G but no ability to recognise thymine, thus forming the basis for the selective binding of MutY to oxoG:A pairs.

As previously mentioned for UNG2, MYH also interacts with PCNA through its consensus binding motif, and with RPA; these interactions may serve to recruit MYH to replication sites (Boldogh et al., 2001; Parker et al., 2001) (see Table 1). Using an elegant *in vivo* plasmid repair system, Hayashi and coworkers have shown that replication-proficient substrates containing A:oxoG lesions were repaired 14-times more efficiently than damaged replication-deficient ones, strongly suggesting that the A:oxoG mispair in the genome is inefficiently repaired without replication (Hayashi et al., 2002). By taking advantage of *Myh*-deficient murine cells they showed that the repair efficiency was higher in *Myh*-proficient cells, but that murine cells lacking Myh could also repair the damaged template to a significant extent. Furthermore, the authors showed that expression of wild-type *Myh* in *Myh*-deficient cells resulted in increased repair efficiency in replicated substrates; no increase was seen with expression of a mutant form of MYH wherein the PCNA consensus binding sequence was disrupted. These results support the involvement of MYH in replication-associated repair of A:oxoG. However, it is not yet clear if MYH (and UNG2) act during replication to couple repair with replication, or if the glycosylases are recruited to replication foci after detection of the A:oxoG in a post-replicative manner.

MYH was recently implicated in human colorectal cancer, more specifically familial adenomatous polyposis or FAP (nicely reviewed in (Cheadle and Sampson, 2003)). *MYH* mutations (but not *OGG1* or *NTH1*) were found in DNA samples obtained from blood of affected members of a family with a history of multiple colorectal adenomas and carcinomas. For these patients no mutations in the *APC* gene were found in DNA from normal tissues but, a high frequency of G:C to T:A transversions was found at the *APC* locus in each of the tumours sequenced (Al-Tassan et al., 2002). By analysis of over 100 FAP cases from apparently unrelated families, Sampson and coworkers have identified biallelic germline mutations in the *MYH* gene in 23% of the cases examined (Sampson et al., 2003). From the mutations already reported in *MYH* in patients with colorectal polyposis, the most common mutations found in affected Caucasian patients are Y165C (36 mutant alleles reported out of a total of 78) and G382D (22 mutant alleles reported out of a total of 78). The corresponding mutations in *E. coli mutY* were found to significantly reduce catalytic activity (Al-Tassan et al., 2002) and the recombinant human Y165C and G382D MYH could not fully complement the activity of MutY in *MutY* deficient *E. coli* (Chmiel et al., 2003). Fromme and coworkers, in their description of the structure of *B. stearothermophilus* MutY, commented on the effects that these polymorphisms would have on the interaction of MutY with DNA. Human Tyr165 and Gly382 correspond to Tyr88 and Gly260 in *B. stearothermophilus* MutY, and both residues lie in the C-terminal 8-oxoG recognition pocket and directly interact with the 8-oxoG containing strand; Tyr88 intercalates into the duplex between 8-oxoG and the nucleoside 5' to it, and Gly260 contributes to the hydrogen bonding with the two phosphates immediately 5' to the 8-oxoG (Fromme et al., 2004a). Thus it looks as if the ability of MYH to faithfully recognise 8-oxoG in DNA is key in maintaining genomic stability.

Myh knock-out mice were generated by insertion of a neomycin-resistance expression cassette into exon 6 of the mouse *Myh* gene by targeted homologous recombination (Xie et al., 2004). The *Myh* deficiency alone has no effect on survival or tumour incidence within 17 months of murine life (similar to the *Ogg1* knock-out to be discussed in more detail later). Xie *et al.* generated double knock-out mutants between *Myh* and *Ogg1*; the double mutant displayed reduced survival and increased tumour incidence, suggesting a synergistic effect in tumour predisposition (Table 2). The double mutant mice had a significant increase in lung adenomas and a striking increase in frequency of G:C to A:T mutations in codon 12 of the *K-ras* gene was found in lung tumours from *Myh*$^{-/-}$ *Ogg1*$^{-/-}$ mutant animals, as was described for about 17% of *MYH* polyposis tumours (Jones et al., 2004). Not surprisingly, 8-oxoG was found to accumulate with age in the liver, lung and small intestine of *Myh Ogg1* double mutant mice (Russo et al., 2004).

3.2 Bifunctional DNA Glycosylases

3.2.1 8-oxoguanine DNA Glycosylase (OGG1)

The human *OGG1* gene expresses two major forms of human OGG1, a 36-kDa polypeptide designated as α-OGG1 and a 40-kDa polypeptide designated as β-OGG1 (Nishioka et al., 1999). Seven different types of human *OGG1* mRNAs have been identified by RT-PCR amplification, and appear to be derived from alternative splicing events (Figure 4C). These transcripts have been classified as two distinct types based on their terminal exons, resulting in polypeptides sharing a common N-terminal region of 316 amino acids and a unique C-terminal region. All seven types of human *OGG1* mRNAs are expressed in all major organs as well as Jurkat and HeLa cells with a 1.7-kb and 2.3-kb transcript being the major transcripts and corresponding to α-OGG1 and β-OGG1, respectively. α-OGG1 was copurified from nuclear extracts, whereas β-OGG1 was determined to be localised on the inner membrane of mitochondria by electron microscopic immunochemistry and subfractionation of the mitochondria (Nishioka et al., 1999). The nuclear α-OGG1 is well conserved and has been characterised in several eukaryotic model organisms; however, the mitochondrial β-OGG1 has only been identified in human cells (Boiteux and Radicella, 1999; Nishioka et al., 1999). Only the nuclear form of OGG1 will be discussed further in this section and will be referred to simply as OGG1. Immunofluorescence studies have shown OGG1 to be associated with the chromatin and the nuclear matrix during interphase and associated with the condensed chromatin during mitosis (Dantzer et al., 2002). Chromatin bound OGG1 is phosphorylated at a serine residue and OGG1 co-immunoprecipitates with protein kinase C (PKC). *In vitro* OGG1 is indeed a substrate for PKC but this phosphorylation does not appear to alter OGG1 enzyme activity. Instead phosphorylation by PKC may result in re-localisation of OGG1 in that chromatin-associated hOGG1 is phosphorylated, whereas the nuclear matrix-associated OGG1 does not appear to be phosphorylated. Because condensed chromatin and the nuclear matrix are often associated with sites of transcription these finding may hint at an undisclosed role for OGG1 as a mediator of transcriptional control (Dantzer et al., 2002).

OGG1 catalyzes the excision of several oxidatively damaged purines, and in addition exhibits lyase activity that cleaves abasic sites in DNA via a β-elimination reaction resulting in cleavage immediately 3' of the abasic site

(Lu et al., 1997c). Damaged bases in duplex DNA that OGG1 has been documented to excise include: 8-oxoG, 2,6-diamino-4-hydroxy-5-formamidopyrimidine (FapyG), 2,6-diamino-4-hydroxy-5-*N*-methylformamidopyrimidine (N^7-meFapyG) and 8-oxo-7,8-dihydroadenine (8-oxoA) (Figure 1) (Dherin et al., 1999; Girard et al., 1998; Zharkov et al., 2000). 8-oxoG is a major mutagenic lesion in the human genome (Hsu et al., 2004) and since most of the information about OGG1 activity refers to this lesion the focus of this discussion will centre on the repair of 8-oxoG. Interestingly, OGG1 excises 8-oxoG regardless of the identity of the opposing base; however, AP lyase activity by OGG1, both subsequent to and independent of glycosylase activity, occurs only when C is the opposing base (Bjoras et al., 1997). This observation led Seeberg and coworkers to suggest that OGG1 possesses a mechanism to insure that strand incision will occur only if completion of repair will result in insertion of the correct base (guanine), while insuring that strand continuity is preserved for subsequent mismatch removal by postreplicational mismatch repair processes (Bjoras et al., 1997).

OGG1 belongs to the structural superfamily of DNA glycosylases that contain a HhH-GPD motif that is terminated by an invariant catalytically essential aspartic acid residue (D268) which appears to play a structural role as well as a role in catalysis (Fromme et al., 2004a; Nash et al., 1996; Norman et al., 2003; Scharer and Jiricny, 2001). The crystal structure of a mutant OGG1 bound to an 8-oxoG-containing oligonucleotide has suggested that an active site lysine (K249) positioned within the HhH-GPD motif serves as the catalytic nucleophile for DNA glycosylase activity. Subsequent to proton abstraction by D268, the ε-NH_2 group of K249 attacks the 8-oxoG containing nucleotide at the C-1' position of the ribose residue, substituting the glycosidic bond with K249. This aminal enzyme-DNA covalent intermediate rearranges to a Schiff's base intermediate that subsequently undergoes β-elimination and strand scission (Norman et al., 2003; Scharer and Jiricny, 2001). Surprisingly, crystal structures from borohydrate-trapped intermediates suggest that OGG1 does not release 8-oxoG, but rather fortuitously retains the base in the base-recognition pocket (Fromme et al., 2003). Crystallographic and biochemical data generated by Fromme *et al.* elegantly demonstrate that the acid/base catalyzed reactions required for β-elimination do not occur as a result of enzyme-catalyzed reactions, but instead are a result of substrate- (8-oxoG) catalyzed reactions, suggesting that 8-oxoG acts as a cofactor to further process the damaged DNA. Therefore, these results have provided the first example of product-assisted catalysis in an enzyme-mediated reaction (Fromme et al., 2003).

The murine Oggl (Oggl) is homologous to human OGG1 and yeast OGG1 and is therefore a member of the HhH-GPD superfamily (Lu et al., 1997a). The mouse protein possesses similar substrate specificity and biochemical properties as human OGG1 (Lu et al., 1997c; Zharkov et al., 2000) and its biological function has been investigated using *Oggl*$^{-/-}$ knockout mice (Klungland et al., 1999c; Minowa et al., 2000). Cell-free extracts from wild-type, heterozygous and *Ogg1*$^{-/-}$ mice were assayed for Ogg1 enzyme activity using radiolabeled 8-oxoG containing oligonucleotides (Klungland et al., 1999c). Ogg1 activity was detected in wild-type extracts from several major organs with the highest activity in testis and brain. Ogg1 activity was reduced by about 50% in heterozygous animals and not detectable in tissues derived from *Ogg1*$^{-/-}$ mice (Klungland et al., 1999c). Despite these phenotypes, no increase in spontaneous tumour incidence was observed for *Ogg1*$^{-/-}$ mice (Klungland et al., 1999c; Minowa et al., 2000). *Ogg1*$^{-/-}$ mice are viable and fertile and remain indistinguishable from wild-type littermates after more than 20 months (Table 2). However, genomic DNA isolated from the livers of 13-15 week-old *Ogg1*$^{-/-}$ animals showed a 1.7- (Klungland et al., 1999c) to 7-fold (Minowa et al., 2000) increase in the level of 8-oxoG compared to wild-type animals. In 10- and 20-week-old *Ogg1*$^{-/-}$ animals a 2- to 3-fold increase in spontaneous mutation frequency was observed only in the liver and testes (Klungland et al., 1999c; Minowa et al., 2000). Consistent with the base pairing properties of 8-oxoG (Wood et al., 1990), these mutations were predominantly G:C to T:A transversions (Klungland et al., 1999c; Minowa et al., 2000). Interestingly, more recent experiments have shown that *Ogg1*$^{-/-}$ mice exposed to chronic oxidative stress accumulate a 70-fold higher level of 8-oxoG in genomic DNA isolated from the kidney compared to wild-type mice (Arai et al., 2002). Sequence analysis of the *gpt* gene from this DNA showed that the predominant mutations were G:C to T:A transversions. In addition deletions and G:C to A:T transitions were also observed (Arai et al., 2002). These results suggest that mOgg1 is required to prevent accumulation of 8-oxoG in the genome during normal oxidative conditions as well as during oxidative stress.

In situ hybridization and Southern blot analysis has mapped the human *OGG1* gene to chromosome 3p25 (Lu et al., 1997b; Radicella et al., 1997; Roldan-Arjona et al., 1997). Interestingly, approximately one third of all lung cancers harbour deletions of the short arm of chromosome 3 (3p) (Brauch et al., 1987; Kok et al., 1987; Naylor et al., 1987; Yokota et al., 1987). Because the majority of all lung cancers have arisen in patients with a history of smoking, Lu et al. (Lu et al., 1997a) speculated that a loss of homozygosity for the *OGG1* allele in these patients could result in an acceleration of the diseased state. One of the components of cigarette smoke are ROS, molecules that are known to produce 8-oxoG in DNA, the major substrate for OGG1. Lu *et al.* reasoned that a reduction in the intracellular OGG1 levels in smokers may lead to increased mutation that in turn could result in accelerating cell transformation (Lu et al., 1997a). This theory is supported by a recent finding, which shows that low OGG1 activity in peripheral blood mononuclear cells of smokers does indeed correlate with an increased risk of lung cancer (Paz-Elizur et al., 2003).

3.2.2 NTH1

NTH1 is another bifunctional DNA glycosylase belonging to the HhH-GPD superfamily of DNA glycosylases (Thayer et al., 1995). The gene encoding the 34.3 kDa NTH1, is located in human chromosome region 16p13.2-.3 and encodes a 1-kb transcript (Aspinwall et al., 1997; Hilbert et al., 1997). Northern blot analysis has demonstrated that while NTH1 is expressed in all major human tissues, its expression varies several fold between different tissues, being highest in the heart and lowest in the lung and kidney (Aspinwall et al., 1997). Expression of NTH1 in human skin keratinocytes (HaCat) is cell cycle regulated with increased expression during early and mid S-phase (Luna et al., 2000). Localisation studies using a green fluorescent protein-tagged hNTH1 fusion protein in HeLa cells showed that the recombinant protein was localised to the nucleus exclusively (Luna et al., 2000). However, other studies have identified sporadic cytoplasmic localisation as well as mitochondrial localisation suggesting that alternate sorting may exist (Ikeda et al., 1998; Takao et al., 1998).

NTH1 is an *E. coli* endonuclease III homolog that removes several oxidatively damaged pyrimidines including 5-hydroxycytosine, thymine glycol (Tg), 5-hydroxy-6-hydrothymine, 5,6-dihydroxycytosine, and 5-hydroxyuracil residues (see Figure 1) (Aspinwall et al., 1997; Dizdaroglu et al., 1999; Hilbert et al., 1997). Interestingly, NTH1 has greater activity when Tg is paired with A than paired with G (Marenstein et al., 2003). Structural homology with endonuclease III is evident in that NTH has spectral properties consistent with an iron/sulfur ([4Fe-4S]) cluster and four conserved cysteine residues at positions 282, 289, 292 and 300 (Aspinwall et al., 1997; Hilbert et al., 1997), which in the *E. coli* homolog are known to be important for binding the iron/sulfur moiety (Kuo et al., 1992; Thayer et al., 1995). Several additional amino acid residues of NTH1 are well conserved throughout phylogeny, many of which have been identified as essential for DNA binding and catalysis for endonuclease III, suggesting that NTH1 is a functional homolog as well. Crystallographic studies with endonuclease III have show that the enzyme consists of two α -helical domains one of which contains the HhH-GPD domain while the other contains the [4Fe-4S] cluster loop which is important for stabilising the protein fold and for orienting the DNA binding residues (Kuo et al., 1992; Thayer et al., 1995). The crystal structure suggests that DNA binds at the cleft between the two domains and that the catalytically essential Lys120 and Asp138 lie at the mouth of the cleft (Thayer et al., 1995). Site-directed mutagenesis has determined Lys212 to be the active site residue in NTH1, performing the requisite nucleophilic attack of the C-1' position of the damaged nucleotide (Ikeda et al., 1998). Similar to OGG1, NTH1 becomes irreversibly cross-linked with a damaged nucleotide after incubation with sodium cyanoborohydride (Hilbert et al., 1997) suggesting a similar reaction mechanism. Indeed a Schiff base intermediate between Lys212 and the deoxyribose of the damaged base most likely does form for NTH1, in that a sodium borohydrate trapped intermediate, much like the OGG1 intermediate, has been identified in a *Bacillus* endonuclease III homolog-DNA cocrystal structure (Fromme et al., 2003). Subsequent to glycosylase activity, the rate-limiting AP lyase activity of NTH1 catalyzes β-elimination reactions to generate a 3' nicked repair intermediate, completing the second step in BER (Marenstein et al., 2001a).

Recently, several nuclear localised proteins have been reported to stimulate NTH1 activity (see Table 1). XPG, a nucleotide excision repair (NER) protein, enhances NTH1 binding to a Tg lesion resulting in a strong stimulation of base excision (Klungland et al., 1999a) and a direct interaction between the two proteins has been identified in human cell free extracts (Oyama et al., 2004). XPG is known as a structure specific nuclease, catalyzing strand cleavage in NER, however its stimulatory effect on BER is not endonuclease dependent as shown by endonuclease deficient mutants of XPG (Bessho, 1999; Klungland et al., 1999a). Similarly, the tumour suppressor p53 has been shown to directly interact with NTH1 and stimulate Tg excision (Oyama et al., 2004). The exact role of p53 in stimulating NTH1 initiated BER is not well understood, but Oyama *et al.* speculate that p53 may stimulate AP lyase activity in addition to its stimulation of 5'-deoxyribose phosphate (5'-dRP) excision by DNA polymerase β (Polβ) (Zhou et al., 2001). Oyama *et al.* further speculate the enticing possibility that p53 may connect oxidative DNA damage and BER with the p53-dependent damage response pathways (Oyama et al., 2004). Addition of the damage-inducible transcription factor Y box-binding protein 1 (YB-1) stimulates the NTH1 DNA glycosylase and AP lyase activities *in vitro* (Marenstein et al., 2001a). YB-1 affects the steady state equilibrium of reaction intermediates, increases the overall rate of enzyme catalysis, and therefore may provide a mechanism to modulate NTH1 activity (Marenstein et al., 2001b). Furthermore, addition of the human AP-endonuclease (APE1) to the YB-1 and NTH1 *in vitro* assay resulted in a greater increase of NTH1 turnover and substrate processing (Marenstein et al., 2003). These results support a role for YB-1 and APE1 in regulation of NTH1 activity (Marenstein et al., 2003). APE1 has been shown to interact with the downstream BER repair proteins Pol β and DNA ligase, suggesting a coordinated processing scheme for BER (Mol et al., 2000; Tom et al., 2001; Wilson and Kunkel, 2000). A direct physical interaction between NTH1 and PCNA was also observed, however PCNA did not stimulate Tg DNA glycosylase/AP lyase activity of NTH1 (Oyama et al., 2004). PCNA is thought to play a role in replication-coupled BER of misincorporated bases, and it is therefore possible that PCNA recruits NTH1 to stalled replication machinery as opposed to modifying NTH1 activity (Oyama et al., 2004). Taken together, these NTH1-initiated interactions suggest that a multi-protein complex most likely forms during the initial steps of BER and that complex formation may be required for efficient processing of the damaged base.

The murine homolog of endonuclease III (Nth1) has substrate specificity similar to NTH1 and several groups have generated mice deficient in Nth1 (Asagoshi et al., 2000; Elder and Dianov, 2002; Ocampo et al., 2002; Takao et al., 2002a). *Nth1*$^{-/-}$ mice are born at the expected Mendelian ratio, are fertile and display no overt abnormalities after almost two years (Table 2). No increase in spontaneous tumour incidence has been reported, and equivalent levels of damage or sensitivity were observed for wild-type and *Nth1*$^{-/-}$ mice exposed to X-rays, menadione or H_2O_2 (Takao et al., 2002b). The absence of a phenotype is further confounding in that a cross of *Nth1*$^{-/-}$ mice and *Ogg1*$^{-/-}$ mice result in progeny that also show no abnormal phenotypes after 15 months (Parsons and Elder, 2003). Experiments using the alkaline comet assay demonstrate that mouse embryonic fibroblasts (MEFs) lacking *Nth1* show altered repair of BER repair intermediates (DNA strand breaks) and abasic sites (Parsons and Elder, 2003). For wild-type cells, the repair process appears to be well coordinated and the apparent overall DNA damage is significantly reduced after 90 minutes. However for *Nth1*$^{-/-}$ cells, even though repair is observed, it appears to be less coordinated and strand breaks are still evident after two hours (Parsons and Elder, 2003). The fact that some level of repair is observed suggest that a back-up enzymatic activity functions to repair oxidised pyrimidines in these cells. Further evidence for a back-up repair activity was demonstrated by Elder and Dianov using a plasmid-based *in vitro* assay to show that 5,6-dihydrouracil was repaired by short patch BER in testes extracts from *Nth1*$^{-/-}$ mice at about 20% of the efficiency of extracts from wild-type animals (Elder and Dianov, 2002). Additionally, Ocampo et al. identified a previously undescribed back-up enzyme activity that may function to repair oxidised pyrimidine residues in DNA (Ocampo et al., 2002). They reported that the back-up activity was present in the brain, liver and thymus, with the thymus having the highest specific activity. Interestingly, tissues of *Nth1*$^{-/-}$ mice possess an enzymatic activity for Tg containing DNA that is greater when the Tg residue is opposite G rather than when it was opposite A (opposite of Nth1 specificity). Migration of the reaction products on a denaturing polyacrylamide gel were consistent with that of an AP endonuclease cleavage activity suggesting that the newly identified repair activity could result from a DNA *N*-glycosylase, DNA *N*-glycosylase/AP lyase, or an endonuclease that nicks 5' to the Tg-containing site (Ocampo et al., 2002). More recently, this backup activity has been identified as the endonuclease VIII-like homolog, NEIL1 (Rosenquist et al., 2003; Takao et al., 2002b).

3.2.3 NEIL DNA Glycosylases

The NEILs were only recently described and detailed information regarding NEIL1 and 2 is just emerging while reports of NEIL3 are limited. NEIL deficient mouse models are not yet available, however ES cells in which NEIL1 expression has been reduced using short hairpin RNA display an increased sensitivity to ionising radiation, suggesting a role for NEIL1 in the repair of oxidised bases (Carmell et al., 2003). Human NEIL1 and 2 map to 15q22.33 and 4q34.2, respectively, and encode proteins of 43.7 and 36.8 kDa, respectively (Hazra et al., 2002a; Morland et al., 2002). It is interesting to note that except for a few conserved motifs, NEIL1 and 2 show little sequence homology. The presence of an N-terminus proline, and absence of a mitochondrial localising sequence, suggest that the NEILs are not likely to be localised to the mitochondria. Indeed, fusion constructs with green fluorescent protein indicate that NEIL1 and 2 localise to the nucleus, with NEIL1 specifically localised with nucleolin, a major nucleolar protein (Morland et al., 2002). NEIL1 expression in human tissues is highest in the liver, pancreas, and thymus, moderate expression is seen for the brain, spleen and prostate, and low levels for the testis and leukocytes (Hazra et al., 2002a). NEIL2 expression in human tissues was highest in the skeletal muscle and testis, moderate in the brain and heart, and very low in other tissues (Hazra et al., 2002b). Importantly the expression level of NEIL1 increases during S phase, while NEIL2 expression is not cell cycle dependent (Hazra et al., 2002a; Hazra et al., 2002b). In contrast, NEIL2 is regulated by posttranslational modification (Bhakat et al., 2004). Mitra and coworkers found that two lysine residues (Lys49 and Lys153) in NEIL2 are acetylated by p300, that NEIL2 and p300 form a stable interaction, and that acetylation of Lys49 leads to inactivation of the enzyme, while modification of Lys153 has no effect on glycosylase or AP lyase activity. These results may suggest a mechanism for fine tuned regulation of this BER pathway (Bhakat et al., 2004).

NEIL1, 2 and 3 all belong to the *E. coli* MutM/Nei family based on structural homology and reaction mechanism (Hazra et al., 2002a; Hazra et al., 2002b). A crystal structure of NEIL1 has been solved and it shows that NEIL1 exhibits the same overall fold as *E. coli* Nei, however instead of a zinc finger, NEIL1 contains a "zincless finger" structural motif which is composed of two antiparallel β-strands that mimic the zinc finger (Doublie et al., 2004). Site-directed mutagenesis of a highly conserved arginine residue within this motif greatly reduces the glycosylase activity of NEIL1 and modelling experiments suggest that this residue is important for DNA binding (Doublie et al., 2004). Interestingly, a recent report has identified a zinc finger domain in NEIL2 by inductively coupled plasma mass spectrometry (Das et al., 2004). Site directed mutagenesis experiments showed that this domain is essential for DNA binding and that the mutant proteins displayed aberrant structural features and did not bind zinc (Das et al., 2004). Unlike the other two bifunctional DNA *N*-glycosylases discussed above, but like their *E. coli* homologs, the NEILs perform βδ-elimination at an AP site in DNA to generate 3'-phosphate and 5'-phosphate termini subsequent to removal of the deoxyribose residue (Figure 2) (Hazra et al., 2002a; Hazra et al., 2002b; Morland et al., 2002; Zharkov et al., 2000; Zharkov et al., 2003). Similar to OGG1 and NTH1, NEIL1 and 2 form a trapped intermediate when sodium cyanoborohydride is added to DNA glycosylase assays (Hazra et al., 2002a; Hazra et al., 2002b). Site directed mutagenesis reveals that the well conserved N-terminus proline of NEIL1 is required for the formation of the trapped intermediate, and that addition of a N-terminus 6-histidine tag abolishes the formation of this trapped complex for both NEIL1 and 2, as well as activity for NEIL1 supporting a role for this residue as the active site nucleophile (Hazra et al., 2002a; Hazra et al., 2002b).

The NEILs recognise a range of oxidatively damaged bases (Figure 1). NEIL1 can excise Fapy-A and Fapy-G from irradiated duplex DNA, but unlike *E. coli* MutM/Fpg, NEIL1 was initially thought to be unable to excise 8-oxoG (Hazra et al., 2002a). More rigorous analysis of NEIL1 activity was performed with damage containing duplex oligonucleotides, revealing that NEIL1 also excises dihydrouracil and to a lesser extent 8-oxoG, and that the reaction was dependent on the identity of the opposing base (Hazra et al., 2002a). Similar experiments showed that NEIL1 also excises 5-hydroxycytosine opposite G residues, in sharp contrast to OGG1 which shows no activity with this substrate (Morland et al., 2002). NEIL1 also has AP lyase activity that appears equivalent to its activity with the Fapy substrates (Katafuchi et al., 2004). The activity of NEIL1 with 8-oxoG is about 5 to 15-fold less than that of OGG1 activity, supporting a backup role for NTH1 in removal of this lesion from DNA (Morland et al., 2002). In addition, experiments using the mouse NEIL1 (Neil1) have shown that this enzyme has similar substrate specificity as the human homolog (Rosenquist et al., 2003). Substrates for NEIL2 include oxidised pyrimidines, with 5-hydroxyuracil (5-OHU), one of the most mutagenic lesions formed by ROS, showing the highest activity followed by 5,6-dihydrouracil and 5-hydroxycytosine (Hazra et al., 2002b). NEIL2 also appears to have activity with AP sites in duplex DNA that are comparable to NEIL1, but NEIL2 shows no detectable activity with Tg or 8-oxoG (Hazra et al., 2002b; Katafuchi et al., 2004). Considering the broad overlap in substrate specificity between the NEILs, OGG1 and NTH1 it is not surprising that *Ogg1* and *Nth1* null mice displayed little or no phenotype, or that the NEILs were not identified until recently. One remarkable difference between the NEILs and NTH1 or OGG1 is that the NEILs have a unique preference for excising lesions from DNA bubble structures (Dou et al., 2003). In DNA bubble structures, both NEIL1 and 2 have higher catalytic specificity in excising 5-OHU than from a G:5-OHU base pair in duplex DNA. Furthermore, NEIL2 but not NEIL1 can excise an 8-oxoG residue from a DNA bubble structure. Taken together, the activity of the NEILs with single stranded DNA, along with the S phase-specific expression of NEIL1, is consistent with a role for the NEILs in transcription coupled repair and replication activated repair (Dou et al., 2003).

4. STEPS POST-BASE REMOVAL

As discussed above, a variety of lesions and substrates are recognised by an extensive array of glycosylases that catalyze the first step of BER (Figure 2). The next step in BER depends largely on the type of glycosylase that acted on the base damage and the type of processing that is needed for completion of repair. The most commonly used sub-pathway is called short-patch BER, depicted in Figure 3. BER initiated by a monofunctional glycosylase results in the generation of an AP-site that is subsequently cleaved by APE1 resulting in a 3'-OH and a 5'-dRP terminus. Bifunctional glycosylases possess an additional intrinsic lyase activity and can cleave the AP sites using two different mechanisms generating either a 3'-dRP that is removed by APE1 or a 3'phosphate, that has to be removed by a phosphatase activity. This role has recently been shown to be fulfilled by polynucleotide kinase (PNK), since as already mentioned, the mammalian AP-endonuclease has no phosphatase activity. On the other hand, a gap greater than one nucleotide is sometimes generated at the base damage site and 2-7 nucleotides need to be resynthesised in what is called the "long-patch" BER pathway (Figure 5). Long-patch BER is PCNA dependent and uses either Pol β or the replicative Pol δ/ε. Because a longer repair patch is made and displacement of the lesion containing stretch is required, the flap-endonuclease FEN1, which recognises and cleaves at the base of the flap structure, and PCNA, which is the loading clamp for the replicative polymerases, are also required. Repair initiated by monofunctional DNA glycosylases can be completed via either short-patch (one nucleotide gaps) or long-patch BER, while repair initiated by bifunctional DNA glycosylase is completed mainly by short patch BER (Dogliotti et al., 2001). Most of the proteins that act downstream of the glycosylase step of BER are discussed below, with the exception of PCNA and Pol δ/ε, which have been well reviewed elsewhere (Hubscher et al., 2002; Hubscher et al., 2000; Maga and Hubscher, 2003; Matsumoto, 2001; Mitra et al., 2001).

4.1 AP-endonuclease

Human AP endonuclease (APE1) is required for the processing of BER intermediates to facilitate the completion of DNA repair by Pol β and DNA ligase IIIα (Klungland et al., 1999b; Pascucci et al., 2002). In addition to DNA *N*-glycosylase activity, enormous amounts of AP sites are generated daily (about 2,000 to 10,000 per cell per day) from spontaneous depurination (Lindahl, 1993). *APE1*, the human homolog of *E. coli* exonuclease III, is thought to process of over 95% of these AP sites in mammalian cells (Demple and Harrison, 1994). APE1 activity at AP sites is catalyzed by a Mg^{2+} stimulated mechanism, resulting in cleavage of the phosphodiester bond 5' of the AP site and generation of a single-strand break. This reaction produces a 3'-hydroxyl group and a 5'-dRP group flanking the break (Figure 3). Product release has been measured to be much faster than the strand cleavage reaction, which is consistent with a Briggs-Haldane reaction mechanism (Beernink et al., 2001; Mol et al., 2000; Strauss et al., 1997). A catalytic mechanism for AP endonuclease activity was proposed based on APE1 complexed with AP site-containing oligonucleotides. In this mechanism two Mg^{2+} ions are required to stabilise the transition-state intermediate and the O3' leaving group and perform the general acid/base chemistry while several essential residues act to orient the two metal ions (Beernink et al., 2001). Recently, the nonspecific interactions of APE1 with double-stranded DNA and specific interactions with AP site-containing double-stranded DNA have been measured using X-ray analysis. Interestingly, this analysis has shown that nonspecific interactions of APE1 with the phosphate groups of 9-10 nucleotides within its binding cleft provide an increase in affinity of enzyme for any DNA of approximately seven orders of magnitude, whereas specific interactions of APE1 with AP sites further increases the affinity one order of magnitude. Therefore it appears that APE1 has a DNA surveillance mechanism that allows it to slide along DNA and scan for specific interactions with AP sites (Beloglazova et al., 2004). APE1 also processes α-unsaturated aldehyde products that result from the AP lyase activity of the bifunctional DNA *N*-glycosylases. Less information is available for the interaction of APE1 with this particular substrate, however, it is clear that APE1 possesses 3' phosphodiesterase activity with α-unsaturated aldehydes, producing a single nucleotide gap flanked by a 3'-hydroxyl group and a 5' phosphate group (Klungland et al., 1999b; Pascucci et al., 2002). Additionally APE1 has been demonstrated to possess 3' phosphodiesterase activity with several other substrates (Chou et al., 2000; Izumi et al., 2000; Suh et al., 1997; Winters et al., 1994) as well as RNase H activity, and endonucleolytic activity against AP sites in single-stranded DNA (Marenstein et al., 2004; Wilson, 2003). While it is clear that the role of the 3' phosphodiesterase activity is to process damaged nucleotides or repair intermediates, the physiological roles of the other activities remain enigmatic. Taken together, the available evidence supports that these BER intermediates are funnelled into an APE1 processing reaction, allowing for the intermediates to become extension substrates for Pol β or Pol δ/ε.

Studies with mouse models that are homozygous null for the gene encoding the murine Ape1 gene, *Apex*, have shown that *Apex* is essential for embryonic development (Table 2). Curran and co-workers found that matings between heterozygous mice yielded litters that had a wild-type to heterozygous ratio that was consistent with the expected 2:1 outcome for embryonic lethality (Xanthoudakis et al., 1996). Analysis of deciduae from heterozygous crosses at 5.5 days showed that 71% contained healthy animals with 23% of the deciduae containing embryos that were severely necrotic, while 6% were empty. It was also observed at 5.5 days that the percentage of degenerating embryos was significantly higher than would be expected from wild type matings, strongly suggesting that *Apex* null mice die *in utero* following implantation (Xanthoudakis et al., 1996). Expression of the murine *Apex* is greatest during the S-phase, suggesting that APE1 expression coincides with a period of oxidative burst that results from active proliferation during embryonic growth (Fung et al., 2001; Xanthoudakis et al., 1996). Other studies have shown that explanted *Apex* null embryos display increased sensitivity to gamma irradiation, supporting a role for APE1 in the repair of oxidative DNA damage (Ludwig et al., 1998). Studies with *Apex* heterozygous null mice have demonstrated that mutant cells and animals are hypersensitive to increased oxidative stress, that the serum from these animals displays elevated levels of stress markers, and that both of these phenotypes can be reversed with dietary supplementation of antioxidants (Meira et al., 2001). Histopathological analysis of deceased heterozygous animals showed that 25% of these animals developed microscopic tumours, whereas no tumours were found in wild type animals. Furthermore, it was shown that heterozygous embryos and pups show reduced survival, which can also be reversed with antioxidant supplementation (Meira et al., 2001). In addition to the wide range of DNA lesions that APE1 processes, APE1 regulates the DNA binding affinity of transcription factors such as p53, Jun/Fos and NF-κB by a reduction/oxidation mechanism (Jayaraman et al., 1997; Xanthoudakis et al., 1992). It thus appears that APE1 may have an additional role as a central component of signal transduction pathways (Evans et al., 2000). APE1 thus has a critical function in two cellular pathways in humans, both of which play an important role in cellular responses to oxidative damage; and it is not clear whether the absence of one or both functions is responsible for the embryonic lethal phenotype of *Apex* null mice.

In contrast, BER events that are initiated by the NEILs are repaired via an alternative pathway that is AP endonuclease independent (see Figure 3). Because excision of damaged bases by the NEILs results in a one-nucleotide gap flanked by a 5' and 3' phosphate, additional enzymatic activity is required to perform the DNA 3' phosphatase reaction. Because the *E. coli* AP endonucleases do possess DNA 3' phosphatase activity, mammalian APE1 was expected to have this activity as well. However, the DNA 3' phosphatase activity of APE1 was shown to be very weak (Demple and Harrison, 1994; Xu et al., 2003). Recently experimental results reported by Wiederhold *et al.* have revealed that the 3' phosphatase reaction is catalyzed by PNK and not APE1 (Wiederhold et al., 2004). Additionally, a DNA repair complex consisting of PNK, NEIL1, Pol β and DNA ligase IIIα was identified, suggesting coordination of APE1-independent BER. Furthermore, it was determined that NEIL1 and PNK contribute to the repair of AP-sites, which suggests a broad role for this repair pathway in mammalian cells (Wiederhold et al., 2004).

4.2 DNA Polymerase β

Subsequent to AP site processing by APE1 or bifunctional DNA *N*-glycosylases, or processing of a 3' phosphate terminus by PNK, Pol β performs a DNA extension reaction to replace the damaged nucleotide (Klungland et al., 1999b; Pascucci et al., 2002; Wiederhold et al., 2004). Pol β is a member of the X family of DNA polymerases and shares many structural, kinetic and mechanistic features with other replicative polymerases, however it does lack exonucleolytic proofreading activity, and consequently is quite small (39 kDa) (Idriss et al., 2002; Krahn et al., 2004). Pol β consists of a 31 kDa polymerase domain that catalyzes the nucleotidyl transferase reaction, and an 8 kDa amino-terminal lyase domain that excises the 5'-dRP intermediate produced from bifunctional DNA glycosylases/AP lyases initiated repair (Idriss et al., 2002). In addition to the 5'-dRP lyase activity, the amino-terminal domain also possesses single strand DNA binding activity. This activity directs Pol β to short gaps that possess a 5' phosphate at the margin (Prasad et al., 1996). Upon binding to gapped DNA Pol β extends the nascent DNA in a processive manner to fill gaps of 6 nucleotides or fewer (Singhal and Wilson, 1993). Structural analysis of Pol β suggests that the mechanism of substrate specificity is consistent with the induced-fit model and that misincorporation of an incorrect base (non Watson-Crick base pair) prevents a conformational change that is thought to be required for catalytic cycling (Idriss et al., 2002; Krahn et al., 2004). Additionally, Pol β has been implicated in double-strand break repair and appears to play a role in synapsis and recombination during meiosis (Idriss et al., 2002).

Several variant forms of Pol β have been found to be mutagenic and associated with cancer and disease (Bhattacharyya et al., 1999a; Bhattacharyya et al., 1999b; Iwanaga et al., 1999; Starcevic et al., 2004). Three mutants of Tyr265 have been isolated and been shown to display a large decrease in fidelity compared to wild-type Pol β. A mutation to histidine for Tyr265 results in a 120-fold reduction in fidelity due to a structural alteration that leads to a rate of polymerisation that is only 8-fold faster for the correct nucleotide versus a non-Watson Crick base pairing nucleotide (Shah et al., 2001). Experiments have shown that extension from nascent DNA by wild-type Pol β is much less efficient for mispaired DNA than base paired DNA (Beard et al., 2004). In contrast, extension experiments with Y265F and Y265W mutants have shown that these mutants extend nascent 3' mismatched DNA with a proper or improper nucleotide end much more efficiently than the wild-type enzyme (Shah et al., 2003). These results have prompted Sweasy and researchers to suggest that mutations of Tyr265 affect structural properties of Pol β that prevent misincorporation and mispair extension (Shah et al., 2003). A genetic screen for the selection of Pol β mutants that confer resistance to AZT has shown that several mutations in the flexible loop domain of Pol β affect polymerase fidelity. This loop appears disordered in crystal structures and is thought to play an indirect role in properly positioning the nascent DNA in the active site and preventing extension from mispaired DNA (Dalal et al., 2004; Kosa and Sweasy, 1999a; Kosa and Sweasy, 1999b). Similarly, a mutation in Pol β has been identified in a colorectal carcinoma in which

Lys289 has been mutated to methionine (Lang et al., 2004). Expression of this mutation in mouse cells was shown to result in a 2.5-fold increase in mutation frequency, with a 16-fold increase in C to G or G to C substitution frequency. Like the Tyr265 mutation this mutation was shown to result in improper positioning or misalignment of the nascent DNA strand. Careful perusal of mutational hotspots sequences revealed that the K289M mutant has a tendency to misalign the nascent strand with the template strand, causing the nucleotide 5' of the proper templating nucleotide to become the new templating nucleotide (Lang et al., 2004). Similar results were observed with the Y265F and Y265W mutants (Shah et al., 2001). Importantly, unpublished observations cited by the Sweasy laboratory have shown that the expression of K289M in mouse cells results in a transformed phenotype, supporting the link between the K289M variant and cancer (Lang et al., 2004).

Evidence is emerging which suggest that in a normal cell context Pol β expression is tightly regulated (Bergoglio et al., 2004). Several years ago it was shown that mice homozygous null for Pol β are inviable and die as embryos after day 10.5 (Table 2)(Gu et al., 1994). Subsequent studies showed that exposing embryonic fibroblasts that are homozygous null for Pol β to DNA alkylating and oxidising agents results in increased cytotoxicity, mutations and chromosomal aberrations, compared to wild-type cells (Sobol et al., 1996). More recent reports have documented that mice heterozygous for Pol β are viable, however they exhibit increased levels of single strand breaks and chromosomal aberrations compared to wild-type littermates (Cabelof et al., 2003). In addition extracts from Pol β heterozygous mice show decreased levels of repair of G:U mismatches and 8-oxoG:C lesions. Furthermore, these mice show an increased level of mutagenicity by dimethyl sulfate, but do not exhibit increased sensitivity to UV and radiation, supporting that the defect is indeed a defect specifically in BER (Cabelof et al., 2003). Conversely, overexpression of Pol β of only 2-fold has been shown to result in an increase in mutation rate, chromosomal instability and accelerated tumorigenesis (Bergoglio et al., 2002; Canitrot et al., 1998). Likewise, high levels of Pol β have been found in several cancer cells and tumours (Canitrot et al., 2000; Scanlon et al., 1989; Srivastava et al., 1999). Taken together, these results suggest that polymorphisms affecting expression levels of Pol β and enzymatic activity may contribute to human diseases.

4.3 Flap-endonuclease 1

The long-patch BER pathway (Figure 5) is initiated subsequent to Pol β catalyzed addition of the first nucleotide into the repair gap in the event that the 5'-sugar phosphate is reduced or oxidised and therefore resistant to Pol β catalyzed lyase activity (Podlutsky et al., 2001). Subsequently, additional DNA synthesis is required to dislodge the damaged 5'-sugar phosphate, resulting in a DNA flap that is processed by FEN1 (Kim et al., 1998; Prasad et al., 2000). The FEN1 enzyme interacts with other BER components including PCNA (Gary et al., 1999) and APE1 (Dianova et al., 2001) (see Table 1). PCNA forms a doughnut shaped homotrimeric clamp around the DNA molecule and, in addition to forming a polymerase clamp, appears to act as a scaffold to coordinate interactions among long-patch repair proteins to facilitate the transition from APE1 excision to strand ligation (Cox, 1997). The FEN1/PCNA interaction was shown to result in a stimulation of FEN1 endonuclease activity *in vitro*; such stimulation was significantly reduced when the respective binding sites were mutated in the interacting proteins, demonstrating direct stimulation of long-patch BER by PCNA (Gary et al., 1999). Similarly, the addition of APE1 to an *in vitro* long-patch BER reaction resulted in increased coordination and stimulation of the enzymatic activities of both FEN1 and DNA ligase I, suggesting a role for APE1 in facilitating long-patch repair (Dianova et al., 2001; Ranalli et al., 2002). Expression of a nuclease-defective FEN1 mutant in human cells results in increased sensitivity to MMS and UV-irradiation, but not ionising radiation, and a prolonged delay of S-phase progression (Shibata and Nakamura, 2002). Importantly, cells expressing the nuclease-defective and PCNA-binding defective FEN1 double mutant displayed an MMS-sensitive phenotype similar to wild-type, therefore demonstrating *in vivo* a direct interaction between FEN1 and PCNA (Shibata and Nakamura, 2002). In addition to long-patch BER, FEN1 has been demonstrated to have a role in replicative DNA synthesis including Okazaki fragment maturation and the prevention of repeat sequence expansion (Liu et al., 2004).

4.4 Polynucleotide Kinase

PNK was initially described as a 5'-DNA kinase, however more recent molecular characterisation of PNK has revealed a 3'-DNA phosphatase activity, suggesting a role for this enzyme in DNA repair (Karimi-Busheri et al., 1999). Further evidence of a role for PNK in DNA repair was gleaned from experiments where expression of PNK was stably down-regulated in human lung adenocarcinoma cells. These cells exhibit a 7-fold increase in spontaneous mutation frequency as well as increased sensitivity to several DNA damaging agents (Rasouli-Nia et al., 2004). In addition to the role PNK plays in BER (see Figure 3), PNK also has a role in both DNA single-strand and double-strand break repair (Chappell et al., 2002; Whitehouse et al., 2001). To determine the repair activity of PNK on a single-strand break Caldecott and workers performed an *in vitro* assay containing recombinant human PNK, Pol β, DNA ligase IIIα, and a double-stranded oligonucleotide containing a single nucleotide gap flanked by a 3'-phosphate and a 5'-OH (Whitehouse et al., 2001). Completion of repair was observed when all enzymes were present, and in the absence of PNK the 3'-phosphatase activity was abolished and repair was not completed, suggesting that both the 3'-DNA phosphatase and 5'-DNA kinase activity of PNK is required for processing of single-strand breaks. Further experimentation showed that the scaffolding protein XRCC1 stimulated the 3'-DNA phosphatase and 5'-DNA kinase activities of PNK, and accelerates the single-strand break repair reaction *in vitro* (Whitehouse et al., 2001). One method cells use to repair double-strand breaks is termed "non-homologous end joining" (NHEJ). Double-stranded DNA breaks that enter the NHEJ pathway often contain non-ligatable end groups that must be converted to ligatable 3'-OH and 5'-phosphate moieties which can be efficiently repaired by NHEJ (Chappell et al., 2002). Using mammalian cell free extracts, West and workers determined that PNK can phosphorylate terminal 5'-OH groups, but that this activity was blocked by depletion of the NHEJ factor XRCC4 or by a PNK-inactivating mutation. These results indicate that the DNA kinase activity of PNK is coupled to active NHEJ processes (Chappell et al., 2002). Taken together, these studies suggest that PNK is an important factor for several pathways that maintain genomic integrity.

4.5 Completion of the BER Pathway

Completion of the BER pathway can be performed by either DNA ligase I or the DNA ligase IIIα/XRCC1 complex (Kubota et al., 1996; Sleeth et al., 2004; Wiederhold et al., 2004). DNA ligase IIIα has been shown to be a molecular nick-sensor while pol β has been shown to interact with XRCC1, and it has been suggested that the role of XRCC1 is to act as a scaffold to bridge polymerase and ligase activity (Table 1) (Caldecott et al., 1996; Cappelli et al., 1997; Kubota et al., 1996). XRCC1 homozygous null mice are inviable and embryos begin to show developmental abnormalities by 7.5 days (Tebbs et al., 1999). XRCC1 mutant cells exhibit an elevated level of spontaneous chromosomal aberrations and deletions (Thompson and West, 2000), and a correlation between elevated somatic mutation and increased cancer risk has been identified for a genetic polymorphism in human XRCC1 (Divine et al., 2001; Shen et al., 1998; Sturgis et al., 1999). Importantly, cells expressing greatly reduced levels of XRCC1 are characteristically deficient in short patch BER specifically, suggesting that the DNA ligase IIIα/XRCC1 complex is the major ligase in short patch BER (Cappelli et al., 1997). Despite this, recent biochemical experiments have shown that DNA ligase I is in fact capable of efficiently completing the short patch BER pathway (Sleeth et al., 2004). Although DNA ligase I plays an essential role in DNA replication (Waga and Stillman, 1998) it has also been implicated in DNA repair, in that DNA ligase I interacts with human pol β *in vitro* (Prasad et al., 1996).

Tainer and coworkers have provided structural and kinetic data with APE1 mutants suggesting that APE1 most likely aids in the displacement of bound glycosylases and retains the nicked dsDNA product thus promoting a coordinated transfer of unstable repair intermediates between the excision and repair steps (Mol et al., 2000). Several reports have provided experimental data to support this. For instance, the α-unsaturated aldehyde repair intermediates generated by AP lyase activity can readily react with an adjacent protein and result in the formation of a potentially toxic covalent DNA-protein cross-link (Klungland et al., 1999b). However, the finding that APE1 stimulates OGG1 glycosylase activity and displaces OGG1 at the AP site (resulting in greater turnover for OGG1) suggests a mechanism for preventing α-unsaturated aldehyde repair intermediates from becoming long-lived (Hill et al., 2001). Therefore, these data provide an excellent example for how the cell remedies the confounding situation of dealing with toxic repair intermediates in addition to exemplifying the elegant coordination present in the processing of BER intermediates. Similarly, APE1 has been shown to physically interact with Pol β at AP sites, resulting in recruitment of the polymerase to the 5'-incised AP site and stimulation of 5'-dRP excision (Table 1) (Bennett et al., 1997; Wong et al., 2003). More recently, APE1 has been shown to display 3' to 5' exonucleolytic activity. Interestingly, this activity shows a preference for 3' mispaired DNA, suggesting that APE1 may provide the proofreading mechanism that is lacking in Pol β (Chou and Cheng, 2002; Wong et al., 2003). Likewise, interactions between DNA polymerase β and DNA ligase I or the DNA ligaseIIIα/XRCC1 complex have been documented (Table 1). Subsequently, the picture emerging from these studies suggests that a dynamic repair complex consisting of several different proteins works in concert to remediate the enormous number of DNA lesions that enter BER pathways.

5. CONCLUDING REMARKS

It is now firmly established that alterations in genes required for the normal processing of DNA damage can result in detrimental consequences to human health. Defective nucleotide excision repair and mismatch repair had been for long associated with the human cancer predisposition syndromes Xeroderma pigmentosum (XP) and hereditary non-polyposis colon cancer (HNPCC), respectively. Until recently, when MYH mutations were found in familial adenomatous polyposis, no such association had been found for defective BER. To complicate matters even further, single knock-out mouse models for the many proteins involved in BER were not terribly informative. Glycosylase-deficient mice were largely devoid of any significant phenotype whereas most of the knock-outs for the subsequent BER steps were reported lethal (summarised in Table 2). This could be a reflection of the inherent versatility and redundancy of the BER pathway since glycosylases recognise and excise a broad range of substrates while downstream enzymes such as APE1 are multi-functional and act not only in BER but also in other cellular pathways. However, recent reports have linked deficiencies in some glycosylases with certain human pathological conditions, increasingly pointing at a very important role of BER in maintaining cellular homeostasis (Table 2). As discussed above, several polymorphisms in BER factors are found associated with increased cancer risk and not long ago, the MYH glycosylase was actually found to be mutated in familial adenomatous polyposis, a type of colorectal cancer. Defects in *UNG2* were linked to hyper-IgM syndrome, or HIGM, where patients suffer from increased susceptibility to bacterial infections and are compromised immunologically, probably due to the role of *UNG2* mediated uracil repair in antibody diversification. Moreover, it is not clear yet if *UNG2* deficiency would be found associated with late-onset lymphomas in humans, mimicking what was seen in the mouse. The same can be said for other glycosylases and future research is required to substantiate the link between polymorphisms in BER genes and increased predisposition to different types of cancer. Finally, the observation that patients with ulcerative colitis show increased expression of *AAG* and *APE1* in inflamed tissues and altered genomic instability as manifested by increased microsatellite instability, gives support to the "Imbalance in Base Excision Repair" paradigm and links inflammation with BER. Since inflammatory processes are also know to affect cancer incidence, this is yet another mechanism by which alterations in BER can lead to increased genomic instability and cancer.

REFERENCES

Al-Tassan, N., N.H. Chmiel, J. Maynard, N. Fleming, A.L. Livingston, G.T. Williams, A.K. Hodges, D.R. Davies, S.S. David, J.R. Sampson, and J.P. Cheadle. 2002. Inherited variants of MYH associated with somatic G:C-->T:A mutations in colorectal tumors. *Nat Genet*. 30:227-32.

Arai, T., V.P. Kelly, O. Minowa, T. Noda, and S. Nishimura. 2002. High accumulation of oxidative DNA damage, 8-hydroxyguanine, in Mmh/Ogg1 deficient mice by chronic oxidative stress. *Carcinogenesis*. 23:2005-10.

Aravind, L., and E.V. Koonin. 2000. The alpha/beta fold uracil DNA glycosylases: a common origin with diverse fates. *Genome Biol*. 1:RESEARCH0007.

Asagoshi, K., H. Odawara, H. Nakano, T. Miyano, H. Terato, Y. Ohyama, S. Seki, and H. Ide. 2000. Comparison of substrate specificities of Escherichia coli endonuclease III and its mouse homologue (mNTH1) using defined oligonucleotide substrates. *Biochemistry*. 39:11389-98.

Aspinwall, R., D.G. Rothwell, T. Roldan-Arjona, C. Anselmino, C.J. Ward, J.P. Cheadle, J.R. Sampson, T. Lindahl, P.C. Harris, and I.D. Hickson. 1997. Cloning and characterization of a functional human homolog of Escherichia coli endonuclease III. *Proc Natl Acad Sci U S A*. 94:109-14.

Barbin, A., R. Wang, P.J. O'Connor, and R.H. Elder. 2003. Increased formation and persistence of 1,N(6)-ethenoadenine in DNA is not associated with higher susceptibility to carcinogenesis in alkylpurine-DNA-N-glycosylase knockout mice treated with vinyl carbamate. *Cancer Res*. 63:7699-703.

Bartsch, H., J. Nair, and R.W. Owen. 2002. Exocyclic DNA adducts as oxidative stress markers in colon carcinogenesis: potential role of lipid peroxidation, dietary fat and antioxidants. *Biol Chem*. 383:915-21.

Basu, A.K., M.L. Wood, L.J. Niedernhofer, L.A. Ramos, and J.M. Essigmann. 1993. Mutagenic and genotoxic effects of three vinyl chloride-induced DNA lesions: 1,N6-ethenoadenine, 3,N4-ethenocytosine, and 4-amino-5-(imidazol-2-yl)imidazole. *Biochemistry*. 32:12793-801.

Beard, W.A., W.P. Osheroff, R. Prasad, M.R. Sawaya, M. Jaju, T.G. Wood, J. Kraut, T.A. Kunkel, and S.H. Wilson. 1996. Enzyme-DNA interactions required for efficient nucleotide incorporation and discrimination in human DNA polymerase beta. *J Biol Chem.* 271:12141-4.

Beard, W.A., D.D. Shock, and S.H. Wilson. 2004. Influence of DNA structure on DNA polymerase beta active site function: extension of mutagenic DNA intermediates. *J Biol Chem.* 279:31921-9.

Beernink, P.T., B.W. Segelke, M.Z. Hadi, J.P. Erzberger, D.M. Wilson, 3rd, and B. Rupp. 2001. Two divalent metal ions in the active site of a new crystal form of human apurinic/apyrimidinic endonuclease, Ape1: implications for the catalytic mechanism. *J Mol Biol.* 307:1023-34.

Begum, N.A., K. Kinoshita, N. Kakazu, M. Muramatsu, H. Nagaoka, R. Shinkura, D. Biniszkiewicz, L.A. Boyer, R. Jaenisch, and T. Honjo. 2004. Uracil DNA glycosylase activity is dispensable for immunoglobulin class switch. *Science.* 305:1160-3.

Bellacosa, A., L. Cicchillitti, F. Schepis, A. Riccio, A.T. Yeung, Y. Matsumoto, E.A. Golemis, M. Genuardi, and G. Neri. 1999. MED1, a novel human methyl-CpG-binding endonuclease, interacts with DNA mismatch repair protein MLH1. *Proc Natl Acad Sci U S A.* 96:3969-74.

Beloglazova, N.G., O.O. Kirpota, K.V. Starostin, A.A. Ishchenko, V.I. Yamkovoy, D.O. Zharkov, K.T. Douglas, and G.A. Nevinsky. 2004. Thermodynamic, kinetic and structural basis for recognition and repair of abasic sites in DNA by apurinic/apyrimidinic endonuclease from human placenta. *Nucleic Acids Res.* 32:5134-46.

Bennett, R.A., D.M. Wilson, 3rd, D. Wong, and B. Demple. 1997. Interaction of human apurinic endonuclease and DNA polymerase beta in the base excision repair pathway. *Proc Natl Acad Sci U S A.* 94:7166-9.

Bergoglio, V., M. Frechet, M. Philippe, A. Bieth, P. Mercier, D. Morello, M. Lacroix-Tricki, G. Delsol, J.S. Hoffmann, and C. Cazaux. 2004. Evidence of finely tuned expression of DNA polymerase beta in vivo using transgenic mice. *FEBS Lett.* 566:147-50.

Bergoglio, V., M.J. Pillaire, M. Lacroix-Triki, B. Raynaud-Messina, Y. Canitrot, A. Bieth, M. Gares, M. Wright, G. Delsol, L.A. Loeb, C. Cazaux, and J.S. Hoffmann. 2002. Deregulated DNA polymerase beta induces chromosome instability and tumorigenesis. *Cancer Res.* 62:3511-4.

Bessho, T. 1999. Nucleotide excision repair 3' endonuclease XPG stimulates the activity of base excision repairenzyme thymine glycol DNA glycosylase. *Nucleic Acids Res.* 27:979-83.

Bhakat, K.K., T.K. Hazra, and S. Mitra. 2004. Acetylation of the human DNA glycosylase NEIL2 and inhibition of its activity. *Nucleic Acids Res.* 32:3033-9.

Bhattacharyya, N., H.C. Chen, S. Comhair, S.C. Erzurum, and S. Banerjee. 1999a. Variant forms of DNA polymerase beta in primary lung carcinomas. *DNA Cell Biol.* 18:549-54.

Bhattacharyya, N., H.C. Chen, S. Grundfest-Broniatowski, and S. Banerjee. 1999b. Alteration of hMSH2 and DNA polymerase beta genes in breast carcinomas and fibroadenomas. *Biochem Biophys Res Commun.* 259:429-35.

Bjoras, M., L. Luna, B. Johnsen, E. Hoff, T. Haug, T. Rognes, and E. Seeberg. 1997. Opposite base-dependent reactions of a human base excision repair enzyme on DNA containing 7,8-dihydro-8-oxoguanine and abasic sites. *EMBO J.* 16:6314-22.

Boiteux, S., O. Huisman, and J. Laval. 1984. 3-Methyladenine residues in DNA induce the SOS function sfiA in Escherichia coli. *EMBO J.* 3:2569-73.

Boiteux, S., and J.P. Radicella. 1999. Base excision repair of 8-hydroxyguanine protects DNA from endogenous oxidative stress. *Biochimie.* 81:59-67.

Boldogh, I., D. Milligan, M.S. Lee, H. Bassett, R.S. Lloyd, and A.K. McCullough. 2001. hMYH cell cycle-dependent expression, subcellular localization and association with replication foci: evidence suggesting replication-coupled repair of adenine:8-oxoguanine mispairs. *Nucleic Acids Res.* 29:2802-9.

Brauch, H., B. Johnson, J. Hovis, T. Yano, A. Gazdar, O.S. Pettengill, S. Graziano, G.D. Sorenson, B.J. Poiesz, J. Minna, and et al. 1987. Molecular analysis of the short arm of chromosome 3 in small-cell and non-small-cell carcinoma of the lung. *N Engl J Med.* 317:1109-13.

Cabelof, D.C., Z. Guo, J.J. Raffoul, R.W. Sobol, S.H. Wilson, A. Richardson, and A.R. Heydari. 2003. Base excision repair deficiency caused by polymerase beta haploinsufficiency: accelerated DNA damage and increased mutational response to carcinogens. *Cancer Res.* 63:5799-807.

Caldecott, K.W., S. Aoufouchi, P. Johnson, and S. Shall. 1996. XRCC1 polypeptide interacts with DNA polymerase beta and possibly poly (ADP-ribose) polymerase, and DNA ligase III is a novel molecular 'nick-sensor' in vitro. *Nucleic Acids Res.* 24:4387-94.

Calmels, S., H. Oshima, M. Crespi, H. Leclerc, C. Cattoen, and H. Bartsch. 1987. *In* The relevance of N-nitroso compounds to human cancer. H. Bartsch, O'Neill, I. & Schulte-Hermann, R., editor. IARC Scientific Publications, Lyon, France. 391-395.

Canitrot, Y., C. Cazaux, M. Frechet, K. Bouayadi, C. Lesca, B. Salles, and J.S. Hoffmann. 1998. Overexpression of DNA polymerase beta in cell results in a mutator phenotype and a decreased sensitivity to anticancer drugs. *Proc Natl Acad Sci U S A.* 95:12586-90.

Canitrot, Y., J.S. Hoffmann, P. Calsou, H. Hayakawa, B. Salles, and C. Cazaux. 2000. Nucleotide excision repair DNA synthesis by excess DNA polymerase beta: a potential source of genetic instability in cancer cells. *Faseb J.* 14:1765-74.

Cappelli, E., R. Taylor, M. Cevasco, A. Abbondandolo, K. Caldecott, and G. Frosina. 1997. Involvement of XRCC1 and DNA ligase III gene products in DNA base excision repair. *J Biol Chem.* 272:23970-5.

Carmell, M.A., L. Zhang, D.S. Conklin, G.J. Hannon, and T.A. Rosenquist. 2003. Germline transmission of RNAi in mice. *Nat Struct Biol.* 10:91-2.

Cerda, S.R., P.W. Turk, A.D. Thor, and S.A. Weitzman. 1998. Altered expression of the DNA repair protein, N-methylpurine-DNA glycosylase (MPG) in breast cancer. *FEBS Lett.* 431:12-8.

Chappell, C., L.A. Hanakahi, F. Karimi-Busheri, M. Weinfeld, and S.C. West. 2002. Involvement of human polynucleotide kinase in double-strand break repair by non-homologous end joining. *EMBO J.* 21:2827-32.

Cheadle, J.P., and J.R. Sampson. 2003. Exposing the MYtH about base excision repair and human inherited disease. *Hum Mol Genet.* 12 Spec No 2:R159-65. Epub 2003 Aug 05.

Chmiel, N.H., A.L. Livingston, and S.S. David. 2003. Insight into the functional consequences of inherited variants of the hMYH adenine glycosylase associated with colorectal cancer: complementation assays with hMYH variants and pre-steady-state kinetics of the corresponding mutated E.coli enzymes. *J Mol Biol.* 327:431-43.

Chou, K.M., and Y.C. Cheng. 2002. An exonucleolytic activity of human apurinic/apyrimidinic endonuclease on 3' mispaired DNA. *Nature.* 415:655-9.

Chou, K.M., M. Kukhanova, and Y.C. Cheng. 2000. A novel action of human apurinic/apyrimidinic endonuclease: excision of L-configuration deoxyribonucleoside analogs from the 3' termini of DNA. *J Biol Chem.* 275:31009-15.

Coulondre, C., J.H. Miller, P.J. Farabaugh, and W. Gilbert. 1978. Molecular basis of base substitution hotspots in Escherichia coli. *Nature.* 274:775-80.

Cox, L.S. 1997. Who bind wins: competition for PCNA rings out cell-cycle changes. *Trends Cell Biol.* 7:493-8.

Dalal, S., J.L. Kosa, and J.B. Sweasy. 2004. The D246V mutant of DNA polymerase beta misincorporates nucleotides: evidence for a role for the flexible loop in DNA positioning within the active site. *J Biol Chem.* 279:577-84.

Dantzer, F., L. Luna, M. Bjoras, and E. Seeberg. 2002. Human OGG1 undergoes serine phosphorylation and associates with the nuclear matrix and mitotic chromatin in vivo. *Nucleic Acids Res.* 30:2349-57.

Das, A., L. Rajagopalan, V.S. Mathura, S.J. Rigby, S. Mitra, and T.K. Hazra. 2004. Identification of a zinc finger domain in the human NEIL (Nei like)-2 protein. *J Biol Chem.*

Demple, B., and L. Harrison. 1994. Repair of oxidative damage to DNA: enzymology and biology. *Annu Rev Biochem.* 63:915-48.

Dherin, C., J.P. Radicella, M. Dizdaroglu, and S. Boiteux. 1999. Excision of oxidatively damaged DNA bases by the human alpha-hOgg1 protein and the polymorphic alpha-hOgg1(Ser326Cys) protein which is frequently found in human populations. *Nucleic Acids Res*. 27:4001-7.

Dianova, II, V.A. Bohr, and G.L. Dianov. 2001. Interaction of human AP endonuclease 1 with flap endonuclease 1 and proliferating cell nuclear antigen involved in long-patch base excision repair. *Biochemistry*. 40:12639-44.

Dickerson, S.K., E. Market, E. Besmer, and F.N. Papavasiliou. 2003. AID mediates hypermutation by deaminating single stranded DNA. *J Exp Med*. 197:1291-6.

Dimitriadis, E.K., R. Prasad, M.K. Vaske, L. Chen, A.E. Tomkinson, M.S. Lewis, and S.H. Wilson. 1998. Thermodynamics of human DNA ligase I trimerization and association with DNA polymerase beta. *J Biol Chem*. 273:20540-50.

Divine, K.K., F.D. Gilliland, R.E. Crowell, C.A. Stidley, T.J. Bocklage, D.L. Cook, and S.A. Belinsky. 2001. The XRCC1 399 glutamine allele is a risk factor for adenocarcinoma of the lung. *Mutat Res*. 461:273-8.

Dizdaroglu, M., B. Karahalil, S. Senturker, T.J. Buckley, and T. Roldan-Arjona. 1999. Excision of products of oxidative DNA base damage by human NTH1 protein. *Biochemistry*. 38:243-6.

Dogliotti, E., P. Fortini, B. Pascucci, and E. Parlanti. 2001. The mechanism of switching among multiple BER pathways. *Prog Nucleic Acid Res Mol Biol*. 68:3-27.

Dosanjh, M.K., R. Roy, S. Mitra, and B. Singer. 1994. 1,N6-ethenoadenine is preferred over 3-methyladenine as substrate by a cloned human N-methylpurine-DNA glycosylase (3-methyladenine-DNA glycosylase). *Biochemistry*. 33:1624-1628.

Dou, H., S. Mitra, and T.K. Hazra. 2003. Repair of oxidized bases in DNA bubble structures by human DNA glycosylases NEIL1 and NEIL2. *J Biol Chem*. 278:49679-84.

Doublie, S., V. Bandaru, J.P. Bond, and S.S. Wallace. 2004. The crystal structure of human endonuclease VIII-like 1 (NEIL1) reveals a zincless finger motif required for glycosylase activity. *Proc Natl Acad Sci U S A*. 101:10284-9.

Doublie, S., S. Tabor, A.M. Long, C.C. Richardson, and T. Ellenberger. 1998. Crystal structure of a bacteriophage T7 DNA replication complex at 2.2 A resolution. *Nature*. 391:251-8.

Duncan, B.K., and J.H. Miller. 1980. Mutagenic deamination of cytosine residues in DNA. *Nature*. 287:560-1.

el Ghissassi, F., A. Barbin, J. Nair, and H. Bartsch. 1995. Formation of 1,N6-ethenoadenine and 3,N4-ethenocytosine by lipid peroxidation products and nucleic acid bases. *Chem Res Toxicol*. 8:278-83.

Elder, R.H., and G.L. Dianov. 2002. Repair of dihydrouracil supported by base excision repair in mNTH1 knock-out cell extracts. *J Biol Chem*. 277:50487-90.

Engelward, B.P., G. Weeda, M.D. Wyatt, J.L.M. Broekhof, J. De Wit, I. Donker, J.M. Allan, B. Gold, J.H.J. Hoeijmakers, and L.D. Samson. 1997. Base excision repair deficient mice lacking the Aag alkyladenine DNA glycosylase. *Proc. Natl. Acad. Sci. USA*. 94:13087-13092.

Evans, A.R., M. Limp-Foster, and M.R. Kelley. 2000. Going APE over ref-1. *Mutat Res*. 461:83-108.

Fortini, P., B. Pascucci, E. Parlanti, M. D'Errico, V. Simonelli, and E. Dogliotti. 2003. The base excision repair: mechanisms and its relevance for cancer susceptibility. *Biochimie*. 85:1053-71.

Fortini, P., B. Pascucci, E. Parlanti, R.W. Sobol, S.H. Wilson, and E. Dogliotti. 1998. Different DNA polymerases are involved in the short- and long-patch base excision repair in mammalian cells. *Biochemistry*. 37:3575-80.

Friedberg, E.C., G.C. Walker, and W. Siede. 1995. DNA Repair and Mutagenesis. ASM Press, Washington, D.C. 698 pp.

Fromme, J.C., A. Banerjee, S.J. Huang, and G.L. Verdine. 2004a. Structural basis for removal of adenine mispaired with 8-oxoguanine by MutY adenine DNA glycosylase. *Nature*. 427:652-6.

Fromme, J.C., A. Banerjee, and G.L. Verdine. 2004b. DNA glycosylase recognition and catalysis. *Curr Opin Struct Biol*. 14:43-9.

Fromme, J.C., S.D. Bruner, W. Yang, M. Karplus, and G.L. Verdine. 2003. Product-assisted catalysis in base-excision DNA repair. *Nat Struct Biol*. 10:204-11.

Fung, H., R.A. Bennett, and B. Demple. 2001. Key role of a downstream specificity protein 1 site in cell cycle-regulated transcription of the AP endonuclease gene APE1/APEX in NIH3T3 cells. *J Biol Chem*. 276:42011-7.

Gary, R., K. Kim, H.L. Cornelius, M.S. Park, and Y. Matsumoto. 1999. Proliferating cell nuclear antigen facilitates excision in long-patch base excision repair. *J Biol Chem*. 274:4354-63.

Girard, P.M., C. D'Ham, J. Cadet, and S. Boiteux. 1998. Opposite base-dependent excision of 7,8-dihydro-8-oxoadenine by the Ogg1 protein of Saccharomyces cerevisiae. *Carcinogenesis*. 19:1299-305.

Glassner, B.J., L.J. Rasmussen, M.T. Najarian, L.M. Posnick, and L.D. Samson. 1998. Generation of a strong mutator phenotype in yeast by imbalanced base excision repair. *Proc Natl Acad Sci U S A*. 95:9997-10002.

Gogos, A., J. Cillo, N.D. Clarke, and A.L. Lu. 1996. Specific recognition of A/G and A/7,8-dihydro-8-oxoguanine (8-oxoG) mismatches by Escherichia coli MutY: removal of the C-terminal domain preferentially affects A/8-oxoG recognition. *Biochemistry*. 35:16665-71.

Gu, H., J.D. Marth, P.C. Orban, H. Mossmann, and K. Rajewsky. 1994. Deletion of a DNA polymerase beta gene segment in T cells using cell type-specific gene targeting. *Science*. 265:103-6.

Ham, A.J., B.P. Engelward, H. Koc, R. Sangaiah, L.B. Meira, L.D. Samson, and J.A. Swenberg. 2004. New immunoaffinity-LC-MS/MS methodology reveals that Aag null mice are deficient in their ability to clear 1,N6-etheno-deoxyadenosine DNA lesions from lung and liver in vivo. *DNA Repair (Amst)*. 3:257-65.

Hang, B., B. Singer, G.P. Margison, and R.H. Elder. 1997. Targeted deletion of alkylpurine-DNA-N-glycosylase in mice eliminates repair of 1,N6-ethenoadenine and hypoxanthine but not of 3,N4-ethenocytosine or 8-oxoguanine. *Proc Natl Acad Sci U S A*. 94:12869-74.

Hardeland, U., M. Bentele, J. Jiricny, and P. Schar. 2000. Separating substrate recognition from base hydrolysis in human thymine DNA glycosylase by mutational analysis. *J Biol Chem*. 275:33449-56.

Hardeland, U., M. Bentele, T. Lettieri, R. Steinacher, J. Jiricny, and P. Schar. 2001a. Thymine DNA glycosylase. *Prog Nucleic Acid Res Mol Biol*. 68:235-53.

Hardeland, U., M. Bentele, T. Lettieri, R. Steinacher, J. Jiricny, and P. Schar. 2001b. Thymine DNA glycosylase. *Prog Nucleic Acid Res Mol Biol*. 68:235-53.

Hardeland, U., R. Steinacher, J. Jiricny, and P. Schar. 2002. Modification of the human thymine-DNA glycosylase by ubiquitin-like proteins facilitates enzymatic turnover. *EMBO J*. 21:1456-64.

Hayashi, H., Y. Tominaga, S. Hirano, A.E. McKenna, Y. Nakabeppu, and Y. Matsumoto. 2002. Replication-associated repair of adenine:8-oxoguanine mispairs by MYH. *Curr Biol*. 12:335-9.

Hazra, T.K., T. Izumi, I. Boldogh, B. Imhoff, Y.W. Kow, P. Jaruga, M. Dizdaroglu, and S. Mitra. 2002a. Identification and characterization of a human DNA glycosylase for repair of modified bases in oxidatively damaged DNA. *Proc Natl Acad Sci U S A*. 99:3523-8.

Hazra, T.K., Y.W. Kow, Z. Hatahet, B. Imhoff, I. Boldogh, S.K. Mokkapati, S. Mitra, and T. Izumi. 2002b. Identification and characterization of a novel human DNA glycosylase for repair of cytosine-derived lesions. *J Biol Chem*. 277:30417-20.

Hendrich, B., U. Hardeland, H.H. Ng, J. Jiricny, and A. Bird. 1999. The thymine glycosylase MBD4 can bind to the product of deamination at methylated CpG sites. *Nature*. 401:301-4.

Hilbert, T.P., W. Chaung, R.J. Boorstein, R.P. Cunningham, and G.W. Teebor. 1997. Cloning and expression of the cDNA encoding the human homologue of the DNA repair enzyme, Escherichia coli endonuclease III. *J Biol Chem*. 272:6733-40.

Hill, J.W., T.K. Hazra, T. Izumi, and S. Mitra. 2001. Stimulation of human 8-oxoguanine-DNA glycosylase by AP-endonuclease: potential coordination of the initial steps in base excision repair. *Nucleic Acids Res.* 29:430-8.

Hill-Perkins, M., M.D. Jones, and P. Karran. 1986. Site-specific mutagenesis in vivo by single methylated or deaminated purine bases. *Mutat Res.* 162:153-63.

Hitchcock, T.M., L. Dong, E.E. Connor, L.B. Meira, L.D. Samson, M.D. Wyatt, and W. Cao. 2004. Oxanine DNA glycosylase activity from mammalian AlkylAdenine glycosylase. *J Biol Chem.*.

Hofseth, L.J., M.A. Khan, M. Ambrose, O. Nikolayeva, M. Xu-Welliver, M. Kartalou, S.P. Hussain, R.B. Roth, X. Zhou, L.E. Mechanic, I. Zurer, V. Rotter, L.D. Samson, and C.C. Harris. 2003. The adaptive imbalance in base excision-repair enzymes generates microsatellite instability in chronic inflammation. *J Clin Invest.* 112:1887-94.

Hsu, G.W., M. Ober, T. Carell, and L.S. Beese. 2004. Error-prone replication of oxidatively damaged DNA by a high-fidelity DNA polymerase. *Nature.* 431:217-21.

Hubscher, U., G. Maga, and S. Spadari. 2002. Eukaryotic DNA polymerases. *Annu Rev Biochem.* 71:133-63.

Hubscher, U., H.P. Nasheuer, and J.E. Syvaoja. 2000. Eukaryotic DNA polymerases, a growing family. *Trends Biochem Sci.* 25:143-7.

Idriss, H.T., O. Al-Assar, and S.H. Wilson. 2002. DNA polymerase beta. *Int J Biochem Cell Biol.* 34:321-4.

Ikeda, S., T. Biswas, R. Roy, T. Izumi, I. Boldogh, A. Kurosky, A.H. Sarker, S. Seki, and S. Mitra. 1998. Purification and characterization of human NTH1, a homolog of Escherichia coli endonuclease III. Direct identification of Lys-212 as the active nucleophilic residue. *J Biol Chem.* 273:21585-93.

Imai, K., G. Slupphaug, W.I. Lee, P. Revy, S. Nonoyama, N. Catalan, L. Yel, M. Forveille, B. Kavli, H.E. Krokan, H.D. Ochs, A. Fischer, and A. Durandy. 2003. Human uracil-DNA glycosylase deficiency associated with profoundly impaired immunoglobulin class-switch recombination. *Nat Immunol.* 4:1023-8. Epub 2003 Sep 7.

Iwanaga, A., M. Ouchida, K. Miyazaki, K. Hori, and T. Mukai. 1999. Functional mutation of DNA polymerase beta found in human gastric cancer--inability of the base excision repair in vitro. *Mutat Res.* 435:121-8.

Izumi, T., T.K. Hazra, I. Boldogh, A.E. Tomkinson, M.S. Park, S. Ikeda, and S. Mitra. 2000. Requirement for human AP endonuclease 1 for repair of 3'-blocking damage at DNA single-strand breaks induced by reactive oxygen species. *Carcinogenesis.* 21:1329-34.

Jayaraman, L., K.G. Murthy, C. Zhu, T. Curran, S. Xanthoudakis, and C. Prives. 1997. Identification of redox/repair protein Ref-1 as a potent activator of p53. *Genes Dev.* 11:558-70.

Jones, S., S. Lambert, G.T. Williams, J.M. Best, J.R. Sampson, and J.P. Cheadle. 2004. Increased frequency of the k-ras G12C mutation in MYH polyposis colorectal adenomas. *Br J Cancer.* 90:1591-3.

Karimi-Busheri, F., G. Daly, P. Robins, B. Canas, D.J. Pappin, J. Sgouros, G.G. Miller, H. Fakhrai, E.M. Davis, M.M. Le Beau, and M. Weinfeld. 1999. Molecular characterization of a human DNA kinase. *J Biol Chem.* 274:24187-94.

Katafuchi, A., T. Nakano, A. Masaoka, H. Terato, S. Iwai, F. Hanaoka, and H. Ide. 2004. Differential specificity of human and Escherichia coli endonuclease III and VIII homologues for oxidative base lesions. *J Biol Chem.* 279:14464-71.

Kavli, B., O. Sundheim, M. Akbari, M. Otterlei, H. Nilsen, F. Skorpen, P.A. Aas, L. Hagen, H.E. Krokan, and G. Slupphaug. 2002. hUNG2 is the major repair enzyme for removal of uracil from U:A matches, U:G mismatches, and U in single-stranded DNA, with hSMUG1 as a broad specificity backup. *J Biol Chem.* 277:39926-36. Epub 2002 Aug 2.

Kim, K., S. Biade, and Y. Matsumoto. 1998. Involvement of flap endonuclease 1 in base excision DNA repair. *J Biol Chem.* 273:8842-8.

Klungland, A., M. Hoss, A. Constantinou, S.G. Clarkson, P.W. Doetch, P.H. Bolton, R.D. Wood, and T. Lindahl. 1999a. Base excision repair of oxidative DNA damage activated by XPG protein. *Molecular Cell.* 3:33-42.

Klungland, A., M. Hoss, D. Gunz, A. Constantinou, S.G. Clarkson, P.W. Doetsch, P.H. Bolton, R.D. Wood, and T. Lindahl. 1999b. Base excision repair of oxidative DNA damage activated by XPG protein. *Mol Cell.* 3:33-42.

Klungland, A., I. Rosewell, S. Hollenbach, E. Larsen, G. Daly, B. Epe, E. Seeberg, T. Lindahl, and D.E. Barnes. 1999c. Accumulation of premutagenic DNA lesions in mice defective in removal of oxidative base damage. *Proc Natl Acad Sci U S A.* 96:13300-5.

Kok, K., J. Osinga, B. Carritt, M.B. Davis, A.H. van der Hout, A.Y. van der Veen, R.M. Landsvater, L.F. de Leij, H.H. Berendsen, P.E. Postmus, and et al. 1987. Deletion of a DNA sequence at the chromosomal region 3p21 in all major types of lung cancer. *Nature.* 330:578-81.

Kosa, J.L., and J.B. Sweasy. 1999a. 3'-Azido-3'-deoxythymidine-resistant mutants of DNA polymerase beta identified by in vivo selection. *J Biol Chem.* 274:3851-8.

Kosa, J.L., and J.B. Sweasy. 1999b. The E249K mutator mutant of DNA polymerase beta extends mispaired termini. *J Biol Chem.* 274:35866-72.

Krahn, J.M., W.A. Beard, and S.H. Wilson. 2004. Structural insights into DNA polymerase Beta deterrents for misincorporation support an induced-fit mechanism for fidelity. *Structure (Camb).* 12:1823-32.

Krokan, H.E., F. Drablos, and G. Slupphaug. 2002. Uracil in DNA--occurrence, consequences and repair. *Oncogene.* 21:8935-48.

Kubota, Y., R.A. Nash, A. Klungland, P. Schar, D.E. Barnes, and T. Lindahl. 1996. Reconstitution of DNA base excision-repair with purified human proteins: interaction between DNA polymerase beta and the XRCC1 protein. *EMBO J.* 15:6662-70.

Kuo, C.F., D.E. McRee, C.L. Fisher, S.F. O'Handley, R.P. Cunningham, and J.A. Tainer. 1992. Atomic structure of the DNA repair [4Fe-4S] enzyme endonuclease III. *Science.* 258:434-40.

Lang, T., M. Maitra, D. Starcevic, S.X. Li, and J.B. Sweasy. 2004. A DNA polymerase beta mutant from colon cancer cells induces mutations. *Proc Natl Acad Sci U S A.* 101:6074-9.

Larson, K., J. Sahm, R. Shenkar, and B. Strauss. 1985. Methylation-induced blocks to in vitro DNA replication. *Mutat Res.* 150:77-84.

Lau, A.Y., O.D. Scharer, L. Samson, G.L. Verdine, and T. Ellenberger. 1998. Crystal structure of a human alkylbase-DNA repair enzyme complexed to DNA: mechanisms for nucleotide flipping and base excision. *Cell.* 95:249-58.

Lau, A.Y., M.D. Wyatt, B.J. Glassner, L.D. Samson, and T. Ellenberger. 2000. Molecular basis for discriminating between normal and damaged bases by the human alkyladenine glycosylase, AAG. *Proc Natl Acad Sci U S A.* 97:13573-8.

Likhite, V.S., E.I. Cass, S.D. Anderson, J.R. Yates, and A.M. Nardulli. 2004. Interaction of estrogen receptor alpha with 3-methyladenine DNA glycosylase modulates transcription and DNA repair. *J Biol Chem.* 279:16875-82.

Lindahl, T. 1993. Instability and decay of the primary structure of DNA. *Nature.* 362:709-15.

Liu, Y., H.I. Kao, and R.A. Bambara. 2004. Flap endonuclease 1: a central component of DNA metabolism. *Annu Rev Biochem.* 73:589-615.

Lu, R., H.M. Nash, and G.L. Verdine. 1997a. A mammalian DNA repair enzyme that excises oxidatively damaged guanines maps to a locus frequently lost in lung cancer. *Current Biology.* 7:397-407.

Lu, R., H.M. Nash, and G.L. Verdine. 1997b. A mammalian DNA repair enzyme that excises oxidatively damaged guanines maps to a locus frequently lost in lung cancer. *Curr Biol.* 7:397-407.

Lu, R., H.M. Nash, and G.L. Verdine. 1997c. A mammalian DNA repair enzyme that excises oxidatively damaged guanines maps to a locus frequently lost in lung cancer. *Curr Biol.* 7:397-407.

Lucas, L.T., D. Gatehouse, and D.E. Shuker. 1999. Efficient nitroso group transfer from N-nitrosoindoles to nucleotides and 2'-deoxyguanosine at physiological pH. A new pathway for N-nitrosocompounds to exert genotoxicity. *J Biol Chem.* 274:18319-26.

Ludwig, D.L., M.A. MacInnes, Y. Takiguchi, P.E. Purtymun, M. Henrie, M. Flannery, J. Meneses, R.A. Pedersen, and D.J. Chen. 1998. A murine AP-endonuclease gene-targeted deficiency with post-implantation embryonic progression and ionizing radiation sensitivity. *Mutat Res.* 409:17-29.

Luna, L., M. Bjoras, E. Hoff, T. Rognes, and E. Seeberg. 2000. Cell-cycle regulation, intracellular sorting and induced overexpression of the human NTH1 DNA glycosylase involved in removal of formamidopyrimidine residues from DNA. *Mutat Res.* 460:95-104.

Maga, G., and U. Hubscher. 2003. Proliferating cell nuclear antigen (PCNA): a dancer with many partners. *J Cell Sci.* 116:3051-60.

Manuel, R.C., E.W. Czerwinski, and R.S. Lloyd. 1996. Identification of the structural and functional domains of MutY, an Escherichia coli DNA mismatch repair enzyme. *J Biol Chem.* 271:16218-26.

Marenstein, D.R., M.K. Chan, A. Altamirano, A.K. Basu, R.J. Boorstein, R.P. Cunningham, and G.W. Teebor. 2003. Substrate specificity of human endonuclease III (hNTH1). Effect of human APE1 on hNTH1 activity. *J Biol Chem.* 278:9005-12.

Marenstein, D.R., M.T. Ocampo, M.K. Chan, A. Altamirano, A.K. Basu, R.J. Boorstein, R.P. Cunningham, and G.W. Teebor. 2001a. Stimulation of human endonuclease III by Y box-binding protein 1 (DNA-binding protein B). Interaction between a base excision repair enzyme and a transcription factor. *J Biol Chem.* 276:21242-9.

Marenstein, D.R., M.T. Ocampo, M.K. Chan, A. Altamirano, A.K. Basu, R.J. Boorstein, R.P. Cunningham, and G.W. Teebor. 2001b. Stimulation of human endonuclease III by Y box-binding protein 1 (DNA-binding protein B). Interaction between a base excision repair enzyme and a transcription factor. *J Biol Chem.* 276:21242-9.

Marenstein, D.R., D.M. Wilson, 3rd, and G.W. Teebor. 2004. Human AP endonuclease (APE1) demonstrates endonucleolytic activity against AP sites in single-stranded DNA. *DNA Repair (Amst).* 3:527-33.

Marnett, L.J. 2000. Oxyradicals and DNA damage. *Carcinogenesis.* 21:361-70.

Matsumoto, Y. 2001. Molecular mechanism of PCNA-dependent base excision repair. *Prog Nucleic Acid Res Mol Biol.* 68:129-38.

Meira, L.B., S. Devaraj, G.E. Kisby, D.K. Burns, R.L. Daniel, R.E. Hammer, S. Grundy, I. Jialal, and E.C. Friedberg. 2001. Heterozygosity for the mouse Apex gene results in phenotypes associated with oxidative stress. *Cancer Res.* 61:5552-7.

Miao, F., M. Bouziane, R. Dammann, C. Masutani, F. Hanaoka, G. Pfeifer, and T.R. O'Connor. 2000. 3-Methyladenine-DNA glycosylase (MPG protein) interacts with human RAD23 proteins. *J Biol Chem.* 275:28433-8.

Millar, C.B., J. Guy, O.J. Sansom, J. Selfridge, E. MacDougall, B. Hendrich, P.D. Keightley, S.M. Bishop, A.R. Clarke, and A. Bird. 2002. Enhanced CpG mutability and tumorigenesis in MBD4-deficient mice. *Science.* 297:403-5.

Minowa, O., T. Arai, M. Hirano, Y. Monden, S. Nakai, M. Fukuda, M. Itoh, H. Takano, Y. Hippou, H. Aburatani, K. Masumura, T. Nohmi, S. Nishimura, and T. Noda. 2000. Mmh/Ogg1 gene inactivation results in accumulation of 8-hydroxyguanine in mice. *Proc Natl Acad Sci U S A.* 97:4156-61.

Mitra, S., I. Boldogh, T. Izumi, and T.K. Hazra. 2001. Complexities of the DNA base excision repair pathway for repair of oxidative DNA damage. *Environ Mol Mutagen.* 38:180-90.

Mol, C.D., A.S. Arvai, G. Slupphaug, B. Kavli, I. Alseth, H.E. Krokan, and J.A. Tainer. 1995. Crystal structure and mutational analysis of human uracil-DNA glycosylase: structural basis for specificity and catalysis. *Cell.* 80:869-78.

Mol, C.D., T. Izumi, S. Mitra, and J.A. Tainer. 2000. DNA-bound structures and mutants reveal abasic DNA binding by APE1 and DNA repair coordination [corrected]. *Nature.* 403:451-6.

Morland, I., V. Rolseth, L. Luna, T. Rognes, M. Bjoras, and E. Seeberg. 2002. Human DNA glycosylases of the bacterial Fpg/MutM superfamily: an alternative pathway for the repair of 8-oxoguanine and other oxidation products in DNA. *Nucleic Acids Res.* 30:4926-36.

Nagelhus, T.A., T. Haug, K.K. Singh, K.F. Keshav, F. Skorpen, M. Otterlei, S. Bharati, T. Lindmo, S. Benichou, R. Benarous, and H.E. Krokan. 1997. A sequence in the N-terminal region of human uracil-DNA glycosylase with homology to XPA interacts with the C-terminal part of the 34-kDa subunit of replication protein A. *J Biol Chem*. 272:6561-6.

Nash, H.M., S.D. Bruner, O.D. Scharer, T. Kawate, T.A. Addona, E. Spooner, W.S. Lane, and G.L. Verdine. 1996. Cloning of a yeast 8-oxoguanine DNA glycosylase reveals the existence of a base-excision DNA-repair protein superfamily. *Curr Biol*. 6:968-80.

Naylor, S.L., B.E. Johnson, J.D. Minna, and A.Y. Sakaguchi. 1987. Loss of heterozygosity of chromosome 3p markers in small-cell lung cancer. *Nature*. 329:451-4.

Nilsen, H., K.A. Haushalter, P. Robins, D.E. Barnes, G.L. Verdine, and T. Lindahl. 2001. Excision of deaminated cytosine from the vertebrate genome: role of the SMUG1 uracil-DNA glycosylase. *EMBO J*. 20:4278-86.

Nilsen, H., M. Otterlei, T. Haug, K. Solum, T.A. Nagelhus, F. Skorpen, and H.E. Krokan. 1997. Nuclear and mitochondrial uracil-DNA glycosylases are generated by alternative splicing and transcription from different positions in the UNG gene. *Nucleic Acids Res*. 25:750-5.

Nilsen, H., I. Rosewell, P. Robins, C.F. Skjelbred, S. Andersen, G. Slupphaug, G. Daly, H.E. Krokan, T. Lindahl, and D.E. Barnes. 2000. Uracil-DNA glycosylase (UNG)-deficient mice reveal a primary role of the enzyme during DNA replication. *Mol Cell*. 5:1059-65.

Nilsen, H., G. Stamp, S. Andersen, G. Hrivnak, H.E. Krokan, T. Lindahl, and D.E. Barnes. 2003. Gene-targeted mice lacking the Ung uracil-DNA glycosylase develop B-cell lymphomas. *Oncogene*. 22:5381-6.

Nishioka, K., T. Ohtsubo, H. Oda, T. Fujiwara, D. Kang, K. Sugimachi, and Y. Nakabeppu. 1999. Expression and Differential Intracellular Localization of Two Major Forms of Human 8-Oxoguanine DNA Glycosylase Encoded by Alternatively Spliced OGG1 mRNAs. *Mol. Biol. Cell*. 10:1637-1652.

Noll, D.M., A. Gogos, J.A. Granek, and N.D. Clarke. 1999. The C-terminal domain of the adenine-DNA glycosylase MutY confers specificity for 8-oxoguanine.adenine mispairs and may have evolved from MutT, an 8-oxo-dGTPase. *Biochemistry*. 38:6374-9.

Norman, D.P., S.J. Chung, and G.L. Verdine. 2003. Structural and biochemical exploration of a critical amino acid in human 8-oxoguanine glycosylase. *Biochemistry*. 42:1564-72.

O'Brien, P.J., and T. Ellenberger. 2004. Dissecting the broad substrate specificity of human 3-methyladenine-DNA glycosylase. *J Biol Chem*. 279:9750-7.

Ocampo, M.T., W. Chaung, D.R. Marenstein, M.K. Chan, A. Altamirano, A.K. Basu, R.J. Boorstein, R.P. Cunningham, and G.W. Teebor. 2002. Targeted deletion of mNth1 reveals a novel DNA repair enzyme activity. *Mol Cell Biol*. 22:6111-21.

O'Connor, T.R., and F. Laval. 1990. Isolation and structure of a cDNA expressing a mammalian 3-methyladenine-DNA glycosylase. *EMBO*. 9:3337-3342.

Ohshima, H., and H. Bartsch. 1994. Chronic infections and inflammatory processes as cancer risk factors: possible role of nitric oxide in carcinogenesis. *Mutat Res*. 305:253-64.

Ohtsubo, T., K. Nishioka, Y. Imaiso, S. Iwai, H. Shimokawa, H. Oda, T. Fujiwara, and Y. Nakabeppu. 2000. Identification of human MutY homolog (hMYH) as a repair enzyme for 2-hydroxyadenine in DNA and detection of multiple forms of hMYH located in nuclei and mitochondria. *Nucleic Acids Res*. 28:1355-64.

Otterlei, M., E. Warbrick, T.A. Nagelhus, T. Haug, G. Slupphaug, M. Akbari, P.A. Aas, K. Steinsbekk, O. Bakke, and H.E. Krokan. 1999. Post-replicative base excision repair in replication foci. *EMBO J*. 18:3834-44.

Oyama, M., M. Wakasugi, T. Hama, H. Hashidume, Y. Iwakami, R. Imai, S. Hoshino, H. Morioka, Y. Ishigaki, O. Nikaido, and T. Matsunaga. 2004. Human NTH1 physically interacts with p53 and proliferating cell nuclear antigen. *Biochem Biophys Res Commun*. 321:183-91.

Pandya, G.A., and M. Moriya. 1996. 1,N6-ethenodeoxyadenosine, a DNA adduct highly mutagenic in mammalian cells. *Biochemistry*. 35:11487-92.

Park, K.K., A. Liem, B.C. Stewart, and J.A. Miller. 1993. Vinyl carbamate epoxide, a major strong electrophilic, mutagenic and carcinogenic metabolite of vinyl carbamate and ethyl carbamate (urethane). *Carcinogenesis*. 14:441-50.

Parker, A., Y. Gu, W. Mahoney, S.H. Lee, K.K. Singh, and A.L. Lu. 2001. Human homolog of the MutY repair protein (hMYH) physically interacts with proteins involved in long patch DNA base excision repair. *J Biol Chem*. 276:5547-55. Epub 2000 Nov 22.

Parsons, J.L., and R.H. Elder. 2003. DNA N-glycosylase deficient mice: a tale of redundancy. *Mutat Res*. 531:165-75.

Pascucci, B., G. Maga, U. Hubscher, M. Bjoras, E. Seeberg, I.D. Hickson, G. Villani, C. Giordano, L. Cellai, and E. Dogliotti. 2002. Reconstitution of the base excision repair pathway for 7,8-dihydro-8-oxoguanine with purified human proteins. *Nucleic Acids Res*. 30:2124-30.

Paz-Elizur, T., M. Krupsky, S. Blumenstein, D. Elinger, E. Schechtman, and Z. Livneh. 2003. DNA repair activity for oxidative damage and risk of lung cancer. *J Natl Cancer Inst*. 95:1312-9.

Petersen-Mahrt, S.K., R.S. Harris, and M.S. Neuberger. 2002. AID mutates E. coli suggesting a DNA deamination mechanism for antibody diversification. *Nature*. 418:99-103.

Podlutsky, A.J., Dianova, II, V.N. Podust, V.A. Bohr, and G.L. Dianov. 2001. Human DNA polymerase beta initiates DNA synthesis during long-patch repair of reduced AP sites in DNA. *EMBO J*. 20:1477-82.

Prasad, R., G.L. Dianov, V.A. Bohr, and S.H. Wilson. 2000. FEN1 stimulation of DNA polymerase beta mediates an excision step in mammalian long patch base excision repair. *J Biol Chem*. 275:4460-6.

Prasad, R., R.K. Singhal, D.K. Srivastava, J.T. Molina, A.E. Tomkinson, and S.H. Wilson. 1996. Specific interaction of DNA polymerase beta and DNA ligase I in a multiprotein base excision repair complex from bovine testis. *J Biol Chem*. 271:16000-7.

Rada, C., J.M. Di Noia, and M.S. Neuberger. 2004. Mismatch recognition and uracil excision provide complementary paths to both Ig switching and the a/t-focused phase of somatic mutation. *Mol Cell*. 16:163-71.

Rada, C., G.T. Williams, H. Nilsen, D.E. Barnes, T. Lindahl, and M.S. Neuberger. 2002. Immunoglobulin isotype switching is inhibited and somatic hypermutation perturbed in UNG-deficient mice. *Curr Biol*. 12:1748-55.

Radicella, J.P., C. Dherin, C. Desmaze, M.S. Fox, and S. Boiteux. 1997. Cloning and characterization of hOGG1, a human homolog of the OGG1 gene of Saccharomyces cerevisiae. *PNAS*. 94:8010-8015.

Ranalli, T.A., S. Tom, and R.A. Bambara. 2002. AP endonuclease 1 coordinates flap endonuclease 1 and DNA ligase I activity in long patch base excision repair. *J Biol Chem*. 277:41715-24.

Rasouli-Nia, A., F. Karimi-Busheri, and M. Weinfeld. 2004. Stable down-regulation of human polynucleotide kinase enhances spontaneous mutation frequency and sensitizes cells to genotoxic agents. *Proc Natl Acad Sci U S A*. 101:6905-10.

Rebeck, G.W., and L. Samson. 1991. Increased spontaneous mutation and alkylation sensitivity of Escherichia coli strains lacking the ogt O6-methylguanine DNA repair methyltransferase. *J Bacteriol*. 173:2068-76.

Rinne, M., D. Caldwell, and M.R. Kelley. 2004. Transient adenoviral N-methylpurine DNA glycosylase overexpression imparts chemotherapeutic sensitivity to human breast cancer cells. *Mol Cancer Ther*. 3:955-67.

Roldan-Arjona, T., Y.F. Wei, K.C. Carter, A. Klungland, C. Anselmino, R.P. Wang, M. Augustus, and T. Lindahl. 1997. Molecular cloning and functional expression of a human cDNA encoding the antimutator enzyme 8-hydroxyguanine-DNA glycosylase. *Proc Natl Acad Sci U S A*. 94:8016-20.

Rosenquist, T.A., E. Zaika, A.S. Fernandes, D.O. Zharkov, H. Miller, and A.P. Grollman. 2003. The novel DNA glycosylase, NEIL1, protects mammalian cells from radiation-mediated cell death. *DNA Repair (Amst)*. 2:581-91.

Roth, R.B., and L.D. Samson. 2002. 3-Methyladenine DNA glycosylase-deficient Aag null mice display unexpected bone marrow alkylation resistance. *Cancer Res*. 62:656-60.

Rouet, P., and J.M. Essigmann. 1985. Possible role for thymine glycol in the selective inhibition of DNA synthesis on oxidized DNA templates. *Cancer Res*. 45:6113-8.

Russo, M.T., G. De Luca, P. Degan, E. Parlanti, E. Dogliotti, D.E. Barnes, T. Lindahl, H. Yang, J.H. Miller, and M. Bignami. 2004. Accumulation of the oxidative base lesion 8-hydroxyguanine in DNA of tumor-prone mice defective in both the Myh and Ogg1 DNA glycosylases. *Cancer Res*. 64:4411-4.

Sampson, J.R., S. Dolwani, S. Jones, D. Eccles, A. Ellis, D.G. Evans, I. Frayling, S. Jordan, E.R. Maher, T. Mak, J. Maynard, F. Pigatto, J. Shaw, and J.P. Cheadle. 2003. Autosomal recessive colorectal adenomatous polyposis due to inherited mutations of MYH. *Lancet*. 362:39-41.

Sansom, O.J., S.M. Bishop, A. Bird, and A.R. Clarke. 2004. MBD4 deficiency does not increase mutation or accelerate tumorigenesis in mice lacking MMR. *Oncogene*. 23:5693-6.

Saparbaev, M., S. Langouet, C.V. Privezentzev, F.P. Guengerich, H. Cai, R.H. Elder, and J. Laval. 2002. 1,N(2)-ethenoguanine, a mutagenic DNA adduct, is a primary substrate of Escherichia coli mismatch-specific uracil-DNA glycosylase and human alkylpurine-DNA-N-glycosylase. *J Biol Chem*. 277:26987-93.

Saparbaev, M., and J. Laval. 1994. Excision of hypoxanthine from DNA containing dIMP residues by the *Eschericia coli*, yeast, rat, and human alkylpurine DNA glycosylsaes. *Proc. Natl. Acad. Sci. USA*. 91:5873-5877.

Savva, R., K. McAuley-Hecht, T. Brown, and L. Pearl. 1995. The structural basis of specific base-excision repair by uracil-DNA glycosylase. *Nature*. 373:487-93.

Scanlon, K.J., M. Kashani-Sabet, and H. Miyachi. 1989. Differential gene expression in human cancer cells resistant to cisplatin. *Cancer Invest*. 7:581-7.

Scharer, O.D., and J. Jiricny. 2001. Recent progress in the biology, chemistry and structural biology of DNA glycosylases. *Bioessays*. 23:270-81.

Schouten, K.A., and B. Weiss. 1999. Endonuclease V protects Escherichia coli against specific mutations caused by nitrous acid. *Mutat Res*. 435:245-54.

Shah, A.M., S.X. Li, K.S. Anderson, and J.B. Sweasy. 2001. Y265H mutator mutant of DNA polymerase beta. Proper teometric alignment is critical for fidelity. *J Biol Chem*. 276:10824-31.

Shah, A.M., M. Maitra, and J.B. Sweasy. 2003. Variants of DNA polymerase Beta extend mispaired DNA due to increased affinity for nucleotide substrate. *Biochemistry*. 42:10709-17.

Shen, M.R., I.M. Jones, and H. Mohrenweiser. 1998. Nonconservative amino acid substitution variants exist at polymorphic frequency in DNA repair genes in healthy humans. *Cancer Res*. 58:604-8.

Shibata, Y., and T. Nakamura. 2002. Defective flap endonuclease 1 activity in mammalian cells is associated with impaired DNA repair and prolonged S phase delay. *J Biol Chem*. 277:746-54.

Shimizu, Y., S. Iwai, F. Hanaoka, and K. Sugasawa. 2003. Xeroderma pigmentosum group C protein interacts physically and functionally with thymine DNA glycosylase. *EMBO J*. 22:164-73.

Singhal, R.K., and S.H. Wilson. 1993. Short gap-filling synthesis by DNA polymerase beta is processive. *J Biol Chem*. 268:15906-11.

Sleeth, K.M., R.L. Robson, and G.L. Dianov. 2004. Exchangeability of Mammalian DNA Ligases between Base Excision Repair Pathways. *Biochemistry*. 43:12924-30.

Slupphaug, G., I. Eftedal, B. Kavli, S. Bharati, N.M. Helle, T. Haug, D.W. Levine, and H.E. Krokan. 1995. Properties of a recombinant human uracil-DNA glycosylase from the UNG gene and evidence that UNG encodes the major uracil-DNA glycosylase. *Biochemistry*. 34:128-38.

Slupphaug, G., B. Kavli, and H.E. Krokan. 2003. The interacting pathways for prevention and repair of oxidative DNA damage. *Mutat Res*. 531:231-51.

Slupphaug, G., C.D. Mol, B. Kavli, A.S. Arvai, H.E. Krokan, and J.A. Tainer. 1996. A nucleotide-flipping mechanism from the structure of human uracil-DNA glycosylase bound to DNA. *Nature*. 384:87-92.

Slupska, M.M., C. Baikalov, W.M. Luther, J.H. Chiang, Y.F. Wei, and J.H. Miller. 1996. Cloning and sequencing a human homolog (hMYH) of the Escherichia coli mutY gene whose function is required for the repair of oxidative DNA damage. *J Bacteriol.* 178:3885-92.

Slupska, M.M., W.M. Luther, J.H. Chiang, H. Yang, and J.H. Miller. 1999. Functional expression of hMYH, a human homolog of the Escherichia coli MutY protein. *J Bacteriol.* 181:6210-3.

Sobol, R.W., J.K. Horton, R. Kuhn, H. Gu, R.K. Singhal, R. Prasad, K. Rajewsky, and S.H. Wilson. 1996. Requirement of mammalian DNA polymerase-beta in base-excision repair. *Nature.* 379:183-6.

Sobol, R.W., M. Kartalou, K.H. Almeida, D.F. Joyce, B.P. Engelward, J.K. Horton, R. Prasad, L.D. Samson, and S.H. Wilson. 2003. Base excision repair intermediates induce p53-independent cytotoxic and genotoxic responses. *J Biol Chem.* 278:39951-9. Epub 2003 Jul 25.

Srivastava, D.K., I. Husain, C.L. Arteaga, and S.H. Wilson. 1999. DNA polymerase beta expression differences in selected human tumors and cell lines. *Carcinogenesis.* 20:1049-54.

Starcevic, D., S. Dalal, and J.B. Sweasy. 2004. Is There a Link Between DNA Polymerase beta and Cancer? *Cell Cycle.* 3.

Strauss, P.R., W.A. Beard, T.A. Patterson, and S.H. Wilson. 1997. Substrate binding by human apurinic/apyrimidinic endonuclease indicates a Briggs-Haldane mechanism. *J Biol Chem.* 272:1302-7.

Stucki, M., B. Pascucci, E. Parlanti, P. Fortini, S.H. Wilson, U. Hubscher, and E. Dogliotti. 1998. Mammalian base excision repair by DNA polymerases delta and epsilon. *Oncogene.* 17:835-43.

Sturgis, E.M., E.J. Castillo, L. Li, R. Zheng, S.A. Eicher, G.L. Clayman, S.S. Strom, M.R. Spitz, and Q. Wei. 1999. Polymorphisms of DNA repair gene XRCC1 in squamous cell carcinoma of the head and neck. *Carcinogenesis.* 20:2125-9.

Suh, D., D.M. Wilson, 3rd, and L.F. Povirk. 1997. 3'-phosphodiesterase activity of human apurinic/apyrimidinic endonuclease at DNA double-strand break ends. *Nucleic Acids Res.* 25:2495-500.

Takao, M., H. Aburatani, K. Kobayashi, and A. Yasui. 1998. Mitochondrial targeting of human DNA glycosylases for repair of oxidative DNA damage. *Nucleic Acids Res.* 26:2917-22.

Takao, M., S. Kanno, K. Kobayashi, Q.M. Zhang, S. Yonei, G.T. van der Horst, and A. Yasui. 2002a. A back-up glycosylase in Nth1 knock-out mice is a functional Nei (endonuclease VIII) homologue. *J Biol Chem.* 277:42205-13.

Takao, M., S. Kanno, T. Shiromoto, R. Hasegawa, H. Ide, S. Ikeda, A.H. Sarker, S. Seki, J.Z. Xing, X.C. Le, M. Weinfeld, K. Kobayashi, J. Miyazaki, M. Muijtjens, J.H. Hoeijmakers, G. van der Horst, and A. Yasui. 2002b. Novel nuclear and mitochondrial glycosylases revealed by disruption of the mouse Nth1 gene encoding an endonuclease III homolog for repair of thymine glycols. *EMBO J.* 21:3486-93.

Tebbs, R.S., M.L. Flannery, J.J. Meneses, A. Hartmann, J.D. Tucker, L.H. Thompson, J.E. Cleaver, and R.A. Pedersen. 1999. Requirement for the Xrcc1 DNA base excision repair gene during early mouse development. *Dev Biol.* 208:513-29.

Thayer, M.M., H. Ahern, D. Xing, R.P. Cunningham, and J.A. Tainer. 1995. Novel DNA binding motifs in the DNA repair enzyme endonuclease III crystal structure. *EMBO J.* 14:4108-20.

Thompson, L.H., and M.G. West. 2000. XRCC1 keeps DNA from getting stranded. *Mutat Res.* 459:1-18.

Tini, M., A. Benecke, S.J. Um, J. Torchia, R.M. Evans, and P. Chambon. 2002. Association of CBP/p300 acetylase and thymine DNA glycosylase links DNA repair and transcription. *Mol Cell.* 9:265-77.

Tom, S., T.A. Ranalli, V.N. Podust, and R.A. Bambara. 2001. Regulatory roles of p21 and apurinic/apyrimidinic endonuclease 1 in base excision repair. *J Biol Chem.* 276:48781-9.

Vidal, A.E., I.D. Hickson, S. Boiteux, and J.P. Radicella. 2001. Mechanism of stimulation of the DNA glycosylase activity of hOGG1 by the major human AP endonuclease: bypass of the AP lyase activity step. *Nucleic Acids Res*. 29:1285-92.

Volk, D.E., P.G. House, V. Thiviyanathan, B.A. Luxon, S. Zhang, R.S. Lloyd, and D.G. Gorenstein. 2000. Structural similarities between MutT and the C-terminal domain of MutY. *Biochemistry*. 39:7331-6.

Waga, S., and B. Stillman. 1998. The DNA replication fork in eukaryotic cells. *Annu Rev Biochem*. 67:721-51.

Waters, T.R., P. Gallinari, J. Jiricny, and P.F. Swann. 1999. Human thymine DNA glycosylase binds to apurinic sites in DNA but is displaced by human apurinic endonuclease 1. *J Biol Chem*. 274:67-74.

Whitehouse, C.J., R.M. Taylor, A. Thistlethwaite, H. Zhang, F. Karimi-Busheri, D.D. Lasko, M. Weinfeld, and K.W. Caldecott. 2001. XRCC1 stimulates human polynucleotide kinase activity at damaged DNA termini and accelerates DNA single-strand break repair. *Cell*. 104:107-17.

Wiederhold, L., J.B. Leppard, P. Kedar, F. Karimi-Busheri, A. Rasouli-Nia, M. Weinfeld, A.E. Tomkinson, T. Izumi, R. Prasad, S.H. Wilson, S. Mitra, and T.K. Hazra. 2004. AP endonuclease-independent DNA base excision repair in human cells. *Mol Cell*. 15:209-20.

Williams, S.D., and S.S. David. 1998. Evidence that MutY is a monofunctional glycosylase capable of forming a covalent Schiff base intermediate with substrate DNA. *Nucleic Acids Res*. 26:5123-33.

Williams, S.D., and S.S. David. 1999. Formation of a Schiff base intermediate is not required for the adenine glycosylase activity of Escherichia coli MutY. *Biochemistry*. 38:15417-24.

Wilson, D.M., 3rd. 2003. Properties of and substrate determinants for the exonuclease activity of human apurinic endonuclease Ape1. *J Mol Biol*. 330:1027-37.

Wilson, S.H., and T.A. Kunkel. 2000. Passing the baton in base excision repair. *Nat Struct Biol*. 7:176-8.

Winters, T.A., W.D. Henner, P.S. Russell, A. McCullough, and T.J. Jorgensen. 1994. Removal of 3'-phosphoglycolate from DNA strand-break damage in an oligonucleotide substrate by recombinant human apurinic/apyrimidinic endonuclease 1. *Nucleic Acids Res*. 22:1866-73.

Wong, D., M.S. DeMott, and B. Demple. 2003. Modulation of the 3'-->5'-exonuclease activity of human apurinic endonuclease (Ape1) by its 5'-incised Abasic DNA product. *J Biol Chem*. 278:36242-9.

Wood, M.L., M. Dizdaroglu, E. Gajewski, and J.M. Essigmann. 1990. Mechanistic studies of ionizing radiation and oxidative mutagenesis: genetic effects of a single 8-hydroxyguanine (7-hydro-8-oxoguanine) residue inserted at a unique site in a viral genome. *Biochemistry*. 29:7024-32.

Wood, R.D., M. Mitchell, J. Sgouros, and T. Lindahl. 2001. Human DNA repair genes. *Science*. 291:1284-9.

Wuenschell, G.E., T.R. O'Connor, and J. Termini. 2003. Stability, miscoding potential, and repair of 2'-deoxyxanthosine in DNA: implications for nitric oxide-induced mutagenesis. *Biochemistry*. 42:3608-16.

Xanthoudakis, S., G. Miao, F. Wang, Y.C. Pan, and T. Curran. 1992. Redox activation of Fos-Jun DNA binding activity is mediated by a DNA repair enzyme. *EMBO J*. 11:3323-35.

Xanthoudakis, S., R.J. Smeyne, J.D. Wallace, and T. Curran. 1996. The redox/DNA repair protein, Ref-1, is essential for early embryonic development in mice. *Proc Natl Acad Sci U S A*. 93:8919-23.

Xie, Y., H. Yang, C. Cunanan, K. Okamoto, D. Shibata, J. Pan, D.E. Barnes, T. Lindahl, M. McIlhatton, R. Fishel, and J.H. Miller. 2004. Deficiencies in mouse Myh and Ogg1 result in tumor predisposition and G to T mutations in codon 12 of the K-ras oncogene in lung tumors. *Cancer Res*. 64:3096-102.

Xu, Y.J., M.S. DeMott, J.T. Hwang, M.M. Greenberg, and B. Demple. 2003. Action of human apurinic endonuclease (Ape1) on C1'-oxidized deoxyribose damage in DNA. *DNA Repair (Amst)*. 2:175-85.

Yang, Y., J. Nair, A. Barbin, and H. Bartsch. 2000. Immunohistochemical detection of 1,N(6)-ethenodeoxyadenosine, a promutagenic DNA adduct, in liver of rats exposed to vinyl chloride or an iron overload. *Carcinogenesis*. 21:777-81.

Yokota, J., M. Wada, Y. Shimosato, M. Terada, and T. Sugimura. 1987. Loss of heterozygosity on chromosomes 3, 13, and 17 in small-cell carcinoma and on chromosome 3 in adenocarcinoma of the lung. *Proc Natl Acad Sci U S A*. 84:9252-6.

Zharkov, D.O., T.A. Rosenquist, S.E. Gerchman, and A.P. Grollman. 2000. Substrate specificity and reaction mechanism of murine 8-oxoguanine-DNA glycosylase. *J Biol Chem*. 275:28607-17.

Zharkov, D.O., G. Shoham, and A.P. Grollman. 2003. Structural characterization of the Fpg family of DNA glycosylases. *DNA Repair (Amst)*. 2:839-62.

Zhou, J., J. Ahn, S.H. Wilson, and C. Prives. 2001. A role for p53 in base excision repair. *EMBO J*. 20:914-23.

Chapter 2.4

GENOMIC INSTABILITY IN CANCER DEVELOPMENT

Penny A. Jeggo
Genome Damage and Stability Centre, University of Sussex, Brighton, UK

1. INTRODUCTION

A DNA double strand break (DSB) can lead to several outcomes. The break may be repaired accurately or small sequence changes at the break site may be tolerated. If unrepaired, a DSB will very likely lead to loss of genomic material with the likely consequence of cell death. However, more importantly, mis-rejoining of a DSB can cause small deletions or genomic rearrangements that can potentially result in the activation of tumour promoters or the inactivation of tumour suppressors. The importance of a DSB is underscored by the fact that cells have evolved several pathways for their repair and that cell lines, mice or human patients deficient in DSB damage response pathways have marked genomic instability frequently associated with cancer predisposition. The aim of this chapter is to discuss the role of the DNA DSB repair pathways as genetic caretakers in the maintenance of genomic stability and cancer avoidance. I will firstly consider the processes and exposures that can cause DSBs. I will then discuss the characterised DSB repair pathways and their contribution to limiting genomic instability and cancer development. Where possible, I will discuss the findings in the context of cancer development in humans.

2. THE ORIGIN AND NATURE OF DSBS

DSBs can be produced following exposure to exogenous DNA damaging agents and represent the most biologically significant lesion induced by ionising radiation. DSBs can also arise from endogenously generated reactive oxygen species and during certain cellular processes. Replication probably represents the most important source of endogenous DSB

E.A. Nigg (ed.), Genome Instability in Cancer Development, 175-197.

generation. Double strand ends can arise at replication forks following attempts to replicate past lesions such as a single strand break (SSB) or various types of DNA adducts. Attempted replication past a SSB can generate a one sided DSB which differs topologically from DSBs introduced by IR since the structure lacks a second end for rejoining (Fig. 1). There is also evidence that stalled replication forks can reverse to generate a chicken foot structure with a double stranded end (Fig. 1). Again, such a structure is topologically distinct from a radiation induced DSB in that it encompasses a single double strand end rather than two double strand ends (see (Thompson and Limoli, 2004) for more detailed overview of DSB generation at a replication forks). Meiotic recombination and V(D)J recombination represent two additional developmental processes during which DSBs are generated (Alt et al., 1992). In both cases, DSBs are introduced by specific nucleases, Spo-11 for meiotic recombination and RAG1/2 for V(D)J recombination, with the aim of generating genetic diversity (Richardson et al., 2004).

Three aspects need to be considered in assessing the impact of DSBs on genomic instability; the topology of the break, which has been considered above in the context of replication associated DSBs, the nature of the DSB termini and the cell cycle stage in which the break is generated. IR introduces complex DSBs, which frequently have associated base damage or damaged sugars (Nikjoo et al., 2001). The complexity of the damage depends upon the linear energy transfer (LET) of the radiation. Around 20-30% of the DSBs generated by low energy electrons are associated with additional breaks (for example, SSBs) which increases to 70% for high energy alpha particle irradiation. Around 90% of such breaks have associated base damage (Nikjoo et al., 2001). Such complex DSBs pose a challenge to the repair machinery and very likely require the co-ordination of different repair pathways. Endogenously produced reactive oxygen species (ROS) predominantly introduce SSBs but DSBs can arise from two overlapping SSBs, during repair processing or following replication. DSBs produced in this way are less complex than IR induced DSBs, but are still likely to require processing prior to rejoining. The double strand ends generated during V(D)J recombination are distinct from IR-induced DSBs but nonetheless require specific processing. During V(D)J recombination the *RAG1* and *2* gene products introduce a single strand nick that undergoes a transesterification reaction to generate a hairpin ended DNA molecule and a blunt double stranded DNA end (Hiom and Gellert, 1997). For successful V(D)J recombination, two hairpin ends are cleaved and subsequently rejoined, frequently with additional processing of the DNA ends to create small nucleotide additions or deletions (Alt et al., 1992). DSBs that arise at stalled replication forks, may or may not be associated with additional damages. However, in addition to the unique topological natures of such breaks, they are further distinct in being closely associated with a sister homologue. The impact of these factors on DNA repair and on genomic instability will be discussed below.

Cell cycle stage possibly plays an important role in influencing the DSB repair pathway availability. This aspect will be discussed below.

3. OVERVIEW OF THE DAMAGE RESPONSE PROCESSES RESPONDING TO DSBS

One line of defence for a cell faced with a DSB is to attempt DSB repair. An additional strategy is to establish a signal transduction pathway that serves to effect cell cycle checkpoint arrest and/or apoptosis. Combining such strategies maximises the opportunity for repair, prevents cell cycle progression prior to the completion of repair and couples failed repair to the prevention of proliferation of the damaged cell either by permanent cell cycle arrest or by the onset of apoptosis. Below, I will briefly overview these processes.

3.1 Non-homologous End-joining

An important DSB repair pathway in mammalian cells is DNA non-homologous end-joining (NHEJ). This process has been adequately covered in reviews and only a brief overview will be presented (Jeggo, 1998; van Gent et al., 2001). Five core components of NHEJ have been identified, namely Ku70, Ku80, the DNA-dependent protein kinase (DNA-PKcs), Xrcc4 and DNA ligase IV. Current evidence suggests that the Ku heterodimer is a basket-shaped molecule with a handle, two pillars and base that together form a loop through which double stranded DNA can pass (Walker et al., 2001). Ku, therefore, encircles the DNA and, moreover, can translocate along the DNA molecule (Blier et al., 1993; de Vries et al., 1989; Smith and Jackson, 1999). The crystal structure demonstrates that Ku needs a DNA end for loading providing an explanation for specificity for binding double strand DNA ends. DNA-bound Ku can then recruit DNA-PKcs forming the DNA-PK complex with the consequent activation of its kinase activity (Dvir et al., 1992; Gottlieb and Jackson, 1993; Smith and Jackson, 1999). DNA-PKcs undergoes autophosphorylation which is essential for NHEJ and likely serves to regulate the process and/or facilitate processing of DNA ends via the recruitment of additional proteins (Kurimasa et al., 1999). The DNA-PK complex also facilitates recruitment of the DNA ligase IV/Xrcc4 complex, which carries out the rejoining step (Calsou et al., 2003). The factors influencing end-processing have still not been fully characterised but likely include polynucleotide kinase (PNK) and Polbeta, which interacts with the DNA ligase IV/Xrcc4 complex (see Jeggo, 2004 for an overview). Artemis, a member of the beta-lactamase family with nuclease activity, which can funtion as a 5' to 3' single strand overhang nuclease, is an additional component of NHEJ that also plays a role in end-processing prior to rejoining (Ma et al., 2002; Riballo et al., 2004).

NHEJ represents the major mechanism for the repair of DSBs in G1 phase. NHEJ also functions to effect rearrangements during V(D)J recombination (Alt et al., 1992). Consequently, cells lacking NHEJ components are exquisitely radiosensitive due to their inability to repair IR-induced DSBs and are also unable to carry out V(D)J recombination (Jeggo, 1998).

3.2 Homologous Recombination

Homologous recombination (HR) uses an undamaged template to rejoin a DSB and/or to restore coding information at a DSB gap. (For reviews see Helleday, 2003; Kanaar et al., 1998; Thompson and Limoli, 2004; Thompson and Schild, 2001). An initial step in HR is the generation of a 3' single-stranded overhang following 5' to 3' resection of the DNA end, a step in which the MRN (Mre11/Rad50/Nbs1) complex has been implicated (Helleday, 2003; Thompson and Schild, 2001). These single strand tails rapidly become coated with RPA, due to its high affinity for single stranded DNA (Wold, 1997). Notwithstanding the high affinity of RPA for single stranded DNA, RAD51 is able to displace RPA on the single strand overhang, a step likely promoted by BRCA2 (Pellegrini et al., 2002; Yu et al., 2003). The loading of RAD51 on DNA may also be facilitated by the RAD51 paralogs, which include RAD51B, RAD51C, RAD51D and XRCC2 (Lio et al., 2003; Masson et al., 2001; Sigurdsson et al., 2001). RAD51 then promotes single strand invasion. RAD54 also appears to play a role in stimulating strand invasion by forming negative supercoils in duplex DNA (Petukhova et al., 1998). Following strand invasion, DNA synthesis elongates the invading strand using an intact strand as template. This causes displacement of the original strand and the formation of a Holliday junction (HJ). Branch migration then ensues. The invading strand may simply be displaced and repaired by single strand annealing (SSA). Alternatively, a double HJ structure may be generated, resolution of which can occur either with or without crossing over. Whilst the proteins required for branch migration have not been fully identified, recent results have shown that RAD51C and XRCC3 are required for HJ resolution and very likely also function in branch migration (Liu et al., 2004).

3.3 Other DSB Rejoining Processes

In addition to these two characterised DSB repair pathways, there is evidence that DSBs can be rejoined by other less well defined mechanisms. Single strand annealing (SSA) represents one alternative pathway. SSA uses short regions of microhomology to facilitate rejoining possibly by two single strand rejoining events. Analysis of cell extracts from cells lacking components of DNA-PK have provided further evidence of a back-up pathway, potentially regulated by DNA-PK (Perrault et al., 2004; Wang et al., 2003). Recent evidence has also shown that DSB ends can be healed by fusion with telomeres, generating telomere-DSB fusion events (Bailey et al., 2004; Latre et al., 2003).

3.4 Damage Response Signalling

As mentioned above, another important response to the presence of DSBs is the establishment of a signal transduction pathway that effects cell cycle checkpoint arrest and/or apoptosis. (For reviews see (Rouse and Jackson, 2002; Shiloh, 2001; Shiloh, 2003; Thompson and Limoli, 2004)). Two phosphatidylinositol-3'kinase related kinases (PIKKs) recognise distinct forms of DNA damage and initiate overlapping signal transduction responses. Ataxia-telangiectasia mutated protein (ATM) is the predominant kinase activated by DSBs. ATM activation results in phosphorylation of H2AX, 53BP1, Chk2, Nbs1, SMC1 and a range of additional proteins involved in the damage response. Via transducer proteins, the signal transduction pathway ultimately effects arrest at one of several cell cycle checkpoints including the G1/S boundary, intra-S phase and the G2/M boundary.

4. THE CONTRIBUTION OF HR TO GENOMIC STABILITY

A major function of HR is to maintain genomic stability during replication. Studies on yeast and bacterial model systems have suggested that replication forks frequently encounter blocks to their progression including lesions such as single strand breaks (for a review see (Cox et al., 2000). There is also evidence that certain chromosomal sites, called fragile sites, are prone to replication fork stalling (Casper et al., 2002). As discussed above and shown in Figure 1, structures such as a one sided DSB or a chicken foot structure generated by fork reversal can arise as a consequence of such replication stalling and the available evidence suggests a major function of HR is to repair or resolve such lesions. Studies have shown that in mammalian cells, HR only utilises a sister homologue, thereby restricting HR function to S/G2 phase (Johnson and Jasin, 2000). This has been consolidated by studies showing the HR does not significantly contribute to IR-induced DSB repair in G1 (Rothkamm et al., 2003).

Mice lacking many of the HR proteins are embryonic lethal demonstrating the importance of HR during development (Deans et al., 2000; Lim and Hasty, 1996; Tsuzuki et al., 1996). More importantly, in the context of this article, cells impaired in HR show very marked genomic instability. This is consistent with the notion that replication fork stalling occurs frequently during each replication cycle and that elevated DNA breakage arising due to a failure to resolve or restore stalled replication leads to elevated spontaneous chromosomal instability.

The importance of HR to the maintenance of genomic stability is most dramatically demonstrated by the fact that *BRCA1* and *BRCA2*, the genes

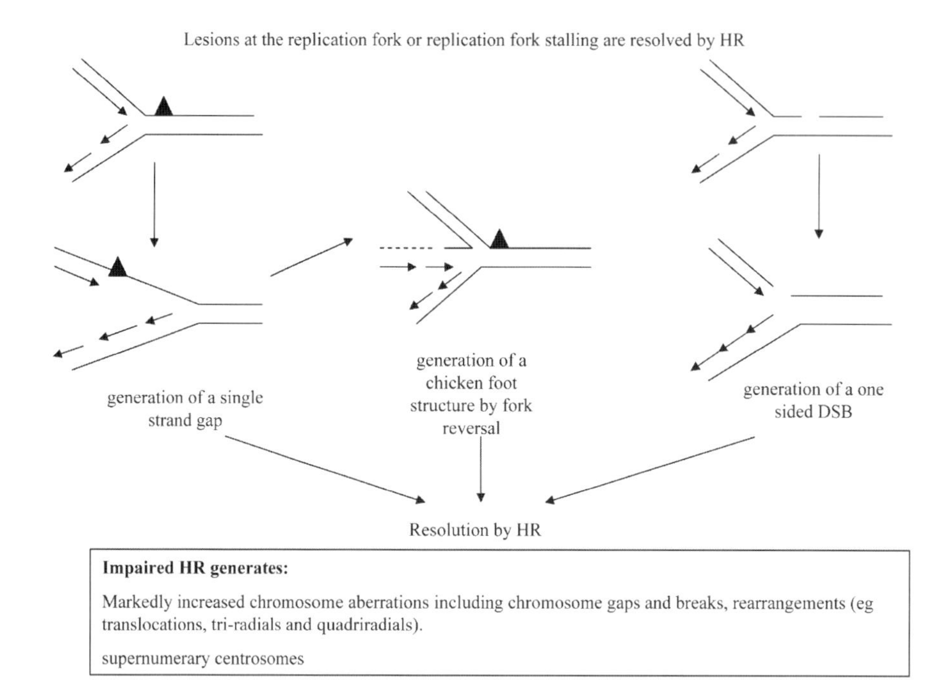

Figure 1. Role of HR in maintaining Genomic Stability.

mutated in human hereditary breast cancer, are required for HR (Moynahan et al., 1999; Moynahan et al., 2001) (see (Shivji and Venkitaraman, 2004; Venkitaraman, 2002) for reviews). Around five percent of all breast cancer incidence can be attributed to mutations in *BRCA1* or *BRCA2*. Strikingly, the lifetime risk of breast cancer amongst heterozygous carriers is greater than 80 % demonstrating the enormous role these genes play as tumour suppressors (King et al., 2003). Predisposition to ovarian and other epithelial cancers is also observed amongst carriers (Struewing et al., 1997; Thompson and Easton, 2002; Thorlacius et al., 1996). Interestingly, specific bi-allelic mutations in *BRCA2* are responsible for FANCB and FANCD1, sub-groups of Fanconi anaemia, which is associated with haematological and developmental abnormalities as well as a marked cancer predisposition (Howlett et al., 2002). The available evidence suggests that the major role of BRCA2 as a tumour suppressor lies in regulating HR via its control of RAD51. A critical and evolutionary conserved feature of BRCA2 is the presence of eight BRC repeat motifs, which bind directly to RAD51. The crystal structure of a RAD51-BRC4 complex has been solved and shows that the BRC repeat mimics a motif in RAD51 required for RAD51 oligomerisation (Pellegrini et al., 2002). Furthermore, expression of the BRC4 region prevents RAD51 monomer-monomer interactions (Davies et al., 2001) and microscopy studies show that damage induced RAD51 nuclear foci fail to form in the absence of BRCA2 and specifically following mutation of the BRC4 interaction domain in RAD51 (Pellegrini et al., 2002). Whilst the precise explanation for these findings is unclear, the evidence points to a role for BRCA2 in facilitating the delivery of RAD51 as nuclear protein filaments to RPA coated single strand regions of DNA. Truncating mutations in BRCA2 result in impaired HR and sensitivity to a range of DNA damaging agents including IR and cross-linking agents (Connor et al., 1997; Patel et al., 1998; Sharan et al., 1997). A further striking phenotype indicative of genetic instability in BRCA2 deficient cells is elevated spontaneous chromosome aberrations, which include chromatid breaks, rearrangement figures and end-to-end chromosome fusions (Hirsch et al., 2004; Patel et al., 1998). Spectral karotyping has provided added insight into such events revealing the presence of gross chromosomal rearrangements and deletions (Yu et al., 2000). A further surprising phenotype, which will be discussed below, is the presence of cells with supernumerary centrosomes (Tutt et al., 1999). Finally, BRCA2 deficiency also confers an elevated spontaneous mutation rate and increased IR-induced mutagenesis (Kraakman-van der Zwet et al., 2003).

Although BRCA1 also appears to regulate HR, both the structure of the protein and its role in HR and the damage response pathway in general is quite distinct to that of BRCA2 (Moynahan et al., 1999; Snouwaert et al., 1999). BRCA1 has an N-terminal RING domain and two BRCT domains in its C-terminus. It interacts with a number of damage response proteins including BARD1, the MRN complex (Mre11/Rad50/Nbs1), RNA polymerase II and CtIP. The role of BRCA1 in HR is much more poorly defined than the role of BRCA2. BRCA1 localises to gamma-H2AX foci, which form at the sites of DNA damage and a range of evidence suggests that it functions in the signalling response that leads to cell cycle checkpoint arrest and/or apoptosis. Hence, BRCA1 deficient cells show impaired S phase and G2/M arrest (Xu et al., 2001). A plausible model is that BRCA1 acts as a docking station for ATM and ATR substrates and facilitates their phosphorylation by the PIKKs after DNA damage. Consistent with this model, the phosphorylation of a range of ATM and ATR-dependent substrates is impaired in BRCA1 deficient cells (Foray et al., 2003; Kim et al., 2002; Yarden et al., 2002). A potential role for BRCA1, therefore, is that it functions to facilitate ATR or ATM phosphorylation events essential for HR. An additional feature of BRCA1 deficiency is observed in mouse embryo fibroblasts with truncations in BRCA1, which show high levels of centrosome amplification leading to unequal chromosome segregation, abnormal nuclear division and aneuploidy (Xu et al., 1999).

Loss of RAD51 results in embryonic lethality making it difficult to assess its contribution to genomic stability. However, studies have been carried out on two RAD51 paralogues, XRCC2 and XRCC3, which are informative for considering the impact of impaired HR on genomic stability. In genetic studies aimed at investigating DNA repair pathways, hamster cell lines were used to isolate mammalian mutants sensitive to DNA damaging agents. After a decade of studying such lines, the defective genes in many have now been identified. Two radiosensitive hamster cell lines, irs1 and irs1SF have mutations in XRCC2 and XRCC3, respectively (Jones et al.,

1987). These two genes are RAD51 paralogues and are required for HR (Johnson et al., 1999; Pierce et al., 1999). The defective lines display marked sensitivity to DNA cross linking agents and milder sensitivity to additional agents including X-rays and UV. The lines also show elevated frequencies of spontaneous and radiation-induced mutations and chromosome aberrations (Thacker et al., 1994; Tucker et al., 1991). Mice knocked out for XRCC2 are embryonic lethal, however XRCC2$^{-/-}$ embryonic cells showed genetic instability with high levels of chromosome aberrations (Deans et al., 2000). XRCC2$^{-/-}$ MEFs also displayed high levels of aneuploidy and complex exchanges evident using spectral karyotyping (Deans et al., 2003). Deficiency of XRCC3 also confers genomic instability with a surprising and novel phenotype of elevated endoreduplication (Yoshihara et al., 2004). This provides the first evidence that impaired HR might confer a defect in the co-ordination of the initiation of replication.

A further striking phenotype of both XRCC2 and XRCC3 deficient cells, similar to that observed in BRCA2 deficient cells, is the presence of elevated numbers of cells with supernumerary centrosomes (Griffin et al., 2000). Multipolar spindle formation is evident in such cells. Interestingly, ATR-Seckel Syndrome cell lines, which are deficient in ATR, the PIKK activated by stalled replication forks, display an elevated frequency of mitotic cells with supernumerary centrosomes (Alderton et al, 2004). These findings could represent a consequence of unresolved DNA damage at the replication fork preventing normal chromosome condensation and segregation or they could be a consequence of prolonged S or G2 phase. Alternatively, it is possible that the proteins involved have dual functions or that HR regulates centrosome duplication. Although the basis underlying this pronounced phenotype remains to be resolved, it demonstrates a further mechanism in which impaired HR can contribute to genomic instability.

5. THE CONTRIBUTION OF NHEJ TO GENOMIC STABILITY

5.1 Fidelity of NHEJ

HR is an exquisitely elegant mechanism that is able to repair a DSB using an undamaged template to restore any coding information lost at the site of the break. It is, therefore, expected to be a high fidelity mechanism of DSB rejoining. Curiously, however, the majority of DSB rejoining outside of S/late G2 phase occurs by NHEJ. Although it is widely stated that NHEJ is an error-prone rejoining mechanism, the level of fidelity achieved by NHEJ is actually unclear. In fact, the blunt ended signal ends generated at the junctions of the non-coding segment created during V(D)J recombination are rejoined with a high degree of fidelity by NHEJ. Additionally, in the absence of NHEJ, elevated chromosomal rearrangements as well as breaks and gaps are observed. Whilst this does not address the issue of NHEJ fidelity, it does demonstrate that there are alternative rejoining pathways that have a higher level of misrejoining. The most likely factor of significance for the fidelity of NHEJ is the nature of the DNA end. For breaks involving loss of nucleotides or multiple damages at the termini, it is difficult to see how the original sequence can be regenerated by NHEJ. It is possible that loss of a few nucleotides at a DSB is not dramatically harmful to a mammalian cell harbouring extensive amounts of intronic DNA. In contrast, mis-rejoining of previously unconnected DNA ends has the potential to result in activation of oncogenes or the inactivation of tumour suppressor genes (see Figure 2).

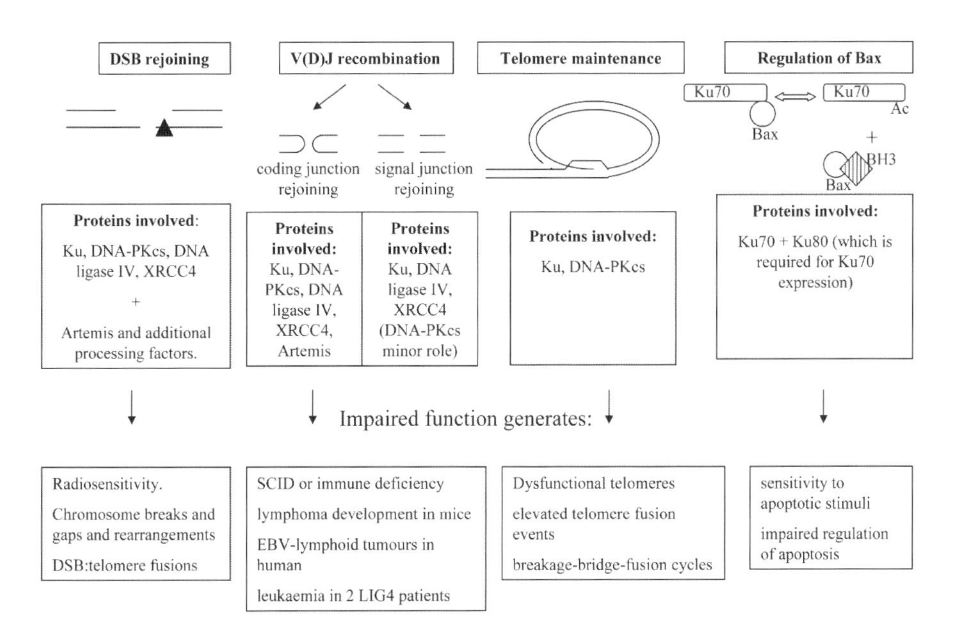

Figure 2. Role of NHEJ proteins and the impact on genomic stability.

5.2 Role of NHEJ Proteins at the Telomere

There is now strong evidence that NHEJ proteins, particularly Ku and DNA-PKcs, have roles that are distinct from their role in NHEJ. Prior to assessing the role of NHEJ in maintaining genomic stability, the impact of these additional roles needs to be considered. One striking example is the role of Ku in the maintenance of telomere length, a role that has been demonstrated in yeast, mice and mammalian cells. Since DNA ligase IV/XRCC4 does not localise to telomeres and does not impact upon telomere shortening, the evidence suggests that it does not function in telomere maintenance (Bailey et al., 1999; Boulton and Jackson, 1998; d'Adda di Fagagna et al., 2001; Herrmann et al., 1998; Teo and Jackson, 1997). Thus, the role for Ku in telomere maintenance is distinct from its role in NHEJ and in evaluating genomic instability in cells and mice lacking Ku, one has to consider the contribution of impaired telomere maintenance as well as impaired NHEJ.

In *S. cerevisiae* and *S. pombe*, Ku is physically associated with telomeres as demonstrated by chromatin immunoprecipitation and immunolocalisation studies and deficient mutants display shortened telomeres (Boulton and Jackson, 1996; Gravel et al., 1998; Laroche et al., 1998). Ku interacts with several proteins that localise to telomeres including TLC1 (hTR), the RNA component of telomerase (Peterson et al., 2001; Stellwagen et al., 2003). On the basis of these yeast studies, one model is that Ku helps to recruit telomerase to telomeres, facilitating the replication of the lagging strand at the telomeres (Gravel and Wellinger, 2002). In addition, Ku also plays a role in transcriptional silencing at telomeres (Tsukamoto et al., 1997).

In addition to a role in telomere length maintenance, Ku also appears to function in the capping of broken chromosomes with telomeric DNA sequences via its interaction with TLC1 (Stellwagen et al., 2003). Separation of function yeast mutants show that this function of Ku in healing chromosome ends is distinct from its role in NHEJ and requires interaction with TLC1. Whether a similar pathway exists in mammalian cells is currently unclear, however.

Ku is also physically associated with telomeres in mammalian cells and interacts with known telomere associated proteins (d'Adda di Fagagna et al., 2001; Hsu et al., 1999). Whether loss of Ku leads to telomere shortening in mammalian cells is more equivocal with some studies reporting shortened (d'Adda di Fagagna et al., 2001) and others lengthened telomeres (Samper et al, 2000; Espejel et al., 2002; Myung et al., 2004). However, Ku appears to function to prevent telomere-telomere fusions when telomeres become critically short, a feature observed in cells doubly defective in Ku80 and telomerase (Espejel et al., 2002). This demonstrates that a pathway generating telomere fusions is Ku independent and distinct from NHEJ.

DNA-PKcs, like Ku also localises to telomeres (d'Adda di Fagagna et al., 2001). Interestingly, loss of DNA-PKcs results in telomere shortening which is observed after several generations in deficient mice (Espejel et al., 2004). Inactivation of DNA-PKcs also leads to telomeric fusion events even when telomere shortening is not observed (Bailey et al., 2001; Gilley et al., 2001; Goytisolo et al., 2001). Defects in DNA-PKcs, therefore, promote telomere dysfunction and misrepair events similar to those described for defects in Ku (Goytisolo et al., 2001). A pathway similar to telomere fusions has been described to operate after introduction of DSBs by IR treatment, when not only telomere-telomere fusions but also telomere-DSB fusions can be observed (Bailey et al., 2004). Telomere-DSB fusions occur at elevated frequency in DNA-PKcs deficient cells suggesting that, like telomere-telomere fusions, they do not arise by NHEJ but rather that the process represents a mis-repair pathway that operates when unrepaired DSBs persist (Bailey et al., 2004). Interestingly, telomerase knock out cell lines also show increased telomere-DSB fusion events (Latre et al., 2003). Thus, two factors can result in elevated telomere-DSB fusion events; elevated DSBs arising from impaired DSB rejoining and elevated dysfunctional telomeres arising from deficiencies in telomere maintenance. Hence, by functioning in DSB repair and in promoting telomere maintenance, DNA-PK plays an important role in limiting telomere-DSB fusion events and hence genetic instability. Significantly, loss of DNA-PKcs was associated with a higher incidence of T lymphomas in mice after several generations (Espejel et al., 2004). Although such cancer predisposition of mice lacking DNA-PKcs has not been reported in other studies on DNA-PKcs deficient mice, it is consistent with the elevated genomic instability observed.

Taken together, therefore, DNA-PK plays a role in telomere maintenance serving to prevent telomere-telomere fusion events. In yeast, Ku can also play a role in capping of broken chromosomes with telomeric sequences although this has not been demonstrated in mammalian cells. Telomere-DSB fusion events, however, appear to be DNA-PK independent and, indeed, are enhanced in DNA-PK defective cells most likely due to elevated unrejoined DSBs. Such events may, therefore, be a route to genomic instability when NHEJ is compromised.

5.3 Role of Ku in Suppressing Apoptosis

A second role for Ku that is independent from its role in NHEJ is the recent demonstration that Ku70 interacts with Bax, thereby sequestering it from mitochondria and thus suppressing apoptosis (Sawada et al., 2003). Interestingly, the Ku70-Bax interaction is regulated by acetylation providing the basis for a novel route to apoptosis (Cohen et al., 2004). This important finding, therefore, provides a role for Ku in maintaining genomic stability that is distinct form its role in NHEJ and in telomere maintenance. Thus, caution should be taken in attributing the phenotype of Ku defective mice and cell lines solely to a defect in NHEJ. Although less dramatic, Ku80 deficient mice also show early tumour onset which could be attributed either to impaired NHEJ or to the decreased Ku70 levels as a result of loss of Ku80.

5.4 Role of NHEJ in the Repair of Endogenous DSBs

As stated previously, most endogenous DSBs arise at the replication fork and HR plays the major role in repairing such DSBs. Although studies with hamster mutants lacking NHEJ components showed that they play a major role in repair of radiation induced DSBs, no evidence of elevated endogenous chromosome gaps or breaks was observed (Kemp and Jeggo, 1986). However, three studies with primary MEFs defective for NHEJ components (Ku80, Ku70, DNA ligase IV and DNA-PKcs) have reported an increased frequency of chromosome breaks compared to wild type MEFs (Difilippantonio et al., 2000a; Ferguson et al., 2000; Karanjawala et al., 1999). Spectral karotype analysis showed that the aberrant events included chromosomal fragmentation as well as non reciprocal translocations. One possibility is that cultured primary MEFs grown are under considerable oxidative stress which may result in enhanced DSB formation and thus a greater dependency on NHEJ repair. However, metaphases prepared from non-cultured cells also showed structural abnormalities (Ferguson et al., 2000). Taken together, these findings provide strong evidence for a role of the NHEJ proteins in the suppression of genomic instability induced by endogenous DNA damage (see Fig. 2).

Mice lacking DNA ligase IV or XRCC4 are embryonic lethal and MEFs lacking these components undergo only a limited number of divisions (Barnes et al., 1998; Frank et al., 2000). Thus, DNA ligase IV/XRCC4 appears to be essential for cell growth. Deficiency in p53 partially overcomes these phenotypes suggesting the DNA ligase IV/XRRC4 handles endogenously arising lesions that, if unrepaired in the presence of p53, lead to cell death (Frank et al., 2000). From the endpoint of considering genomic instability in humans, the impact of impaired but not abolished DSB repair function should also be considered. LIG4 syndrome is a rare, inherited disorder conferred by hypomorphic mutations in DNA ligase IV (O'Driscoll et al., 2001). Of six characterised patients, two developed leukaemias, providing evidence that LIG4 syndrome is associated with some level of cancer predisposition (O'Driscoll et al., 2001)(Ben-Owran and Concannon, personal communication). The cancer predisposition appears, however, to be less marked than that observed in the clinically similar disorder, Nijmegen Breakage Syndrome (International Nijmegen Breakage Syndrome Study Group, 2000). Interestingly, elevated spontaneous chromosome instability was not observed in cell lines from LIG4 syndrome patients potentially due to their hypomorphic mutations (O'Driscoll et al., 2001). In contrast, however, a striking feature was the decreased fidelity of signal joint formation during V(D)J recombination (O'Driscoll et al., 2001). This might suggest that whilst residual rejoining is retained in these hypomorphic cell lines, the fidelity of rejoining is compromised. This is important in considering the impact of polymorphisms in genes that function in these DSB repair pathways. Small changes in the proteins may not significantly impact upon the frequency of DSB rejoining but could confer a small change in the fidelity of rejoining, which may be of significance in considering cancer predisposition. This may be of particular importance for the development of certain lymphoid malignancies, the origin of which might be influenced by erroneous V(D)J recombination events (see below and discussions in (Bassing and Alt, 2004; Mills et al., 2003)). Interestingly in this context, two LIG4 syndrome patients displayed the same homozygous mutational change in the active site of the protein (O'Driscoll et al., 2001). One patient harbouring this mutational change alone which reduced DNA ligase IV activity to around 10% of the wild type activity was apparently clinically normal until he developed leukaemia (Girard et al., 2004). A second patient had two linked polymorphisms in addition to the homozygous active site mutation. These combined mutational changes decreased ligation activity to around 2% of the wild type activity. This patient had more severe clinical features with significantly reduced T and B cells due to impaired V(D)J recombination. This patient has not yet developed any tumours. Although no conclusions can be derived from the comparison of just two patients, it nonetheless raises the possibility that severely diminished V(D)J rejoining may be preferable from the standpoint of cancer predisposition than a higher level of error prone rejoining.

5.5 Role of Artemis in the Maintenance of Genomic Stability

Currently, the role of Artemis in NHEJ is unclear. It does, however, function to cleave the hairpin during V(D)J recombination (Ma et al., 2002). As a consequence, Artemis patients display a SCID phenotype. Interestingly, patients with hypomorphic Artemis mutations have been described and such patients have elevated EBV-associated lymphomas (Moshous et al., 2003). One possibility is that the compromised immune response in the Artemis patients enhances the possibility of EBV infection and the development of an EBV-associated malignancy. This model would suggest a novel role for Artemis in maintaining genomic stability, not in a direct caretaker role, but rather by facilitating an efficient immune response to prevent cancer predisposing EBV infection. However, it is also possible that Artemis plays a more direct role in maintaining genomic stability, which is discussed further below. In this context, Artemis deficient MEFs displayed chromosomal instability suggesting a genomic caretaker role (Rooney et al., 2002).

6. THE CONTRIBUTION OF SIGNALLING RESPONSES TO DSBS

An unrepaired DSB will likely result in cell death either due to attempted replication past a DSB or due to ATM-dependent activation of cell cycle checkpoint arrest and/or apoptosis. Thus, from the endpoint of genomic instability and cancer predisposition, mis-rejoining events may be potentially more harmful than unrejoined DSBs. However, it has recently become clear that the propagation of genetically altered cells can be readily facilitated if accompanied by mutations in genes such as p53 which suppress the normal gate keeper functions (see section below). Several studies have shown that double mutant mice defective in components of the DSB repair machinery and p53 display dramatically increased genetic instability. For example, Ku80$^{-/-}$ mice have only a slightly earlier onset of cancer compared to control mice, but Ku80$^{-/-}$ p53$^{-/-}$ mice succumb to lymphomas at a very early age (Lim et al., 2000). Similarly, LIG4$^{-/-}$ mice are non viable but LIG4$^{-/-}$p53$^{-/-}$ mice are born, die early and display elevated chromosomal abnormalities (Frank et al., 2000). Interestingly, there is evidence that p53 mutations are more frequent in the tumours of BRCA mutation carriers compared to the level found in other breast cancers. This would be consistent with the notion that homozygosity of BRCA1 or 2 results in lethality due to highly elevated genetic instability unless coupled with additional mutation changes that allow the propagation of genetically damaged cells. Consistent with such a model, tumours in mice with a truncated form of BRCA2, frequently have mutations in p53 as well Bub1 and Mad3L, components of the spindle assembly checkpoint machinery (Lee et al., 1999). These findings demonstrate that mutations in genes that affect apoptosis (eg p53) as well as in genes that function in the mitotic checkpoint (e.g. Bub1 and BubR1 (Mad3)) co-operate with mutations in DNA repair genes to allow tumour development.

In this context, the response of ATM to DSBs is relevant. To date, the role of ATM in DNA repair as opposed to its role in signalling is unclear although indirect evidence suggests that cells from ataxia-telangiectasia (A-T) patients, which carry mutations in ATM, have a DSB repair defect that is distinct from their cell cycle checkpoint defect (Jeggo et al., 1998; Kuhne et al., 2004). A-T, therefore, may represent an important situation where a repair defect is coupled to a defect that allows the propagation of damaged cells. This dual defect may contribute to the high cancer predisposition that is characteristic of the A-T phenotype (Taylor, 1992). Nbs1 and Mre11, defective in Nijmegen Breakage Syndrome and Ataxia-telangiectasia Like Disorder (ATLD), respectively, have overlapping functions with ATM (Carney et al., 1998; Stewart et al., 1999; Varon et al., 1998). Mounting evidence suggests that both proteins are required for ATM phosphorylation events (Girard et al., 2002; Lee and Paull, 2004). Indirect evidence also points to a role in DNA DSB repair (Petrini, 1999). Again, this dual defect may be an important aspect of the high cancer incidence observed at least in NBS.

7. CANCER INCIDENCE ASSOCIATED WITH V(D)J RECOMBINATION

In the sections above, instances of elevated cancer incidence in mice and humans conferred by impaired HR or NHEJ have been described (overviewed in Table 1). From studies in mice, it is clear that under appropriate conditions, mis-repair of V(D)J induced DSBs can lead to lymphoma development (Rooney et al., 2004). This association will be discussed in this section. As mentioned earlier, NHEJ-deficient mice fail to carry out V(D)J recombination and, thus, exhibit a severe combined immunodeficiency (SCID) phenotype with dramatically decreased numbers of T and B cells (see (Mills et al., 2003) for a review). This is largely due to the inability of progenitor lymphocytes to proliferate due to the presence of unrejoined DSBs. However, NHEJ/p53$^{-/-}$ double mutant mice show early onset development of pro-B lymphomas in a manner dependent upon DSB formation by the RAG proteins (reviewed in (Bassing and Alt, 2004; Mills et al., 2003)). Most lymphomas exhibit a characteristic non-reciprocal der(12)t(12;15) translocation with co-amplification of c-Myc/*IgH* (Mills et al., 2003). The model proposed is that unrepaired DSBs generated during V(D)J recombination in G1 fuse downstream of c-myc in S phase, which

Table 1. Elevated cancer incidence associated with impaired DSB responses

Genotype	Human/Mice	Tumour Type	Reference
BRCA1/2 heterozygosity	Human	Breast cancer also ovary, pancreas and prostrate	Venkitaraman, 2002
BRCA2/FANCD1 heterozygosity	Human	solid tumours	Hirsch et al., 2004
Ku80$^{-/-}$p53$^{-/-}$	mice	pro-B cell lymphomas	Difilippantonio et al., 2000b; Lim et al., 2000
Ku70$^{-/-}$	mice	T cell lymphomas	Gu et al., 1997; Li et al., 1998
DNA-PKcs$^{-/-}$	mice	lymphomas	Custer et al., 1985; Espejel et al., 2004; Jhappan et al., 1997
DNA-PKcs$^{-/-}$p53$^{-/-}$	mice	lymphoblastic leukaemia	Gladdy et al., 2003
DNA ligase IV$^{-/-}$p53$^{-/-}$	mice	pro-B cell lymphomas	Frank et al., 2000
XRCC4$^{-/-}$p53$^{-/-}$	mice	pro-B cell lymphomas	Gao et al., 2000
DNA ligase IV (hypomorphic mutations)	humans	leukaemia	O'Driscoll et al., 2001; Riballo et al., 1999
Artemis (hypomorphic mutations)	humans	EBV-associated lymphoma	Moshous et al., 2003
Artemis$^{-/-}$p53$^{-/-}$	mice	pro-B cell lymphomas	Rooney et al., 2004
AT M$^{-/-}$	humans	mainly leukaemias and lymphomas	Taylor, 1992
ATM$^{-/-}$	mice	lymphomas	Barlow et al., 1996; Xu et al., 1996
Nbs1 mutations	humans	mainly leukaemias and lymphomas	International Nijmegen Breakage Syndrome Study Group, 2000

leads to c-myc amplification via cycles of breakage-bridge-fusion. Artemis defective mice, like NHEJ defective mice, exhibit a SCID phenotype with few T and B lymphocytes (Rooney et al., 2002). However Artemis/p53 deficient mice succumb to early onset progenitor B cell tumours (Rooney et al., 2004). Importantly, in contrast, to the situation with loss of the core NHEJ proteins, the Artemis/p53 deficient tumours have co-amplified *IgH*/N-

Myc rather than c-Myc. These results demonstrate that RAG induced DSBs can undergo erroneous rejoining in the absence of NHEJ components to generate translocation events that are precursor events for lymphoma development. Although the reason by which Artemis deficiency versus deficiency of other NHEJ components leads to c-Myc versus N-Myc amplification is unclear, the result demonstrate an important link between Myc amplification and lymphoma development and importantly provide strong evidence that translocations generated during aberrant V(D)J recombination can lead to malignancy.

8. CONCLUDING REMARKS

DSBs arise frequently at stalled replication forks and HR represents the major mechanism for the repair of such lesions. As a consequence, cells deficient in HR display dramatically elevated genomic instability. In humans, this likely underlies the highly elevated cancer predisposition observed in carriers of BRCA1 and BRCA2, two proteins required for efficient HR. HR, therefore, represents an important DSB repair mechanism, that contributes to genomic stability and cancer avoidance in humans. Ku and DNA-PKcs, which function in NHEJ, also function in telomere maintenance and shortened or uncapped telomeres confer genomic instability due to the formation of telomere fusion events. However, additionally, impaired NHEJ itself, as seen for example in LIG4 defective cell lines, results in elevated chromosome breaks and rearrangements. Thus, NHEJ also functions to protect against endogenously arising DSBs and promote genomic stability. Defects in DSB repair, however, results predominantly in broken chromosomes and the propagation of cells harbouring such lesions is facilitated by additional mutations in checkpoint proteins, such as p53. ATM is particularly interesting in this context, since it appears to function in both a DSB repair and a damage response checkpoint pathway and is, possibly as a consequence, associated with very high cancer predisposition.

ACKNOWLEDGEMENTS

I would like to thank Dr. M. O'Driscoll and A. Parker for helpful discussion and comments. The PAJ laboratory is supported by grants from the Medical Research Council, the Leukaemia Research Fund, the Primary Immunodeficiency Association, the Human Frontiers Science Programme, the Department of Health and the European Union.

REFERENCES

Alderton, G.K., Joenje, H., Varon, R., Borglum, A.D., Jeggo, P.A. and O'Driscoll, M. 2004 Seckel syndrome exhibits cellular features demonstrating defects in the ATR signalling pathway. *Hum. Mol. Genet.* in press.

Alt, F.W., E.M. Oltz, F. Young, J. Gorman, G. Taccioli, and J. Chen. 1992. VDJ recombination. *Immunology Today*. 13:306-314.

Bailey, S.M., M.N. Cornforth, A. Kurimasa, D.J. Chen, and E.H. Goodwin. 2001. Strand-specific postreplicative processing of mammalian telomeres. *Science*. 293:2462-5.

Bailey, S.M., M.N. Cornforth, R.L. Ullrich, and E.H. Goodwin. 2004. Dysfunctional mammalian telomeres join with DNA double-strand breaks. *DNA Repair (Amst)*. 3:349-57.

Bailey, S.M., J. Meyne, D.J. Chen, A. Kurimasa, G.C. Li, B.E. Lehnert, and E.H. Goodwin. 1999. DNA double-strand break repair proteins are required to cap the ends of mammalian chromosomes. *Proceedings of the National Academy of Sciences of the United States of America*. 96:14899-14904.

Barlow, C., S. Hirotsune, R. Paylor, M. Liyanage, M. Eckhaus, F. Collins, Y. Shiloh, J.N. Crawley, T. Ried, D. Tagle, and A. Wynshaw-Boris. 1996. Atm-deficient mice: a paradigm of ataxia-telangiectasia. *Cell*. 86:159-171.

Barnes, D.E., G. Stamp, I. Rosewell, A. Denzel, and T. Lindahl. 1998. Targeted disruption of the gene encoding DNA ligase IV leads to lethality in embryonic mice. *Current Biology*. 8:1395-1398.

Bassing, C.H., and F.W. Alt. 2004. The cellular response to general and programmed DNA double strand breaks. *DNA Repair (Amst)*. 3:781-96.

Blier, P.R., A.J. Griffith, J. Craft, and J.A. Hardin. 1993. Binding of Ku protein to DNA. Measurement of affinity for ends and demonstration of binding to nicks. *Journal of Biological Chemistry*. 268:7594-7601.

Boulton, S.J., and S.P. Jackson. 1996. Identification of a *Saccharomyces cerevisiae* Ku80 homologue: roles in DNA double-strand break repair and in telomeric maintenance. *Nucleic Acids Research*. 24:4639-4648.

Boulton, S.J., and S.P. Jackson. 1998. Components of the Ku-dependent non-homologous end-joining pathway are involved in telomeric length maintenance and telomeric silencing. *EMBO Journal*. 17:1819-1828.

Calsou, P., C. Delteil, P. Frit, J. Drouet, and B. Salles. 2003. Coordinated assembly of Ku and p460 subunits of the DNA-dependent protein kinase on DNA ends is necessary for XRCC4-ligase IV recruitment. *J Mol Biol*. 326:93-103.

Carney, J.P., R.S. Maser, H. Olivares, E.M. Davis, M. Le Beau, J.R. Yates III, L. Hays, W.F. Morgan, and J.H.J. Petrini. 1998. The hMre11/hRad50 protein complex and Nijmegen breakage-syndrome: linkage of double-strand break repair to the cellular DNA damage response. *Cell*. 93:477-486.

Casper, A.M., P. Nghiem, M.F. Arlt, and T.W. Glover. 2002. ATR regulates fragile site stability. *Cell*. 111:779-789.

Cohen, H.Y., S. Lavu, K.J. Bitterman, B. Hekking, T.A. Imahiyerobo, C. Miller, R. Frye, H. Ploegh, B.M. Kessler, and D.A. Sinclair. 2004. Acetylation of the C terminus of Ku70 by CBP and PCAF controls Bax-mediated apoptosis. *Mol Cell*. 13:627-38.

Connor, F., D. Bertwistle, P.J. Mee, G.M. Ross, S. Swift, E. Grigorieva, V.L. Tybulewicz, and A. Ashworth. 1997. Tumorigenesis and a DNA repair defect in mice with a truncating *Brca2* mutation. *Nature Genetics*. 17:423-430.

Cox, M.M., M.F. Goodman, K.N. Kreuzer, D.J. Sherratt, S.J. Sandler, and M.K. J. 2000. The importance of repairing stalled replication forks. *Nature*. 404:37-41.

Custer, R.P., G.C. Bosma, and M.J. Bosma. 1985. Severe combined immunodeficiency (SCID) in the mouse: Pathology, reconstitution, neoplasms. *American Journal of Pathology*. 120:464-477.

d'Adda di Fagagna, F., M.P. Hande, W.M. Tong, D. Roth, P.M. Lansdorp, Z.Q. Wang, and S.P. Jackson. 2001. Effects of DNA nonhomologous end-joining factors on telomere length and chromosomal stability in mammalian cells. *Curr Biol*. 11:1192-6.

Davies, A.A., J.Y. Masson, M.J. McIlwraith, A.Z. Stasiak, A. Stasiak, A.R. Venkitaraman, and S.C. West. 2001. Role of BRCA2 in control of the RAD51 recombination and DNA repair protein. *Mol Cell*. 7:273-282.

de Vries, E., W. van Driel, W.G. Bergsma, A.C. Arnberg, and P.C. van der Vliet. 1989. HeLa nuclear protein recognizing DNA termini and translocating on DNA forming a regular DNA-multimeric protein complex. *Journal of Molecular Biology*. 208:65-78.

Deans, B., C.S. Griffin, M. Maconochie, and J. Thacker. 2000. Xrcc2 is required for genetic stability, embryonic neurogenesis and viability in mice. *EMBO J*. 19:6675-6685.

Deans, B., C.S. Griffin, P. O'Regan, M. Jasin, and J. Thacker. 2003. Homologous recombination deficiency leads to profound genetic instability in cells derived from Xrcc2-knockout mice. *Cancer Res*. 63:8181-7.

Difilippantonio, M.J., J. Zhu, H.T. Chen, E. Meffre, M.C. Nussenzweig, E.E. Max, T. Ried, and A. Nussenzweig. 2000a. DNA repair protein Ku80 suppresses chromosomal aberrations and malignant transformation. *Nature*. 404:510-514.

Difilippantonio, M.J., J. Zhu, H.T. Chen, E. Meffre, M.C. Nussenzweig, E.E. Max, T. Ried, and A. Nussenzweig. 2000b. DNA repair protein Ku80 suppresses chromosomal aberrations and malignant transformation [In Process Citation]. *Nature*. 404:510-514.

Dvir, A., S.R. Peterson, M.W. Knuth, H. Lu, and W.S. Dynan. 1992. Ku autoantigen is the regulatory component of a template-associated protein kinase that phosphorylates RNA polymerase II. *Proceedings of the National Academy of Sciences of the United States of America*. 89:11920-11924.

Espejel, S., S. Franco, S. Rodriguez-Perales, S.D. Bouffler, J.C. Cigudosa, and M.A. Blasco. 2002. Mammalian Ku86 mediates chromosomal fusions and apoptosis caused by critically short telomeres. *EMBO J*. 21:2207-19.

Espejel, S., M. Martin, P. Klatt, J. Martin-Caballero, J.M. Flores, and M.A. Blasco. 2004. Shorter telomeres, accelerated ageing and increased lymphoma in DNA-PKcs-deficient mice. *Embo Rep*. 5:503-9.

Ferguson, D.O., J.M. Sekiguchi, S. Chang, K.M. Frank, Y. Gao, R.A. DePinho, and F.W. Alt. 2000. The nonhomologous end-joining pathway of DNA repair is required for genomic stability and the suppression of translocations. *Proc Natl Acad Sci U S A*. 97:6630-3.

Foray, N., D. Marot, A. Gabriel, V. Randrianarison, A. Carr, M. Perricaudet, A. Ashworth, and P. Jeggo. 2003. A subset of ATM and ATR-dependent phosphorylation events requires the BRCA1 protein. *EMBO*. 22:2860-2871.

Frank, K.M., N.E. Sharpless, Y. Gao, J.M. Sekiguchi, D.O. Ferguson, C. Zhu, J.P. Manis, J. Horner, R.A. DePinho, and F.W. Alt. 2000. DNA ligase IV deficiency in mice leads to defective neurogenesis and embryonic lethality via the p53 pathway. *Mol Cell*. 5:993-1002.

Gao, Y., D.O. Ferguson, W. Xie, J.P. Manis, J. Sekiguchi, K.M. Frank, J. Chaudhuri, J. Horner, R.A. DePinho, and F.W. Alt. 2000. Interplay of p53 and DNA-repair protein XRCC4 in tumorigenesis, genomic stability and development. *Nature*. 404:897-900.

Gilley, D., H. Tanaka, M.P. Hande, A. Kurimasa, G.C. Li, M. Oshimura, and D.J. Chen. 2001. DNA-PKcs is critical for telomere capping. *Proc Natl Acad Sci U S A*. 98:15084-8.

Girard, P.-M., B. Kysela, C.J. Harer, A.J. Doherty, and P.A. Jeggo. 2004. Analysis of DNA ligase IV mutations found in LIG4 syndrome patients: the impact of two linked polymorphisms. *Human Molecular Genetics*:In press.

Girard, P.-M., E. Riballo, A. Begg, A. Waugh, and P.A. Jeggo. 2002. Nbs1 promotes ATM dependent phosphorylation events including those required for G1/S arrest. *Oncogene*. 21:4191-4199.

Gladdy, R.A., M.D. Taylor, C.J. Williams, I. Grandal, J. Karaskova, J.A. Squire, J.T. Rutka, C.J. Guidos, and J.S. Danska. 2003. The RAG-1/2 endonuclease causes genomic instability and controls CNS complications of lymphoblastic leukemia in p53/Prkdc-deficient mice. *Cancer Cell*. 3:37-50.

Gottlieb, T.M., and S.P. Jackson. 1993. The DNA-dependent protein kinase: requirement of DNA ends and association with Ku Antigen. *Cell*. 72:131-142.

Goytisolo, F.A., E. Samper, S. Edmonson, G.E. Taccioli, and M.A. Blasco. 2001. The absence of the dna-dependent protein kinase catalytic subunit in mice results in anaphase bridges and in increased telomeric fusions with normal telomere length and G-strand overhang. *Mol Cell Biol*. 21:3642-51.

Gravel, S., M. Larrivee, P. Labrecque, and R.J. Wellinger. 1998. Yeast Ku as a regulator of chromosomal DNA end structure. *Science*. 280:741-744.

Gravel, S., and R.J. Wellinger. 2002. Maintenance of double-stranded telomeric repeats as the critical determinant for cell viability in yeast cells lacking Ku. *Mol Cell Biol*. 22:2182-93.

Griffin, C.S., P.J. Simpson, C.R. Wilson, and J. Thacker. 2000. Mammalian recombination-repair genes XRCC2 and XRCC3 promote correct chromosome segregation. *Nat Cell Biol*. 2:757-61.

Gu, Y.S., K.J. Seidl, G.A. Rathbun, C.M. Zhu, J.P. Manis, N. van der Stoep, L. Davidson, H.L. Cheng, J.M. Sekiguchi, K. Frank, P. StanhopeBaker, M.S. Schlissel, D.B. Roth, and F.W. Alt. 1997. Growth retardation and leaky SCID phenotype of Ku70-deficient mice. *Immunity*. 7:653-665.

Helleday, T. 2003. Pathways for mitotic homologous recombination in mammalian cells. *Mutat Res*. 532:103-115.

Herrmann, G., T. Lindahl, and P. Schar. 1998. *Saccharomyces cerevisiae* LIF1: a function involved in DNA double-strand break repair related to mammalian XRCC4. *EMBO Journal*. 17:4188-4198.

Hiom, K., and M. Gellert. 1997. A stable RAG1-RAG2-DNA complex that is active in V(D)J cleavage. *Cell*. 88:65-72.

Hirsch, B., A. Shimamura, L. Moreau, S. Baldinger, M. Hag-alshiekh, B. Bostrom, S. Sencer, and A.D. D'Andrea. 2004. Association of biallelic BRCA2/FANCD1 mutations with spontaneous chromosomal instability and solid tumors of childhood. *Blood*. 103:2554-9.

Howlett, N.G., T. Taniguchi, S. Olson, B. Cox, Q. Waisfisz, C. De Die-Smulders, N. Persky, M. Grompe, H. Joenje, G. Pals, H. Ikeda, E.A. Fox, and A.D. D'Andrea. 2002. Biallelic Inactivation of BRCA2 in Fanconi Anemia. *Science*. 13:13.

Hsu, H.L., D. Gilley, E.H. Blackburn, and D.J. Chen. 1999. Ku is associated with the telomere in mammals. *Proc Natl Acad Sci U S A*. 96:12454-8.

International Nijmegen Breakage Syndrome Study Group. 2000. Nijmegen breakage syndrome. The International Nijmegen Breakage Syndrome Study Group. *Arch Dis Child*. 82:400-406.

Jeggo, P.A. 1998. DNA breakage and repair. *Advances in Genetics*. 38:185-211.

Jeggo, P.A. 2004. The mechanism of DNA non-homologous end-joining: Lessons learned from biophysical, biochemical and cellular studies. *In* Eukaryotic DNA damage surveillance and repair. K.W. Caldecott, editor. Eurekah.com and Kluwer Academic/Plenum Publishers. 146-158.

Jeggo, P.A., A.M. Carr, and A.R. Lehmann. 1998. Splitting the ATM: distinct repair and checkpoint defects in ataxia-telangiectasia. *Trends in Genetics*. 14:312-316.

Jhappan, C., H.C. Morse, 3rd, R.D. Fleischmann, M.M. Gottesman, and G. Merlino. 1997. DNA-PKcs: a T-cell tumour suppressor encoded at the mouse scid locus. *Nat Genet*. 17:483-6.

Johnson, R.D., and M. Jasin. 2000. Sister chromatid gene conversion is a prominent double-strand break repair pathway in mammalian cells. *EMBO J*. 19:3398-3407.

Johnson, R.D., N. Liu, and M. Jasin. 1999. Mammalian XRCC2 promotes the repair of DNA double-strand breaks by homologous recombination. *Nature*. 401:397-399.

Jones, N.J., R. Cox, and J. Thacker. 1987. Isolation and cross-sensitivity of X-ray-sensitive mutants of V79-4 hamster cells. *Mutation Research*. 183:279-286.

Kanaar, R., J.H. Hoeijmakers, and D.C. van Gent. 1998. Molecular mechanisms of DNA double strand break repair. *Trends Cell Biol*. 8:483-489.

Karanjawala, Z.E., U. Grawunder, C.L. Hsieh, and M.R. Lieber. 1999. The nonhomologous DNA end joining pathway is important for chromosome stability in primary fibroblasts. *Curr Biol.* 9:1501-1504.

Kemp, L.M., and P.A. Jeggo. 1986. Radiation-induced chromosome damage in X-ray-sensitive mutants (xrs) of the Chinese hamster ovary cell line. *Mutat Res.* 166:255-263.

Kim, S.T., B. Xu, and M.B. Kastan. 2002. Involvement of the cohesin protein, Smc1, in Atm-dependent and independent responses to DNA damage. *Genes Dev.* 16:560-570.

King, M.C., J.H. Marks, and J.B. Mandell. 2003. Breast and ovarian cancer risks due to inherited mutations in BRCA1 and BRCA2. *Science.* 302:643-6.

Kraakman-van der Zwet, M., W.W. Wiegant, and M.Z. Zdzienicka. 2003. Brca2 (XRCC11) deficiency results in enhanced mutagenesis. *Mutagenesis.* 18:521-5.

Kuhne, M., E. Riballo, N. Rief, K. Rothkamm, P.A. Jeggo, and M. Lobrich. 2004. A double-strand break repair defect in ATM-deficient cells contributes to radiosensitivity. *Cancer Res.* 64:500-508.

Kurimasa, A., S. Kumano, N. Boubnov, M.D. Story, C. Tung, S.R. Peterson, and D.J. Chen. 1999. Requirement for the kinase activity of human DNA-dependent protein kinase catalytic subunit in DNA strand break rejoining. *Molecular and Cellular Biology.* 19:3877-3884.

Laroche, T., S.G. Martin, M. Gotta, H.C. Gorham, F.E. Pryde, E.J. Louis, and S.M. Gasser. 1998. Mutation of yeast Ku genes disrupts the subnuclear organization of telomeres. *Current Biology.* 8:653-656.

Latre, L., L. Tusell, M. Martin, R. Miro, J. Egozcue, M.A. Blasco, and A. Genesca. 2003. Shortened telomeres join to DNA breaks interfering with their correct repair. *Exp Cell Res.* 287:282-288.

Lee, H., A.H. Trainer, L.S. Friedman, F.C. Thistlethwaite, M.J. Evans, B.A. Ponder, and A.R. Venkitaraman. 1999. Mitotic checkpoint inactivation fosters transformation in cells lacking the breast cancer susceptibility gene, Brca2. *Mol Cell.* 4:1-10.

Lee, J.H., and T.T. Paull. 2004. Direct activation of the ATM protein kinase by the Mre11/Rad50/Nbs1 complex. *Science.* 304:93-6.

Li, G.C., H. Ouyang, X. Li, H. Nagasawa, J.B. Little, D.J. Chen, C.C. Ling, Z. Fuks, and C. Cordon-Cardo. 1998. Ku70: a candidate tumor suppressor gene for murine T cell lymphoma. *Molecular and Cellular Biology.* 2:1-8.

Lim, D.-S., and P. Hasty. 1996. A mutation in mouse *rad51*results in an early embryonic lethal that is suppressed by a mutation in *p53*. *Molecular and Cellular Biology.* 16:7133-7143.

Lim, D.S., H. Vogel, D.M. Willerford, A.T. Sands, K.A. Platt, and P. Hasty. 2000. Analysis of ku80-mutant mice and cells with deficient levels of p53. *Mol Cell Biol.* 20:3772-80.

Lio, Y.C., A.V. Mazin, S.C. Kowalczykowski, and D.J. Chen. 2003. Complex formation by the human Rad51B and Rad51C DNA repair proteins and their activities in vitro. *J Biol Chem.* 278:2469-78.

Liu, Y., J.Y. Masson, R. Shah, P. O'Regan, and S.C. West. 2004. RAD51C is required for Holliday junction processing in mammalian cells. *Science.* 303:243-246.

Ma, Y., U. Pannicke, K. Schwarz, and M.R. Lieber. 2002. Hairpin Opening and Overhang Processing by an Artemis/DNA-Dependent Protein Kinase Complex in Nonhomologous End Joining and V(D)J Recombination. *Cell.* 108:781-794.

Masson, J.Y., M.C. Tarsounas, A.Z. Stasiak, A. Stasiak, R. Shah, M.J. McIlwraith, F.E. Benson, and S.C. West. 2001. Identification and purification of two distinct complexes containing the five RAD51 paralogs. *Genes Dev.* 15:3296-307.

Mills, K.D., D.O. Ferguson, and F.W. Alt. 2003. The role of DNA breaks in genomic instability and tumorigenesis. *Immunol Rev.* 194:77-95.

Moshous, D., C. Pannetier, R. Chasseval Rd, F. Deist Fl, M. Cavazzana-Calvo, S. Romana, E. Macintyre, D. Canioni, N. Brousse, A. Fischer, J.L. Casanova, and J.P. Villartay. 2003. Partial T and B lymphocyte immunodeficiency and predisposition to lymphoma in patients with hypomorphic mutations in Artemis. *J Clin Invest.* 111:381-387.

Moynahan, M.E., J.W. Chiu, B.H. Koller, and M. Jasin. 1999. Brca1 controls homology-directed DNA repair. *Mol Cell*. 4:511-518.

Moynahan, M.E., A.J. Pierce, and M. Jasin. 2001. BRCA2 is required for homology-directed repair of chromosomal breaks. *Mol Cell*. 7:263-272.

Myung, K., G. Ghosh, F.J. Fattah, G. Li, H. Kim, A. Dutia, E. Pak, S. Smith, and E.A. Hendrickson. 2004. Regulation of telomere length and suppression of genomic instability in human somatic cells by Ku86. *Mol Cell Biol*. 24:5050-9.

Nikjoo, H., P. O'Neill, W.E. Wilson, and D.T. Goodhead. 2001. Computational approach for determining the spectrum of DNA damage induced by ionizing radiation. *Radiat Res*. 156:577-583.

O'Driscoll, M., K.M. Cerosaletti, P.-M. Girard, Y. Dai, M. Stumm, B. Kysela, B. Hirsch, A. Gennery, S.E. Palmer, J. Seidel, R.A. Gatti, R. Varon, M.A. Oettinger, K. Sperling, P.A. Jeggo, and P. Concannon. 2001. DNA Ligase IV mutations identified in patients exhibiting development delay and immunodeficiency. *Molecular Cell*. 8:1175-1185.

Patel, K.J., V.P. Yu, H. Lee, A. Corcoran, F.C. Thistlethwaite, M.J. Evans, W.H. Colledge, L.S. Friedman, B.A. Ponder, and A.R. Venkitaraman. 1998. Involvement of Brca2 in DNA repair. *Mol Cell*. 1:347-57.

Pellegrini, L., D.S. Yu, T. Lo, S. Anand, M. Lee, T.L. Blundell, and A.R. Venkitaraman. 2002. Insights into DNA recombination from the structure of a RAD51-BRCA2 complex. *Nature*. 420:287-293.

Perrault, R., H. Wang, M. Wang, B. Rosidi, and G. Iliakis. 2004. Backup pathways of NHEJ are suppressed by DNA-PK. *J Cell Biochem*. 92:781-94.

Peterson, S.E., A.E. Stellwagen, S.J. Diede, M.S. Singer, Z.W. Haimberger, C.O. Johnson, M. Tzoneva, and D.E. Gottschling. 2001. The function of a stem-loop in telomerase RNA is linked to the DNA repair protein Ku. *Nat Genet*. 27:64-7.

Petrini, J.H. 1999. The Mammalian Mre11-Rad50-nbs1 Protein Complex: Integration of functions in the cellular DNA damage response. *Am J Hum Genet*. 64:1264-1269.

Petukhova, G., S. Stratton, and P. Sung. 1998. Catalysis of homologous dna pairing by yeast rad51 and rad 54 proteins. *Nature*. 393:91-94.

Pierce, A.J., R.D. Johnson, L.H. Thompson, and M. Jasin. 1999. XRCC3 promotes homology-directed repair of DNA damage in mammalian cells. *Genes Dev*. 13:2633-2638.

Riballo, E., S.E. Critchlow, S.H. Teo, A.J. Doherty, A. Priestley, B. Broughton, B. Kysela, H. Beamish, N. Plowman, C.F. Arlett, A.R. Lehmann, S.P. Jackson, and P.A. Jeggo. 1999. Identification of a defect in DNA ligase IV in a radiosensitive leukaemia patient. *Current Biology*. 19:699-702.

Richardson, C., N. Horikoshi, and T.K. Pandita. 2004. The role of the DNA double-strand break response network in meiosis. *DNA Repair (Amst)*. 3:1149-64.

Rooney, S., J. Sekiguchi, S. Whitlow, M. Eckersdorff, J.P. Manis, C. Lee, D.O. Ferguson, and F.W. Alt. 2004. Artemis and p53 cooperate to suppress oncogenic N-myc amplification in progenitor B cells. *Proc Natl Acad Sci U S A*. 101:2410-5.

Rooney, S., J. Sekiguchi, C. Zhu, H.L. Cheng, J. Manis, S. Whitlow, J. DeVido, D. Foy, J. Chaudhuri, D. Lombard, and F.W. Alt. 2002. Leaky Scid phenotype associated with defective V(D)J coding end processing in Artemis-deficient mice. *Mol Cell*. 10:1379-1390.

Rothkamm, K., I. Kruger, L.H. Thompson, and M. Lobrich. 2003. Pathways of DNA double-strand break repair during the mammalian cell cycle. *Mol Cell Biol*. 23:5706-5715.

Rouse, J., and S.P. Jackson. 2002. Interfaces between the detection, signaling, and repair of DNA damage. *Science*. 297:547-551.

Samper, E., Goytisolo, F.A., Slijepcevic, P., van Buul, P.P. and M.A. Blasco. 2000 Mammalian Ku86 protein prevents telomeric fusions independently of the length of TTAGGG repeats and the G-strand overhang. *EMBO Rep,* 1: 244-52.

Sawada, M., W. Sun, P. Hayes, K. Leskov, D.A. Boothman, and S. Matsuyama. 2003. Ku70 suppresses the apoptotic translocation of Bax to mitochondria. *Nat Cell Biol*. 5:320-9.

Sharan, S.K., M. Morimatsu, U. Albrecht, D.-S. Lim, E. Regel, C. Dinh, A. Sands, G. Eichele, P. Hasty, and A. Bradley. 1997. Embryonic lethality and radiation hypersensitivity mediated by Rad51 in mice lacking *Brca2*. *Nature*. 386:804-810.

Shiloh, Y. 2001. ATM and ATR: networking cellular responses to DNA damage. *Curr Opin Genet Dev*. 11:71-77.

Shiloh, Y. 2003. ATM and related protein kinases: safeguarding genome integrity. *Nat Rev Cancer*. 3(155-168.

Shivji, M.K., and A.R. Venkitaraman. 2004. DNA recombination, chromosomal stability and carcinogenesis: insights into the role of BRCA2. *DNA Repair (Amst)*. 3:835-43.

Sigurdsson, S., S. Van Komen, W. Bussen, D. Schild, J.S. Albala, and P. Sung. 2001. Mediator function of the human Rad51B-Rad51C complex in Rad51/RPA-catalyzed DNA strand exchange. *Genes Dev*. 15:3308-18.

Smith, G.C., and S.P. Jackson. 1999. The DNA-dependent protein kinase. *Genes Dev*. 13:916-934.

Snouwaert, J.N., L.C. Gowen, A.M. Latour, A.R. Mohn, A. Xiao, L. DiBiase, and B.H. Koller. 1999. BRCA1 deficient embryonic stem cells display a decreased homologous recombination frequency and an increased frequency of non-homologous recombination that is corrected by expression of a brca1 transgene. *Oncogene*. 18:7900-7.

Stellwagen, A.E., Z.W. Haimberger, J.R. Veatch, and D.E. Gottschling. 2003. Ku interacts with telomerase RNA to promote telomere addition at native and broken chromosome ends. *Genes Dev*. 17:2384-95.

Stewart, G.S., R.S. Maser, T. Stankovic, D.A. Bressan, M.I. Kaplan, N.G. Jaspers, A. Raams, P.J. Byrd, J.H. Petrini, and A.M. Taylor. 1999. The DNA double-strand break repair gene hMRE11 is mutated in individuals with an ataxia-telangiectasia-like disorder. *Cell*. 99:577-587.

Struewing, J.P., P. Hartge, S. Wacholder, S.M. Baker, M. Berlin, M. McAdams, M.M. Timmerman, L.C. Brody, and M.A. Tucker. 1997. The risk of cancer associated with specific mutations of BRCA1 and BRCA2 among Ashkenazi Jews. *N Engl J Med*. 336:1401-8.

Taylor, A.M.R. 1992. Aataxia-telangiectasia genes and predisposition to leukaemia, lymphoma and breast cancer. *British Journal of Cancer*. 66:5-9.

Teo, S.-H., and S.P. Jackson. 1997. Identification of *Saccharomyces cerevisiae* DNA ligase IV: involvement in DNA double-strand break repair. *EMBO Journal*. 16:4788-4795.

Thacker, J., A.N. Ganesh, A. Stretch, D.M. Benjamin, A.J. Zahalsky, and E.A. Hendrickson. 1994. Gene mutation and V(D)J recombination in the radiosensitive *irs* lines. *Mutagenesis*. 9:163-168.

Thompson, D., and D.F. Easton. 2002. Cancer Incidence in BRCA1 mutation carriers. *J Natl Cancer Inst*. 94:1358-65.

Thompson, L.H., and C.L. Limoli. 2004. Origin, recognition, signaling and repair of DNA double-strand breaks in mammalian cells. *In* Eukaryotic DNA damage surveillance and repair. K.W. Caldecott, editor. Landes Bioscience/Eurekah.com, Texas. 107-145.

Thompson, L.H., and D. Schild. 2001. Homologous recombinational repair of DNA ensures mammalian chromosome stability. *Mutat Res*. 477:131-153.

Thorlacius, S., G. Olafsdottir, L. Tryggvadottir, S. Neuhausen, J.G. Jonasson, S.V. Tavtigian, H. Tulinius, H.M. Ogmundsdottir, and J.E. Eyfjord. 1996. A single BRCA2 mutation in male and female breast cancer families from Iceland with varied cancer phenotypes. *Nat Genet*. 13:117-9.

Tsukamoto, Y., J. Kato, and H. Ikeda. 1997. Silencing factors participate in DNA repair and recombination in *Saccharomyces cerevisiae*. *Nature*. 388:900-903.

Tsuzuki, T., Y. Fujii, K. Sakumi, Y. Tominaga, K. Nakao, M. Sekiguchi, A. Matsushiro, Y. Yoshimura, and T. Morita. 1996. Targeted disruption of the Rad51 gene leads to lethality in embryonic mice. *Proceedings of the National Academy of Sciences of the United States of America*. 93:6236-6240.

Tucker, J.D., N.J. Jones, N.A. Allen, J.L. Minkler, L.H. Thompson, and A.V. Carrano. 1991. Cytogenetic characterisation of the ionizing radiation-sensitive Chinese hamster mutant irs1. *Mutation Research*. 254:143-152.

Tutt, A., A. Gabriel, D. Bertwistle, F. Connor, H. Paterson, J. Peacock, G. Ross, and A. Ashworth. 1999. Absence of Brca2 causes genome instability by chromosome breakage and loss associated with centrosome amplification. *Curr Biol*. 9:1107-10.

van Gent, D.C., J.H. Hoeijmakers, and R. Kanaar. 2001. Chromosomal stability and the DNA double-stranded break connection. *Nat Rev Genet*. 2:196-206.

Varon, R., C. Vissinga, M. Platzer, K.M. Cerosaletti, K.H. Chrzanowska, K. Saar, G. Beckmann, E. Seemanova, P.R. Cooper, N.J. Nowak, M. Stumm, C.M.R. Weemaes, R.A. Gatti, R.K. Wilson, M. Digweed, A. Rosenthal, K. Sperling, P. Concannon, and A. Reis. 1998. Nibrin, a novel DNA double-strand break repair protein, is mutated in Nijmegen breakage syndrome. *Cell*. 93:467-476.

Venkitaraman, A.R. 2002. Cancer susceptibility and the functions of BRCA1 and BRCA2. *Cell*. 108:171-182.

Walker, J.R., R.A. Corpina, and J. Goldberg. 2001. Structure of the Ku heterodimer bound to DNA and its implications for double-strand break repair. *Nature*. 412:607-614.

Wang, H., A.R. Perrault, Y. Takeda, W. Qin, and G. Iliakis. 2003. Biochemical evidence for Ku-independent backup pathways of NHEJ. *Nucleic Acids Res*. 31:5377-88.

Wold, M.S. 1997. Replication protein A: a heterotrimeric, single-stranded DNA-binding protein required for eukaryotic DNA metabolism. *Annu Rev Biochem*. 66:61-92.

Xu, B., S. Kim, and M.B. Kastan. 2001. Involvement of brca1 in s-phase and g(2)-phase checkpoints after ionizing irradiation. *Mol Cell Biol*. 21:3445-3450.

Xu, X., Z. Weaver, S.P. Linke, C. Li, J. Gotay, X.W. Wang, C.C. Harris, T. Ried, and C.X. Deng. 1999. Centrosome amplification and a defective G2-M cell cycle checkpoint induce genetic instability in BRCA1 exon 11 isoform-deficient cells. *Mol Cell*. 3:389-395.

Xu, Y., T. Ashley, E.E. Brainerd, R.T. Bronson, M.S. Meyn, and D. Baltimore. 1996. Targeted disruption of ATM leads to growth retardation, chromosomal fragmentation during meiosis, immune defects and thymic lymphoma. *Genes and Development*. 10:2411-2422.

Yarden, R.I., S. Pardo-Reoyo, M. Sgagias, K.H. Cowan, and L.C. Brody. 2002. BRCA1 regulates the G2/M checkpoint by activating Chk1 kinase upon DNA damage. *Nat Genet*. 30:285-289.

Yoshihara, T., M. Ishida, A. Kinomura, M. Katsura, T. Tsuruga, S. Tashiro, T. Asahara, and K. Miyagawa. 2004. XRCC3 deficiency results in a defect in recombination and increased endoreduplication in human cells. *EMBO J*. 23:670-80.

Yu, D.S., E. Sonoda, S. Takeda, C.L. Huang, L. Pellegrini, T.L. Blundell, and A.R. Venkitaraman. 2003. Dynamic control of Rad51 recombinase by self-association and interaction with BRCA2. *Mol Cell*. 12:1029-1041.

Yu, V.P., M. Koehler, C. Steinlein, M. Schmid, L.A. Hanakahi, A.J. van Gool, S.C. West, and A.R. Venkitaraman. 2000. Gross chromosomal rearrangements and genetic exchange between nonhomologous chromosomes following BRCA2 inactivation. *Genes Dev*. 14:1400-6.

Chapter 2.5

TRANSLESION SYNTHESIS AND ERROR-PRONE POLYMERASES

Catherine M. Green and Alan R. Lehmann
Genome Damage and Stability Centre, University of Sussex, Brighton, UK

1. INTRODUCTION

The previous chapters have described the many ways in which cells are able to remove almost all kinds of lesions from their genomes. Furthermore, as discussed later in the book, when the DNA is damaged, signals are generated which lead to cell cycle delays. This provides the cell with the time required to repair the damage before critical events, such as S phase or mitosis, occur. One might reasonably expect that all these processes were sufficient to defend the cell against the consequences of genome damage. In fact, however, some of the repair processes are relatively slow and incomplete, and the cell cycle checkpoints take some time to mediate their effects. They also cause slowing down rather than complete blockage of cell cycle progression. As a consequence, the DNA replication machinery will encounter lesions, which the replicative DNA polymerases are unable to recognise. The replication apparatus is a superb tailor-made machine designed to replicate the DNA at high speed and with high fidelity. The price to be paid is that the replicating polymerases cannot accommodate damaged bases in their active sites, and consequently most lesions block the progression of the replication fork. These blocks could potentially be catastrophic. In order to overcome or tolerate such blocks to replication, the cell is able to utilise several different mechanisms. One of these involves a group of specialised DNA polymerases that are capable of using damaged DNA as a template, but are, as a consequence, often highly error-prone. These polymerases are known as translesion synthesis (TLS) polymerases. Genetic studies in bacteria and yeast have demonstrated that TLS polymerases are a major source of spontaneous genome instability at the nucleotide level. However, TLS pathways can also protect against cancer development in humans as demonstrated by the cancer-prone nature of

E.A. Nigg (ed.), Genome Instability in Cancer Development, 199-223.

patients suffering from the variant form of xeroderma pigmentosum (XP-V), which results from a lack of the major UV TLS polymerase polη (eta). A large number of TLS polymerases have now been identified, conserved from bacteria to man. *In vitro* and *in vivo* studies have demonstrated that these polymerases have the capacity to perform TLS past many different lesions. It is however, only recently that mechanisms by which the cell can coordinate and control their activity have begun to be elucidated. Given the error-prone nature of these enzymes, it seems likely that failure or perturbation of their control mechanisms may well play a part in the generation of mutations that can eventually lead to cancer development.

2. TOLERANCE OF DNA LESIONS (POSTREPLICATION REPAIR)

The ways in which the cell is able to overcome DNA damage during replication are collectively termed postreplication repair (PRR) (Figure 1).

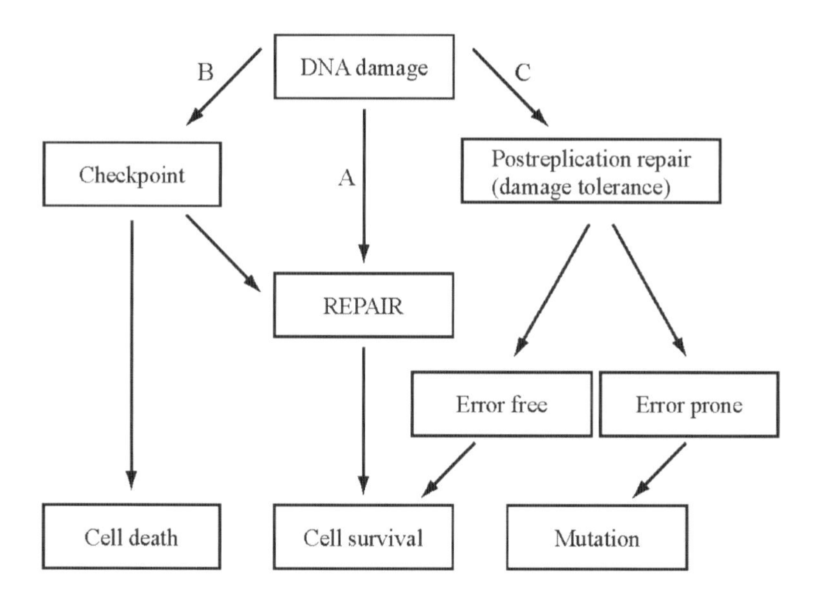

Figure 1. Cellular responses to DNA damage. As discussed in previous chapters, the recognition of DNA damage leads to a multitude of cellular responses including activation of repair (A) and checkpoint pathways (B). However, because lesions present in the genome at the time of replication will impede the progression of replication forks, the cell has also evolved mechanisms to maintain replicative capacity when damage is present. These are termed damage tolerance, or postreplication repair (C). The blocking lesion is not removed by the process, but the block is overcome in either an error-free or error-prone manner. The error-prone subset of the lesion tolerance pathway is therefore mutagenic.

To some extent this term is misleading. On the one hand, overcoming the replication block can be considered a repair process, on the other hand the lesion remains in the DNA and, in this sense, it is not repaired. Likewise, the processes may occur at the replication fork as well as behind it. Putting aside these niceties of nomenclature, we will use the term merely for convenience. The first evidence for PRR or damage tolerance came from the observation that *E. coli uvrA* mutants that were completely deficient in nucleotide excision repair (NER) were still able to tolerate about 50 UV photoproducts in their DNA. However, in double mutants defective in both NER and the RecA protein, the presence of only 1 or 2 photoproducts was lethal. The implication was that there was a PRR process in *E. coli* that absolutely required the RecA protein. Since RecA is required for recombination, it was inferred that PRR was mediated by recombinational exchanges. This was partially correct, but as we shall see, RecA plays three roles in PRR and only one of these is in recombination. The model proposed from early studies of Rupp, Howard-Flanders and colleagues (Rupp and Howard-Flanders, 1968; Rupp et al., 1971) was that in NER-deficient *E. coli,* following blockage of replication fork progression (Figure 2i), gaps were left in the daughter DNA strands opposite photoproducts. The replication machinery re-initiated DNA synthesis beyond the damage, probably at the start of the next Okazaki fragment. The genetic information lost in the daughter duplex at the damage-gap site was recovered using a sister-strand exchange mechanism (Figure 2ii), for which direct physical evidence was obtained (Rupp et al., 1971). An alternative mechanism for PRR by damage avoidance is indicated in Figure 2iii. Here, when the replication fork is stalled at a damaged site, the fork reverses and the newly synthesised daughter-strands anneal. This provides a template for the blocked daughter strand to continue synthesis. Reverse branch migration then restores the fork to its original configuration and replication can proceed.

3. ERROR PRONE PRR, TRANSLESION SYNTHESIS AND MUTATION IN *ESCHERICHIA COLI*

UV light, like most DNA-damaging agents, generates mutations, and, in all organisms, the number of UV-induced mutations is substantially increased in mutants defective in NER. The implication of this is that these mutations are generated during replication of the unrepaired damage, i.e. in a PRR process. There is no reason to expect that the sister-strand exchange or fork reversal mechanisms mentioned above would be error-prone. It was therefore hypothesised that in a minority of cases, the replication machinery could somehow be altered to permit synthesis of DNA past damage in the template, in an error- prone manner which results in mutations. This process is termed translesion synthesis (TLS) (Figure 2iv).

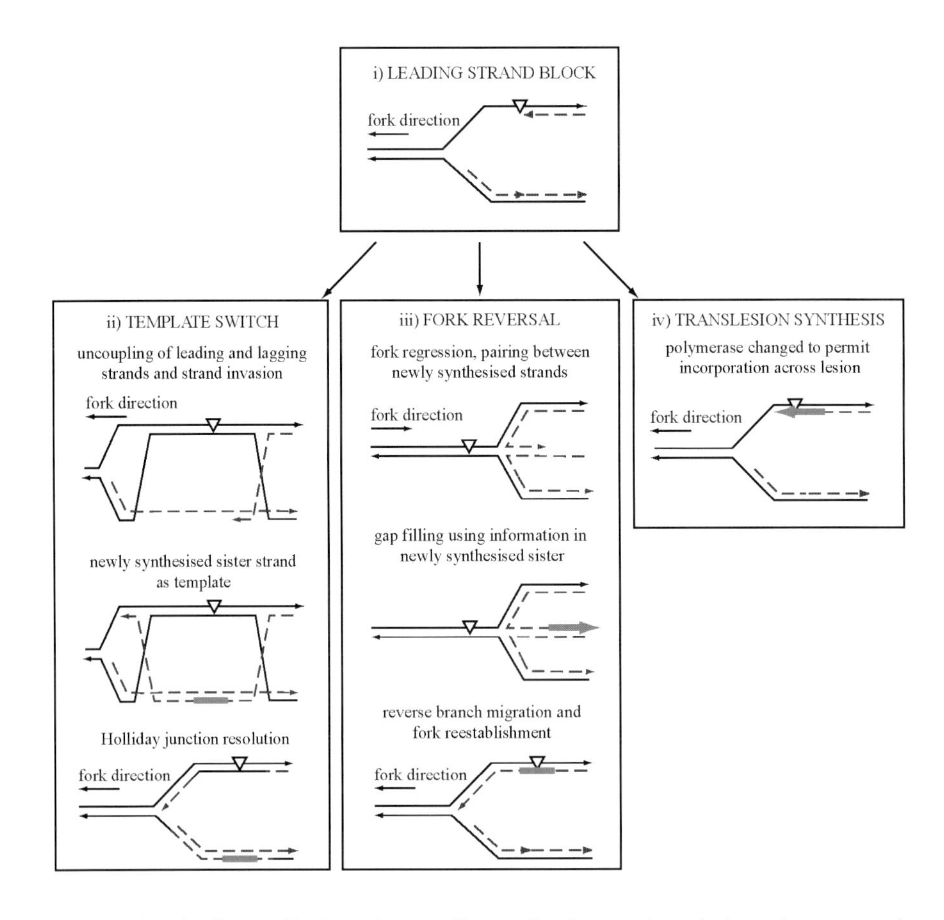

Figure 2. Mechanisms of lesion tolerance. Postreplication repair or lesion tolerance can be divided into sub-pathways, some of which are error-free and others error-prone. Both can be initiated when the replicative polymerase stalls at a replication-blocking lesion, here shown on the template for leading strand synthesis (i). The error-free pathways are likely to involve a template switch "damage avoidance" mechanism, to utilise the information in the undamaged DNA strand. This inherently implies that leading and lagging strand synthesis must become uncoupled to allow daughter strand synthesis on the undamaged parental strand. Two possibilities for how lesion bypass can be achieved after this step are depicted here: ii) a strand invasion mechanism, with Holliday junction intermediates, that is likely to involve recombination proteins such as Rad51, and iii) a fork reversal mechanism leading to formation of a "chicken foot" structure, perhaps involving the RecQ-like helicases. A third mechanism of lesion bypass (iv) is potentially error-prone and involves the TLS polymerases, such as polη, which can directly insert nucleotides opposite the damaged site. Each of these three methods allows the leading strand to bypass the lesion, and subsequent restoration of processive replication. Replication blocking lesions that halt lagging strand progression will lead to daughters containing single-stranded gapped regions containing the lesion, which can also be filled in by similar mechanisms.

Mutations in four genes, *recA, lexA, umuC* and *umuD* abolish UV mutagenesis in *E. coli.* These genes are all involved in the SOS response first proposed by Radman in 1974 (Radman, 1974), and now understood in great detail (Friedberg et al., 1995). Some 40-50 genes in the *E. coli* chromosome are under the control of the LexA repressor (Fernandez de

Henestrosa et al., 2000), including the *recA* gene, the *lexA* gene itself and the *umuC-umuD* operon. Blockage of the replication fork exposes single-stranded regions of DNA, on which RecA binds to form a nucleoprotein filament. This activates RecA, which in turn catalyses the autodigestion of the LexA repressor and induction of the SOS response genes. Increased synthesis of RecA protein provides positive feedback, whereas LexA synthesis provides negative feedback. As well as inducing the synthesis of UmuD and C proteins, activated RecA mediates the autolytic cleavage of a $UmuD_2$ dimer, removing the 24 N-terminal amino acids, to generate the active form $UmuD'_2$. UmuC and $UmuD'_2$ are absolutely required for UV mutagenesis, but what are their roles? The dogma that was widely accepted for almost 30 years was that these proteins somehow lower the stringency and fidelity of the replicative polymerase, DNA polymerase III, and enable it to carry out TLS past UV photoproducts. Neither the crystal structure of UmuD' (Peat et al., 1996) nor the sequence of UmuC gave any clue as to how they might carry out this function. The principal obstacle was the failure to purify UmuC protein.

4. 1999, A MOMENTOUS YEAR FOR TLS

After almost 20 years with little progress in understanding the mechanism of TLS, 1999 proved to be a turning point, with several more or less simultaneous independent major discoveries. UmuC was finally purified in 1999 with dramatic results. The accepted dogma was proved wrong – a heterotrimer of $UmuD'_2UmuC$ was itself a DNA polymerase and could carry out TLS past abasic sites and UV photoproducts with the assistance of activated RecA protein (Reuven et al., 1999; Tang et al., 2000; Tang et al., 1999).

The *dinB* gene of *E. coli* is required for mutagenesis of phage λ and confers a mutator phenotype to cells when over-expressed. DinB was purified and found also to have DNA polymerase activity. It was designated DNA polymerase IV (polIV) (Wagner et al., 1999), whilst $UmuD'_2UmuC$ became polV. It was already apparent from their sequences that UmuC and DinB were related, and a homolog in *Saccharomyces cerevisiae* had been identified and designated Rad30 (McDonald et al., 1997; Roush et al., 1998). Rad30 was also found to have DNA polymerase activity and was designated polη (eta) (Johnson et al., 1999b).

The genetic disorder xeroderma pigmentosum (XP) has been discussed in detail elsewhere in the book. To recap briefly, XP is an autosomal recessive genetic disorder characterised by severe sensitivity of the skin to sunlight and multiple sunlight-induced pigmentation abnormalities. Eventually the patients develop multiple skin cancers in sun-exposed areas, the incidence being up to 1000-fold higher than that in normal individuals (Kraemer et al., 1987). Most XP patients are defective in NER, but about 20%, designated XP variants (XP-V), have normal NER and are only mildly sensitive to killing by UV-irradiation. However UV mutagenesis is elevated in XP-V and NER-defective XP cells to similar extents (Maher et al., 1976). At the cellular level, it was shown that XP-V cells were defective in PRR. They had a reduced ability to make intact daughter DNA strands after UV-irradiation (Lehmann et al., 1975). The precise nature of the molecular defect remained elusive for almost 25 years, until two groups cloned the XPV gene in 1999, by completely different routes. Johnson *et al.* searched the sequence databases and identified human homologs of Rad30 (Johnson et al., 1999a). Masutani *et al.* developed a cell-free assay for TLS, which was defective in XP-V extracts. They purified the missing protein by complementation of the defective extracts and identified a DNA polymerase capable of carrying out TLS past cyclobutane pyrimidine dimers (CPD) (Masutani et al., 1999a). It encoded a human homolog of polη. Both groups identified mutations in the *POLH* gene in several XP-V patients (Johnson et al., 1999a; Masutani et al., 1999b). In the same year, polι (iota), a second human homolog of Rad30 (McDonald et al., 1999) and polκ (kappa), a human homolog of DinB (PolIV) (Gerlach et al., 1999; Ogi et al., 1999) were also discovered.

5. TLS POLYMERASES: THE Y-FAMILY

DNA polymerases have been assigned to several families based on their amino acid sequences. These families have been designated A,B, C and X (Rattray and Strathern, 2003). The seminal findings in the area of TLS led to the discovery of a plethora of new DNA polymerases. Although related to each other, most of these polymerases were not obviously related to previously known polymerases. This led to the designation of a new DNA polymerase family, the Y-family, which includes *E. coli* polIV, polV and yeast and human polη (Ohmori et al., 2001) (Table 1). PolIV and V are the only Y-family members in *E. coli.* In *S. cerevisiae*, there are also two members, polη and Rev1. Rev1 together with Rev3 and Rev7 has been extensively characterised genetically and shown to be essential for UV-induced mutagenesis. Rev1 is a member of the Y-family, but as we shall see below is not a DNA polymerase, but a deoxycytidine nucleotidyl transferase. The Rev3-Rev7 heterodimer is also a DNA polymerase, polζ (zeta), but is a member of the B-family. *Schizosaccharomyces pombe* contains Polζ, Rev1, polη and polκ. Curiously, in this organism, polη is part of a fusion protein, Eso1. The N-terminal 600 amino acids (aa) of Eso1 comprise polη, whereas the C-terminal 300 aa encode an ortholog of *S. cerevisiae* Eco1, a protein required to establish sister chromatid cohesion (Tanaka et al., 2000). Whether there is any biological significance to the existence of this fusion protein is not known.

Table 1. Properties of TLS polymerases.

E. coli	*S. cerevisiae*	*S. pombe*	Human	Family	TLS *in vitro*[a]	*in vivo* mutant phenotype	Comments[b]
PolII				B	AP, AAF (i+e)	Spontaneous and UV-induced mutations reduced.	Required for replication restart after UV. Also involved in HN2 tolerance.
PolIV (DinB)		Polκ (DinB)	Polκ	Y	AP, 8-Og, BaP (i+e)	Reduced adaptive and induced mutation. Knockout mice viable, BaP sensitive.	Associates with β-clamp. Overproduction increases mutation rate. Crystal structure solved.
PolV (UmuC)				Y	6-4, CPD, AP, AAF (i+e) BaP (e)	Drastic reduction in induced mutagenesis.	Associates with β-clamp.
	Polζ (Rev3-Rev7)	Polζ (Rev3-Rev7)	Polζ (Rev3-Rev7)	B	CPD (e) AAF (e)	Drastic reduction in induced and spontaneous mutagenesis. Essential gene in mouse.	Likely functions as an extender of mispaired termini after TLS insertion by a different polymerase.
	Rev1	Rev1	hRev1	Y	AP, 8-Og, BaP (i)	Drastic reduction in induced and spontaneous mutagenesis. Required for IgG gene maturation in DT40	Physical interaction with many TLS polymerases.
	Polη (Rad30)	Eso1	Polη	Y	CPD, TG, 8-Og (i+e)	Yeast are slightly UV-sensitive. Humans have XPV.	Interacts with ubiquitinated-PCNA. Present at replication foci. Possibly involved in SHM. Crystal structure solved.
			Polι	Y	6-4, AP, AAF (i)	No phenotype detected in mouse.	Interacts with PCNA. Present at replication foci. Possibly involved in SHM. Crystal structure available.
			Polλ	X	AP	n/c	Interacts with PCNA. May be involved in BER and NHEJ. Crystal structure solved.
			Polμ	X	AAF, 8-Og, AP, BaP (e) CPD (i+e)	n/c	Creates frameshift mutations on undamaged templates May be involved in NHEJ.

An overview of the *in vitro* TLS properties and possible *in vivo* phenotypes of TLS polymerases from *E. coli*, yeast and humans.

[a]Abbreviations are: AP, abasic site; AAF, acetylaminofluorene; i, insertion; e, extension; HN2, nitrogen mustard, 8-Og, 8-oxoguanine; BaP, benzo[a]pyrene; 6-4, 6-4 photoproduct; CPD, cyclobutane pyrimidine dimer.

[b]TLS, translesion synthesis; SHM, somatic hypermutation; BER, base excision repair; NHEJ, non homologous end joining; n/c, not characterised. Indicated crystal structures are of catalytic domains only.

Human cells contain four Y-family proteins, polη, polκ, hRev1 (ortholog of Rev1) and polι. polι is a paralog of polη, both proteins being homologs of *S. cerevisiae* Rad30 (polη). In addition there is a human homolog of Polζ. Two other newly discovered DNA polymerases are polλ (lambda) and polμ (mu), both of which belong to the X-family. With the exception of polη, all these human polymerases were initially identified by scanning the human genome database for homologs either to Y-family polymerases in other organisms, or for sequence similarity to other human polymerases. Although much work has been done on their activities *in vitro*, there have been relatively few biological studies, and in most cases their proposed functions are based on speculative extrapolations rather than on experimental evidence. The properties of the polymerases have been the topics of several reviews (see Table 1) (Friedberg et al., 2002; Goodman, 2002; Lehmann, 2002; Pages and Fuchs, 2002; Prakash and Prakash, 2002).

The Y-family polymerases have very different properties from replicative polymerases. As mentioned above, the latter are highly processive, have very low error-rates and are very stringent in the bases that can be accommodated in their active sites. In contrast, the Y-family polymerases are distributive (i.e. they dissociate after incorporating a few nucleotides), they have high error rates when using undamaged DNA as template and they have low stringency. These are precisely the properties needed for a TLS polymerase. Their low stringency allows them to incorporate abnormal bases in their active sites. This makes them more error-prone when copying normal bases. It is thus vital that they dissociate after they have by-passed the damage, so that the error-free replicative polymerase can take over as soon as possible.

Many studies have been carried out in which defined oligonucleotide templates containing single lesions have been used to analyse the ability of different polymerases to bypass these lesions. Such studies are very informative in characterising the properties of the polymerases, telling us what the polymerases can do and pointing to possible roles *in vivo*. What they do not tell us however, is what the polymerases actually do in the cell. The substrate is different, the conformation of the DNA in chromatin is different and there are many accessory proteins whose interactions may have profound effects. The conclusions from these studies are regrettably often overstated, and biological data to back up the biochemistry is in many cases still lacking.

5.1 Insertion and Extension

The TLS process consists of two different stages. The first is the insertion of a nucleotide opposite the damaged base(s). The product of the insertion step is not however a substrate for replicative polymerases, because the newly inserted nucleotide at the 3' end of the primer strand is not correctly base-paired with the template. The second stage of the process is the extension of the new strand beyond the site of the damage to a position where template and primer can again correctly base-pair and replication can restart. The insertion and extension steps might, with some lesions, require the consecutive action of different polymerases. There is biochemical and some genetic evidence to support this theory.

5.2 *E. coli* TLS Polymerases

The extensive genetic studies on induced mutagenesis in *E. coli* suggested that polV was the principal polymerase involved in TLS. This was backed up by *in vitro* studies showing that polV could bypass abasic (AP) sites and the major UV photoproducts, the cyclobutane pyrimidine dimer (CPD) and the pyrimidine(6-4)pyrimidone photoproduct (6-4). A's were usually inserted opposite AP sites. With T-T CPDs, again A's were inserted in most cases. With T-T 6-4s, G was preferentially inserted opposite the first (3') base and A opposite the second (5') base (Reuven et al., 1999; Tang et al., 2000; Tang et al., 1999). These data corresponded with known mutation spectra. However a series of elegant studies by Fuchs and colleagues showed that TLS could also employ polIV and the B-family polymerase polII, depending on the exact nature of the lesion and its sequence context. This work showed, for example, that for 6-4's, polV was absolutely required for both error-free and error-prone TLS (Napolitano et al., 2000), whereas bypass of benzo[a]pyrene (BaP) adducts required polIV and polV. For a 2-acetylaminofluorene-guanine (AAF-G) adduct in a particular sequence context, error-free TLS was absolutely dependent only on polV, whereas error-prone TLS required polII and partially required polV (Napolitano et al., 2000; Wagner et al., 2002). *In vitro*, polII is particularly efficient at extension reactions from some template-primer structures (Becherel and Fuchs, 2001).

5.3 Polη

The DNA polymerase activity of polη is contained within the N-terminal 430 aa. A wealth of both biochemical and biological data suggest that the major function of polη is to carry out TLS past a CPD, in most cases inserting the "correct nucleotides" opposite the damaged pyrimidines. For example, *S. cerevisiae rad30* mutants are UV sensitive, although they do not show a defect in UV induced mutagenesis (McDonald et al., 1997; Roush et al., 1998). Similarly, in *S. pombe,* removal of the Rad30 domain of Eso1 results in UV sensitivity (Tanaka et al., 2000). Cells from patients with XP-V (which lack polη) are not significantly sensitive to killing by UV, but become sensitised by treatment with caffeine (Arlett et al., 1975). This implies that an alternative UV survival pathway to TLS exists in human cells, that is abrogated by caffeine. As with NER-defective XP cells, XP-V cells are highly mutable by UV light.

Using *in vitro* systems, polη is able to replicate a template containing a CPD very efficiently. Indeed, there is evidence that it may actually be more efficient with the CPD template than with an undamaged template (McCulloch et al., 2004). Using a T-T CPD-containing template, polη preferentially inserts adenines opposite both damaged T's (Masutani et al., 2000). Furthermore, polη can only extend efficiently when A's are inserted, providing an additional check for "correct" insertion. Thus polη is capable of carrying out both insertion and extension steps in an error-free manner with a CPD template. This property of polη provides a satisfying explanation for the XP-V phenotype. In normal human cells the major PRR pathway is TLS by polη, which is relatively error-free. In XP-V cells, polη is defective, PRR is less efficient, presumably because there is a less efficient substitute, and UV mutagenesis is increased, presumably because the substitute is more error-prone. Thus, polη protects against UV-induced mutations *in vivo.*

A further characteristic of TLS by polη, at least in a simple *in vitro* system, is a marked tendency to dissociate from the template two bases beyond the CPD (McCulloch et al., 2004). This would make sense *in vivo*, since polη is relatively error-prone on undamaged templates and it is therefore desirable for it to be replaced by a replicative polymerase, as soon as a suitable primer is available.

In simple *in vitro* systems polη can carry out TLS past a variety of other lesions with varying efficiencies. These include 8-oxoguanine, O-6-methylguanine and AAF-G adducts. In the absence of supporting biological data, it is not clear which of these is relevant inside cells.

5.4 Polι

Polι is a paralog of polη, discovered by screening the sequence databases. It has about 20-25% sequence identity both to human polη and to *S. cerevisiae* Rad30 (McDonald et al., 1999). Its properties *in vitro* and *in vivo* are however quite different from those of polη. Using undamaged templates, it has a curious base-specific error-proneness. Opposite template A or C it misincorporates with a fairly low frequency (approx. 10^{-4}). In contrast, opposite template T, it incorporates dGMP three times more efficiently than the correct dAMP (Tissier et al., 2000b). Polι is able to insert nucleotides opposite a variety of damaged bases, but it is unable to extend from these incorporated nucleotides (Vaisman et al., 2002). Thus polι cannot carry out TLS on its own, but it is able to complete TLS when partnered with an "extender" polymerase like polζ (Johnson et al., 2000; Tissier et al., 2000a). Neither *S. pombe* nor *S. cerevisiae* have polι orthologs. Serendipitously, the widely used mouse 129 strain contains a nonsense mutation at codon 27 in the polι gene rendering these mice polι deficient. 129 mice are fertile, not UV sensitive and have no immune system defects, suggesting that redundant pathways can perform the role of polι when it is lacking (McDonald et al., 2003).

5.5 Polκ

Polκ was identified as a homolog of *E.coli* PolIV (DinB) (Gerlach et al., 1999; Ogi et al., 1999). Overexpression of polκ/DinB in mammalian cells leads to an increased rate of spontaneous mutagenesis. This suggests that unbalancing the polymerase pools is detrimental, and that polκ has an error prone activity *in vivo* (Ogi et al., 2001). Polκ-deficient mice are viable, fertile and have normal immune system development. The deficient mouse ES cells are slightly UV sensitive but very sensitive to BaP (Ogi et al., 2002; Schenten et al., 2002). *In vitro*, Polκ is unable to carry out TLS past UV photoproducts but can bypass a variety of lesions with differing efficiencies. The most studied lesions are BaP adducts. Polκ is able to bypass the major adduct efficiently (Suzuki et al., 2002; Zhang et al., 2000), and the sensitivity of polκ-deficient mouse embryonic fibroblasts to BaP supports the biological relevance of this finding (Ogi et al., 2002). It has been proposed that polκ is an extender rather than an inserter but convincing evidence to support this contention is lacking.

5.6 Rev1,3,7

The *REV1, 3* and *7* genes of *S. cerevisiae* were originally isolated because they are absolutely required for damage-induced mutagenesis (Lemontt, 1971). Rev3 and Rev7 form a heterodimer designated polζ. Rev3 has the structure of a B-family DNA polymerase and Rev7 is a regulatory subunit (Nelson et al., 1996b). The precise role of polζ in TLS is not yet understood, but it has been proposed that it might act as an extender (see above) (Guo et al., 2001; Johnson et al., 2000; Tissier et al., 2000a). Rev3$^{-/-}$ mice are not viable. However human fibroblast lines expressing antisense REV3 transcripts can divide although they are slightly UV sensitive and have low UV mutagenesis rates (Diaz et al., 2003; Gibbs et al., 1998; Li et al., 2002).

The catalytic domain of Rev1 is characteristic of the Y-family but it does not have full-blown polymerase activity. *In vitro* both the yeast Rev1 and its human ortholog have dCMP transferase activity with which dCMP can be incorporated opposite template G or AP sites (Lin et al., 1999; Nelson et al., 1996a). Curiously the role of Rev1 in mutagenesis can be separated from this transferase function (Nelson et al., 2000). In addition to the catalytic domain Rev1 orthologs from all species have an N-terminal BRCT domain, probably involved in protein-protein interactions. Mammalian Rev1 interacts with polη, ι, κ and Rev7 via the same 150 aa domain at the extreme C-terminus (Guo et al., 2003; Ohashi et al., 2004; Tissier et al., 2004). It has therefore been proposed that Rev1 might provide a TLS platform to assist in switching between TLS polymerases (see below). Rev1 has been deleted in the chicken DT40 system. The resulting cells are UV sensitive as well as being deficient in immunoglobulin gene maturation (Simpson and Sale, 2003).

5.7 Crystal Structures

Elucidation of the three-dimensional structures of TLS polymerases is vital to understanding how they are able to carry out TLS, whereas replicative polymerases are not. The amino acid sequences of the Y-family polymerases show conservation between family members of 250-400 aa close to the N-termini that form the catalytic domain, but this sequence bears no resemblance to the sequences of classical DNA polymerases. X-ray crystal structures of the catalytic domains of several of these polymerases have now been solved and they reveal a common pattern. Despite the lack of sequence similarity to classical polymerases, Y-family polymerases have a hand-like configuration with the same three domains of "thumb", "palm" and "fingers" found in other polymerases. In addition they have another domain, which has been variously designated as little finger, PAD or wrist. The key feature, which enables these polymerases to accommodate altered bases in their active sites, is a more open configuration than classical polymerases, with shorter thumb and finger domains (Ling et al., 2001; Trincao et al., 2001). Most detailed information has come from the Dpo4 protein from *Sulfolobus sulfataricus* P2, which is an archaeal homolog of polIV, but has features similar to polη in that it can carry out TLS past CPDs, albeit less efficiently than polη. In the ternary structure with DNA, the thumb and little finger grip the ds DNA across both minor and major grooves (Ling et al., 2001). The limited and non-specific interactions between Dpo4 and the replicating base pair provide the structural basis for low fidelity replication on undamaged templates. Unlike classical polymerases, which can only accommodate a single template base in their active sites, in Dpo4 two adjacent bases are admitted into the active site simultaneously. This provides a means for the enzyme to accommodate both bases of a CPD and carry out TLS past this lesion. In crystal structures of Dpo4 with DNA containing a CPD, the 3' thymine of the CPD forms a Watson-Crick base pair with incoming ddATP, whereas the 5'thymine forms a Hoogsteen base pair with ddATP in the *syn* conformation (Ling et al., 2003). The structure of the catalytic domain of yeast polη has also been solved (Trincao et al., 2001), and modelling of this structure complexed with a CPD suggests a similar base-pairing mechanism to that determined for Dpo4 (Ling et al., 2003).

Polι and polκ have also been crystallised. In the structure of polι the template A was constrained into the *syn* position and formed a Hoogsteen base-pair with the incoming dTTP (Nair et al., 2004). The authors proposed that similar Hoogsteen base-pairing could account for the unusual base-pairing specificity of polι (see above). The active site of polκ is more tightly restrained with respect to template base than is polη and it has an N-terminal extension of the thumb domain not found in any of the other Y-family polymerases. The little finger domain in the structure crystallised in the absence of DNA is in a completely different position relative to the other domains, when compared with structures of other Y-family polymerases

(Uljon et al., 2004). It is likely that, when bound to substrate, the little finger will swing into a position similar to that in the other polymerases, suggesting that the little finger domain is much more flexible in polκ.

6. THE X-FAMILY OF DNA POLYMERASES: POSSIBLE ROLES IN TLS

Y-family of DNA polymerases are implicated in TLS not only from their activities *in vitro*, but also, for many of them, from the phenotypes of mutants deficient in one or other of the polymerases. The X-family includes polβ as well as the recently discovered polλ and polμ. Polβ has a well-characterised role in base excision repair, but there is also some evidence for a possible role in TLS . *In vitro* studies have shown that both pol λ and polμ can carry out TLS (Covo et al., 2004; Havener et al., 2003; Zhang et al., 2002), raising the possibility that they can fulfil such a role *in vivo*. However as yet there is no biological evidence in support this possibility, and we will not discuss these polymerases further.

7. TLS POLYMERASES AND SOMATIC HYPERMUTATION

Somatic hypermutation (SHM) is one of the processes by which diversity is generated during development of the immune response. The mutation rates that occur during this process are far higher than in other proliferating cells. The discovery of polymerases with high error-frequencies therefore immediately raised the possibility of their involvement in SHM (Goodman and Tippin, 2000). A detailed discussion of TLS polymerases and SHM is outside the scope of this chapter. Suffice it to say that evidence has been produced to implicate many of the polymerases in SHM, but much of the evidence is controversial, and it is likely that the polymerases have redundant functions (Li et al., 2004).

8. TLS POLYMERASES FUNCTION AT SITES OF REPLICATION

In vitro analysis gave us some idea of what these TLS polymerases are capable of and *in vivo* analysis of knockouts and mutants answered some of the questions regarding what they might actually be used for, but the question remained how to connect the two. Insights have been gained from microscopic analysis of human cells exogenously expressing fluorescent-tagged versions of these polymerases. Visualisation of GFP-tagged polη immediately led to a surprising observation. The fluorescent signal of this protein was nuclear as expected but in about 15% of the cells it was seen as discrete bright dots or foci within the nucleus (Kannouche et al., 2001). The proportion of cells with such foci increased dramatically after UV irradiation or treatment of cells with the replication inhibitor hydroxyurea (HU). The co-localisation of tagged polη with PCNA or incorporated bromodeoxyuridine (both markers of replication factories) led to the suggestion that this polymerase is associated with replication forks after UV damage or replication inhibition. The fact that polη foci were also observed in undamaged cells suggests either that some forks stall throughout a normal S phase and polη is recruited, or that the polymerase is always associated with ongoing replication and is ready to be called up when required. Further studies showed that not only polη but also exogenously expressed polι and Rev1 co-localise at replication factories (Kannouche et al., 2003; Tissier et al., 2004). In the case of polι, a reduced focal localisation in cells derived from XP-V individuals lacking polη suggests that polη plays a role in targeting polι to such sites. Curiously, whereas polη, ι and Rev1 can be detected in all replication factories, as judged by co-localisation with PCNA (see below), polκ only colocalises with PCNA in factories in a small proportion of replicating cells (Ogi et al., 2005).

Both pol η and Rev1 molecules that are found within the foci are resistant to extraction with detergent, suggesting that they are tightly associated with chromatin or the nuclear matrix (Kannouche et al., 2004; Tissier et al., 2004). In contrast polι and κ did not show this behaviour, being released upon Triton treatment (Kannouche and Lehmann, 2004, Ogi et al., 2005). Thus polη and Rev1 may be more central components of a putative multi-polymerase complex associated with replication.

8.1 Polymerases and Clamps – Keeping Replication on Track

Sliding clamps are a highly conserved family of structural proteins required for replication in bacteria, archaea, yeasts and higher eukaryotes. They perform an essential role as processivity factors for DNA synthesis during chromosomal replication, acting to tether the polymerase to the template to allow rapid and processive synthesis by preventing polymerase disengagement. They achieve this by virtue of their ring structure, formed, depending on the organism, of two or three subunits that can encircle a DNA duplex and slide along it. Once loaded onto a DNA strand by an ATP-dependent clamp loader, these clamps act as a moving platform with which the polymerase can stably associate as it tracks along its template (Bruck and O'Donnell, 2001). In *E. coli* this role is performed by the homodimeric β clamp, which assists polIII holoenzyme in chromosomal replication of leading and lagging strands. The β clamp has also been shown to associate with the *E. coli* TLS polymerases IV and V (Lenne-Samuel et al., 2002) and to increase the processivity of PolIV (Wagner et al., 2000). Mutations of polII, IV and V predicted to abolish clamp-binding result in greatly reduced error-free and error-prone TLS *in vivo*, suggesting that clamp-association is necessary for the function of TLS polymerases (Becherel et al., 2002; Lenne-Samuel et al., 2002). All five DNA polymerases of *E. coli* bind to the β clamp at the same site, suggesting that clamp binding might be competitive and regulated (Lopez de Saro et al., 2003). A crystal structure of the complex formed between the β clamp and a domain of pol IV has recently been solved (Bunting et al., 2003). In this structure the contacts between clamp and TLS polymerase are similar to those made by the replicative polymerases, but a secondary interface allows pol IV to also interact with the clamp in an inactive conformation. If this is true for the eukaryotic homologs it would provide a mechanistic explanation for how TLS polymerases can associate with active replication sites even when they are not engaged in DNA synthesis themselves. In yeast and mammals the trimeric PCNA processivity clamps are also essential for replicative polymerases. PCNA is localised in replication factories and is resistant to Triton extraction (Bravo and Macdonald-Bravo, 1987). *In vitro* investigations have shown that many of the TLS polymerases, including polη can associate physically and directly with PCNA (Haracska et al., 2001a; Haracska et al., 2001b; Haracska et al., 2002; Vidal et al., 2004), and they contain the conserved sequence motif (Q-I/L-FF) that is characteristic of PCNA-binding proteins (Warbrick, 2000). The addition of PCNA to primer extension or lesion bypass assays has often demonstrated that PCNA can enhance the activity or processivity of these polymerases in such simple systems. Whether such systems are really representative of the far more complex situation within a cell remains an open issue. Given that these polymerases are so error prone, it is perhaps not immediately intuitive that enhancing their processivity would be a desirable outcome in a cell. Hence it remains to be determined whether this *in vitro* effect can be extrapolated to the *in vivo* situation.

The association of TLS polymerases with PCNA immediately suggested a mechanism by which they could find their target sites: once stalled at a lesion, the PCNA clamp of the replicative polymerase could be utilised to recruit the appropriate TLS polymerase to allow bypass, and subsequently processive replication could resume. Indeed mutation analysis of polη and ι shows that the PCNA-binding motif is required for their localisation to the replication foci suggesting that an interaction with PCNA is likely to be important for this targeting (P. Kannouche and ARL, manuscript in preparation; Vidal et al., 2004).

A second trimeric protein complex that structurally resembles PCNA has been identified in eukaryotes (Venclovas and Thelen, 2000). This Rad9, Rad1, Hus1 (9-1-1) complex is involved in the checkpoint responses to DNA damage. In *S. pombe* there is an increase in chromatin-associated polκ (DinB) following checkpoint arrest, resulting from either a temperature-sensitive mutation in polα or treatment with methyl methanesulfonate, and polκ becomes associated with the 9-1-1 complex. The clamp loader for the 9-1-1 complex, Rad17, is also required for TLS-dependent mutagenesis (Kai and Wang, 2003). Hence it is possible that in this organism, the checkpoint-related sliding clamp acts to recruit polκ to lesions, resulting in increased mutagenesis.

8.2 Clamps as Choreographers of Protein Exchanges

Many elegant studies in bacterial systems have analysed the role of the β clamp in coordination of sequential events at the replication fork. Alterations in binding affinities between the clamp and its partners provide one controlling mechanism by which protein traffic on a clamp can be regulated. One good example of this is the lagging strand Okazaki fragment synthesis cycle in *E. coli.* (Lopez de Saro et al., 2004). In this cycle the sequential associations of the β clamp with the clamp loader, the polymerase and the clamp unloader are dependent upon binding affinities mediated by the status of the clamp (Naktinis et al., 1996). Can a simple hierarchy of affinities explain the regulation of the polymerase switch to TLS upon encountering a lesion? Shingo and Fuchs have been able to reconstitute the TLS process in vitro using a primed single-stranded plasmid containing a single AAF-G lesion and polIII holoenzyme (which includes the β-clamp), polV and the RecA protein. With this system, PolIII synthesised DNA up to the lesion. A switch to polV then took place and polV synthesised a TLS patch of 1-60 nucleotides. If the patch was greater than 5 nucleotides, polIII was able to resume synthesis beyond the lesion (Fujii and Fuchs, 2004). Based on these results it appears that different affinities and mass action are sufficient to bring about TLS in this simple system. In eukaryotes, given the number of proteins now shown to be able to physically interact with PCNA, other levels of regulation might be necessary. Indeed, recent insights from *S. cerevisiae* and humans have led us to believe that additional levels of control are provided by the ability of eukaryotic cells to modify the clamp post-translationally.

8.3 Modified Clamps during PRR

The first hint that post-translational modifications play a role in TLS came from genetic analyses in *S. cerevisiae* (Game, 2000). The PRR pathway is controlled by the product of the *RAD6* gene, an E2 ubiquitin-conjugating enzyme, which can catalyse the attachment of ubiquitin to histones *in vitro* and *in vivo* (Jentsch et al., 1987). Both the error-free and error-prone sub-pathways of PRR are controlled by *RAD6*, in cooperation with the E3 ubiquitin ligase *RAD18*, which suggests that a ubiquitination event is likely to be involved in both pathways (Bailly et al., 1997). Genetic analysis showed that the ubiquitin E2 Ubc13/Mms2, which synthesises ubiquitin chains joined via lysine 63 residues (Hofmann and Pickart, 1999), and the E3 *RAD5*, are also involved in PRR. Extensive genetic analyses suggested that *UBC13/MMS2* and *RAD5* control an error-free branch of PRR, whereas the error prone, mutagenic branch contains *REV1*, *REV3* and

REV7 (Xiao et al., 2000). Protein interaction studies showed that Rad5 could physically interact with the Mms2-Ubc13 heterodimer and also with Rad18 (Ulrich and Jentsch, 2000). Taken together, this work emphasised the importance of ubiquitination processes in PRR, but gave no clues as to which proteins were being ubiquitinated.

The first biochemical insights into how ubiquitination connects with PRR came with the discovery that PCNA is a critical substrate for the Rad6/Rad18 ubiquitin-conjugation machinery, and that the Rad6-dependent ubiquitination of PCNA on lysine 164 (K164) occurs in a damage-dependent manner (Hoege et al., 2002). Mutants of PCNA had previously been shown to be in the *RAD6* epistasis group (Torres-Ramos et al., 1996). Hoege *et al.* also demonstrated that the same residue of PCNA, K164, is modified in a cell cycle- and DNA damage-dependent manner by the attachment of the small ubiquitin-like modifier SUMO (Hoege et al., 2002). However, mutation of Siz1, which catalyses this attachment, does not confer sensitivity to damaging agents, suggesting that it is not essential for repair (Johnson and Gupta, 2001). In contrast, an *S. cerevisiae* strain that only contains PCNA that cannot be modified (because K164 is mutated to arginine(R)) is sensitive to the DNA damaging agent methyl methanesulfonate. This mutation is in the same epistasis group as *MMS2*, highlighting the importance of PCNA-ubiquitination for PRR (Hoege et al., 2002). These data, and further elegant genetic analysis of the K164R mutation showing that TLS by Rad30 (*S. cerevisiae* pol η) or Rev3 (pol ζ) depends upon mono-ubiquitination at K164 (Stelter and Ulrich, 2003) led to the model depicted in Figure 3. In this model, detection of DNA damage in *S. cerevisiae* leads to the mono-ubiquitination of PCNA on K164. After this mono-ubiquitination event, a choice is made either to undertake TLS with polη (error-free) or polζ (error-prone) to deal with the damage in a possibly mutagenic way, or to catalyse lysine 63-linked poly-ubiquitination via Ubc13, Mms2 and Rad5. The genetic studies suggest that this poly-ubiquitination will channel lesions into an error-free pathway, probably utilising one of the damage avoidance mechanisms indicated in Figure 2ii or 2iii. Many details of how this choice is made still remain to be clarified, but at least one aspect of how PCNA ubiquitination controls subsequent PRR events has recently been elucidated from studies in human cells (see below).

8.4 A Ubiquitinated Clamp for Recruitment of polη

Human cells contain single orthologs of Rad18 and Ubc13, and two orthologs of Rad6 and Mms2. No clear human ortholog of Rad5 has yet been identified. In human cells the modification of PCNA appears to be simpler than in *S. cerevisiae*, as to date neither SUMO-modified nor poly-ubiquitinated forms of PCNA have been detected. However, a mono-ubiquitinated form is detected after UV damage. This was shown to be dependent upon the hRad6 and hRad18 proteins, as in *S. cerevisiae*

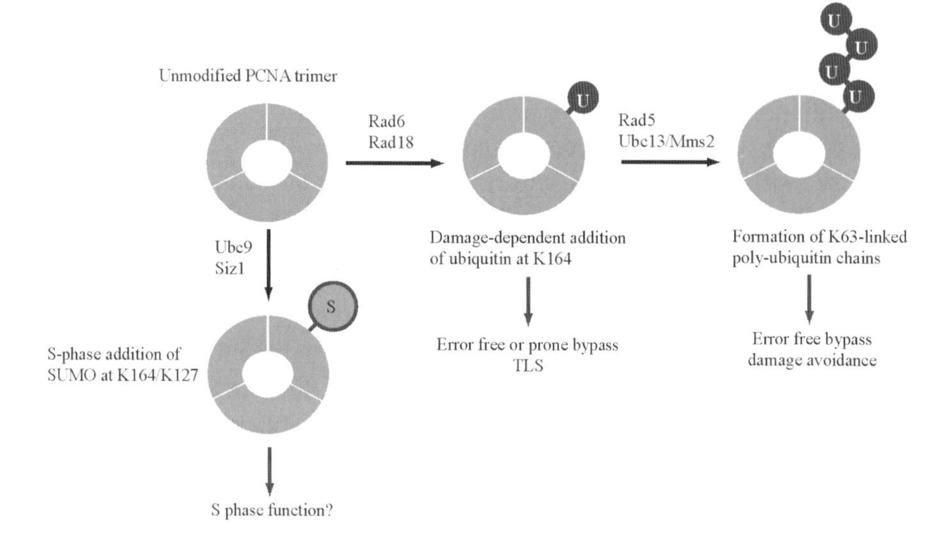

Figure 3. Modification of PCNA in *S. cerevisiae*. Three types of modification of the replicative sliding clamp, PCNA, have been detected in *S. cerevisiae*, and each seems to lead to a different biological endpoint. Sumoylation of residues K127 and K164 occurs during S phase, and acts antagonistically to the ubiquitination that occurs at K164. The Rad18/Rad6-dependent addition of a single ubiquitin residue allows the cell to perform lesion bypass via TLS. The subsequent formation of a K63-linked poly-ubiquitin chain by Mms2/Ubc13 with Rad5 channels PRR down an error-free pathway dependent on recombination proteins.

(Kannouche et al., 2004; Watanabe et al., 2004). This relative simplicity has perhaps facilitated study of the function of this modification in higher eukaryotes. Transient transfection of epitope-tagged PCNA into human cells and subsequent UV irradiation demonstrated that such expressed PCNA can be mono-ubiquitinated *in vivo* in a damage-dependent manner. Subsequent crosslinking and purification of the modified PCNA from these cells showed that mono-ubiquitinated PCNA was specifically able to interact physically with polη (Kannouche et al., 2004). This leads to an elegant model for how the polymerase switch is regulated in human cells: following UV damage, forks stalled at damaged sites result in the mono-ubiquitination of PCNA. This mono-ubiquitinated PCNA is then bound by the TLS polymerase to achieve bypass (Figure 4). Clearly there are many details that remain to be resolved in order for this to be a complete picture of events that occur at stalled forks, not least how the signals are initiated that lead to the activation of Rad18 and the choice of PCNA as a substrate. It will also be crucial to determine the mechanism by which the replicative polymerase is displaced, and then re-engaged beyond the damage site, and the fate of the ubiquitinated PCNA (Kannouche and Lehmann, 2004).

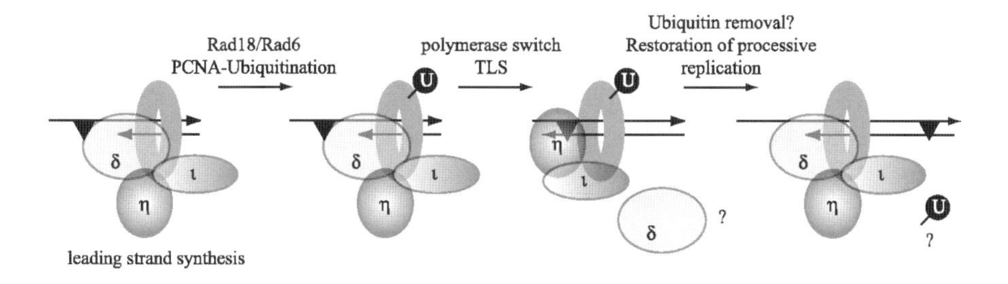

Figure 4. A mechanism for the polymerase switch in human cells. TLS polymerases may be associated with processive replication forks, in order to be available should the fork encounter a lesion. As in *S. cerevisiae,* DNA damage leads to modification of PCNA, but to date, in human cells, only a mono-ubiquitination event has been detected. This is observed after cell treatments expected to result in replication fork stalling, such as UV irradiation or hydroxyurea, and is dependent upon hRad6A/B and hRad18. The TLS polymerase, polη, specifically interacts with mono-ubiquitinated PCNA. Hence a simple mechanism can be proposed for the polymerase switch, in which the ubiquitination alters the affinity of PCNA for the different polymerases, such that at a stalled fork, polη is preferred to the replicative polδ. Once lesion bypass is achieved there must be a switch back to processive replication, although the mechanism by which this occurs remains to be determined.

9. FUTURE DIRECTIONS – THE QUESTION OF CHOICE/CONTROL

As discussed above, we now have a working model for how polη might be recruited to a UV damaged site. However this only begins to address the difficulties of decision-making inherent when the presence of all the other TLS polymerases are taken into account. What allows a cell to target a specific polymerase to a specific lesion? Is it merely a case that all polymerases are tested until one can do the job? For example we know that polη is recruited to replication foci when cells are treated with HU though it is unlikely that the presence of polη at such stalled forks is productive for a cell in this case. Such a "suck it and see" strategy does not seem to be an intrinsically sensible choice given the fact that some polymerases are far more likely to insert a mutational nucleotide across from a particular lesion than others. Does the cell step through an ordered series of polymerases starting with the least promiscuous, until it finds one that will do the job? The fact that in higher eukaryotes the Rev1 protein appears to interact with many TLS polymerases raises the possibility that it may act as a toolbelt, from which polymerases can be alternatively selected. An in depth study of the temporal recruitment of polymerases to damage sites *in vivo* may well be currently technically unreachable, but is certainly something to which we shall look in the future. In *S. cerevisiae,* the decision to attempt TLS as opposed to a recombinational repair mechanism appears to be determined by whether PCNA is mono- or poly-ubiquitinated. Does the fact that poly-ubiquitination has not been identified to date in human cells mean that higher eukaryotes regulate this decision in an unrelated manner? The complexity of these issues leaves us certain that there is much still to be learned in the future.

10. TLS POLYMERASES AND CANCER

Studies demonstrating that the spontaneous mutation rates of *E. coli* and yeast are due in large part to the activity of TLS polymerases suggest that the underlying mutations that lead to carcinogenesis in human cells may also derive from the activity of these polymerases. Indeed, XP-V patients give us a dramatic example of the problems for genomic stability that can arise when a single TLS polymerase is perturbed. It is certainly possible that the phenotypes of other diseases are to some extent a result of deregulation of other polymerases. There are some reports of overexpression of TLS polymerases in certain types of cancer, but as yet a cause-effect relationship has not been established (e.g. O-Wang et al., 2001). Furthermore the toxicity of many drugs used to treat cancer may be due in part to their mutagenic nature. Understanding the biochemical basis of mutagenesis resulting from bypass of lesions induced by chemotherapeutic agents may allow their rational design, to ensure that secondary mutagenic effects can be minimised in non target tissues while toxicity against target tumours remains high. Given the limited nature of *in vivo* mammalian studies of these proteins to date, we are sure that work in this area over the next few years will produce many insights into the precise roles that this plethora of polymerases plays within living cells.

REFERENCES

Arlett, C.F., S.A. Harcourt, and B.C. Broughton. 1975. The influence of caffeine on cell survival in excision- proficient and excision-deficient xeroderma pigmentosum and normal human cell strains following ultraviolet light irradiation. *Mutation Res.* 33:341-346.

Bailly, V., S. Lauder, S. Prakash, and L. Prakash. 1997. Yeast DNA repair proteins Rad6 and Rad18 form a heterodimer that has ubiquitin conjugating, DNA binding, and ATP hydrolytic activities. *J Biol Chem.* 272:23360-23365.

Becherel, O.J., and R.P. Fuchs. 2001. Mechanism of DNA polymerase II-mediated frameshift mutagenesis. *Proc Natl Acad Sci U S A.* 98:8566-71.

Becherel, O.J., R.P.P. Fuchs, and J. Wagner. 2002. Pivotal role of the beta-clamp in translesion DNA synthesis and mutagenesis in *E. coli* cells. *DNA Repair.* 1:703-708.

Bravo, R., and H. Macdonald-Bravo. 1987. Existence of two populations of cyclin/proliferating cell nuclear antigen during the cell cycle: association with DNA replication sites. *J Cell Biol.* 105:1549-1554.

Bruck, I., and M. O'Donnell. 2001. The ring-type polymerase sliding clamp family. *Genome Biol.* 2: reviews3001.1-3001.3.

Bunting, K.A., S.M. Roe, and L.H. Pearl. 2003. Structural basis for recruitment of translesion DNA polymerase Pol IV/DinB to the beta-clamp. *EMBO J.* 22:5883-92.

Covo, S., L. Blanco, and Z. Livneh. 2004. Lesion bypass by human DNA polymerase mu reveals a template-dependent, sequence-independent nucleotidyl transferase activity. *J Biol Chem.* 279:859-65.

Diaz, M., N.B. Watson, G. Turkington, L.K. Verkoczy, N.R. Klinman, and W.G. McGregor. 2003. Decreased frequency and highly aberrant spectrum of ultraviolet-induced mutations in the hprt gene of mouse fibroblasts expressing antisense RNA to DNA polymerase zeta. *Mol Cancer Res.* 1:836-47.

Fernandez de Henestrosa, A.R., T. Ogi, S. Aoyagi, D. Chafin, J.J. Hayes, H. Ohmori, and R. R. Woodgate. 2000. Identification of additional genes belonging to the LexA-regulon in *Escherichia coli. Molecular Microbiology.* 35:1560-1572.

Friedberg, E.C., R. Wagner, and M. Radman. 2002. Specialized DNA polymerases, cellular survival, and the genesis of mutations. *Science.* 296:1627-1630.

Friedberg, E.C., G.C. Walker, and W. Siede. 1995. DNA Repair and Mutagenesis. ASM Press, Washington, USA.

Fujii, S. and Fuchs, R.P. 2004. Defining the position of the switches between replicative and bypass DNA polymerases. *EMBO J.* 23:4342-4352.

Game, J.C. 2000. The *Saccharomyces* repair genes at the end of the century. *Mutat Res.* 451:277-93.

Gerlach, V.L., L. Aravind, G. Gotway, R.A. Schultz, E.V. Koonin, and E.C. Friedberg. 1999. Human and mouse homologs of *Escherichia coli* DinB (DNA polymerase IV), members of the UmuC/DinB superfamily. *Proc Natl Acad Sci U S A.* 96:11922-11927.

Gibbs, P.E.M., W.G. McGregor, V.M. Maher, P. Nisson, and C.W. Lawrence. 1998. A human homolog of the *Saccharomyces cerevisiae REV3* gene, which encodes the catalytic subunit of DNA polymerase ζ. *Proc Natl AcadSci., USA.* 95:6876-6880..

Goodman, M.F. 2002. Error-prone repair DNA polymerases in prokaryotes and eukaryotes. *Annu Rev Biochem.* 71:17-50.

Goodman, M.F., and B. Tippin. 2000. Sloppier copier DNA polymerases involved in genome repair. *Curr Opin Genet Dev.* 10:162-168.

Guo, C., P.L. Fischhaber, M.J. Luk-Paszyc, Y. Masuda, J. Zhou, K. Kamiya, C. Kisker, and E.C. Friedberg. 2003. Mouse Rev1 protein interacts with multiple DNA polymerases involved in translesion DNA synthesis. *EMBO J.* 22:6621-30.

Guo, D., X. Wu, D.K. Rajpal, J.S. Taylor, and Z. Wang. 2001. Translesion synthesis by yeast DNA polymerase zeta from templates containing lesions of ultraviolet radiation and acetylaminofluorene. *Nucleic Acids Res.* 29:2875-2883.

Haracska, L., R.E. Johnson, I. Unk, B. Phillips, J. Hurwitz, L. Prakash, and S. Prakash. 2001a. Physical and Functional Interactions of Human DNA Polymerase η with PCNA. *Mol Cell Biol.* 21:7199-7206.

Haracska, L., R.E. Johnson, I. Unk, B.B. Phillips, J. Hurwitz, L. Prakash, and S. Prakash. 2001b. Targeting of human DNA polymerase ι to the replication machinery via interaction with PCNA. *Proc Natl Acad Sci U S A.* 98:14256-14261.

Haracska, L., I. Unk, R.E. Johnson, B.B. Phillips, J. Hurwitz, L. Prakash, and S. Prakash. 2002. Stimulation of DNA synthesis activity of human DNA polymerase κ by PCNA. *Mol Cell Biol.* 22:784-791.

Havener, J.M., S.A. McElhinny, E. Bassett, M. Gauger, D.A. Ramsden, and S.G. Chaney. 2003. Translesion synthesis past platinum DNA adducts by human DNA polymerase μ. *Biochemistry.* 42:1777-88.

Hoege, C., B. Pfander, G.-L. Moldovan, G. Pyrolowakis, and S. Jentsch. 2002. *RAD6*-dependent DNA repair is linked to modification of PCNA by ubiquitin and SUMO. *Nature.* 419:135-141.

Hofmann, R.M., and C.M. Pickart. 1999. Noncanonical MMS2-encoded ubiquitin-conjugating enzyme functions in assembly of novel polyubiquitin chains for DNA repair. *Cell.* 96:645-653.

Jentsch, S., J.P. McGrath, and A. Varshavsky. 1987. The yeast DNA repair gene *RAD6* encodes a ubiquitin-conjugating enzyme. *Nature*. 329:131-134.

Johnson, E.S., and A.A. Gupta. 2001. An E3-like factor that promotes SUMO conjugation to the yeast septins. *Cell*. 106:735-744.

Johnson, R.E., C.M. Kondratick, S. Prakash, and L. Prakash. 1999a. *hRAD30* mutations in the variant form of xeroderma pigmentosum. *Science*. 285:263-265.

Johnson, R.E., S. Prakash, and L. Prakash. 1999b. Efficient bypass of a thymine-thymine dimer by yeast DNA polymerase, Polη. *Science*. 283:1001-1004.

Johnson, R.E., M.T. Washington, L. Haracska, S. Prakash, and L. Prakash. 2000. Eukaryotic polymerases ι and ζ act sequentially to bypass DNA lesions. *Nature*. 406:1015-1019.

Kai, M., and T. Wang. 2003. Checkpoint activation regulates mutagenic translesion synthesis. *Genes Dev*. 17:64-76.

Kannouche, P., B.C. Broughton, M. Volker, F. Hanaoka, L.H.F. Mullenders, and A.R. Lehmann. 2001. Domain structure, localization and function of DNA polymerase η, defective in xeroderma pigmentosum variant cells. *Genes Dev*. 15:158-172.

Kannouche, P., A.R. Fernandez de Henestrosa, B. Coull, A.E. Vidal, C. Gray, D. Zicha, R. Woodgate, and A.R. Lehmann. 2003. Localization of DNA polymerases η and ι to the replication machinery is tightly co-ordinated in human cells. *EMBO J*. 22:1223-1233.

Kannouche, P.L., and A.R. Lehmann. 2004. Ubiquitination of PCNA and the Polymerase Switch in Human Cells. *Cell Cycle*. 3: 1011-1013.

Kannouche, P.L., J. Wing, and A.R. Lehmann. 2004. Interaction of Human DNA Polymerase η with Monoubiquitinated PCNA; A Possible Mechanism for the Polymerase Switch in Response to DNA Damage. *Mol Cell*. 14:491-500.

Kraemer, K.H., M.M. Lee, and J. Scotto. 1987. Xeroderma Pigmentosum. Cutaneous, ocular and neurologic abnormalities in 830 published cases. *Arch Dermatol*. 123:241-250.

Lehmann, A.R. 2002. Replication of damaged DNA in mammalian cells: new solutions to an old problem. *Mutat Res*. 509:23-34.

Lehmann, A.R., S. Kirk-Bell, C.F. Arlett, M.C. Paterson, P.H.M. Lohman, E.A. de Weerd-Kastelein, and D. Bootsma. 1975. Xeroderma pigmentosum cells with normal levels of excision repair have a defect in DNA synthesis after UV-irradiation. *Proc Natl Acad Sci, USA*. 72:219-223.

Lemontt, J.F. 1971. Mutants of yeast defective in mutation induced by ultraviolet light. *Adv. Genet*. 68:21-33.

Lenne-Samuel, N., J. Wagner, H. Etienne, and R.P. Fuchs. 2002. The processivity factor beta controls DNA polymerase IV traffic during spontaneous mutagenesis and translesion synthesis in vivo. *Embo Rep*. 3:45-9.

Li, Z., C.J. Woo, M.D. Iglesias-Ussel, D. Ronai, and M.D. Scharff. 2004. The generation of antibody diversity through somatic hypermutation and class switch recombination. *Genes Dev*. 18:1-11.

Li, Z., H. Zhang, T.P. McManus, J.J. McCormick, C.W. Lawrence, and V.M. Maher. 2002. hREV3 is essential for error-prone translesion synthesis past UV or benzo[a]pyrene diol epoxide-induced DNA lesions in human fibroblasts. *Mutat Res*. 510:71-80.

Lin, W., H. Xin, Y. Zhang, X. Wu, F. Yuan, and Z. Wang. 1999. The human *REV1* gene codes for a DNA template-dependent dCMP transferase. *Nucleic Acids Res*. 27:4468-4475.

Ling, H., F. Boudsocq, B.S. Plosky, R. Woodgate, and W. Yang. 2003. Replication of a cis-syn thymine dimer at atomic resolution. *Nature*. 424:1083-7.

Ling, H., F. Boudsocq, R. Woodgate, and W. Yang. 2001. Crystal structure of a Y-family DNA polymerase in action: a mechanism for error-prone and lesion-bypass replication. *Cell*. 107:91-102.

Lopez de Saro, F., R.E. Georgescu, F. Leu, and M. O'Donnell. 2004. Protein trafficking on sliding clamps. *Philos Trans R Soc Lond B Biol Sci*. 359:25-30.

Lopez de Saro, F.J., R.E. Georgescu, M.F. Goodman, and M. O'Donnell. 2003. Competitive processivity-clamp usage by DNA polymerases during DNA replication and repair. *EMBO J*. 22:6408-18.

Maher, V.M., L.M. Ouellette, R.D. Curren, and J.J. McCormick. 1976. Frequency of ultraviolet light-induced mutations is higher in xeroderma pigmentosum variant cells than in normal human cells. *Nature*. 261:593-595.

Masutani, C., M. Araki, A. Yamada, R. Kusumoto, T. Nogimori, T. Maekawa, S. Iwai, and F. Hanaoka. 1999a. Xeroderma pigmentosum variant (XP-V) correcting protein from HeLa cells has a thymine dimer bypass DNA polymerase activity. *EMBO J*. 18:3491-3501.

Masutani, C., R. Kusumoto, S. Iwai, and F. Hanaoka. 2000. Accurate translesion synthesis by human DNA polymerase η. *EMBO J*. 19:3100-3109.

Masutani, C., R. Kusumoto, A. Yamada, N. Dohmae, M. Yokoi, M. Yuasa, M. Araki, S. Iwai, K. Takio, and F. Hanaoka. 1999b. The XPV (xeroderma pigmentosum variant) gene encodes human DNA polymerase eta. *Nature*. 399:700-4.

McCulloch, S.D., R.J. Kokoska, C. Masutani, S. Iwai, F. Hanaoka, and T.A. Kunkel. 2004. Preferential cis-syn thymine dimer bypass by DNA polymerase η occurs with biased fidelity. *Nature*. 428:97-100.

McDonald, J.P., E.G. Frank, B.S. Plosky, I.B. Rogozin, C. Masutani, F. Hanaoka, R. Woodgate, and P.J. Gearhart. 2003. 129-derived strains of mice are deficient in DNA polymerase ι and have normal immunoglobulin hypermutation. *J Exp Med*. 198:635-43.

McDonald, J.P., A.S. Levine, and R. Woodgate. 1997. The *Saccharomyces cerevisiae RAD30* gene, a homologue of *Escherichia coli dinB* and *umuC*, is DNA damage inducible and functions in a novel error-free postreplication repair mechanism. *Genetics*. 147:1557-1568.

McDonald, J.P., V. Rapic-Otrin, J.A. Epstein, B.C. Broughton, X. Wang, A.R. Lehmann, D.J. Wolgemuth, and R. Woodgate. 1999. Novel human and mouse homologs of *Saccharomyces cerevisiae* DNA polymerase η. *Genomics*. 60:20-30.

Nair, D.T., R.E. Johnson, S. Prakash, L. Prakash, and A.K. Aggarwal. 2004. Replication by human DNA polymerase-iota occurs by Hoogsteen base-pairing. *Nature*. 430:377-80.

Naktinis, V., J. Turner, and M. O'Donnell. 1996. A molecular switch in a replication machine defined by an internal competition for protein rings. *Cell*. 84:137-45.

Napolitano, R., R. Janel-Bintz, J. Wagner, and R.P. Fuchs. 2000. All three SOS-inducible DNA polymerases (Pol II, Pol IV and Pol V) are involved in induced mutagenesis. *EMBO J*. 19:6259-65.

Nelson, J.R., P.E. Gibbs, A.M. Nowicka, D.C. Hinkle, and C.W. Lawrence. 2000. Evidence for a second function for Saccharomyces cerevisiae Rev1p. *Mol Microbiol*. 37:549-554.

Nelson, J.R., C.W. Lawrence, and D.C. Hinkle. 1996a. Deoxycytidyl transferase activity of yeast *REV1* protein. *Nature*. 382:729-31.

Nelson, J.R., C.W. Lawrence, and D.C. Hinkle. 1996b. Thymine-thymine dimer bypass by yeast DNA polymerase ζ. *Science*. 272:1646-1649.

Ogi, T., T. Kato, and H. Ohmori. 1999. Mutation enhancement by DINB1, a mammalian homologue of the *Escherichia coli* mutagenesis protein dinB. *Genes Cells*. 4:607-618.

Ogi, T., J. Mimura, M. Hikida, H. Fujimoto, Y. Fujii-Kuriyama, and H. Ohmori. 2001. Expression of human and mouse genes encoding polκ: testis-specific developmental regulation and AhR-dependent inducible transcription. *Genes Cells*. 6:943-953.

Ogi, T., Y. Shinkai, K. Tanaka, and H. Ohmori. 2002. Pol κ protects mammalian cells against the lethal and mutagenic effects of benzo[a]pyrene. *Proc Natl Acad Sci U S A*. 99:15548-15553.

Ogi, T., P. Kannouche, and A. R. Lehmann. 2005. Localization of human DNA polymerase κ (polκ), a Y-family DNA polymerase: relationship to PCNA foci. *J Cell Sci*.118:129-136.

Ohashi, E., Y. Murakumo, N. Kanjo, J. Akagi, C. Masutani, F. Hanaoka, and H. Ohmori. 2004. Interaction of hREV1 with three human Y-family DNA polymerases. *Genes Cells*. 9:523-31.

Ohmori, H., E.C. Friedberg, R.P.P. Fuchs, M.F. Goodman, F. Hanaoka, D. Hinkle, T.A. Kunkel, C.W. Lawrence, Z. Livneh, T. Nohmi, L. Prakash, S. Prakash, T. Todo, G.C. Walker, Z. Wang, and R. Woodgate. 2001. The Y-family of DNA polymerases. *Molecular Cell*. 8:7-8.

O-Wang, J., Kawamura, K., Tada, Y., Ohmori, H., Kimura, H., Sakiyama, S., and Tagawa, M. 2001. DNA polymerase kappa, implicated in spontaneous and DNA damage-induced mutagenesis, is overexpressed in lung cancer. *Cancer Res* 61:5366-5369.

Pages, V., and R.P. Fuchs. 2002. How DNA lesions are turned into mutations within cells? *Oncogene*. 21:8957-8966.

Peat, T.S., E.G. Frank, J.P. McDonald, A.S. Levine, R. Woodgate, and W.A. Hendrickson. 1996. Structure of the UmuD' protein and its regulation in response to DNA damage. *Nature*. 380:727-30.

Prakash, S., and L. Prakash. 2002. Translesion DNA synthesis in eukaryotes: a one- or two-polymerase affair. *Genes Dev*. 16:1872-1883.

Radman, M. 1974. Phenomenology of an inducible mutagenic DNA repair pathway in Escherichia coli: SOS repair hypothesis. *In* Molecular and environmental aspects of mutagenesis. P. L, F. Sherman, M. Miller, C. Lawrence, and W.H. Tabor, editors. Charles C. Thomas, Springfield Ill. 128-142.

Rattray, A.J., and J.N. Strathern. 2003. Error-prone DNA polymerases: when making a mistake is the only way to get ahead. *Annu Rev Genet*. 37:31-66.

Reuven, N.B., G. Arad, A. Maor-Shoshani, and Z. Livneh. 1999. The mutagenesis protein UmuC is a DNA polymerase activated by UmuD', RecA, and SSB and Is specialized for translesion replication. *J Biol Chem*. 274:31763-31766.

Roush, A.A., M. Suarez, E.C. Friedberg, M. Radman, and W. Siede. 1998. Deletion of the *Saccharomyces cerevisiae* gene *RAD30* encoding an *Escherichia coli* DinB homolog confers UV radiation sensitivity and altered mutability. *Mol Gen Genet*. 257:686-692.

Rupp, W.D., and P. Howard-Flanders. 1968. Discontinuities in the DNA synthesized in an excision-defective strain of *Escherichia coli* following ultraviolet irradiation. *J Mol Biol*. 31:291-304.

Rupp, W.D., C.E. Wilde, D.L. Reno, and P. Howard-Flanders. 1971. Exchanges between DNA strands in ultraviolet irradiated *Escherichia coli*. *J. Mol Biol.* 61:25-44.

Schenten, D., V.L. Gerlach, C. Guo, S. Velasco-Miguel, C.L. Hladik, C.L. White, E.C. Friedberg, K. Rajewsky, and G. Esposito. 2002. DNA polymerase κ deficiency does not affect somatic hypermutation in mice. *Eur J Immunol*. 32:3152-3160.

Simpson, L.J., and J.E. Sale. 2003. Rev1 is essential for DNA damage tolerance and non-templated immunoglobulin gene mutation in a vertebrate cell line. *EMBO J*. 22:1654-64.

Stelter, P., and H.D. Ulrich. 2003. Control of spontaneous and damage-induced mutagenesis by SUMO and ubiquitin conjugation. *Nature*. 425:188-191.

Suzuki, N., E. Ohashi, A. Kolbanovskiy, N.E. Geacintov, A.P. Grollman, H. Ohmori, and S. Shibutani. 2002. Translesion synthesis by human DNA polymerase καππα on a DNA template containing a single stereoisomer of dG-(+)- or dG-(-)-anti-N(2)-BPDE (7,8-dihydroxy-anti-9,10-epoxy-7,8,9,10-tetrahydrobenzo[a]pyrene). *Biochemistry*. 41:6100-6.

Tanaka, K., T. Yonekawa, Y. Kawasaki, M. Kai, K. Furuya, M. Iwasaki, H. Murakami, M. Yanagida, and H. Okayama. 2000. Fission yeast eso1p is required for establishing sister chromatid cohesion during S phase. *Mol Cell Biol*. 20:3459-3469.

Tang, M., P. Pham, X. Shen, J.-S. Taylor, M. O'Donnell, R. Woodgate, and M. Goodman. 2000. Roles of *E. coli* DNA polymerases IV and V in lesion-targeted and untargeted SOS mutagenesis. *Nature*. 404:1014-1018.

Tang, M., X. Shen, E.G. Frank, M. O'Donnell, R. Woodgate, and M.F. Goodman. 1999. UmuD'(2)C is an error-prone DNA polymerase, *Escherichia coli* pol V. *Proc Natl Acad Sci U S A*. 96:8919-8924.

Tissier, A., E.G. Frank, J.P. McDonald, S. Iwai, F. Hanaoka, and R. Woodgate. 2000a. Misinsertion and bypass of thymine-thymine dimers by human DNA polymerase ι. *EMBO J*. 19:5259-5266.

Tissier, A., P. Kannouche, M.-P. Reck, A.R. Lehmann, R.P.P. Fuchs, and A. Cordonnier. 2004. Co-localization in replication foci and interaction of human Y-family members, DNA polymerase polη and Rev1 protein. *DNA repair*. 3:1503-1514.

Tissier, A., J.P. McDonald, E.G. Frank, and R. Woodgate. 2000b. Polι, a remarkably error-prone human DNA polymerase. *Genes Dev*. 14:1642-1650.

Torres-Ramos, C.A., B.L. Yoder, P.M. Burgers, S. Prakash, and L. Prakash. 1996. Requirement of proliferating cell nuclear antigen in *RAD6*-dependent postreplicational DNA repair. *Proc Natl Acad Sci U S A*. 93:9676-9681.

Trincao, J., R.E. Johnson, C.R. Escalante, S. Prakash, L. Prakash, and A.K. Aggarwal. 2001. Structure of the catalytic core of *S. cerevisiae* DNA polymerase η: implications for translesion DNA synthesis. *Mol Cell*. 8:417-426.

Uljon, S.N., R.E. Johnson, T.A. Edwards, S. Prakash, L. Prakash, and A.K. Aggarwal. 2004. Crystal structure of the catalytic core of human DNA polymerase κ *Structure (Camb)*. 12:1395-404.

Ulrich, H.D., and S. Jentsch. 2000. Two RING finger proteins mediate cooperation between ubiquitin-conjugating enzymes in DNA repair. *EMBO J*. 19:3388-3397.

Vaisman, A., E.G. Frank, J.P. McDonald, A. Tissier, and R. Woodgate. 2002. poliota-dependent lesion bypass in vitro. *Mutat Res*. 510:9-22.

Venclovas, C., and M.P. Thelen. 2000. Structure-based predictions of rad1, rad9, hus1 and rad17 participation in sliding clamp and clamp-loading complexes. *Nucleic Acids Res*. 28:2481-2493.

Vidal, A.E., P.P. Kannouche, V.N. Podust, W. Yang, A.R. Lehmann, and R. Woodgate. 2004. PCNA-dependent coordination of the biological functions of human DNA polymerase ι. *J Biol Chem*. 279: 48360-48368

Wagner, J., H. Etienne, R. Janel-Bintz, and R.P.P. Fuchs. 2002. Genetics of mutagenesis in E. coli: various combinations of translesion polymerases (Pol II, IV, and V) deal with lesion/sequence diversity. *DNA Repair*. 1:159-167.

Wagner, J., S. Fujii, P. Gruz, T. Nohmi, and R.P. Fuchs. 2000. The beta clamp targets DNA polymerase IV to DNA and strongly increases its processivity. *Embo Rep*. 1:484-488.

Wagner, J., P. Gruz, S.R. Kim, M. Yamada, K. Matsui, R.P. Fuchs, and T. Nohmi. 1999. The dinB gene encodes a novel E. coli DNA polymerase, DNA pol IV, involved in mutagenesis. *Mol Cell*. 4:281-286.

Warbrick, E. 2000. The puzzle of PCNA's many partners. *Bioessays*. 22:997-1006.

Watanabe, K., S. Tateishi, M. Kawasuji, T. Tsurimoto, H. Inoue, and M. Yamaizumi. 2004. Rad18 guides poleta to replication stalling sites through physical interaction and PCNA monoubiquitination. *EMBO J*. 23.

Xiao, W., B.L. Chow, S. Broomfield, and M. Hanna. 2000. The *Saccharomyces cerevisiae* RAD6 group is composed of an error-prone and two error-free postreplication repair pathways. *Genetics*. 155:1633-41.

Zhang, Y., X. Wu, D. Guo, O. Rechkoblit, J.S. Taylor, N.E. Geacintov, and Z. Wang. 2002. Lesion bypass activities of human DNA polymerase μ. *J Biol Chem*. 277:44582-7.

Zhang, Y., F. Yuan, X. Wu, M. Wang, O. Rechkoblit, J.S. Taylor, N.E. Geacintov, and Z. Wang. 2000. Error-free and error-prone lesion bypass by human DNA polymerase κ in vitro. *Nucleic Acids Res*. 28:4138-4146.

Part 3

Cell Cycle Progression and Chromosome Aberration

Chapter 3.1

THE INK4A/ARF NETWORK – CELL CYCLE CHECKPOINT OR EMERGENCY BRAKE?

Ana Gutierrez del Arroyo and Gordon Peters
Cancer Research UK, London Research Institute, London, UK

1. INTRODUCTION

Most contemporary reviews on cell cycle regulation in mammalian cells include some discussion of the *Ink4a/Arf* locus. The reasons are rather obvious. The proteins encoded by the locus impact significantly on key regulatory networks centred on the retinoblastoma (*RB1*) and p53 (*TP53*) tumour suppressor genes, and *Ink4a/Arf* itself is an accredited tumour suppressor, fostering the idea that it has role in controlling cell division and proliferation. However, a cogent case can be made that *Ink4a/Arf* may have more to do with emergencies than the day to day business of cell division and checkpoint control. In this article, we will summarise the current understanding of the regulation and function of *Ink4a/Arf* and present the evidence for and against a role in genome instability.

2. ORGANISATION OF THE *INK4A/ARF* LOCUS

One of the unusual features of the *Ink4a/Arf* locus is that it specifies two structurally and functionally distinct proteins from transcripts that initiate at separate promoters but share a common exon which is translated in alternative reading frames (Figure 1a). For more detailed information, the reader can refer to a number of previous reviews (Drayton and Peters, 2002; Ruas and Peters, 1998; Sharpless and DePinho, 1999; Sherr, 2001). To our knowledge, such a gene organisation is unique in the human genome and without the need to economise, as in the condensed genomes of viruses, it is hard to envisage the evolutionary pressures that led to such an arrangement. The p16^{Ink4a} protein, like other members of the Ink4 (inhibitors of Cdk4) family, comprises a set of ankyrin-like repeats and little else (Figure 2).

E.A. Nigg (ed.), Genome Instability in Cancer Development, 227-247.

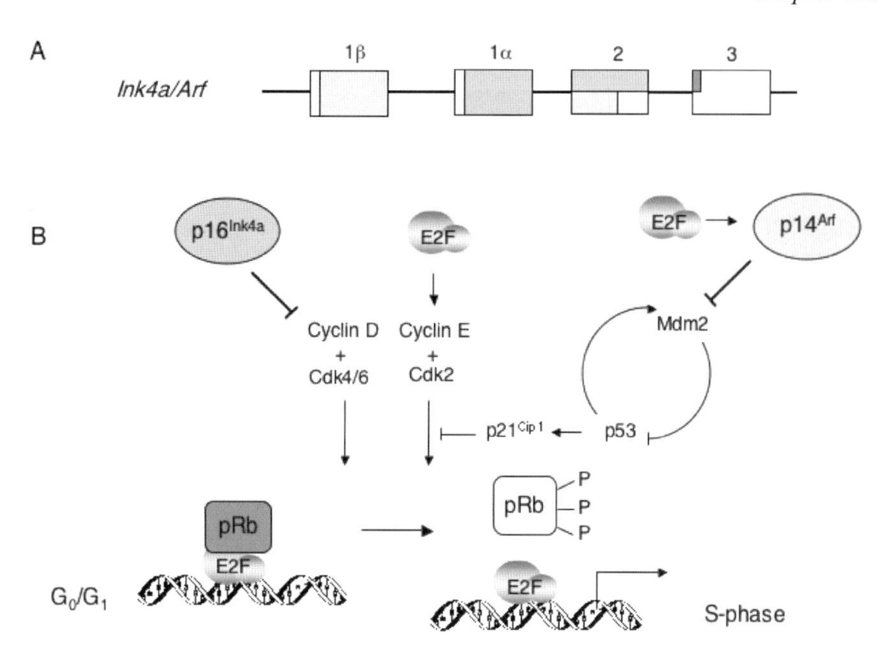

Figure 1. **A.** Organisation of the *Ink4a/Arf* locus. The *Ink4a/Arf* locus on human chromosome 9p21 (mouse chromosome 4) has two promoters and alternative first exons (1α and 1β) that are spliced to the same second and third exons. The α transcript encodes Ink4a (darker shading) and the β transcript encodes Arf (lighter shading). The Arf sequences specified by exon 2 are translated in the –1 reading frame relative to that of Ink4a. **B.** The G_0/G_1 to S phase transition in mammalian cells. Current understanding of G1 progression envisages sequential phosphorylation of the retinoblastoma protein (pRb) and its relatives by the cyclin D-dependent kinases Cdk4 and Cdk6 followed by cyclin E-Cdk2. This releases the repressive influence of pRb on the E2F family of transcription factors and allows the temporally ordered expression of genes required for S-phase. Cyclin E expression is E2F-dependent and the activity of Cdk2 is also modulated by the extent to which the $p21^{Cip1}/p27^{Kip1}$ Cdk inhibitors are sequestered in cyclin D-dependent complexes. By binding directly to Cdk4 and 6, Ink4a prevents them from associating with D cyclins and causes redistribution of $p21^{Cip1}/p27^{Kip1}$ onto cyclin E-Cdk2. Levels of $p21^{Cip1}$ are also regulated by the p53 transcription factor. The activity of p53 is in turn controlled by Mdm2, an E3 ubiquitin ligase that promotes the proteasome-mediated destruction of p53 as well as directly blocking its transcription activity. As the expression of Mdm2 is p53 dependent, there is a feedback loop through which p53 activity is tightly constrained. Arf interferes with this balance by binding to Mdm2 and inhibiting its ubiquitin ligase activity, thereby stabilising p53. Inactivation of pRb also impacts on this pathway because expression of Arf is activated by E2F. In a further feedback loop, p53 appears to repress the transcription of Arf.

Available crystal structures and the distribution of missense mutations found in human cancers suggest that contacts between Ink4a and its targets, the cyclin dependent kinases Cdk4 and Cdk6, requires maintenance of all four ankyrin repeats (Ruas and Peters, 1998). In contrast, the Arf (alternative reading frame) protein is rather variable in length in different species because of the different extents to which the relevant reading frame in exon 2 remains open (Figure 3). Most of the experimental evidence favours the idea that the amino acids encoded by exon 2 are dispensable for Arf function

(Quelle et al., 1997; Rizos et al., 2000; Stott et al., 1998; Zhang et al., 1998) and indeed the chicken Arf protein has no contribution at all from exon 2 because the splice from exon 1β into exon 2 occurs in a different register to that used in mammals (Kim et al., 2003). Apart from being very rich in arginine, the known forms of Arf in different species are poorly conserved at the primary sequence level (Figure 3) and the only structural information available thus far suggests that the region of the protein specified by exon 1β is unlikely to have an ordered conformation unless associated with other proteins (Bothner et al., 2001). We would argue, therefore, that if there is an evolutionary rationale for the location of *Arf* next to *Ink4a* in the genome, it must reflect their regulation rather than their composition. While the sharing of an exon could be pure coincidence, perhaps providing a way in which the *Arf* transcript can be processed and polyadenylated, a more attractive possibility is that this unique arrangement reflects some common purpose for the encoded products.

Continuing the evolutionary theme, it is important to point out that although *Ink4* homologues have been identified in fish (Gilley and Fried, 2001; Kazianis et al., 1999), the earliest known *Arf* ancestor is in chickens (Kim et al., 2003) and there are no obvious relatives in the genome databases. This clearly sets *Ink4a/Arf* apart from many of the other genes discussed in this volume, where functional analogies can be traced from yeasts to humans.

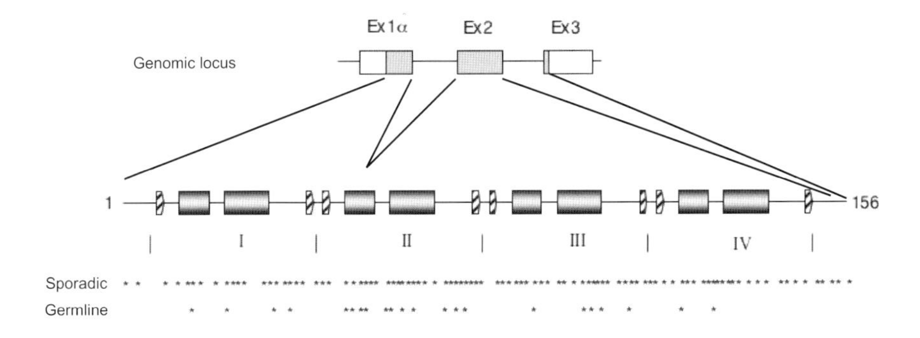

Figure 2. Location of missense mutations relative to the organisation and structure of Ink4a. The figure shows how the three exons that specify $p16^{Ink4a}$ relate to the ankyrin repeats within the protein. Each repeat, numbered I-IV, comprises β-hairpin (striped arrows) and α helical (cylinders) modules as indicated. Below this linear depiction of the primary sequence are the codons that have been reported to sustain missense mutations in sporadic human cancers or in the germline of familial melanoma kindreds. The germline mutations are known to inactivate the protein to some degree but not all the sporadic mutations have been functionally evaluated. Note that the locus also sustains nonsense and frameshift mutations, deletions and insertions, splicing defects and promoter methylation in human cancers, that are not depicted in this diagram.

			Contribution from exon 2
Human	MVRRFLVTLRIRR--ACGPPRVRVFVVHIPRLTGEWAAPGAPAAVALVLMLLRSQRLGQQPLPRRP	64	68
Pig	MVRRLLITVRIRR--SCGPPRVRAFVVQIARPAGEWAAPGVRAAAARVLLLVRSQRRAQQPHPRP	63	68
Mouse	MGRRFLVTVRIQR--AGRPLQERVFLVKFVRSRRPRTASCALAFVNMLLRLERILRRGPHRNPGP	63	106
Rat	MGRRFVVTVRIRR--TGRSPQVRVFLVQFLGSSRPRSANGTRGFVALVLRPERIARRGPQPHPGP	63	97
Hamster	MGRRFVVTVRIRRRRADRPPRVRAFVVQFPRSSRHRSASRARAVVALLLMLARSQRQRGPRLHSGA	66	57
Opossum	MIRRVRVTVRVSR--ACRPHHVRIFVAKIVQALCRASASINQGTPFQVLLIVRKKRHRGRS	59	96
Chicken	MTSRIRCTVCLRR--ARSRPLSFSLLRRILRGVAAVLRRSGTLRRILRRVLRRRHRGSRRPR	60	0

Figure 3. Comparison of Arf sequences from different species. The region of Arf specified by exon 1β is shown in single letter amino acid code. The figure aligns available information for the human, pig, mouse, rat, hamster, opossum and chicken homologues. The numbers refer to the amino acid residues specified by exon 1β, followed by the number specified by the alternative reading frame in exon 2. Residues shared by all seven homologues are highlighted. Note that there are other ways of maximising the alignment and several residues that are common to most but not all versions of the protein.

3. CELL CYCLE ARREST BY INK4A

The most obvious attribute shared by Ink4a and Arf is their ability to cause a cell cycle arrest. As its name implies, $p16^{Ink4a}$ achieves this by binding directly to the cyclin-dependent kinases Cdk4 and Cdk6 (Serrano et al., 1993), preventing them from associating with their regulatory subunits, the D-type cyclins (Figure 1). The accepted wisdom is that following induction of the D-type cyclins by extracellular growth factors, these cyclin-Cdk complexes are responsible for initiating the phosphorylation and functional inactivation of the retinoblastoma protein (pRb) and the related p107 and p130 proteins (Sherr and McCormick, 2002). This releases their inhibitory influence on the E2F family of transcription factors, enabling E2F-dependent transcription of a set of genes that are required for DNA replication and S phase. Among the target genes is cyclin E which in conjunction with Cdk2 is thought to complete or maintain the phosphorylation of pRb. As well as creating an amplification loop for inactivating pRb, an obvious attraction of this model is that it provides an explanation for the G1 restriction point, because the phosphorylation of pRb switches from being growth factor dependent (via D cyclins) to growth factor independent (via cyclin E). In addition to inhibiting Cdk4 and Cdk6, excess $p16^{Ink4a}$ can also inhibit Cdk2 by causing the redistribution of the $p21^{Cip1}$ family of Cdk inhibitors from cyclin D-Cdk complexes onto cyclin E-containing complexes (Jiang et al., 1998; McConnell et al., 1999). The net result is an arrest in the G1 phase. Despite the wide acceptance of this model and a wealth of supporting evidence, it is important to point out that it faces challenges from the recently reported phenotypes of mice lacking the various cyclins and Cdks (Kozar et al., 2004; Malumbres et al., 2004; Ortega et al., 2003).

4. CELL CYCLE ARREST BY ARF

In the case of Arf, cell cycle arrest is for the most part achieved by its ability to bind to Mdm2 and inhibit its associated E3 ubiquitin ligase activity (Honda and Yasuda, 1999; Kamijo et al., 1998; Midgley et al., 2000; Pomerantz et al., 1998; Stott et al., 1998). Mdm2 serves as a critical modulator of p53 activity by directly blocking its transcriptional activation functions and by promoting its degradation by the proteasome (Michael and Oren, 2003). The Mdm2 gene is itself activated by p53, setting up a feedback loop through which p53 function is tightly controlled (Figure 1). Although there has been some debate about the mechanistic details (Kashuba et al., 2004; Llanos et al., 2001; Lohrum et al., 2000a; Rizos et al., 2000; Tao and Levine, 1999; Weber et al., 1999; Zhang and Xiong, 1999), expression of Arf causes the stabilisation of p53 and consequent up-regulation of p53 target genes. These include the $p21^{Cip1}$ gene, which presumably underlies the ability of Arf to cause a cell cycle arrest in G1 and G2 (Quelle et al., 1997; Stott et al., 1998) in line with the ability of $p21^{Cip1}$ to inhibit multiple Cdks (Medema et al., 1998; Niculescu III et al., 1998). In general terms, therefore, Ink4a-mediated arrest is pRb-dependent whereas Arf-mediated arrest is p53-dependent (Figure 1). Although there is much experimental support for this conjecture, it is still a subject of debate (Korgaonkar et al., 2002; Weber et al., 2002; Yarbrough et al., 2002) and Arf has the capacity to arrest mouse embryo fibroblasts (MEFs) that lack $p21^{Cip1}$ (Modestou et al., 2001) or both p53 and Mdm2 (Weber et al., 2000a), prompting a great deal of interest in the possible mechanisms.

5. INK4A/ARF IN THE NORMAL CELL CYCLE

The type of linear pathways drawn in Figure 1 may be helpful in formulating functional connections but they can also give a false impression. They do not take account of the relative amounts of the various components or their rates of synthesis and turnover. For example, Ink4a is a relatively stable protein, with a half life of several hours (Parry et al., 1995) and in normal, i.e. primary cells, its levels are rarely if ever high enough to restrict the functions of Cdk4 and Cdk6 during G1 progression. Moreover, the levels of Ink4a do not fluctuate appreciably during cell cycle progression and only marginal changes have been reported when quiescent cells are stimulated with growth factors (Hara et al., 1996; Soucek et al., 1995; Tam et al., 1994). Although it has been shown that $p16^{Ink4a}$ levels are increased when pRb is completely inactivated by viral proteins such as SV40 large T-antigen, adenovirus E1A or HPV E7, or by naturally occurring mutations (Hara et al., 1996; Li et al., 1994; Parry et al., 1995), this does not equate with the cyclical inactivation of pRb by Cdk-mediated phosphorylation. Thus, it is probably incorrect to classify $p16^{Ink4a}$ as a "cell cycle regulator" although other members of the Ink4 family, such as $p18^{Ink4c}$ or $p19^{Ink4d}$ may perform this function (Hirai et al., 1995). It is also wrong to conclude from diagrams such as Figure 1 that loss of Ink4a will inevitably result in unrestricted phosphorylation of pRb, release of E2F and up-regulation of Arf.

Similar considerations apply to Arf. Despite consistent reports that it is activated by members of the E2F family of transcription factors (Bates et al., 1998; DeGregori et al., 1997; Dimri et al., 2000; Lomazzi et al., 2002; Parisi et al., 2002; Robertson and Jones, 1998; Zindy et al., 1998), evidence for cell cycle regulation remains equivocal (Aslanian et al., 2004; Buschmann et al., 2000; Quelle et al., 1995; Stone et al., 1995; Stott et al., 1998). Nevertheless, agents that incapacitate pRb, such as viral oncoproteins, do activate Arf and elicit a p53 response (De Stanchina et al., 1998) and viruses generally compensate for this by encoding proteins that inactivate p53. To complicate matters, p53 also exerts a negative influence on Arf expression by an as yet unknown mechanism (Kamijo et al., 1998; Stott et al., 1998). As a result, Arf is more readily detected in cells that have no or mutant p53 (Quelle et al., 1995; Stott et al., 1998). In general terms, it has proved very difficult to detect Arf protein in primary cells or normal tissues unless they are subjected to some activating stimulus (Llanos et al., 2001; Wei et al., 2001; Zindy et al., 1997; Zindy et al., 2003).

6. SENESCENCE AND STASIS

The predominant impression, therefore, is that both Ink4a and Arf are activated by aberrant signals rather than the normal chain of events that dictate cell cycle progression. But what are these signals and do Ink4a and Arf respond to the same signals and in what circumstances? Paradoxically, attempts to answer these questions using cultured cells have revealed that the very process of placing cells in standard tissue culture conditions is enough to activate the locus.

Somatic cells generally have a limited capacity to proliferate in tissue culture before undergoing an irreversible growth arrest termed senescence in which they remain metabolically viable and adopt a characteristic flat cell phenotype (Campisi, 1997). The phenomenon was first observed in human diploid fibroblasts (Hayflick, 1965) and it is now recognised that there are at least two contributory factors (Sherr and DePinho, 2000; Wright and Shay, 2002). The first is the erosion of the telomeres (see elsewhere in this book). Although there are conflicting views (Masutomi et al., 2003), human fibroblasts, like most somatic cell types, either fail to express or have insufficient amounts of the enzyme telomerase to maintain the telomeric structures at the ends of the chromosomes. The inexorable loss of telomeric DNA with each cell division eventually registers as a form of DNA damage invoking a p53-mediated response that engages many of the components and mechanisms discussed in other chapters of this volume (d'Adda di Fagagna et al., 2003; Harley, 1991).

The second factor is what is vaguely referred to as "culture shock", a catch-all term to describe the reaction of a cell to the non-physiological milieu of tissue culture, such as high serum, high oxygen tension and plastic substratum (Sherr and DePinho, 2000; Wright and Shay, 2002). Thus, whereas most strains of primary human fibroblasts can be immortalised simply by supplying hTERT, the catalytic component of telomerase (Bodnar et al., 1998; Vaziri and Benchimol, 1998), primary human epithelial cells often arrest after only a few population doublings, long before telomere erosion has reached critical proportions (Brenner et al., 1998; Foster et al., 1998; Huschtscha et al., 1998). Although in some cases the stress response can be alleviated by optimising the culture conditions (Ramirez et al., 2001), these cells will not be immortalised by hTERT unless the arrest mechanism is turned off or over-ridden (Dickson et al., 2000; Kiyono et al., 1998). An obvious source of confusion is that the relative severity and timing of culture stress can vary in different cell types grown under different conditions. A clear distinction must also be drawn between mouse and human cells. As laboratory mice have exceptionally long telomeres and telomerase is more widely expressed, the lifespan of mouse cells in culture is solely determined by culture stress (Wright and Shay, 2000). In addition, mouse cells seem be more sensitive to particular stresses (Parrinello et al., 2003; Rangarajan and Weinberg, 2003).

In addition to the unavoidable consequences of tissue culture, deliberate forms of cellular stress also elicit a senescence-like phenotype. A classic example is the introduction of the constitutively active G12V allele of H-Ras which at sufficiently high doses will cause primary cells to growth arrest (Newbold and Overell, 1983; Serrano et al., 1997). This occurs irrespective of the age of the cell or the presence or absence of hTERT (Morales et al., 1999; Wei et al., 1999). Similarly, a variety of non-specific agents such as UV and γ irradiation, hydrogen peroxide, and histone deacetylase inhibitors can elicit a senescence-like arrest in primary fibroblasts (Chen et al., 1995; Gorbunova et al., 2002; Naka et al., 2004; Ogryzko et al., 1996). As it is difficult to justify a distinction between a "natural" type of senescence influenced by the vagaries of tissue culture and "premature senescence" induced by oncogenic Ras or other insults, we have suggested the term "stasis" (stress and aberrant signalling induced senescence) to describe the phenomenon (Drayton and Peters, 2002).

7. ROLE OF INK4A IN SENESCENCE AND STASIS

Irrespective of the causes, it is clear that the pRb and p53 pathways play key roles in the implementation of senescence and stasis, and that the *Ink4a/Arf* locus is intimately involved. For example, the levels of $p16^{Ink4a}$ increase substantially in senescent human fibroblasts (Alcorta et al., 1996; Hara et al., 1996) and the early arrest of human epithelial cells is almost entirely attributable to the up-regulation of Ink4a. Cells spontaneously bypass this arrest by transcriptional silencing of Ink4a (Brenner et al., 1998; Dickson et al., 2000; Foster et al., 1998; Huschtscha et al., 1998; Kiyono et al., 1998) and fibroblasts in which $p16^{Ink4a}$ activity has been reduced or ablated in one way or another show an extended lifespan (Brookes et al., 2004; Morris et al., 2002; Wei et al., 2003). Similarly, oncogenes such as Ras and Myc induce expression of Ink4a (Drayton et al., 2003; Serrano et al., 1997) and the basal and induced levels of the protein clearly have an impact on the ensuing growth arrest (Benanti and Galloway, 2004; Brookes et al., 2002; Drayton et al., 2003; Huot et al., 2002; Serrano et al., 1997; Voorhoeve and Agami, 2003). Stasis caused by histone deacetylase inhibitors is also attributable to Ink4a (Munro et al., 2004). Significantly, in the context of genome stability, deliberate interference with telomere integrity using a dominant negative form of TRF2 has been shown to activate $p16^{Ink4a}$ (Smogorzewska and de Lange, 2002) and there have been sporadic reports that DNA double strand breaks and UVC irradiation can induce $p16^{Ink4a}$ accumulation, albeit over widely different time scales (Robles and Adami, 1998; Wang et al., 1996).

Although there are some dissenting views (Herbig et al., 2004), most of the evidence would be consistent with the idea that in human cells at least, senescence and stasis are implemented by the combined actions of $p16^{Ink4a}$ and $p21^{Cip1}$ (Alcorta et al., 1996; Stein et al., 1999). Given that Ink4 proteins can displace $p21^{Cip1}$ (and $p27^{Kip1}$) from Cdk4/6 complexes onto Cdk2 complexes, logic suggests that the amount of $p16^{Ink4a}$ needed to cause cell cycle arrest will depend on how much $p21^{Cip1}$ is available to inhibit Cdk2. Conversely, cells will be more sensitive to $p21^{Cip1}$-mediated arrest if they have elevated levels of $p16^{Ink4a}$. Indeed, there seems little doubt that the main conduit for the cell's response to telomere erosion and oxidative stress is actually via p53 and $p21^{Cip1}$ rather than Ink4a (Itahana et al., 2003b; Wei et al., 2003). Ink4a-deficient cells still succumb to telomere exhaustion (Brookes et al., 2004), albeit after some delay, and in normal fibroblasts, the accumulation of $p16^{Ink4a}$ becomes more obvious after the cells have stopped proliferating (Alcorta et al., 1996; Stein et al., 1999). In the context of genome instability, therefore, the open question is whether there is a direct signalling pathway that links DNA damage in one form or another to the up-regulation of Ink4a, or whether Ink4a is wired to respond indirectly or to chronic rather than acute forms of cellular stress. The same question can be posed for oncogene induced expression of $p16^{Ink4a}$ which generally takes several days to become apparent. This is much longer than would be expected for a simple kinase cascade culminating in, for example, the modification and activation of a transcription factor.

In seeking an explanation, two intriguing possibilities come to mind. The first is that the induction of $p16^{Ink4a}$ by Ras may be mediated indirectly by the activation of the p38 MAP kinase (Figure 4 and (Deng et al., 2004; Wang et al., 2002). This idea has several attractions. In addition to Ras, the p38 pathway is activated by a variety of stresses, including DNA damage, oxidative stress and inflammatory cytokines, many of which have also been shown to induce stasis or senescence in primary cells. Moreover, constitutive activation of p38 can elicit a senescence-like arrest in primary cells that can in part be explained by effects on $p16^{Ink4a}$ (Haq et al., 2002; Iwasa et al., 2003). As p38 is also known to activate p53, via phosphorylation of Ser33 and Ser46 (Bulavin et al., 1999; Wang et al., 2000), it has the potential to engage both branches of the senescence mechanism. The p38-mediated activation of p53 is regulated by a feedback loop involving the p53-inducible phosphatase Wip1 (Takekawa et al., 2000) and it has recently been shown that the gene encoding Wip1 is amplified in human cancers (Bulavin et al., 2002).

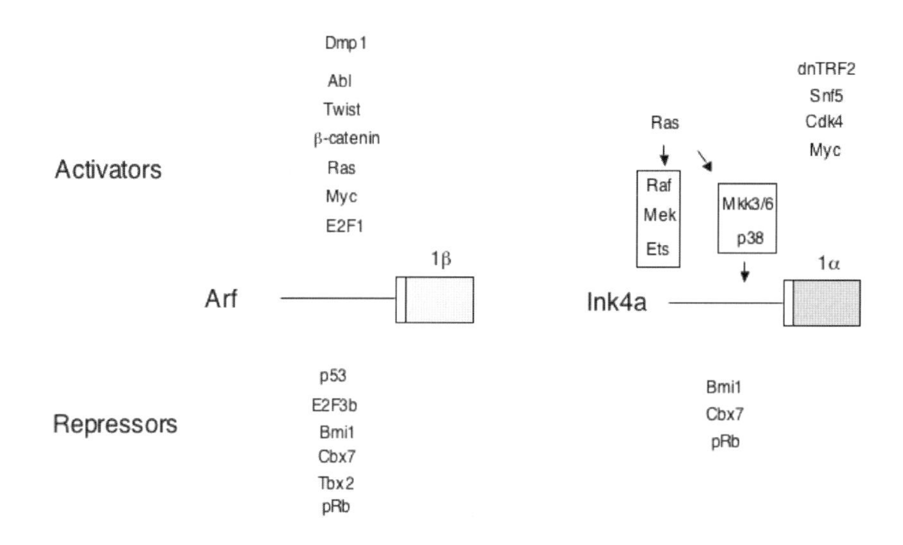

Figure 4. Positive and negative regulators of Ink4a and Arf. The figure shows the agents and/or pathways that have been reported to increase or decrease transcription from the Ink4 or Arf promoters. Note that we have made no distinction between the human and mouse loci and there are several instances in which the supporting evidence is confined to only one or other species.

The alternative proposition is that Ink4a transcription requires the remodelling of heterochromatin. It is now established that Ink4a expression is negatively regulated by two members of the so-called Polycomb group (PcG) of transcriptional repressors, Bmi1 and Cbx7 (Figure 4 and (Gil et al., 2004; Jacobs et al., 1999). Bmi1 is a relative of the Posterior sex combs gene of *Drosophila* and is a *bona fide* oncogene that was originally identified in virally induced leukaemias in mice (van Lohuizen et al., 1991). Cbx7 is related to the *Drosophila* Polycomb gene and was identified in a screen for cDNAs that can extend the lifespan of human epithelial cells (Gil et al., 2004). Both Bmi1 and Cbx7 are thought to participate in the multi-protein complexes that bind to specific methylated lysines in the amino terminal tails of histones H3 and H4 (Lund and van Lohuizen, 2004). These epigenetic marks are established by a separate complex, but the net effect is to shut down transcription of adjacent genes via the formation of heterochromatin. Interestingly, Bmi1 and Cbx7 appear to act independently of one another, but both can extend the lifespan of primary cells and their levels decline with increasing numbers of population doublings (Gil et al., 2004; Itahana et al., 2003b). The details of the underlying mechanisms remain obscure at this point, but the slow response of Ink4a to activating agents could in part reflect the need to erase the epigenetic marks in the surrounding chromatin.

8. ROLE OF ARF IN SENESCENCE AND STASIS

The Arf promoter is also subject to repression by Bmi1 and Cbx7 (Gil et al., 2004; Jacobs et al., 1999), as well as other factors such as Tbx2 and E2F3b (Figure 4 and (Aslanian et al., 2004; Jacobs et al., 2000). However, most studies conclude that in human cells at least, Arf expression is not significantly affected by telomere erosion or in response to oncogenic Ras (Brookes et al., 2002; Dimri et al., 2000; Ferbeyre et al., 2000; Munro et al., 1999; Wei et al., 2001). In striking contrast, $p19^{Arf}$ clearly plays a central role in regulating the lifespan of mouse embryo fibroblasts (MEFs), where senescence is thought to be largely determined by oxidative stress (Parrinello et al., 2003; Zindy et al., 1998). Thus, Arf-null MEFs are immortal and resemble p53-null MEFs in this respect (Kamijo et al., 1997). They are also resistant to Ras-induced arrest (Kamijo et al., 1997, Groth, 2000 #2642) in keeping with the idea that Arf is the major sensor of oncogenic stress in mouse cells.

In contrast, Arf-null MEFs show a normal response to many DNA damaging agents, as do human cells in which the locus has been deleted or silenced, indicating that p53 can register the upstream signalling pathways independently of Arf (Kamijo et al., 1999b; Kamijo et al., 1997; Stott et al., 1998). Indeed, the prevailing view is that Arf is not up-regulated in response to DNA damage. However, even at low but physiological levels, Arf can presumably influence the sensitivity of the p53-Mdm2 loop, and there are indications that Arf can modulate the intensity or duration of the response in some contexts (Khan et al., 2004; Khan et al., 2000). Conversely, induction of Arf by other factors may engage facets of the DNA damage response, such as the ATM/ATR kinases, to reinforce its effects on p53 (Li et al., 2004). Whatever its role in normal cell physiology, it is clear that knocking down Arf with siRNA provides a proliferative advantage (Voorhoeve and Agami, 2003).

Part of the difficulty in constructing plausible models for Arf function is in deciding whether the growing list of proteins with which it allegedly interacts are its regulators or its targets. In addition to Mdm2 and p53, the list now includes E2F1 (Eymin et al., 2001; Martelli et al., 2001), DP1 (Datta et al., 2002), Myc (Datta et al., 2004; Qi et al., 2004), B23/nucleophosmin (Bertwistle et al., 2004; Itahana et al., 2003a), HIF1α

(Fatyol and Szalay, 2001), topoisomerase 1 (Karayan et al., 2001), spinophilin (Vivo et al., 2001), TBP-1 (Pollice et al.), and Pex19 (Sugihara et al., 2001). With striking similarity to the Arf-p53 and Mdm2-p53 feedback loops depicted in Figure 1, Arf is transcriptionally activated by Myc and by E2F1, but appears capable of inhibiting the functions of both of these prominent transcription factors (Datta et al., 2004; Datta et al., 2002; Eymin et al., 2001; Martelli et al., 2001; Qi et al., 2004; Russell et al., 2002). Given that Myc is thought to cause genomic instability (Felsher and Bishop, 1999; Vafa et al., 2002), and that E2F1 and Mdm2 are functionally modulated by DNA damage (Blattner et al., 1999; Lin et al., 2001; Maya et al., 2001), there are numerous ways in which Arf can be implicated in these events as well as considerable scope for confusion.

Another source of debate relates to the predominant localisation of Arf in the nucleolus (Lindström et al., 2000; Quelle et al., 1995; Stott et al., 1998). A prominent school of thought holds that Arf executes its effects on cellular physiology by modulating nucleolar function or by bringing client proteins into the nucleolus (Bertwistle et al., 2004; Datta et al., 2004; Datta et al., 2002; Fatyol and Szalay, 2001; Itahana et al., 2003a; Karayan et al., 2001; Lohrum et al., 2000a; Lohrum et al., 2000b; Martelli et al., 2001; Rizos et al., 2000; Sugimoto et al., 2003; Weber et al., 2000b; Weber et al., 1999). Even in situations where Arf has been maximally induced, it is not clear that cells can express enough endogenous Arf to cause the physical sequestration of all these proteins, not to mention its proposed effects on the processing of ribosomal RNA (Sugimoto et al., 2003). As well as concerns over the validity of some of these targets, given that Arf is such an odd and poorly conserved protein, the stoichiometry would be easier to rationalise if the localization of Arf in the nucleolus was in fact a mechanism through which Arf activity is controlled (Kuo et al., 2004). There are clearly situations in which Arf performs its known functions without nucleolar sequestration (Clark et al., 2002; Kashuba et al., 2004; Korgaonkar et al., 2002; Lin and Lowe, 2001; Llanos et al., 2001; Qi et al., 2004; Zhang and Xiong, 1999) and forms of human Arf that are excluded from the nucleolus are inherently unstable but can be stabilised if redirected to the nucleolus by addition of a basic motif (Rodway et al., 2004). Interestingly, nucleolar disruption appears to be a common factor among the many agents that are known to activate p53 (Rubbi and Milner, 2003) and one of the potential consequences of nucleolar breakdown would be the release of Arf and transient stabilisation of p53.

9. CONTRIBUTION OF INK4A/ARF TO GENOME STABILITY AND CANCER SUSCEPTIBILITY

From what we know about the *Ink4a/Arf* locus, it is clear that there are many ways in which the absence of $p16^{Ink4a}$ or $p14^{Arf}$ could impact on the cellular response to genotoxic insults or telomere erosion. For example, the ability of p21^{Cip1} to implement a G1 arrest, the effectiveness of Cdk inhibition by ATM/ATR signalling to Cdc25A (Falck et al., 2001), as well as other effects on D cyclin levels (Agami and Bernards, 2000) could all be influenced by the basal levels of Ink4 proteins within the cell. Similarly, as discussed above, the absence of Arf could have a substantial effect on the sensitivity of p53-MDM2 dependent responses. Nevertheless one is left with the impression that *Ink4a/Arf* is not part of the cell's front line defences against DNA damage and genome instability. Cells that lack one or both gene products do not have obvious defects in their checkpoint mechanisms and do not show evidence for widespread genomic instability. What they do show is an inability to protect themselves from what might be vaguely termed "oncogenic stress".

This presumably underlies the role of the locus in tumour suppression and here the evidence is irrefutable. For example, germline mutations in Ink4a/Arf are associated with familial predisposition to melanoma and certain other cancers, and the locus is affected by missense mutations, homozygous deletions and promoter methylation in a wide variety of human cancers (Ruas and Peters, 1998). The only contentious issue is whether Ink4a or Arf plays the more prominent role in different settings, as the genes are often co-deleted and mutations in exon 2 have the potential to affect both proteins. However, as discussed above, the known functions of Arf are entirely attributable to the sequences encoded by exon 1β suggesting that mutations in exons 1α, 2 and 3 specifically target p16^{Ink4a}. This would agree with the bias towards deletion and methylation of Ink4a in sporadic cancers but it by no means excludes a contribution from Arf. For example, rare germline alterations have been reported that exclusively affect exon 1β (Hewitt et al., 2002) and there are indications that they may be specifically associated with the combined occurrence of melanoma and neural system tumours (Randerson-Moor et al., 2001). Arf is also specifically deleted or methylated in some tumours (Esteller et al., 2000) but in these contexts there has been no formal proof that the regulation of Ink4a has not been affected in some perhaps subtle way.

Curiously, the mouse locus shows almost the inverse bias in terms of tumour suppression. Thus, mice that are specifically nullizygous for Arf are tumour prone, developing mostly leukaemias and lymphomas within the first few months of age (Kamijo et al., 1999a; Kamijo et al., 1997). In contrast, specific deletion of Ink4a does not increase the incidence of spontaneous tumours unless associated with Arf heterozygosity or incorporated into a chemical carcinogenesis protocol (Krimpenfort et al., 2001; Sharpless et al., 2001).

10. CONCLUDING REMARKS

In conclusion, a consensus view of the Ink4a/Arf locus is that it operates as an intrinsic defence mechanism that enables a cell to shut down when subjected to aberrant proliferative signals. Whereas a classical checkpoint monitors the cells credentials for passage to the next phase of the cell cycle, the transgressions that activate Ink4a/Arf may not warrant the death penalty (apoptosis) but are serious enough to commit the cell to life imprisonment in a senescent state.

REFERENCES

Agami, R., and R. Bernards. 2000. Distinct initiation and maintenance mechanisms cooperate to induce G1 cell cycle arrest in response to DNA damage. *Cell.* 102:55-66.

Alcorta, D.A., Y. Xiong, D. Phelps, G. Hannon, D. Beach, and J.C. Barrett. 1996. Involvement of the cyclin-dependent kinase inhibitor p16 (INK4a) in replicative senescence. *Proc. Natl. Acad. Sci. USA.* 93:13742-13747.

Aslanian, A., P.J. Iaquinta, R. Verona, and J.A. Lees. 2004. Repression of the *Arf* tumor suppressor by E2F3 is required for normal cel cycle kinetics. *Genes & Dev.* 18:1413-1422.

Bates, S., A.C. Phillips, P.A. Clark, F. Stott, G. Peters, R.L. Ludwig, and K.H. Vousden. 1998. $p14^{ARF}$ links the tumour suppressors RB and p53. *Nature.* 395:124-125.

Benanti, J.A., and D.A. Galloway. 2004. Normal human fibroblasts are resistant to RAS-induced senescence. *Mol. Cell. Biol.* 24:2842-2852.

Bertwistle, D., M. Sugimoto, and C.J. Sherr. 2004. Physical and functional interactions of the Arf tumor suppressor protein with nucleophosmin/B23. *Mol. Cell. Biol.* 24:985-996.

Blattner, C., A. Sparks, and D. Lane. 1999. Transcription factor E2F-1 is upregulated in response to DNA damage in a manner analogous to that of p53. *Mol. Cell. Biol.* 19:3704-3713.

Bodnar, A.G., M. Ouellette, M. Frolkis, S.E. Holt, C.-P. Chiu, G.B. Morin, C.B. Harley, J.W. Shay, S. Lichsteiner, and W.E. Wright. 1998. Extension of life-span by introduction of telomerase into normal human cells. *Science.* 279:349-352.

Bothner, B., W.S. Lewis, E.L. DiGiammarino, J.D. Weber, S.J. Bothner, and R.W. Kriwacki. 2001. Defining the molecular basis of Arf and Hdm2 interactions. *J. Mol. Biol.* 314:263-277.

Brenner, A.J., M.R. Stampfer, and C.M. Aldaz. 1998. Increased *p16* expression with first senescence arrest in human mammary epithelial cells and extended growth capacity with *p16* inactivation. *Oncogene.* 17:199-205.

Brookes, S., J. Rowe, A. Gutierrez del Arroyo, J. Bond, and G. Peters. 2004. Contribution of $p16^{INK4a}$ to replicative senescence of human fibroblasts. *Exp. Cell Res.* 298:549-559.

Brookes, S., J. Rowe, M. Ruas, S. Llanos, P.A. Clark, M. Lomax, M.C. James, R. Vatcheva, S. Bates, K.H. Vousden, D. Parry, N. Gruis, N. Smit, W. Bergman, and G. Peters. 2002. INK4a-deficient human diploid fibroblasts are resistant to RAS-induced senescence. *EMBO J.* 21:2936-2945.

Bulavin, D.V., O.N. Demidov, S. Saito, P. Kauraniemi, C. Phillips, S.A. Amundson, C. Ambrosino, G. Sauter, A.R. Nebreda, C.W. Anderson, A. Kallioniemi, A.J. Fornace Jr., and E. Appella. 2002. Amplification of *PPM1D* in human tumors abrogates p53 tumor-suppressor activity. *Nat. Genet.* 31:210-215.

Bulavin, D.V., S. Saito, M.C. Hollander, K. Sakaguchi, C.W. Anderson, E. Appella, and A.J. Fornace Jr. 1999. Phosphorylation of human p53 by p38 kinase coordinates N-terminal phosphorylation and apoptosis in response to UV radiation. *EMBO J.* 18:6845-6854.

Buschmann, T., V. Adler, E. Matusevich, S.Y. Fuchs, and Z. Ronai. 2000. p53 phosphorylation and association wih murine double minute 2, c-Jun NH_2-terminal kinase, $p14^{ARF}$, and p300/CBP during the cell cycle and after exposure to ultraviolet irradiation. *Cancer Res.* 60:896-900.

Campisi, J. 1997. The biology of replicative senescence. *Eur. J. Cancer.* 33:703-709.

Chen, Q., A. Fischer, J.D. Reagan, L.J. Yan, and B.N. Ames. 1995. Oxidative DNA damage and senescence of human diploid fibroblast cells. *Proc. Natl. Acad. Sci. USA*. 92:4337-4341.

Clark, P.A., S. Llanos, and G. Peters. 2002. Multiple interacting domains contribute to $p14^{ARF}$ mediated inhibition of MDM2. *Oncogene*. 21:4498-4507.

d'Adda di Fagagna, F., P.M. Reaper, L. Clay-Farrace, H. Fiegler, P. Carr, T. von Zglinicki, G. Saretzki, N.P. Carter, and S.P. Jackson. 2003. A DNA damage checkpoint response in telomere-initiated senescence. *Nature*. 426:194-198.

Datta, A., A. Nag, W. Pan, N. Hay, A.L. Gartel, O. Colamonici, Y. Mori, and P. Raychaudhuri. 2004. Myc-ARF (alternative reading frame) interaction inhibits the functions of Myc. *J. Biol. Chem.* 279:36698-36707.

Datta, B., A. Nag, and P. Raychaudhuri. 2002. Differential regulation of E2F1, DP1, and the E2F1/DP1 complex by ARF. *Mol. Cell. Biol.* 22:8398-8408.

De Stanchina, E., M.E. McCurrach, F. Zindy, S.-Y. Shieh, G. Ferbeyre, A.V. Samuelson, C. Prives, M.F. Roussel, C.J. Sherr, and S.W. Lowe. 1998. E1A signaling to p53 involves the $p19^{ARF}$ tumor suppressor. *Genes Dev.* 12:2434-2442.

DeGregori, J., G. Leone, A. Miron, L. Jakoi, and J.R. Nevins. 1997. Distinct roles for E2F proteins in cell growth control. *Proc. Natl. Acad. Sci. USA*. 94:7245-7250.

Deng, Q., R. Liao, B.-L. Wu, and P. Sun. 2004. High intensity *ras* signaling induces premature senescence by activating p38 pathway in primary human fibroblasts. *J. Biol. Chem.* 279:1050-1059.

Dickson, M.A., W.C. Hahn, Y. Ino, V. Ronfard, J.Y. Wu, R.A. Weinberg, D.N. Louis, F.P. Li, and J.G. Rheinwald. 2000. Human keratinocytes that express hTERT and also bypass a $p16^{INK4a}$-enforced mechanism that limits life span become immortal yet retain normal growth and differentiation characteristics. *Mol. Cell. Biol.* 20:1436-1447.

Dimri, G.P., K. Itahana, M. Acosta, and J. Campisi. 2000. Regulation of a senescence checkpoint response by the E2F1 transcription factor and $p14^{ARF}$ tumor suppressor. *Mol. Cell. Biol.* 20:273-285.

Drayton, S., and G. Peters. 2002. Immortalisation and transformation revisited. *Curr. Opin. Genet. Dev.* 12:98-104.

Drayton, S., J. Rowe, R. Jones, R. Vatcheva, D. Cuthbert-Heavens, J. Marshall, M. Fried, and G. Peters. 2003. Tumor suppressor $p16^{INK4a}$ determines sensitivity of human cells to transformation by cooperating cellular oncogenes. *Cancer Cell*. 4:301-310.

Esteller, M., S. Tortola, M. Toyota, G. Capella, M.A. Peinado, S.B. Baylin, and J.G. Herman. 2000. Hypermethylation-associated inactivation of $p14^{ARF}$ is independent of $p16^{INK4a}$ methylation and *p53* mutational status. *Cancer Res.* 60:129-133.

Eymin, B., L. Karayan, P. Séité, C. Brambilla, E. Brambilla, C.-J. Larsen, and S. Gazzéri. 2001. Human ARF binds E2F1 and inhibits its transcriptional activity. *Oncogene*. 20:1033-1041.

Falck, J., N. Mailand, R.G. Syljuåsen, J. Bartek, and J. Lukas. 2001. The ATM-Chk2-Cdc25A checkpoint pathway guards against radioresistant DNA synthesis. *Nature*. 410:842-847.

Fatyol, K., and A.A. Szalay. 2001. The $p14^{ARF}$ tumor suppressor protein facilitates nucleolar sequestration of hypoxia-inducible factor-1α (HIF-1α) and inhibits HIF-1-mediated transcription. *J. Biol. Chem.* 276:28421-28429.

Felsher, D.W., and J.M. Bishop. 1999. Transient excess of *MYC* activity can elicit genomic instability and tumorigenesis. *Proc. Natl. Acad. Sci. USA*. 96:3940-3944.

Ferbeyre, G., E. de Stanchina, E. Querido, N. Baptiste, C. Prives, and S.W. Lowe. 2000. PML is induced by oncogenic ras and promotes premature senescence. *Genes Dev.* 14:2015-2027.

Foster, S.A., D.J. Wong, M.T. Barrett, and D.A. Galloway. 1998. Inactivation of p16 in human mammary epithelial cells by CpG island methylation. *Mol. Cell. Biol.* 18:1793-1801.

Gil, J., D. Bernard, D. Martinez, and D. Beach. 2004. Polycomb CBX7 has a unifying role in cellular lifespan. *Nature Cell Biol.* 6:67-72.

Gilley, J., and M. Fried. 2001. One INK4 gene and no ARF at the Fugu equivalent of the human INK4A/ARF/INK4B tumour suppressor locus. *Oncogene*. 20:7447-7452.

Gorbunova, V., A. Seluanov, and O.M. Pereira-Smith. 2002. Expression of human telomerase (hTERT) does not prevent stress-induced senescence in normal human fibroblasts but protects the cells from stress-induced apoptosis and necrosis. *J. Biol. Chem.* 277:38540-38549.

Haq, R., J.D. Brenton, M. Takahashi, D. Finan, R. Rottapel, and B. Zanke. 2002. Constitutive p38HOG mitogen-activated protein kinase activation induces permanent cell cycle arrest and senescence. *Cancer Res.* 62:5067-5082.

Hara, E., R. Smith, D. Parry, H. Tahara, S. Stone, and G. Peters. 1996. Regulation of p16^{CDKN2} expression and its implications for cell immortalization and senescence. *Mol. Cell. Biol.* 16:859-867.

Harley, C.B. 1991. Telomere loss: mitotic clock or genetic time bomb? *Mutat. Res.* 256:271-282.

Hayflick, L. 1965. The limited *in vitro* lifespan of human diploid cell strains. *Exp. Cell Res.* 37:614-636.

Herbig, U., W.A. Jobling, B.P.C. Chen, D.J. Chen, and J.M. Sedivy. 2004. Telomere shortening triggers senescence of human cells through a pathway involving ATM, p53, and 21^{CIP1}, but not p16^{INK4a}. *Mol. Cell.* 14:501-513.

Hewitt, C., C.L. Wu, G. Evans, A. Howell, R.G. Elles, R. Jordan, P. Sloan, A.P. Read, and N. Thakker. 2002. Germline mutation of *ARF* in a melanoma kindred. *Hum. Mol. Genet.* 11:1273-1279.

Hirai, H., M.F. Roussel, J.-Y. Kato, R.A. Ashmun, and C.J. Sherr. 1995. Novel INK4 proteins, p19 and p18, are specific inhibitors of the cyclin D-dependent kinases CDK4 and CDK6. *Mol. Cell. Biol.* 15:2672-2681.

Honda, R., and H. Yasuda. 1999. Association of p19^{ARF} with Mdm2 inhibits ubiquitin ligase activity of Mdm2 for tumour suppressor p53. *EMBO J.* 18:22-27.

Huot, T.J., J. Rowe, M. Harland, S. Drayton, S. Brookes, C. Goopta, P. Purkis, M. Fried, V. Bataille, E. Hara, J. Newton-Bishop, and G. Peters. 2002. Biallelic mutations in p16^{INK4a} confer resistance to Ras- and Ets-induced senescence in human diploid fibroblasts. *Mol. Cell. Biol.* 22:8135-8143.

Huschtscha, L.I., J.R. Noble, A.A. Neumann, E.L. Moy, P. Barry, J.R. Melki, S.J. Clark, and R.R. Reddel. 1998. Loss of p16^{INK4} expression by methylation is associated with lifespan extension of human mammary epithelial cells. *Cancer Res.* 58:3508-3512.

Itahana, K., K.P. Bhat, A. Jin, Y. Itahana, D. Hawke, R. Kobayashi, and Y. Zhang. 2003a. Tumor suppressor ARF degrades B23, a nucleolar protein involved in ribosome biogenesis and cell proliferation. *Mol. Cell.* 12:1151-1164.

Itahana, K., Y. Zou, Y. Itahana, J.-L. Martinez, C. Beausejour, J.J.L. Jacobs, M. van Lohuizen, V. Band, J. Campisi, and G.P. Dimri. 2003b. Control of the replicative life span of human fibroblasts by p16 and the polycomb protein Bmi-1. *Mol. Cell. Biol.* 23:389-401.

Iwasa, H., J. Han, and F. Ishikawa. 2003. Mitogen-activated protein kinase p38 defines the common senescence-signalling pathway. *Genes to Cells*. 8:131-144.

Jacobs, J.J.L., P. Keblusek, E. Robanus-Maandag, P. Kristel, M. Lingbeek, P.M. Nederlof, T. van Welsem, M.J. van de Vijver, E.Y. Koh, G.Q. Daley, and M. van Lohuizen. 2000. Senescence bypass screen identifies *TBX2*, which represses *Cdkn2a* (*p19^{ARF}*) and is amplified in a subset of human breast cancers. *Nature Genet.* 291:291-299.

Jacobs, J.J.L., K. Kieboom, S. Marino, R.A. DePinho, and M. van Lohuizen. 1999. The oncogene and Polycomb-group gene *bmi-1* regulates cell proliferation and senescence through the *ink4a* locus. *Nature*. 397:164-168.

Jiang, H., H.S. Chou, and L. Zhu. 1998. Requirement of cyclin E-Cdk2 inhibition in p16^{INK4a}-mediated growth suppression. *Mol. Cell. Biol.* 18:5284-5290.

Kamijo, T., S. Bodner, E. van de Kamp, D.H. Randle, and C.J. Sherr. 1999a. Tumor spectrum in *ARF*-deficient mice. *Cancer Res.* 59:2217-2222.

Kamijo, T., E. van de Kamp, M.J. Chong, F. Zindy, J.A. Diehl, C.J. Sherr, and P.J. McKinnon. 1999b. Loss of ARF tumor suppressor reverses premature replicative arrest but not radiation hypersensitivity arising from disabled Atm function. *Cancer Res.* 59:2464-2469.

Kamijo, T., J.D. Weber, G. Zambetti, F. Zindy, M.F. Roussel, and C.J. Sherr. 1998. Functional and physical interactions of the ARF tumor suppressor with p53 and Mdm2. *Proc. Natl. Acad. Sci. USA.* 95:8292-8297.

Kamijo, T., F. Zindy, M.F. Roussel, D.E. Quelle, J.R. Downing, R.A. Ashmun, G. Grosveld, and C.J. Sherr. 1997. Tumor suppression at the mouse *INK4a* locus mediated by the alternative reading frame product $p19^{ARF}$. *Cell.* 91:649-659.

Karayan, L., J.-F. Riou, P. Séité, J. Migeon, A. Cantereau, and C.-J. Larsen. 2001. Human ARF protein interacts with topoisomerase I and stimulates its activity. *Oncogene.* 20:836-848.

Kashuba, E., K. Mattson, G. Klein, and L. Szekely. 2004. p14ARF induces the relocation of HDM2 and p53 to extranucleolar sites that are targeted by PML bodies and proteasomes. *Mol. Cancer.* 2:18.

Kazianis, S., D.C. Morizot, L. Della Coletta, D.A. Johnston, B. Woolcock, J.R. Vielkind, and R.S. Nairn. 1999. Comparative structure and characterization of a *CDKN2* gene in a *Xiphophorus* fish melanoma model. *Oncogene.* 18:5088-5099.

Khan, S., C. Guevara, G. Fujii, and D. Parry. 2004. p14ARF is a component of the p53 response following ionizing irradiation of normal human fibroblasts. *Oncogene.* 23:6040-6046.

Khan, S.H., J. Moritsugu, and G.M. Wahl. 2000. Differential requirement for p19ARF in the p53-dependent arrest induced by DNA damage, microtubule disruption, and ribonucloetide depletion. *Proc. Natl. Acad. Sci. USA.* 97:3266-3271.

Kim, S.-H., M. Mitchell, H. Fujii, S. Llanos, and G. Peters. 2003. Absence of $p16^{INK4a}$ and truncation of ARF tumor suppressors in chickens. *Proc. Natl. Acad. Sci. USA.* 100:211-216.

Kiyono, T., S.A. Foster, J.I. Koop, J.K. McDougall, D.A. Galloway, and A.J. Klingelhutz. 1998. Both Rb/$p16^{INK4a}$ inactivation and telomerase activity are required to immortalize human epithelial cells. *Nature.* 396:84-88.

Korgaonkar, C., L. Zhao, M. Modestou, and D.E. Quelle. 2002. ARF function does not require p53 stabilization or Mdm2 relocalization. *Mol. Cell. Biol.* 22:196-206.

Kozar, K., M.A. Ciemerych, V.I. Rebel, H. Shigematsu, A. Zagozdzon, E. Sicinska, Y. Geng, Q. Yu, S. Batthacharya, R.T. Bronson, K. Akashi, and P. Sicinski. 2004. Mouse development and cell proliferation in the absence of D-cyclins. *Cell.* 118:477-491.

Krimpenfort, P., K.C. Quon, W.J. Mool, A. Loonstra, and A. Berns. 2001. Loss of $p16^{Ink4a}$ confers susceptibility to metastatic melanoma in mice. *Nature.* 413:83-86.

Kuo, M.-L., W. den Besten, D. Bertwistle, M.F. Roussel, and C.J. Sherr. 2004. N-terminal polyubiquitination and degradation of the Arf tumor suppressor. *Genes & Dev.* 18:1862-1874.

Li, Y., M.A. Nichols, J.W. Shay, and Y. Xiong. 1994. Transcriptional repression of the D-type cyclin-dependent kinase inhibitor p16 by the retinoblastoma susceptibility gene product pRb. *Cancer Res.* 54:6078-6082.

Li, Y., D. Wu, B. Chen, A. Ingram, L. He, L. Liu, D. Zhu, A. Kapoor, and D. Tang. 2004. ATM activity contributes to the tumor-suppressing functions of $p14^{ARF}$. *Oncogene.* 23:7355-7365.

Lin, A.W., and S.W. Lowe. 2001. Oncogenic *ras* activates the ARF-p53 pathway to suppress epthelial cell transformation. *Proc. Natl. Acad. Sci. USA.* 98:5025-5030.

Lin, W.-C., F.-T. Lin, and J.R. Nevins. 2001. Selective induction of E2F1 in response to DNA damage, mediated by ATM-dependent phosphorylation. *Genes Dev.* 15:1833-1844.

Lindström, M.S., U. Klangby, R. Inoue, P. Pisa, K.G. Wiman, and C.E. Asker. 2000. Immunolocalization of human $p14^{ARF}$ to the granular component of the interphase nucleus. *Exp. Cell Res.* 256:400-410.

Llanos, S., P.A. Clark, J. Rowe, and G. Peters. 2001. Stabilisation of p53 by p14^{ARF} without relocation of MDM2 to the nucleolus. *Nature Cell Biol.* 3:445-452.

Lohrum, M.A.E., M. Ashcroft, M.H.G. Kubbutat, and K.H. Vousden. 2000a. Contribution of two independent MDM2-binding domains in p14^{ARF} to p53 stabilization. *Curr. Biol.* 10:539-542.

Lohrum, M.A.E., M. Ashcroft, M.H.G. Kubbutat, and K.H. Vousden. 2000b. Identification of a cryptic nucleolar-localization signal in MDM2. *Nature Cell Biol.* 2:179-181.

Lomazzi, M., M.C. Moroni, M.R. Jensen, E. Frittoli, and K. Helin. 2002. Suppression of the p53- or pRB-mediated G1 checkpoint is required for E2F-induced S-phase entry. *Nat. Genet.* 31:190-194.

Lund, A.H., and M. van Lohuizen. 2004. Polycomb complexes and silencing mechanisms. *Curr. Opin. Cell Biol.* 16:239-246.

Malumbres, M., R. Sotillo, D. Santamaria, J. Galan, A. Cerezo, S. Ortega, P. Dubus, and M. Barbacid. 2004. Mammalian cells cycle without the D-type cyclin-dependent kinases Cdk4 and Cdk6. *Cell.* 118:493-504.

Martelli, F., T. Hamilton, D.P. Silver, N.E. Sharpless, N. Bardeesy, M. Rokas, R.A. DePinho, D.M. Livingston, and S.R. Grossman. 2001. p19^{ARF} targets certain E2F species for degradation. *Proc. Natl. Acad. Sci. USA.* 8:4455-4460.

Masutomi, K., E.Y. Yu, S. Khurts, I. Ben-Porath, J.L. Currier, G.B. Metz, M.W. Brooks, S. Kaneko, S. Murakami, J.A. DeCaprio, R.A. Weinberg, S.A. Stewart, and W.C. Hahn. 2003. Telomerase maintains telomere structure in normal human cells. *Cell.* 114:241-253.

Maya, R., M. Balass, S.-T. Kim, D. Shkedy, J.-F.M. Leal, O. Shifman, M. Moas, T. Buschmann, Z. Ronai, Y. Shiloh, M.B. Kastan, E. Katzir, and M. Oren. 2001. ATM-dependent phosphorylation of Mdm2 on serine 395: role in p53 activation by DNA damage. *Genes & Dev.* 15:1067-1077.

McConnell, B.B., F.J. Gregory, F.J. Stott, E. Hara, and G. Peters. 1999. Induced expression of p16^{INK4a} inhibits both CDK4- and CDK2-associated kinase activity by reassortment of cyclin-CDK-inhibitor complexes. *Mol. Cell. Biol.* 19:1981-1989.

Medema, R.H., R. Klompmaker, V.A.J. Smits, and G. Rijksen. 1998. p21^{waf1} can block cells at two points in the cell cycle, but does not interfere with processive DNA-replication or stress-activated kinases. *Oncogene.* 16:431-441.

Michael, D., and M. Oren. 2003. The p53-Mdm2 module and the ubiquitin system. *Sem. Cancer Biol.* 13:49-58.

Midgley, C.A., J.M.P. Desterro, M.K. Saville, S. Howard, A. Sparks, R.T. Hay, and D.P. Lane. 2000. An N-terminal p14^{ARF}peptide blocks Mdm2-dependent ubiquitination *in vitro* and can activate p53 *in vivo*. *Oncogene.* 19:2312-2323.

Modestou, M., V. Puig-Antich, C. Korgaonkar, A. Eapen, and D.E. Quelle. 2001. The alternative reading frame tumor suppressor inhibits growth through p21-dependent and p21-independent pathways. *Cancer Res.* 61:3145-3150.

Morales, C.P., S.E. Holt, M. Ouellette, K.J. Kaur, Y. Yan, K.S. Wilson, M.A. White, W.E. Wright, and J.W. Shay. 1999. Absence of cancer-associated changes in human fibroblasts immortalized with telomerase. *Nat. Genet.* 21:115-118.

Morris, M., P. Hepburn, and D. Wynford-Thomas. 2002. Sequential extension of proliferative lifespan in human fibroblasts induced by over-expression of CDK4 or 6 and loss of p53 function. *Oncogene.* 21:4277-4288.

Munro, J., N.I. Barr, H. Ireland, V. Morrison, and E.K. Parkinson. 2004. Histone deacetylase inhibitors induce a senescence-like state in human cells by a p16-dependent mechanism that is independent of a mitotic clock. *Exp. Cell Res.* 295:528-538.

Munro, J., F.J. Stott, K.H. Vousden, G. Peters, and E.K. Parkinson. 1999. Role of the alternative *INK4A* proteins in human keratinocyte senescence: evidence for the specific inactivation of *p16*^{INK4A} upon immortalization. *Cancer Res.* 59:2516-2521.

Naka, K., A. Tachibana, K. Ikeda, and N. Motoyama. 2004. Stress-induced premature senescence in hTERT-expressing Ataxia-telangiectasia fibroblasts. *J. Biol. Chem.* 279:2030-2037.

Newbold, R.F., and R.W. Overell. 1983. Fibroblast immortality is a prerequisite for transformation by EJ c-Ha-*ras* oncogene. *Nature*. 304:648-651.

Niculescu III, A.B., X. Chen, M. Smeets, L. Hengst, C. Prives, and S.I. Reed. 1998. Effects of $p21^{Cip1/Waf1}$ at both the G_1/S and the G_2/M cell cycle transitions: pRb is a critical determinant in blocking DNA replication and in preventing endoreduplication. *Mol. Cell. Biol.* 18:629-643.

Ogryzko, V.V., T.H. Hirai, V.R. Russanova, D.A. Barbie, and B.H. Howard. 1996. Human fibroblast commitment to a senescence-like state in response to histone deacetylase inhibitors is cell cycle dependent. *Mol. Cell. Biol.* 16:5210-5218.

Ortega, S., I. Prieto, J. Odajima, A. Martin, P. Dubus, R. Sotillo, J.L. Barbero, M. Malumbres, and M. Barbacid. 2003. Cyclin-dependent kinase 2 is essential for meiosis but not for mitotic cell division in mice. *Nat. Genet.* 35:25-31.

Parisi, T., A. Pollice, A. Di Cristofano, V. Calabrò, and G. La Mantia. 2002. Transcriptional regulation of the human tumor suppressor $p14^{ARF}$ by E2F1, E2F2, E2F3, and Sp1-like factors. *Biochem. Biophys. Res. Commun.* 291:1138-1145.

Parrinello, S., E. Samper, A. Krtolica, J. Goldstein, S. Melov, and J. Campisi. 2003. Oxygen sensitivity severely limits the replicative lifespan of murine fibroblasts. *Nat. Cell. Biol.* 5:741-747.

Parry, D., S. Bates, D.J. Mann, and G. Peters. 1995. Lack of cyclin D-Cdk complexes in Rb-negative cells correlates with high levels of $p16^{INK4/MTS1}$ tumour suppressor gene product. *EMBO J.* 14:503-511.

Pollice, A., V. Nasti, R. Ronca, M. Vivo, M. Lo Iacono, R. Calogero, V. Calabrò, and G. La Mantia. Functional and physical interaction of the human ARF tumor suppressor with Tat-binding protein-1. *J. Biol. Chem.* 279:6345-6353.

Pomerantz, J., N. Schreiber-Agus, N.J. Liégeois, A. Silverman, L. Alland, L. Chin, J. Potes, K. Chen, I. Orlow, H.-W. Lee, C. Cordon-Cardo, and R.A. DePinho. 1998. The *Ink4a* tumor suppressor gene product, $p19^{Arf}$, interacts with MDM2 and neutralizes MDM2's inhibition of p53. *Cell*. 92:713-723.

Qi, Y., M.A. Gregory, Z. Li, J.P. Brousal, K. West, and S.R. Hann. 2004. $p19^{ARF}$ directly and differentially controls the functions of c-Myc independently of p53. *Nature*.

Quelle, D.E., M. Cheng, R.A. Ashmun, and C.J. Sherr. 1997. Cancer-associated mutations at the *INK4a* locus cancel cell cycle arrest by $p16^{INK4a}$ but not by the alternative reading frame protein $p19^{ARF}$. *Proc. Natl. Acad. Sci. USA*. 94:669-673.

Quelle, D.E., F. Zindy, R.A. Ashmun, and C.J. Sherr. 1995. Alternative reading frames of the *INK4a* tumor suppressor gene encode two unrelated proteins capable of inducing cell cycle arrest. *Cell*. 83:993-1000.

Ramirez, R.D., C.P. Morales, B.-S. Herbert, J.M. Rohde, C. Passons, J.W. Shay, and W.E. Wright. 2001. Putative telomere-independent mechanisms of replicative aging reflect inadequate growth conditions. *Genes Dev.* 15:398-403.

Randerson-Moor, J.A., M. Harland, S. Williams, D. Cuthbert-Heavens, E. Sheridan, J. Aveyard, K. Sibley, L. Whitaker, M. Knowles, J. Newton Bishop, and D.T. Bishop. 2001. A germline deletion of $p14^{ARF}$ but not *CDKN2A* in a melanoma-neural system tumour syndrome family. *Hum. Mol. Genet.* 10:55-62.

Rangarajan, A., and R.A. Weinberg. 2003. Comparative biology of mouse versus human cells: modelling human cancer in mice. *Nat Rev. Cancer*. 3:952-959.

Rizos, H., A.P. Darmanian, G.J. Mann, and R.F. Kefford. 2000. Two arginine rich domains in the p14ARF tumour suppressor mediate nucleolar localization. *Oncogene*. 19:2978-2985.

Robertson, K.D., and P.A. Jones. 1998. The human ARF cell cycle regulatory gene promoter is a CpG island which can be silenced by DNA methylation and down regulated by wild-type p53. *Mol. Cell. Biol.* 18:6457-6473.

Robles, S.J., and G.R. Adami. 1998. Agents that cause DNA double strand breaks lead to $p16^{INK4a}$ enrichment and the premature senescence of normal fibroblasts. *Oncogene*. 16:1113-1123.

Rodway, H., S. Llanos, J. Rowe, and G. Peters. 2004. Stability of nucleolar *versus* non-nucleolar forms of human $p14^{ARF}$. *Oncogene*. 23:6186-6192.

Ruas, M., and G. Peters. 1998. The p16^{INK4a}/CDKN2A tumor suppressor and its relatives. *Biochim. Biophys. Acta*. 1378:115-177.

Rubbi, C.P., and J. Milner. 2003. Disruption of the nucleolus mediates stabilization of p53 in response to DNA damage and other stresses. *EMBO J.* 22:6068-6077.

Russell, J.L., J.T. Powers, R.J. Rounbehler, P.M. Rogers, C.J. Conti, and D.G. Johnson. 2002. ARF differentially modulates apoptosis induced by E2F1 and Myc. *Mol. Cell. Biol.* 22:1360-1368.

Serrano, M., G.J. Hannon, and D. Beach. 1993. A new regulatory motif in cell-cycle control causing specific inhibition of cyclin D/CDK4. *Nature*. 366:704-707.

Serrano, M., A.W. Lin, M.E. McCurrach, D. Beach, and S.W. Lowe. 1997. Oncogenic *ras* provokes premature cell senescence associated with accumulation of p53 and p16^{INK4a}. *Cell*. 88:593-602.

Sharpless, N.E., N. Bardeesy, K.-H. Lee, D. Carrasco, D.H. Castrillon, A.J. Aguirre, E.A. Wu, J.W. Horner, and R.A. DePinho. 2001. Loss of p16^{Ink4a} with retention of p19^{Arf} predisposes mice to tumorigenesis. *Nature*. 413:86-91.

Sharpless, N.E., and R.A. DePinho. 1999. The *INK4A/ARF* locus and its two gene products. *Curr. Opin. Genet. Dev.* 9:22-30.

Sherr, C.J. 2001. The INK4a/ARF network in tumour suppression. *Nature Reviews*. 2:731-737.

Sherr, C.J., and R.A. DePinho. 2000. Cellular senescence: mitotic clock or culture shock? *Cell*. 102:407-410.

Sherr, C.J., and F. McCormick. 2002. The RB and p53 pathways in cancer. *Cancer Cell*. 2:103-112.

Smogorzewska, A., and T. de Lange. 2002. Different telomere damage signaling pathways in human and mouse cells. *EMBO J.* 21:4338-4348.

Soucek, T., O. Pusch, E. Hengstschläger-Ottnad, E. Wawra, G. Bernaschek, and M. Hengstschläger. 1995. Expression of the cyclin-dependent kinase inhibitor p16 during the ongoing cell cycle. *FEBS Lett*. 373:164-169.

Stein, G.H., L.F. Drullinger, A. Soulard, and V. Dulic. 1999. Differential roles for cyclin-dependent kinase inhibitors p21 and p16 in the mechanisms of senescence and differentiation in human fibroblasts. *Mol. Cell. Biol.* 19:2109-2117.

Stone, S., P. Jiang, P. Dayanath, S.V. Tavtigian, H. Katcher, D. Parry, G. Peters, and A. Kamb. 1995. Complex structure and regulation of the *P16* (*MTS1*) locus. *Cancer Res.* 55:2988-2994.

Stott, F.J., S. Bates, M.C. James, B.B. McConnell, M. Starborg, S. Brookes, I. Palmero, E. Hara, K.H. Vousden, and G. Peters. 1998. The alternative product from the human *CDKN2A* locus, p14^{ARF}, participates in a regulatory feedback loop with p53 and MDM2. *EMBO J.* 17:5001-5014.

Sugihara, T., S.C. Kaul, J. Kato, R.R. Reddel, H. Nomura, and R. Wadhwa. 2001. Pex19p dampens the p19ARF-p53-p21WAF1 tumor suppressor pathway. *J. Biol. Chem.* 276:18649-18652.

Sugimoto, M., M.-L. Kuo, M.F. Roussel, and C.J. Sherr. 2003. Nucleolar Arf tumor suppressor inhibits ribosomal RNA processing. *Mol. Cell*. 11:415-424.

Takekawa, M., M. Adachi, A. Nakahata, I. Nakayama, F. Itoh, H. Tsukuda, Y. Taya, and K. Imai. 2000. p53-inducible Wip1 phosphatase mediates a negative feedback regulation of p38 MAPK-p53 signaling in response to UV radiation. *EMBO J.* 19:6517-6526.

Tam, S.W., J.W. Shay, and M. Pagano. 1994. Differential expression and cell cycle regulation of the cyclin-dependent kinase 4 inhibitor p16^{Ink4}. *Cancer Res.* 54:5816-5820.

Tao, W., and A.J. Levine. 1999. P19^{ARF} stabilizes p53 by blocking nucleo-cytoplasmic shuttling of Mdm2. *Proc. Natl. Acad. Sci. USA*. 96:6937-6941.

Vafa, O., M. Wade, S. Kern, M. Beeche, T.K. Pandita, G.M. Hampton, and G.M. Wahl. 2002. c-Myc can induce DNA damage, increase reactive oxygen species, and mitigate p53 function: a mechanism for oncogene-induced genetic instability. *Mol. Cell*. 9:1031-1044.

van Lohuizen, M., S. Verbeek, B. Scheijen, E. Wientjens, H. van der Gulden, and B. A. 1991. Identification of cooperating oncogenes in Eμ-myc transgenic mice. *Cell*. 65:737-752.

Vaziri, H., and S. Benchimol. 1998. Reconstitution of telomerase activity in normal human cells leads to elongation of telomeres and extended replicative lifespan. *Curr. Biol.* 8:279-282.

Vivo, M., R.A. Calogero, F. Sansone, V. Calabro, T. Parisi, L. Borrelli, S. Saviozzi, and G. La Mantia. 2001. The human tumor suppressor ARF interacts with spinophilin/neurabin II, a type 1 protein-phosphatase-binding protein. *J. Biol. Chem.* 276:14161-14169.

Voorhoeve, M., and R. Agami. 2003. The tumor-suppressive functions of the human *INK4A* locus. *Cancer Cell.* 4:311-319.

Wang, W., J.X. Chen, R. Liao, Q. Deng, J.J. Zhou, S. Huang, and P. Sun. 2002. Sequential activation of the MEK-extracellular signal-regulated kinase and MKK3/6-p38 mitogen-activated protein kinase pathways mediates oncogenic *ras*-induced premature senescence. *Mol. Cell. Biol.* 22:3389-3403.

Wang, X.-F., C.H. McGowan, M. Zhao, L. He, J.S. Downey, C. Fearns, Y. Wang, S. Huang, and J. Han. 2000. Involvement of the MKK6-p38γ cascade in γ-radiation-induced cell cycle arrest. *Mol. Cell. Biol.* 20:4543-4552.

Wang, X.Q., B.G. Gabrielli, A. Milligan, J.L. Dickinson, T.M. Antalis, and K.A.O. Ellem. 1996. Accumulation of p16^{CDKN2A} in response to ultraviolet irradiation correlates with a late S-G_2-phase cell cycle delay. *Cancer Res.* 56:2510-2514.

Weber, H.O., T. Samuel, P. Rauch, and J.O. Funk. 2002. Human p14^{ARF}-mediated cell cycle arrest strictly depends on intact p53 signaling pathways. *Oncogene.* 21:3207-3212.

Weber, J.D., J.R. Jeffers, J.E. Rehg, D.H. Randle, G. Lozano, M.F. Roussel, C.J. Sherr, and G.P. Zambetti. 2000a. p53-independent functions of the p19^{ARF} tumor suppressor. *Genes Dev.* 14:2358-2365.

Weber, J.D., M.-L. Kuo, B. Bothner, E.L. DiGiammarino, R.W. Kriwacki, M.F. Roussel, and C.J. Sherr. 2000b. Cooperative signals governing ARF-MDM2 interaction and nucleolar localization of the complex. *Mol. Cell. Biol.* 20:2517-2528.

Weber, J.D., L.J. Taylor, M.F. Roussel, C.J. Sherr, and D. Bar-Sagi. 1999. Nucleolar Arf sequesters Mdm2 and activates p53. *Nature Cell Biol.* 1:20-26.

Wei, S., W. Wei, and J.M. Sedivy. 1999. Expression of catalytically active telomerase does not prevent premature senescence caused by overexpression of oncogenic Ha-Ras in normal human fibroblasts. *Cancer Res.* 59:1539-1543.

Wei, W., R.M. Hemmer, and J.M. Sedivy. 2001. The role of p14^{ARF} in replicative and induced senescence of human fibroblasts. *Mol. Cell. Biol.* 21:6748-6757.

Wei, W., U. Herbig, A. Dutriaux, and J.M. Sedivy. 2003. Loss of retinoblastoma but not p16 function allows bypass of replicative senescence in human fibroblasts. *Embo Rep.* 4:1061-1066.

Wright, W., and J.W. Shay. 2000. Telomere dynamics in cancer progression and prevention: fundamental differences in human and mouse telomere biology. *Nature Med.* 6:849-851.

Wright, W.E., and J.W. Shay. 2002. Historical claims and current interpretations of replicative aging. *Nat. Biotechnol.* 20:682-688.

Yarbrough, W.G., M. Bessho, A. Zanation, J.E. Bisi, and Y. Xiong. 2002. Human tumor suppressor ARF impedes S-phase progression independent of p53. *Cancer Res.* 62:1171-1177.

Zhang, Y., and Y. Xiong. 1999. Mutations in human ARF exon 2 disrupt its nucleolar localization and impair its ability to block nuclear export of MDM2 and p53. *Molec. Cell.* 3:579--591.

Zhang, Y., Y. Xiong, and W.G. Yarbrough. 1998. ARF promotes MDM2 degradation and stabilizes p53: *ARF-INK4a* locus deletion impairs both the Rb and p53 tumor suppression pathways. *Cell.* 92:725-734.

Zindy, F., C.M. Eischen, D.H. Randle, T. Kamijo, J.L. Cleveland, C.J. Sherr, and M.F. Roussel. 1998. Myc signaling via the ARF tumor suppressor regulates p53-dependent apoptosis and immortalization. *Genes Dev.* 12:2424-2433.

Zindy, F., D.E. Quelle, M.F. Roussel, and C.J. Sherr. 1997. Expression of the p16^{INK4A} tumor suppressor *versus* other INK4 family members during mouse development and aging. *Oncogene.* 15:203-211.

Zindy, F., R.T. Williams, T.A. Baudino, J.E. Rehg, S.X. Skapek, J.L. Cleveland, M.F. Roussel, and C.J. Sherr. 2003. *Arf* tumor suppressor promoter monitors latent oncogenic signals *in vivo*. *Proc. Natl. Acad. Sci. USA*. 100:15930-15935.

Chapter 3.2

DNA REPLICATION AND GENOMIC INSTABILITY

Wenge Zhu, Tarek Abbas and Anindya Dutta
Department of Biochemistry and Molecular Genetics, University of Virginia School of Medicine, Charlottesville, USA

1. INTRODUCTION

The ability of cells to faithfully and fully replicate their genetic material is of utmost significance to maintain genomic integrity, but of equal significance is the ability of cells to limit this highly orchestrated activity to once per cell cycle. Cells have evolved multiple mechanisms to cope with the various tasks that continuously threaten both the integrity and the fidelity of the replication process. These mechanisms can be divided into two major classes. (i) Mechanisms that respond to intrinsic replication perturbations as well as to extrinsic factors that interfere with the replication process to cause various forms of DNA damage, and (ii) Mechanisms that prevent premature firing of origins of replication, and that ensure that DNA is replicated only once during each cell cycle. The former mechanisms include DNA repair and checkpoint mechanisms, both of which have been described in depth elsewhere and are only mentioned here for classification purposes. In this chapter, we focus on DNA replication, and the relationship between DNA replication and genomic instability. We begin this chapter with a review of origins of replication as well as a summary of the current knowledge of the various replication initiation factors known to regulate origin firing. This is followed by description of the various steps involving the assembly and activation of initiation complexes, and of regulatory mechanisms that control initiation of DNA replication with a focus on the role of CDKs (cyclin-dependent kinases) and DDKs (Dbf4-dependent kinases) in regulating this process. We conclude this chapter with a description of the various mechanisms that prevent re-replication, and how cells respond to replication perturbations when these mechanisms are derailed.

E.A. Nigg (ed.), Genome Instability in Cancer Development, 249-279.

In eukaryotes, DNA replication initiates at multiple sites known as origins of DNA replication. Trans-acting replication-initiation factors recognise these sites, assemble replication initiation, and activate the replication initiation process.

2. ORIGINS OF REPLICATION

In both prokaryotic and eukaryotic cells, initiation of DNA replication occurs at origins of replication (Baker and Kornberg, 1988; Jacob, 1963) The well-established eukaryotic origins of replication are those from the yeast *Saccharomyces cerevisiae,* originally identified as autonomously replicating sequences (ARS), with the capacity to promote high frequency transformation (Hsiao and Carbon, 1979; Kingsman et al., 1979; Stinchcomb et al., 1979; Struhl et al., 1979). ARSs have high (73-83%) A+T nucleic acid content, and contain a consensus sequence 5'-(A/T)TTTAT(A/G)TTT(A/T)-3', known as ARS consensus sequence (ACS) (Broach et al., 1983). The ACS elements serve as the binding site for the origin recognition complex (ORC) which is discussed in the following section, and plays an essential role in the initiation of DNA replication (Bell and Stillman, 1992). The budding yeast replication origins contain elements that are distributed over a region of 200 bp, including an A element, as well as two or three B elements (Marahrens and Stillman, 1992).

Unlike replication origins in the budding yeast, origins from higher eukaryotes are not as simply structured. One of the best-studied human replication origins is the Dihydrofolate reductase (DHFR) locus, which is located within a 55 kb spacer region between the 3' end of the DHFR gene and the 5' end of the next adjacent genes. This large replication region contains three sites termed ori-β, ori-β', and ori-γ, from which replication initiation is preferred (DePamphilis, 1999; Hamlin and Dijkwel, 1995; Kobayashi et al., 1998; Linskens and Huberman, 1990; Wang et al., 1998). Deletion of ori-β at the endogenous locus does not alter replication of the locus, while deletion of the 3' end of the DHFR gene eliminates ori-β activity, suggesting that initiation of DNA replication in higher eukaryotes can be controlled by both local sequences, as well as by sequences distant from origins of replications (Kalejta et al., 1998). This conclusion is further supported by studies on the human β-globin origins encompassing an 8 kb DNA sequence, where locus control region (LCR) 40 kb distant from the initiation region, is required for firing from the β-globin origin (Aladjem and Fanning, 2004; Aladjem et al., 1995; Aladjem et al., 2002). Adding to the complexity is the possibility that transacting transcriptional factors may also regulate origin firing in higher eukaryotes. For example, deletion of the transcriptional promoter at the DHFR locus prevents initiation in the downstream initiation zone. This deficiency in replication initiation is restored by replacement with a wild-type Chinese hamster promoter or a *Drosophila*-based construct (Saha et al., 2004). Other factors such as histone acetylation at replication origins have been shown to be critical for origin specificity and activity in *Drosophila* (Aggarwal and Calvi, 2004), indicating that chromatin structure may serve as a modulator of origin activity.

3. REPLICATION INITIATION FACTORS

A number of replication initiation factors have been identified and characterised. The major components are summarised below.

3.1 The Origin Recognition Complex (ORC)

The ORC (origin recognition complex) was originally identified as a six-protein subunit complex that associates with replication origins in *S. cerevisiae* (Bell and Stillman, 1992). The ORC is highly conserved in all eukaryotic organisms and plays a key role in initiation of DNA replication (Bell and Dutta, 2002). High resolution *in vivo* footprinting shows that the ORC can bind to both ACS and B1 elements simultaneously (Diffley and Cocker, 1992; Rao and Stillman, 1995; Rowley et al., 1995).

In *S. cerevisiae*, the ORC protein complex binds specifically to replication origins throughout the cell cycle in an ATP-dependent manner (Bell and Stillman, 1992). Orc1 and Orc5 are two subunits with ATP binding sites, but only Orc1 has an ATP hydrolysis motif (Klemm et al., 1997). Interestingly, the binding of ORC to single stranded DNA induces conformational changes in ORC and stimulates its ATPase activity. In striking contrast, however, the binding of ORC to double-stranded origin DNA inhibits the same activity *in vitro*, suggesting a dual state of ORC; an ATP-bound state, and an ATP hydrolyzed state (Lee et al., 2000). Given the fact that ORC is exposed to ssDNA during DNA melting that accompanies initiation of DNA replication, it is possible that initiation of DNA replication induces the hydrolysis of ORC-bound ATP, contributing to the control of replication initiation. How ORC interacts with replication origins remains under extensive investigation, but protein-DNA cross-linking experiments suggest that five of the six ORC subunits coordinate this interaction, and that Orc1, Orc2, and Orc4 are major subunits mediating the interaction with ACS DNA (Lee and Bell, 1997).

Although ORC proteins display high homology, the regulation of ORC protein interactions with chromatin seems to differ widely among the various eukaryotes (Bell and Dutta, 2002). For instance, in *Drosophila melanogaster, Xenopus laevis*, and *Homo sapiens*, the association of one or more ORC subunits with chromatin is cell cycle regulated. *Xenopus* ORC was shown to dissociate from metaphase chromatin (Romanowski et al., 1996). In human cells, whereas the levels of ORC subunits Orc2-5 remain constant throughout the cell cycle, the Orc1 protein is stable in G1, but is degraded in S phase by the SCF^{Skp2} ubiquitin ligase-mediated proteolytic pathway (Mendez et al., 2002; Tatsumi et al., 2003). Moreover, Orc2-5 proteins were shown to appear in the nuclease-insoluble, non-chromatin structure in a manner that parallels the increased levels of Orc1 associating with the non-chromatin nuclear fractions (Ohta et al., 2003). In hamster cells, Orc1 protein is mono-ubiquitinated during S phase, and mono-ubiquitinated Orc1 is detected in the non-chromatin bound fraction, suggesting that mono-ubiquitination of Orc1 may regulate its activity (Li and DePamphilis, 2002). Depletion of Orc1 protein by short-interfering RNAi allows Orc2 protein to be detected in the nuclease-soluble fraction, and blocks the association of MCM (minichromosome maintenance) proteins with chromatin (Ohta et al., 2003). These results indicate that Orc1 may regulate the association of the ORC complex with replication origins, and regulate the loading of MCM proteins onto chromatin (DePamphilis, 2003). Thus, downregulation of Orc1 protein in S-phase of the cell cycle provides a mechanism by which human cells avoid reinitiation of DNA replication during the same cell cycle. Depletion of Orc2 by siRNA on the other hand, significantly decreases the levels of Orc1 and Orc3-6, suggesting a key role of Orc2 in the maintenance or stability of the ORC complex. Interestingly, however, depletion of Orc2 stabilises $p27^{Kip1}$ protein, an inhibitor of cyclin E-Cdk2 (Y. Machida and A. Dutta, unpublished data), suggesting that inadequate levels of Orc2 or ORC complex may activate a $p27^{Kip1}$-dependent checkpoint, which prevents genomic instability by inhibiting cell cycle progression. Moreover, depletion of Orc2 by siRNA in human cancer cells causes abnormally condensed chromosomes, failed chromosome congression, and multiple centrosomes, implicating Orc2 protein as a key player regulating chromosome duplication, chromosome structure and centrosome copy number control (Prasanth et al., 2004). In addition to its apparent role in the initiation of DNA replication, Orc6 protein has also been shown to localise at the site of the cleavage furrow, where septin protein forms rings, suggesting that Orc6 protein may play a role in chromosome segregation and cytokinesis (Prasanth et al., 2002). It appears that different ORC subunits may incorporate into different complexes mediating different biological functions.

3.2 Cdc6

The Cdc6 protein was first identified in a screen for mutants that affect the cell division cycle in *S. cerevisiae* (Hartwell 1973), and was later shown to play important roles in cell cycle-specific activation of DNA replication origins (Dutta and Bell, 1997; Stillman, 1996). Cdc6 is a member of the large family of AAA+ ATPases. This group of proteins functions in the assembly, operation, and disassembly of diverse protein complexes (Neuwald et al., 1999). Many other proteins involved in the initiation of DNA replication also have AAA+ containing modules, including Orc1, Orc4, and Orc5, MCM proteins, and RFC family members. AAA+ proteins contain two protein domains: an ATP binding domain, and a smaller domain composed of a three-helix bundle that is involved directly in protein-substrate remodelling (Neuwald et al., 1999). ATP binding and hydrolysis may regulate the physical relationship of the two domains, which in turn may mediate the assembly, remodelling, or disassembly of protein complexes (Neuwald et al., 1999). In addition to the two domains that are conserved in AAA+ proteins, Cdc6 has an N-terminal region, which interacts with CDKs and contains the major sites for CDK-dependent phosphorylation (Delmolino et al., 2001; Elsasser et al., 1996). The function of Cdc6's C-terminal region is still unknown, but it may interact with other proteins in the initiation complex or with DNA. The C-terminal region of *Archaea* Cdc6 forms a helix-turn-helix domain that could be involved in protein-protein or protein-DNA interactions (Grabowski and Kelman, 2001).

While yeast Cdc6 is required for Mcm2-7 protein loading onto replication origins as part of the assembly of pre-replicative complexes (pre-RCs) in early G1 phase of the cell cycle, it is degraded at the onset of S phase by the SCF^{Cdc4}-dependent proteolytic pathway (Aparicio et al., 1997; Cocker et al., 1996; Liang and Stillman, 1997; Perkins and Diffley, 1998; Tanaka et al., 1997; Weinreich et al., 1999). *In vitro*, Cdc6 increases the DNA binding specificity of ORC by inhibiting non-specific DNA binding of ORC, due to a conformational change in ORC, suggesting that Cdc6 may facilitate the recruitment of MCM proteins to replication origins by changing ORCs conformation (Mizushima et al., 2000). At the end of mitosis, yeast Cdc6 also cooperates with Sic1 to inactivate mitotic cyclin-dependent kinase, thus promoting pre-RC formation (Calzada et al., 2001). Although the level of human Cdc6 protein is not changed throughout the cell cycle, upon phosphorylation by CDKs at the G1/S transition of the cell cycle, Cdc6 is exported from the nucleus to the cytoplasm (Fujita et al., 1999; Jiang et al., 1999; Petersen et al., 1999; Saha et al., 1998a; Thome et al., 2000). However, a recent study showed that even non-chromatin bound Cdc6 in Chinese Hamster Ovary cells remains nuclear throughout S phase (Alexandrow and Hamlin, 2004). Despite these conflicting results, the persistence of chromatin-bound Cdc6 in S phase does not promote replication, because Cdc6, at least in *Xenopus,* is no longer required for efficient DNA replication once DNA has been licensed (Rowles et al., 1999). In quiescent human cells, Cdc6 protein is degraded via an APC^{Cdh1}-dependent proteolytic pathway (Petersen et al., 1999). Overexpression of human Cdc6 in cells during the G2 phase of the cell cycle activates a Chk1-mediated G2 phase checkpoint, and prevents entry into mitosis (Clay-Farrace et al., 2003). Additionally, a study from *Schizosaccharomcyes pombe* revealed a role of Cdc6 in S phase checkpoint activation (Murakami and Nurse, 1999). In this study, the activation of Cds1 (Chk2) checkpoint kinase was found to be abolished in *cdc6 (cdc18)* mutant cells treated with hydroxyurea (Murakami and Nurse, 1999). During early stages of programmed cell death, mammalian Cdc6 is destroyed by a p53-independent, ubiquitin-mediated pathway, indicating that Cdc6 may play a role in the uncoupling of DNA replication from the cell cycle in cells undergoing apoptosis (Blanchard et al., 2002; Guo and Hay, 1999).

3.3 Minichromosome Maintenance (MCM) Proteins2-7

The MCM proteins consist of a family of six related proteins (Mcm2-7), most of which were first identified in genetic screens for mutants defective in plasmid maintenance or cell cycle progression (Maine et al., 1984; Moir et al., 1982). All six MCM proteins are essential for cell viability, highly conserved among eukaryotes, and required for the initiation of DNA replication (Tye, 1999). Each MCM protein is a member of the AAA+ class of ATPase, and all six proteins show a high sequence similarity to each other, especially in a 240 amino acid conserved region containing an ATP binding motif (Koonin, 1993). Biochemical studies suggest that the six MCM proteins interact with one another to form a hexameric complex composed of stoichiometric amounts of each protein, with a molecular weight of about 600 kDa (Thommes et al., 1997). Thus, the six MCM proteins probably function together in a large multiprotein complex. In *S. cerevisiae*, the total levels of MCM proteins do not vary significantly during the cell cycle. However, the localisation of yeast MCM proteins is cell cycle-regulated, with a localisation in the nucleus during G1, and a cytoplasmic localisation as cells enter S phase (Dalton and Whitbread, 1995; Hennessy et al., 1990; Yan et al., 1993). Mcm2 and Mcm3 exist with different phosphorylation states in different phases of the cell cycle, suggesting that the localisation of MCM proteins may be regulated by phosphorylation (Young and Tye, 1997).

Biochemical studies suggest that the MCM complex acts as a replicative helicase (Kelly and Brown, 2000; Nishitani and Lygerou, 2002). Mutational analyses indicate that there are two distinct MCM subgroups: Mcm4, Mcm6, and Mcm7 contribute the ATP binding motif, whereas the subgroup composed of Mcm2, Mcm3, and Mcm5 regulates the activity of this motif (Schwacha and Bell, 2001). The Mcm4/6/7 trimeric complex has weak processive DNA helicase activity, although the MCM heterohexamer lacks helicase activity *in vitro* (Adachi et al., 1997; Ishimi, 1997; Lee and Hurwitz, 2000; Lee and Hurwitz, 2001). These data suggest that Mcm4/6/7 is essential for helicase activity, whereas Mcm2/3/5 may negatively regulate this activity. Given that Mcm4/6/7 helicase activity can be inhibited by the simple addition of Mcm2, it is likely that Mcm2 is part of a control mechanism that prevents initiation of DNA replication from occurring during the assembly of initiation complexes (Ishimi et al., 1998). Alternatively, it is possible that MCM proteins are loaded onto replication origins as an inactive heterohexameric complex containing all six members (Mcm2/3/4/5/6/7). At the time of initiation of DNA replication, the Mcm4/6/7 helicase is activated upon removal of Mcm2/3/5. However, there is no conclusive evidence that the MCM complexes dissociate into subcomplexes during DNA replication. Mcm2 is phosphorylated by Cdc7, a kinase that is required for the activation of the initiation complex (Lei et al., 1997). Whether the phosphorylation of Mcm2 by Cdc7 can remove Mcm2, or Mcm2/3/5 from Mcm4/6/7 is not yet clear. In HeLa cells the phosphorylation of Mcm4 results in a loss of the Mcm4,6,7 helicase activity, suggesting a potential mechanism by which cells inhibit Mcm4/6/7 helicase activity later in S phase to prevent re-replication (Ishimi and Komamura-Kohno, 2001). Structural studies on *Archaea* MCM complexes suggest that MCM proteins form a bi-lobed double hexameric ring with a large central channel lined with positive charge that could accommodate the DNA double helix (Fletcher et al., 2003; Pape et al., 2003). Head to head interactions of the N-terminal lobes would position the C-terminal lobes containing the helicase domain in two rings at the outermost edge of the double hexameric ring. Structural studies with the SV40 T antigen (Li et al., 2003a) suggest that such a double hexameric helicase could act as a DNA pump that draws in double stranded DNA from the two ends and extrudes single stranded DNA, one stand from the gap between the two hexamers and the other strand through gaps between the subunits of the hexamer.

3.4 Mcm10

Although identified in the same screen as other MCM proteins, Mcm10 is not an AAA+ ATPase, but is required for DNA replication from yeast to humans (Homesley et al., 2000; Izumi et al., 2000; Wohlschlegel et al., 2002). The Mcm10 protein does not share significant sequence similarity with Mcm2-7, and is an abundant chromatin-binding protein, which interacts with all six subunits of the Mcm2-7 complex (Homesley et al., 2000; Tye, 1999). In yeast, Mcm10 specifically binds to replication origins, and this binding is essential for the chromatin association of Mcm2-7 proteins, suggesting that Mcm10 is a component of the pre-RC (Homesley et al., 2000). Interestingly, however, *Xenopus* and human Mcm10 has been shown to bind origins of DNA replication subsequent to Mcm2-7 binding, and to further stimulate origin binding of Cdc45 (Wohlschlegel et al., 2002). These observations may be explained by evolutionary differences between *X. laevis* and *S. cerevisiae* in the order of events during replication initiation. The Mcm10 protein seems to be also required for the completion of S phase after release from a hydroxyurea block, indicating a role of Mcm10 in replication elongation (Kawasaki et al., 2000). The potential role of Mcm10 in promoting replication elongation is supported by the pausing of elongation forks in *mcm10-1* mutants, and the interactions between Mcm10 protein and elongation factors in *S. cerevisiae* (Homesley et al., 2000). In *S. pombe*, the Mcm10 protein has been shown to interact specifically with the catalytic p180 subunit of DNA polymerase α and to stimulate DNA synthesis catalyzed by the polymerase-α-primase complex *in vitro* (Fien et al., 2004). In human cells, the level of Mcm10 protein decreases in late M phase and remains low during G_1 phase of the cell cycle. It then begins to accumulate and binds chromatin at the onset of S phase of the cycle, suggesting that mammalian Mcm10 is involved in the activation of pre-RCs (Izumi et al., 2001; Izumi et al., 2004). Recently, in *S. pombe*, Mcm10 was shown to be required for the phosphorylation of MCM proteins by the Dfp1-Hsk1 kinase (the fission yeast homolog the Dfb4-Cdc7 kinase of *S. cerevisiae*), indicating that Mcm10 may participate in the activation of pre-RCs by recruiting the Dfp1-Hsk1 kinase, and stimulating the phosphorylation of the MCM complex (Lee et al., 2003).

3.5 Cdt1

The Cdt1 (Cdc10 dependent transcript 1) protein was first identified as a target of Cdc10, a G1/S specific transcription factor, and as an essential protein for the initiation of DNA replication in *S. pombe* (Hofmann and Beach, 1994). Later studies showed that Cdt1 is required for recruiting Mcm4 to chromatin, and that Cdc18/Cdc6, and Cdt1 bind to chromatin independently of each other (Nishitani et al., 2000). Homologues of Cdt1 have been identified in *S. cerevisiae, Xenopus*, *Drosophila,* as well as in humans. In most species, Cdt1 is negatively regulated by proteolysis, with levels peaking in G1 phase of the cell cycle, and declining *via* degradation in S phase. While Cdt1 protein from *C. elegans* is degraded by the Cul4-mediated proteolytic pathway (Zhong et al., 2003), the human Cdt1 is phosphorylated by cyclin A-dependent kinases in S phase, resulting in the binding of Cdt1 to the F-box protein Skp2, and its subsequent proteolysis (Sugimoto et al., 2004). Studies in *Drosophila* showed that phosphorylation and degradation of Cdt1(Dup) at G_1/S requires cyclin E/Cdk2 (Thomer et al., 2004). In response to UV-induced DNA damage, human Cdt1 is rapidly degraded through either SCF^{Skp2} or Cul4-mediated proteolysis pathways (Higa et al., 2003; Kondo et al., 2004), indicating that Cdt1 acts as one of the checkpoint targets that prevent cells from replicating damaged DNA. In sharp contrast to Cdt1 from most organisms, the Cdt1 protein from *S. cerevisiae* is associated with the Mcm2-7 complex and is excluded from the nucleus after G1 at the same time that Mcm2-7 is exported (Nishitani et al., 2000; Tanaka and Diffley, 2002).

3.6 CDC45 (Sld4)

Cdc45 is a protein essential for initiation of DNA replication (Dutta and Bell, 1997; Kelly and Brown, 2000). The yeast Cdc45 protein associates with chromatin in late G1, and during S phase of the cell cycle (Zou and Stillman, 1998). The human Cdc45 protein associates with Orc2 protein (Saha et al., 1998b) and, like its yeast homolog, associates with chromatin in G_1, but progressively loses association with nuclei as S phase proceeds. The chromatin binding activity of yeast Cdc45 protein is dependent on both Clb-Cdc28 and Cdc7-Dbf4 kinase activities, as well as on functional Cdc6 and Mcm2 proteins (Zou and Stillman, 1998). Cdc45 and the single strand binding factor RPA associate with origins of replication in a mutually dependent manner (Zou and Stillman, 2000). In both *S. cerevisiae* and *X. laevis*, Cdc45 protein is required for loading DNA polymerase α onto chromatin, and both of these proteins colocalise throughout S phase (Mimura and Takisawa, 1998; Zou and Stillman, 2000). Using chromatin immunoprecipitation assays, it has been shown that Cdc45, RPA, MCMs and DNA polymerase ε associate with replication forks that move away from replication origins after initiation of DNA replication (Aparicio et al., 1997; Tanaka and Nasmyth, 1998). Since Cdc45, RPA, and MCMs move together with replication forks, it is possible that Cdc45, together with RPA, may stimulate the MCM helicase activity during replication elongation. Support for this hypothesis comes from a study showing that, in *X. laevis,* the Cdc45 protein is required for chromosome unwinding during elongation, by functioning as a helicase co-factor, because depletion of Cdc45 abolishes unwinding by uncoupled helicases (Pacek and Walter, 2004). In addition, the degradation of Cdc45 in yeast prevents completion of S phase and fork progression (Tercero et al., 2000). Activation of the S phase checkpoint by treating cells with hydroxurea inhibits the association of Cdc45 with late replication origins in wild type, but not in *rad53* mutant cells (Aparicio et al., 1999). Given the fact that Rad53 plays a key role in a checkpoint controlling late–firing origins of DNA replication (Santocanale and Diffley, 1998), it has been suggested that Rad53-dependent checkpoint inhibition of late origin firing occurs at the level of Cdc45 (Aparicio et al., 1999).

3.7 Dpb11, Sld2, Sld3, and GINS

Dpb11 was isolated as a multi-copy suppressor of mutations in the *POL2* and *DPB2* genes, which encode the catalytic, and the second-largest subunits of DNA polymerase ε, respectively (Araki et al., 1995). Dpb11 and Polε simultaneously associate with replication origins in a mutually dependent manner and Dpb11-Pol2 complexes accumulate during S phase (Masumoto et al., 2000), suggesting a role of Dbp11 in elongation. Additionally, Dpb11 may be involved in S phase checkpoint control because the association of Pol2 with late-firing-origins is inhibited by HU in Dpb11 wild-type cells but not in *dpb11* mutant cells (Masumoto et al., 2000). Homologs of Dpb11 in mammals (TopBP1) and *Drosophila* (Mus101) have also been recently identified and shown to have similar function as Dpb11 in yeast (Makiniemi et al., 2001; Van Hatten et al., 2002).

Sld2 and Sld3 are two proteins identified in a screen for mutants that are synthetically lethal with *dpb11.* The Sld2 and its counterpart in *S. pombe*, Drc1, were subsequently shown to be required for initiation of DNA replication, and the activation of the S phase checkpoint (Kamimura et al., 1998; Wang and Elledge, 1999). The Sld2 protein is phosphorylated by S-CDKs, and this phosphorylation is required for its interaction with Dpb11 (Masumoto et al., 2002; Noguchi et al., 2002). Thus, an important role for S-CDK in the regulation of DNA replication is to promote the interaction between Sld2 and Dpb11, and therefore facilitate the association of polymerases with replication origins. The Sld3 protein associates with Cdc45 throughout the cell cycle, and both proteins bind to replication origins simultaneously in a mutually dependent manner (Masumoto et al., 2002). These data suggest that Sld3 may function together with Cdc45 for polymerase α/primase loading and origin unwinding in DNA replication (Kamimura et al., 2001). Unlike observations in budding yeast, recent studies in the fission yeast *S. pombe* suggest that Sld3-loading is independent of Cdc45-loading, and that Sld3 forms complexes with MCM proteins without Cdc45 (Yamada et al., 2004). The functional significance of Sld3-MCM protein interaction in DNA replication, however, remains to be elucidated. Despite the essential role of Sld3 in the initiation of DNA replication in yeast, the homolog of Sld3 in mammals is not yet identified.

GINS is a protein complex containing Sld5, PSF1 (partner of SLD five), PSF2, and PSF3 and was identified in a screen for proteins that interact with Sld5. All of these subunits are highly conserved in eukaryotic cells (Kubota et al., 2003; Takayama et al., 2003). Studies in both *S. cerevisae* and *X. laevis* demonstrate that GINS is required for initiation of DNA replication, and that its association with replication origins is mutually dependent on Dpb11, Sld3, and Cdc45. GINS associates with replication origins and then with neighbouring fragments as S phase progresses, indicating a possible role in DNA elongation (Takayama et al., 2003). A ring-like structure of GINS, revealed by electron microscopy (Kubota et al., 2003), suggest that GINS could function as a clamp for polymerases in a manner reminiscent to that of PCNA. It is, however, not clear how GINS regulates DNA replication elongation by DNA polymerases, or how GINS cooperates with PCNA in mediating DNA replication elongation.

4. ASSEMBLY OF REPLICATION INITIATION COMPLEXES

The initiation of eukaryotic replication at origins of replication can be divided into two steps. The first step is the sequential assembly of ORC complex, Cdc6, Cdt1, and MCMs onto replication origins to form pre-RCs, a process that is called "licensing" of replication origins. This model of pre-RC assembly was initially based on chromatin footprinting assays, in which ARS1 DNA is protected by protein complexes throughout the cell cycle, but protection becomes more extensive from late mitosis till the onset of S phase (Diffley, 1994). Diffley and colleagues referred to this more extensively protected structure as the pre-replicative complex (pre-RC), and to the other structure from early S phase to the end of mitosis as the post-replicative complex (post-RC). The second step in replication initiation involves the firing of "licensed" origins at different times during S phase of the cell cycle. The first step occurs from late mitosis to late G1 phase of the cell cycle, during which CDK activity is maintained at low levels. Cells have developed two mechanisms by which they maintain low CDK activity to allow for pre-RC assembly. One mechanism is to degrade G1, S, and G2-specific cyclins via the APC/C (anaphase promoting complex/cyclosome)-mediated proteolytic pathway. The second mechanism involves the inhibition of CDK activity via the induction of CDK inhibitors, such as Sic1, an inhibitor of Cdc28-Clb-5/6 in *S. cerevisiae*, and p27 ^{Kip1}, an inhibitor of cyclin E/Cdk2 in mammalian cells.

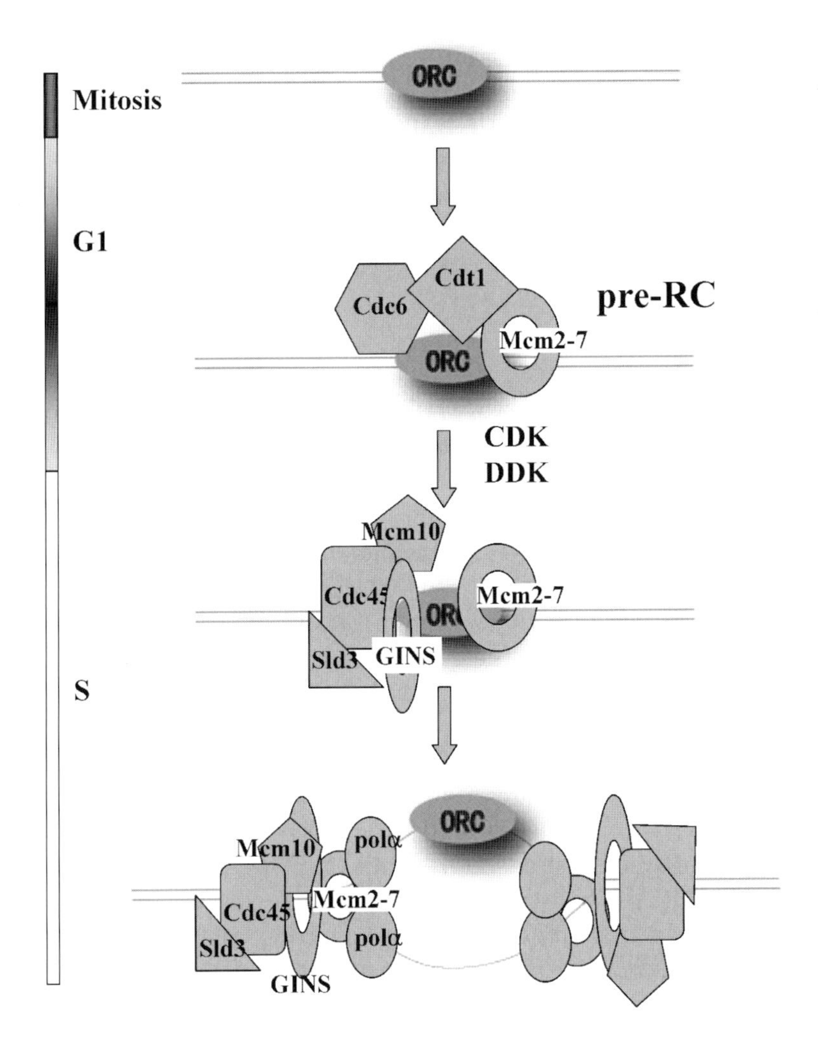

Figure 1. The assembly and activation of pre-replicative complexes. Pre-RCs are assembled at late G1 phase and activated by CDK and DDK. See text for detail.

A relatively simple model of pre-RC assembly and origin firing is depicted in Figure 1. In this model, the ORC complex, which is constitutively bound to replication origins throughout the cell cycle (Aparicio et al., 1997; Diffley et al., 1994; Liang and Stillman, 1997; Lygerou and Nurse, 1999), acts as a landing pad for Cdc6 and Cdt1 from late mitosis to early G1 phase of the cell cycle (Liang et al., 1995; Tanaka and Diffley, 2002). Pre-RCs are then fully assembled when the Mcm2-7 complex is recruited to origins (Coleman et al., 1996; Maiorano et al., 2000; Tanaka et al., 1997). As mentioned above, the Mcm2-7 complex is required throughout S phase for both initiation of DNA replication and elongation (Kelly and Brown, 2000). In yeast, Cln-CDK phosphorylates Cdc6 in late G1, and the phosphorylated Cdc6 is targeted for ubiquitination and proteolysis (Drury et al., 1997; Jallepalli et al., 1997; Sanchez et al., 1999). In human cells, however, Cdc6 is exported from the nucleus to the cytoplasm after phosphorylation by CDKs at the G1/S transition. At the onset of S phase, the Mcm10 protein is recruited to origins of replication, and both Clb5,6-Cdc28 (Cyclin E–Cdk2 in human cells) and Cdc7-Dbf4 kinases are thought to phosphorylate the MCM complex, triggering its helicase activity. Active MCM helicase initiates the melting of origin DNA, and the DNA replication proteins Sld3, Cdc45, RPA, GINS and DNA polymerase α/primase are subsequently recruited to origins where DNA synthesis initiates (Diffley and Labib, 2002).

5. CDKS AND DDKS: MASTER REGULATORS OF INITIATION OF DNA REPLICATION

Two classes of protein kinases are required for initiation of DNA replication: Cdc7/Dbf4 (DDKs) and cyclin dependent kinases (CDKs) (Bell and Dutta, 2002; Kelly and Brown, 2000; Newlon, 1997). Cdc7 is a kinase required for initiation of DNA replication, (Hartwell, 1973; Hollingsworth and Sclafani, 1990; Patterson et al., 1986; Yoon and Campbell, 1991), and homologs of Cdc7 have been identified in a number of organisms ranging from yeast to human (Diffley et al., 1995; Hardy, 1996; Jares et al., 2000; Johnston et al., 1999; Lei and Tye, 2001; Masai et al., 2000; Sclafani, 2000). The activation of Cdc7 requires its association with a regulatory protein, Dbf4 in *S. cerevisiae*, Dfp1/Him1 in *S. pombe*, or ASK in mammals, through a conserved Dbf4-motif-M, and Dbf4-motif-C (Dowell et al., 1994; Masai and Arai, 2000; Masai and Arai, 2002). In budding yeast, Cdc7 is required for the initiation of DNA replication from both early, and late firing origins (Bousset and Diffley, 1998; Donaldson et al., 1998). Thus, the Cdc7-Dbf4 complex appears to be not a trigger of S phase, but instead, an important player in mediating firing from individual origins of replication.

Several lines of evidence indicate that Cdc7 kinase acts at replication origins, presumably to phosphorylate some components of the pre-initiation complex. In yeast, a mutant allele of *mcm5* (*mcm5-bob1*) was initially identified as a suppressor of the *cdc7* mutant, and the *mcm5-bob1* mutation suppresses all mutations in *CDC7* or *DBF4*, suggesting that *mcm5-bob1* is able to bypass the essential function of Cdc7-Dbf4 (Hardy et al., 1997; Jackson et al., 1993). Additionally, Mcm2 physically interacts with, and is phosphorylated by Cdc7-Dbf4 both *in vitro* and *in vivo* (Lei et al., 1997) Interestingly, the efficiency of Mcm2 phosphorylation by Cdc7 is significantly increased when the Mcm2 is pre-phosphorylated by CDKs (Masai et al., 2000), suggesting that CDKs and Cdc7-Dbf4 may collaborate to produce the efficient phosphorylation of Mcm2 in the initiation complex for initiation of DNA replication. In yeast, CDK activity is required prior to DDK, although the opposite order was observed in *Xenopus* (Nougarede et al., 2000; Walter, 2000). Whether DDK activity precedes that of CDK activity in the activation of replication initiation in mammals is not yet clear. In addition to MCM proteins, the p180 subunit of DNA polymerase α/primase is also specifically phosphorylated by Cdc7-Dbf4 *in vitro* (Weinreich and Stillman, 1999), but the functional significance of this phosphorylation is yet to be determined.

Both the budding yeast Dbf4, and its homologue in fission yeast (Dfp1/Him1) are hyperphosphorylated in cells arrested in S phase in response to hydroxyurea, and this hyperphosphorylation has been shown to be Rad53-dependent (Brown and Kelly, 1999; Weinreich and Stillman, 1999), suggesting that the Cdc7-Dbf4 complex is a target of an intra-S phase checkpoint pathway. The phosphorylation of Dbf4 is associated with a decrease in the Cdc7-Dbf4 kinase activity (Elledge, 1996). Furthermore, in *X. laevis*, single-stranded DNA gaps generated following treatment with etoposide or exonuclease induce an ATR-dependent checkpoint that downregulates Cdc7-Dbf4 protein kinase activity by impeding the association of Cdc7 with Dbf4, resulting in inhibition of further origin activation (Costanzo et al., 2003). Whether the inactivation of DDK following these insults to DNA serve as a safeguard against the replication of damaged DNA remains to be confirmed.

The second class of protein kinases required for initiation of DNA replication are the CDKs. These kinases are required for activation of pre-RCs, accounting for the fact that the chromatin association of Cdc45 is dependent on CDK activity (Walter and Newport, 2000; Zou and Stillman, 2000). How CDKs contribute to pre-RC activation is not entirely clear, although it is well established that a large number of replication factors are substrates for CDKs *in vitro* and/or *in vivo* (Kelly and Brown, 2000). One of the few examples where the role of phosphorlyation seems to be clear involves the Sld2 protein discussed above. In both budding and fission yeast, the Sld2 protein is phosphorylated by CDKs, and this phosphorylation is essential for the association of DNA polymerase with the replication origins (Masumoto et al., 2002; Noguchi et al., 2002). The positive role of CDKs in regulating DNA replication initiation and its significant negative role in preventing re-replication is discussed in detail in the following section.

6. RE-REPLICATION AND GENOMIC STABILITY

Timely assembly, disassembly, and reassembly of pre-RCs at replication origins is tightly controlled to ensure that genomic DNA is replicated once per cell cycle. Multiple overlapping mechanisms have evolved to ensure that cells replicate their genetic material accurately. A common characteristic of these mechanisms is that they are inherent to the replication initiation machinery itself in that origins that have already fired during S phase of a cell cycle are prevented from firing again until the next replication cycle. Overriding this replication control leads to abnormal DNA replication and genomic instability. One form of abnormal DNA replication is re-replication, which can be classified into two main types. (i). Re-replication after the completion of S phase, thereby bypassing cell division. This type of re-replication is associated with complete duplication of the genome within the same cell. Cyclin A-CDK activity is required for suppressing this kind of re-replication, because depletion of cyclin A leads to duplication of the whole genome (Mihaylov et al., 2002). (ii). Re-replication from origins of replication that have already fired before the completion of S phase of the same cell cycle, resulting in DNA amplification only from origins that have already fired during that cycle. This type of re-replication has been extensively studied, and in this section we focus on this type of re-replication. We will first discuss CDK-dependent pathways that are implicated in preventing pre-RC assembly until the completion of S and M phase, and then discuss possible CDK-independent pathways that are mainly dependent on a carefully maintained balance between Cdt1 and protein known as geminin.

6.1 CDKs and Re-replication Control

In addition to the positive role that CDKs play in the initiation of DNA replication, CDKs also play a role in preventing re-replication to ensure that DNA replication occurs once, and only once, per cell cycle (Bell and Dutta, 2002). The activity of CDKs is not constant during the cell cycle. It is maintained at low levels in early G1 phase, during which pre-RCs can assemble at origins, but initiation cannot occur. On the other hand, increased CDK activity from late G1 phase to late M phase allows pre-assembled complexes to be activated to initiate DNA replication, while new pre-RCs are prevented from forming (Jallepalli et al., 1997). Several lines of evidence support this conclusion. First, in *S. cerevisiae*, inhibition of CDK activity in M phase cells leads to reversion of the post-RC pattern to a pre-RC pattern, suggesting that high CDK activity suppresses the assembly of initiation complexes (Dahmann et al., 1995). Second, overexpression of Clb2 in cells in early G1 phase of the cell cycle inhibits pre-RC formation (Detweiler and Li, 1998). Third, chromatin immunoprecipitation (ChIP) experiments demonstrate that high CDK activity in G2/M phase prevents the association of MCM proteins with origins of replication (Tanaka et al., 1997), and finally, as mentioned above, depletion of cyclin A in *Drosophila* cells results in the full duplication of the genome (Mihaylov et al., 2002). More recent work from yeast further supports this notion. Yeast cells lacking the CDK inhibitor Sic1 exhibit precocious CDK activation, resulting in fewer origin firing and prolonged S phase, and show a rate of gross chromosomal rearrangements, which can be rescued by delaying S phase CDK activation (Lengronne and Schwob, 2002). This finding demonstrates that a window of low CDK activity is critical for cells to reset origins before the onset of S phase, and that precocious CDK activation inhibits replication origin licensing in late G1, and causes genomic instability.

How do CDKs block pre-RC re-assembly in the same cell cycle? As mentioned above, several components of the pre-RC (ORC, Cdt1, Cdc6, and the MCMs) have been shown to be substrates of CDKs. In *S. cerevisiae*, both Orc6 and Orc2 are phosphorylated by Clb-Cdc28, resulting in inhibition of ORC function for assembly of pre-initiation complexes (Nguyen et al., 2001). The fission yeast Orc2 protein is phosphorylated and inactivated by rising CDK activity as cells enter S phase (Vas et al., 2001). Furthermore, in *Xenopus* egg extracts, high cyclin A-associated kinase activity promotes the release of ORC along with Cdc6 and MCM from sperm chromatin, whereas high cyclin E/CDK activity promotes chromatin binding (Findeisen et al., 1999). Additionally, as mammalian cells proceed through S to M phase, the Orc1 protein is selectively released from chromatin (Li and DePamphilis, 2002). Chromatin-unbound human Orc1 protein is polyubiquitinated by the SCF^{Skp2} ubiquitin ligase pathway, and degraded by the 26S proteasome (Fujita et al., 2002; Mendez et al., 2002). Although S phase CDK (S-CDK) phosporylates Orc1, it is still unclear if this phosphorylation is required for targeting Orc1 for degradation (Mendez et al., 2002). More recently, Orc1 protein was shown to be phosphorylated by CyclinA/Cdk1 in mitosis in cells where it is not degraded in S phase, and this phosphorylation prevents Orc1 protein from binding to chromatin, suggesting that Orc1 is targeted by CDK to block the assembly of pre-RC at mitosis (Li et al., 2004). In yeast cells, Orc6 protein interacts with the S-phase cyclin Clb5 at origins of replication only after initiation has occurred, and this interaction is maintained at origins during the remainder of S phase (Wilmes et al., 2004). Interestingly, eliminating the Clb5-Orc6 interaction by mutating the CDK phosphorylation site in Orc6 results in re-replication (Wilmes et al., 2004). Taken together, these findings indicate that phosphorylation of ORC proteins by S-CDKs inhibit ORC activity after initiation, and that inhibition of ORC activity is important for cells to prevent the re-assembly of pre-RCs until the beginning of the next cycle.

Cyclin-dependent protein kinases also phosphorylate the replication initiation factor Cdc6 (Cdc18 in fission yeast) at the G1/S transition of the cell cycle. In fission as well as budding yeast, the phosphorylation of Cdc6 by CDKs leads to its ubiquitin-dependent proteolytic degradation (Dietrich et al., 1997; Jallepalli et al., 1997; Kelly and Brown, 2000). In budding yeast, phosphorylated Cdc6 protein is targeted for ubiquitination by the SCF^{CDC4} E3 ubiquitin ligase, and polyubiquitinated Cdc6 protein is subsequently degraded by the 26S proteasome (Calzada et al., 2000; Drury et al., 2000; Elsasser et al., 1999). Furthermore, Clb-Cdc28 not only triggers the ubiquitination and degradation of Cdc6 protein, but also phosphorylates the transcriptional activator Swi5 and prevents it from entering the nucleus (Moll et al., 1991), thereby blocking the expression of Cdc6, and maintaining low levels of Cdc6 until cells complete mitosis. In higher eukaryotes, Cdc6 interacts specifically with active cyclin A/Cdk2 both *in vivo* and *in vitro* (Petersen et al., 1999; Saha et al., 1998a), and this interaction leads to the phosphorylation and subsequent cytoplasmic translocation of Cdc6 at the onset of S phase (Jiang et al., 1999; Petersen et al., 1999; Saha et al., 1998a). The significance of the role CDKs play in down-regulating Cdc6 is underscored by the finding that *S. pombe* cells expressing a mutant form of Cdc18, lacking CDK phosphorylation sites, exhibit significant re-replication (Gopalakrishnan et al., 2001; Jallepalli et al., 1997). Interestingly, however, when this mutant protein was expressed from a Cdc18 promoter, cells did not rereplicate their genomes unless Cdt1 was also coexpressed (Gopalakrishnan et al., 2001). This suggests that, at least in fission yeast, Cdc18 and Cdt1 work synergistically to allow initiation of DNA replication, and that deregulation of both of these proteins is required if a cell is to re-replicate its genetic material.

Phosphorylation of other replication proteins also plays a major role in preventing re-replication. The budding yeast MCM proteins, for example, are phosphorylated and exported from the nucleus during the G2 to M transition (Labib et al., 1999; Nguyen et al., 2000; Pasion and Forsburg, 1999). Although fission yeast and metazoan MCM proteins are not subject to such translocation effect after phosphorylation, their chromatin binding affinities are dramatically diminished (Lei and Tye, 2001). In *X. laevis,* Mcm4 is phosphorylated by mitotic cyclin-CDK, resulting in release of MCM proteins from chromatin (Coue et al., 1996; Pereverzeva et al., 2000). Interestingly, CDKs also regulate the helicase activity of MCMs proteins apart from their effects on replication initiation described above. *In vitro* studies, for example, have shown that the amino-terminal region of Mcm4 protein in the Mcm4,6,7 helicase complex is phosphorylated by cyclin A/Cdk2, resulting in inhibition of its helicase activity (Ishimi and Komamura-Kohno, 2001). The phosphorylation and inactivation of MCM proteins, which takes place in the S and G2/M phases of the cell cycle (Kimura et al., 1994; Musahl et al., 1995; Thommes et al., 1997; Todorov et al., 1995) is thought to be an important factor preventing genomic instability in eukaryotic cells.

A yeast strain carrying mutations in ORC, CDC6 and MCM that prevent the inhibition of these proteins by CDKs, demonstrates re-replication (Nguyen et al., 2001). Interestingly, all three inactivating mechanisms had to be annulled before the re-replication was observed, suggesting redundancy in the control extended by CDKs. Even in this strain, however, the cells did not enter into run away re-replication, and did not continue proliferation, suggesting the presence of additional mechanisms that allow a cell to resist re-replication.

Cdt1, another replication initiation factor, is targeted for degradation upon phosphorylation by CDKs. The budding yeast Cdt1 accumulates in the nucleus during G1 phase, but is excluded from the nucleus in late G1 in a CDK-dependent manner (Tanaka and Diffley, 2002). As mentioned above, the level of human Cdt1 protein is high in G1 phase, but rapidly declines as cells enter S phase of the cycle (Nishitani et al., 2001). In mammalian cells, Cdt1 is targeted for degradation via the SCF^{Skp2} ubiquitination pathway (Li et al., 2003b). Over-expression of the CDK inhibitors $p21^{Cip1}$ and $p27^{Kip1}$ suppresses the phosphorylation of Cdt1 and disrupts the association between Cdt1 and the F box protein Skp2, thereby stabilising Cdt1 (Liu et al., 2004). In fact, an unphosphorylatable mutant of Cdt1, which lacks the cyclin/CDK consensus phosphorylation sites is more stable than wild type Cdt1, and fails to interact with Skp2 protein (Liu et al., 2004). Recently, *Drosophila* Cdt1 (Dup) was shown to be phosphorylated by cyclin E/Cdk2 for degradation at the G1/S transition and mutation of Dup CDK sites increased genomic re-replication (Thomer et al., 2004). The fact that CDKs target multiple components of the pre-RC complexes, as shown in Figure 2, demonstrates that eukaryotic cells utilise various avenues to ensure that DNA replication initiates only once per cell cycle.

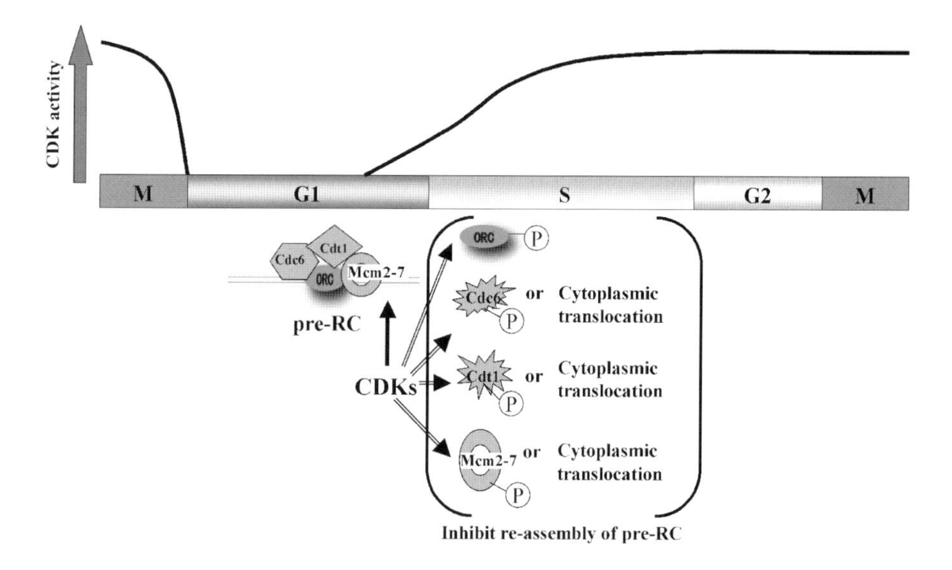

Figure 2. CDKs inhibits re-replication by blocking re-assembly of pre-RCs. In late G1 phase, CDKs play a positive role by activating pre-RCs. Once pre-RCs are activated, CDKs block re-assembly of pre-RCs by phosphorylating ORC proteins, MCM proteins, Cdc6 or Cdt1, which results in their inactivation by degradation or cytoplasmic translocation. See text for detail.

6.2 Cdt1- Geminin Balance and Re-replication

Cdt1 promotes the assembly of pre-RC complexes, and this function sets up its positive role in the initiation of DNA replication. Overexpression of Cdt1 in human cells results in re-replication (Vaziri et al., 2003). Similarly, increased expression of *Drosophila* Cdt1(Dup) in diploid cells is sufficient to induce polyploidizaion and cell death in developing tissues (Thomer et al., 2004). Given the evidence that loading of MCM proteins is the most important event in the formation of pre-RCs, and that Cdt1 is required for loading of MCM proteins, it is likely that excess Cdt1 may promote the re-assembly of pre-RCs despite the presence of active cyclin A-Cdk2. Thus, the degradation of Cdt1 at the onset of S phase is an important event that cells must execute very carefully if it were to prevent re-replication. This conclusion is supported by the observation that inactivation of the CUL-4 ubiquitin ligase, which results in the accumulation of Cdt1, causes re-replication (Zhong et al., 2003), suggesting that Cdt1 may be the limiting factor in the assembly of pre-RCs. Since Cdt1 activity is critical for genomic stability, and since it may serve as the limiting factor for pre-RC assembly, cells have evolved an additional layer of complexity to insure proper control of this important protein. This layer of complexity depends on geminin, an inhibitor of Cdt1 protein.

Geminin was first identified in a screen for proteins that are degraded in mitosis by anaphase-promoting complex in *Xenopus* (McGarry and Kirschner, 1998). Geminin protein levels are cell cycle regulated, accumulating in S, G2, and M phases of the cell cycle, and disappearing in G1 phase. *In vitro*, geminin was shown to inhibit initiation of DNA replication by interfering with the assembly of pre-RCs at the point between the binding of Cdc6 to chromatin, and loading of MCMs. Later, two groups independently reported that geminin inhibits DNA replication by inhibiting Cdt1 activity (Tada et al., 2001; Wohlschlegel et al., 2000). The recently described crystal structure of geminin-Cdt1 reveals that the middle region of geminin forms a parallel coiled-coil homodimer, that interacts with both the N-and middle domains of Cdt1. Disruption of geminin dimerisation by point mutation abolishes its interaction with Cdt1, and its ability to inhibit replication (Lee et al., 2004; Saxena et al., 2004). Additionally, in *Xenopus* embryos, geminin induces uncommitted embryonic cells to differentiate as neurons (Kroll et al., 1998), which may account for the stable geminin level throughout the cell cycle in *Xenopus* early embryonic cells (Hodgson et al., 2002; Quinn et al., 2001). In *Xenopus* early embryonic cells, instead of being degraded, geminin is ubquitinated, and this ubiquitination triggers geminin inactivation without ubiquitin-dependent proteolysis, and this is essential for replication origins to become licensed (Li and Blow, 2004). More recently, geminin was shown to associate with the transcription factors Hox and Six 3, and these interactions regulate the Hox and Six transcriptional activity (Del Bene et al., 2004; Luo et al., 2004). Whether the ability of geminin to regulate the transcriptional activity of Hox or Six proteins has anything to do with its role in suppressing replication is yet to be determined. Since geminin inhibits Cdt1, a decrease in geminin protein levels is predicted to increase Cdt1 activity, and therefore result in re-replication. This prediction proved to be true in both human and *Drosophila* cells, in which depletion of geminin by siRNA caused re-replication

(Melixetian et al., 2004; Mihaylov et al., 2002; Zhu et al., 2004). Re-replication induced by the depletion of geminin is dependent on its ability to inhibit Cdt1 activity, because co-silencing of Cdt1 suppresses this re-replication (Ballabeni et al., 2004; McGarry, 2002; Melixetian et al., 2004; Zhu et al., 2004). Surprisingly, inhibition of CDK in M phase causes replication re-licensing (Ballabeni et al., 2004). Thus mitotic CDK, rather than geminin, may play a dominant role in blocking re-assembly of pre-RCs in mitosis. It was shown recently that geminin plays positive roles in pre-RC assembly by binding and stabilising Cdt1 in mitosis (Ballabeni et al., 2004). Although both geminin and Cdt1 protein levels are cell cycle regulated, there is an intervening period at the onset of S phase when both proteins are present. During this short period, the appearance of geminin seems to inhibit the activity of Cdt1 and block the re-assembly of pre-RC. This is confirmed by the fact that co-overexpression of geminin suppresses re-replication induced by overexpression of Cdt1 (Vaziri et al., 2003).

Despite a role similar to Cdt1 in the assembly of pre-RCs, Cdc6 does not seem to play a major role in the control of re-replication. Overexpression of Cdc6 does not induce robust re-replication in human cells, although re-replication is robust when Cdc18/Cdc6 is over-expressed in *S. pombe* (Muzi-Falconi and Kelly, 1995; Nishitani and Nurse, 1995; Vaziri et al., 2003). These results suggest that higher eukaryotic cell may have more vigilant pathways to suppress re-replication. Rather than inducing re-replication, overexpression of Cdc6 by adenovirus in human cells results in double-stranded DNA breaks (Wagle and Dutta, unpublished observation). This damaged DNA activates the ATM/ATR/Chk2 DNA damage checkpoint, which subsequently triggers an apoptotic cascade (N. Wagle and A. Dutta, unpublished). It appears that Cdc6 and Cdt1 may have different roles in the assembly and activation of initiation complexes even though both are required for loading MCM proteins. It is likely that over-expression of Cdt1 triggers multiple rounds of assembly and activation of MCM proteins, resulting in re-replication, whereas over-expression of Cdc6 may just promote the assembly of incompletely active pre-RCs because of the absence of sufficient Cdt1 protein. These incomplete pre-RCs may generate damaged DNA, which may in turn activate a DNA damage response culminating in apoptosis. This model is supported by the fact that co-overexpression of both Cdc6 and Cdt1 caused more over-replicated DNA than the mere overexpression of Cdt1 (Vaziri et al., 2003). Additionally, a higher endogenous level of Cdt1 was observed in human megakaryocytic cells transiently expressing ectopic Cdc6 (Bermejo et al., 2002), again suggesting that Cdt1, but not Cdc6, is the limiting factor in initiation of DNA replication. Interestingly, overexpression of Cdc6 in G2 phase cells activates Chk1 protein, and prevents cells from entering mitosis, indicating that Cdc6 can regulate entry into mitosis in addition to its role in the initiation of DNA replication (Clay-Farrace et al., 2003). Alternatively, overexpressed Cdc6 may cause DNA damage which may activate a G2/M checkpoint.

In addition to CDKs and geminin, chromatin structure has been shown to be an important factor controlling re-replication. In *metazoa*, mutation of the histone deacetylase (HDAC) Rpd3 induced genome-wide hyperacetylation, a redistribution of the origin-binding protein Orc2, and increased DNA replication. Tethering Rpd3 or Polycomb proteins to the origin decreased its activity, whereas tethering the *Chameau* acetyltransferase increased origin activity. These results suggest that nucleosome acetylation could be important for controlling DNA replication and genomic stability (Aggarwal and Calvi, 2004).

6.3 Cellular Response to Re-replication and Checkpoint Activation

Checkpoint pathways are activated in cells with re-replicated DNA. Re-replication by overexpression of both Cdc6 and Cdt1 is seen only in cells lacking the tumour suppressor protein p53 but not in wild type cells. Furthermore, co-overexpression of Mdm2 induces re-replication in wild type p53 containing cells, suggesting that p53 plays a major role in suppressing re-replication. Indeed, p53 is activated through an ATM/ATR/Chk2 DNA damage checkpoint pathway, resulting in the induction of the CDK inhibitor $p21^{Cip1}$, which prevents replication. Paradoxically, unlike re-replication induced by the overexpression of Cdt1 and Cdc6, re-replication induced by the depletion of geminin occurs irrespective of the p53 status of the cell. This activates a Chk1- and Chk2-mediated G2/M checkpoint that inhibits Cdc25 and prevents cells from entering mitosis, the abrogation of which causes apoptosis (Zhu et al., 2004). These observations raise an interesting question: why is p53 activated in re-replication induced by the overexpression of Cdt1, but not in that induced by the depletion of geminin? One explanation is that elevated levels of Cdc6, which do not occur in cells depleted of geminin, are most critical for activating a p53 response. This notion is supported by the observations that overexpression of Cdc6 alone induces the accumulation of p53 (Wagle and Dutta, unpublished data). Alternatively, overexpression of Cdt1 may activate an alternative checkpoint pathway that is different from that activated in geminin-depleted cells. This later possibility is supported by the fact that caffeine, which blocks the activation of ATM and ATR, suppresses re-replication in geminin-depleted cells, whereas it increases that induced by the overexpression of Cdt1 (Zhu et al., 2004); Zhu and Dutta unpublished data). Another possibility is that geminin has a role in directing the Chk1 and Chk2 kinases to p53, so that the p53 pathway is activated upon Cdt1 overexpression but not in cells deficient in geminin. Further investigations are required to elucidate the mechanisms by which DNA damage checkpoint pathways are activated in cells undergoing re-replication. Re-replication by depletion of geminin in human cells results in the formation of γH2AX foci, indicating that DNA damage occurs in cells with re-replicated DNA (Melixetian et al., 2004; Zhu et al., 2004). However, what kind of DNA damage is generated, and how cells recognise this type of damage in geminin-depleted cells is still unclear. Given the fact that geminin is critical to suppress re-replication, it will be interesting to investigate whether geminin operates as a tumour suppressor protein, and whether it is mutated in human cancer cells. It appears that the balance between Cdt1 and geminin is important for cells to prevent re-replication, and loss of this balance leads to genomic instability (Figure 3).

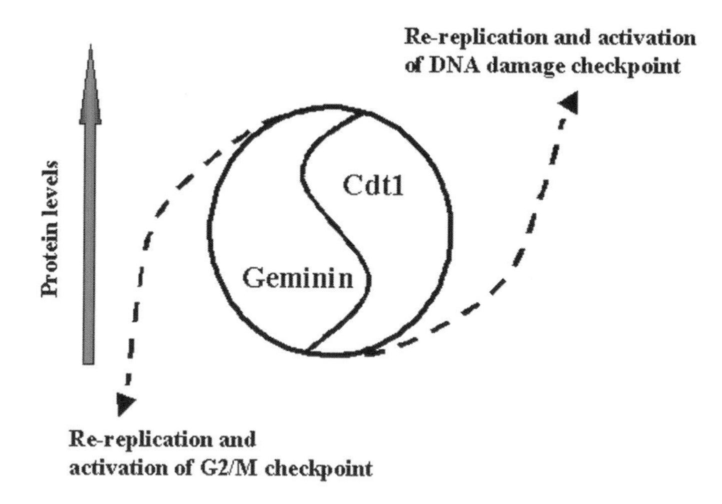

Figure 3. The geminin-Cdt1 balance is important for the inhibition of re-replication. Abrogation of the geminin-Cdt1 balance by stabilisation of Cdt1or lowering of geminin leads to re-replication, resulting in the activation of DNA damage checkpoint pathways. See text for detail.

7. CONCLUSIONS

DNA replication is perhaps the most important event governing cellular proliferation. Even a slight perturbation in the molecular events that govern the replication apparatus may have deleterious effects, which may ultimately lead to an unstable genome, the hallmark of cancer. In fact, cancer cells invariably exhibit a variety of genomic anomalies, including chromosomal rearrangements, gains and losses of chromosomes, and amplification or deletion of genomic material. In face of the extrinsic and intrinsic insults that continuously threaten the integrity of the replication machinery, cells have evolved complicated mechanisms to ensure that DNA is faithfully replicated. When necessary, checkpoints are activated, so that cells either pause to fix the error, or die when the damage is irreparable. When these checkpoints are abrogated (as seen in some human genetic disorders such as ataxia-telangiectasia (ATM loss), Fanconi's anemia (FA)(BRCA2 mutations), Nijmegen breakage syndrome (NBS) (NBS1 mutated), Bloom's syndrome (BLM protein defective), genomic instability inevitably ensues. It will be interesting to investigate whether mutations in clinically relevant genes allow cells to escape from the checkpoints induced by re-replication.

In this review, we have highlighted the molecular mechanisms of re-replication. Both CDK-dependent and CDK-independent pathways are essential to block re-replication. Pre-RC components (ORC, Cdc6, MCMs, Cdt1) are targets for CDKs to inhibit re-replication. However, the balance of Cdt1 and its inhibitor geminin plays a key role in CDK-independent pathways. Eukaryotes use multiple inhibitory mechanisms to block re-replication to ensure that DNA replicates once per cell cycle.

ACKNOWLEDGEMENT

Work in the Dutta Laboratory is supported by grants CA60499 and CA89406 from the National Cancer Institute.

REFERENCES

Adachi, Y., J. Usukura, and M. Yanagida. 1997. A globular complex formation by Nda1 and the other five members of the MCM protein family in fission yeast. *Genes Cells*. 2:467-79.

Aggarwal, B.D., and B.R. Calvi. 2004. Chromatin regulates origin activity in Drosophila follicle cells. *Nature*. 430:372-6.

Aladjem, M.I., and E. Fanning. 2004. The replicon revisited: an old model learns new tricks in metazoan chromosomes. *EMBO Rep*. 5:686-91.

Aladjem, M.I., M. Groudine, L.L. Brody, E.S. Dieken, R.E. Fournier, G.M. Wahl, and E.M. Epner. 1995. Participation of the human beta-globin locus control region in initiation of DNA replication. *Science*. 270:815-9.

Aladjem, M.I., L.W. Rodewald, C.M. Lin, S. Bowman, D.M. Cimbora, L.L. Brody, E.M. Epner, M. Groudine, and G.M. Wahl. 2002. Replication initiation patterns in the beta-globin loci of totipotent and differentiated murine cells: evidence for multiple initiation regions. *Mol Cell Biol*. 22:442-52.

Alexandrow, M.G., and J.L. Hamlin. 2004. Cdc6 chromatin affinity is unaffected by serine-54 phosphorylation, S-phase progression, and overexpression of cyclin A. *Mol Cell Biol*. 24:1614-27.

Aparicio, O.M., A.M. Stout, and S.P. Bell. 1999. Differential assembly of Cdc45p and DNA polymerases at early and late origins of DNA replication. *Proc Natl Acad Sci U S A*. 96:9130-5.

Aparicio, O.M., D.M. Weinstein, and S.P. Bell. 1997. Components and dynamics of DNA replication complexes in S. cerevisiae: redistribution of MCM proteins and Cdc45p during S phase. *Cell*. 91:59-69.

Araki, H., S.H. Leem, A. Phongdara, and A. Sugino. 1995. Dpb11, which interacts with DNA polymerase II(epsilon) in Saccharomyces cerevisiae, has a dual role in S-phase progression and at a cell cycle checkpoint. *Proc Natl Acad Sci U S A*. 92:11791-5.

Baker, T.A., and A. Kornberg. 1988. Transcriptional activation of initiation of replication from the E. coli chromosomal origin: an RNA-DNA hybrid near oriC. *Cell*. 55:113-23.

Ballabeni, A., M. Melixetian, R. Zamponi, L. Masiero, F. Marinoni, and K. Helin. 2004. Human Geminin promotes pre-RC formation and DNA replication by stabilizing CDT1 in mitosis. *EMBO J*. 23:3122-32.

Bell, S.P., and A. Dutta. 2002. DNA replication in eukaryotic cells. *Annu Rev Biochem*. 71:333-74.

Bell, S.P., and B. Stillman. 1992. ATP-dependent recognition of eukaryotic origins of DNA replication by a multiprotein complex. *Nature*. 357:128-34.

Bermejo, R., N. Vilaboa, and C. Cales. 2002. Regulation of CDC6, geminin, and CDT1 in human cells that undergo polyploidization. *Mol Biol Cell*. 13:3989-4000.

Blanchard, F., M.E. Rusiniak, K. Sharma, X. Sun, I. Todorov, M.M. Castellano, C. Gutierrez, H. Baumann, and W.C. Burhans. 2002. Targeted destruction of DNA replication protein Cdc6 by cell death pathways in mammals and yeast. *Mol Biol Cell*. 13:1536-49.

Bousset, K., and J.F. Diffley. 1998. The Cdc7 protein kinase is required for origin firing during S phase. *Genes Dev*. 12:480-90.

Broach, J.R., Y.Y. Li, J. Feldman, M. Jayaram, J. Abraham, K.A. Nasmyth, and J.B. Hicks. 1983. Localization and sequence analysis of yeast origins of DNA replication. *Cold Spring Harb Symp Quant Biol*. 47 Pt 2:1165-73.

Brown, G.W., and T.J. Kelly. 1999. Cell cycle regulation of Dfp1, an activator of the Hsk1 protein kinase. *Proc Natl Acad Sci U S A*. 96:8443-8.

Calzada, A., M. Sacristan, E. Sanchez, and A. Bueno. 2001. Cdc6 cooperates with Sic1 and Hct1 to inactivate mitotic cyclin-dependent kinases. *Nature*. 412:355-8.

Calzada, A., M. Sanchez, E. Sanchez, and A. Bueno. 2000. The stability of the Cdc6 protein is regulated by cyclin-dependent kinase/cyclin B complexes in Saccharomyces cerevisiae. *J Biol Chem*. 275:9734-41.

Clay-Farrace, L., C. Pelizon, D. Santamaria, J. Pines, and R.A. Laskey. 2003. Human replication protein Cdc6 prevents mitosis through a checkpoint mechanism that implicates Chk1. *EMBO J*. 22:704-12.

Cocker, J.H., S. Piatti, C. Santocanale, K. Nasmyth, and J.F. Diffley. 1996. An essential role for the Cdc6 protein in forming the pre-replicative complexes of budding yeast. *Nature*. 379:180-2.

Coleman, T.R., P.B. Carpenter, and W.G. Dunphy. 1996. The Xenopus Cdc6 protein is essential for the initiation of a single round of DNA replication in cell-free extracts. *Cell*. 87:53-63.

Costanzo, V., D. Shechter, P.J. Lupardus, K.A. Cimprich, M. Gottesman, and J. Gautier. 2003. An ATR- and Cdc7-dependent DNA damage checkpoint that inhibits initiation of DNA replication. *Mol Cell*. 11:203-13.

Coue, M., S.E. Kearsey, and M. Mechali. 1996. Chromotin binding, nuclear localization and phosphorylation of Xenopus cdc21 are cell-cycle dependent and associated with the control of initiation of DNA replication. *EMBO J*. 15:1085-97.

Dahmann, C., J.F. Diffley, and K.A. Nasmyth. 1995. S-phase-promoting cyclin-dependent kinases prevent re-replication by inhibiting the transition of replication origins to a pre-replicative state. *Current Biology*. 5:1257-69.

Dalton, S., and L. Whitbread. 1995. Cell cycle-regulated nuclear import and export of Cdc47, a protein essential for initiation of DNA replication in budding yeast. *Proc Natl Acad Sci U S A*. 92:2514-8.

Del Bene, F., K. Tessmar-Raible, and J. Wittbrodt. 2004. Direct interaction of geminin and Six3 in eye development. *Nature*. 427:745-9.

Delmolino, L.M., P. Saha, and A. Dutta. 2001. Multiple mechanisms regulate subcellular localization of human CDC6. *Journal of Biological Chemistry*. 276:26947-54.

DePamphilis, M.L. 1999. Replication origins in metazoan chromosomes: fact or fiction? *Bioessays*. 21:5-16.

DePamphilis, M.L. 2003. Eukaryotic DNA replication origins: reconciling disparate data. *Cell*. 114:274-5.

Detweiler, C.S., and J.J. Li. 1998. Ectopic induction of Clb2 in early G1 phase is sufficient to block prereplicative complex formation in Saccharomyces cerevisiae. *Proc Natl Acad Sci U S A*. 95:2384-9.

Dietrich, C., K. Wallenfang, F. Oesch, and R. Wieser. 1997. Translocation of cdk2 to the nucleus during G1-phase in PDGF-stimulated human fibroblasts. *Exp Cell Res*. 232:72-8.

Diffley, J.F. 1994. Eukaryotic DNA replication. *Curr Opin Cell Biol*. 6:368-72.

Diffley, J.F., and J.H. Cocker. 1992. Protein-DNA interactions at a yeast replication origin. *Nature*. 357:169-72.

Diffley, J.F., J.H. Cocker, S.J. Dowell, J. Harwood, and A. Rowley. 1995. Stepwise assembly of initiation complexes at budding yeast replication origins during the cell cycle. *J Cell Sci Suppl*. 19:67-72.

Diffley, J.F., J.H. Cocker, S.J. Dowell, and A. Rowley. 1994. Two steps in the assembly of complexes at yeast replication origins in vivo. *Cell*. 78:303-16.

Diffley, J.F., and K. Labib. 2002. The chromosome replication cycle. *J Cell Sci*. 115:869-72.

Donaldson, A.D., W.L. Fangman, and B.J. Brewer. 1998. Cdc7 is required throughout the yeast S phase to activate replication origins. *Genes Dev*. 12:491-501.

Dowell, S.J., P. Romanowski, and J.F. Diffley. 1994. Interaction of Dbf4, the Cdc7 protein kinase regulatory subunit, with yeast replication origins in vivo. *Science*. 265:1243-6.

Drury, L.S., G. Perkins, and J.F. Diffley. 1997. The Cdc4/34/53 pathway targets Cdc6p for proteolysis in budding yeast. *EMBO J*. 16:5966-76.

Drury, L.S., G. Perkins, and J.F. Diffley. 2000. The cyclin-dependent kinase Cdc28p regulates distinct modes of Cdc6p proteolysis during the budding yeast cell cycle. *Curr Biol*. 10:231-40.

Dutta, A., and S.P. Bell. 1997. Initiation of DNA replication in eukaryotic cells. *Annu Rev Cell Dev Biol*. 13:293-332.

Elledge, S.J. 1996. Cell cycle checkpoints: preventing an identity crisis. *Science*. 274:1664-72.

Elsasser, S., Y. Chi, P. Yang, and J.L. Campbell. 1999. Phosphorylation controls timing of Cdc6p destruction: A biochemical analysis. *Mol Biol Cell*. 10:3263-77.

Elsasser, S., F. Lou, B. Wang, J.L. Campbell, and A. Jong. 1996. Interaction between yeast Cdc6 protein and B-type cyclin/Cdc28 kinases. *Mol Biol Cell*. 7:1723-35.

Fien, K., Y.S. Cho, J.K. Lee, S. Raychaudhuri, I. Tappin, and J. Hurwitz. 2004. Primer utilization by DNA polymerase alpha-primase is influenced by its interaction with Mcm10p. *J Biol Chem*. 279:16144-53.

Findeisen, M., M. El-Denary, T. Kapitza, R. Graf, and U. Strausfeld. 1999. Cyclin A-dependent kinase activity affects chromatin binding of ORC, Cdc6, and MCM in egg extracts of Xenopus laevis. *Eur J Biochem*. 264:415-26.

Fletcher, R.J., B.E. Bishop, R.P. Leon, R.A. Sclafani, C.M. Ogata, and X.S. Chen. 2003. The structure and function of MCM from archaeal M. Thermoautotrophicum.[see comment]. *Nature Structural Biology*. 10:160-7.

Fujita, M., Y. Ishimi, H. Nakamura, T. Kiyono, and T. Tsurumi. 2002. Nuclear organization of DNA replication initiation proteins in mammalian cells. *J Biol Chem*. 277:10354-61.

Fujita, M., C. Yamada, H. Goto, N. Yokoyama, K. Kuzushima, M. Inagaki, and T. Tsurumi. 1999. Cell cycle regulation of human CDC6 protein. Intracellular localization, interaction with the human mcm complex, and CDC2 kinase-mediated hyperphosphorylation. *Journal of Biological Chemistry*. 274:25927-32.

Gopalakrishnan, V., P. Simancek, C. Houchens, H.A. Snaith, M.G. Frattini, S. Sazer, and T.J. Kelly. 2001. Redundant control of re-replication in fission yeast. *Proc Natl Acad Sci U S A*. 98:13114-9.

Grabowski, B., and Z. Kelman. 2001. Autophosphorylation of archaeal Cdc6 homologues is regulated by DNA. *Journal of Bacteriology*. 183:5459-64.

Guo, M., and B.A. Hay. 1999. Cell proliferation and apoptosis. *Curr Opin Cell Biol*. 11:745-52.

Hamlin, J.L., and P.A. Dijkwel. 1995. On the nature of replication origins in higher eukaryotes. *Curr Opin Genet Dev*. 5:153-61.

Hardy, C.F. 1996. Characterization of an essential Orc2p-associated factor that plays a role in DNA replication. *Mol Cell Biol*. 16:1832-41.

Hardy, C.F., O. Dryga, S. Seematter, P.M. Pahl, and R.A. Sclafani. 1997. mcm5/cdc46-bob1 bypasses the requirement for the S phase activator Cdc7p. *Proc Natl Acad Sci U S A*. 94:3151-5.

Hartwell, L.H., Mortimer, R.K., Culotti, J., and Culotti, M. 1973. Genetic control of the cell division cycle in yeast. *Genetics*. 74:267-86.

Hennessy, K.M., C.D. Clark, and D. Botstein. 1990. Subcellular localization of yeast CDC46 varies with the cell cycle. *Genes Dev*. 4:2252-63.

Higa, L.A., I.S. Mihaylov, D.P. Banks, J. Zheng, and H. Zhang. 2003. Radiation-mediated proteolysis of CDT1 by CUL4-ROC1 and CSN complexes constitutes a new checkpoint. *Nat Cell Biol*. 5:1008-15.

Hodgson, B., A. Li, S. Tada, and J.J. Blow. 2002. Geminin becomes activated as an inhibitor of Cdt1/RLF-B following nuclear import. *Curr Biol*. 12:678-83.

Hofmann, J.F., and D. Beach. 1994. cdt1 is an essential target of the Cdc10/Sct1 transcription factor: requirement for DNA replication and inhibition of mitosis. *EMBO J*. 13:425-34.

Hollingsworth, R.E., Jr., and R.A. Sclafani. 1990. DNA metabolism gene CDC7 from yeast encodes a serine (threonine) protein kinase. *Proc Natl Acad Sci U S A*. 87:6272-6.

Homesley, L., M. Lei, Y. Kawasaki, S. Sawyer, T. Christensen, and B.K. Tye. 2000. Mcm10 and the MCM2-7 complex interact to initiate DNA synthesis and to release replication factors from origins. *Genes Dev*. 14:913-26.

Hsiao, C.L., and J. Carbon. 1979. High-frequency transformation of yeast by plasmids containing the cloned yeast ARG4 gene. *Proc Natl Acad Sci U S A*. 76:3829-33.

Ishimi, Y. 1997. A DNA helicase activity is associated with an MCM4, -6, and -7 protein complex. *J Biol Chem*. 272:24508-13.

Ishimi, Y., Y. Komamura, Z. You, and H. Kimura. 1998. Biochemical function of mouse minichromosome maintenance 2 protein. *J Biol Chem*. 273:8369-75.

Ishimi, Y., and Y. Komamura-Kohno. 2001. Phosphorylation of Mcm4 at specific sites by cyclin-dependent kinase leads to loss of Mcm4,6,7 helicase activity. *J Biol Chem*. 276:34428-33.

Izumi, M., K. Yanagi, T. Mizuno, M. Yokoi, Y. Kawasaki, K.Y. Moon, J. Hurwitz, F. Yatagai, and F. Hanaoka. 2000. The human homolog of Saccharomyces cerevisiae Mcm10 interacts with replication factors and dissociates from nuclease-resistant nuclear structures in G(2) phase. *Nucleic Acids Res*. 28:4769-77.

Izumi, M., F. Yatagai, and F. Hanaoka. 2001. Cell cycle-dependent proteolysis and phosphorylation of human Mcm10. *J Biol Chem*. 276:48526-31.

Izumi, M., F. Yatagai, and F. Hanaoka. 2004. Localization of human Mcm10 is spatially and temporally regulated during the S phase. *J Biol Chem*. 279:32569-77.

Jackson, A.L., P.M. Pahl, K. Harrison, J. Rosamond, and R.A. Sclafani. 1993. Cell cycle regulation of the yeast Cdc7 protein kinase by association with the Dbf4 protein. *Mol Cell Biol*. 13:2899-908.

Jacob, F., Brenner, S., and Cuzin, F. 1963. On the regulation of DNA replication in bacteria. *Cold Spring Habor Symp. Quant. Biol.* 28:329-348.

Jallepalli, P.V., G.W. Brown, M. Muzi-Falconi, D. Tien, and T.J. Kelly. 1997. Regulation of the replication initiator protein p65cdc18 by CDK phosphorylation. *Genes Dev*. 11:2767-79.

Jares, P., A. Donaldson, and J.J. Blow. 2000. The Cdc7/Dbf4 protein kinase: target of the S phase checkpoint? *Embo Rep*. 1:319-22.

Jiang, W., N.J. Wells, and T. Hunter. 1999. Multistep regulation of DNA replication by Cdk phosphorylation of HsCdc6. *Proc Natl Acad Sci U S A*. 96:6193-8.

Johnston, L.H., H. Masai, and A. Sugino. 1999. First the CDKs, now the DDKs. *Trends Cell Biol*. 9:249-52.

Kalejta, R.F., X. Li, L.D. Mesner, P.A. Dijkwel, H.B. Lin, and J.L. Hamlin. 1998. Distal sequences, but not ori-beta/OBR-1, are essential for initiation of DNA replication in the Chinese hamster DHFR origin. *Mol Cell*. 2:797-806.

Kamimura, Y., H. Masumoto, A. Sugino, and H. Araki. 1998. Sld2, which interacts with Dpb11 in Saccharomyces cerevisiae, is required for chromosomal DNA replication. *Mol Cell Biol*. 18:6102-9.

Kamimura, Y., Y.S. Tak, A. Sugino, and H. Araki. 2001. Sld3, which interacts with Cdc45 (Sld4), functions for chromosomal DNA replication in Saccharomyces cerevisiae. *EMBO J.* 20:2097-107.

Kawasaki, Y., S. Hiraga, and A. Sugino. 2000. Interactions between Mcm10p and other replication factors are required for proper initiation and elongation of chromosomal DNA replication in Saccharomyces cerevisiae. *Genes Cells*. 5:975-89.

Kelly, T.J., and G.W. Brown. 2000. Regulation of chromosome replication. *Annu Rev Biochem*. 69:829-80.

Kimura, H., N. Nozaki, and K. Sugimoto. 1994. DNA polymerase alpha associated protein P1, a murine homolog of yeast MCM3, changes its intranuclear distribution during the DNA synthetic period. *EMBO J.* 13:4311-20.

Kingsman, A.J., L. Clarke, R.K. Mortimer, and J. Carbon. 1979. Replication in Saccharomyces cerevisiae of plasmid pBR313 carrying DNA from the yeast trpl region. *Gene*. 7:141-52.

Klemm, R.D., R.J. Austin, and S.P. Bell. 1997. Coordinate binding of ATP and origin DNA regulates the ATPase activity of the origin recognition complex. *Cell*. 88:493-502.

Kobayashi, T., T. Rein, and M.L. DePamphilis. 1998. Identification of primary initiation sites for DNA replication in the hamster dihydrofolate reductase gene initiation zone. *Mol Cell Biol*. 18:3266-77.

Kondo, T., M. Kobayashi, J. Tanaka, A. Yokoyama, S. Suzuki, N. Kato, M. Onozawa, K. Chiba, S. Hashino, M. Imamura, Y. Minami, N. Minamino, and M. Asaka. 2004. Rapid degradation of Cdt1 upon UV-induced DNA damage is mediated by SCFSkp2 complex. *J Biol Chem*. 279:27315-9.

Koonin, E.V. 1993. A common set of conserved motifs in a vast variety of putative nucleic acid-dependent ATPases including MCM proteins involved in the initiation of eukaryotic DNA replication. *Nucleic Acids Res*. 21:2541-7.

Kroll, K.L., A.N. Salic, L.M. Evans, and M.W. Kirschner. 1998. Geminin, a neuralizing molecule that demarcates the future neural plate at the onset of gastrulation. *Development*. 125:3247-58.

Kubota, Y., Y. Takase, Y. Komori, Y. Hashimoto, T. Arata, Y. Kamimura, H. Araki, and H. Takisawa. 2003. A novel ring-like complex of Xenopus proteins essential for the initiation of DNA replication. *Genes Dev*. 17:1141-52.

Labib, K., J.F. Diffley, and S.E. Kearsey. 1999. G1-phase and B-type cyclins exclude the DNA-replication factor Mcm4 from the nucleus. *Nat Cell Biol*. 1:415-22.

Lee, C., B. Hong, J.M. Choi, Y. Kim, S. Watanabe, Y. Ishimi, T. Enomoto, S. Tada, and Y. Cho. 2004. Structural basis for inhibition of the replication licensing factor Cdt1 by geminin. *Nature*. 430:913-7.

Lee, D.G., and S.P. Bell. 1997. Architecture of the yeast origin recognition complex bound to origins of DNA replication. *Mol Cell Biol*. 17:7159-68.

Lee, D.G., A.M. Makhov, R.D. Klemm, J.D. Griffith, and S.P. Bell. 2000. Regulation of origin recognition complex conformation and ATPase activity: differential effects of single-stranded and double-stranded DNA binding. *EMBO J.* 19:4774-82.

Lee, J.K., and J. Hurwitz. 2000. Isolation and characterization of various complexes of the minichromosome maintenance proteins of Schizosaccharomyces pombe. *J Biol Chem*. 275:18871-8.

Lee, J.K., and J. Hurwitz. 2001. Processive DNA helicase activity of the minichromosome maintenance proteins 4, 6, and 7 complex requires forked DNA structures. *Proc Natl Acad Sci U S A*. 98:54-9.

Lee, J.K., Y.S. Seo, and J. Hurwitz. 2003. The Cdc23 (Mcm10) protein is required for the phosphorylation of minichromosome maintenance complex by the Dfp1-Hsk1 kinase. *Proc Natl Acad Sci U S A*. 100:2334-9.

Lei, M., Y. Kawasaki, M.R. Young, M. Kihara, A. Sugino, and B.K. Tye. 1997. Mcm2 is a target of regulation by Cdc7-Dbf4 during the initiation of DNA synthesis. *Genes Dev*. 11:3365-74.

Lei, M., and B.K. Tye. 2001. Initiating DNA synthesis: from recruiting to activating the MCM complex. *J Cell Sci.* 114:1447-54.
Lengronne, A., and E. Schwob. 2002. The yeast CDK inhibitor Sic1 prevents genomic instability by promoting replication origin licensing in late G(1). *Mol Cell.* 9:1067-78.
Li, A., and J.J. Blow. 2004. Non-proteolytic inactivation of geminin requires CDK-dependent ubiquitination. *Nat Cell Biol.* 6:260-7.
Li, C.J., and M.L. DePamphilis. 2002. Mammalian Orc1 protein is selectively released from chromatin and ubiquitinated during the S-to-M transition in the cell division cycle. *Mol Cell Biol.* 22:105-16.
Li, C.J., A. Vassilev, and M.L. DePamphilis. 2004. Role for Cdk1 (Cdc2)/cyclin A in preventing the mammalian origin recognition complex's largest subunit (Orc1) from binding to chromatin during mitosis. *Mol Cell Biol.* 24:5875-86.
Li, D., R. Zhao, W. Lilyestrom, D. Gai, R. Zhang, J.A. DeCaprio, E. Fanning, A. Jochimiak, G. Szakonyi, and X.S. Chen. 2003a. Structure of the replicative helicase of the oncoprotein SV40 large tumour antigen. *Nature.* 423:512-8.
Li, X., Q. Zhao, R. Liao, P. Sun, and X. Wu. 2003b. The SCF(Skp2) ubiquitin ligase complex interacts with the human replication licensing factor Cdt1 and regulates Cdt1 degradation. *J Biol Chem.* 278:30854-8.
Liang, C., and B. Stillman. 1997. Persistent initiation of DNA replication and chromatin-bound MCM proteins during the cell cycle in cdc6 mutants. *Genes Dev.* 11:3375-86.
Liang, C., M. Weinreich, and B. Stillman. 1995. ORC and Cdc6p interact and determine the frequency of initiation of DNA replication in the genome. *Cell.* 81:667-76.
Linskens, M.H., and J.A. Huberman. 1990. The two faces of higher eukaryotic DNA replication origins. *Cell.* 62:845-7.
Liu, E., X. Li, F. Yan, Q. Zhao, and X. Wu. 2004. Cyclin-dependent kinases phosphorylate human Cdt1 and induce its degradation. *J Biol Chem.* 279:17283-8.
Luo, L., X. Yang, Y. Takihara, H. Knoetgen, and M. Kessel. 2004. The cell-cycle regulator geminin inhibits Hox function through direct and polycomb-mediated interactions. *Nature.* 427:749-53.
Lygerou, Z., and P. Nurse. 1999. The fission yeast origin recognition complex is constitutively associated with chromatin and is differentially modified through the cell cycle. *J Cell Sci.* 112 (Pt 21):3703-12.
Maine, G.T., P. Sinha, and B.K. Tye. 1984. Mutants of S. cerevisiae defective in the maintenance of minichromosomes. *Genetics.* 106:365-85.
Maiorano, D., J.M. Lemaitre, and M. Mechali. 2000. Stepwise regulated chromatin assembly of MCM2-7 proteins. *J Biol Chem.* 275:8426-31.
Makiniemi, M., T. Hillukkala, J. Tuusa, K. Reini, M. Vaara, D. Huang, H. Pospiech, I. Majuri, T. Westerling, T.P. Makela, and J.E. Syvaoja. 2001. BRCT domain-containing protein TopBP1 functions in DNA replication and damage response. *J Biol Chem.* 276:30399-406.
Marahrens, Y., and B. Stillman. 1992. A yeast chromosomal origin of DNA replication defined by multiple functional elements. *Science.* 255:817-23.
Masai, H., and K. Arai. 2000. Dbf4 motifs: conserved motifs in activation subunits for Cdc7 kinases essential for S-phase. *Biochem Biophys Res Commun.* 275:228-32.
Masai, H., and K. Arai. 2002. Cdc7 kinase complex: a key regulator in the initiation of DNA replication. *J Cell Physiol.* 190:287-96.
Masai, H., E. Matsui, Z. You, Y. Ishimi, K. Tamai, and K. Arai. 2000. Human Cdc7-related kinase complex. In vitro phosphorylation of MCM by concerted actions of Cdks and Cdc7 and that of a criticial threonine residue of Cdc7 bY Cdks. *J Biol Chem.* 275:29042-52.
Masumoto, H., S. Muramatsu, Y. Kamimura, and H. Araki. 2002. S-Cdk-dependent phosphorylation of Sld2 essential for chromosomal DNA replication in budding yeast. *Nature.* 415:651-5.
Masumoto, H., A. Sugino, and H. Araki. 2000. Dpb11 controls the association between DNA polymerases alpha and epsilon and the autonomously replicating sequence region of budding yeast. *Mol Cell Biol.* 20:2809-17.

McGarry, T.J. 2002. Geminin deficiency causes a Chk1-dependent G2 arrest in Xenopus. *Mol Biol Cell*. 13:3662-71.

McGarry, T.J., and M.W. Kirschner. 1998. Geminin, an inhibitor of DNA replication, is degraded during mitosis. *Cell*. 93:1043-53.

Melixetian, M., A. Ballabeni, L. Masiero, P. Gasparini, R. Zamponi, J. Bartek, J. Lukas, and K. Helin. 2004. Loss of Geminin induces re-replication in the presence of functional p53. *J Cell Biol*. 165:473-82.

Mendez, J., X.H. Zou-Yang, S.Y. Kim, M. Hidaka, W.P. Tansey, and B. Stillman. 2002. Human origin recognition complex large subunit is degraded by ubiquitin-mediated proteolysis after initiation of DNA replication. *Mol Cell*. 9:481-91.

Mihaylov, I.S., T. Kondo, L. Jones, S. Ryzhikov, J. Tanaka, J. Zheng, L.A. Higa, N. Minamino, L. Cooley, and H. Zhang. 2002. Control of DNA replication and chromosome ploidy by geminin and cyclin A. *Mol Cell Biol*. 22:1868-80.

Mimura, S., and H. Takisawa. 1998. Xenopus Cdc45-dependent loading of DNA polymerase alpha onto chromatin under the control of S-phase Cdk. *EMBO J*. 17:5699-707.

Mizushima, T., N. Takahashi, and B. Stillman. 2000. Cdc6p modulates the structure and DNA binding activity of the origin recognition complex in vitro. *Genes Dev*. 14:1631-41.

Moir, D., S.E. Stewart, B.C. Osmond, and D. Botstein. 1982. Cold-sensitive cell-division-cycle mutants of yeast: isolation, properties, and pseudoreversion studies. *Genetics*. 100:547-63.

Moll, T., G. Tebb, U. Surana, H. Robitsch, and K. Nasmyth. 1991. The role of phosphorylation and the CDC28 protein kinase in cell cycle-regulated nuclear import of the S. cerevisiae transcription factor SWI5. *Cell*. 66:743-58.

Murakami, H., and P. Nurse. 1999. Meiotic DNA replication checkpoint control in fission yeast. *Genes Dev*. 13:2581-93.

Musahl, C., D. Schulte, R. Burkhart, and R. Knippers. 1995. A human homologue of the yeast replication protein Cdc21. Interactions with other Mcm proteins. *Eur J Biochem*. 230:1096-101.

Muzi-Falconi, M., and T.J. Kelly. 1995. Orp1, a member of the Cdc18/Cdc6 family of S-phase regulators, is homologous to a component of the origin recognition complex. *Proceedings of the National Academy of Sciences of the United States of America*. 92:12475-9.

Neuwald, A.F., L. Aravind, J.L. Spouge, and E.V. Koonin. 1999. AAA+: A class of chaperone-like ATPases associated with the assembly, operation, and disassembly of protein complexes. *Genome Res*. 9:27-43.

Newlon, C.S. 1997. Putting it all together: building a prereplicative complex. *Cell*. 91:717-20.

Nguyen, V.Q., C. Co, K. Irie, and J.J. Li. 2000. Clb/Cdc28 kinases promote nuclear export of the replication initiator proteins Mcm2-7. *Curr Biol*. 10:195-205.

Nguyen, V.Q., C. Co, and J.J. Li. 2001. Cyclin-dependent kinases prevent DNA re-replication through multiple mechanisms. *Nature*. 411:1068-73.

Nishitani, H., and Z. Lygerou. 2002. Control of DNA replication licensing in a cell cycle. *Genes Cells*. 7:523-34.

Nishitani, H., Z. Lygerou, T. Nishimoto, and P. Nurse. 2000. The Cdt1 protein is required to license DNA for replication in fission yeast. *Nature*. 404:625-8.

Nishitani, H., and P. Nurse. 1995. p65cdc18 plays a major role controlling the initiation of DNA replication in fission yeast. *Cell*. 83:397-405.

Nishitani, H., S. Taraviras, Z. Lygerou, and T. Nishimoto. 2001. The human licensing factor for DNA replication Cdt1 accumulates in G1 and is destabilized after initiation of S-phase. *J Biol Chem*. 276:44905-11.

Noguchi, E., P. Shanahan, C. Noguchi, and P. Russell. 2002. CDK phosphorylation of Drc1 regulates DNA replication in fission yeast. *Curr Biol*. 12:599-605.

Nougarede, R., F. Della Seta, P. Zarzov, and E. Schwob. 2000. Hierarchy of S-phase-promoting factors: yeast Dbf4-Cdc7 kinase requires prior S-phase cyclin-dependent kinase activation. *Mol Cell Biol*. 20:3795-806.

Ohta, S., Y. Tatsumi, M. Fujita, T. Tsurimoto, and C. Obuse. 2003. The ORC1 cycle in human cells: II. Dynamic changes in the human ORC complex during the cell cycle. *J Biol Chem*. 278:41535-40.

Pacek, M., and J.C. Walter. 2004. A requirement for MCM7 and Cdc45 in chromosome unwinding during eukaryotic DNA replication. *EMBO J*. 23:3667-76.

Pape, T., H. Meka, S. Chen, G. Vicentini, M. van Heel, and S. Onesti. 2003. Hexameric ring structure of the full-length archaeal MCM protein complex. *Embo Reports*. 4:1079-83.

Pasion, S.G., and S.L. Forsburg. 1999. Nuclear localization of Schizosaccharomyces pombe Mcm2/Cdc19p requires MCM complex assembly. *Mol Biol Cell*. 10:4043-57.

Patterson, M., R.A. Sclafani, W.L. Fangman, and J. Rosamond. 1986. Molecular characterization of cell cycle gene CDC7 from Saccharomyces cerevisiae. *Mol Cell Biol*. 6:1590-8.

Pereverzeva, I., E. Whitmire, B. Khan, and M. Coue. 2000. Distinct phosphoisoforms of the Xenopus Mcm4 protein regulate the function of the Mcm complex. *Mol Cell Biol*. 20:3667-76.

Perkins, G., and J.F. Diffley. 1998. Nucleotide-dependent prereplicative complex assembly by Cdc6p, a homolog of eukaryotic and prokaryotic clamp-loaders. *Mol Cell*. 2:23-32.

Petersen, B.O., J. Lukas, C.S. Sorensen, J. Bartek, and K. Helin. 1999. Phosphorylation of mammalian CDC6 by cyclin A/CDK2 regulates its subcellular localization. *EMBO J*. 18:396-410.

Prasanth, S.G., K.V. Prasanth, K. Siddiqui, D.L. Spector, and B. Stillman. 2004. Human Orc2 localizes to centrosomes, centromeres and heterochromatin during chromosome inheritance. *EMBO J*. 23:2651-63.

Prasanth, S.G., K.V. Prasanth, and B. Stillman. 2002. Orc6 involved in DNA replication, chromosome segregation, and cytokinesis. *Science*. 297:1026-31.

Quinn, L.M., A. Herr, T.J. McGarry, and H. Richardson. 2001. The Drosophila Geminin homolog: roles for Geminin in limiting DNA replication, in anaphase and in neurogenesis. *Genes Dev*. 15:2741-54.

Rao, H., and B. Stillman. 1995. The origin recognition complex interacts with a bipartite DNA binding site within yeast replicators. *Proc Natl Acad Sci U S A*. 92:2224-8.

Romanowski, P., M.A. Madine, A. Rowles, J.J. Blow, and R.A. Laskey. 1996. The Xenopus origin recognition complex is essential for DNA replication and MCM binding to chromatin. *Curr Biol*. 6:1416-25.

Rowles, A., S. Tada, and J.J. Blow. 1999. Changes in association of the Xenopus origin recognition complex with chromatin on licensing of replication origins. *J Cell Sci*. 112 (Pt 12):2011-8.

Rowley, A., J.H. Cocker, J. Harwood, and J.F. Diffley. 1995. Initiation complex assembly at budding yeast replication origins begins with the recognition of a bipartite sequence by limiting amounts of the initiator, ORC. *EMBO J*. 14:2631-41.

Saha, P., J. Chen, K.C. Thome, S.J. Lawlis, Z.H. Hou, M. Hendricks, J.D. Parvin, and A. Dutta. 1998a. Human CDC6/Cdc18 associates with Orc1 and cyclin-cdk and is selectively eliminated from the nucleus at the onset of S phase. *Mol Cell Biol*. 18:2758-67.

Saha, P., K.C. Thome, R. Yamaguchi, Z. Hou, S. Weremowicz, and A. Dutta. 1998b. The human homolog of Saccharomyces cerevisiae CDC45. *J Biol Chem*. 273:18205-9.

Saha, S., Y. Shan, L.D. Mesner, and J.L. Hamlin. 2004. The promoter of the Chinese hamster ovary dihydrofolate reductase gene regulates the activity of the local origin and helps define its boundaries. *Genes Dev*. 18:397-410.

Sanchez, M., A. Calzada, and A. Bueno. 1999. The Cdc6 protein is ubiquitinated in vivo for proteolysis in Saccharomyces cerevisiae. *J Biol Chem*. 274:9092-7.

Santocanale, C., and J.F. Diffley. 1998. A Mec1- and Rad53-dependent checkpoint controls late-firing origins of DNA replication. *Nature*. 395:615-8.

Saxena, S., P. Yuan, S.K. Dhar, T. Senga, D. Takeda, H. Robinson, S. Kornbluth, K. Swaminathan, and A. Dutta. 2004. A dimerized coiled-coil domain and an adjoining part of geminin interact with two sites on Cdt1 for replication inhibition. *Mol Cell*. 15:245-58.

Schwacha, A., and S.P. Bell. 2001. Interactions between two catalytically distinct MCM subgroups are essential for coordinated ATP hydrolysis and DNA replication. *Mol Cell.* 8:1093-104.

Sclafani, R.A. 2000. Cdc7p-Dbf4p becomes famous in the cell cycle. *J Cell Sci.* 113 (Pt 12):2111-7.

Stillman, B. 1996. Cell cycle control of DNA replication. *Science.* 274:1659-64.

Stinchcomb, D.T., K. Struhl, and R.W. Davis. 1979. Isolation and characterisation of a yeast chromosomal replicator. *Nature.* 282:39-43.

Struhl, K., D.T. Stinchcomb, S. Scherer, and R.W. Davis. 1979. High-frequency transformation of yeast: autonomous replication of hybrid DNA molecules. *Proc Natl Acad Sci U S A.* 76:1035-9.

Sugimoto, N., Y. Tatsumi, T. Tsurumi, A. Matsukage, T. Kiyono, H. Nishitani, and M. Fujita. 2004. Cdt1 phosphorylation by cyclin A-dependent kinases negatively regulates its function without affecting geminin binding. *J Biol Chem.* 279:19691-7.

Tada, S., A. Li, D. Maiorano, M. Mechali, and J.J. Blow. 2001. Repression of origin assembly in metaphase depends on inhibition of RLF-B/Cdt1 by geminin. *Nat Cell Biol.* 3:107-13.

Takayama, Y., Y. Kamimura, M. Okawa, S. Muramatsu, A. Sugino, and H. Araki. 2003. GINS, a novel multiprotein complex required for chromosomal DNA replication in budding yeast. *Genes Dev.* 17:1153-65.

Tanaka, S., and J.F. Diffley. 2002. Interdependent nuclear accumulation of budding yeast Cdt1 and Mcm2-7 during G1 phase. *Nat Cell Biol.* 4:198-207.

Tanaka, T., D. Knapp, and K. Nasmyth. 1997. Loading of an Mcm protein onto DNA replication origins is regulated by Cdc6p and CDKs. *Cell.* 90:649-60.

Tanaka, T., and K. Nasmyth. 1998. Association of RPA with chromosomal replication origins requires an Mcm protein, and is regulated by Rad53, and cyclin- and Dbf4-dependent kinases. *EMBO J.* 17:5182-91.

Tatsumi, Y., S. Ohta, H. Kimura, T. Tsurimoto, and C. Obuse. 2003. The ORC1 cycle in human cells: I. cell cycle-regulated oscillation of human ORC1. *J Biol Chem.* 278:41528-34.

Tercero, J.A., K. Labib, and J.F. Diffley. 2000. DNA synthesis at individual replication forks requires the essential initiation factor Cdc45p. *EMBO Journal.* 19:2082-93.

Thome, K.C., S.K. Dhar, D.G. Quintana, L. Delmolino, A. Shahsafaei, and A. Dutta. 2000. Subsets of human origin recognition complex (ORC) subunits are expressed in non-proliferating cells and associate with non-ORC proteins. *J Biol Chem.* 275:35233-41.

Thomer, M., N.R. May, B.D. Aggarwal, G. Kwok, and B.R. Calvi. 2004. Drosophila double-parked is sufficient to induce re-replication during development and is regulated by cyclin E/CDK2. *Development.* 131:4807-18.

Thommes, P., Y. Kubota, H. Takisawa, and J.J. Blow. 1997. The RLF-M component of the replication licensing system forms complexes containing all six MCM/P1 polypeptides. *EMBO J.* 16:3312-9.

Todorov, I.T., A. Attaran, and S.E. Kearsey. 1995. BM28, a human member of the MCM2-3-5 family, is displaced from chromatin during DNA replication. *J Cell Biol.* 129:1433-45.

Tye, B.K. 1999. MCM proteins in DNA replication. *Annu Rev Biochem.* 68:649-86.

Van Hatten, R.A., A.V. Tutter, A.H. Holway, A.M. Khederian, J.C. Walter, and W.M. Michael. 2002. The Xenopus Xmus101 protein is required for the recruitment of Cdc45 to origins of DNA replication. *J Cell Biol.* 159:541-7.

Vas, A., W. Mok, and J. Leatherwood. 2001. Control of DNA re-replication via Cdc2 phosphorylation sites in the origin recognition complex. *Mol Cell Biol.* 21:5767-77.

Vaziri, C., S. Saxena, Y. Jeon, C. Lee, K. Murata, Y. Machida, N. Wagle, D.S. Hwang, and A. Dutta. 2003. A p53-dependent checkpoint pathway prevents re-replication. *Mol Cell.* 11:997-1008.

Walter, J., and J. Newport. 2000. Initiation of eukaryotic DNA replication: origin unwinding and sequential chromatin association of Cdc45, RPA, and DNA polymerase alpha. *Mol Cell.* 5:617-27.

Walter, J.C. 2000. Evidence for sequential action of cdc7 and cdk2 protein kinases during initiation of DNA replication in Xenopus egg extracts. *J Biol Chem*. 275:39773-8.

Wang, H., and S.J. Elledge. 1999. DRC1, DNA replication and checkpoint protein 1, functions with DPB11 to control DNA replication and the S-phase checkpoint in Saccharomyces cerevisiae. *Proc Natl Acad Sci U S A*. 96:3824-9.

Wang, S., P.A. Dijkwel, and J.L. Hamlin. 1998. Lagging-strand, early-labelling, and two-dimensional gel assays suggest multiple potential initiation sites in the Chinese hamster dihydrofolate reductase origin. *Mol Cell Biol*. 18:39-50.

Weinreich, M., C. Liang, and B. Stillman. 1999. The Cdc6p nucleotide-binding motif is required for loading mcm proteins onto chromatin. *Proc Natl Acad Sci U S A*. 96:441-6.

Weinreich, M., and B. Stillman. 1999. Cdc7p-Dbf4p kinase binds to chromatin during S phase and is regulated by both the APC and the RAD53 checkpoint pathway. *EMBO J*. 18:5334-46.

Wilmes, G.M., V. Archambault, R.J. Austin, M.D. Jacobson, S.P. Bell, and F.R. Cross. 2004. Interaction of the S-phase cyclin Clb5 with an "RXL" docking sequence in the initiator protein Orc6 provides an origin-localized replication control switch. *Genes Dev*. 18:981-91.

Wohlschlegel, J.A., S.K. Dhar, T.A. Prokhorova, A. Dutta, and J.C. Walter. 2002. Xenopus Mcm10 binds to origins of DNA replication after Mcm2-7 and stimulates origin binding of Cdc45. *Mol Cell*. 9:233-40.

Wohlschlegel, J.A., B.T. Dwyer, S.K. Dhar, C. Cvetic, J.C. Walter, and A. Dutta. 2000. Inhibition of eukaryotic DNA replication by geminin binding to Cdt1. *Science*. 290:2309-12.

Yamada, Y., T. Nakagawa, and H. Masukata. 2004. A novel intermediate in initiation complex assembly for fission yeast DNA replication. *Mol Biol Cell*. 15:3740-50.

Yan, H., A.M. Merchant, and B.K. Tye. 1993. Cell cycle-regulated nuclear localization of MCM2 and MCM3, which are required for the initiation of DNA synthesis at chromosomal replication origins in yeast. *Genes Dev*. 7:2149-60.

Yoon, H.J., and J.L. Campbell. 1991. The CDC7 protein of Saccharomyces cerevisiae is a phosphoprotein that contains protein kinase activity. *Proc Natl Acad Sci U S A*. 88:3574-8.

Young, M.R., and B.K. Tye. 1997. Mcm2 and Mcm3 are constitutive nuclear proteins that exhibit distinct isoforms and bind chromatin during specific cell cycle stages of Saccharomyces cerevisiae. *Mol Biol Cell*. 8:1587-601.

Zhong, W., H. Feng, F.E. Santiago, and E.T. Kipreos. 2003. CUL-4 ubiquitin ligase maintains genome stability by restraining DNA-replication licensing. *Nature*. 423:885-9.

Zhu, W., Y. Chen, and A. Dutta. 2004. Re-replication by depletion of geminin is seen regardless of p53 status and activates a G2/M checkpoint. *Mol Cell Biol*. 24:7140-50.

Zou, L., and B. Stillman. 1998. Formation of a preinitiation complex by S-phase cyclin CDK-dependent loading of Cdc45p onto chromatin. *Science*. 280:593-6.

Zou, L., and B. Stillman. 2000. Assembly of a complex containing Cdc45p, replication protein A, and Mcm2p at replication origins controlled by S-phase cyclin-dependent kinases and Cdc7p-Dbf4p kinase. *Mol Cell Biol*. 20:3086-96.

Chapter 3.3

THE DREAM OF EVERY CHROMOSOME: EQUAL SEGREGATION FOR A HEALTHY LIFE OF THE HOST

Tomohiro Matsumoto [1,2] and Mitsuhiro Yanagida[1]
[1:] *Graduate School of Biostudies, Kyoto University, Kyoto, Japan* [2:] *Radiation Biology Center, Kyoto University, Kyoto, Japan*

1. INTRODUCTION

Mitosis is the culmination of the cell cycle in which the genetic material duplicated in the preceding S phase is segregated equally to two daughter cells. At this stage of the cell cycle, sister chromatids must be condensed, interact with the spindle that correctly positions them, and separate upon resolution of cohesion, a linkage between sister chromatids. In this chapter, we will first review the structure and components of the kinetochore, an important chromosomal site for equal segregation. The major topics of this chapter also include the two chromosomal processes, condensation and cohesion in mitosis, both of which have recently been described at a molecular level. Finally, we will consider how chromosomal processes gone awry could result in chromosome instability and become a cause of human diseases.

2. OVERVIEW OF CHROMOSOME DYNAMICS IN MITOSIS

Each chromosome is duplicated in S phase and the resulting sister chromatids remain attached to each other throughout the rest of interphase. The visible sign of the onset of mitosis is chromosome condensation in prophase. Sister chromatids become rod-shape structures as the condensation process goes on. During this first stage of mitosis, the cytoskeletal microtubules begin to disassemble and the mitotic spindle begins to form outside the nucleus between the two centrosomes, which

E.A. Nigg (ed.), Genome Instability in Cancer Development, 281-310.

have replicated and moved apart. The following stage of mitosis, prometaphase, is marked by breakdown of the nuclear membrane in higher eukaryotes. This event allows active interaction between the mitotic spindle and condensed chromosomes. The chromosomes are aligned at the equator of the spindle. By the end of metaphase, all sister chromatids are connected to the opposite poles, centrosomes, via the mitotic spindle. Anaphase begins abruptly. The sister chromatids separate synchronously and move toward the pole. During telophase, the two sets of chromosomes arrive to the daughter cells and decondense. Nuclear membrane reforms around the set of chromosomes. In lower eukaryotes such as fission and budding yeasts, the nuclear membrane persists throughout the cell cycle. In these organisms, the spindle pole body (SPB), a structure equivalent to the centrosome, is embedded in the nuclear membrane prior to mitosis. The mitotic spindle can form inside the nucleus and thereby the persisting nuclear membrane does not interfere with the interaction between the spindle and chromosomes. In addition to the dynamic change of shape and position of chromosomes in mitosis, the composition of proteins that are found specifically at/around centromeric DNA is also changed during mitosis for formation of the functional kinetochore.

3. REQUIREMENT FOR EQUAL SEGREGATION

As illustrated in Figure 1, the mitotic spindle plays an important role in placing chromosomes at the position ideal for equal segregation by metaphase. It also serves as a pilot that guides each chromosome to a daughter cell after separation of sister chromatids at anaphase.

Equal segregation of chromosomes largely depends on a symmetrical arrangement of the spindle apparatus and chromosomes, which is established by the end of metaphase. To be more precise, all sister chromatids must biorient (i.e., they attach to the spindle via two kinetochores each of which interacts with the spindle radiated from one of the two poles) and are positioned on the spindle equator (Figure 1f), a mid point between the two poles. In mitosis, each sister chromatid moves in two modes, oscillation and congression.

A sister chromatid initially interacts with the spindle radiated from one pole via one kinetochore, leaving the sister kinetochore unattached (Figure 1a). This monooriented sister chromatid is pulled toward the pole by the attached spindle to the leading kinetochore. On the other hand, its arm is pushed away from the attached pole likely by interaction between kinesin-related proteins, Kid (Funabiki and Murray, 2000; Levesque and Compton, 2001) and microtubules (Figure 1b). The leading kinetochore, as a result, experiences the two opposing forces, the poleward force and the polar ejection force. As it moves closer to the attached pole, the polar ejection force increases. It has been proposed that the leading kinetochore switches

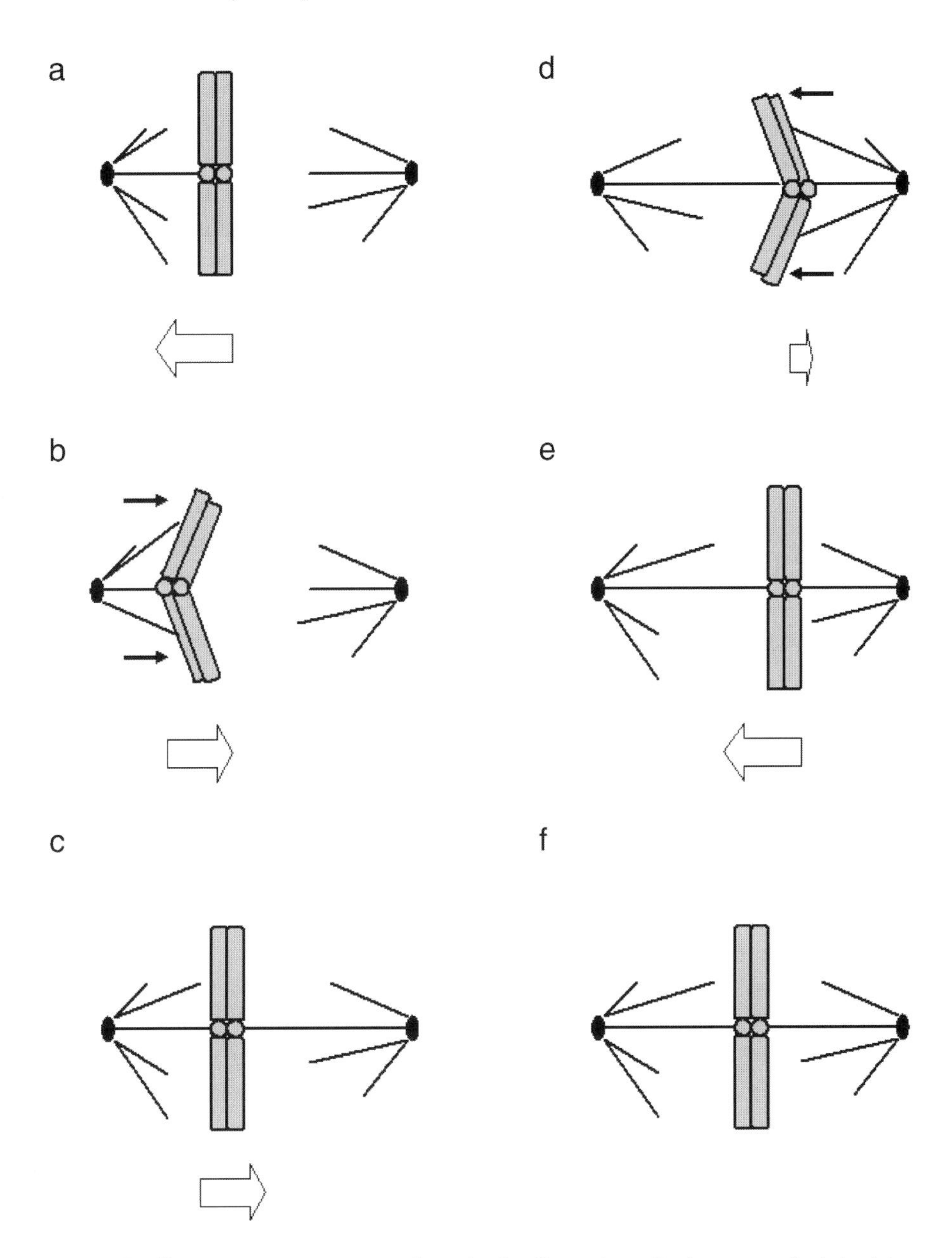

Figure 1. Chromosome movements in mitosis: In early mitosis monooriented sister chromatids oscillate (a and b). When they are bioriented, they start congression (c, d and e) to establish the symmetrical arrangment (f). The thin solid arrows (b and d) indicate the polar ejection force and the open arrows (a to e) indicate the total net force, which contributes to sister chromatid movement. Only one sister chromatid is shown for simplicity.

off the pole ward force when it senses the increasing polar ejection force, and thereby allows movement away from the pole (Skibbens et al., 1993). When the sister chromatid moves away from the pole and the polar ejection force decreases, the leading kinetochore switches on the poleward force again and moves toward the pole. The leading kinetochore switches on and off the poleward force and allows oscillation of the sister chromatid.

During the period of oscillation, monooriented sister chromatids are attached to the spindle radiated from the other pole and establish biorientation (Figure 1c) and the bioriented sister chromatids start congression, movement toward the spindle equator. The kinetochore closer to its attached pole encounters the poleward force through the attached spindle and polar ejection force through both the arm and the sister kinetochore. As the polar ejection force increases, the kinetochore switches off the poleward force and lets the sister chromatids move toward the other pole. The sister kinetochore then becomes the leading kinetochore and moves the sister chromatid toward its attached pole (Figure 1d). When the sister chromatid passes the spindle equator, the new leading kinetochore switches off the poleward force due to an increase in the ejection force. The sister kinetochore then switches on the poleward force and pulls the sister chromatid back to its attached pole (Figure 1e). Repeating these switches results in congression and the sister chromatid is eventually positioned at the spindle equator at which the polar ejection forces are equal or minimal (Figure 1f; Skibbens et al., 1993; Kapoor and Compton, 2002).

Attachment of the spindle to kinetochores, oscillation and congression are three major events required for the establishment of the symmetrical arrangement. They are initiated and progress independently at each sister chromatid. Thereby, some sister chromatid pairs arrive at the spindle equator while others are yet unattached or monooriented. Sister chromatids, which have arrived at the spindle equator earlier, do not separate until all sister chromatids are bioriented. A surveillance mechanism termed the spindle checkpoint is responsible for preventing premature sister chromatid separation. Kinetochores not attached to the spindle or attached abnormally activate the spindle checkpoint.

Obviously, chromosomes not only passively travel inside the cell, but also play critical roles for the establishment of the symmetrical arrangement by the end of metaphase. The kinetochore, in particular, must perform "intelligent" and regulatory functions in mitosis. It senses its own position for appropriate congression. Moreover, it must remain attached to the spindle during oscillation and congression, which are processes involving polymerisation/de-polymerisation of microtubules. The kinetochore also emits a signal for activation of the spindle checkpoint if the spindle is not attached. Finally, the linkage between the sister kinetochores needs to be stiff and yet elastic to resist the antagonistic forces during oscillation and congression.

4. KINETOCHORE

4.1 Centromere DNA

The kinetochore is a specialised assembly of proteins that bind, directly or indirectly, to the centromere DNA. The budding yeast centromere is probably the simplest one. The functional centromere DNAs were first identified as DNA sequences which increase the mitotic stability of circular plasmids with a replication origin, ARS (autonomously replicating sequence) (Clarke and Carbon, 1980). These DNA segments (approximately 125 bp), when combined with ARS and telomere DNA in an appropriate order, are able to behave as linear artificial chromosomes, which can be transmitted stably during mitosis as well as meiosis (Murray and Szostak, 1983). These studies demonstrated that the functional centromere is defined primarily by the context of DNA sequence in budding yeast, *Saccharomyces cerevisiae*. Comparison of the centromere DNAs isolated from multiple chromosomes of budding yeast revealed that the centromere DNA consists of three distinct DNA elements, CDE I, II and III. While CDE I (~15 bp) and CDE III (~25 bp) are well-conserved at all chromosomes, CDE II (~85 bp), exhibiting high A-T content, is less-conserved among the budding yeast chromosomes (Figure 2a).

In fission yeast, *Schizosaccharomyces pombe*, an attempt to clone the centromere DNA as a short DNA segment was not successful, suggesting that the functional centromere is much larger than that of budding yeast and can not be cloned in a conventional bacterial plasmid. An alterative approach was to isolate genomic DNAs from the centromere regions by chromosome walking and characterise the DNA sequences (Nakaseko et al., 1986, Clarke et al., 1986; Chikashige et al., 1989; Murakami et al., 1991; Takahashi et al., 1992). Artificial mini-chromosomes were also constructed by removing arm domains from the native chromosome III in an aneuploid disomic for chromosome III (Niwa et al., 1989). Determination of their sizes and end points assisted in estimating the size of the functional centromere. These efforts allowed understanding of the structural basis of the fission yeast centromeres. As shown in figure 2b (Fishel et al., 1988; Takahashi et al., 1992; Baum et al., 1994), each of the three centromeres ranges in size from 40 kb to 100 kb and is composed of a pair of inverted DNA repeats (otr) surrounding a non-repetitive central core domain (imr and cnt).

In larger eukaryotes such as fly and human, the centromere DNA is more complex and less defined (Figure 2c). Human centromeres contain tandemly repeated arrays of a short (171 bp) DNA segment termed α-satellite. Their sizes range from 0.3 to 5 Mbp DNA consisting of 1,500 - 30,000 copies of the α-satellite. Similarly, fly centromeres are located in highly repetitive DNA regions, consisting of 5 ~ 7 bp-satellite sequences and transposons. Deletion analysis of human X chromosome (Schueler et al., 2001) and fly X-

a: Budding Yeast

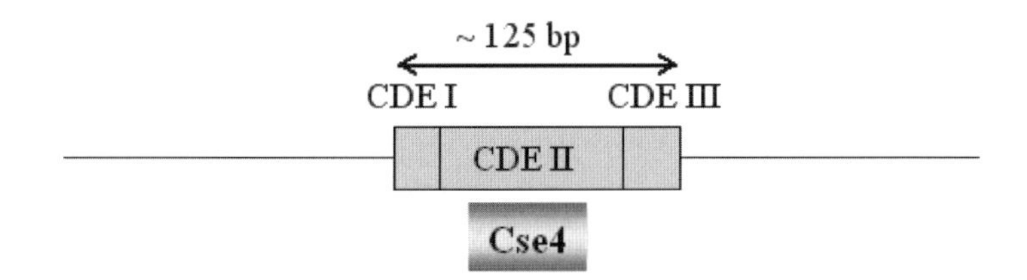

b: Fission Yeast

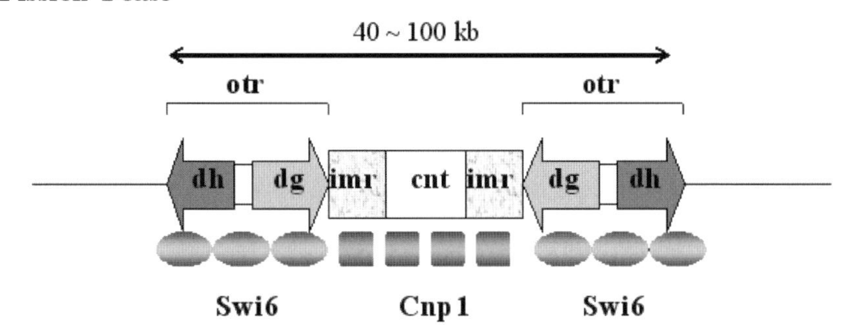

c: Human

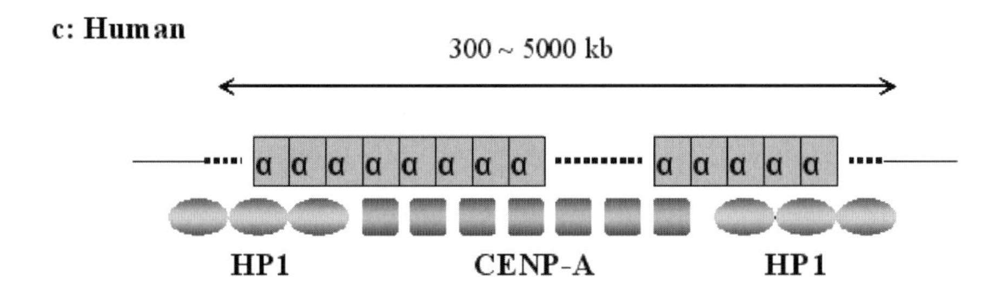

Figure 2. Centromere structure: The budding yeast centromere (a), which is only ~125 bp in length, is probably a single nucleosome centromere. The structures of fission yeast centromeres are slightly different among the three chromosomes; the structure of Cen I is shown in (b). For human centromeres (c), HP1 may also spread into the region in which CENP-A is found.

derived mini-chromosome, Dp1187 (Murphy et al., 1995; Sun et al., 1997) indicated that centromere function maps to a satellite DNA array. Although centromere function is conferred by these large satellite DNA segments, it is unlikely that the entire satellite arrays are required for function. Supporting this view, artificial human chromosomes were constructed by introducing approximately 100 kb-DNA containing α-satellite DNA arrays combined with telomeres (Harrington et al., 1997; Ikeno et al., 1998). The seeding activity must be regulated strictly. Once a kinetochore is assembled on a chromosome, formation of another one is deleterious. While one kinetochore is attached to the spindle radiated from one pole, another one on the same chromatid could be attached to the spindle from the other pole. Such a dicentric chromatid would be broken because the two kinetochores move to opposite directions after the onset of anaphase.

4.2 Kinetochore Proteins

Remarkably, most of the proteins involved in centromere/kinetochore assembly are well-conserved from yeast to human, though we cannot find any significant homology among the centromere DNA sequences in eukaryotes. As we discuss below, while some proteins are the founding members to build up the kinetochore, others play roles in chromosome movement and/or cell cycle regulation. The overall structure of the kinetochore can be dissected into the following three parts; inner kinetochore, central kinetochore and outer kinetochore.

4.2.1 Inner Kinetochore

The human CENP-A protein is a variant of histone H3 that is assembled into the nucleosome. The CENP-A containing nucleosome is found specifically in the centromere. By binding to the centromere DNA as the nucleosome, CENP-A with other proteins builds up the inner kinetochore during mitosis. It is also present in the centromere during interphase. The budding yeast homolog of CENP-A, Cse4, associates with the A-T rich CDE II (Figure 1a, Stoler et al., 1995). Similarly, fission yeast CENP-A, Cnp1, is found at the highly A-T rich imr and cnt regions (Takahashi et al., 2000).

As CENP-A and its homologs are found exclusively at the centromere, it is particularly interesting to investigate how CENP-A is loaded on the centromere for assembly of only one kinetochore per chromosome. In budding yeast, localisation of Cse4 requires a chromatin assembly factor, CAF-1, or Hir proteins (Sharp et al., 2002). A protein complex, CBF3, binds to CDE III. Loss of this complex abolishes localisation of Cse4 (Ortiz et al., 1999). Loading of the fission yeast CENP-A (Cnp1) onto the centromere is dependent on Mis6 (Takahashi et al., 2000) and Mis16-Mis18, two other kinetochore proteins (Hayashi et al, 2004). Mis16 and Mis18 are evolutionarily conserved. They form a complex and act as upstream CENP-A loading factors. RbAp46 and RbAp48, human homologues of Mis16, are also essential for CENP-A loading. In a temperature sensitive *mis6* mutant, chromosomes segregate unequally when the mutant is incubated at the restrictive temperature in the preceding G1/S phase (Saitoh et al., 1997). Mis6 plays an important role in establishment of the inner kinetochore before mitosis. In vertebrate cells, however, a weak homolog of Mis6, CENP-I, is not required for CENP-A loading (Liu et al., 2003; Goshima et al., 2003), but the requirement of the Mis16-Mis18 complex for CENP-A loading is conserved. CENP-A loading could be regulated by multiple mechanisms and the dependency on a particular mechanism may have been modified through evolution. The vertebrate CENP-C is also localised at the inner kinetochore (Saitoh et al., 1992). The CENP-A nucleosome interacts with CENP-C and another protein, CENP-B, through the α-satellite DNA array (Ando et al., 2002). Null mutant mice for CENP-A fail to survive beyond 6.5 days postconception. The null embryos show mitotic defects, including formation of micronuclei and macronuclei, nuclear bridging and chromatin fragmentation. In the cells lacking CENP-A, CENP-B and C disperse throughout the nucleus (Howman et al., 2000). The CENP-B gene seems to be non-essential, since knockout mice were viable without any apparent defects (Hudson et al., 1998; Kapoor et al., 1998; Perez-Castro et al., 1998). Deletion of the CENP-C gene results in embryonic lethality in mouse and causes defects in chromosome segregation (Kalitsis et al., 1998).

4.2.2 Central Kinetochore

The direct interaction between the kinetochore and the spindle is mediated by a set of proteins localised at the "outer kinetochore" (see below). We like to refer to some of the kinetochore proteins, which belong to neither the "inner" nor the "outer" kinetochore proteins, as "central kinetochore" proteins (Cheeseman et al., 2002). This classification is tentative and may be redefined by future studies.

Human hMis12 is localised at centromeres in interphase and at kinetochores in mitosis. This localisation does not require CENP-A. Conversely, CENP-A is still localised at kinetochores in cells lacking Mis12. Thus, the two proteins are recruited to kinetochores independently. In human cells, depletion of Mis12 causes misaligned chromosomes in metaphase, lagging chromosomes in anaphase and micronuclei in interphase. In addition, the length of the metaphase spindle is approximately 60 % longer (Goshima et al., 2003). In fission yeast, a temperature sensitive mutant, mis12-537, exhibits phenotypes similar to those seen upon depletion of Mis12 in human cells (Goshima, et al., 1999). The longer mitotic spindle and abnormal segregation of chromosomes would suggest that Mis12 plays an important role in maintenance/regulation of stable interactions between the kinetochores and the spindle. Recent studies indicated that budding yeast Mtw1, a homolog of fission yeast Mis12, forms a complex designated MIND and serves as a bridge between the inner kinetochore and the outer kinetochore (De Wulf et al., 2003). In nematodes, Mis12 was copurified with a number of proteins with a role at the kinetochore-microtuble interface. Based on the phenotype observed upon its depletion, it is proposed that Mis12 regulates the rate and extent of the outer kinetochore assembly (Cheeseman et al., 2004). Human hMis12 was copurified with an outer kinetochore protein, Zwint-1 (Cheeseman et al., 2004; Obuse et al., 2004), as well as with HP1, a protein localised at the centromeric heterochromatin (Obuse et al., 2004), suggesting an important role of hMis12 as a kinetochore skeleton.

Ndc80/Hec1/Tid3 (hereafter referred to as Ndc80) forms a complex with Spc24, Spc25 and Nuf2 (Janke et al., 2001, McCleland et al., 2003, 2004).

The vertebrate Ndc80-complex appears at kinetochores from prometaphase to anaphase. Depletion of the complex affects chromosome congression in early mitosis (Martin-Lluesma et al., 2002) and movement in anaphase, suggesting a role of the complex in the establishment/maintenance of interactions between the spindle and kinetochores.

The Aurora kinases are a family of cell-cycle kinases whose activity peaks in mitosis. The family can be sub-divided into three classes, A, B and C based on the sequence similarity in higher eukaryotes (Nigg, 2001). Aurora B kinases are so called "chromosome passenger" proteins that form a complex with INCENP and survivin (Cooke et al., 1987; Adams et al., 2000; Kaitna et al., 2000; Adams et al., 2001). At prophase, they are localised along the length of chromosomes and then found at the central kinetochore at prometaphase. During anaphase, these proteins dissociate from chromosomes and are localised at the central spindle. Localisation of Aurora requires an appropriate chromosome configuration. If, for example, cohesion between sister chromatids is impaired, fission yeast Aurora (Ark1) cannot localise at the kinetochore (Morishita et al., 2001). There is an accumulating body of evidence that Aurora B kinases regulate chromosome segregation as well as cytokinesis. In a number of organisms, Aurora B has been shown to promote chromosome condensation through phosphorylation of histone H3 at serine 10. Loss of Aurora B correlates with a reduced level of phosphorylation of histone H3 and results in a failure in proper chromosome condensation in *S. cerevisiae*, *C. elegans* and *Drosophila* (Hsu et al., 2000; Adams et al., 2001; Giet and Glover, 2001). Another study demonstrated that fission yeast Aurora B (Ark1) and survivin (Bir1) are required for recruitment of condensin (a protein complex necessary for chromosome condensation, see below) to chromatin in mitosis (Morishita et al., 2001). Overexpression of a dominant negative mutant of Aurora B results in disruption of cleavage furrow formation in mammalian cells. Consequently, cytokinesis frequently fails and cell polyploidy and cell death are apparent (Terada et al., 1998). Inhibition of Aurora B kinase in mammalian cells results in disorganisation of the kinetochore. CENP-E and dynein, which normally target to the kinetochore outer plate (see below), are no longer found at kinetochores. Furthermore, most of the kinetochores remain unattached to the spindle (Murata-Hori and Wang, 2002). Budding yeast studies demonstrated that Ipl1, the yeast homolog of Aurora, is required for activation of the spindle checkpoint when kinetochores are not under tension (Biggins and Murray, 2001). More recently, it has been shown that Aurora B kinase is responsible for preventing and correcting syntelic attachment (attachment of sister kinetochores to microtubules extended from the same pole). It has also been shown that Aurora kinases promote selective disassembly of microtubules involved in syntelic attachment (Tanaka et al., 2002; Dewar et al., 2004; Lampson et al., 2004). Another recent study indicated that novel chromosome passenger proteins, termed Dasra A/B (Sampath et al., 2004) or Borealin (Gassmann et al., 2004) form a complex with Aurora B and INCENP and contribute to the stabilisation of chromatin-associated microtubules. Obviously, Aurora kinases perform multiple functions in mitosis. They likely select appropriate substrates depending on the stage of mitosis as well as their location (Vagnarelli and Earnshaw, 2004).

4.2.3 Outer Kinetochore

A central role of the kinetochore is to connect chromosomes to the mitotic spindle. Most of the proteins in this category mediate, regulate and/or monitor attachment of the spindle to kinetochores.

CENP-F first appears at centromeres in early G2 (Rattner et al., 1993; Liao et al., 1995). The level of CENP-F reaches a peak at G2/M and the protein is rapidly degraded after mitosis (Liao et al., 1995). CENP-F ends with a CAAX-motif, a signature for farnesylation, and the biological significance of this modification has been explored. Both the localisation of CENP-F to kinetochores and its degradation are dependent on the CAAX-motif (Hussein and Taylor, 2002).

MCAK, a kinesin-related protein, follows CENP-F and is localised at kinetochores by prophase (Wordeman, 1995). During prometaphase, two motor proteins, CENP-E and dynein (and its associating protein complex, dynactin) arrive at kinetochores (Yen et al., 1992, Pfarr et al., 1990; Steuer et al., 1990; Echeverri et al., 1996). The temporal recruitment of these proteins would indicate that a specialised structure of the outer kinetochore is assembled for specific functions from G2 to mid-mitosis. MCAK has a unique activity to stimulate disassembly of microtubules in an ATP-dependent manner (Desai et al., 1999; Hunter et al., 2003). Inhibition of MCAK *in vivo* affects chromosome alignment as well as poleward movement in anaphase (Maney et al., 1998; Walczak et al., 2002). These results would indicate that MCAK is an important motor protein required for congression and delivery of chromosomes after separation of sister chromatids. When the function of CENP-E is compromised, biorientation and the consecutive congression are inhibited (Schaar et al., 1997). The two proteins, dynein and CENP-E, are also involved in regulation of the spindle checkpoint. Injection of an anti-dynein antibody does not block congression but causes an arrested at metaphase. Inactivation of the spindle checkpoint releases the arrest and allows segregation of chromosomes, indicating that a major role of dynein may be in negative regulation of the spindle checkpoint (Howell et al., 2001). Depletion of CENP-E in frog egg extracts causes a defect in checkpoint signalling (Abrieu et al., 2000). Placed at an appropriate position to monitor the attachment of the spindle to kinetochores, these proteins regulate the activity of the spindle checkpoint.

A group of checkpoint proteins including Mad1 (Jin et al., 1998) and Mad2 (Chen et al., 1996; Li and Benezra, 1996) assemble onto unattached kinetochores between prophase and prometaphase. A target of the spindle checkpoint, CDC20/p55CDC, also arrives at the kinetochore at this stage (Kallio et al., 1998). CDC20/p55CDC stimulates proteolysis necessary for sister chromatid separation (see below) and its activity is inhibited by binding of Mad2 (Kim et al., 1998; Hwang et al., 1998; Fang et al., 1998). The half-life of Mad2 at unattached kinetochores is approximately 30 seconds (Howell et al., 2000). Although the precise mechanism is still to be explored, it has been speculated that unattached kinetochores serve as a factory to assemble Mad2 and its target, CDC20, into a complex. Depletion of central kinetochore proteins, Nuf2 or Hec1, causes a reduction in the levels of Mad1 and Mad2 at kinetochores, suggesting that they are required for retention of Mad1 and Mad2 (Martin-Lluesma et al., 2002; DeLuca et al., 2003). In addition, CENP-I, a human homolog of the fission yeast Mis6, is required for localisation of CENP-F, Mad1 and Mad2 to kinetochores (Liu et al., 2003). These central kinetochore proteins may provide a docking site for Mad1 and Mad2, which is presumably active only when kinetochores are unattached. Finally, three protein kinases, Bub1, BubR1, and Mps1 also contribute to spindle checkpoint signalling at the kinetochore (see chapter by Yen and Kao in this volume). However, the regulation of these kinases remains poorly understood and most their physiological substrates remain to be identified.

5. CHROMOSOME COHESION

After replication in S phase, duplicated sisters are linked together until mitosis. The linkage holding sister chromatids together, termed cohesion, plays an important role in establishment of the symmetrical arrangement of chromosomes (Figure 1f). By keeping sister kinetochores attached back to back, cohesion maintains a centromere geometry that favours the attachment of sister kinetochores to the spindles radiated from the two opposite poles. Tying up sister chromatids together, cohesion also provides a memory of which chromatids are to be separated and delivered to two daughter cells.

5.1 Molecular Mechanism of Cohesion

Genetic approaches toward understanding chromosome structure led to the identification of proteins responsible for cohesion in yeast. Through a screen for mutants exhibiting a higher rate of chromosome loss, a *smc1* mutant was isolated (Strunnikov et al., 1993). Another screen (Michaelis et al., 1997; Guacci et al., 1997; Toth et al., 1999), in which a chromosome locus marked by fluorescence was examined microscopically, was also conducted. This screen was based on an assumption that the fluorescent signal would be observed as two separate dots after S phase in a mutant defective in cohesion (or its related process). In the wild type background, the signal appears as a single dot before mitosis because the fluorescently marked sister chromatids are linked closely by cohesion. Through these genetic screens, four genes, namely, Scc1/Mcd1, Scc3, Smc1 and Smc3, were identified as essential components for chromosome cohesion in budding yeast. Homologs of these genes were identified in frog, fission yeast (Losada et al., 1998; Tomonaga et al., 2000) as well as other eukaryotes (Haering and Nasmyth, 2003).

Smc1 and Smc3 belong to the SMC (Structural Maintenance of Chromosomes) protein family. They are large (1,000-1,500 amino acids) proteins with a characteristic structure. The globular amino- and carboxy-terminal domains flank the central domain of SMC proteins, which form coiled-coil and hinge structures. The central domain forms an intra-molecular anti-parallel coiled coil and keeps the amino- and carboxy-terminal globular domains together (Melby et al., 1998). The amino-terminal globular domain contains an ATPase motif. Smc1 and Smc3 proteins dimerize via the central hinge domains to form a V-shape molecule. To the resulting heterodimer of Smc1 and Smc3, two non-SMC components, Scc1/Mcd1 and Scc3 bind to build a protein complex, called cohesin. Biochemical analysis and direct observation of fine structure (Melby et al., 1998; Anderson et al., 2002; Gruber et al., 2003) suggest that cohesin may from a hinge ring, in which the opening of the V-shaped Smc1-Smc3 heterodimer is closed by Scc1-Scc3. Although it is speculated that cohesin maintains sister chromatids together by encircling them, this model remains to be proven experimentally (Uhlmann, 2004).

5.2 Loading of Cohesin onto DNA

Loading of cohesin onto DNA must be regulated so that it maintains linkage between sister chromatids, but not other chromatids. In this regard, it is perhaps not surprising that cohesion between sister chromatids is established during S phase, a cell cycle stage at which sister strands would be in the closest proximity. If synthesis of Scc1, one of the subunits of cohesin, is limited to only after S phase, the cohesin assembled in G2 still binds to DNA, but fails to link sister chromatids together (Uhlmann and Nasmyth, 1998). It is likely that the establishment of cohesion and DNA replication are tightly coupled. Two prime questions, how cohesin links only sister chromatids, and how the action of cohesin is coupled with DNA replication, remain to be answered. In budding yeast, a mutant of Smc1 protein, which lacks the ATPase activity, is still assembled into the cohesin complex but fails to associate with DNA *in vivo* (Weitzer et al., 2003; Arumugam et al., 2003). On the other hand, a purified cohesin complex associates with DNA *in vitro* without a requirement for ATP hydrolysis (Losada and Hirano, 2001; Kagansky et al., 2004), though association *in vitro* may not be in the same configuration as *in vivo*. We should consider the possibility that other proteins, such as adherin (budding yeast Scc2 and fission yeast Mis4;Michaelis et al., 1997; Furuya et al., 1998) and/or proteins assembled in the replication folk, may be involved in the loading process of cohesin.

5.3 Pericentric Heterochromatin and Cohesin

In most eukaryotes, the centromere/kinetochore is flanked by heterochromatic regions, specialised chromatin found predominantly, but not exclusively, at pericentric regions. It is rich in repeating sequences and transposons, and poor in expressed genes. Epigenetic control determines heterochromatic state of the pericentric region. Modification of the N-terminal tail of histone H3 is a major target of this control. The tail of histone H3 is hypoacetylated and di- or trimethylated at lysine 9 (H3-K9), and monomethylated at lysine 27 (H3-K27). The heterochromatin-associated protein, Swi6/HP1, is recruited to methylated histone H3 (Nakayama et al., 2001; Bannister et al, 2001). The biological significance of pericentric heterochromatin is evident from the observation that a loss of heterochromatin components results in abnormal chromosome segregation (Allshire et al., 1995; Bernard et al., 2001; Peters et al., 2001). Swi6/HP1 recruits cohesin to the pericentric heterochromatic regions (Bernard et al, 2001; Nonaka et al., 2002). Furthermore, fission yeast Hsk1, a conserved protein kinase that regulates initiation of DNA replication, interacts with and phosphorylates the heterochromatin protein, Swi6 (Bailis et al., 2003). These results not only demonstrate the importance of Swi6 for loading of cohesin at the pericentric regions, but also suggest that an analogous mechanism may be responsible for loading cohesin onto chromosome arms.

It has also been demonstrated that the heterochromatic state of the pericentric chromatin is established/maintained by the RNA-mediated interference (RNAi) pathway, a mechanism responsible for post-transcriptional gene silencing in a wide range of biological processes (Hannon, 2002; Denli and Hannon, 2003). In fission yeast, several genes encoding components for RNAi, Ago1, Dcr1 and Rdp1, were deleted and shown to be essential for faithful chromosome segregation in mitosis as well as meiosis (Volpe et al., 2002; Provost et al., 2002; Hall et al., 2003; Volpe et al., 2003). A part of centric inverted DNA repeats (otr, Figure 2b) is transcribed and converted into double stranded RNA (dsRNA). It has been proposed that dsRNA derived from the otr regions are cleaved by Dcr1 and assembled into an RNAi effector complex. This complex may promote association of the heterochromatin protein, Swi6, by recruiting a methyltransferase for lysine 9 of histone H3. Supporting this model, it has been demonstrated that deletion of RNAi control results in loss of Swi6 as well as cohesin at the centromeric region (Hall et al., 2003; Volpe et al., 2003). A recent study has also indicated that vertebrate cells employ a similar mechanism to maintain heterochromatic state (Fukagawa et al., 2004).

5.4 Unloading of Cohesin

At the onset of anaphase, cohesin is unloaded from chromosomes for sister chromatid separation. Budding yeast studies revealed that the majority of Scc1, a subunit of the cohesin complex is cleaved by a specific protease, separase, at anaphase onset. If amino-acid motifs of Scc1 recognised by separase are mutated, Scc1 remains chromatin-bound and sister chromatids cannot be separated at anaphase. On the other hand, if a recognition site by an exogenous protease (TEV) is generated in Scc1, expression of TEV allows both unloading of Scc1 and sister chromatid separation without the endogenous separase. In budding yeast, cleavage of Scc1, which is sufficient for sister chromatid separation, is tightly coupled with unloading of the cohesin complex (Uhlmann et al., 1999, 2000).

Fission yeast studies indicated that although cleavage of Rad21 (the fission yeast homolog of Scc1) is necessary for sister chromatid separation, the amount of cleaved Rad21 is very small (less than 5%). Furthermore, Rad21 remains in the nucleus throughout the cell cycle (Tomonaga et al., 2000). In vertebrates most cohesin, in contrast to yeast, is removed from chromosome arms at prophase, well before anaphase (Losada et al., 1998; Schmiesing et al., 1998; Darwiche et al., 1999; Waizenegger et al., 2000). The removal of cohesin at early mitosis is seemingly not accompanied by cleavage of Scc1. Residual amounts of Scc1 are cleaved at the onset of anaphase by separase. Higher eukaryotes may use two distinct mechanisms to remove cohesin from chromosomes. The cohesin complexes at the arm regions, which can be removed without separase, may bind to DNA in a mode different from the ones, which require the activity of separase.

Nonetheless, separase is a key enzyme that allows dissociation of the cohesin complex at anaphase. For most time of the cell cycle, separase is kept inactive and thereby premature sister chromatid separation is prevented (Figure 3). Securin (budding yeast Pds1/fission yeast Cut2), a protein degraded at the onset of anaphase, is responsible for the regulation of separase (Cohen-Fix et al., 1996, Funabiki et al., 1996). Separase, confined in a complex with its inhibitor, securin, is liberated upon destruction of securin at anaphase and promotes sister chromatid separation (Ciosk et al., 1998, Kumada et al., 1998). The timing of destruction of securin, therefore, determines the timing of sister chromatid separation. A large protein complex, APC/Cyclosome, together with Cdc20/p55CDC, ubiquitinates securin for its destruction (King et al., 1995; Sudakin et al., 1995; Visintin et al., 1997).

Sister chromatids must separate only after all sister chromatids are correctly attached to the spindle and are bioriented at the spindle equator. If anaphase occurs prematurely, some chromatids may not interact with the spindle, a pilot to guide them to daughter cells. Such chromatids would be a major cause of aneuploidy, which is highly harmful to the host body. To prevent this deleterious event, Mad2, a component of the spindle checkpoint, binds to Cdc20/p55CDC and inhibits its activity to promote ubiquitination of securin (Kim et al., 1998; Hwang et al., 1998; Fang et al., 1998).

Although activation of separase determines the timing of sister chromatid separation primarily, a recent report (Gerlich et al., 2003) indicated that each

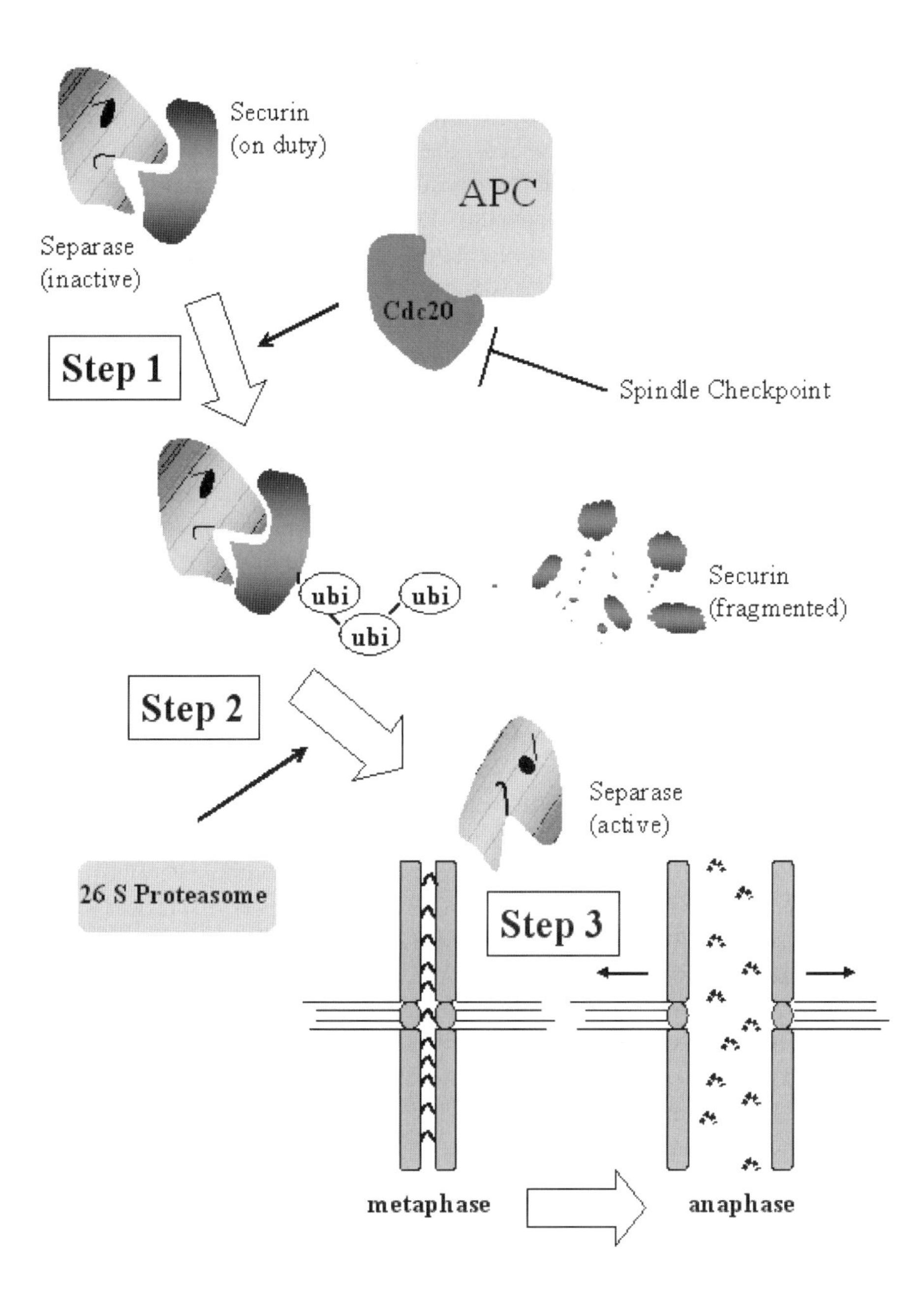

Figure 3. Separase activation and sister chromatid separation at anaphase: A separase inhibitor, securin, is ubiquitinated by APC-Cdc20 (Step 1) and consecutively degraded by 26 S proteasome (Step 2). Securin, once fragmented, can no longer inhibit separase and separase in its active form then cleaves the subunit of cohesin, Scc1, for sister chromatid separation (Step 3).

sister chromatid separates with a slightly different timing during anaphase. An early separating chromatid migrates further than a late separating one and thereby can reach a more distal domain of the daughter nucleus. It has been proposed that this difference in the timing of separation loosely determines 3D-positioning of each chromosome in the interphase nucleus. Interestingly, disruption of heterochromatin formation by treatment with a drug, Hoechst3325, abolishes the temporal order of sister chromatid separation. As mentioned earlier, the centromere is surrounded by heterochromatin in most eukaryotes and Swi6/HP1 is responsible for recruiting cohesin to heterochromatin. Perhaps the quality or quantity of cohesin at heterochromatin of each centromere differs for inheritance of the positioning information of chromosomes.

Cleavage of cohesin is not limited to mitosis. A recent study (Nagao et al., 2004) demonstrated that upon introduction of DNA damage fission yeast Scc1 (Rad21) is cleaved in a separase-dependent fashion. Furthermore expression of a Rad21-mutant, which cannot be cleaved by separase, as well as a strain with a partially defective separase are impaired in DNA damage repair. Based on these and other results, it has been proposed that removal of cohesin at a damage site may facilitate a recombination-based repair process. Separase is active presumably only in the vicinity of a damage site. It will be of considerable interest to address the question of how separase is activated locally.

6. CHROMOSOME CONDENSATION

Shortly after the onset of mitosis, chromosomes undergo a dramatic change in shape during chromosome condensation. This chromosome process is essential for accurate chromosome segregation in several aspects. Fist of all, chromosome condensation shortens the length of chromosomes and allows efficient partition of sister chromatids in the limited space of the nucleus. The human genome, for example, consists of approximately 3.2×10^9 nucleotides, and, without any compaction, is about 2 meters in total length. (Thus, an average DNA length/chromosome would be about 80,000 micro-meters). In interphase, human chromosomes are packed in an approximate ratio of 1000 ~ 2000 fold, thereby resulting in a length of 80 micro-meters/chromosome. Considering that the diameter of the nucleus of human cells is less than 10 micro-meters, a chromosome in the interphase configuration, whose length exceeds the diameter of the nucleus, would not move its entirety into the daughter nucleus. Mitotic chromosome condensation is required to make chromosomes more compact. Secondly, condensation is expected to make chromosomes stiffer or more plastic. As mentioned earlier, the arms of chromosomes encounter a pole-ejection force in mitosis that regulates oscillation and congression. If chromosomes were flabby, they could not sense the force and thus mitotic chromosome movement would be out of control. Finally, chromosome condensation resolves tangles between sister chromatids or different chromosomes. This process allows individualisation of chromosomes in early mitosis and requires Topo II (Maeshima and Laemmli, 2003).

6.1 Condensin Complex

Chromosome biochemistry was a powerful approach to elucidate the molecules responsible for chromosome condensation. Two components (XCAP-C/Smc4 and XCAP-E/Smc2) required for condensation of mitotic chromosomes were identified in frog egg extracts (Hirano and Mitchison, 1994). Subsequent biochemical studies identified 3 additional components, CAP-D2, CAP-G and CAP-H, which are not related to Smc proteins (Hirano et al., 1997). Like the cohesin complex, the two Smc-components form a V-shape heterodimer, to which the three non-Smc components bind to form the condensin (Anderson et al., 2002). Studies in yeast and fly also identified homologous condensin complexes (Saka et al., 1994; Saitoh et al., 1994; Strunnikov et al., 1995; Sutani and Yanagida, 1997; Sutani et al., 1999). In fission yeast condensin mutants, chromosomes fail to condense but are forced to segregate. As a result, chromosomes, which are not compact enough, remain in the middle of the cell and are eventually "cut" by cytokinesis. Unlike cohesin that is loaded onto chromosomes in S phase, condensin is recruited to the chromatin in mitosis. In fission yeast, where the nuclear membrane persists during mitosis, condensin is transported into the nucleus in a Cdc2-dependent manner (Sutani et al., 1999) and concentrates at the central cnt region of centromeres during mitosis (Aono et al., 2002).

6.2 Biochemical Activity of Condensin

Studies *in vitro* demonstrated that condensin cooperates with other enzymes to alter the topology of DNA. In the presence of type I topoisomerase, condensin introduces positive supercoils into close circular DNA depending on ATP-hydrolysis by the ATPase at the amino-terminal globular domains of XCAP-C/Smc4 and XCAP-E/Smc2 (Kimura and Hirano, 1997). It also exhibits a knotting activity toward nicked circular DNA with type II topoisomerase (Kimura et al., 1999). While type I topoisomerase is not essential for viability in yeast (implying that the importance of its cooperation with condensin is unclear at present), type II topoisomerase, an essential enzyme for viability, may play an important role in chromosome condensation with condensin *in vivo*. It was also shown that a single condensin complex can introduce two (or more) supercoils into closed circular DNA in an ATP-dependent manner. Observation of the condensin-induced supercoiled DNA by ESI (electron spectroscopic imaging) suggested that a single condensin complex introduces two supercoils to its bound region, and the torsion is cancelled by supercoils in the condensin-free region (Bazett-Jones et al., 2002). If a single condensin complex is responsible for compaction of local chromosomal DNA *in vivo*, it is likely that DNA is trapped within the individual condensin complex. As the complex is assembled in a manner similar to the cohesin complex, DNA might be trapped within the postulated ring structure of the condensin.

The condensin complex also exhibits an annealing activity *in vivo* (Sutani and Yanagida, 1997). The activity of promoting ssDNA reannealing *in vitro* has also been demonstrated for the mammalian recombination complex RC-1 (Jessberger et al., 1996) and for the *Bacillus subtilis* SMC homodimeric complex (Hirano and Hirano, 1998). Although the cohesin is similar in the content and structure of the complex, it does not exhibit this activity (Sakai et al., 2003). Chromatin DNA isolated from fission yeast condensin mutants (*cut3* and *cut14*) is sensitive to S1 nuclease, an enzyme that preferentially digests ssDNA (Sutani and Yanagida, 1997). Although it remains to be elucidated, this annealing activity may be required for a hidden biological process in chromosome condensation. It is equally possible that the condensin plays an important role in interphase. Supporting this notion, a fission yeast study indicated that a mutant of a condensin subunit (Cnd2) fails to recover from S phase interference or activate the replication checkpoint kinase Cds1/Chk2 (Aono et al., 2002).

7. CHROMOSOME INSTABILITY AND HUMAN DISEASES

7.1 Cancer Development/Progression

An abnormal karyotype is a hallmark of cancer cells. It has been reported that about 60-80% of the cases of cancers show clear evidence of an aneuploid DNA content (Fiegl et al., 1995; Shackney et al., 1995). Furthermore, most of the solid tumour cells exhibit considerable variability in chromosome number from cell to cell in the same tumour tissue (Shackney et al., 1989). It has therefore been postulated that 1) a normal cell looses the ability to maintain the normal chromosome number, 2) such a cell would produce a variety of cells with extra and/or missing chromosomes and 3) a cell with a particular chromosome content acquires the ability to grow aggressively and becomes the germ of cancer. The abnormal karyotype found in cancer cells may result from failure in faithful chromosome segregation. Supporting this notion, a number of components required for faithful segregation are found mutated or abnormally expressed in human cancer cells.

Mad2, a component of the spindle checkpoint, was poorly expressed in some breast cancer cells (Li and Benezra, 1996). More directly, removal of a single copy of the Mad2 gene from the diploid mouse genome, which causes a reduction of the protein level to about 70%, results in development of lung cancer as well as lymphoma (Michel et al., 2001). Mad1 was also reported to be a target of Tax, a gene product of human T cell leukemia virus (Jin et al., 1998). Expression of either Tax or a transdominant-negative Mad1 results in multinucleated cells, a phenotype consistent with a loss of the functional spindle checkpoint. BubR1, a kinase to play an important role in the spindle checkpoint, has been found mutated in some colon cancers, which exhibit instability of chromosome number (Cahill et al., 1998).

In some colon cancer cells, the level of CENP-A is elevated (up to 32-fold compared to normal cells) whereas that of CENP-B seems to be normal (Tomonaga et al., 2003). Cytological observations indicated that some of the CENP-A foci do not coincide with the CENP-B foci, suggesting that a fraction of CENP-A is mislocalised in these cells. If the elevated level of CENP-A results in formation of additional kinetochores, it may cause abnormal chromosome segregation and account for chromosome instability in cancer cells.

All Aurora kinases (A, B and C) are often found overexpressed in cancers. Aurora A, in particular, has been identified as a tumour-susceptibility gene in both mouse and human (Ewart-Toland et al., 2003), thereby demonstrating a direct contribution to carcinogenesis. Although they are still elusive, some mechanisms by which a high dosage of Aurora kinases leads to tumour development, have been proposed. Aurora A, when overexpressed, may cause abnormal centrosome amplification, which, in turn, induces aneuploidy/polyploidy (Meraldi et al., 2002; Goepfert et al., 2002). It has also been shown that a high dosage expression of Aurora A disables the spindle checkpoint (Anand, 2003). These models are not mutually exclusive, and Aurora A may induce instability of chromosome number by disregulating multiple control mechanisms.

Human securin is identical to the product of the gene called pituitary tumour-transforming gene (PTTG), which is overexpressed in some tumours and exhibits transforming activity in NIH 3T3 cells (Zou et al., 1999). It has been reported that p53 interacts specifically with securin both *in vitro* and *in vivo*. This interaction blocks the specific binding of p53 to DNA and inhibits its transcriptional activity. Securin also inhibits the ability of p53 to induce cell death (Bernal et al., 2002). An inhibitory activity of securin toward p53 could account for its transforming activity.

7.2 Chromosome Diseases

As discussed above, the structure of the mitotic chromosome is maintained by multiple mechanisms, notably post-translational modifications, DNA-protein and protein-protein interactions. A defect in any one of these mechanisms could result in an aberrant mitotic process and cause a severe human disease. If a defect is caused by a mutation of a gene, it could be inherited as a genetic disease.

Neocentromeres are new centromeres that emerge at chromosomal loci, which are normally non-centromeric. The first human neocentromere was identified through a karyotying of a patient with learning difficulties (Voullaire et al., 1993). This new centromere was found on a small chromosome termed mardel(10), which is formed by a *de novo* rearrangement of chromosome 10. To date, 60 cases of human neocentromeres have been reported. Like mardel(10), these neocentromeres are found on marker chromosomes generated through rearrangement and typically associated with developmental delay or congenital abnormalities. Some neocentromeres have also been found in patients with cancers, such as lipomatous tumours and acute myeloid leukemia (Amor and Choo, 2002). Mitotic stability of a marker chromosome is generally high, though not fully comparable to that of authentic chromosomes, indicating that the neocetromere is functional. Although most of the neocentromeres lack α-satellite DNA arrays, kinetochores built on the neocentromeres contain most of the kinetochore proteins, including CENP-A, C, E, F, spindle checkpoint proteins, dynein, INCENP as well as HP1 (Saffery et al., 2000). CENP-B is an exception in that it has so far not been detectable on neocentromeres. As demonstrated in the mouse studies, CENP-B is not essential, implying that it is not absolutely required for kinetochore formation.

Townes-Brocks syndrome (TBS) is an autosomal dominantly transmitted syndrome characterised by congenital abnormalities such as imperforate anus, triphalangeal and supernumerary thumbs, malformed ears and sensorineural hearing loss. The TBS-predisposing gene encodes a zinc-finger protein, SALL1 (Kohlhase et al., 1998). SALL1 is localised at chromocentres including pericentromeric heterochromatin and telomeres in mouse NIH-3T3 cells (Netzer et al., 2001). The mouse SALL1 exhibits activity as a transcriptional repressor. It physically interacts with components of chromatin remodelling complexes, HDAC1, HDAC2, RbAp46/48, MTA-1, and MTA-2 (Kiefer et al., 2002). Taken together, these results would suggest that SALL1 is a component involved in a silencing process at heterochromatin.

Roberts syndrome is an autosomal recessive disorder characterised by symmetric reduction of all limbs and growth retardation. Cells isolated from patients with this syndrome exhibit premature separation of heterochromatin regions of many chromosomes (Van den Berg and Francke, 1993). In addition, these cells are abnormal in metaphase and anaphase progression. An elevated level of lagging chromosomes or chromosomes prematurely advancing toward the poles was found. They also show aneuploidy, micronuclei formation, poor growth, reduced plating efficiency and lower density for confluent cultures (Jabs et al., 1991). These observations would suggest that Roberts syndrome is a mitotic mutation syndrome, leading to secondary developmental defects. Recent study has demonstrated that inhibition of INCENP, ZW10 or ZWINT-1 results in a phenocopy of the syndrome, suggesting that these genes (or a gene closely interacting with them) may be predisposing to Roberts syndrome (Musio et al., 2004).

Cornelia de Lange syndrome (CdLS) is a multiple malformation disorder characterised by dysmorphic facial features, mental retardation, growth delay and limb reduction defects. The mutations responsible for CdLS have recently been identified on a gene encoding NIPBL (Krantz et al., 2004; Tonkin et al., 2004). NIPBL exhibits a significant homology to adherins, budding yeast Scc2 and fission yeast Mis4 (Michaelis et al., 1997; Furuya et al., 1998), both of which are required for sister chromatid cohesion, probably acting for cohesin loading (Tomonaga et al., 2000). Nipped-B, a fly homolog of NIPBL, is required for transactivation of cut and Ultrabithorax genes by remote promoter. It has recently been shown that Nipped-B mutants display an increased rate of precocious sister chromatid separation (Rollins et al., 2004). These results tempt us to speculate that fly Nipped-B and human NIPBL may regulate gene expression perhaps by regulating local chromosome cohesion.

8. CONCLUSION

Once, François Jacob wrote "The dream of every cell is to become two cells". Following Jacob's aphorism, we would say that the dream of every chromosome is to segregate equally in mitosis for a healthy life of the host". Chromosomes are equipped with the kinetochore, a critical control centre to regulate their movement as well as cell cycle progression. Recent advance in biochemical technology has greatly contributed to the identification of novel kinetochore components. While we are able to deduce some biological functions for each component by depleting them one by one from cells, we still do not understand the underlying mechanisms. For example, what kind of biochemical (or physical) change occurs on a kinetochore component when the pole-ejecting force is generated? How does the kinetochore know that the spindle is correctly attached to it? How does the kinetochore determine its position on a chromosome? Besides the kinetochore, a number of components considerably contribute to equal segregation of chromosomes as well. How does cohesin tie up only the sister chromatids, but not others? What determines the timing of cohesin-unloading? Similarly, condensin should promote chromosome condensation by interacting with DNA of only one sister chromatid. How does condensin avoid mediating interactions between different chromosomes? Apparently we need to build a clever model and design elegant experiments to answer these questions.

ACKNOWLEDGEMENT

The work from the authors' laboratories was supported by grants (Specially Promoted COE Research to M. Y. and Scientific Research to T. M.) from the Ministry of Education, Culture, Sports, Science and Technology of Japan.

REFERENCES

Abrieu, A., J.A. Kahana, K.W. Wood, and D.W. Cleveland. 2000. CENP-E as an essential component of the mitotic checkpoint in vitro. *Cell*. 102:817-26.

Adams, R.R., M. Carmena, and W.C. Earnshaw. 2001. Chromosomal passengers and the (aurora) ABCs of mitosis. *Trends Cell Biol.* 11:49-54.

Adams, R.R., H. Maiato, W.C. Earnshaw, and M. Carmena. 2001. Essential roles of Drosophila inner centromere protein (INCENP) and aurora B in histone H3 phosphorylation, metaphase chromosome alignment, kinetochore disjunction, and chromosome segregation. *J Cell Biol.* 153:865-80.

Adams, R.R., S.P. Wheatley, A.M. Gouldsworthy, S.E. Kandels-Lewis, M. Carmena, C. Smythe, D.L. Gerloff, and W.C. Earnshaw. 2000. INCENP binds the Aurora-related kinase AIRK2 and is required to target it to chromosomes, the central spindle and cleavage furrow. *Curr Biol.* 10:1075-8.

Allshire, R.C., E.R. Nimmo, K. Ekwall, J.P. Javerzat, and G. Cranston. 1995. Mutations derepressing silent centromeric domains in fission yeast disrupt chromosome segregation. *Genes Dev.* 9:218-33.

Amor, D.J., and K.H. Choo. 2002. Neocentromeres: role in human disease, evolution, and centromere study. *Am J Hum Genet.* 71:695-714.

Anand, S., S. Penrhyn-Lowe, and A.R. Venkitaraman. 2003. AURORA-A amplification overrides the mitotic spindle assembly checkpoint, inducing resistance to Taxol. *Cancer Cell.* 3:51-62.

Anderson, D.E., A. Losada, H.P. Erickson, and T. Hirano. 2002. Condensin and cohesin display different arm conformations with characteristic hinge angles. *J Cell Biol.* 156:419-24.

Ando, S., H. Yang, N. Nozaki, T. Okazaki, and K. Yoda. 2002. CENP-A, -B, and -C chromatin complex that contains the I-type alpha-satellite array constitutes the prekinetochore in HeLa cells. *Mol Cell Biol.* 22:2229-41.

Aono, N., T. Sutani, T. Tomonaga, S. Mochida, and M. Yanagida. 2002. Cnd2 has dual roles in mitotic condensation and interphase. *Nature.* 417:197-202.

Arumugam, P., S. Gruber, K. Tanaka, C.H. Haering, K. Mechtler, and K. Nasmyth. 2003. ATP hydrolysis is required for cohesin's association with chromosomes. *Curr Biol.* 13:1941-53.

Bailis, J.M., P. Bernard, R. Antonelli, R.C. Allshire, and S.L. Forsburg. 2003. Hsk1-Dfp1 is required for heterochromatin-mediated cohesion at centromeres. *Nat Cell Biol.* 5:1111-6.

Bannister, A.J., P. Zegerman, J.F. Partridge, E.A. Miska, J.O. Thomas, R.C. Allshire, and T. Kouzarides. 2001. Selective recognition of methylated lysine 9 on histone H3 by the HP1 chromo domain. *Nature.* 410:120-4.

Baum, M., V.K. Ngan, and L. Clarke. 1994. The centromeric K-type repeat and the central core are together sufficient to establish a functional Schizosaccharomyces pombe centromere. *Mol Biol Cell.* 5:747-61.

Bazett-Jones, D.P., K. Kimura, and T. Hirano. 2002. Efficient supercoiling of DNA by a single condensin complex as revealed by electron spectroscopic imaging. *Mol Cell.* 9:1183-90.

Bernal, J.A., R. Luna, A. Espina, I. Lazaro, F. Ramos-Morales, F. Romero, C. Arias, A. Silva, M. Tortolero, and J.A. Pintor-Toro. 2002. Human securin interacts with p53 and modulates p53-mediated transcriptional activity and apoptosis. *Nat Genet.* 32:306-11.

Bernard, P., J.F. Maure, J.F. Partridge, S. Genier, J.P. Javerzat, and R.C. Allshire. 2001. Requirement of heterochromatin for cohesion at centromeres. *Science.* 294:2539-42.

Biggins, S., and A.W. Murray. 2001. The budding yeast protein kinase Ipl1/Aurora allows the absence of tension to activate the spindle checkpoint. *Genes Dev.* 15:3118-29.

Cahill, D.P., C. Lengauer, J. Yu, G.J. Riggins, J.K. Willson, S.D. Markowitz, K.W. Kinzler, and B. Vogelstein. 1998. Mutations of mitotic checkpoint genes in human cancers. *Nature.* 392:300-3.

Cheeseman, I.M., D.G. Drubin, and G. Barnes. 2002. Simple centromere, complex kinetochore: linking spindle microtubules and centromeric DNA in budding yeast. *J Cell Biol.* 157:199-203.

Cheeseman, I.M., S. Niessen, S. Anderson, F. Hyndman, J.R. Yates, 3rd, K. Oegema, and A. Desai. 2004. A conserved protein network controls assembly of the outer kinetochore and its ability to sustain tension. *Genes Dev.* 18:2255-68.
Chen, R.H., J.C. Waters, E.D. Salmon, and A.W. Murray. 1996. Association of spindle assembly checkpoint component XMAD2 with unattached kinetochores. *Science.* 274:242-6.
Chikashige, Y., N. Kinoshita, Y. Nakaseko, T. Matsumoto, S. Murakami, O. Niwa, and M. Yanagida. 1989. Composite motifs and repeat symmetry in S. pombe centromeres: direct analysis by integration of NotI restriction sites. *Cell.* 57:739-51.
Ciosk, R., W. Zachariae, C. Michaelis, A. Shevchenko, M. Mann, and K. Nasmyth. 1998. An ESP1/PDS1 complex regulates loss of sister chromatid cohesion at the metaphase to anaphase transition in yeast. *Cell.* 93:1067-76.
Clarke, L., H. Amstutz, B. Fishel, and J. Carbon. 1986. Analysis of centromeric DNA in the fission yeast Schizosaccharomyces pombe. *Proc Natl Acad Sci U S A.* 83:8253-7.
Clarke, L., and J. Carbon. 1980. Isolation of a yeast centromere and construction of functional small circular chromosomes. *Nature.* 287:504-9.
Cohen-Fix, O., J.M. Peters, M.W. Kirschner, and D. Koshland. 1996. Anaphase initiation in Saccharomyces cerevisiae is controlled by the APC-dependent degradation of the anaphase inhibitor Pds1p. *Genes Dev.* 10:3081-93.
Cooke, C.A., M.M. Heck, and W.C. Earnshaw. 1987. The inner centromere protein (INCENP) antigens: movement from inner centromere to midbody during mitosis. *J Cell Biol.* 105:2053-67.
Darwiche, N., L.A. Freeman, and A. Strunnikov. 1999. Characterization of the components of the putative mammalian sister chromatid cohesion complex. *Gene.* 233:39-47.
De Wulf, P., A.D. McAinsh, and P.K. Sorger. 2003. Hierarchical assembly of the budding yeast kinetochore from multiple subcomplexes. *Genes Dev.* 17:2902-21.
DeLuca, J.G., B.J. Howell, J.C. Canman, J.M. Hickey, G. Fang, and E.D. Salmon. 2003. Nuf2 and Hec1 are required for retention of the checkpoint proteins Mad1 and Mad2 to kinetochores. *Curr Biol.* 13:2103-9.
Denli, A.M., and G.J. Hannon. 2003. RNAi: an ever-growing puzzle. *Trends Biochem Sci.* 28:196-201.
Desai, A., S. Verma, T.J. Mitchison, and C.E. Walczak. 1999. Kin I kinesins are microtubule-destabilizing enzymes. *Cell.* 96:69-78.
Dewar, H., K. Tanaka, K. Nasmyth, and T.U. Tanaka. 2004. Tension between two kinetochores suffices for their bi-orientation on the mitotic spindle. *Nature.* 428:93-7.
Echeverri, C.J., B.M. Paschal, K.T. Vaughan, and R.B. Vallee. 1996. Molecular characterization of the 50-kD subunit of dynactin reveals function for the complex in chromosome alignment and spindle organization during mitosis. *J Cell Biol.* 132:617-33.
Ewart-Toland, A., P. Briassouli, J.P. de Koning, J.H. Mao, J. Yuan, F. Chan, L. MacCarthy-Morrogh, B.A. Ponder, H. Nagase, J. Burn, S. Ball, M. Almeida, S. Linardopoulos, and A. Balmain. 2003. Identification of Stk6/STK15 as a candidate low-penetrance tumor-susceptibility gene in mouse and human. *Nat Genet.* 34:403-12.
Fang, G., H. Yu, and M.W. Kirschner. 1998. The checkpoint protein MAD2 and the mitotic regulator CDC20 form a ternary complex with the anaphase-promoting complex to control anaphase initiation. *Genes Dev.* 12:1871-83.
Fiegl, M., C. Tueni, T. Schenk, R. Jakesz, M. Gnant, A. Reiner, M. Rudas, H. Pirc-Danoewinata, C. Marosi, H. Huber, and et al. 1995. Interphase cytogenetics reveals a high incidence of aneuploidy and intra-tumour heterogeneity in breast cancer. *Br J Cancer.* 72:51-5.
Fishel, B., H. Amstutz, M. Baum, J. Carbon, and L. Clarke. 1988. Structural organization and functional analysis of centromeric DNA in the fission yeast Schizosaccharomyces pombe. *Mol Cell Biol.* 8:754-63.
Fukagawa, T., M. Nogami, M. Yoshikawa, M. Ikeno, T. Okazaki, Y. Takami, T. Nakayama, and M. Oshimura. 2004. Dicer is essential for formation of the heterochromatin structure in vertebrate cells. *Nat Cell Biol.* 6:784-91.

Funabiki, H., and A.W. Murray. 2000. The Xenopus chromokinesin Xkid is essential for metaphase chromosome alignment and must be degraded to allow anaphase chromosome movement. *Cell*. 102:411-24.

Funabiki, H., H. Yamano, K. Kumada, K. Nagao, T. Hunt, and M. Yanagida. 1996. Cut2 proteolysis required for sister-chromatid seperation in fission yeast. *Nature*. 381:438-41.

Furuya, K., K. Takahashi, and M. Yanagida. 1998. Faithful anaphase is ensured by Mis4, a sister chromatid cohesion molecule required in S phase and not destroyed in G1 phase. *Genes Dev*. 12:3408-18.

Gassmann, R., A. Carvalho, A.J. Henzing, S. Ruchaud, D.F. Hudson, R. Honda, E.A. Nigg, D.L. Gerloff, and W.C. Earnshaw. 2004. Borealin: a novel chromosomal passenger required for stability of the bipolar mitotic spindle. *J Cell Biol*. 166:179-91

Gerlich, D., J. Beaudouin, B. Kalbfuss, N. Daigle, R. Eils, and J. Ellenberg. 2003. Global chromosome positions are transmitted through mitosis in mammalian cells. *Cell*. 112:751-64.

Giet, R., and D.M. Glover. 2001. Drosophila aurora B kinase is required for histone H3 phosphorylation and condensin recruitment during chromosome condensation and to organize the central spindle during cytokinesis. *J Cell Biol*. 152:669-82.

Goepfert, T.M., Y.E. Adigun, L. Zhong, J. Gay, D. Medina, and W.R. Brinkley. 2002. Centrosome amplification and overexpression of aurora A are early events in rat mammary carcinogenesis. *Cancer Res*. 62:4115-22.

Goshima, G., T. Kiyomitsu, K. Yoda, and M. Yanagida. 2003. Human centromere chromatin protein hMis12, essential for equal segregation, is independent of CENP-A loading pathway. *J Cell Biol*. 160:25-39.

Goshima, G., S. Saitoh, and M. Yanagida. 1999. Proper metaphase spindle length is determined by centromere proteins Mis12 and Mis6 required for faithful chromosome segregation. *Genes Dev*. 13:1664-77.

Gruber, S., C.H. Haering, and K. Nasmyth. 2003. Chromosomal cohesin forms a ring. *Cell*. 112:765-77.

Guacci, V., D. Koshland, and A. Strunnikov. 1997. A direct link between sister chromatid cohesion and chromosome condensation revealed through the analysis of MCD1 in S. cerevisiae. *Cell*. 91:47-57.

Haering, C.H., and K. Nasmyth. 2003. Building and breaking bridges between sister chromatids. *Bioessays*. 25:1178-91.

Hall, I.M., K. Noma, and S.I. Grewal. 2003. RNA interference machinery regulates chromosome dynamics during mitosis and meiosis in fission yeast. *Proc Natl Acad Sci U S A*. 100:193-8.

Hannon, G.J. 2002. RNA interference. *Nature*. 418:244-51.

Harrington, J.J., G. Van Bokkelen, R.W. Mays, K. Gustashaw, and H.F. Willard. 1997. Formation of de novo centromeres and construction of first-generation human artificial microchromosomes. *Nat Genet*. 15:345-55.

Hayashi, T., Y. Fujita, O. Iwasaki, Y. Adachi, K. Takahashi, and M. Yanagida. 2004. Mis16 and Mis18 are required for CENP-A loading and histone deacetylation at centromeres. *Cell*. 118:715-29.

Hirano, M., and T. Hirano. 1998. ATP-dependent aggregation of single-stranded DNA by a bacterial SMC homodimer. *EMBO J*. 17:7139-48.

Hirano, T., R. Kobayashi, and M. Hirano. 1997. Condensins, chromosome condensation protein complexes containing XCAP-C, XCAP-E and a Xenopus homolog of the Drosophila Barren protein. *Cell*. 89:511-21.

Hirano, T., and T.J. Mitchison. 1994. A heterodimeric coiled-coil protein required for mitotic chromosome condensation in vitro. *Cell*. 79:449-58.

Howell, B.J., D.B. Hoffman, G. Fang, A.W. Murray, and E.D. Salmon. 2000. Visualization of Mad2 dynamics at kinetochores, along spindle fibers, and at spindle poles in living cells. *J Cell Biol*. 150:1233-50.

Howell, B.J., B.F. McEwen, J.C. Canman, D.B. Hoffman, E.M. Farrar, C.L. Rieder, and E.D. Salmon. 2001. Cytoplasmic dynein/dynactin drives kinetochore protein transport to the spindle poles and has a role in mitotic spindle checkpoint inactivation. *J Cell Biol.* 155:1159-72.

Howman, E.V., K.J. Fowler, A.J. Newson, S. Redward, A.C. MacDonald, P. Kalitsis, and K.H. Choo. 2000. Early disruption of centromeric chromatin organization in centromere protein A (Cenpa) null mice. *Proc Natl Acad Sci U S A.* 97:1148-53.

Hsu, J.Y., Z.W. Sun, X. Li, M. Reuben, K. Tatchell, D.K. Bishop, J.M. Grushcow, C.J. Brame, J.A. Caldwell, D.F. Hunt, R. Lin, M.M. Smith, and C.D. Allis. 2000. Mitotic phosphorylation of histone H3 is governed by Ipl1/aurora kinase and Glc7/PP1 phosphatase in budding yeast and nematodes. *Cell.* 102:279-91.

Hudson, D.F., K.J. Fowler, E. Earle, R. Saffery, P. Kalitsis, H. Trowell, J. Hill, N.G. Wreford, D.M. de Kretser, M.R. Cancilla, E. Howman, L. Hii, S.M. Cutts, D.V. Irvine, and K.H. Choo. 1998. Centromere protein B null mice are mitotically and meiotically normal but have lower body and testis weights. *J Cell Biol.* 141:309-19.

Hunter, A.W., M. Caplow, D.L. Coy, W.O. Hancock, S. Diez, L. Wordeman, and J. Howard. 2003. The kinesin-related protein MCAK is a microtubule depolymerase that forms an ATP-hydrolyzing complex at microtubule ends. *Mol Cell.* 11:445-57.

Hussein, D., and S.S. Taylor. 2002. Farnesylation of Cenp-F is required for G2/M progression and degradation after mitosis. *J Cell Sci.* 115:3403-14.

Hwang, L.H., L.F. Lau, D.L. Smith, C.A. Mistrot, K.G. Hardwick, E.S. Hwang, A. Amon, and A.W. Murray. 1998. Budding yeast Cdc20: a target of the spindle checkpoint. *Science.* 279:1041-4.

Ikeno, M., B. Grimes, T. Okazaki, M. Nakano, K. Saitoh, H. Hoshino, N.I. McGill, H. Cooke, and H. Masumoto. 1998. Construction of YAC-based mammalian artificial chromosomes. *Nat Biotechnol.* 16:431-9.

Jabs, E.W., C.M. Tuck-Muller, R. Cusano, and J.B. Rattner. 1991. Studies of mitotic and centromeric abnormalities in Roberts syndrome: implications for a defect in the mitotic mechanism. *Chromosoma.* 100:251-61.

Janke, C., J. Ortiz, J. Lechner, A. Shevchenko, A. Shevchenko, M.M. Magiera, C. Schramm, and E. Schiebel. 2001. The budding yeast proteins Spc24p and Spc25p interact with Ndc80p and Nuf2p at the kinetochore and are important for kinetochore clustering and checkpoint control. *EMBO J.* 20:777-91.

Jessberger, R., B. Riwar, H. Baechtold, and A.T. Akhmedov. 1996. SMC proteins constitute two subunits of the mammalian recombination complex RC-1. *EMBO J.* 15:4061-8.

Jin, D.Y., F. Spencer, and K.T. Jeang. 1998. Human T cell leukemia virus type 1 oncoprotein Tax targets the human mitotic checkpoint protein MAD1. *Cell.* 93:81-91.

Kagansky, A., L. Freeman, D. Lukyanov, and A. Strunnikov. 2004. Histone tail-independent chromatin binding activity of recombinant cohesin holocomplex. *J Biol Chem.* 279:3382-8.

Kaitna, S., M. Mendoza, V. Jantsch-Plunger, and M. Glotzer. 2000. Incenp and an aurora-like kinase form a complex essential for chromosome segregation and efficient completion of cytokinesis. *Curr Biol.* 10:1172-81.

Kalitsis, P., K.J. Fowler, E. Earle, J. Hill, and K.H. Choo. 1998. Targeted disruption of mouse centromere protein C gene leads to mitotic disarray and early embryo death. *Proc Natl Acad Sci U S A.* 95:1136-41.

Kallio, M., J. Weinstein, J.R. Daum, D.J. Burke, and G.J. Gorbsky. 1998. Mammalian p55CDC mediates association of the spindle checkpoint protein Mad2 with the cyclosome/anaphase-promoting complex, and is involved in regulating anaphase onset and late mitotic events. *J Cell Biol.* 141:1393-406.

Kapoor, M., R. Montes de Oca Luna, G. Liu, G. Lozano, C. Cummings, M. Mancini, I. Ouspenski, B.R. Brinkley, and G.S. May. 1998. The cenpB gene is not essential in mice. *Chromosoma.* 107:570-6.

Kapoor, T.M., and D.A. Compton. 2002. Searching for the middle ground: mechanisms of chromosome alignment during mitosis. *J Cell Biol.* 157:551-6.

Kiefer, S.M., B.W. McDill, J. Yang, and M. Rauchman. 2002. Murine Sall1 represses transcription by recruiting a histone deacetylase complex. *J Biol Chem.* 277:14869-76.

Kim, S.H., D.P. Lin, S. Matsumoto, A. Kitazono, and T. Matsumoto. 1998. Fission yeast Slp1: an effector of the Mad2-dependent spindle checkpoint. *Science*. 279:1045-7.

Kimura, K., and T. Hirano. 1997. ATP-dependent positive supercoiling of DNA by 13S condensin: a biochemical implication for chromosome condensation. *Cell*. 90:625-34.

Kimura, K., V.V. Rybenkov, N.J. Crisona, T. Hirano, and N.R. Cozzarelli. 1999. 13S condensin actively reconfigures DNA by introducing global positive writhe: implications for chromosome condensation. *Cell*. 98:239-48.

King, R.W., J.M. Peters, S. Tugendreich, M. Rolfe, P. Hieter, and M.W. Kirschner. 1995. A 20S complex containing CDC27 and CDC16 catalyzes the mitosis-specific conjugation of ubiquitin to cyclin B. *Cell*. 81:279-88.

Kohlhase, J., A. Wischermann, H. Reichenbach, U. Froster, and W. Engel. 1998. Mutations in the SALL1 putative transcription factor gene cause Townes-Brocks syndrome. *Nat Genet*. 18:81-3.

Krantz, I.D., J. McCallum, C. DeScipio, M. Kaur, L.A. Gillis, D. Yaeger, L. Jukofsky, N. Wasserman, A. Bottani, C.A. Morris, M.J. Nowaczyk, H. Toriello, M.J. Bamshad, J.C. Carey, E. Rappaport, S. Kawauchi, A.D. Lander, A.L. Calof, H.H. Li, M. Devoto, and L.G. Jackson. 2004. Cornelia de Lange syndrome is caused by mutations in NIPBL, the human homolog of Drosophila melanogaster Nipped-B. *Nat Genet*. 36:631-5.

Kumada, K., T. Nakamura, K. Nagao, H. Funabiki, T. Nakagawa, and M. Yanagida. 1998. Cut1 is loaded onto the spindle by binding to Cut2 and promotes anaphase spindle movement upon Cut2 proteolysis. *Curr Biol*. 8:633-41.

Lampson, M.A., K. Renduchitala, A. Khodjakov, and T.M. Kapoor. 2004. Correcting improper chromosome-spindle attachments during cell division. *Nat Cell Biol*. 6:232-7.

Levesque, A.A., and D.A. Compton. 2001. The chromokinesin Kid is necessary for chromosome arm orientation and oscillation, but not congression, on mitotic spindles. *J Cell Biol*. 154:1135-46.

Li, Y., and R. Benezra. 1996. Identification of a human mitotic checkpoint gene: hsMAD2. *Science*. 274:246-8.

Liao, H., R.J. Winkfein, G. Mack, J.B. Rattner, and T.J. Yen. 1995. CENP-F is a protein of the nuclear matrix that assembles onto kinetochores at late G2 and is rapidly degraded after mitosis. *J Cell Biol*. 130:507-18.

Liu, S.T., J.C. Hittle, S.A. Jablonski, M.S. Campbell, K. Yoda, and T.J. Yen. 2003. Human CENP-I specifies localization of CENP-F, MAD1 and MAD2 to kinetochores and is essential for mitosis. *Nat Cell Biol*. 5:341-5.

Losada, A., M. Hirano, and T. Hirano. 1998. Identification of Xenopus SMC protein complexes required for sister chromatid cohesion. *Genes Dev*. 12:1986-97.

Losada, A., and T. Hirano. 2001. Intermolecular DNA interactions stimulated by the cohesin complex in vitro: implications for sister chromatid cohesion. *Curr Biol*. 11:268-72.

Maeshima, K., and U.K. Laemmli. 2003. A two-step scaffolding model for mitotic chromosome assembly. *Dev Cell*. 4:467-80.

Maney, T., A.W. Hunter, M. Wagenbach, and L. Wordeman. 1998. Mitotic centromere-associated kinesin is important for anaphase chromosome segregation. *J Cell Biol*. 142:787-801.

Martin-Lluesma, S., V.M. Stucke, and E.A. Nigg. 2002. Role of Hec1 in spindle checkpoint signaling and kinetochore recruitment of Mad1/Mad2. *Science*. 297:2267-70.

McCleland, M.L., R.D. Gardner, M.J. Kallio, J.R. Daum, G.J. Gorbsky, D.J. Burke, and P.T. Stukenberg. 2003. The highly conserved Ndc80 complex is required for kinetochore assembly, chromosome congression, and spindle checkpoint activity. *Genes Dev*. 17:101-14.

McCleland, M.L., M.J. Kallio, G.A. Barrett-Wilt, C.A. Kestner, J. Shabanowitz, D.F. Hunt, G.J. Gorbsky, and P.T. Stukenberg. 2004. The vertebrate Ndc80 complex contains Spc24 and Spc25 homologs, which are required to establish and maintain kinetochore-microtubule attachment. *Curr Biol*. 14:131-7.

Melby, T.E., C.N. Ciampaglio, G. Briscoe, and H.P. Erickson. 1998. The symmetrical structure of structural maintenance of chromosomes (SMC) and MukB proteins: long, antiparallel coiled coils, folded at a flexible hinge. *J Cell Biol.* 142:1595-604.

Meraldi. P., R. Honda, and E.A. Nigg. 2002. Aurora-A overexpression reveals tetraploidization as a major route to centrosome amplification in p53-/- cells. *EMBO J.* 21:483-92.

Michaelis, C., R. Ciosk, and K. Nasmyth. 1997. Cohesins: chromosomal proteins that prevent premature separation of sister chromatids. *Cell.* 91:35-45.

Michel, L.S., V. Liberal, A. Chatterjee, R. Kirchwegger, B. Pasche, W. Gerald, M. Dobles, P.K. Sorger, V.V. Murty, and R. Benezra. 2001. MAD2 haplo-insufficiency causes premature anaphase and chromosome instability in mammalian cells. *Nature.* 409:355-9.

Morishita, J., T. Matsusaka, G. Goshima, T. Nakamura, H. Tatebe, and M. Yanagida. 2001. Bir1/Cut17 moving from chromosome to spindle upon the loss of cohesion is required for condensation, spindle elongation and repair. *Genes Cells.* 6:743-63.

Murakami, S., T. Matsumoto, O. Niwa, and M. Yanagida. 1991. Structure of the fission yeast centromere cen3: direct analysis of the reiterated inverted region. *Chromosoma.* 101:214-21.

Murata-Hori, M., and Y.L. Wang. 2002. The kinase activity of aurora B is required for kinetochore-microtubule interactions during mitosis. *Curr Biol.* 12:894-9.

Murphy, T.D., and G.H. Karpen. 1995. Localization of centromere function in a Drosophila minichromosome. *Cell.* 82:599-609.

Murray, A.W., and J.W. Szostak. 1983. Construction of artificial chromosomes in yeast. *Nature.* 305:189-93.

Musio, A., T. Mariani, C. Montagna, D. Zambroni, C. Ascoli, T. Ried, and P. Vezzoni. 2004. Recapitulation of the Roberts syndrome cellular phenotype by inhibition of INCENP, ZWINT-1 and ZW10 genes. *Gene.* 331:33-40.

Nagao, K., Y. Adachi, and M. Yanagida. 2004. Separase-mediated cleavage of cohesin at interphase is required for DNA repair. *Nature.* 430:1044-8.

Nakaseko, Y., Adachi, Y., Funahashi, S., Niwa, O. and Yanagida, M. Chromosome Walking shows a highly repetitive sequence present in all the centromere regions of fission yeast. *EMBO J.* 5:1011-21.

Nakayama, J., J.C. Rice, B.D. Strahl, C.D. Allis, and S.I. Grewal. 2001. Role of histone H3 lysine 9 methylation in epigenetic control of heterochromatin assembly. *Science.* 292:110-3.

Netzer, C., L. Rieger, A. Brero, C.D. Zhang, M. Hinzke, J. Kohlhase, and S.K. Bohlander. 2001. SALL1, the gene mutated in Townes-Brocks syndrome, encodes a transcriptional repressor which interacts with TRF1/PIN2 and localizes to pericentromeric heterochromatin. *Hum Mol Genet.* 10:3017-24.

Nigg, E. A. 2001 Mitotic kinases as regulators of cell division and its checkpoints. *Nat Rev Mol Cell Biol.* 2:21-32.

Niwa, O., T. Matsumoto, Y. Chikashige, and M. Yanagida. 1989. Characterization of Schizosaccharomyces pombe minichromosome deletion derivatives and a functional allocation of their centromere. *EMBO J.* 8:3045-52.

Nonaka, N., T. Kitajima, S. Yokobayashi, G. Xiao, M. Yamamoto, S.I. Grewal, and Y. Watanabe. 2002. Recruitment of cohesin to heterochromatic regions by Swi6/HP1 in fission yeast. *Nat Cell Biol.* 4:89-93.

Obuse, C., O. Iwasaki, T. Kiyomitsu, G. Goshima, Y. Toyoda, and M. Yanagida. 2004. A conserved Mis12 centromere complex is linked to heterochromatic HP1 and outer kinetochore protein Zwint-1. *Nat Cell Biol.* 6:1135-41.

Ortiz, J., O. Stemmann, S. Rank, and J. Lechner. 1999. A putative protein complex consisting of Ctf19, Mcm21, and Okp1 represents a missing link in the budding yeast kinetochore. *Genes Dev.* 13:1140-55.

Perez-Castro, A.V., F.L. Shamanski, J.J. Meneses, T.L. Lovato, K.G. Vogel, R.K. Moyzis, and R. Pedersen. 1998. Centromeric protein B null mice are viable with no apparent abnormalities. *Dev Biol.* 201:135-43.

Peters, A.H., D. O'Carroll, H. Scherthan, K. Mechtler, S. Sauer, C. Schofer, K. Weipoltshammer, M. Pagani, M. Lachner, A. Kohlmaier, S. Opravil, M. Doyle, M. Sibilia, and T. Jenuwein. 2001. Loss of the Suv39h histone methyltransferases impairs mammalian heterochromatin and genome stability. *Cell*. 107:323-37.

Pfarr, C.M., M. Coue, P.M. Grissom, T.S. Hays, M.E. Porter, and J.R. McIntosh. 1990. Cytoplasmic dynein is localized to kinetochores during mitosis. *Nature*. 345:263-5.

Provost, P., R.A. Silverstein, D. Dishart, J. Walfridsson, I. Djupedal, B. Kniola, A. Wright, B. Samuelsson, O. Radmark, and K. Ekwall. 2002. Dicer is required for chromosome segregation and gene silencing in fission yeast cells. *Proc Natl Acad Sci U S A*. 99:16648-53.

Rattner, J.B., A. Rao, M.J. Fritzler, D.W. Valencia, and T.J. Yen. 1993. CENP-F is a.ca 400 kDa kinetochore protein that exhibits a cell-cycle dependent localization. *Cell Motil Cytoskeleton*. 26:214-26.

Rollins, R.A., M. Korom, N. Aulner, A. Martens, and D. Dorsett. 2004. Drosophila nipped-B protein supports sister chromatid cohesion and opposes the stromalin/Scc3 cohesion factor to facilitate long-range activation of the cut gene. *Mol Cell Biol*. 24:3100-11.

Saffery, R., D.V. Irvine, B. Griffiths, P. Kalitsis, L. Wordeman, and K.H. Choo. 2000. Human centromeres and neocentromeres show identical distribution patterns of >20 functionally important kinetochore-associated proteins. *Hum Mol Genet*. 9:175-85.

Saitoh, H., J. Tomkiel, C.A. Cooke, H. Ratrie, 3rd, M. Maurer, N.F. Rothfield, and W.C. Earnshaw. 1992. CENP-C, an autoantigen in scleroderma, is a component of the human inner kinetochore plate. *Cell*. 70:115-25.

Saitoh, S., K. Takahashi, and M. Yanagida. 1997. Mis6, a fission yeast inner centromere protein, acts during G1/S and forms specialized chromatin required for equal segregation. *Cell*. 90:131-43.

Saka, Y., T. Sutani, Y. Yamashita, S. Saitoh, M. Takeuchi, Y. Nakaseko, and M. Yanagida. 1994. Fission yeast cut3 and cut14, members of a ubiquitous protein family, are required for chromosome condensation and segregation in mitosis. *EMBO J*. 13:4938-52.

Sakai, A., K. Hizume, T. Sutani, K. Takeyasu, and M. Yanagida. 2003. Condensin but not cohesin SMC heterodimer induces DNA reannealing through protein-protein assembly. *EMBO J*. 22:2764-75.

Sampath, S.C., R. Ohi, O. Leismann, A. Salic, A. Pozniakovski, and H. Funabiki. 2004. The chromosomal passenger complex is required for chromatin-induced microtubule stabilization and spindle assembly. *Cell*. 118:187-202.

Schaar, B.T., G.K. Chan, P. Maddox, E.D. Salmon, and T.J. Yen. 1997. CENP-E function at kinetochores is essential for chromosome alignment. *J Cell Biol*. 139:1373-82.

Schmiesing, J.A., A.R. Ball, Jr., H.C. Gregson, J.M. Alderton, S. Zhou, and K. Yokomori. 1998. Identification of two distinct human SMC protein complexes involved in mitotic chromosome dynamics. *Proc Natl Acad Sci U S A*. 95:12906-11.

Schueler, M.G., A.W. Higgins, M.K. Rudd, K. Gustashaw, and H.F. Willard. 2001. Genomic and genetic definition of a functional human centromere. *Science*. 294:109-15.

Shackney, S.E., S.G. Singh, R. Yakulis, C.A. Smith, A.A. Pollice, S. Petruolo, A. Waggoner, and R.J. Hartsock. 1995. Aneuploidy in breast cancer: a fluorescence in situ hybridization study. *Cytometry*. 22:282-91.

Shackney, S.E., C.A. Smith, B.W. Miller, D.R. Burholt, K. Murtha, H.R. Giles, D.M. Ketterer, and A.A. Pollice. 1989. Model for the genetic evolution of human solid tumours. *Cancer Res*. 49:3344-54.

Sharp, J.A., A.A. Franco, M.A. Osley, and P.D. Kaufman. 2002. Chromatin assembly factor I and Hir proteins contribute to building functional kinetochores in S. cerevisiae. *Genes Dev*. 16:85-100.

Skibbens, R.V., V.P. Skeen, and E.D. Salmon. 1993. Directional instability of kinetochore motility during chromosome congression and segregation in mitotic newt lung cells: a push-pull mechanism. *J Cell Biol*. 122:859-75.

Steuer, E.R., L. Wordeman, T.A. Schroer, and M.P. Sheetz. 1990. Localization of cytoplasmic dynein to mitotic spindles and kinetochores. *Nature*. 345:266-8.

Stoler, S., K.C. Keith, K.E. Curnick, and M. Fitzgerald-Hayes. 1995. A mutation in CSE4, an essential gene encoding a novel chromatin-associated protein in yeast, causes chromosome nondisjunction and cell cycle arrest at mitosis. *Genes Dev.* 9:573-86.

Strunnikov, A.V., E. Hogan, and D. Koshland. 1995. SMC2, a Saccharomyces cerevisiae gene essential for chromosome segregation and condensation, defines a subgroup within the SMC family. *Genes Dev.* 9:587-99.

Strunnikov, A.V., V.L. Larionov, and D. Koshland. 1993. SMC1: an essential yeast gene encoding a putative head-rod-tail protein is required for nuclear division and defines a new ubiquitous protein family. *J Cell Biol.* 123:1635-48.

Sudakin, V., D. Ganoth, A. Dahan, H. Heller, J. Hershko, F.C. Luca, J.V. Ruderman, and A. Hershko. 1995. The cyclosome, a large complex containing cyclin-selective ubiquitin ligase activity, targets cyclins for destruction at the end of mitosis. *Mol Biol Cell.* 6:185-97.

Sun, X., J. Wahlstrom, and G. Karpen. 1997. Molecular structure of a functional Drosophila centromere. *Cell.* 91:1007-19.

Sutani, T., and M. Yanagida. 1997. DNA renaturation activity of the SMC complex implicated in chromosome condensation. *Nature.* 388:798-801.

Sutani, T., T. Yuasa, T. Tomonaga, N. Dohmae, K. Takio, and M. Yanagida. 1999. Fission yeast condensin complex: essential roles of non-SMC subunits for condensation and Cdc2 phosphorylation of Cut3/SMC4. *Genes Dev.* 13:2271-83.

Takahashi, K., E.S. Chen, and M. Yanagida. 2000. Requirement of Mis6 centromere connector for localizing a CENP-A-like protein in fission yeast. *Science.* 288:2215-9.

Takahashi, K., S. Murakami, Y. Chikashige, H. Funabiki, O. Niwa, and M. Yanagida. 1992. A low copy number central sequence with strict symmetry and unusual chromatin structure in fission yeast centromere. *Mol Biol Cell.* 3:819-35.

Tanaka, T.U., N. Rachidi, C. Janke, G. Pereira, M. Galova, E. Schiebel, M.J. Stark, and K. Nasmyth. 2002. Evidence that the Ipl1-Sli15 (Aurora kinase-INCENP) complex promotes chromosome bi-orientation by altering kinetochore-spindle pole connections. *Cell.* 108:317-29.

Terada, Y., M. Tatsuka, F. Suzuki, Y. Yasuda, S. Fujita, and M. Otsu. 1998. AIM-1: a mammalian midbody-associated protein required for cytokinesis. *EMBO J.* 17:667-76.

Tomonaga, T., K. Matsushita, S. Yamaguchi, T. Oohashi, H. Shimada, T. Ochiai, K. Yoda, and F. Nomura. 2003. Overexpression and mistargeting of centromere protein-A in human primary colorectal cancer. *Cancer Res.* 63:3511-6.

Tomonaga, T., K. Nagao, Y. Kawasaki, K. Furuya, A. Murakami, J. Morishita, T. Yuasa, T. Sutani, S.E. Kearsey, F. Uhlmann, K. Nasmyth, and M. Yanagida. 2000. Characterization of fission yeast cohesin: essential anaphase proteolysis of Rad21 phosphorylated in the S phase. *Genes Dev.* 14:2757-70.

Tonkin, E.T., T.J. Wang, S. Lisgo, M.J. Bamshad, and T. Strachan. 2004. NIPBL, encoding a homolog of fungal Scc2-type sister chromatid cohesion proteins and fly Nipped-B, is mutated in Cornelia de Lange syndrome. *Nat Genet.* 36:636-41.

Toth, A., R. Ciosk, F. Uhlmann, M. Galova, A. Schleiffer, and K. Nasmyth. 1999. Yeast cohesin complex requires a conserved protein, Eco1p(Ctf7), to establish cohesion between sister chromatids during DNA replication. *Genes Dev.* 13:320-33.

Uhlmann, F. 2004. The mechanism of sister chromatid cohesion. *Exp Cell Res.* 296:80-5.

Uhlmann, F., F. Lottspeich, and K. Nasmyth. 1999. Sister-chromatid separation at anaphase onset is promoted by cleavage of the cohesin subunit Scc1. *Nature.* 400:37-42.

Uhlmann, F., and K. Nasmyth. 1998. Cohesion between sister chromatids must be established during DNA replication. *Curr Biol.* 8:1095-101.

Uhlmann, F., D. Wernic, M.A. Poupart, E.V. Koonin, and K. Nasmyth. 2000. Cleavage of cohesin by the CD clan protease separin triggers anaphase in yeast. *Cell.* 103:375-86.

Van Den Berg, D.J., and U. Francke. 1993. Roberts syndrome: a review of 100 cases and a new rating system for severity. *Am J Med Genet.* 47:1104-23.

Vagnarelli, P., and W.C. Earnshaw. 2004. Chromosomal passengers: the four-dimensional regulation of mitotic events. *Chromosoma* 113:211-22.

Visintin, R., S. Prinz, and A. Amon. 1997. CDC20 and CDH1: a family of substrate-specific activators of APC-dependent proteolysis. *Science*. 278:460-3.

Volpe, T., V. Schramke, G.L. Hamilton, S.A. White, G. Teng, R.A. Martienssen, and R.C. Allshire. 2003. RNA interference is required for normal centromere function in fission yeast. *Chromosome Res*. 11:137-46.

Volpe, T.A., C. Kidner, I.M. Hall, G. Teng, S.I. Grewal, and R.A. Martienssen. 2002. Regulation of heterochromatic silencing and histone H3 lysine-9 methylation by RNAi. *Science*. 297:1833-7.

Voullaire, L.E., H.R. Slater, V. Petrovic, and K.H. Choo. 1993. A functional marker centromere with no detectable alpha-satellite, satellite III, or CENP-B protein: activation of a l atent centromere? *Am J Hum Genet*. 52:1153-63.

Waizenegger, I.C., S. Hauf, A. Meinke, and J.M. Peters. 2000. Two distinct pathways remove mammalian cohesin from chromosome arms in prophase and from centromeres in anaphase. *Cell*. 103:399-410.

Walczak, C.E., E.C. Gan, A. Desai, T.J. Mitchison, and S.L. Kline-Smith. 2002. The microtubule-destabilizing kinesin XKCM1 is required for chromosome positioning during spindle assembly. *Curr Biol*. 12:1885-9.

Weitzer, S., C. Lehane, and F. Uhlmann. 2003. A model for ATP hydrolysis-dependent binding of cohesin to DNA. *Curr Biol*. 13:1930-40.

Wordeman, L., and T.J. Mitchison. 1995. Identification and partial characterization of mitotic centromere-associated kinesin, a kinesin-related protein that associates with centromeres during mitosis. *J Cell Biol*. 128:95-104.

Yen, T.J., G. Li, B.T. Schaar, I. Szilak, and D.W. Cleveland. 1992. CENP-E is a putative kinetochore motor that accumulates just before mitosis. *Nature*. 359:536-9.

Zou, H., T.J. McGarry, T. Bernal, and M.W. Kirschner. 1999. Identification of a vertebrate sister-chromatid separation inhibitor involved in transformation and tumorigenesis. *Science*. 285:418-22.

Chapter 3.4

TELOMERE STRUCTURAL DYNAMICS IN GENOME INTEGRITY CONTROL AND CARCINOGENESIS

Roger A. Greenberg[1] and K. Lenhard Rudolph[2]
Department of Cancer Biology, Dana Farber Cancer Institute, Boston, Massasuchsetts, USA[1], Department of Gastroenterology, Hepatology, and Endocrinology, Medical School Hannover, Germany[2]

1. INTRODUCTION

Telomeres cap the ends of linear eukaryotic chromosomes. In addition to telomere length, proper maintenance of telomere structure by telomere binding proteins is necessary for telomere capping function. Telomere binding proteins include both single and double strand DNA-binding proteins and interact with a variety of DNA damage recognition, signalling and repair proteins. Due to the end replication problem of DNA-polymerase telomeres loose 50-100 base pairs during each cell division. When telomeres reach a critically short length they lose capping function possibly because the tertiary telomere structure can no longer be maintained. Critically short, dysfunctional telomeres generate DNA damage like signals that induce a permanent proliferation arrest known as replicative senescence. Senescence limits proliferation of primary cells to a finite number of divisions *in vitro* and the regenerative capacity of tissues and organs *in vivo* during aging and chronic disease. Senescence has been proposed to function as a tumour suppressor mechanism by imposing a barrier to cellular immortalisation. When components of the DNA damage pathway are deleted or inhibited, cells can bypass the senescence checkpoint and continue to proliferate, eventually reaching a stage of extreme telomere dysfunction - crisis. This "crisis" stage appears to induce carcinogenesis since it fosters chromosomal instability – the hallmark of human cancer in the elderly. In addition, there is growing evidence that telomere binding proteins have a direct role in DNA damage recognition and repair. The following chapter summarises our current knowledge of telomere structure, senescence, and crisis with a

E.A. Nigg (ed.), Genome Instability in Cancer Development, 311-341.

special emphasis on the role of telomere dysfunction in accelerating chromosomal instability and cancer.

2. TELOMERE STRUCTURE

Telomeres form the ends of linear chromosomes in a variety of different species. In eukaryotes telomeres are composed of simple tandem repeats – $(TTAGGG)_n$ in vertebrates (Blackburn, 1991). The length of telomeres varies substantially between different species, e.g. ~1kb in yeast, ~10kb in human cells, ~50 kb in cells from laboratory mice. The main function of telomeres is to cap chromosomal ends thus preventing chromosomal fusions and induction of DNA damage responses that would lead to chromosomal instability, cell cycle arrest or apoptosis (Blackburn, 2001). It appears that for proper telomere function a minimum length of telomere repeats is required, although it is not yet defined what this minimum length is. However, studies in telomerase deficient cell lines (Allsopp and Harley, 1995), knockout mice (Blasco *et al.*, 1997), and human disease (Dokal, 2001) have clearly demonstrated a correlation between telomere shortening and loss of telomere capping function. In addition to the length of telomere repeats, telomere structure appears to be essential for its function. Telomeres are highly heterochromatic with pronounced DNA folding. Telomeres are DNA-double strand repetitive elements but terminate with a single strand, G-rich 3' overhang (Henderson and Blackburn, 1989; Makarov *et al.*, 1997; Wright *et al.*, 1997). Telomeres form tertiary structures such as G-quadruplex (Rezler *et al.*, 2003) and loop-structures (Griffith *et al.*, 1999). The "T-loop model" predicts that the G-strand telomeric overhang folds back and invades the more proximal double stranded telomere sequence (Stansel *et al.*, 2001).

To maintain the higher order tertiary telomere structure specific telomere binding proteins are required. According to the telomere nucleotide structure these proteins are grouped into single-strand and double-strand DNA binding proteins. Single strand telomere binding proteins bind specifically to the telomere 3'single stranded region thus forming a protein-nucleotide complex at chromosomal termini (Table 1). These proteins include *Cdc13* in budding yeast (Hughes *et al.*, 2000; Nugent *et al.*, 1996), *hypotrichous proteins* in ciliates (Hicke *et al.*, 1990), and POT1 in *Schizosaccharomyces pombe* and human cells (Baumann and Cech, 2001). POT1 is similar to Cdc13 since both proteins contain an OB-fold binding single stranded DNA and appear to have a similar function (Colgin *et al.*, 2003). Experimental mutations of these single strand binding proteins induce rapid degradation of the telomeric C-strand by exonuclease indicating that they are essential for telomere integrity (Baumann and Cech, 2001). In addition, the single strand telomere binding proteins play a role in telomere length regulation by controlling the accessibility of the telomeres

Table 1. Human telomere binding proteins

A) single strand telomere binding proteins		
POT1	Similar to cdc13 in yeast; protects the G-strand from degradation by exonuclease 1, recruits telomerase to the telomere, inhibits telomere elongation of long telomeres	Colgin *et al.*, 2003; Ye *et al.*, 2004
Telomerase: TERC TERT	Necessary for telomere elongation, telomere capping function, not necessary for cell for organsimal viability over several generations.	Bryan *et al.*, 1998; Collins and Gandhi, 1998; Greenberg *et al.*, 1998; Lingner *et al.*, 1997
EST1a-c	Three homologs of yeast EST1. EST1a is necessary for unfolding of telomeres and telomere elongation by telomerase	Reichenbach *et al.*, 2003
B) double strand telomere binding proteins		
PIP1	Mediates binding of POT1 to the TRF1/TIN2 complex, negative regulator of telomere length	Ye *et al.*, 2004
TRF1	Negative regulator of telomere length, telomere independent function necessary for organismal viability	Karlseder *et al.*, 2002; Karlseder *et al.*, 2003
Tankyrase1, Tankyrase2	poly (ADP-ribose) polymerases that inhibit TRF1 binding to the telomere, positive regulators of telomere length, involved in sister chromatid separation during anaphase	Smith and de Lange, 2000; Smith *et al.*, 1998; Dyneck *et al.*, 2004
TIN2	Protects TRF1 from Tankyrase action, negative regulator of telomere length	Smogorzewska and De Lange, 2004
PinX2	Negative regulator of telomerase	Zhou and Lu, 2001
TRF2	Stabilises the T-Loop, necessary for telomere function, inhibits activation of ATM	Karlseder *et al.*, 2004
RAP1	Binds to TRF2, negative regulator of telomere length	Li and de Lange, 2003; Li *et al.*, 2000
RIF1	Binds to dysfunctional telomeres and activates ATM and p53bp	Silverman *et al.*, 2004

for telomerase – the enzyme that synthesises telomeres *de novo* (see below and Colgin *et al.*, 2003). Telomerase, the protein-RNA complex itself is binding at the telomere overhang in order to synthesise telomere sequence. The telomerase reverse transcriptase (TERT in human cells, Est2p in *Saccharomyces cerevisiae*) is the catalytic subunit of the holoenzyme complex (Bryan *et al.*, 1998; Collins and Gandhi, 1998; Greenberg *et al.*, 1998; Lingner *et al.*, 1997). The telomerase RNA component (TERC in human cell, Tlc1 in *S. cerevisiae*) serves as a template for telomere sequence synthesis (Blasco *et al.*, 1995; Shippen-Lentz and Blackburn, 1990). Deletion of any of these two essential components leads to complete disruption of telomerase activity and telomere shortening in cells (Niida *et al.*, 1998) and in different organs of an organism (Blasco *et al.*, 1997; Yuan *et al.*, 1999). In addition to the two essential components of telomerase (TERT & TERC), a number of other proteins associate with telomerase and telomere binding proteins to control recruitment and activity of telomerase at the telomere (Smogorzewska and De Lange, 2004). For instance, telomerase interacting proteins in *S. cerevisiae* include Est1p and Est3p, which are required for telomere maintenance by telomerase (Blackburn *et al.*, 2000). In humans three Est1 homologs have been identified and Est1a was shown to be associated with telomerase and to uncap telomeres possibly to facilitate telomere elongation by telomerase (Reichenbach *et al.*, 2003). In addition to the function of telomerase in telomere length maintenance, there is also evidence for a direct telomere capping function of telomerase, since over expression of TERT prolongs the lifespan of human cells without net telomere elongation (Zhu *et al.*, 1999).The importance of single strand telomere binding proteins for telomere integrity was demonstrated by experiments using a mutant form of the telomerase RNA component. Mutation in the template region of TERC (see below) resulted in the incorporation of erratic bases in telomere repeats, thus interfering with the binding of telomere binding proteins to the single strand overhang, and consequently provoking rapid induction of telomere dysfunction and cell death (Kirk *et al.*, 1997).

Along with the telomere single strand binding proteins a growing number of telomere double-strand binding proteins have been identified (Table 1), which function to ensure telomere structure integrity and telomere length homeostasis. In mammalian cells a double strand telomere binding protein necessary for telomere structure is TRF2; inhibition of this protein results in rapid induction of telomere dysfunction, G-strand degradation, and chromosomal fusions (van Steensel *et al.*, 1998). Conversely, wildtype TRF2 promotes T-loop formation *in vitro* (Stansel *et al.*, 2001) and over-expression of wildtype TRF2 has been shown to stabilise telomere structure thus allowing cell growth beyond the telomere length limit that halts cell proliferation at the senescence stage (Karlseder *et al.*, 2002). A variety of double strand telomere binding proteins have been implicated in telomere length control in different species: In *S. cerevisiae* these proteins include RAP1 and its interaction partners Rif1 and Rif2, which are negative regulators of telomere length (Conrad *et al.*, 1990; Hardy *et al.*, 1992; Kyrion *et al.*, 1992; Lustig *et al.*, 1990). The length of telomere repeats determines the number of Rap1 proteins bound to the telomere and this counting mechanism inhibits further elongation of long telomeres, thus setting a cellular telomere length equilibrium (Grossi *et al.*, 2001; Marcand *et al.*, 1999; Marcand *et al.*, 1997; Ray and Runge, 1998). In mammalian cells inhibition of telomere elongation is exerted by TRF1 (Karlseder *et al.*, 2002) possibly by limiting the accessibility of telomeres for telomerase binding (Ancelin *et al.*, 2002; van Steensel and de Lange, 1997). Similar to Rap1 in yeast, a telomere length counting mechanism has been proposed for TRF1 (Smogorzewska and De Lange, 2004). Various proteins interact with TRF1 thereby modifying the binding of TRF1 to the telomere and telomere length control. The poly (ADP-ribose) polymerases Tankyrase-1 and Tankyrase-2 bind and ADP-ribosylate TRF1 leading to diminished TRF1 binding at the telomere, telomerase de-repression and telomere elongation (Smith and de Lange, 2000; Smith *et al.*, 1998). Another TRF1 binding protein is PinX2, which appears to inhibit telomerase (Zhou and Lu, 2001). TRF1 also associates with TIN2, which affects telomere length (Kim *et al.*, 1999) by protecting TRF1 from modification by Tankyrase (Smogorzewska and De Lange, 2004). There are significant differences in telomere binding proteins comparing yeast and mammalian cells, complicating the research aiming to understand their role in telomere structure and length control. A detailed understanding of the functions of these proteins might identify new targets for cancer therapies and regenerative medicine.

To summarise, telomeres represent a special form of chromatin that differs from the rest of the chromosome in several aspects; (1) telomere specific binding proteins (Table 1) instead of octameric histone containing nucleosomes, (2) a 3' G-rich single stranded overhang that is generated by telomerase during S-phase (Makarov *et al.*, 1997; Masutomi *et al.*, 2003), (3) formation of G-quadruplexes (Rezler *et al.*, 2003), and (4) invasion of single stranded overhang into duplex telomeric DNA to create a loop structure known as a T-loop (Griffith *et al.*, 1999). The unique structure of the telomere is thought to contribute to its ability to evade detection by cellular DNA damage response proteins and disruption of telomere structure elicits DNA damage responses and arrests cell proliferation.

3. TELOMERE MAINTENANCE AND THE DNA DAMAGE RESPONSE

The cellular DNA damage response (DDR) machinery efficiently recognises and responds to the presence of single DNA double strand breaks (DSB) (Bakkenist and Kastan, 2003). The eukaryotic genome, however, presents this same set of signalling pathways with the intuitively complex task of distinguishing a single DSB from the presence of multiple chromosomal termini known as telomeres (several thousand in the case of *Tetrahymena thermophila*). The molecular responses needed to distinguish telomeres from DNA damage induced DSBs have yet to be defined. Observations from lower eukaryotic and mammalian systems indicate that there occurs considerable interaction between DDR proteins and telomeric DNA. Moreover dynamic association exists even in the absence of exogenous damaging agents or telomere dysfunction.

Among the earliest detectable responses to DSBs is activation of the checkpoint kinase ATM (*Ataxia-telangiectasia Mutated*). ATM is a large protein related to the phosphoinositide 3-kinases (PIKKs) and is mutated in a human syndrome known as Ataxia-telangiectasia, that is characterised by vascular and CNS abnormalities as well as a predisposition to malignancy. ATM phosphorylates a myriad of DNA repair proteins in response to DSBs such as BRCA1, NBS1, Rad17, SMC1 and many others (Shiloh, 2000; Shiloh, 2003). In addition, phosphorylation of ATM substrates is essential for execution of cell cycle checkpoints in response to DNA damage including targets such as p53 and Chk2. Almost the entire repertoire of

ATM protein in a cell is activated via intramolecular phosphorylation within minutes following a dose of IR (0.1-0.5 Gy) that is predicted to generate only 4-18 DSBs per human genome (Bakkenist and Kastan, 2003). It appears that ATM activation occurs in part via sensing of an undefined change in global chromatin structure that occurs after DSBs occur. A critical substrate of ATM is the Rad50/NBS1/Mre11 complex (R/M/N) (Wu *et al.*, 2000), which also plays a role in activation of ATM. Phosphorylation of the proteins of the R/M/N complex is essential for the S-phase and G2/M checkpoints as well as the repair of DSBs and sensitivity to ionising radiation. Human syndromes are known to occur as a result of inherited biallelic mutation in either NBS1 or Mre11. These syndromes, Nijmegen Breakage Syndrome (NBS1) and Ataxia-like-disorder (ATL-D), respectively, bear a strong similarity to patients carrying homozygous mutations in the ATM gene in terms of sensitivity to ionising radiation, cytogenetic abnormalities and cancer predisposition (Petrini, 2000). Though both NBS1 and Mre11 are substrates of ATM, they also play a key role in activation of ATM kinase activity (Carson *et al.*, 2003) and this is thought to occur through direct physical interaction (Lee and Paull, 2004). Thus ATM and the R/M/N complex represent intertwined systems to sense and transduce growth inhibitory signals following DNA damage.

In addition to ATM and the R/M/N complex, a second PIKK, ATR, is involved in the early recognition of DNA lesions. ATR is a large kinase related to ATM that phosphorylates many of the known ATM substrates at the ATM/ATR consensus SQ site. While substrates are overlapping for ATM and ATR, the stimulatory signals appear to be somewhat distinct. ATM is activated primarily by DSBs and consistent with this observation, AT cells are hypersensitive to ionising radiation (IR), but not UV, alkylating agents or inhibitors of DNA replication. Conversely, ATR phosphorylates its substrates in response to UV rays induced damage and replication stress. This suggests that distinct DNA repair intermediates may target either ATM or ATR for activation. ATR along with its binding partner ATRIP are recruited to single stranded DNA following DNA damage (Zou and Elledge, 2003). A model has been proposed that DNA is resected to long single stranded intermediates following DNA damage and that this forms a stimulus for recruitment of ATR and other damage response proteins (Zou and Elledge, 2003). Phosphorylation of ATR targets such as BRCA1, Rad17 etc. occur at the site of DNA damage, allowing propagation of the damage response.

At first glance, telomeric DNA would appear to be an ideal substrate to initiate a DDR. As mentioned previously, single stranded DNA is a potent inducer of the DNA damage response and the presence of DNA termini containing long single stranded 3'tails mimics 5'resected DNA that occurs during DSBR. *In vitro* assays revealed this to be the preferred substrate of the early DSB sensors Replication Protein A (RPA) and Rad17 (Ellison and Stillman, 2003; Zou *et al.*, 2002). In order to evade these responses telomeres have developed an elaborate system of telomere-specific binding proteins (see above, Table 1). Consistent with this is the observation that disruption of telomere binding by expression of dominant negative forms of TRF2 or Pot-1 results in a recruitment of DNA repair proteins to the site of the dysfunctional telomere (Takai *et al.*, 2003). Moreover, telomere erosion that occurs during normal replicative limits induces all of the hallmarks of a DNA damage response including phosphorylation of ATM and ATR substrates and activation of the downstream checkpoint kinases Chk1 and Chk2 (d'Adda di Fagagna *et al.*, 2003). Mutation of ATM also attenuates apoptotic responses to telomere dysfunction caused by dominant negative TRF2 expression (Karlseder *et al.*, 1999). Thus disruption of native telomere structure triggers an active DNA damage response that is largely indistinguishable from the response to IR induced DSBs (Reaper *et al.*, 2004). A reasonable model would be that telomere-binding proteins protect telomeric DNA by excluding DDR proteins from access to telomeric DNA. In line with this hypothesis, the telomere binding protein TRF2 inhibits activation of ATM by directly interacting with this PIKK (Karlseder *et al.*, 2004). Additional evidence for a crosstalk of telomere binding and DDR proteins comes from the observation that the human ortholog of a telomere binding protein in yeast, RIF1, does not interact with functional telomeres but localises to dysfunctional telomeres and plays a role in the S-phase checkpoint activated by ATM and p53BP1 in response to DNA damage (Silverman *et al.*, 2004). Moreover, it has recently been shown that the telomere binding protein TRF2 transiently localizes to laser-induced double-strand breaks within seconds after damage induction – faster than ATM (Bradshaw *et al.* 2005). Transient localization of TRF2 to double strand break possibly inhibits premature DNA repair necessary for correct assembly of the multimolecular DDR complex (Wright and Shay, 2005). Together, the current data indicate that the interaction between telomere binding proteins and DDR proteins is necessary for chromosome capping and for control of DDR following DNA damage.

There is also evidence that many DNA repair proteins that interact with telomeres are required to properly maintain telomere length (Table 2). Components of the DNA damage pathway needed for telomere maintenance include the Rad50/Mre11/Xrs2 (yeast)/Nbs (human) proteins (Boulton and Jackson, 1998; Gallego and White, 2001; Nugent *et al.*, 1998; Ranganathan *et al.*, 2001; Ritchie and Petes, 2000), which are involved in activating ATM in response to DNA damage (see above). Similarly RPA has been implicated in telomere length maintenance (Mallory *et al.*, 2003; Smith *et al.*, 2000). RPA facilitates recruitment and activation of ATR in response to DNA damage (see above). At the telomere RPA provides EST1p access to telomere sequences thus facilitating telomere lengthening (Schramke *et al.*, 2004). Both PIKKs, ATM/ATR, in human and their corresponding homologs Tel1/Mec1 in *S. cerevisiae* and Tel1/Rad3 in *S. pombe* appear to have a role in telomere maintenance. In humans and *S. cerevisiae,* mutation or deletion of ATM/Tel1 leads to telomere shortening (Greenwell *et al.*, 1995; Hande *et al.*, 2001; Lustig and Petes, 1986). Similarly, in *S. pombe* mutation of Rad3 was associated with telomere shortening (Matsuura *et al.*, 1999; Naito *et al.*, 1998). Whereas inactivation of one of the two PIKKs leads to moderate telomere shortening, inactivation of both PIKKs in double mutant yeast strains induced massive loss of telomere sequences and senescence, indicating that there is some potential of compensation between

Table 2. DDR proteins with reported functions at the telomere

Protein	Reported function at the Telomere	Reference
ATM/ATR	Checkpoint signalling kinases. Associate with telomeres in a cell cycle dependent manner. ATM is required for activation of an apoptotic pathway downstream of TRF2 dysfunction. Telomeres are completely eliminated in yeast deleted for ATM and ATR homologues.	Bi *et al.*, 2004; Ciapponi *et al.*, 2004; Silva *et al.*, 2004; Takata *et al.*, 2004; Greenwell *et al.*, 1995; Hande *et al.*, 2001; Lustig and Petes, 1986; Matsuura *et al.*, 1999; Naito *et al.*, 1998
Ku	Forms a heterodimer with Ku70/Ku80 subunits and is required for NHEJ. Yeast display telomere defects that appear to be independent of the Ku DNA repair function	Gravel *et al.* 1998; d'Adda di Fagagna *et al.*, 2001; Samper *et al.*, 2000; Bertuch *et al.*, 2003
MSH2	Necessary for the recognition and repair of base mismatches in DNA. Deletion in yeast allows for recombination-based maintenance of chromosomal termini in the absence of telomerase.	Rizki and Lundblad 2001
Pif1	DNA helicase reported to inhibit telomerase activity at the telomere and at DSBs. Telomere healing of terminal DSBs in yeast is dramatically increased after Pif1 mutation.	Bessler *et al.*, 2001; Myung *et al.*, 2001
Rad50/Mre 11/NBS1	This checkpoint and repair complex associates with the telomere through TRF2. Cells containing biallelic mutations in NBS1 display reduced telomere length and cytogenetic evidence of telomere dysfunction. Mre11 mutants in D. melanogaster exhibit telomere-telomere fusions in approximately 30% of metaphase chromosomes.	Boulton and Jackson, 1998; Gallego and White, 2001; Nugent *et al.*, 1998; Ranganathan *et al.*, 2001; Ritchie and Petes, 2000
RPA1	Composed of 2 subunits, this single stranded DNA binding protein is required for DNA replication and telomere length maintenance.	Schramke, *et al.* 2004
Rad51D	Rad51 homolog involved in the strand invasion reaction during HR. Reported to be important for T-loop formation.	Tarsounas, *et al.* 2004
Rad52	Required for HR in yeast and the ALT pathway of telomere maintenance in the absence of telomerase.	McEachern and Blackburn, 1996

the two PIKK-pathways (Naito *et al.*, 1998; Nakamura *et al.*, 2002; Ritchie *et al.*, 1999). Both PIKKs show an alternating cell cycle dependent association with telomeres in *S. cerevisiae* (Takata *et al.*, 2004). Mutation of the ATM in humans is associated with premature aging and increased cancer development. In mice ATM and telomere shortening cooperate to induce premature aging (Wong *et al.*, 2003), whereas ATR deletion is cell lethal (Brown and Baltimore, 2000). Other upstream factors of the DDR-pathway have been implicated in telomere length homeostasis and telomere capping function, including members of the 9-1-1 complex and the RF-C-like checkpoint (Ahmed and Hodgkin, 2000; Corda *et al.*, 1999; Grandin *et al.*, 2001; Kanellis *et al.*, 2003; Longhese *et al.*, 2000; Mieczkowski *et al.*, 2003; Nakamura *et al.*, 2002; Smolikov *et al.*, 2004).

In contrast to the upstream members of the DDR pathways, the downstream components of these pathways do not have a major effect on telomere length control in yeast and mammalian cells (Chin *et al.*, 1999; Longhese *et al.*, 2000; Mallory *et al.*, 2003). The exact function of the DDR pathway in telomere maintenance has yet to be determined. A possible model indicates that the DDR components are needed to detect exposed telomeres and to recruit and activate telomerase at the exposed telomere (Blackburn, 2001; Smogorzewska and De Lange, 2004). Possible targets of the DDR pathway in telomere maintenance concern the loading of Cdc13/Pot1 on telomere overhangs (Diede and Gottschling, 2001) and the inhibition of telomerase repressors like Rap1, Rif1, and Rif2 (Craven and Petes, 1999; Ray and Runge, 1999). There is experimental evidence that components of the DDR pathway directly target telomerase activity (Kharbanda *et al.*, 2000). Telomerase activation and distinct changes in telomerase nuclear localisation have been detected in response to irradiation (Hande *et al.*, 1997; Hande *et al.*, 1998). Evaluating the role of DNA damage proteins at functional and dysfunctional telomeres and its interaction with telomere binding proteins appears as one of the important areas to further understand the molecular mechanisms underlying senescence signalling.

In addition to components of the DNA damage recognition and signalling pathway, components of the DNA repair machinery are associated with telomeres and have a role in telomere homeostasis (Table 2). In yeast and mammalian cells the non-homologous end-joining (NHEJ) protein Ku is found at telomeres (d'Adda di Fagagna *et al.*, 2001; Gravel *et al.*, 1998; Hsu *et al.*, 2000; Nakamura *et al.*, 2002). Ku also interacts with the RNA component of telomerase in *S. cerevisiae* (Peterson *et al.*, 2001; Stellwagen *et al.*, 2003) and the telomere binding proteins TRF1 and TRF2 in mammals (Hsu *et al.*, 2000; Peterson *et al.*, 2001). Together these reports suggest that Ku assists telomerase to bind at the telomere. In agreement with this hypothesis Ku mutants/deletions result in telomere shortening in yeast cells (Baumann and Cech, 2000; Boulton and Jackson, 1996; Boulton and Jackson, 1998; Porter *et al.*, 1996; Stellwagen *et al.*, 2003) inducing a short telomere-length equilibrium (Baumann and Cech, 2000; Boulton and Jackson, 1998; Nugent *et al.*, 1998). Similarly inactivation of a single Ku allele in human cells induces telomere shortening (Myung *et al.*, 2004), however, deletion of both Ku alleles is cell lethal (Li *et al.*, 2002). In contrast, deletion of Ku is not lethal in mice and there are conflicting reports on the impact of Ku on telomere length in $Ku^{-/-}$ mice (d'Adda di Fagagna *et al.*, 2001; Samper *et al.*, 2000). Both studies found that Ku deletion induced telomere dysfunction and chromosomal fusions. In yeast cells Ku has been implicated in maintenance of the telomeric G-strand overhang (Gravel *et al.*, 1998), however, Ku deletion did not induce changes in G-strand stability in knockout mice (d'Adda di Fagagna *et al.*, 2001; Samper *et al.*, 2000). Experiments with a variety of different Ku mutations in *S. cerevisiae* have shown that Ku's function at the telomere is separate from its function at the site of DNA damage and that Ku performs distinct activities at subtelomeric chromatin *versus* the end of the chromosome (Bertuch *et al.*, 2003). According to these studies, it appears that Ku participates in telomerase-mediated G-strand synthesis, thereby contributing to telomere length regulation, and it separately protects against resection of the C-strand, thereby contributing to the protection at chromosome termini. In addition to its role in telomere homeostasis and function, Ku is necessary for transcriptional silencing at telomeres in yeast possibly depending on its interaction with Sir-proteins (Boulton and Jackson, 1998; Tsukamoto *et al.*, 1997). Beside Ku, the Ku interacting protein, DNA-PK interacts with human telomeres (d'Adda di Fagagna *et al.*, 2001) and appears to be important for telomere function and length homeostasis (Bailey *et al.*, 2004; Bailey *et al.*, 2001; Espejel *et al.*, 2002; Gilley *et al.*, 2001)

The role of proteins of the homologous recombination (HR) DNA repair pathway in telomere structure and length control is still under debate. Mutations of Rad51 and Rad52 appears to have an effect on senescence and the telomeric G-strand (Le *et al.*, 1999; Lundblad and Blackburn, 1993; Wei *et al.*, 2002). The human Rad51 related protein Rad51D localises at the telomere and its inhibition results in cell death (Tarsounas *et al.*, 2004). Moreover, Rad54 deletion induces telomere shortening in mice (Jaco *et al.*, 2003) and PARP-1 (D'Amours *et al.*, 1999; Tong *et al.*, 2001) and XPF/XRCC1 (Zhu *et al.*, 2003), which are involved in nucleotide excision repair, appear to be involved in telomere structure maintenance

The mechanistic basis for telomere maintenance by DNA repair proteins is unclear. It has been proposed that ATM, ATR and the R/M/N facilitate access of the telomerase holoenzyme (hTERT and hTERC for human telomerase) to chromosomal termini, allowing efficient telomere synthesis (see above). However, recent findings suggest a more complex relationship. For example, *Drosophila melanogaster* lacks the telomerase enzyme and is thought to maintain telomeres via recombination mechanisms. *Drosophila* atm and mre11 mutants display severe defects in telomere integrity with nearly a third of chromosomal termini engaged in telomere-telomere fusions (Bi *et al.*, 2004; Ciapponi *et al.*, 2004; Silva *et al.*, 2004). These fusion events were partially reduced by mutation of NHEJ proteins suggesting that both NHEJ and HR mechanisms of repair contribute to telomere fusion reactions. Together, these findings provide compelling arguments that DNA repair proteins also play a role in telomere maintenance independent of telomere synthesis by telomerase. Rather a role for ATM and R/M/N in telomere maintenance by inhibition of illegitimate recombination and end-joining events has been proposed (Bi *et al.*, 2004; Ciapponi *et al.*, 2004; Silva *et al.*, 2004). Furthermore, expression of ectopic NBS1 containing serine to alanine mutations at ATM phosphorylation sites resulted in increased incidence of complete telomere loss, consistent with a role for these proteins in preventing recombination of telomeric elements as opposed to TERT accessibility (Bai and Murnane, 2003)

Telomeres are highly repetitive sequences in the genome that are ideal substrates for recombination with telomeric sequences at non-homologous chromosomes. T-loop formation illustrates that intra-molecular strand invasion into homologous sequences on the same telomere can readily occur. A potential mechanism for the catastrophic telomere fusions that occur in the absence of TRF2 is that telomeric single stranded overhangs become available to invade homologous telomere sequences on nonhomologous chromosomes. TRF2 may be necessary at the telomere to prevent illegitimate strand invasion reactions from occurring between interchromosomal telomere elements. Tight regulation by DNA repair proteins is enacted to prevent inappropriate recombination between homologous sequences throughout the genome. In support of this, homologous recombination between sister chromatids is favoured by several orders of magnitude over recombination between homologous, non-sister chromosomes (Jasin, 2000). Furthermore, illegitimate recombination is a common cytogenetic abnormality in genetic instability syndromes stemming from mutation in BRCA1, BRCA2, FANCD2, Mre11, NBS1 and Blooms syndrome genes (Venkitaraman, 2002). Although not rigorously proven, it appears that repair proteins inhibit telomere-telomere reactions and this may instead favour telomere synthesis by telomerase-based mechanisms. Consistent with this, correction of telomere length in NBS1 fibroblasts requires both NBS1 and hTERT, suggesting that both factors cooperate for proper telomere maintenance (Ranganathan *et al.*, 2001).

Telomere length also appears to play a role in cellular sensitivity and response to DSBs. The telomerase RNA component nullizygous mouse mTERC$^{-/-}$ displays shortened telomeres in late generations and concomitant phenotypes in highly proliferative organ systems (Blasco *et al.*, 1997). Cells derived from late generation, but not early generation knockouts display hypersensitivity to IR and DNA damaging agents such as doxorubicin (Lee *et al.*, 2001; Wong *et al.*, 2000). They also display a decreased ability to repair DSBs in the presence of short telomeres, suggesting that telomere length itself could influence the ability to repair subtelomeric DSBs (Wong *et al.*, 2000). Consistent with this notion, recent reports document the presence of DSBs throughout the genome in senescent cells but not in early passage human diploid fibroblasts (Sedelnikova *et al.*, 2004). The location of these DSBs was primarily non-telomeric, suggesting that DDR at both the telomere and non-telomeric regions of the genome were responsible for inducing senescence programs. This may also be viewed in light of reports that reducing the O_2 concentration in which cells are cultured, significantly increases the replicative lifespan (Hayflick limit: (Balin *et al.*, 1984; Forsyth *et al.*, 2003). Interestingly, TERT immortalised fibroblasts also contained γ-H2A.X foci suggesting that maintenance of telomere length did not affect the prevalence of these lesions (Sedelnikova *et al.*, 2004). This also suggests that TERT overexpression may be able to override checkpoint activation by accumulated DSBs throughout the genome. A second possibility is that telomere length influences the cellular response to DSBs. Reports that ectopic hTERT reduces the sensitivity of cells to DNA damaging agents may be related to this phenomenon (Sharma *et al.*, 2003), although it is unclear whether this is related to telomere length or the presence of telomerase holoenzyme. As mentioned previously, telomeres interact with many DDR proteins seemingly in a manner that does not result in their activation. Moreover in yeast, Sir proteins and Ku70/80 migrate from telomeres to the site of a DSB for efficient NHEJ repair (Martin *et al.*, 1999; Mills *et al.*, 1999). Thus telomeres may act as a depot for DNA repair factors. Insufficient telomere reserve may thus trigger DDR in part through an inability to sequester such factors.

A second potential role for telomerase and the DDR may be in direct healing of DSBs by addition of telomere repeats to DSB termini. Telomerase displays a rather promiscuous affinity for DNA termini and can synthesise telomeric repeats at the ends of broadly divergent G rich sequences (Kim *et al.*, 1994; Morin, 1989). The generation of genomic DSBs could potentially be healed by telomere addition by telomerase. However, few reports of this phenomenon have occurred (Harrington and Greider, 1991; Mangahas *et al.*, 2001; Morin, 1991; Myung *et al.*, 2001). It appears that eukaryotes have developed multiple mechanisms to prevent telomere healing, a potential mechanism to promote genomic instability by inappropriate repair of DSBs. Yeast contain a DNA helicase Pif1 that both controls telomere length and the ability of telomerase to heal DSBs (Mangahas *et al.*, 2001; Myung *et al.*, 2001). Deletion of Pif1 results in elongated telomeres. It also causes a dramatic increase in the rate of telomere addition to DSBs (Bessler *et al.*, 2001; Myung *et al.*, 2001). A mammalian homolog of Pif1 exists, although it is unclear if it plays a conserved role in inhibiting telomerase in mammalian cells (Bessler *et al.*, 2001). However, mammalian cells do have at least one mechanism to prevent telomere addition to DSBs. Utilising hTERT-GFP fusions expressed at low levels under the direction of a weak promoter, it was demonstrated that the telomerase enzyme migrates from the nucleoplasm to the nucleolus upon DNA damage, presumably preventing inappropriate telomere addition to DSB termini (Wong *et al.*, 2002). The signalling pathway responsible for this migration is currently unknown. Disruption of these mechanisms in cancer cells expressing strong telomerase activity could be a contributing mechanism for genomic instability via stabilisation of chromosomal fragments.

4. TELOMERE INDUCED SENESCENCE AND ITS ROLE IN REGENERATION, AGING, AND CANCER SUPPRESSION

At the end of DNA replication, lagging strand synthesis results in a loss of chromosomal termini of 50-100 base pairs in human cells (Olovnikov, 1996). This shortening of telomere sequence is more than the size of the RNA segment used for priming the most telomeric okazaki fragment indicating that telomere loss results in part from processing of telomeric termini during S-Phase. The telomerase enzyme synthesises telomere sequence *de novo* and thus can prevent telomere attrition. However, in humans, telomerase activity is tightly regulated (Meyerson *et al.*, 1997; Nakamura *et al.*, 1997). Telomerase is active in immature germ cells and during embryogenesis but is suppressed in adult somatic cells by repression of TERT expression (Schaetzlein *et al.*, 2004). Therefore, the growth of primary human cells is limited to a finite number of cell divisions (50-70 divisions for human fibroblasts) as first described by Hayflick (Hayflick, 1979). Cessation of cell proliferation at the Hayflick limit has been named replicative senescence and correlates with telomere shortening (Allsopp *et al.*, 1992). The proof for the concept that telomere shortening limits cell proliferation at the senescence stage came from studies showing that over-expression of TERT induced telomerase activation, telomere stabilisation and, most importantly, immortalisation of primary human cells (Bodnar *et al.*, 1998).

In human cells telomere shortening and replicative senescence have been linked to the exposure (van Steensel *et al.*, 1998) and degradation (Stewart *et al.*, 2003) of the telomeric 3'end overhang leading to activation of DNA damage pathways involving p53. In agreement with this model, inactivation of p53 allows cell proliferation beyond the senescence stage (Vaziri and Benchimol, 1996; Wright and Shay, 1992). It appears that at the senescence stage critically short telomeres lose capping function, inducing DDR involving the PIKKs ATM/ATR and downstream targets Chk2 and p53 (d'Adda di Fagagna *et al.*, 2003; Karlseder *et al.*, 1999; Takai *et al.*, 2003). Therefore, the same DDR proteins that transiently interact with functional telomeres (see above) induce the senescence program in response to telomere dysfunction, possibly via a prolonged interaction with the dysfunctional telomere. There is experimental evidence that exposure of the single strand telomeric overhang (see above) initiates the DNA damage signal inducing senescence (Eller *et al.*, 2002; Li *et al.*, 2003) a process which seems to require exonuclease 1 (Exo1) activity in yeast cells (Maringele and Lydall, 2002). A possible explanation is that Exo1 activity resects double strand telomeric sequences thereby elongating the telomeric G-strand overhang and facilitating RPA binding and ATR activation (d'Adda di Fagagna *et al.*, 2004). The exact mechanisms by which a dysfunctional telomere triggers a DDR have yet to be determined.

In agreement with the DNA damage hypothesis of senescence induction, it was shown that p53 mediates the adverse effects of telomere dysfunction on organ homeostasis in mTERC$^{-/-}$ mice (Chin *et al.*, 1999). Activation of the p53 pathway increases the expression of the CDK inhibitor p21, which appears to be an important negative regulator of cell proliferation at senescence (Brown *et al.*, 1997). Senescent cells fail to phosphorylate Rb in response to mitogen stimulation (Futreal and Barrett, 1991), and this is in part mediated by p21 (Sherr and Roberts, 1999). In addition, there is another CDK-inhibitor, p16, that also controls Rb activity and cessation of cell proliferation at the senescence stage (Alcorta *et al.*, 1996; Hara *et al.*, 1996; Stein *et al.*, 1999). However, it appears that senescence induced by telomere dysfunction primarily involves the p53-p21-pathway, whereas up-regulation of p16 is a downstream effect regulated by other yet undefined mechanisms (Beausejour *et al.*, 2003; Herbig *et al.*, 2004; Itahana *et al.*, 2003; Stein *et al.*, 1999). Beside the induction of senescence by telomere shortening, over-stimulation of the Ras/Raf/MEK/ mitogen-activated protein kinase (MAPK) pathway induces a premature senescence arrest in cells with functional telomeres (Lin *et al.*, 1998; Serrano *et al.*, 1997; Zhu *et al.*, 1998). Similarly the constant growth factor signalling induced by serum during *in vitro* culture of primary cells induces senescence in a variety of human and mouse cells. Mouse cells that escape this "cell culture shock" induced senescence often show deletion of p19^{ARF} (Sherr and DePinho, 2000). There is also a connection between telomere shortening and mitogen stimulation during initiation of senescence since telomere dysfunction does not induce a DNA damage signalling response in mitogen-deprived cells (Satyanarayana *et al.*, 2004a). In addition, there are other cellular stresses that induce a senescence response including oxidative stress and DNA damage (Matuoka and Chen, 2002; Parrinello *et al.*, 2003) and there is some connection between telomere shortening and oxidative stress (von Zglinicki *et al.*, 1995). The interaction between these different conditions that induce senescence and its relevance to *in vivo* aging has yet to be determined.

There is mounting evidence that telomere shortening not only limits cell proliferation *in vitro* but also the regeneration of human organs and tissues during aging and chronic disease. A variety of studies have demonstrated telomere shortening in different human tissues during aging (Djojosubroto *et al.*, 2003). In addition, there is accelerated telomere shortening in chronic diseases that elevate the rate of cell turnover (Djojosubroto *et al.*, 2003). Moreover, short telomeres are significantly associated with reduced survival in humans at an age of 60 years and older (Cawthon *et al.*, 2003). Direct evidence for an influence of telomere shortening on human aging and organ homeostasis has been revealed in a rare disease, Dyskeratosis congenita, which is characterised by telomere shortening, premature aging, bone marrow failure, liver cirrhosis, skin lesions, increased frequency of malignancies, and reduced survival. A mutation of hTERC was linked to the autosomal dominant form of this disease (Vulliamy *et al.*, 2001). In addition, a mutation of the Dyskerin gene has been linked to the autosomal recessive form of the disease. Dyskerin has a role in maturation of small ribosomal RNA but also interacts with TERC, which seems to be important for TERC-stability (Mitchell *et al.*, 1999). Experimental evidence for a role of telomere shortening in aging and regeneration comes from studies in mTERC$^{-/-}$ mice. Telomere shortening in mTERC$^{-/-}$ mice impaired organ homeostasis, induced premature aging, inhibited organ regeneration and regenerative stress responses, and reduced survival (Lee *et al.*, 1998; Rudolph *et al.*, 1999). Impaired organ regeneration in mTERC$^{-/-}$ mice was due to critical telomere shortening at cellular level resulting in induction of cellular senescence and a decreased population of proliferative organ cells with sufficient telomere reserve (Satyanarayana *et al.*, 2003). Reactivation of telomerase by genetic or adenoviral gene delivery rescued defects in organ regeneration in mTERC$^{-/-}$ mice (Rudolph *et al.*, 2000; Samper *et al.*, 2001).

The postnatal suppression of telomerase in humans and the senescence limit to cell proliferation are believed to function as a tumour suppressor mechanism. In agreement with this hypothesis telomere shortening in mTERC$^{-/-}$ mice limits the growth of tumours in different genetic mouse models of lymphoma, sarcoma, liver tumour, and intestinal tumour formation (Farazi *et al.*, 2003; Greenberg *et al.*, 1999; Rudolph *et al.*, 2001; Wong *et al.*, 2003) as well in response to carcinogen treatment (Farazi *et al.*, 2003; Gonzalez-Suarez *et al.*, 2000). Similarly, the transformation of primary human cells with short telomeres requires telomerase reactivation (Hahn *et al.*, 1999a). Impaired tumour progression in mTERC$^{-/-}$ mice was associated with increased level of p53, increased apoptosis and decreased proliferation of tumour cells indicating that the senescence signalling pathway mediates tumour suppression in response to telomere shortening (Rudolph *et al.*, 2001). In line with these experimental mouse data on impairment of tumour progression by telomere shortening, a variety of studies have shown that inhibition of telomerase in telomerase positive human cancer cell lines results in loss of cell viability (Feng *et al.*, 1995; Hahn *et al.*, 1999b). In accordance with known telomere biology there was a lag period between the start of telomerase inhibition and the onset of reduced cell viability; the length of this lag phase depended on the initial telomere length in tumour cells before initiating telomerase inhibition (Hahn *et al.*, 1999b). According to these data telomere shortening and senescence appear to have a dual role during aging – on the one hand protecting the organism against cancer formation, and on the other contributing to regenerative exhaustion in later life and chronic disease. Therefore, the idea of exploiting telomerase-activation for the treatment of regenerative disorders during aging and chronic disease could be limited by increasing the risk of neoplasia. In this context, it will be important to analyze which of the effects of telomere dysfunction is dominantly affecting survival of the organism. Recent data suggest that there are disease stages where a telomere-stabilizing therapy could be beneficial (Wiemann *et al.* 2005).

5. TELOMERE INDUCED CRISIS AND ITS ROLE IN CHROMOSOMAL INSTABILITY AND CARCINOGENESIS

The pioneering work of Shay and Wright established a biphasic response to telomere attrition. Senescence induced checkpoints occur following attainment of replicative (Hayflick) limits and this checkpoint could be surpassed by ectopic expression of viral oncoproteins (Shay and Wright, 1989). Inactivation of tumour suppressor proteins p53 and pRb produce similar extensions in replicative capacity. Viral oncoprotein expression permits an additional 20-30 population doublings until a second checkpoint known as crisis occurs. As opposed to senescence, crisis is characterised by cell death rather than arrest. Crisis is also characterised by cytogenetic abnormalities such as telomere fusions (Counter *et al.*, 1992). These chromosomal aberrations are thought to result from rampant telomere dysfunction and to be the initiating stimulus for cell death during crisis (Shay and Wright, 1989).

From crisis, rare clones may emerge that are able to maintain telomere length by reactivation of telomerase or via an alternative pathway of telomere maintenance known as ALT. Immortalisation of viral oncoprotein expressing human diploid fibroblasts occurs by hTERT reactivation in 60-90% of cases and ALT pathway telomere maintenance in the other 10-40% (Bryan *et al.*, 1995). In yeast, ALT is dependent on Rad52 mediated recombination (McEachern and Blackburn, 1996). Double mutants for telomerase and Rad52 are not viable beyond crisis. It appears that cancer cells also require these mechanisms as the majority of human tumours have reactivated telomerase (Kim *et al.*, 1994). Consistent with the need for telomere maintenance at crisis, telomerase activity positively correlates with advanced stages of malignancy (Chadeneau *et al.*, 1995; Satyanarayana *et al.*, 2004b; Tang *et al.*, 2001; Yan *et al.*, 2003; Yan *et al.*, 2002). ALT has been reported in a small percentage of tumours (Bryan *et al.*, 1997), although the evidence for long heterogeneous telomeres in primary tumours is lacking. ALT is also seen in mouse cells that lack telomerase (Chang *et al.*, 2003; Hande *et al.*, 1999). Passage of myc and ras V12G together transformed mTERC$^{-/-}$; INK4a$^{-/-}$ MEFs results in telomere shortening and reduced ability to form tumours in SCID mice (Chang *et al.*, 2003; Greenberg *et al.*, 1999). The tumours that did arise were unable to form lung metastasis, however transfection with a genomic fragment of mTERC produced a rescue of this phenotype, allowing lung metastasis after intravenous injection. Passage of these cells in culture revealed a stabilisation of telomere length by ALT mechanisms (Chang *et al.*, 2003). Together, these data indicate that ALT might represent an alternative mechanism beside telomerase reactivation allowing immortal growth of tumour cells. However, the data also indicate that ALT appears to be a weaker mechanism in terms of tumour progression and metastasis as compared to telomerase reactivation. Experimental data on transformation of ALT-positive immortalised primary human cells suggested that telomerase might have additional functions during transformation that are not linked to telomere stabilisation (Stewart *et al.*, 2002)

The cloning of the telomerase catalytic subunit hTERT provided direct experimental evidence for the telomere shortening hypothesis of senescence and crisis. Forced expression of exogenous hTERT in primary human cells bypassed senescence and crisis limits (Bodnar *et al.*, 1998). Moreover, unambiguous demonstration for telomere maintenance as a requisite pathway in cellular transformation has been revealed in mouse cells with short telomeres (Greenberg *et al.*, 1999) and in primary human diploid cells (Hahn *et al.*, 1999a). The important role of telomerase for telomere stabilisation was also demonstrated in experiments showing that telomeres can be maintained at lengths shorter than crisis levels if hTERT is expressed (Zhu *et al.*, 1999). It may be that the presence of even short telomeric sequences can provide both chromosomal protection and avoidance of DDR in the presence of telomerase

As mentioned previously ALT is thought to occur by HR. Disruption in the fidelity of HR results in illegitimate recombination between similar but not identical (homologous) sequences in the genome. The mismatch repair (MMR) proteins are involved in preventing replication errors and the accumulation of point mutations throughout the genome (Modrich and Lahue, 1996). Disruption of MMR activity by mutation results in instability of repetitive elements or microsatellite instability (MSI). This leads to a predisposition to gastrointestinal and other malignancies. MMR genes also ensure the fidelity of HR. Cell lines lacking the full complement of MMR activity display an approximately 10 fold increase in recombination between homologous sequences (Modrich and Lahue, 1996). This also appears to affect telomere maintenance by ALT. Yeast deficient in the MMR gene MSH2 were able to proliferate in the absence of telomerase (Rizki and Lundblad, 2001). These strains were able to maintain telomere length by recombination of subtelomeric repetitive elements. Some evidence that this is relevant to human cancer has been reported since inhibition of telomerase in MMR deficient colon cancer cell activated ALT mechanisms to maintain telomere length (Bechter *et al.*, 2004). Examination of MMR pathway status may therefore be necessary prior to patient selection during clinical trials for telomerase inhibition in human cancers.

Though telomere length maintenance is necessary for cellular immortalisation, the accrual of chromosomal abnormalities prior to telomerase reactivation may actually be a mechanism of generating genetic alterations required for tumorigenesis (Maser and DePinho, 2003). Initial experimental evidence is derived from studies in aged telomerase deficient mice (Rudolph *et al.*, 1999). Mice carrying a genetic deletion in the mTERC gene displayed accelerated aging and an increased susceptibility to lymphomas and teratocarcinoma. These mice similarly displayed hallmarks of telomere dysfunction and crisis such as chromosomal end-to-end fusions and anaphase bridges. Considerable evidence indicates that telomere dysfunction may play a role in the early stages of human malignancy (Romanov *et al.*, 2001). Most compelling however is the discovery that in the human syndrome Dyskeratosis Congenita (see above) hypomorphic function of telomerase recapitulates many of the findings in the late generation aged mTERC$^{-/-}$ mice including a predisposition to cancer formation. Furthermore, recent evidence reveals that telomere dysfunction may be more widespread as a tumour initiating mechanism. In situ analysis of sporadic breast cancers was performed for analysis of telomere dysfunction (Chin *et al.*, 2004). This study revealed telomere shortening and a sharp increase in subtelomeric chromosomal fusions during breast cancer formation at the transition from ductal hyperplasia to ductal carcinoma *in situ*. Similarly, significant telomere shortening and a drastic increase in anaphase bridges – a sign of telomere dysfunction – were observed at the adenoma-carcinoma transition in human colorectal carcinogenesis (Plentz *et al.*, 2003; Rudolph *et al.*, 2001). In addition, liver nodules in cirrhotic liver – a precancerous disease stage – showed shortened telomeres specifically in hepatocellular carcinoma but not in benign regenerative nodules (Plentz *et al.*, 2004). These observations in mice and humans provide clear evidence that telomere dysfunction enhances tumour initiation. It will be interesting to examine the tumours that arise in Dyskeratosis Congenita to determine what eventual mechanisms they utilise to maintain telomeres. In addition, it appears to be of great interest to identify genetic lesions that cooperate with telomere shortening to induce chromosomal instability and cancer initiation. A good candidate in this regard is the tumour suppressor p53, which is a major component of the senescence pathway (see above) and is deleted in many human cancers. The concomitant deletion of the p53 tumour suppressor in mTERC$^{-/-}$ mice greatly enhanced the cancer-prone phenotype in cell culture based oncogene transformation assays (Chin *et al.*, 1999). Homozygous deletion of p53 in mice results in lymphoma and sarcoma formation (Donehower *et al.*, 1992; Jacks *et al.*, 1994). Telomere shortening in mTERC$^{-/-}$, p53$^{-/-}$ double mutant mice resulted in premature death as compared to p53$^{-/-}$ mTERC$^{+/+}$ mice (Artandi *et al.*, 2000). A possible explanation is that telomere shortening and loss of p53 cooperate to induce malignancies *in vivo* thus resulting in decreased mouse survival. An alternative explanation is that lymphoma formation reduced survival in mTERC$^{-/-}$ mice due to the impact of telomere shortening on decreasing the fitness of the aging organism to cope with the stress induced by tumour growth, e.g. diminished hematopoiesis by invading lymphoma cells. In mice carrying a heterozygous deletion of one p53 allele (p53$^{+/-}$) telomere shortening provoked a striking change in the tumour spectrum with a high incidence of epithelial cancers (Artandi *et al.*, 2000). Notably, epithelial cancers are uncommon in mice with wildtype telomerase suggesting that telomere dysfunction may contribute to the emergence of epithelial cancers (Artandi and DePinho, 2000). The tumours in mTERC$^{-/-}$, p53$^{+/-}$ mice showed LOH at the remaining p53 allele (Artandi *et al.*, 2000). It remains to be tested whether telomere shortening contributed to tumorigenesis in epithelial cell compartments by increasing the rate of LOH of p53 or whether loss of p53 and telomere shortening cooperated to foster chromosomal instability to allow additional genetic lesions necessary for cancer formation. Interestingly, the tumours of $p53^{+/-}$, $mTERC^{-/-}$ double mutant mice showed high rates of chromosomal instability including non-reciprocal translocations – a hallmark of epithelial cancers in aged humans. In addition, the combination of shortened telomeres and loss of p53 function is often observed in human cancer. Studies in conditional p53 deficient mice may further clarify the role of p53 during telomere dysfunction induced carcinogenesis.

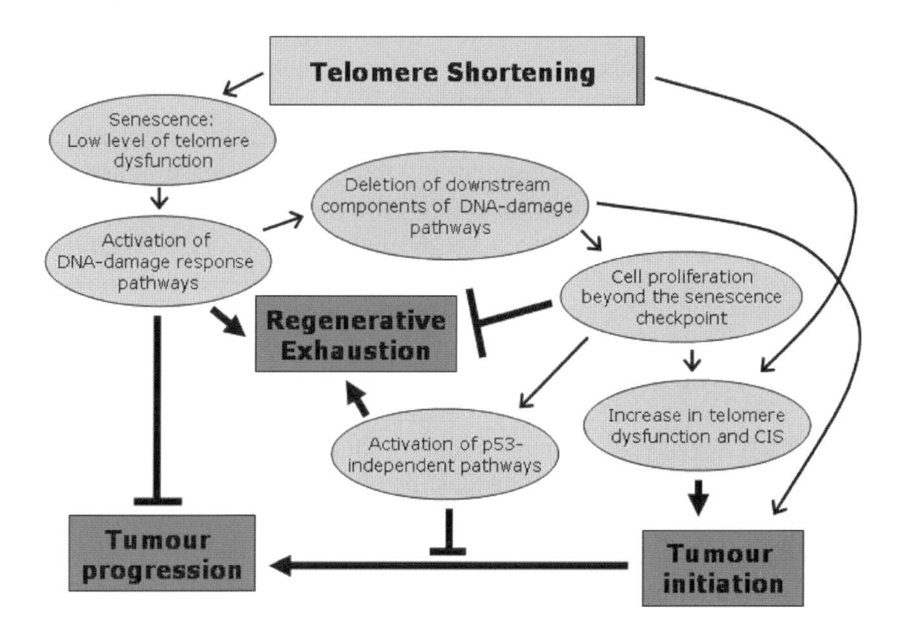

Figure 1. Diverse effects of telomere dysfunction on regeneration and carcinogenesis. Telomere shortening leads to induction of senescence when a subset of telomeres has lost telomere function resulting in chromosomal instability (CIS). Senescence signalling is similar to DNA damage signalling and involves the p53-pathway. Senescence functions as a tumour suppressor mechanism but at the same time limits the regenerative capacity during aging and chronic disease. When components of the DNA damage pathway are deleted/mutated cells proliferate beyond the senescence limit, allowing clonal expansion of such cells during organ regeneration. Further telomere shortening increases the rate of telomere dysfunction and CIS in these cells eventually leading to tumour initiation. At the same time, severe telomere dysfunction induces p53-independent pathways leading to cell cycle arrest or apoptosis thus inhibiting the progression of tumours initiated by telomere dysfunction.

An emerging area of future research is to analyze whether other members of the DNA damage response pathway involved in senescence signalling contribute to tumour initiation when they are deleted. However, lymphoma formation in $ATM^{-/-}$, $mTERC^{-/-}$ double knockout mice was reduced compared to ATM$^{-/-}$, mTERC$^{+/+}$ single mutant mice. One possible explanation is that ATM itself has a crucial role for telomere integrity (see above) and that severe telomere dysfunction induced by loss of ATM in combination with telomere shortening impaired tumour formation. ATM deficiency induces hypersensitivity to ionising radiation by a failure of ATM mutant cells to repair DSBs, while p53 deficiency results in resistance to apoptosis after DSBs (Lowe *et al.*, 1994). As telomere dysfunction is recognised in a similar fashion to DSBs, ATM deficiency may sensitise cells containing dysfunctional telomeres to p53-dependent apoptotic responses due to a persistence of unrepaired and uncapped chromosomal termini. In addition, it seems possible that severe telomere dysfunction induces apoptosis by p53-independent signals. It has been suggested from studies in mTERC$^{-/-}$, p53-/ double knockout mice that p53-independent mechanisms induce defects in organ homeostasis in response to severe telomere dysfunction (Chin *et al.*, 1999). In line with this hypothesis it was observed that high cellular levels of telomere dysfunction induce p53-independent apoptosis in mouse liver cells (Lechel *et al.* 2005).

Another important area of current research is to delineate the mechanisms inducing chromosomal instability in response to telomere dysfunction. Uncapping of telomeres results in chromosomal fusions that are easily visualised in metaphase chromosome spreads or by the presence of anaphase bridges. Failure of chromosome separation during anaphase produces anaphase bridges. This results in both non-disjunction and aneuploidy as well as chromosome breakage. Fragmented chromosomes are then repaired by either NHEJ or recombination into homologous sequences in the genome. NEHJ results in non-reciprocal translocations and inappropriate repair – a common finding in human cancers and in tumours derived from mTERC$^{-/-}$ mice (Artandi *et al.*, 2000b).

In summation telomere dysfunction appears to enhance tumour initiation via increasing genetic instability, however stabilisation of telomeres is eventually necessary for tumour progression (DePinho, 2000). The potential therapeutic use of telomerase inhibitors for cancer or, conversely, activators of telomerase to facilitate regenerative processes will ultimately depend on which of the divergent effects of telomere shortening has a dominant effect on organism survival. The answer to this question will likely be influenced by the disease state in question and the age of the individual being treated.

ACKNOWLEDGMENTS

We thank J. Karlseder, A. Satyanarayana, N. K. Ramireddy, M. Djojosubroto and S. Schaetzlein for critical reading of the manuscript. KLR is supported by the Deutsche Forschungsgemeinschaft (Emmy-Noether-Program: Ru745/2-3, KFO119/1-1, and Ru745/4-1) and by the Deutsche Krebshilfe e.V. (10-2236-Ru2).

REFERENCES

Ahmed, S., and J. Hodgkin. 2000. MRT-2 checkpoint protein is required for germline immortality and telomere replication in C. elegans. *Nature*. 403:159-64.

Alcorta, D.A., Y. Xiong, D. Phelps, G. Hannon, D. Beach, and J.C. Barrett. 1996. Involvement of the cyclin-dependent kinase inhibitor p16 (INK4a) in replicative senescence of normal human fibroblasts. *Proc Natl Acad Sci U S A*. 93:13742-7.

Allsopp, R.C., and C.B. Harley. 1995. Evidence for a critical telomere length in senescent human fibroblasts. *Exp Cell Res*. 219:130-6.

Allsopp, R.C., H. Vaziri, C. Patterson, S. Goldstein, E.V. Younglai, A.B. Futcher, C.W. Greider, and C.B. Harley. 1992. Telomere length predicts replicative capacity of human fibroblasts. *Proc Natl Acad Sci U S A*. 89:10114-8.

Ancelin, K., M. Brunori, S. Bauwens, C.E. Koering, C. Brun, M. Ricoul, J.P. Pommier, L. Sabatier, and E. Gilson. 2002. Targeting assay to study the cis functions of human telomeric proteins: evidence for inhibition of telomerase by TRF1 and for activation of telomere degradation by TRF2. *Mol Cell Biol*. 22:3474-87.

Artandi, S.E., S. Chang, S.L. Lee, S. Alson, G.J. Gottlieb, L. Chin, and R.A. DePinho. 2000. Telomere dysfunction promotes non-reciprocal translocations and epithelial cancers in mice. *Nature*. 406:641-5.

Artandi, S.E., and R.A. DePinho. 2000. A critical role for telomeres in suppressing and facilitating carcinogenesis. *Curr Opin Genet Dev*. 10:39-46.

Bai, Y., and J.P. Murnane. 2003. Telomere instability in a human tumour cell line expressing NBS1 with mutations at sites phosphorylated by ATM. *Mol Cancer Res*. 1:1058-69.

Bailey, S.M., M.A. Brenneman, J. Halbrook, J.A. Nickoloff, R.L. Ullrich, and E.H. Goodwin. 2004. The kinase activity of DNA-PK is required to protect mammalian telomeres. *DNA Repair (Amst)*. 3:225-33.

Bailey, S.M., M.N. Cornforth, A. Kurimasa, D.J. Chen, and E.H. Goodwin. 2001. Strand-specific postreplicative processing of mammalian telomeres. *Science*. 293:2462-5.

Bakkenist, C.J., and M.B. Kastan. 2003. DNA damage activates ATM through intermolecular autophosphorylation and dimer dissociation. *Nature*. 421:499-506.

Balin, A.K., A.J. Fisher, and D.M. Carter. 1984. Oxygen modulates growth of human cells at physiologic partial pressures. *J Exp Med*. 160:152-66.

Baumann, P., and T.R. Cech. 2000. Protection of telomeres by the Ku protein in fission yeast. *Mol Biol Cell*. 11:3265-75.

Baumann, P., and T.R. Cech. 2001. Pot1, the putative telomere end-binding protein in fission yeast and humans. *Science*. 292:1171-5.

Beausejour, C.M., A. Krtolica, F. Galimi, M. Narita, S.W. Lowe, P. Yaswen, and J. Campisi. 2003. Reversal of human cellular senescence: roles of the p53 and p16 pathways. *EMBO J*. 22:4212-22.

Bechter, O.E., Y. Zou, W. Walker, W.E. Wright, and J.W. Shay. 2004. Telomeric recombination in mismatch repair deficient human colon cancer cells after telomerase inhibition. *Cancer Res*. 64:3444-51.

Bertuch, A.A., K. Buckley, and V. Lundblad. 2003. The way to the end matters--the role of telomerase in tumour progression. *Cell Cycle*. 2:36-8.

Bessler, J.B., J.Z. Torredagger, and V.A. Zakian. 2001. The Pif1p subfamily of helicases: region-specific DNA helicases? *Trends Cell Biol*. 11:60-5.

Bi, X., S.C. Wei, and Y.S. Rong. 2004. Telomere protection without a telomerase; the role of ATM and Mre11 in Drosophila telomere maintenance. *Curr Biol*. 14:1348-53.

Blackburn, E.H. 1991. Structure and function of telomeres. *Nature*. 350:569-73.

Blackburn, E.H. 2001. Switching and signalling at the telomere. *Cell*. 106:661-73.

Blackburn, E.H., S. Chan, J. Chang, T.B. Fulton, A. Krauskopf, M. McEachern, J. Prescott, J. Roy, C. Smith, and H. Wang. 2000. Molecular manifestations and molecular determinants of telomere capping. *Cold Spring Harb Symp Quant Biol*. 65:253-63.

Blasco, M.A., W. Funk, B. Villeponteau, and C.W. Greider. 1995. Functional characterization and developmental regulation of mouse telomerase RNA. *Science*. 269:1267-70.

Blasco, M.A., H.W. Lee, M.P. Hande, E. Samper, P.M. Lansdorp, R.A. DePinho, and C.W. Greider. 1997. Telomere shortening and tumour formation by mouse cells lacking telomerase RNA. *Cell*. 91:25-34.

Bodnar, A.G., M. Ouellette, M. Frolkis, S.E. Holt, C.P. Chiu, G.B. Morin, C.B. Harley, J.W. Shay, S. Lichtsteiner, and W.E. Wright. 1998. Extension of life-span by introduction of telomerase into normal human cells. *Science*. 279:349-52.

Boulton, S.J., and S.P. Jackson. 1996. Identification of a Saccharomyces cerevisiae Ku80 homologue: roles in DNA double strand break rejoining and in telomeric maintenance. *Nucleic Acids Res*. 24:4639-48.

Boulton, S.J., and S.P. Jackson. 1998. Components of the Ku-dependent non-homologous end-joining pathway are involved in telomeric length maintenance and telomeric silencing. *EMBO J*. 17:1819-28.

Bradshaw, P.S., D.J. Stavropoulus, and M.S. Meyn. 2005. Human telomeric protein TRF2 associates with genomic double-strand breaks as an early response to DNA damage. *Nat. Genet*. 37:193-197

Brown, E.J., and D. Baltimore. 2000. ATR disruption leads to chromosomal fragmentation and early embryonic lethality. *Genes Dev*. 14:397-402.

Brown, J.P., W. Wei, and J.M. Sedivy. 1997. Bypass of senescence after disruption of p21CIP1/WAF1 gene in normal diploid human fibroblasts. *Science*. 277:831-4.

Bryan, T.M., A. Englezou, L. Dalla-Pozza, M.A. Dunham, and R.R. Reddel. 1997. Evidence for an alternative mechanism for maintaining telomere length in human tumours and tumour-derived cell lines. *Nat Med*. 3:1271-4.

Bryan, T.M., A. Englezou, J. Gupta, S. Bacchetti, and R.R. Reddel. 1995. Telomere elongation in immortal human cells without detectable telomerase activity. *EMBO J*. 14:4240-8.

Bryan, T.M., J.M. Sperger, K.B. Chapman, and T.R. Cech. 1998. Telomerase reverse transcriptase genes identified in Tetrahymena thermophila and Oxytricha trifallax. *Proc Natl Acad Sci U S A*. 95:8479-84.

Carson, C.T., R.A. Schwartz, T.H. Stracker, C.E. Lilley, D.V. Lee, and M.D. Weitzman. 2003. The Mre11 complex is required for ATM activation and the G2/M checkpoint. *EMBO J*. 22:6610-20.

Cawthon, R.M., K.R. Smith, E. O'Brien, A. Sivatchenko, and R.A. Kerber. 2003. Association between telomere length in blood and mortality in people aged 60 years or older. *Lancet*. 361:393-5.

Chadeneau, C., K. Hay, H.W. Hirte, S. Gallinger, and S. Bacchetti. 1995. Telomerase activity associated with acquisition of malignancy in human colorectal cancer. *Cancer Res*. 55:2533-6.

Chang, S., C.M. Khoo, M.L. Naylor, R.S. Maser, and R.A. DePinho. 2003. Telomere-based crisis: functional differences between telomerase activation and ALT in tumour progression. *Genes Dev*. 17:88-100.

Chin, K., C.O. de Solorzano, D. Knowles, A. Jones, W. Chou, E.G. Rodriguez, W.L. Kuo, B.M. Ljung, K. Chew, K. Myambo, M. Miranda, S. Krig, J. Garbe, M. Stampfer, P. Yaswen, J.W. Gray, and S.J. Lockett. 2004. In situ analyses of genome instability in breast cancer. *Nat Genet*. 36:984-8.

Chin, L., S.E. Artandi, Q. Shen, A. Tam, S.L. Lee, G.J. Gottlieb, C.W. Greider, and R.A. DePinho. 1999. p53 deficiency rescues the adverse effects of telomere loss and cooperates with telomere dysfunction to accelerate carcinogenesis. *Cell*. 97:527-38.

Ciapponi, L., G. Cenci, J. Ducau, C. Flores, D. Johnson-Schlitz, M.M. Gorski, W.R. Engels, and M. Gatti. 2004. The Drosophila Mre11/Rad50 complex is required to prevent both telomeric fusion and chromosome breakage. *Curr Biol*. 14:1360-6.

Colgin, L.M., K. Baran, P. Baumann, T.R. Cech, and R.R. Reddel. 2003. Human POT1 facilitates telomere elongation by telomerase. *Curr Biol*. 13:942-6.

Collins, K., and L. Gandhi. 1998. The reverse transcriptase component of the Tetrahymena telomerase ribonucleoprotein complex. *Proc Natl Acad Sci U S A*. 95:8485-90.

Conrad, M.N., J.H. Wright, A.J. Wolf, and V.A. Zakian. 1990. RAP1 protein interacts with yeast telomeres in vivo: overproduction alters telomere structure and decreases chromosome stability. *Cell*. 63:739-50.

Corda, Y., V. Schramke, M.P. Longhese, T. Smokvina, V. Paciotti, V. Brevet, E. Gilson, and V. Geli. 1999. Interaction between Set1p and checkpoint protein Mec3p in DNA repair and telomere functions. *Nat Genet*. 21:204-8.

Counter, C.M., A.A. Avilion, C.E. LeFeuvre, N.G. Stewart, C.W. Greider, C.B. Harley, and S. Bacchetti. 1992. Telomere shortening associated with chromosome instability is arrested in immortal cells which express telomerase activity. *EMBO J*. 11:1921-9.

Craven, R.J., and T.D. Petes. 1999. Dependence of the regulation of telomere length on the type of subtelomeric repeat in the yeast Saccharomyces cerevisiae. *Genetics*. 152:1531-41.

d'Adda di Fagagna, F., M.P. Hande, W.M. Tong, D. Roth, P.M. Lansdorp, Z.Q. Wang, and S.P. Jackson. 2001. Effects of DNA nonhomologous end-joining factors on telomere length and chromosomal stability in mammalian cells. *Curr Biol*. 11:1192-6.

d'Adda di Fagagna, F., P.M. Reaper, L. Clay-Farrace, H. Fiegler, P. Carr, T. Von Zglinicki, G. Saretzki, N.P. Carter, and S.P. Jackson. 2003. A DNA damage checkpoint response in telomere-initiated senescence. *Nature*. 426:194-8.

d'Adda di Fagagna, F., S.H. Teo, and S.P. Jackson. 2004. Functional links between telomeres and proteins of the DNA damage response. *Genes Dev*. 18:1781-99.

D'Amours, D., S. Desnoyers, I. D'Silva, and G.G. Poirier. 1999. Poly(ADP-ribosyl)ation reactions in the regulation of nuclear functions. *Biochem J*. 342 (Pt 2):249-68.

DePinho, R.A. 2000. The age of cancer. *Nature*. 408:248-54.

Diede, S.J., and D.E. Gottschling. 2001. Exonuclease activity is required for sequence addition and Cdc13p loading at a de novo telomere. *Curr Biol*. 11:1336-40.

Djojosubroto, M.W., Y.S. Choi, H.W. Lee, and K.L. Rudolph. 2003. Telomeres and telomerase in aging, regeneration and cancer. *Mol Cells*. 15:164-75.

Dokal, I. 2001. Dyskeratosis congenita. A disease of premature ageing. *Lancet*. 358 Suppl:S27.

Donehower, L.A., M. Harvey, B.L. Slagle, M.J. McArthur, C.A. Montgomery, Jr., J.S. Butel, and A. Bradley. 1992. Mice deficient for p53 are developmentally normal but susceptible to spontaneous tumours. *Nature*. 356:215-21.

Dynek, J.N., and S. Smith. 2004. Resolution of sister telomere association is required for progression through mitosis. *Science*. 304:97-100.

Eller, M.S., N. Puri, I.M. Hadshiew, S.S. Venna, and B.A. Gilchrest. 2002. Induction of apoptosis by telomere 3' overhang-specific DNA. *Exp Cell Res*. 276:185-93.

Ellison, V., and B. Stillman. 2003. Biochemical Characterization of DNA Damage Checkpoint Complexes: Clamp Loader and Clamp Complexes with Specificity for 5' Recessed DNA. *PLoS Biol*. 1:E33

Espejel, S., S. Franco, A. Sgura, D. Gae, S.M. Bailey, G.E. Taccioli, and M.A. Blasco. 2002. Functional interaction between DNA-PKcs and telomerase in telomere length maintenance. *EMBO J*. 21:6275-87.

Farazi, P.A., J. Glickman, S. Jiang, A. Yu, K.L. Rudolph, and R.A. DePinho. 2003. Differential impact of telomere dysfunction on initiation and progression of hepatocellular carcinoma. *Cancer Res*. 63:5021-7.

Feng, J., W.D. Funk, S.S. Wang, S.L. Weinrich, A.A. Avilion, C.P. Chiu, R.R. Adams, E. Chang, R.C. Allsopp, J. Yu, and et al. 1995. The RNA component of human telomerase. *Science*. 269:1236-41.

Forsyth, N.R., A.P. Evans, J.W. Shay, and W.E. Wright. 2003. Developmental differences in the immortalization of lung fibroblasts by telomerase. *Aging Cell*. 2:235-43.

Futreal, P.A., and J.C. Barrett. 1991. Failure of senescent cells to phosphorylate the RB protein. *Oncogene*. 6:1109-13.

Gallego, M.E., and C.I. White. 2001. RAD50 function is essential for telomere maintenance in Arabidopsis. *Proc Natl Acad Sci U S A*. 98:1711-6.

Gilley, D., H. Tanaka, M.P. Hande, A. Kurimasa, G.C. Li, M. Oshimura, and D.J. Chen. 2001. DNA-PKcs is critical for telomere capping. *Proc Natl Acad Sci U S A*. 98:15084-8.

Gonzalez-Suarez, E., E. Samper, J.M. Flores, and M.A. Blasco. 2000. Telomerase-deficient mice with short telomeres are resistant to skin tumorigenesis. *Nat Genet.* 2:114-7.

Grandin, N., C. Damon, and M. Charbonneau. 2001. Ten1 functions in telomere end protection and length regulation in association with Stn1 and Cdc13. *EMBO J.* 20:1173-83.

Gravel, S., M. Larrivee, P. Labrecque, and R.J. Wellinger. 1998. Yeast Ku as a regulator of chromosomal DNA end structure. *Science*. 280:741-4.

Greenberg, R.A., R.C. Allsopp, L. Chin, G.B. Morin, and R.A. DePinho. 1998. Expression of mouse telomerase reverse transcriptase during development, differentiation and proliferation. *Oncogene*. 16:1723-30.

Greenberg, R.A., L. Chin, A. Femino, K.H. Lee, G.J. Gottlieb, R.H. Singer, C.W. Greider, and R.A. DePinho. 1999. Short dysfunctional telomeres impair tumorigenesis in the INK4a(delta2/3) cancer-prone mouse. *Cell*. 97:515-25.

Greenwell, P.W., S.L. Kronmal, S.E. Porter, J. Gassenhuber, B. Obermaier, and T.D. Petes. 1995. TEL1, a gene involved in controlling telomere length in S. cerevisiae, is homologous to the human ataxia-telangiectasia gene. *Cell*. 82:823-9.

Griffith, J.D., L. Comeau, S. Rosenfield, R.M. Stansel, A. Bianchi, H. Moss, and T. de Lange. 1999. Mammalian telomeres end in a large duplex loop. *Cell*. 97:503-14.

Grossi, S., A. Bianchi, P. Damay, and D. Shore. 2001. Telomere formation by rap1p binding site arrays reveals end-specific length regulation requirements and active telomeric recombination. *Mol Cell Biol*. 21:8117-28.

Hahn, W.C., C.M. Counter, A.S. Lundberg, R.L. Beijersbergen, M.W. Brooks, and R.A. Weinberg. 1999a. Creation of human tumour cells with defined genetic elements. *Nature*. 400:464-8.

Hahn, W.C., S.A. Stewart, M.W. Brooks, S.G. York, E. Eaton, A. Kurachi, R.L. Beijersbergen, J.H. Knoll, M. Meyerson, and R.A. Weinberg. 1999b. Inhibition of telomerase limits the growth of human cancer cells. *Nat Med.* 5:1164-70.

Hande, M.P., A.S. Balajee, and A.T. Natarajan. 1997. Induction of telomerase activity by UV-irradiation in Chinese hamster cells. *Oncogene*. 15:1747-52.

Hande, M.P., A.S. Balajee, A. Tchirkov, A. Wynshaw-Boris, and P.M. Lansdorp. 2001. Extra-chromosomal telomeric DNA in cells from Atm(-/-) mice and patients with ataxia-telangiectasia. *Hum Mol Genet*. 10:519-28.

Hande, M.P., P.M. Lansdorp, and A.T. Natarajan. 1998. Induction of telomerase activity by in vivo X-irradiation of mouse splenocytes and its possible role in chromosome healing. *Mutat Res*. 404:205-14.

Hande, M.P., E. Samper, P. Lansdorp, and M.A. Blasco. 1999. Telomere length dynamics and chromosomal instability in cells derived from telomerase null mice. *J Cell Biol*. 144:589-601.

Hara, E., R. Smith, D. Parry, H. Tahara, S. Stone, and G. Peters. 1996. Regulation of p16CDKN2 expression and its implications for cell immortalization and senescence. *Mol Cell Biol*. 16:859-67.

Hardy, C.F., L. Sussel, and D. Shore. 1992. A RAP1-interacting protein involved in transcriptional silencing and telomere length regulation. *Genes Dev.* 6:801-14.

Harrington, L.A., and C.W. Greider. 1991. Telomerase primer specificity and chromosome healing. *Nature*. 353:451-4.

Hayflick, L. 1979. The cell biology of aging. *J Invest Dermatol.* 73:8-14.

Henderson, E.R., and E.H. Blackburn. 1989. An overhanging 3' terminus is a conserved feature of telomeres. *Mol Cell Biol*. 9:345-8.

Herbig, U., W.A. Jobling, B.P. Chen, D.J. Chen, and J.M. Sedivy. 2004. Telomere shortening triggers senescence of human cells through a pathway involving ATM, p53, and p21(CIP1), but not p16(INK4a). *Mol Cell*. 14:501-13.

Hicke, B.J., D.W. Celander, G.H. MacDonald, C.M. Price, and T.R. Cech. 1990. Two versions of the gene encoding the 41-kilodalton subunit of the telomere binding protein of Oxytricha nova. *Proc Natl Acad Sci U S A*. 87:1481-5.

Hsu, H.L., D. Gilley, S.A. Galande, M.P. Hande, B. Allen, S.H. Kim, G.C. Li, J. Campisi, T. Kohwi-Shigematsu, and D.J. Chen. 2000. Ku acts in a unique way at the mammalian telomere to prevent end joining. *Genes Dev.* 14:2807-12.

Hughes, T.R., R.G. Weilbaecher, M. Walterscheid, and V. Lundblad. 2000. Identification of the single-strand telomeric DNA binding domain of the Saccharomyces cerevisiae Cdc13 protein. *Proc Natl Acad Sci U S A.* 97:6457-62.

Itahana, K., Y. Zou, Y. Itahana, J.L. Martinez, C. Beausejour, J.J. Jacobs, M. Van Lohuizen, V. Band, J. Campisi, and G.P. Dimri. 2003. Control of the replicative life span of human fibroblasts by p16 and the polycomb protein Bmi-1. *Mol Cell Biol.* 23:389-401.

Jacks, T., L. Remington, B.O. Williams, E.M. Schmitt, S. Halachmi, R.T. Bronson, and R.A. Weinberg. 1994. Tumour spectrum analysis in p53-mutant mice. *Curr Biol.* 4:1-7.

Jaco, I., P. Munoz, F. Goytisolo, J. Wesoly, S. Bailey, G. Taccioli, and M.A. Blasco. 2003. Role of mammalian Rad54 in telomere length maintenance. *Mol Cell Biol.* 23:5572-80.

Jasin, M. 2000. Chromosome breaks and genomic instability. *Cancer Invest.* 18:78-86.

Kanellis, P., R. Agyei, and D. Durocher. 2003. Elg1 forms an alternative PCNA-interacting RFC complex required to maintain genome stability. *Curr Biol.* 13:1583-95.

Karlseder, J., D. Broccoli, Y. Dai, S. Hardy, and T. de Lange. 1999. p53- and ATM-dependent apoptosis induced by telomeres lacking TRF2. *Science.* 283:1321-5.

Karlseder, J., K. Hoke, O.K. Mirzoeva, C. Bakkenist, M.B. Kastan, J.H. Petrini, and T. de Lange. 2004. The Telomeric Protein TRF2 Binds the ATM Kinase and Can Inhibit the ATM-Dependent DNA Damage Response. *PLoS Biol.* 2:E240.

Karlseder, J., L. Kachatrian, H. Takai, K. Mercer, S. Hingorani, T. Jacks, and T. de Lange. 2003. Targeted deletion reveals an essential function for the telomere length regulator Trf1. *Mol Cell Biol.* 23:6533-41.

Karlseder, J., A. Smogorzewska, and T. de Lange. 2002. Senescence induced by altered telomere state, not telomere loss. *Science.* 295:2446-9.

Kharbanda, S., V. Kumar, S. Dhar, P. Pandey, C. Chen, P. Majumder, Z.M. Yuan, Y. Whang, W. Strauss, T.K. Pandita, D. Weaver, and D. Kufe. 2000. Regulation of the hTERT telomerase catalytic subunit by the c-Abl tyrosine kinase. *Curr Biol.* 10:568-75.

Kim, N.W., M.A. Piatyszek, K.R. Prowse, C.B. Harley, M.D. West, P.L. Ho, G.M. Coviello, W.E. Wright, S.L. Weinrich, and J.W. Shay. 1994. Specific association of human telomerase activity with immortal cells and cancer. *Science.* 266:2011-5.

Kim, S.H., P. Kaminker, and J. Campisi. 1999. TIN2, a new regulator of telomere length in human cells. *Nat Genet.* 23:405-12.

Kirk, K.E., B.P. Harmon, I.K. Reichardt, J.W. Sedat, and E.H. Blackburn. 1997. Block in anaphase chromosome separation caused by a telomerase template mutation. *Science.* 275:1478-81.

Kyrion, G., K.A. Boakye, and A.J. Lustig. 1992. C-terminal truncation of RAP1 results in the deregulation of telomere size, stability, and function in Saccharomyces cerevisiae. *Mol Cell Biol.* 12:5159-73.

Le, S., J.K. Moore, J.E. Haber, and C.W. Greider. 1999. RAD50 and RAD51 define two pathways that collaborate to maintain telomeres in the absence of telomerase. *Genetics.* 152:143-52.

Lechel A., A. Satyanarayana, Z. Ju, R.R. Plentz, S. Schaetzlein, C. Rudolph, L. Wilkens, S.U. Wiemann, G. Saretzki, N.P. Malek, M.P. Manns, J. Buer, and K.L. Rudolph. 2005. The Cellular Level of Telomere Dysfunction Determines Induction of Senescence or Apoptosis *in vivo. EMBO Rep.* Epub ahead of print

Lee, H.W., M.A. Blasco, G.J. Gottlieb, J.W. Horner, 2nd, C.W. Greider, and R.A. DePinho. 1998. Essential role of mouse telomerase in highly proliferative organs. *Nature.* 392:569-74.

Lee, J.H., and T.T. Paull. 2004. Direct activation of the ATM protein kinase by the Mre11/Rad50/Nbs1 complex. *Science.* 304:93-6.

Lee, K.H., K.L. Rudolph, Y.J. Ju, R.A. Greenberg, L. Cannizzaro, L. Chin, S.R. Weiler, and R.A. DePinho. 2001. Telomere dysfunction alters the chemotherapeutic profile of transformed cells. *Proc Natl Acad Sci U S A.* 98:3381-6.

Li, B., and T. de Lange. 2003. Rap1 affects the length and heterogeneity of human telomeres. *Mol Biol Cell.* 14:5060-8.

Li, B., S. Oestreich, and T. de Lange. 2000. Identification of human Rap1: implications for telomere evolution. *Cell.* 101:471-83.

Li, G., C. Nelsen, and E.A. Hendrickson. 2002. Ku86 is essential in human somatic cells. *Proc Natl Acad Sci U S A.* 99:832-7.

Li, G.Z., M.S. Eller, R. Firoozabadi, and B.A. Gilchrest. 2003. Evidence that exposure of the telomere 3' overhang sequence induces senescence. *Proc Natl Acad Sci U S A.* 100:527-31.

Lin, A.W., M. Barradas, J.C. Stone, L. van Aelst, M. Serrano, and S.W. Lowe. 1998. Premature senescence involving p53 and p16 is activated in response to constitutive MEK/MAPK mitogenic signalling. *Genes Dev.* 12:3008-19.

Lingner, J., T.R. Hughes, A. Shevchenko, M. Mann, V. Lundblad, and T.R. Cech. 1997. Reverse transcriptase motifs in the catalytic subunit of telomerase. *Science.* 276:561-7.

Longhese, M.P., V. Paciotti, H. Neecke, and G. Lucchini. 2000. Checkpoint proteins influence telomeric silencing and length maintenance in budding yeast. *Genetics.* 155:1577-91.

Lowe, S.W., T. Jacks, D.E. Housman, and H.E. Ruley. 1994. Abrogation of oncogene-associated apoptosis allows transformation of p53-deficient cells. *Proc Natl Acad Sci U S A.* 91:2026-30.

Lundblad, V., and E.H. Blackburn. 1993. An alternative pathway for yeast telomere maintenance rescues est1- senescence. *Cell.* 73:347-60.

Lustig, A.J., S. Kurtz, and D. Shore. 1990. Involvement of the silencer and UAS binding protein RAP1 in regulation of telomere length. *Science.* 250:549-53.

Lustig, A.J., and T.D. Petes. 1986. Identification of yeast mutants with altered telomere structure. *Proc Natl Acad Sci U S A.* 83:1398-402.

Makarov, V.L., Y. Hirose, and J.P. Langmore. 1997. Long G tails at both ends of human chromosomes suggest a C strand degradation mechanism for telomere shortening. *Cell.* 88:657-66.

Mallory, J.C., V.I. Bashkirov, K.M. Trujillo, J.A. Solinger, M. Dominska, P. Sung, W.D. Heyer, and T.D. Petes. 2003. Amino acid changes in Xrs2p, Dun1p, and Rfa2p that remove the preferred targets of the ATM family of protein kinases do not affect DNA repair or telomere length in Saccharomyces cerevisiae. *DNA Repair (Amst).* 2:1041-64.

Mangahas, J.L., M.K. Alexander, L.L. Sandell, and V.A. Zakian. 2001. Repair of chromosome ends after telomere loss in Saccharomyces. *Mol Biol Cell.* 12:4078-89.

Marcand, S., V. Brevet, and E. Gilson. 1999. Progressive cis-inhibition of telomerase upon telomere elongation. *EMBO J.* 18:3509-19.

Marcand, S., E. Gilson, and D. Shore. 1997. A protein-counting mechanism for telomere length regulation in yeast. *Science.* 275:986-90.

Maringele, L., and D. Lydall. 2002. EXO1-dependent single-stranded DNA at telomeres activates subsets of DNA damage and spindle checkpoint pathways in budding yeast yku70Delta mutants. *Genes Dev.* 16:1919-33.

Martin, S.G., T. Laroche, N. Suka, M. Grunstein, and S.M. Gasser. 1999. Relocalization of telomeric Ku and SIR proteins in response to DNA strand breaks in yeast. *Cell.* 9:621-33.

Maser, R.S., and R.A. DePinho. 2003. Take care of your chromosomes lest cancer take care of you. *Cancer Cell.* 3:4-6.

Masutomi, K., E.Y. Yu, S. Khurts, I. Ben-Porath, J.L. Currier, G.B. Metz, M.W. Brooks, S. Kaneko, S. Murakami, J.A. DeCaprio, R.A. Weinberg, S.A. Stewart, and W.C. Hahn. 2003. Telomerase maintains telomere structure in normal human cells. *Cell.* 114:241-53.

Matsuura, A., T. Naito, and F. Ishikawa. 1999. Genetic control of telomere integrity in Schizosaccharomyces pombe: rad3(+) and tel1(+) are parts of two regulatory networks independent of the downstream protein kinases chk1(+) and cds1(+). *Genetics.* 152:1501-12.

Matuoka, K., and K.Y. Chen. 2002. Telomerase positive human diploid fibroblasts are resistant to replicative senescence but not premature senescence induced by chemical reagents. *Biogerontology*. 3:365-72.

McEachern, M.J., and E.H. Blackburn. 1996. Cap-prevented recombination between terminal telomeric repeat arrays (telomere CPR) maintains telomeres in Kluyveromyces lactis lacking telomerase. *Genes Dev*. 10:1822-34.

Meyerson, M., C. Counter, E. Eaton, L. Ellisen, P. Steiner, S. Caddle, L. Ziaugra, R. Beijersbergen, M. Davidoff, Q. Liu, S. Bacchetti, D. Haber, and R. Weinberg. 1997. hEST2, the putative human telomerase catalytic subunit gene, is up-regulated in tumour cells and during immortalization. *Cell*. 90:785-795.

Mieczkowski, P.A., J.O. Mieczkowska, M. Dominska, and T.D. Petes. 2003. Genetic regulation of telomere-telomere fusions in the yeast Saccharomyces cerevisae. *Proc Natl Acad Sci U S A*. 100:10854-9.

Mills, K.D., D.A. Sinclair, and L. Guarente. 1999. MEC1-dependent redistribution of the Sir3 silencing protein from telomeres to DNA double-strand breaks. *Cell*. 97:609-20.

Mitchell, J.R., E. Wood, and K. Collins. 1999. A telomerase component is defective in the human disease dyskeratosis congenita. *Nature*. 402:551-5.

Modrich, P., and R. Lahue. 1996. Mismatch repair in replication fidelity, genetic recombination, and cancer biology. *Annu Rev Biochem*. 65:101-33.

Morin, G.B. 1989. The human telomere terminal transferase enzyme is a ribonucleoprotein that synthesizes TTAGGG repeats. *Cell*. 59:521-9.

Morin, G.B. 1991. Recognition of a chromosome truncation site associated with alpha-thalassaemia by human telomerase. *Nature*. 353:454-6.

Myung, K., C. Chen, and R.D. Kolodner. 2001. Multiple pathways cooperate in the suppression of genome instability in Saccharomyces cerevisiae. *Nature*. 411:1073-6.

Myung, K., G. Ghosh, F.J. Fattah, G. Li, H. Kim, A. Dutia, E. Pak, S. Smith, and E.A. Hendrickson. 2004. Regulation of telomere length and suppression of genomic instability in human somatic cells by Ku86. *Mol Cell Biol*. 24:5050-9.

Naito, T., A. Matsuura, and F. Ishikawa. 1998. Circular chromosome formation in a fission yeast mutant defective in two ATM homologues. *Nat Genet*. 20:203-6.

Nakamura, T., G. Morin, K. Chapman, S. Weinrich, W. Andrews, J. Lingner, C. Harley, and T. Cech. 1997. Telomerase catalytic subunit homologs from fission yeast and humans. *Science*. 15:955-959.

Nakamura, T.M., B.A. Moser, and P. Russell. 2002. Telomere binding of checkpoint sensor and DNA repair proteins contributes to maintenance of functional fission yeast telomeres. *Genetics*. 161:1437-52.

Niida, H., T. Matsumoto, H. Satoh, M. Shiwa, Y. Tokutake, Y. Furuichi, and Y. Shinkai. 1998. Severe growth defect in mouse cells lacking the telomerase RNA component. *Nat Genet*. 19:203-6.

Nugent, C.I., G. Bosco, L.O. Ross, S.K. Evans, A.P. Salinger, J.K. Moore, J.E. Haber, and V. Lundblad. 1998. Telomere maintenance is dependent on activities required for end repair of double-strand breaks. *Curr Biol*. 8:657-60.

Nugent, C.I., T.R. Hughes, N.F. Lue, and V. Lundblad. 1996. Cdc13p: a single-strand telomeric DNA-binding protein with a dual role in yeast telomere maintenance. *Science*. 274:249-52.

Olovnikov, A.M. 1996. Telomeres, telomerase, and aging: origin of the theory. *Exp Gerontol*. 31:443-8.

Parrinello, S., E. Samper, A. Krtolica, J. Goldstein, S. Melov, and J. Campisi. 2003. Oxygen sensitivity severely limits the replicative lifespan of murine fibroblasts. *Nat Cell Biol*. 5:741-7.

Peterson, S.E., A.E. Stellwagen, S.J. Diede, M.S. Singer, Z.W. Haimberger, C.O. Johnson, M. Tzoneva, and D.E. Gottschling. 2001. The function of a stem-loop in telomerase RNA is linked to the DNA repair protein Ku. *Nat Genet*. 27:64-7.

Petrini, J.H. 2000. The Mre11 complex and ATM: collaborating to navigate S phase. *Curr Opin Cell Biol*. 12:293-6.

Plentz, R.R., M. Caselitz, J.S. Bleck, M. Gebel, P. Flemming, S. Kubicka, M.P. Manns, and K.L. Rudolph. 2004. Hepatocellular telomere shortening correlates with chromosomal instability and the development of human hepatoma. *Hepatology*. 40:80-6.

Plentz, R.R., S.U. Wiemann, P. Flemming, P.N. Meier, S. Kubicka, H. Kreipe, M.P. Manns, and K.L. Rudolph. 2003. Telomere shortening of epithelial cells characterises the adenoma-carcinoma transition of human colorectal cancer. *Gut*. 52:1304-7.

Porter, S.E., P.W. Greenwell, K.B. Ritchie, and T.D. Petes. 1996. The DNA-binding protein Hdf1p (a putative Ku homologue) is required for maintaining normal telomere length in Saccharomyces cerevisiae. *Nucleic Acids Res*. 24:582-5.

Ranganathan, V., W.F. Heine, D.N. Ciccone, K.L. Rudolph, X. Wu, S. Chang, H. Hai, I.M. Ahearn, D.M. Livingston, I. Resnick, F. Rosen, E. Seemanova, P. Jarolim, R.A. DePinho, and D.T. Weaver. 2001. Rescue of a telomere length defect of Nijmegen breakage syndrome cells requires NBS and telomerase catalytic subunit. *Curr Biol*. 11:962-6.

Ray, A., and K.W. Runge. 1998. The C terminus of the major yeast telomere binding protein Rap1p enhances telomere formation. *Mol Cell Biol*. 18:1284-95.

Ray, A., and K.W. Runge. 1999. Varying the number of telomere-bound proteins does not alter telomere length in tel1Delta cells. *Proc Natl Acad Sci U S A*. 96:15044-9.

Reaper, P.M., F. di Fagagna, and S.P. Jackson. 2004. Activation of the DNA damage response by telomere attrition: a passage to cellular senescence. *Cell Cycle*. 3:543-6.

Reichenbach, P., M. Hoss, C.M. Azzalin, M. Nabholz, P. Bucher, and J. Lingner. 2003. A human homolog of yeast Est1 associates with telomerase and uncaps chromosome ends when overexpressed. *Curr Biol*. 13:568-74.

Rezler, E.M., D.J. Bearss, and L.H. Hurley. 2003. Telomere inhibition and telomere disruption as processes for drug targeting. *Annu Rev Pharmacol Toxicol*. 43:359-79.

Ritchie, K.B., J.C. Mallory, and T.D. Petes. 1999. Interactions of TLC1 (which encodes the RNA subunit of telomerase), TEL1, and MEC1 in regulating telomere length in the yeast Saccharomyces cerevisiae. *Mol Cell Biol*. 19:6065-75.

Ritchie, K.B., and T.D. Petes. 2000. The Mre11p/Rad50p/Xrs2p complex and the Tel1p function in a single pathway for telomere maintenance in yeast. *Genetics*. 155:475-9.

Rizki, A., and V. Lundblad. 2001. Defects in mismatch repair promote telomerase-independent proliferation. *Nature*. 411:713-6.

Romanov, S.R., B.K. Kozakiewicz, C.R. Holst, M.R. Stampfer, L.M. Haupt, and T.D. Tlsty. 2001. Normal human mammary epithelial cells spontaneously escape senescence and acquire genomic changes. *Nature*. 409:633-7.

Rudolph, K.L., S. Chang, H.W. Lee, M. Blasco, G.J. Gottlieb, C. Greider, and R.A. DePinho. 1999. Longevity, stress response, and cancer in aging telomerase-deficient mice. *Cell*. 96:701-12.

Rudolph, K.L., S. Chang, M. Millard, N. Schreiber-Agus, and R.A. DePinho. 2000. Inhibition of experimental liver cirrhosis in mice by telomerase gene delivery. *Science*. 287:1253-8.

Rudolph, K.L., M. Millard, M.W. Bosenberg, and R.A. DePinho. 2001. Telomere dysfunction and evolution of intestinal carcinoma in mice and humans. *Nat Genet*. 28:155-9.

Samper, E., J. Flores, and M. Blasco. 2001. Restoration of telomerase activity rescues chromosomal instability and premature aging in Terc-/- mice with short telomeres. *EMBO Rep*. 2:800-7.

Samper, E., F.A. Goytisolo, P. Slijepcevic, P.P. van Buul, and M.A. Blasco. 2000. Mammalian Ku86 protein prevents telomeric fusions independently of the length of TTAGGG repeats and the G-strand overhang. *EMBO Rep*. 1:244-52.

Satyanarayana, A., R.A. Greenberg, S. Schaetzlein, J. Buer, K. Masutomi, W.C. Hahn, S. Zimmermann, U. Martens, M.P. Manns, and K.L. Rudolph. 2004a. Mitogen stimulation cooperates with telomere shortening to activate DNA damage responses and senescence signalling. *Mol Cell Biol*. 24:5459-74.

Satyanarayana, A., M. Manns, and K. Rudolph. 2004b. Telomeres and Telomerase: A dual role in hepatocarcinogenesis. *Hepatology*. 40:276-283.

Satyanarayana, A., S.U. Wiemann, J. Buer, J. Lauber, K.E. Dittmar, T. Wustefeld, M.A. Blasco, M.P. Manns, and K.L. Rudolph. 2003. Telomere shortening impairs organ regeneration by inhibiting cell cycle re-entry of a subpopulation of cells. *EMBO J.* 22:4003-13.

Schaetzlein, S., A. Lucas-Hahn, E. Lemme, W.A. Kues, M. Dorsch, M.P. Manns, H. Niemann, and K.L. Rudolph. 2004. Telomere length is reset during early mammalian embryogenesis. *Proc Natl Acad Sci U S A.* 101:8034-8.

Schramke, V., P. Luciano, V. Brevet, S. Guillot, Y. Corda, M.P. Longhese, E. Gilson, and V. Geli. 2004. RPA regulates telomerase action by providing Est1p access to chromosome ends. *Nat Genet.* 36:46-54.

Sedelnikova, O.A., I. Horikawa, D.B. Zimonjic, N.C. Popescu, W.M. Bonner, and J.C. Barrett. 2004. Senescing human cells and ageing mice accumulate DNA lesions with unrepairable double-strand breaks. *Nat Cell Biol.* 6:168-70.

Serrano, M., A.W. Lin, M.E. McCurrach, D. Beach, and S.W. Lowe. 1997. Oncogenic ras provokes premature cell senescence associated with accumulation of p53 and p16INK4a. *Cell.* 88:593-602.

Sharma, G.G., A. Gupta, H. Wang, H. Scherthan, S. Dhar, V. Gandhi, G. Iliakis, J.W. Shay, C.S. Young, and T.K. Pandita. 2003. hTERT associates with human telomeres and enhances genomic stability and DNA repair. *Oncogene.* 22:131-46.

Shay, J.W., and W.E. Wright. 1989. Quantitation of the frequency of immortalization of normal human diploid fibroblasts by SV40 large T-antigen. *Exp Cell Res.* 184:109-18.

Shay, J.W., and W.E. Wright. 2005. Telomere-binding factors and general DNA repair. *Nat Genet.* 37:116-8

Sherr, C.J., and J.M. Roberts. 1999. CDK inhibitors: positive and negative regulators of G1-phase progression. *Genes Dev.* 13:1501-12.

Sherr, C.J., and R.A. DePinho. Cellular senescence: mitotic clock or culture shock? *Cell.* 102:407-10.

Shiloh, Y. 2000. ATM: sounding the double-strand break alarm. *Cold Spring Harb Symp Quant Biol.* 65:527-33.

Shiloh, Y. 2003. ATM and related protein kinases: safeguarding genome integrity. *Nat Rev Cancer.* 3:155-68.

Shippen-Lentz, D., and E.H. Blackburn. 1990. Functional evidence for an RNA template in telomerase. *Science.* 247:546-52.

Silva, E., S. Tiong, M. Pedersen, E. Homola, A. Royou, B. Fasulo, G. Siriaco, and S.D. Campbell. 2004. ATM is required for telomere maintenance and chromosome stability during Drosophila development. *Curr Biol.* 14:1341-7.

Silverman, J., H. Takai, S.B. Buonomo, F. Eisenhaber, and T. de Lange. 2004. Human Rif1, ortholog of a yeast telomeric protein, is regulated by ATM and 53BP1 and functions in the S-phase checkpoint. *Genes Dev.* 18:2108-19.

Smith, J., H. Zou, and R. Rothstein. 2000. Characterization of genetic interactions with RFA1: the role of RPA in DNA replication and telomere maintenance. *Biochimie.* 82:71-8.

Smith, S., and T. de Lange. 2000. Tankyrase promotes telomere elongation in human cells. *Curr Biol.* 10:1299-302.

Smith, S., I. Giriat, A. Schmitt, and T. de Lange. 1998. Tankyrase, a poly(ADP-ribose) polymerase at human telomeres. *Science.* 282:1484-7.

Smogorzewska, A., and T. De Lange. 2004. Regulation of telomerase by telomeric proteins. *Annu Rev Biochem.* 73:177-208.

Smolikov, S., Y. Mazor, and A. Krauskopf. 2004. ELG1, a regulator of genome stability, has a role in telomere length regulation and in silencing. *Proc Natl Acad Sci U S A.* 101:1656-61.

Stansel, R.M., T. de Lange, and J.D. Griffith. 2001. T-loop assembly in vitro involves binding of TRF2 near the 3' telomeric overhang. *EMBO J.* 20:5532-40.

Stein, G.H., L.F. Drullinger, A. Soulard, and V. Dulic. 1999. Differential roles for cyclin-dependent kinase inhibitors p21 and p16 in the mechanisms of senescence and differentiation in human fibroblasts. *Mol Cell Biol.* 19:2109-17.

Stellwagen, A.E., Z.W. Haimberger, J.R. Veatch, and D.E. Gottschling. 2003. Ku interacts with telomerase RNA to promote telomere addition at native and broken chromosome ends. *Genes Dev.* 17:2384-95.

Stewart, S.A., I. Ben-Porath, V.J. Carey, B.F. O'Connor, W.C. Hahn, and R.A. Weinberg. 2003. Erosion of the telomeric single-strand overhang at replicative senescence. *Nat Genet.* 33. 492-6.

Stewart, S.A., W.C. Hahn, B.F. O'Connor, E.N. Banner, A.S. Lundberg, P. Modha, H. Mizuno, M.W. Brooks, M. Fleming, D.B. Zimonjic, N.C. Popescu, and R.A. Weinberg. 2002. Telomerase contributes to tumorigenesis by a telomere length-independent mechanism. *Proc Natl Acad Sci U S A.* 99:12606-11.

Takai, H., A. Smogorzewska, and T. de Lange. 2003. DNA damage foci at dysfunctional telomeres. *Curr Biol.* 13:1549-56.

Takata, H., Y. Kanoh, N. Gunge, K. Shirahige, and A. Matsuura. 2004. Reciprocal association of the budding yeast ATM-related proteins Tel1 and Mec1 with telomeres in vivo. *Mol Cell.* 14:515-22.

Tang, Z.Y., H. Yi, and B.L. Chen. 2001. [Telomerase activity in bladder cancer tissues]. *Hunan Yi Ke Da Xue Xue Bao.* 26:167-8.

Tarsounas, M., P. Munoz, A. Claas, P.G. Smiraldo, D.L. Pittman, M.A. Blasco, and S.C. West. 2004. Telomere maintenance requires the RAD51D recombination/repair protein. *Cell.* 117:337-47.

Tong, W.M., M.P. Hande, P.M. Lansdorp, and Z.Q. Wang. 2001. DNA strand break-sensing molecule poly(ADP-Ribose) polymerase cooperates with p53 in telomere function, chromosome stability, and tumour suppression. *Mol Cell Biol.* 21:4046-54.

Tsukamoto, Y., J. Kato, and H. Ikeda. 1997. Silencing factors participate in DNA repair and recombination in Saccharomyces cerevisiae. *Nature.* 388:900-3.

van Steensel, B., and T. de Lange. 1997. Control of telomere length by the human telomeric protein TRF1. *Nature.* 385:740-3.

van Steensel, B., A. Smogorzewska, and T. de Lange. 1998. TRF2 protects human telomeres from end-to-end fusions. *Cell.* 92:401-13.

Vaziri, H., and S. Benchimol. 1996. From telomere loss to p53 induction and activation of a DNA damage pathway at senescence: the telomere loss/DNA damage model of cell aging. *Exp Gerontol.* 31:295-301.

Venkitaraman, A.R. 2002. Cancer susceptibility and the functions of BRCA1 and BRCA2. *Cell.* 108:171-82.

von Zglinicki, T., G. Saretzki, W. Docke, and C. Lotze. 1995. Mild hyperoxia shortens telomeres and inhibits proliferation of fibroblasts: a model for senescence? *Exp Cell Res.* 220:186-93.

Vulliamy, T., A. Marrone, F. Goldman, A. Dearlove, M. Bessler, P.J. Mason, and I. Dokal. 2001. The RNA component of telomerase is mutated in autosomal dominant dyskeratosis congenita. *Nature.* 413:432-5.

Wei, C., R. Skopp, M. Takata, S. Takeda, and C.M. Price. 2002. Effects of double-strand break repair proteins on vertebrate telomere structure. *Nucleic Acids Res.* 30:2862-70.

Wiemann S.U., A. Satyanarayana, J. Buer, K. Kamino, M.P. Manns, K.L. Rudolph. 2005. Contrasting effects of telomere shortening on organ homeostasis, tumor suppression, and survival during chronic organ damage. *Oncogene..* 24:1501-9.

Wong, J.M., L. Kusdra, and K. Collins. 2002. Subnuclear shuttling of human telomerase induced by transformation and DNA damage. *Nat Cell Biol.* 4:731-6.

Wong, K.K., S. Chang, S.R. Weiler, S. Ganesan, J. Chaudhuri, C. Zhu, S.E. Artandi, K.L. Rudolph, G.J. Gottlieb, L. Chin, F.W. Alt, and R.A. DePinho. 2000. Telomere dysfunction impairs DNA repair and enhances sensitivity to ionizing radiation. *Nat Genet.* 26:85-8.

Wong, K.K., R.S. Maser, R.M. Bachoo, J. Menon, D.R. Carrasco, Y. Gu, F.W. Alt, and R.A. DePinho. 2003. Telomere dysfunction and Atm deficiency compromises organ homeostasis and accelerates ageing. *Nature.* 421:643-8.

Wright, W.E., and J.W. Shay. 1992. The two-stage mechanism controlling cellular senescence and immortalization. *Exp Gerontol.* 27:383-9.

Wright, W.E., V.M. Tesmer, K.E. Huffman, S.D. Levene, and J.W. Shay. 1997. Normal human chromosomes have long G-rich telomeric overhangs at one end. *Genes Dev.* 11:2801-9.

Wu, X., V. Ranganathan, D.S. Weisman, W.F. Heine, D.N. Ciccone, T.B. O'Neill, K.E. Crick, K.A. Pierce, W.S. Lane, G. Rathbun, D.M. Livingston, and D.T. Weaver. 2000. ATM phosphorylation of Nijmegen breakage syndrome protein is required in a DNA damage response. *Nature.* 405:477-82.

Yan, S.N., B. Deng, and Z.J. Gong. 2003. [Effect of cell cycle on telomerase activity of hepatoma cells and its relationship with replication of hepatitis B virus]. *Ai Zheng.* 22:504-7.

Yan, Y.L., J.X. Zheng, X. Wang, Y. Wang, and J.Y. Yang. 2002. [Detection of telomerase activity in bronchoscopic brush-off samples in patients with lung cancer]. *Ai Zheng.* 21:768-71.

Ye, J.Z., D. Hockemeyer, A.N. Krutchinsky, D. Loayza, S.M. Hooper, B.T. Chait, and T. de Lange. 2004. POT1-interacting protein PIP1: a telomere length regulator that recruits POT1 to the TIN2/TRF1 complex. *Genes Dev.* 18:1649-54.

Yuan, X., S. Ishibashi, S. Hatakeyama, M. Saito, J. Nakayama, R. Nikaido, T. Haruyama, Y. Watanabe, H. Iwata, M. Iida, H. Sugimura, N. Yamada, and F. Ishikawa. 1999. Presence of telomeric G-strand tails in the telomerase catalytic subunit TERT knockout mice. *Genes Cells.* 4:563-72.

Zhou, X.Z., and K.P. Lu. 2001. The Pin2/TRF1-interacting protein PinX1 is a potent telomerase inhibitor. *Cell.* 107:347-59.

Zhu, J., H. Wang, J.M. Bishop, and E.H. Blackburn. 1999. Telomerase extends the lifespan of virus-transformed human cells without net telomere lengthening. *Proc Natl Acad Sci U S A.* 96:3723-8.

Zhu, J., D. Woods, M. McMahon, and J.M. Bishop. 1998. Senescence of human fibroblasts induced by oncogenic Raf. *Genes Dev.* 12:2997-3007.

Zhu, X.D., L. Niedernhofer, B. Kuster, M. Mann, J.H. Hoeijmakers, and T. de Lange. 2003. ERCC1/XPF removes the 3' overhang from uncapped telomeres and represses formation of telomeric DNA-containing double minute chromosomes. *Mol Cell.* 12:1489-98.

Zou, L., D. Cortez, and S.J. Elledge. 2002. Regulation of ATR substrate selection by Rad17-dependent loading of Rad9 complexes onto chromatin. *Genes Dev.* 16:198-208.

Zou, L., and S.J. Elledge. 2003. Sensing DNA damage through ATRIP recognition of RPA-ssDNA complexes. *Science.* 300:1542-8.

Chapter 3.5

GENE AMPLIFICATION MECHANISMS

Michelle Debatisse[1] and Bernard Malfoy[2]
[1]UMR 7147, Institut Curie, CNRS, Université Pierre et Marie Curie and [2]Institut Curie, Section de Recherche, 26 Rue d'ULM, Paris Cedex5, France

1. INTRODUCTION

Gene amplification is one of the most frequent genomic alterations found in cancer. This phenomenon has been first characterised in mammalian cells by Alt et al., who studied mutants selected *in vitro* to resist metothrexate (MTX), an inhibitor of the dihydrofolate reductase (DHFR) (Alt et al., 1978). Resistance was often due to an increase in the copy number of the DHFR gene, which leads to overproduce the protein. Cytogenetic analysis of mutants showed that amplified copies lie on abnormal chromosomal structures. Biedler and Spengler first reported on the presence of chromosome expansions, called HSRs (homogeneously staining regions), in Chinese hamster cells amplified for the DHFR gene (Biedler and Spengler, 1976). The amplified copies may alternatively accumulate as centromere-deficient extra-chromosomal circular elements, called double-minutes (DMs) (Kaufman et al., 1979). HSRs and DMs were also found in mutant cells amplified for different loci following selection by various cytotoxic drugs (Stark and Wahl, 1984). A large number of reports stressed the presence of the same types of abnormal structures in cells of tumours and established tumour cell lines. Molecular screening of collections of tumours for candidate oncogenes established that genes belonging to the MYC, ERBB, FGFR and RAS families are recurrently amplified in these cells (Brison, 1993; Schwab and Amler, 1990). Global searches for genes amplified in tumours by genomic approaches now suggest that most genes coding for proteins that favour cell proliferation, including those involved in cell cycle progression and some house keeping genes, may be selected for in growing tumours (Knuutila et al., 1998; Schwab, 1999).

Gene amplification is commonly found in tumour but not in normal cells. Indeed, when both tumour and normal cells of the same patient were studied,

E.A. Nigg (ed.), Genome Instability in Cancer Development, 343-361.

it appeared that oncogene amplification is strictly limited to tumour cells (Brison, 1993). In good agreement with these observations, normal human or rodent cells failed to give rise to amplified mutants resistant to cytotoxic drugs. In contrast, upon selection with the same drugs, mutants were readily obtained from cells presenting a deficiency for p53 (Livingstone et al., 1992; Yin et al., 1992). This was the first demonstration that amplification relies on loss of cell cycle checkpoint (s). Ishizaka et al. engineered rat cells that express a thermosensitive SV40 T antigen (Tag). Because Tag binds to and inhibits p53, these cells are deficient for p53 activity at permissive temperature but not at non-permissive temperature. This model permitted to show that p53 prevents cell survival as long as the amplification process is operating (Ishizaka et al., 1995). It was later established that p53 deficient cells undergo amplification only when they are capable of cycling under deleterious conditions (Chernova et al., 1998; Paulson et al., 1998; Poupon et al., 1996). Amplification also relies on the p53 status *in vivo*, as shown in mouse models (see below) and in at least some human cancers. This emphasises the existence of general mechanisms controlling the permissivity for amplification.

In cells from advanced tumours, the reconstitution of the amplification mechanisms from the study of the organisation of repeats is often jeopardised by the frequent occurrence of secondary rearrangements. This difficulty has been by-passed by the development of protocols that permit the analysis of mutant cells selected *in vitro* only a few generations after initiation of the amplification process (Smith et al., 1990). The organisation of the repeats accumulated at early stages was studied in cells resistant to coformycin, PALA (N-(phosphonacetyl)-L-aspartate) and MTX, which bear amplified copies of the AMPD2 (adenylate deaminase 2) (Coquelle et al., 1997; Toledo et al., 1993; Toledo et al., 1992b), the CAD (Carbamyl-P-synthetase, Aspartate transcarbamylase, Dihydro-orotase) (Poupon et al., 1996; Smith et al., 1990; Smith et al., 1992) and the DHFR (Ma et al., 1993; Windle et al., 1991) genes, respectively. Mutants selected with drugs such as vinblastine, adriamycin or actinomycin D, bearing amplified copies of the MDR1 (multidrug resistance 1) gene were also studied. This latter model permitted to compare the effect of different selective agents on the frequency and the type of mutants recovered for the same locus in the same genetic background (Coquelle et al., 1997). In all early mutants, the amplified copies were found on DMs or within HSRs, as previously described for established resistant lines and cancer cells. Moreover, the results indicated that at least two different initial amplification mechanisms accumulate intra-chromosomal or extra-chromosomal copies of a same gene in the same cell type (Toledo et al., 1992b; Toledo et al., 1993; Coquelle et al., 1997).

2. HSRS AND THE BFB-CYCLES

In many cases, amplification proceeds through an entirely intra-chromosomal pathway, relying on breakage-fusion-bridge (BFB) cycles (Figure 1), a mechanism long identified by the pioneering work of B. McClintock (McClintock, 1951). This mechanism accumulates large inverted repeats on the chromosome arm that bears the locus in non-amplified cells. The operation of this mechanism has been associated with a deletion of the distal part of the amplified chromosome arm. Fused sister chromatids and anaphase bridges were observed in cell populations undergoing amplification, which definitely established that the BFB-cycles operate during *in vitro* selection (Coquelle et al., 1997; Ma et al., 1993; Toledo et al., 1993; Toledo et al., 1992b). This mechanism turned out to be of wide occurrence, since it also operates *in vivo*, in mouse and human cancers (see below).

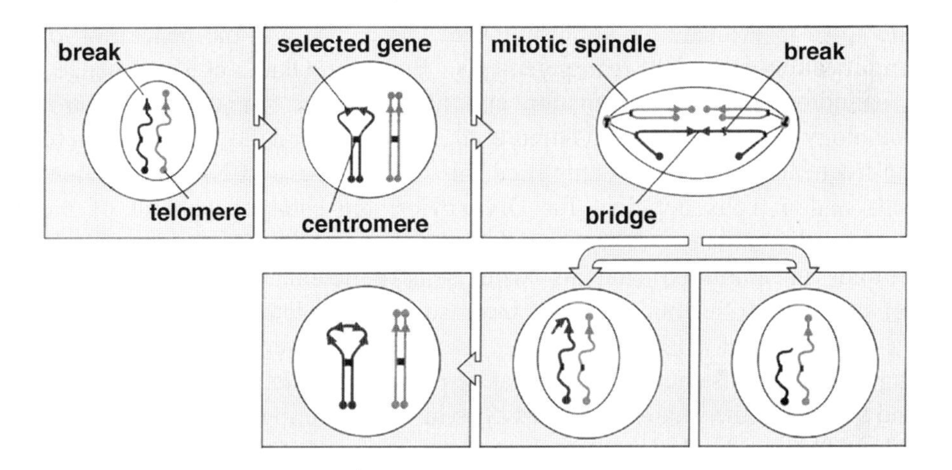

Figure 1. BFB-cycles and intrachromosomal amplification. The process is supposed to arise either from a chromosome break followed by the fusion of the two broken chromatids, or directly from the fusion of sister chromatids with inactive telomeres. At mitosis, the fused chromatids form a bridge that is later broken. One daughter cell presents a terminal deletion of the chromosome and dies in selective medium. The second one gains a third copy of the selected gene, two of them lying in inverted orientation on the broken chromosome arm. After replication the broken chromatids can fuse again, perpetuating the amplification cycles.

2.1 Initiation of BFB-cycles in *in vitro* Models

Triggering of the BFB cycles was proposed to result either from chromosome breakage, or from telomere erosion leading to fusion of unprotected chromosome ends (Ma et al., 1993; Poupon et al., 1996; Smith et al., 1992; Toledo et al., 1993; Toledo et al., 1992b). A growing number of experiments showed that the mechanistic consequences of telomere shortening are roughly similar to those of DSBs (Maser and DePinho, 2004).

Hence, both types of feature might contribute to initiate the BFB-cycles, depending on the localisation of the targeted gene on the chromosome, on the stress initiating the process and on the cell type.

The contribution of DSBs to the triggering of BFB-cycles has been thoroughly explored. Protocols permitting activation of site-specific breaks were developed, notably through expression of the mega-nuclease I-Sce1 (Dujon, 1989) in mammalian cells bearing a transgenic I-Sce1 site (Jasin, 1996). Pipiras et al. showed that I-Sce1 expression induces DHFR amplification in cells with a single copy of the I-Sce1 site integrated close to and telomeric to the DHFR gene. Analysis of the organisation of the repeats indicated that amplification results from BFB-cycles initiated at the I-sce1 site, which demonstrates that a double strand break is able to trigger the whole process of intra-chromosomal amplification (Pipiras et al., 1998). However, it remained to be shown that DSBs contribute efficiently to amplification in cells not designed for having highly recurrent site-specific breaks. Since DSBs are mainly repaired by non-homologous end joining (NHEJ) in mammalian cells (see elsewhere in this book), several groups attempted to evaluate the contribution of this pathway to the generation of amplification-dependent rearrangements. Recently, Okuno et al. sequenced a palindromic amplicon junction and identified the signature of micro-homology-mediated NHEJ (Okuno et al., 2004). Mondello et al. compared the frequencies of mutants amplified for the CAD gene in NHEJ proficient cells and in cells deficient for DNA-PKcs, the catalytic subunit of the complex that initiates NHEJ. They found an increased frequency, by at least tenfold, in amplified mutants with cells deficient for DNA-PKcs as compared to proficient cells. This increase was completely accounted for by the operation of BFB-cycle-driven amplification events. These results suggested the existence of an alternative pathway for sister chromatid fusion and supported the involvement of DSBs in the triggering of the BFB-cycles (Mondello et al., 2001), though it does not exclude a contribution of telomere fusion since components of the NHEJ complex may also participate in telomere maintenance (Maser and DePinho, 2004).

The contribution of hotspots of breakage/recombination to the triggering of BFB-cycles has also been investigated, notably the role of common fragile sites (CFSs). CFSs are chromosomal loci where breaks occur recurrently when cells are grown under stress conditions (Sutherland et al., 1998). Cytogenetic correlations between their locations and those of cancer-associated translocations and deletions have long suggested that they play a role in the development of cancers (Richards, 2001; Smith et al., 1998). Kuo et al. first suggested that CFSs are also involved in gene amplification (Kuo et al., 1994). Since most agents used to select for amplified mutants are potent DNA damaging drugs, Coquelle et al further explored their role at early stages of the amplification process. They showed that CFS-activating drugs specifically induce the BFB-cycles. More extensive analysis of the structures formed upon amplification of three different loci established that fragile sites telomeric to and centromeric to the selected gene frame the initial repeats. This indicated that CFSs play a major role in determining the genetic content of the early repeats (Coquelle et al., 1997) (Figure 2). Whether CFSs contribute to the amplification process in the absence of drugs is still unclear (see below).

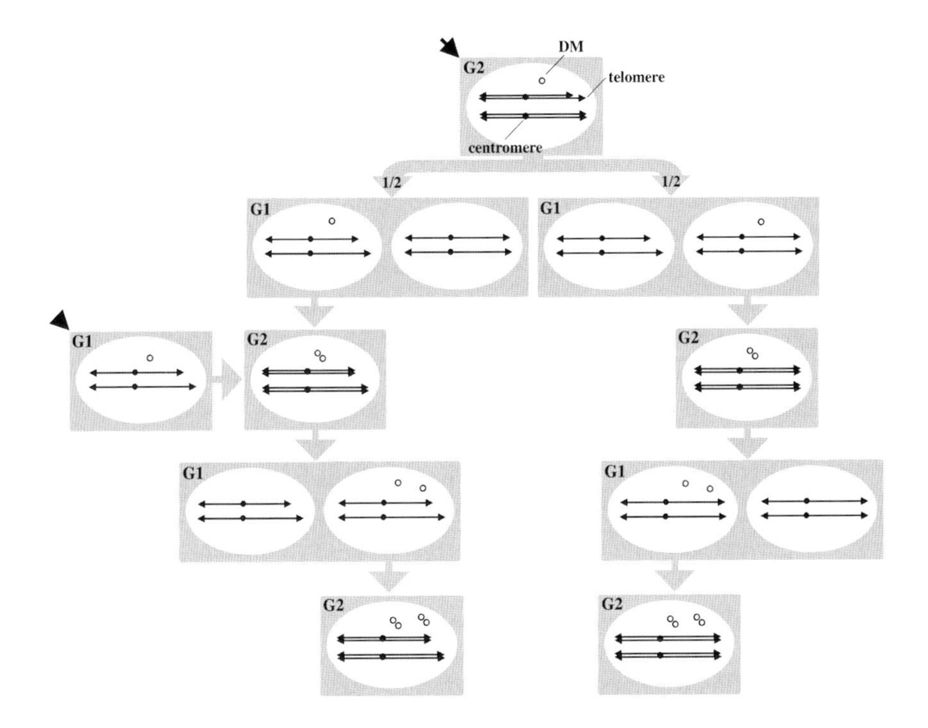

Figure 2. Mechanism of extrachromosomal amplification. Black arrowhead: excision of a sequence bearing the selected gene in a cell in phase G1. After one step of replication and mitosis, one daughter cell (left) has a single copy of the selected gene on the unrearranged chromosome (lethal in selective medium) while three copies are present in the other daughter cell (right). The uneven segregation of the extrachromosomal acentric molecules during subsequent cell cycles could lead to higher levels of amplification. The cells of the resulting amplified clone are characterised by an interstitial deletion corresponding to the excised sequence. Black arrow: excision in a cell in S or G2 phase. Among the four chromatids, only one is deleted. At mitosis, in half the cases, the acentric molecule segregates within a cell with two normal chromosomes (right). Such a cell can later give rise to an amplified clone devoid of chromosomal abnomalies. In the other cases, the acentric molecule segregates within a cell with one normal and one deleted chromosome (left), a situation indistinguishable from the one depicted above.

2.2 Initiation of BFB-cycles in Mouse Models

Even though a large body of evidence indicated that DSBs and possibly telomere dysfunction trigger the BFB-cycles *in vitro*, it remained to be demonstrated that they also trigger amplification *in vivo*. Several groups studied the consequences of DSBs in NHEJ deficient mice. The study of p53 deficient mice in which either the XRCC4, Lig4 or DNA-PKc gene was inactivated revealed that most animals succumb to pro-B cell lymphomas. Amplification of a specific translocation linking the c-myc gene to the IgH locus was recurrently found in cells of these tumours, and further analysis of the amplified structures revealed the signature of the BFB-cycles. It was also established that the generation of these complex rearrangements, named complicons, requires both a functional RAG1/2 endonuclease and p53 deficiency (Difilippantonio et al., 2002; Gladdy et al., 2003; Zhu et al., 2002). Double-mutant mice deficient for p53 and Artemis, another component of the NHEJ pathway, similarly develop pro-B lymphomas, the cells of which also display complex amplified structures. The role of H2AX in the formation of complicons was also examined. In response to DSBs, this histone variant is rapidly phosphorylated (gamma-H2AX) in an ATM and DNA-PKc-dependent manner. Gamma-H2AX forms nuclear foci, in which proteins involved in the repair of DSBs are recruited and accumulated, facilitating the repair process (Fernandez-Capetillo et al., 2003). Gamma-H2AX may also directly control the accuracy of the repair process by tethering the two broken chromosomal ends (Bassing and Alt, 2004). Mice deficient for p53 and H2AX develop the same types of tumours as NHEJ/p53 mice, and also display complicons (Bassing et al., 2003; Celeste et al., 2003). Altogether these studies established a direct link between inaccurate repair of DSBs and BFB-cycle-mediated oncogene amplification *in vivo*.

A still unexplained feature is the very high frequency of rearrangements involving the c-myc region in tumours formed in XRCC4, LIG4, DNA-PKc and H2AX deficient animals as compared to rearrangements involving the N-myc locus in Artemis deficient mice. Rearrangements leading to over-expression of either oncogene probably give similar growth advantages to mutant cells; hence, the choice of the member of the myc family selected for fusion may reflect differences in the frequency of DSBs occurring in the vicinity of these loci. Such a finding suggests the attractive hypothesis that the step at which the DSB repair process is impaired modulates the relative fragility of different chromosomal regions. Finally, the observation, in mice deficient for NHEJ and p53, of chromosome rearrangements and amplifications in tumours from tissues that do not express the RAG genes, highlights a general contribution of DSBs to the control of chromosome stability (references above and (Sharpless et al., 2001)).

The contribution of telomere fusion to the triggering of BFB-cycles is also well documented in mouse models. Artandi et al. studied animals deficient for both p53 and telomerase activity and found that they display a pattern of epithelial cancers mimicking the one found in humans. The cells of these tumours presented features typical of the operation of BFB-cycles, such as fused chromosomes and anaphase bridges (Artandi et al., 2002). Global analysis of these tumours revealed amplification and deletion of chromosomal regions that bear oncogenes or suppressor genes, respectively, the homologues of which are involved in human cancer progression (O'Hagan et al., 2002).

2.3 Intra-chromosomal Amplification and Human Cancers

Cytogenetic signatures of BFB-cycles have been repeatedly observed in tumour cells. These include amplified copies lying on the chromosome arm that bears one copy of the locus in normal cells (Lese et al., 1995, Jin et al., 1998 ; Roelofs et al., 1993), palindromic organisation of the repeats (Ciullo et al., 2002; Hellman et al., 2002; Pedeutour et al., 1994, Gisselsson et al., 2000; Saunders et al., 2000; Shuster et al., 2000), anaphase bridges and fused sister chromatids (Gisselsson et al., 2000; Saunders et al., 2000). Whether CFSs frame the repeats accumulated *in vivo* is still debated. Indeed, although cytogenetic correlations exist between the location of CFSs and the limits of amplicons in cells of human cancers (Popescu, 2003), their presence at the boundaries of amplified units has been unambiguously demonstrated in only two cases (Ciullo et al., 2002; Hellman et al., 2002).

The nature of the stresses that activate CFSs in the absence of drugs and trigger chromosome rearrangements *in vivo* has been intensively studied. Recent results suggest that CFSs are regions of unusual replication programs. In yeast, replication slow zones have been identified, that behave as genetically programmed hotspots of breakage, the fragility of which is enhanced by inactivation of Mec1 (ATR) (Cha and Kleckner, 2002). Casper et al. have established that the frequency of breaks at FRA3B, the most active CFS of the human genome, is strongly enhanced in ATR-deficient cells (Casper et al., 2002). This was confirmed by the finding that instability at common fragile sites occurs *in vivo* in cells of patients bearing a heterozygous mutation for ATR (Seckel syndrome) (Casper et al., 2004). Cells deficient in Brca1, the phosphorylation of which is ATR-dependent, also displayed enhanced frequencies of breaks at CFSs (Arlt et al., 2004). Moreover, replication delays were observed along CFSs (Hellman et al., 2000). Altogether, these results suggested that CFSs might be replication slow zones of the mammalian genomes, prone to spontaneous breakage in S phase (Arlt et al., 2003). Besides occasional induction of DSBs at CFSs by ongoing replication, some constraints directly related to tumour microenvironment may activate CFSs *in vivo*. For example, hypoxia and re-oxygenation are important parameters of solid tumour environment that have been repeatedly correlated with tumour progression (Leo et al., 2004). It was long demonstrated *in vitro* that pretreatment of cells by hypoxia enhances the frequency of amplification of different genes (Luk et al., 1990; Rice et al., 1986). More recently, Coquelle et al. showed that oxygen deprivation and/or re-oxygenation trigger BFB-cycles and generate the same organisation of amplicons as drug treatments. Some clue as to these observations came from the demonstration that hypoxia and/or re-oxygenation induce chromosome breaks at specific loci, most of which co-localise with drug-activated CFSs (Coquelle et al., 1998). A rationale for these findings recently emerged from the demonstration that hypoxia induces a rapid replication arrest and that sensitivity to that stress depends on ATR (Hammond et al., 2004), a cellular response similar to the one induced by aphidicoline, the archetype of CFS-activating drug (Arlt et al., 2003).

The contribution of telomere-based crisis to end-to-end chromosome fusion and genome instability has been also intensively studied ((Gisselsson, 2003), and elsewhere in this book). Indeed, several studies strongly suggest that telomere dysfunction triggers gross genome remodelling through reiterated cycles of BFB, and promotes progression of various types of human cancers from benign to malignant forms (Chin et al., 2004; Gisselsson et al., 2004; Hoglund et al., 2003; O'Sullivan et al., 2002). It is now well established that telomere lengths are heterogeneous among chromosome extremities, but similar in all the cells of a given individual (Londono-Vallejo, 2004). Der-Sarkissian et al. showed that the chromosome arms carrying the shortest telomeres preferentially undergo fusion and subsequent BFB-cycles. Since instability affects almost exclusively the chromosomes with shorter telomeres, the initial distribution of telomere lengths in an individual may, in part, determine the karyopypes of tumour cells (der-Sarkissian et al., 2004).

3. MECHANISM(S) OF EXTRA-CHROMOSOMAL AMPLIFICATION

3.1 Initiation of Extra-chromosomal Amplification in Model Systems

Analyses of amplified mutants at very early stages suggested that chromosome breakage is involved in the generation of DMs bearing the CAD or the DHFR genes. When sub-microscopic, the extra-chromosomal molecules were called "episomes" (Carroll et al., 1987; Maurer et al., 1987; Windle et al., 1991). Increase in copy number of DMs or episomes is acquired through unequal segregation at mitosis, allowing one daughter cell to gain more copies of the gene, while the other one looses copies and dies in selective medium. The same group proposed that breakage occurring across replication bubbles might account for the palindromic organisation of some episomes and the associated gross rearrangements of the chromosome that normally bears the gene (Nonet et al., 1993; Windle and Wahl, 1992).

Early mutants amplified for the AMPD2 or the MDR1 genes bearing extra copies of the selected gene on DMs were also analyzed, and two situations have been encountered - in some clones, one of the homologues was deleted for the amplified sequence while the other one was normal - in other clones, cells had two normal sets of chromosomes. To explain these features, it has been proposed that DMs are formed without chromosome breakage but rather by looping out of a megabase-long sequence (Coquelle et al., 1997; Toledo et al., 1993; Toledo et al., 1992b) (Figure 2). If such an event occurs in a cell in G1 phase, a chromosome deleted for this sequence on both sister chromatids is generated after replication, the second homologue remaining normal. At mitosis, each daughter cell receives a normal and a deleted chromosome. Only those cells that receive at least one copy of the extra-chromosomal element enter the amplification process. This accounts for the first category of clones. If looping out of the same sequence occurs in a cell in S or G2 phase, a single chromatid of one homologue is deleted. After mitosis, one daughter cell has two normal copies of the considered chromosome and the other, as above, contains one normal and one deleted homologue. The DM has a 50 % chance of segregating to the daughter cell that received two normal chromosomes, accounting for the second category of DM-containing clones.

Whether there is more than one mechanism of extra-chromosomal amplification or whether the rather complex situations found in the former set of experiments result from the rapid occurrence of secondary events that masked the initial situation is still unclear.

3.2 Segregation of DMs in Model Systems

It is generally agreed that, when devoid of a centromere, extra-chromosomal elements segregate at random and are frequently lost. However, some acentric molecules, including DMs, appear capable of binding the ends of mitotic chromosomes. Hence, they segregate as passengers of the chromosomes, which leads to a relative mitotic stability. This was observed only when both an origin of DNA replication and a matrix attachment region were present on the acentric molecules (Jenke et al., 2004; Shimizu et al., 2001). Our knowledge of the mechanisms involved in this binding process mainly came from the study of plasmids derived from bovine papillomavirus type1 (BPV) and Epstein-Barr virus (EBV). E2 and EBNA-1 proteins, that respectively initiate replication from BPV and EBV origins, mediate the attachment to the host chromosomes (Ilves et al., 1999; Kanda et al., 2001). In addition, association of EBNA-1 with telomeric proteins such as TRF2, hRAP1 and Tankyrase controls episomal maintenance of EBV-based plasmids (Deng et al., 2002). Analysis of the localisation of DMs in cells of the neuroendocrine colon tumour line COLO 320DM at different phases of the cell cycle suggested a relationship between replication of DMs and chromosome attachment. DMs were preferentially found at the nuclear periphery all through G1 phase. Then, they moved towards the nuclear interior, a re-localisation spatially and temporally coupled to DNA replication. Finally DMs appeared attached to the condensed chromosomes during M phase (Itoh and Shimizu, 1998; Shimizu et al., 1998; Tanaka and Shimizu, 2000).

Elimination of DMs resulting from their capture into micronuclei has also been observed. This process is stimulated *in vitro* by treatment of cells with replication inhibitors or ionising radiation (Canute et al., 1998; Raymond et al., 2001; Schoenlein et al., 2003), probably in response to breakage of DMs, which may inhibit their anchorage to the chromosomes and favour their entrapment in micronuclei (Tanaka and Shimizu, 2000). Hence, the understanding of this mechanism may offer new molecular targets for the treatment of cancers, the progression of which relies on extra-chromosomal amplification of oncogenes.

3.3 Extra-chromosomal Amplification in Cancer Cells

Several studies took advantage of the completed human genome sequence to determine the genetic content and the structure of extra-chromosomal elements found in cells of human cancers. Gene amplification is frequent in gliomas for example. DMs were observed in up to 50% of the glioblastomas, whereas HSRs were rarely found. Vogt et al. have studied a series of 7 gliomas in which the epidermal growth factor receptor (EGFR) gene is amplified. They established that all amplicons of a given tumour derived from a single founding extra-chromosomal DNA molecule. In each tumour, the founding molecule was generated by a simple event that circularised a chromosome segment overlapping the EGFR gene. The signature of micro-homology-based NHEJ has been observed at all sequenced junctions. The corresponding chromosomal loci were not rearranged, suggesting that a post-replicative event was responsible for the formation of each initial amplicon (Vogt et al., 2004).

In hematologic malignancies DMs are present at frequencies below 10%, but a few cases were precisely analyzed. In about 6 % of individuals with T-cell acute lymphoblastic leukaemia (T-ALL), a 500 kb-long segment from chromosome 9q containing the NUP214 and ABL1 genes was circularised, which generates a gene fusion coding a constitutively activated tyrosine kinase. In the only analyzed case, a deletion larger than the amplified region was found in one of the 2 chromosomes 9 (Graux et al., 2004). Analysis of 6 cases of acute myeloid leukemias (AML) displaying amplification of the C-MYC gene on DMs revealed amplicons of more than 4 Mb harbouring up to 8 known genes. In all the cases, the region corresponding to the amplicon was deleted from one homologue of chromosome 8 and in 2 cases, the chromosomal deletions extended beyond the amplified segment on both sides (Storlazzi et al., 2004; Thomas et al., 2004).

Thus, at least in these cases, DM formation is not associated with gross chromosomal rearrangements. The formation of the initial extra-chromosomal element appears coupled to chromosomal deletions in leukemias, but not in gliomas. Whether these observations reflect the existence of preferential excision mechanisms or differences in selection stringency (the cell without deletion has one more copy at the initial stage, Figure 2) from cell type to cell type and/or from locus to locus is unknown. However, the finding that the intra-chromosomal deletion may be larger than the amplified segment in leukemias and that the junctions of DMs found in gliomas rely on NHEJ suggest that the mechanisms forming the founding extra-chromosomal element are more complex than the simple recombination-driven looping-out model previously proposed.

4. SECONDARY REARRANGEMENTS AND EVOLVING AMPLICONS

4.1 Interconversion of DMs into HSRs

It has been long suggested that episomes and DMs may fuse and eventually reintegrate the chromosomes, giving rise to secondary HSRs (Carroll et al., 1988; Cowell, 1982; Wahl, 1989). More recently, Coquelle et al. have shown that fragile site activation in cells containing two different populations of DMs, each bearing a CFS, promotes the fusion of extra-chromosomal molecules, giving rise to large molecules containing multiple copies of both amplified sequences. Reintegration of these large elements was also observed, often targeted to chromosomal CFSs, which strikingly mimics the cytogenetic characteristics of some complex HSRs containing amplified non-syntenic sequences (Coquelle et al., 1998). Shimizu et al. studied by FISH and molecular approaches the evolution of plasmids transfected in mammalian cells, which revealed that head to tail fusion of plasmid copies generates large circular molecules resembling DMs. These elements may remain extra-chromosomal or may recombine with DMs present in the cells before transfection. In some cases, both types of extra-chromosomal elements reintegrate the chromosomes at various loci and initiate BFB-cycles. The authors proposed that, in this case, a conflict between replication and transcription generates the breaks that trigger the processes (Shimizu et al., 2003; Shimizu et al., 2001). Hence, fusion and reintegration appear as a general pathway for the evolution of extra-chromosomal elements that may account for the frequent observation of complex DMs and ectopic HSRs in cells of human cancers (Bernardino et al., 1998; Corvi et al., 1995; Fukumoto et al., 1993; Lafage et al., 1992; Muleris et al., 1995). This pathway may also account for the formation of ectopic HSRs bearing amplified sequences organised as direct repeats in cells of some cancers and cancer cell lines (Kuwahara et al., 2004; Schwab, 2004).

4.2 Interconversion of HSRs into DMs

The formation of DMs following HSR breakdown has also been proposed (Balaban-Mallenbaum et al., 1981; Singer et al., 2000). In order to analyze this putative pathway, Coquelle et al. engineered a cell line bearing an HSR containing numerous copies of the I-SceI site. I-Sce1 expression in these cells gave rise to HSR breakdown and DMs formation, which demonstrates that DMs may originate from an HSR and points to the role of DSBs in triggering this phenomenon (Coquelle et al., 2002).

4.3 Shortening of Intra-chromosomal Amplicons

During expansion of mutant clones selected *in vitro*, or upon re-amplification, it has been repeatedly shown that the large regular units accumulated at early steps of the process evolve rapidly, giving rise to shorter units (Figure 3A, B)(Ma et al., 1993; Smith et al., 1992; Toledo et al., 1992b; Toledo et al., 1992a). Toledo et al. have shown that two sequences co-amplified in early amplicons, which alternate in the large inverted repeats observed on condensed chromosome, often cluster in distinct regions of interphase nuclei. This observation, together with the frequent occurrence of nuclear blebs and micronucleation in cells undergoing BFB-cycles, suggested a model accounting for the shortening of the early units (Fig. 3C) (Toledo et al., 1992b). This phenomenon would, in one step, generate interstitial deletions within most initial repeats and restart cycles of BFB that would re-amplify the rearranged units.

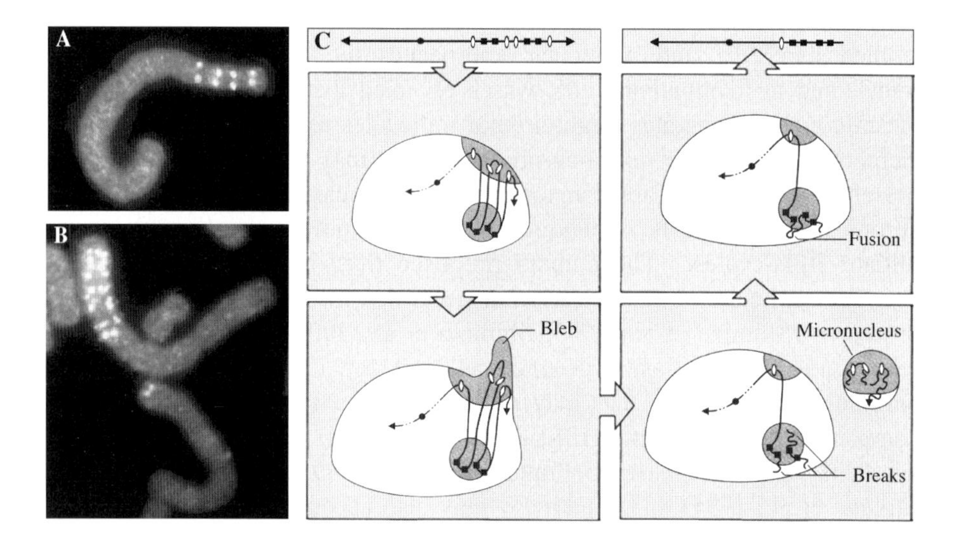

Figure 3. Shortening of intrachromosomal repeats. A: Example of large and regular repeats accumulated at early steps of the process. B: Example of heterogenous and shorter units observed following clonal expansion. C: Segregation-Blebbing-Breakage model: Two-colours FISH revealing chromosome markers co-amplified in the initial repeats showed that while co-amplified sequences alternate along the amplified chromosomes, they tend to cluster within different domains in interphase nuclei. Blebs and micronuclei containing multiple copies of only one marker were commonly observed (see Toledo et al., 1992b). This suggested that internal deletions may occur in a single step in a series of initial units, as presented here. Note that the process generating a broken chromatid, new rounds of BFB-cycles may lead to reamplification of the rearranged repeats.

These observations may explain why CFSs, that were recurrently involved in the generation of translocations and deletions favouring tumour progression (Smith et al., 1998), were not frequently found at the boundaries of amplified units in cells of human cancers. Recently, the development of microarray-based methods permitted to determine more precisely the limits of the amplicons accumulated in various types of tumours and tumour cell lines and the copy number of different sequences along these amplicons. Unexpected variations in copy number were often found for sequences co-linear on the physical map of the normal locus, as examplified for amplification of loci 11q13, 20q and 17q23 in breast cancers (Albertson, 2003; Hodgson et al., 2003; Ormandy et al., 2003; Sinclair et al., 2003). These findings strongly suggest that internal rearrangements occurred in the early amplicons and that regions maintained at high copy number are those bearing oncogenes involved in tumour progression. The mechanism described above and subsequent selection pressure may indeed account for these features.

ACKNOWLEDGEMENTS

We thank Dr G. Buttin for careful reading of the manuscript and helpful discussion. This work was supported in part by the Association pour la Recherche sur le Cancer (ARC) and the Ligue Nationale Française contre le Cancer (Comité d'Ile de France).

REFERENCES

Albertson, D.G. 2003. Profiling breast cancer by array CGH. *Breast Cancer Res Treat.* 78:289-98.

Alt, F., R. Kellems, J.R. Bertino, and R.T. Schimke. 1978. Selective multiplications of dihydrofolate reductase genes in methotrexate-resistant variants of cultured murine cells. *Journal of Biological Chemistry.* 253:1357-1370.

Arlt, M.F., A.M. Casper, and T.W. Glover. 2003. Common fragile sites. *Cytogenet Genome Res.* 100:92-100.

Arlt, M.F., B. Xu, S.G. Durkin, A.M. Casper, M.B. Kastan, and T.W. Glover. 2004. BRCA1 is required for common-fragile-site stability via its G2/M checkpoint function. *Mol Cell Biol.* 24:6701-9.

Artandi, S.E., S. Alson, M.K. Tietze, N.E. Sharpless, S. Ye, R.A. Greenberg, D.H. Castrillon, J.W. Horner, S.R. Weiler, R.D. Carrasco, and R.A. DePinho. 2002. Constitutive telomerase expression promotes mammary carcinomas in aging mice. *Proc Natl Acad Sci U S A.* 99:8191-6.

Balaban-Mallenbaum, G., G. Grove, and F.W. Gilbert. 1981. The proposed origin of double minutes from homogeneously staining regions (HSR)-marker chromosomes in human neuroblastoma hybrid cell lines. *Cancer Genet. Cytogenet.* 2:339-348.

Bassing, C.H., and F.W. Alt. 2004. H2AX may function as an anchor to hold broken chromosomal DNA ends in close proximity. *Cell Cycle.* 3:149-53.

Bassing, C.H., H. Suh, D.O. Ferguson, K.F. Chua, J. Manis, M. Eckersdorff, M. Gleason, R. Bronson, C. Lee, and F.W. Alt. 2003. Histone H2AX: a dosage-dependent suppressor of oncogenic translocations and tumours. *Cell*. 114:359-70.

Bernardino, J., F. Apiou, M. Gerbault-Seureau, B. Malfoy, and B. Dutrillaux. 1998. Characterization of recurrent homogeneously staining regions in 72 breast cancers. *Genes Chromosomes and Cancer*. 23:100-108.

Biedler, J.L., and B.A. Spengler. 1976. Metaphase chromosome anomaly: association with drug resistance and cell-specific products. *Science*. 191:185-187.

Brison, O. 1993. Gene amplification and tumour progression. *Biochim. Biophys. Acta*. 1155:25-41.

Canute, G.W., S.L. Longo, J.A. Longo, M.M. Shetler, T.E. Coyle, J.A. Winfield, and P.J. Hahn. 1998. The hydroxyurea-induced loss of double-minute chromosomes containing amplified epidermal growth factor receptor genes reduces the tumorigenicity and growth of human glioblastoma multiforme. *Neurosurgery*. 42:609-16.

Carroll, S.M., M.L. DeRose, P. Gaudray, C.M. Moore, D.R. Needham-Vandevanter, D.D. Von Hoff, and G.M. Wahl. 1988. Double minute chromosomes can be produced from precursors derived from a chromosomal deletion. *Mol. Cell. Biol.* 8:1525-1533.

Carroll, S.M., P. Gaudray, M.L. DeRose, J.F. Emery, J.L. Meinkoth, E. Nakkim, M. Subler, D.D. VonHoff, and G.M. Wahl. 1987. Characterization of an episome produced in hamster cells that amplify a transfected CAD gene at high frequency: functional evidence for a mammalian replication origin. *Molecular and Cellular Biology*. 7:1740-1750.

Casper, A.M., S.G. Durkin, M.F. Arlt, and T.W. Glover. 2004. Chromosomal Instability at Common Fragile Sites in Seckel Syndrome. *Am J Hum Genet*. 75.

Casper, A.M., P. Nghiem, M.F. Arlt, and T.W. Glover. 2002. ATR regulates fragile site stability. *Cell*. 111:779-89.

Celeste, A., S. Difilippantonio, M.J. Difilippantonio, O. Fernandez-Capetillo, D.R. Pilch, O.A. Sedelnikova, M. Eckhaus, T. Ried, W.M. Bonner, and A. Nussenzweig. 2003. H2AX haploinsufficiency modifies genomic stability and tumour susceptibility. *Cell*. 114:371-83.

Cha, R.S., and N. Kleckner. 2002. ATR homolog Mec1 promotes fork progression, thus averting breaks in replication slow zones. *Science*. 297:602-6.

Chernova, O.B., M.V. Chernov, Y. Ishizaka, M.L. Agarwal, and G.R. Stark. 1998. MYC abrogates p53-mediated cell cycle arrest in N-(phosphonacetyl)-L-aspartate-treated cells, permitting CAD gene amplification. *Mol Cell Biol*. 18:536-45.

Chin, K., C.O. de Solorzano, D. Knowles, A. Jones, W. Chou, E.G. Rodriguez, W.L. Kuo, B.M. Ljung, K. Chew, K. Myambo, M. Miranda, S. Krig, J. Garbe, M. Stampfer, P. Yaswen, J.W. Gray, and S.J. Lockett. 2004. In situ analyses of genome instability in breast cancer. *Nat Genet*. 36:984-8.

Ciullo, M., M.A. Debily, L. Rozier, M. Autiero, A. Billault, V. Mayau, S. El Marhomy, J. Guardiola, A. Bernheim, P. Coullin, D. Piatier-Tonneau, and M. Debatisse. 2002. Initiation of the breakage-fusion-bridge mechanism through common fragile site activation in human breast cancer cells: the model of PIP gene duplication from a break at FRA7I. *Hum Mol Genet*. 11:2887-94.

Coquelle, A., E. Pipiras, F. Toledo, G. Buttin, and M. Debatisse. 1997. Expression of fragile sites triggers intrachromosomal mammalian gene amplification and sets boundaries to early amplicons. *Cell*. 89:215-25.

Coquelle, A., L. Rozier, B. Dutrillaux, and M. Debatisse. 2002. Induction of multiple double-strand breaks within an hsr by meganucleaseI-SceI expression or fragile site activation leads to formation of double minutes and other chromosomal rearrangements. *Oncogene*. 21:7671-9.

Coquelle, A., F. Toledo, S. Stern, A. Bieth, and M. Debatisse. 1998. A new role for hypoxia in tumour progression: induction of fragile sites triggering genomic rearrangements and formation of complex DMs and HSRs. *Mol. Cell*. 2:259-265.

Corvi, R., L. Savelyeva, L. Amler, R. Handgretinger, and M. Schwab. 1995. Cytogenetic evolution of MYCN and MDM2 amplification in the neuroblastoma LS tumour and its cell line. *Eur. J. Cancer*. 31A:520-3.

Cowell, J.K. 1982. Double minutes and homogeneously staining regions : gene amplification in mammalian cells. *Ann. Rev. Genet.* 16:21-59.

Deng, Z., L. Lezina, C.J. Chen, S. Shtivelband, W. So, and P.M. Lieberman. 2002. Telomeric proteins regulate episomal maintenance of Epstein-Barr virus origin of plasmid replication. *Mol Cell.* 9:493-503.

der-Sarkissian, H., S. Bacchetti, L. Cazes, and J.A. Londono-Vallejo. 2004. The shortest telomeres drive karyotype evolution in transformed cells. *Oncogene.* 23:1221-8.

Difilippantonio, M.J., S. Petersen, H.T. Chen, R. Johnson, M. Jasin, R. Kanaar, T. Ried, and A. Nussenzweig. 2002. Evidence for replicative repair of DNA double-strand breaks leading to oncogenic translocation and gene amplification. *J Exp Med.* 196:469-80.

Dujon, B. 1989. Group I introns as mobile genetic elements: facts and mechanistic speculations-a review. *Gene.* 82:91-114.

Fernandez-Capetillo, O., A. Celeste, and A. Nussenzweig. 2003. Focusing on foci: H2AX and the recruitment of DNA damage response factors. *Cell Cycle.* 2:426-7.

Fukumoto, M., A. Suzuki, J. Inazawa, T. Yoshimura, S.Arao, T. Takahashi, H. Nomura, and H. Hiai. 1993. Chromosomal location and structure of amplicons in two human cell lines with coamplification of c-*myc* and Ki-*ras* oncogenes. *Somatic Cell Mol. Genet.* 19:21-28.

Gisselsson, D. 2003. Chromosome instability in cancer: how, when, and why? *Adv Cancer Res.* 87:1-29.

Gisselsson, D., L. Gorunova, M. Hoglund, N. Mandahl, and P. Elfving. 2004. Telomere shortening and mitotic dysfunction generate cytogenetic heterogeneity in a subgroup of renal cell carcinomas. *Br J Cancer.* 91:327-32.

Gisselsson, D., L. Pettersson, M. Hoglund, M. Heidenblad, L. Gorunova, J. Wiegant, F. Mertens, P. Dal Cin, F. Mitelman, and N. Mandahl. 2000. Chromosomal breakage-fusion-bridge events cause genetic intratumor heterogeneity. *Proc Natl Acad Sci U S A.* 97:5357-62.

Gladdy, R.A., M.D. Taylor, C.J. Williams, I. Grandal, J. Karaskova, J.A. Squire, J.T. Rutka, C.J. Guidos, and J.S. Danska. 2003. The RAG-1/2 endonuclease causes genomic instability and controls CNS complications of lymphoblastic leukemia in p53/Prkdc-deficient mice. *Cancer Cell.* 3:37-50.

Graux, C., J. Cools, C. Melotte, H. Quentmeier, A. Ferrando, R. Levine, J.R. Vermeesch, M. Stul, B. Dutta, N. Boeckx, A. Bosly, P. Heimann, A. Uyttebroeck, N. Mentens, R. Somers, R.A. MacLeod, H.G. Drexler, A.T. Look, D.G. Gilliland, L. Michaux, P. Vandenberghe, I. Wlodarska, P. Marynen, and A. Hagemeijer. 2004. Fusion of NUP214 to ABL1 on amplified episomes in T-cell acute lymphoblastic leukemia. *Nat Genet.* 36:1084-9.

Hammond, E.M., M.J. Dorie, and A.J. Giaccia. 2004. Inhibition of ATR leads to increased sensitivity to hypoxia/reoxygenation. *Cancer Res.* 64:6556-62.

Hellman, A., A. Rahat, S.W. Scherer, A. Darvasi, L.C. Tsui, and B. Kerem. 2000. Replication delay along FRA7H, a common fragile site on human chromosome 7, leads to chromosomal instability. *Mol Cell Biol.* 20:4420-7.

Hellman, A., E. Zlotorynski, S.W. Scherer, J. Cheung, J.B. Vincent, D.I. Smith, L. Trakhtenbrot, and B. Kerem. 2002. A role for common fragile site induction in amplification of human oncogenes. *Cancer Cell.* 1:89-97.

Hodgson, J.G., K. Chin, C. Collins, and J.W. Gray. 2003. Genome amplification of chromosome 20 in breast cancer. *Breast Cancer Res Treat.* 78:337-45.

Hoglund, M., D. Gisselsson, G.B. Hansen, T. Sall, and F. Mitelman. 2003. Ovarian carcinoma develops through multiple modes of chromosomal evolution. *Cancer Res.* 63:3378-85.

Ilves, I., S. Kivi, and M. Ustav. 1999. Long-term episomal maintenance of bovine papillomavirus type 1 plasmids is determined by attachment to host chromosomes, which Is mediated by the viral E2 protein and its binding sites. *J Virol.* 73:4404-12.

Ishizaka, Y., M.V. Chernov, C.M. Burns, and G.R. Stark. 1995. p53-dependent growth arrest of REF52 cells containing newly amplified DNA. *Proc Natl Acad Sci U S A.* 92:3224-8.

Itoh, N., and N. Shimizu. 1998. DNA replication-dependent intranuclear relocation of double minute chromatin. *J Cell Sci.* 111 (Pt 22):3275-85.

Jasin, M. 1996. Genetic manipulation of genomes with rare-cutting endonucleases. *Trends Genet.* 12:224-8.

Jenke, A.C., I.M. Stehle, F. Herrmann, T. Eisenberger, A. Baiker, J. Bode, F.O. Fackelmayer, and H.J. Lipps. 2004. Nuclear scaffold/matrix attached region modules linked to a transcription unit are sufficient for replication and maintenance of a mammalian episome. *Proc Natl Acad Sci U S A.* 101:11322-7.

Kanda, T., M. Otter, and G.M. Wahl. 2001. Coupling of mitotic chromosome tethering and replication competence in epstein-barr virus-based plasmids. *Mol Cell Biol.* 21:3576-88.

Kaufman, R.J., P.C. Brown, and R.T. Schimke. 1979. Amplified dihydrofolate reductase genes in unstable methotrexate-resistant cells are associated with double minute chromosomes. *Proc. Natl. Acad. Sci. USA.* 76:5669-5673.

Knuutila, S., A.M. Bjorkqvist, K. Autio, M. Tarkkanen, M. Wolf, O. Monni, J. Szymanska, M.L. Larramendy, J. Tapper, H. Pere, W. El-Rifai, S. Hemmer, V.M. Wasenius, V. Vidgren, and Y. Zhu. 1998. DNA copy number amplifications in human neoplasms: review of comparative genomic hybridization studies. *Am. J. Pathol.* 152:1107-23.

Kuo, M.T., R.C. Vyas, L.X. Jiang, and W.N. Hittelman. 1994. Chromosome breakage at a major fragile site associated with P-glycoprotein gene amplification in multidrug-resistant CHO cells. *Mol. Cell. Biol.* 14:5202-11.

Kuwahara, Y., C. Tanabe, T. Ikeuchi, K. Aoyagi, M. Nishigaki, H. Sakamoto, K. Hoshinaga, T. Yoshida, H. Sasaki, and M. Terada. 2004. Alternative mechanisms of gene amplification in human cancers. *Genes Chromosomes Cancer.* 41:125-32.

Lafage, M., F. Pedeutour, S. Marchetto, J. Simonetti, M.T. Prosperi, P. Gaudray, and D. Birnbaum. 1992. Fusion and amplification of two originally non-syntenic chromosomal regions in a mammary carcinoma cell line. *Genes Chromosomes and Cancer.* 5:40-9.

Leo, C., A.J. Giaccia, and N.C. Denko. 2004. The hypoxic tumour microenvironment and gene expression. *Semin Radiat Oncol.* 14:207-14.

Lese, C.M., K.M. Rossie, B.N. Appel, J.K. Reddy, J.T. Johnson, E.N. Myers, and S.M. Gollin. 1995. Visualization of INT2 and HST1 amplification in oral squamous cell carcinomas. *Genes Chromosomes Cancer.* 12:288-95.

Livingstone, L.R., A. White, J. Sprouse, E. Livanos, T. Jacks, and T.D. Tlsty. 1992. Cell cycle arrest and gene amplification potential accompany loss of wild-type p53. *Cell.* 70:923-935.

Londono-Vallejo, J.A. 2004. Telomere length heterogeneity and chromosome instability. *Cancer Lett.* 212:135-44.

Luk, C.K., L. Veinot-Drebot, E. Tjan, and I.F. Tannock. 1990. Effect of transient hypoxia on sensitivity to doxorubicin in human and murine cell lines. *J. Natl. Cancer Inst.* 82:684-92.

Ma, C., S. Martin, B. Trask, and J.L. Hamlin. 1993. Sister chromatid fusion initiates amplification of the dihydrofolate reductase gene in Chinese hamster cells. *Genes Dev.* 7:605-20.

Maser, R.S., and R.A. DePinho. 2004. Telomeres and the DNA damage response: why the fox is guarding the henhouse. *DNA Repair (Amst).* 3:979-88.

Maurer, B.J., E. Lai, B.A. Hamkalo, L. Hood, and G. Attardi. 1987. Novel submicroscopic extrachromosomic elements containing amplified genes in human cells. *Nature.* 327:434-437.

McClintock, B. 1951. Chromosome organization and genic expression. *Cold Spring Harbor Symp. Quant. Biol.* 16:13-47.

Mondello, C., P. Rebuzzini, M. Dolzan, S. Edmonson, G.E. Taccioli, and E. Giulotto. 2001. Increased gene amplification in immortal rodent cells deficient for the DNA-dependent protein kinase catalytic subunit. *Cancer Res.* 61:4520-5.

Muleris, M., A. Almeida, M. Gerbault-Seureau, B. Malfoy, and B. Dutrillaux. 1995. Identification of amplified DNA sequences in breast cancer and their organization within homogeneously staining regions. *Genes, Chromosomes & Cancer.* 14:155-63.

Nonet, G.H., S.M. Carroll, M.L. DeRose, and G.M. Wahl. 1993. Molecular dissection of an extrachromosomal amplicon reveals a circular structure consisting of an imperfect inverted duplication. *Genomics.* 15:543-58.

O'Hagan, R.C., S. Chang, R.S. Maser, R. Mohan, S.E. Artandi, L. Chin, and R.A. DePinho. 2002. Telomere dysfunction provokes regional amplification and deletion in cancer genomes. *Cancer Cell*. 2:149-55.

O'Sullivan, J.N., M.P. Bronner, T.A. Brentnall, J.C. Finley, W.T. Shen, S. Emerson, M.J. Emond, K.A. Gollahon, A.H. Moskovitz, D.A. Crispin, J.D. Potter, and P.S. Rabinovitch. 2002. Chromosomal instability in ulcerative colitis is related to telomere shortening. *Nat Genet*. 32:280-4.

Okuno, Y., P.J. Hahn, and D.M. Gilbert. 2004. Structure of a palindromic amplicon junction implicates microhomology-mediated end joining as a mechanism of sister chromatid fusion during gene amplification. *Nucleic Acids Res*. 32:749-56.

Ormandy, C.J., E.A. Musgrove, R. Hui, R.J. Daly, and R.L. Sutherland. 2003. Cyclin D1, EMS1 and 11q13 amplification in breast cancer. *Breast Cancer Res Treat*. 78:323-35.

Paulson, T.G., A. Almasan, L.L. Brody, and G.M. Wahl. 1998. Gene amplification in a p53-deficient cell line requires cell cycle progression under conditions that generate DNA breakage. *Mol Cell Biol*. 18:3089-100.

Pedeutour, F., R.F. Suijkerbuijk, A. Forus, J. Van Gaal, W. Van de Klundert, J.M. Coindre, G. Nicolo, F. Collin, U. Van Haelst, K. Huffermann, and et al. 1994. Complex composition and co-amplification of SAS and MDM2 in ring and giant rod marker chromosomes in well-differentiated liposarcoma. *Genes Chromosomes Cancer*. 10:85-94.

Pipiras, E., A. Coquelle, A. Bieth, and M. Debatisse. 1998. Interstitial deletions and intrachromosomal amplification initiated from a double-strand break targeted to a mammalian chromosome. *EMBO J*. 17:325-333.

Popescu, N.C. 2003. Genetic alterations in cancer as a result of breakage at fragile sites. *Cancer Lett*. 192:1-17.

Poupon, M.F., K.A. Smith, O.B. Chernova, C. Gilbert, and G.R. Stark. 1996. Inefficient growth arrest in response to dNTP starvation stimulates gene amplification through bridge-breakage-fusion cycles. *Mol. Biol. Cell*. 7:345-354.

Raymond, E., S. Faivre, G. Weiss, J. McGill, K. Davidson, E. Izbicka, J.G. Kuhn, C. Allred, G.M. Clark, and D.D. Von Hoff. 2001. Effects of hydroxyurea on extrachromosomal DNA in patients with advanced ovarian carcinomas. *Clin Cancer Res*. 7:1171-80.

Rice, G.C., C. Hoy, and R.T. Schimke. 1986. Transient hypoxia enhances the frequency of dihydrofolate reductase gene amplification in Chinese hamster ovary cells. *Proc. Natl. Acad. Sci. USA*. 83:5978-5982.

Richards, R.I. 2001. Fragile and unstable chromosomes in cancer: causes and consequences. *Trends Genet*. 17:339-45.

Roelofs, H., E. Schuuring, J. Wiegant, R. Michalides, and M. Giphart-Gassler. 1993. Amplification of the 11q13 region in human carcinoma cell lines: a mechanistic view. *Genes Chromosomes Cancer*. 7:74-84.

Saunders, W.S., M. Shuster, X. Huang, B. Gharaibeh, A.H. Enyenihi, I. Petersen, and S.M. Gollin. 2000. Chromosomal instability and cytoskeletal defects in oral cancer cells. *Proc Natl Acad Sci U S A*. 97:303-8.

Schoenlein, P.V., J.T. Barrett, A. Kulharya, M.R. Dohn, A. Sanchez, D.Y. Hou, and J. McCoy. 2003. Radiation therapy depletes extrachromosomally amplified drug resistance genes and oncogenes from tumour cells via micronuclear capture of episomes and double minute chromosomes. *Int J Radiat Oncol Biol Phys*. 55:1051-65.

Schwab, M. 1999. Oncogene amplification in solid tumours. *Semin. cancer Biol*. 9:319-325.

Schwab, M. 2004. MYCN in neuronal tumours. *Cancer Lett*. 204:179-87.

Schwab, M., and L.C. Amler. 1990. Amplification of cellular oncogenes: a predictor of clinical outcome in human cancer. *Genes Chromosomes Cancer*. 1:181-193.

Sharpless, N.E., D.O. Ferguson, R.C. O'Hagan, D.H. Castrillon, C. Lee, P.A. Farazi, S. Alson, J. Fleming, C.C. Morton, K. Frank, L. Chin, F.W. Alt, and R.A. DePinho. 2001. Impaired nonhomologous end-joining provokes soft tissue sarcomas harbouring chromosomal translocations, amplifications, and deletions. *Mol Cell*. 8:1187-96.

Shimizu, N., T. Hashizume, K. Shingaki, and J.K. Kawamoto. 2003. Amplification of plasmids containing a mammalian replication initiation region is mediated by controllable conflict between replication and transcription. *Cancer Res.* 63:5281-90.

Shimizu, N., N. Itoh, H. Utiyama, and G.M. Wahl. 1998. Selective entrapment of extrachromosomally amplified DNA by nuclear budding and micronucleation during S phase. *1998*. 140:1307-1320.

Shimizu, N., Y. Miura, Y. Sakamoto, and K. Tsutsui. 2001. Plasmids with a mammalian replication origin and a matrix attachment region initiate the event similar to gene amplification. *Cancer Res.* 61:6987-90.

Shuster, M.I., L. Han, M.M. Le Beau, E. Davis, M. Sawicki, C.M. Lese, N.H. Park, J. Colicelli, and S.M. Gollin. 2000. A consistent pattern of RIN1 rearrangements in oral squamous cell carcinoma cell lines supports a breakage-fusion-bridge cycle model for 11q13 amplification. *Genes Chromosomes Cancer*. 28:153-63.

Sinclair, C.S., M. Rowley, A. Naderi, and F.J. Couch. 2003. The 17q23 amplicon and breast cancer. *Breast Cancer Res Treat.* 78:313-22.

Singer, M.J., L.D. Mesner, C.L. Friedman, B.J. Trask, and J.L. Hamlin. 2000. Amplification of the human dihydrofolate reductase gene via double minutes is initiated by chromosome breaks. *Proc Natl Acad Sci U S A*. 97:7921-6.

Smith, D.I., H. Huang, and L. Wang. 1998. Common fragile sites and cancer. *International Journal of Oncology*. 12:187-196.

Smith, K.A., P.A. Gorman, M.B. Stark, R.P. Groves, and G.R. Stark. 1990. Distinctive chromosomal structures are formed very early in the amplification of CAD genes in Syrian hamster cells. *Cell*. 63:1219-1227.

Smith, K.A., M.B. Stark, P.A. Gorman, and G.R. Stark. 1992. Fusion near telomeres occur very early in the amplification of CAD genes in Syrian hamster cells. *Proc. Natl. Acad. Sci. U.S.A.* 89:5427-5431.

Stark, G.R., and G.M. Wahl. 1984. Gene amplification. *Ann. Rev. Biochem.* 53:447-491.

Storlazzi, C.T., T. Fioretos, K. Paulsson, B. Strombeck, C. Lassen, T. Ahlgren, G. Juliusson, F. Mitelman, M. Rocchi, and B. Johansson. 2004. Identification of a commonly amplified 4.3 Mb region with overexpression of C8FW, but not MYC in MYC-containing double minutes in myeloid malignancies. *Hum Mol Genet.* 13:1479-85.

Sutherland, G.R., E. Baker, and R.I. Richards. 1998. Fragile Sites Still Breaking. *Trends in Genetics*. 14:501-506.

Tanaka, T., and N. Shimizu. 2000. Induced detachment of acentric chromatin from mitotic chromosomes leads to their cytoplasmic localization at G(1) and the micronucleation by lamin reorganization at S phase. *J Cell Sci.* 113 (Pt 4):697-707.

Thomas, L., J. Stamberg, I. Gojo, Y. Ning, and A.P. Rapoport. 2004. Double minute chromosomes in monoblastic (M5) and myeloblastic (M2) acute myeloid leukemia: two case reports and a review of literature. *Am J Hematol.* 77:55-61.

Toledo, F., G. Buttin, and M. Debatisse. 1993. The origin of chromosome rearrangements at early stages of *AMPD2* gene amplification in Chinese hamster cells. *Current Biol.* 3:255-264.

Toledo, F., D. LeRoscouet, G. Buttin, and M. Debatisse. 1992b. Co-amplified markers alternate in megabase long chromosomal inverted repeats and cluster independently in interphase nuclei at early steps of mammalian gene amplification. *EMBO J.* 11:2665-2673.

Toledo, F., K.A. Smith, G. Buttin, and M. Debatisse. 1992a. The evolution of the amplified adenylate-deaminase-2 domains in Chinese hamster cells suggests the sequential intervention of different mechanisms of DNA amplification. *Mutation Res.* 276:261-273.

Vogt, N., S.H. Lefevre, F. Apiou, A.M. Dutrillaux, A. Cor, P. Leuraud, M.F. Poupon, B. Dutrillaux, M. Debatisse, and B. Malfoy. 2004. Molecular structure of double-minute chromosomes bearing amplified copies of the epidermal growth factor receptor gene in gliomas. *Proc Natl Acad Sci U S A*. 101:11368-73.

Wahl, G.M. 1989. The importance of circular DNA in mammalian gene amplification. *Cancer Res.* 49:1333-1340.

Windle, B., B.W. Draper, Y. Yin, S. O'Gorman, and G.M. Wahl. 1991. A central role for chromosome breakage in gene amplification, deletion formation and amplicon integration. *Genes Dev.* 5:160-174.

Windle, B.E., and G.M. Wahl. 1992. Molecular dissection of mammalian gene amplification: new mechanistic insights revealed by analyses of very early events. *Mutat Res.* 276:199-224.

Yin, Y., M.A. Tainsky, F.Z. Bischoff, L.C. Strong, and G.M. Wahl. 1992. Wild-type p53 restores cell cycle control and inhibits gene amplification in cells with mutant p53 alleles. *Cell.* 70:937-948.

Zhu, C., K.D. Mills, D.O. Ferguson, C. Lee, J. Manis, J. Fleming, Y. Gao, C.C. Morton, and F.W. Alt. 2002. Unrepaired DNA breaks in p53-deficient cells lead to oncogenic gene amplification subsequent to translocations. *Cell.* 109:811-21.

Chapter 3.6

DNA METHYLATION AND CANCER-ASSOCIATED GENETIC INSTABILITY

Melanie Ehrlich
Tulane Cancer Center, Human Genetics Program, Department of Biochemistry, Tulane Medical School, New Orleans, Louisiana, USA

1. INTRODUCTION

Epigenetic changes at the DNA level, *i.e.*, alterations in the distribution of 5-methylcytosine (m^5C), play a critical role in carcinogenesis. These changes consist of both increases and decreases in cytosine methylation in different genomic sequences (Baylin and Herman, 2000; Dutrillaux, 2000; Ehrlich, 2002; Issa, 2000; Mitelman et al., 1997). Transmission of the oncogenic phenotype from cell to progeny cell also relies on epigenetic changes at the chromatin level, which may be mechanistically associated with or independent of DNA methylation changes (Momparler, 2003). During carcinogenesis, these epigenetic modifications supplement genetic changes (point mutagenesis, chromosomal rearrangements, aneuploidy, and polyploidy), the classic source of carcinogenic alterations inherited at the cellular, and sometimes, also the individual level. Both genetic and epigenetic changes associated with carcinogenesis occur at much higher frequency in cancer cells than in normal cells and appear early in many types of tumorigenesis. Furthermore, both type of changes tend to evolve in association with tumour progression (Heim and Teixeira, 2000; Hoglund et al., 2003; Itano et al., 2002; Salem et al., 2000; Widschwendter et al., 2004). At the gross chromosomal level, karyotypic instability involves a progressive alteration of the karyotype affecting a cell population (Dutrillaux, 2000). At the level of local mutagenesis, tumour progression is associated with an accumulation of mutations often linked to a mutator phenotype (Albor and Kulesz-Martin, 2000; Fearon et al., 1990). Tumour progression can be accompanied by the spreading in *cis* of DNA methylation changes or an increased frequency of these changes (Itano et al., 2002; Laird, 2003; Widschwendter et al., 2004). Moreover, cancers can display a

E.A. Nigg (ed.), Genome Instability in Cancer Development, 363-392.

hypermethylator phenotype (Issa, 2000) analogous to the mutator phenotype. The high frequency of inherited changes found in cancers seems to result from an acquired tumour-related inability to maintain the stable genotype and epigenotype. This instability results in continuous changes in gene expression patterns that can favour tumour formation, increasingly aggressive tumour behaviour, and metastasis.

It has been proposed that cancer-linked changes in DNA methylation influence genetic instability (Kokalj-Vokac et al., 1993; Qu et al., 1999b; Veigl et al., 1998). Cancer-associated hypermethylation of CpG islands overlapping promoters (Baylin and Herman, 2000) could alter expression of genes involved in maintaining gross chromosomal stability or in minimising point mutagenesis. Hypomethylation of DNA linked to cancer may also play a role in genetic instability. Decreases in DNA methylation in cancer are often much more numerous than increases in this methylation, which leads to a net deficiency in genomic m^5C (Gama-Sosa et al., 1983). The often-observed hypomethylation of tandem DNA repeats, especially in the vicinity of the centromere, in a wide variety of cancers (Ehrlich, 2002; Narayan et al., 1998) could favour chromosomal rearrangements or might interfere with proper chromosome segregation. There is also hypomethylation at interspersed repeats (Florl et al., 1999) and non-repeated DNA sequences (Cho et al., 2001; Gupta et al., 2003; Sato et al., 2003; Scelfo et al., 2002), which might make these regions prone to DNA rearrangements by *cis* effects. Such effects could involve interactions of DNA methylation, histone modification, and non-histone chromatin proteins. Evidence for and against the relationships of cancer-linked methylation changes to genetic instability in tumours will be considered below.

2. CANCER-ASSOCIATED PROMOTER HYPERMETHYLATION AND GENOMIC INSTABILITY

2.1 Epigenetic Silencing of *MLH1* and Microsatellite Instability

As described in an earlier chapter, one of the sources of the hypermutator phenotype characteristic of hereditary non-polyposis colorectal carcinoma (HNPCC) is the inheritance of one mutationally inactivated allele of a mismatch repair (MMR) gene. Usually the affected gene that is mutated in the germline is *hMLH1* or *hMSH2* (Bocker et al., 1999). These genes are involved in repair of spontaneous DNA replication errors. In HNPCC, the remaining, single wild-type allele of *hMLH1* or *hMSH2* is usually genetically inactivated somatically in the tumour, often as the result of loss of heterozygosity (LOH) (Hemminki et al., 1994; Wheeler et al., 2000). The ensuing decrease in MMR results in changes in the length of microsatellite sequences (microsatellite instability, MSI) and the introduction of high frequencies of frameshift mutations, including in the transforming growth factor-β type II receptor gene in the tumours. The latter is a tumour suppressor gene that serves as a hotspot for mutations in HNPCC (Kinzler and Vogelstein, 1996; Markowitz et al., 1995). In contrast to HNPCC colon cancer in which one MMR allele typically has an inherited mutation and the other a somatically acquired mutational loss of function, sporadic colon cancer, endometrial cancer, and gastric cancer often display the MSI hypermutator phenotype as a result of epigenetic MMR gene alterations (Esteller et al., 1998; Fleisher et al., 1999; Kane et al., 1997). Most of these tumours have *hMLH1* promoter hypermethylation, which is often biallelic, and is strongly associated with an MSI phenotype (Esteller et al., 1998; Fleisher et al., 1999; Veigl et al., 1998). Treatment of these cancer cells *in vitro* with a DNA methylation inhibitor, 5-azacytidine, induces re-expression of the previously silenced *hMLH1* gene (Veigl et al., 1998). Therefore, depending upon the type of cancer, including, its familiar or sporadic origin, inactivation of a gene that helps maintain genetic integrity can occur genetically or epigenetically.

Recently, evidence for the opposite type of relationship between MMR and DNA methylation or DNA methyltransferase was presented in a mouse model. Guo *et al.* (Guo et al., 2004) found that a deficiency in DNA methyltransferase 1 (*Dnmt1*) in murine embryonal stem cells, which should result in DNA hypomethylation, was linked to the MMR phenotype in a screen for MMR cells using a revertable gene trap retrovirus. While mechanistic associations are unclear, the authors note that other studies show that mice with *Dnmt1* mutations are predisposed to certain types of tumorigenesis and chromosome instability (Chen et al., 1998; Eden et al., 2003) (see below).

2.2 Epigenetic Silencing of *MGMT* and Transition Mutagenesis

Another DNA repair gene, *MGMT,* has been shown to be hypermethylated in its promoter in a wide variety of cancers (Eads et al., 2001; Esteller et al., 2001; Kang et al., 2004). *MGMT* encodes O^6-methylguanine-DNA methyltransferase and functions in repair of alkylated guanine formed by mutagenic drugs. Consistent with the function of this enzyme and the mispairing of O^6-methylguanine, the promoter hypermethylation is associated with G-to-A mutations in the *K-ras* oncogene in colorectal tumours, but not with tumours having other types of mutations in this gene (Esteller et al., 2000b). Furthermore, in brain tumours with a methylated *MGMT* promoter there was a significantly increased frequency of G-to-A mutations in the *TP53 (p53)* gene compared with tumours having an unmethylated *MGMT* promoter (Yin et al., 2003). Given the lack of *MGMT* RNA in cultured tumour cells with this hypermethylation, the reactivation of the hypermethylated gene by 5-azacytidine, and the association of reduced histone H4 acetylation, reduced H3 K9 acetylation, and increased H3 K9 dimethylation with promoter hypermethylation (Bhakat and Mitra, 2003; Kondo et al., 2003; Nakagawachi et al., 2003; Qian and Brent, 1997), it is likely that hypermethylation of the *MGMT* promoter in colon cancer is responsible for inadequate repair of abnormally methylated G residues in *K-ras* leading to the prevalent observed transition mutations in this oncogene in the cancers.

2.3 Epigenetic Silencing of *BRCA1*, DNA Repair and Cell-Cycle Progression

Like *hMLH1*, *BRCA1* is a tumour suppressor gene that is involved in preserving genetic stability and can be mutant in the germline or become somatically inactivated. Germline mutations in this gene, which has been implicated in double-strand DNA break repair, transcriptional regulation, and cell-cycle progression (Sato et al., 2004; Speit and Trenz, 2004; Westermark et al., 2003; Zhang et al., 2004), are found in a large fraction of familial breast cancer and ovarian cancer patients (Jhanwar-Uniyal, 2003; Wang et al., 2004). Hypermethylation has been observed in appreciable percentages of sporadic breast and ovarian cancers, but not in colon cancer or leukemias (Bianco et al., 2000; Esteller et al., 2000a; Wang et al., 2004). Hypermethylation of the *BRCA1* promoter, as well as *BRCA1* LOH, is associated with decreased expression of BRCA1, at the RNA and protein levels.

2.4 Epigenetic Silencing of Genes and Increased Cell-Cycle Progression or Decreased Apoptosis

In addition to DNA repair genes, cell cycle-progression or pro-apoptotic genes are often down-regulated by cancer-associated DNA hypermethylation. Inhibition of expression of these genes can interfere with normal cell cycle checkpoints to repair cell damage or to eliminate cells with irreparable DNA damage, as described in earlier chapters. The cell cycle- or apoptosis-related genes that are susceptible to cancer-associated 5' gene region hypermethylation include *p16CDKN2A*, 14-3-3 σ, caspase-8, *TMS1*, DAP-kinase, *p14ARF*, and *p21CIPI/WAF1/SDI1* (Ferguson et al., 2000; Mhawech et al., 2004; Momparler, 2003; Roman-Gomez et al., 2002). Cancer-linked *de novo* methylation of this class of genes can contribute to cancer-associated genetic instability.

2.5 Mechanisms of Silencing by Increased Methylation of Transcription Control Sequences

Cancer-associated downregulation of expression of tumour suppressor genes by *de novo* methylation may involve changes in chromatin structure, usually of the promoter or 5' gene region, in response to DNA methylation changes or altered binding of transcription factors to CpG-containing binding sites. Decreases or increases in DNA methylation can affect chromatin structure by altering binding of sequence-nonspecific methylated DNA binding proteins (Wade, 2001), which, in turn, recruit histone deacetylases, corepressors, or other proteins to regulate transcription (Muegge et al., 2003; Robertson, 2002). Alternatively, changing methylation of DNA sequences can affect their interactions with sequence-specific DNA-binding proteins that bind either less or more avidly to their CpG-containing recognition sites in promoters or other transcription regulatory sequences when those sites are methylated and can act as transcription activators or repressors (De Smet et al., 1999; Filippova et al., 2001; Huang et al., 1984; Kanduri et al., 2002; Plass and Soloway, 2002; Sengupta et al., 1999; Sengupta et al., 2002; Takizawa et al., 2001; Zhang et al., 1993).

Furthermore, DNA methyltransferases not only catalyze DNA methylation, but also interact with proliferating cell nuclear antigen (PCNA), histone deacetylases, histone methylases, chromatin remodelling enzymes, heterochromatin protein 1 (HP1α), RB, and other nuclear proteins (Chuang et al., 1997; Fuks et al., 2003a; Geiman et al., 2004; Pradhan and Kim, 2002; Robertson, 2002). Therefore, cancer-associated changes in DNA methylation might affect recruitment of these proteins and thereby control gene expression by changing protein-protein interactions. Although in most normal tissues, one of their three main DNA methyltransferases, DNMT3B, is present at low levels (Robertson, 2002), some cancers have elevated levels of DNMT3B mRNA isoforms, even relative to other cell cycle-regulated genes (Kimura et al., 2003; Robertson, 2002). Increased DNMT RNA levels showed a correlation with hypermethylation of *p15/INK4B* in a study of AML (Mizuno et al., 2001), but in two other cancer studies, DNMT RNA levels displayed little or no correlation with downregulation or hypermethylation of genes with CpG islands overlapping their promoters (Eads et al., 1999; Girault et al., 2003). For hepatocellular carcinomas, it was reported that levels of RNA corresponding to a DNMT3B isoform that encodes a presumably catalytically inactive form of DNMT3B was associated with satellite 2 hypomethylation (Saito et al., 2002). However, we saw no significant association between these in a study of ovarian carcinomas (M. Ehrlich, P. Laird, and M. Yu, unpub. data). Levels of the RNA encoding DNMT1, the main DNA methyltransferase, analyzed in the above studies and others (Lee et al., 1996; Yakushiji et al., 2003) generally have been reported to show no change or to increase in cancers. How cancer-associated DNA hypomethylation (especially combined with hypermethylation in other DNA sequences) is established is unclear. For example, the frequency of mutations in *DNMT1* coding sequences in colorectal cancers was found to be negligible (Kanai et al., 2003). However, in an immunohistochemical study of colorectal cancers, carcinoma cells with very high levels of DNMT1 and other cells in the same cancer with low levels were detected (De Marzo et al., 1999). Moreover, evidence for selective proteolysis of DNMT1 by sulfonate-derived methylating agents has been reported (Chuang et al., 2002).

The interactions of methylated DNA-specific proteins (Billard et al., 2002; Fujita et al., 2003; Fuks et al., 2003b; Reese et al., 2003; Wade, 2001) as well as DNA methyltransferases with histone modifying enzymes, chromatin remodelling complexes, and heterochromatin-associated proteins show that epigenetic changes at the DNA level and at the chromatin level can impact and reinforce one another. In mouse embryonal stem cells and embryos, evidence was provided that one of the H3 K9 and K27 methyltransferases (G9a) is necessary for the maintenance of CpG methylation in an imprinting centre (Xin et al., 2003). Furthermore, in mouse embryonal stem cells exhibiting a greater loss of H3 K9 methylation due to knockout of two other genes encoding H3 K9 methyltransferases (*Suv39h1,Suv39h2* double null cells), juxtacentromeric satellite DNA had decreased cytosine methylation (Lehnertz et al., 2003). While DNA methylation and histone modifications can affect one another, there is only partial overlap of these phenomena in vertebrates. For example, in the latter study, decreased H3 K9 trimethylation did not cause decreases in methylation of centromeric or endogenous C-type retroviral DNAs, and reduced DNA methylation did not decrease H3 K9 methylation in juxtacentromeric heterochromatin (Lehnertz et al., 2003).

3. CANCER-ASSOCIATED HYPOMETHYLATION OF CENTROMERIC AND JUXTACENTROMERIC HETEROCHROMATIN

3.1 Pericentromeric Heterochromatin: Distinct Epigenetic Characteristics and Relevance to Cancer

The epigenetic status of the constitutive heterochromatin in juxtacentromeric or centromeric regions in postnatal somatic human cells is distinctive, with enrichment in H3 K9 trimethylation, a low amount of H4 acetylation, and a high percentage of C methylation (Jiang et al., 2004; Lehnertz et al., 2003; Narayan et al., 1998; Yang et al., 2004). Because of the overrepresentation of rearrangements in the pericentromeric (juxtacentromeric or centromeric) heterochromatin in cancer cells (Mertens et al., 1997; Mitelman et al., 1997) and the frequent aneuploidy, which might partially involve centromere dysfunction at mitosis, epigenetic changes in the pericentromeric heterochromatin in cancer is of much interest. Especially noteworthy are pericentromeric rearrangements involving chromosomes 1 and 16 (Chr1 and Chr16). These rearrangements generally lead to gains of 1q and losses of 16q, respectively (Brito-Babapulle and Atkin, 1981; Le Baccon et al., 2001; Mitelman et al., 1997). Pericentromeric rearrangements of Chr1 and Chr16 can favour tumorigenesis or tumour progression by the resulting gene imbalances in tumour suppressor genes or proto-oncogenes. Furthermore, such rearrangements are sometimes the sole detected cytogenetic abnormality in cancer (Pandis et al., 1994).

3.2 Pericentromeric Instability in the ICF Syndrome

DNA hypomethylation in the juxtacentromeric region of chromosomes 1 and 16 (1qh and 16qh) has been implicated in pericentromeric rearrangements from studies of a rare chromosome breakage disease called the immunodeficiency, centromeric region instability, and facial anomalies syndrome (ICF) (Ehrlich, 2003). ICF is usually caused by mutations in *DNMT3B* (Gowher and Jeltsch, 2002; Hansen et al., 1999) and involves almost exclusively instability of chromosomes 1 and 16, and sometimes 9. Chromosomes 1, 16, 2, and 10 have satellite 2 DNA (Sat2; Figure 1) in their juxtacentromeric heterochromatin (Tagarro et al., 1994). This qh region is much larger for Chr1 than for Chr16, which is in turn much larger than that of Chr2 or Chr10. Chr9 has a large juxtacentromeric heterochromatin region predominantly containing satellite 3 (Sat3), which is distantly related to Sat2. Both Sat2 and Sat3 are hypomethylated in ICF tissues and cell cultures (Jeanpierre et al., 1993; Tuck-Muller et al., 2000). ICF is a recessive disease that has been described in fewer than 50 patients world-wide in the last several decades. The two invariant clinical characteristics of the disease are agammaglobulinemia with B-cells and cytologically detectable rearrangements targeted to the pericentromeric region of chromosomes 1 and/or 16 and sometimes 9 in mitogen-stimulated lymphocytes (Ehrlich, 2003; Smeets et al., 1994). The hypomethylation of satellite DNA sequences in ICF cells is usually, but not always, limited to the juxtacentromeric DNA without extending to centromeric DNA (Miniou et al., 1997; Tuck-Muller et al., 2000). By HPLC analysis of DNA digests, we demonstrated that the hypomethylation of the genome in ICF involved only a rather small percentage of the 5-methylcytosine residues, 7% hypomethylation in brain DNA (Tuck-Muller et al., 2000). We also confirmed that the methylation abnormality of ICF is confined to a small percentage of the genome by two-dimensional electrophoresis on DNA from four ICF *vs.* four control lymphoblastoid cell lines (LCLs) that was digested with two restriction endonucleases, including one that was sensitive to CpG methylation (Kondo et al., 2000). Only 13 of the approximately one thousand spots displayed consistent ICF-specific differences, and all but one of these was derived from tandem copies of two unrelated repeats present in several chromosomal locations. Importantly, these results indicate that the ICF syndrome is a good model implicating DNA hypomethylation in *cis* in certain DNA sequences favouring chromosome instability because the DNA in ICF cells is restricted to a rather small portion of the genome, especially tandem DNA repeats.

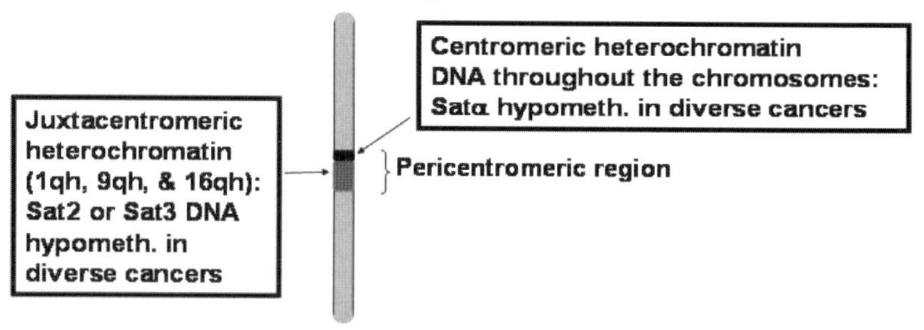

Figure 1. Chromosomal locations of tandem DNA repeats that are frequently hypomethylated in cancer. This cartoon depicts human chromosome 1. In addition to Chr1 having a long juxtacentromeric heterochromatin region with satellite 2 (Sat2) as the major component of the DNA, Chr9 has almost as long a region with satellite 3 (Sat3) instead of Sat2, and Chr16 has a shorter qh region rich in Sat2 sequences that are highly homologous to those of Chr1 but distinguishable under high-stringency hybridization conditions. There are also high frequencies of hypomethylation of the non-satellite DNA tandem arrays in cancer, as indicated in the lower portion of the figure (Itano et al., 2002; Nagai et al., 1999; Thoraval et al., 1996) (Kesmic Jackson and M. Ehrlich, unpub. data). HCC, hepatocellular carcinomas.

The ICF-diagnostic rearrangements, which are targeted to the pericentromeric regions of chromosomes 1 and 16 in ICF LCLs and mitogen-stimulated ICF lymphocytes, usually occur in the juxtacentromeric rather than the centromeric region (Sumner et al., 1998; Tuck-Muller et al., 2000). These rearrangements are predominantly chromosome breaks, whole-arm deletions, multibranched chromosomes, translocations, and isochromosomes usually containing two 1q arms fused in the pericentromeric region (Figure 2). The multibranched chromosomes (multiradials) contain 3 to 12 arms, e.g., dodeciradial(1)(p,p,q,q,q,q,q,q,q,q,q,q) (Sawyer et al., 1995b), derived from Chr1 or Chr16 and occasionally with an additional attached Chr9 (Fryns et al., 1981; Tiepolo et al., 1979). Also, there is frequent decondensation in the pericentromeric region of Chr1 in these cells in the juxtacentromeric, rather than the centromeric region (Sumner et al., 1998; Tuck-Muller et al., 2000). Although ICF patients display no increased cancer incidence, fewer than 50 patients (mostly children) have been identified, and their usually very short average lifespan would preclude detection of a cancer predisposition that was not very high and did not result in tumours rather quickly. The relevance of ICF-specific chromosome rearrangements as a factor contributing to, but not sufficient for, carcinogenesis is suggested by the finding of ICF-like Chr1/Chr16 multiradial chromosomes, which are expected to be very short-lived structures (Tuck-Muller et al., 2000), and 1qh decondensation in multiple myeloma and hepatocellular carcinomas (Sawyer et al., 1995a; Wong et al., 2001). Moreover, unbalanced Chr1 and Chr16 pericentromeric rearrangements are overrepresented in a wide variety of cancers (Mitelman et al., 1997).

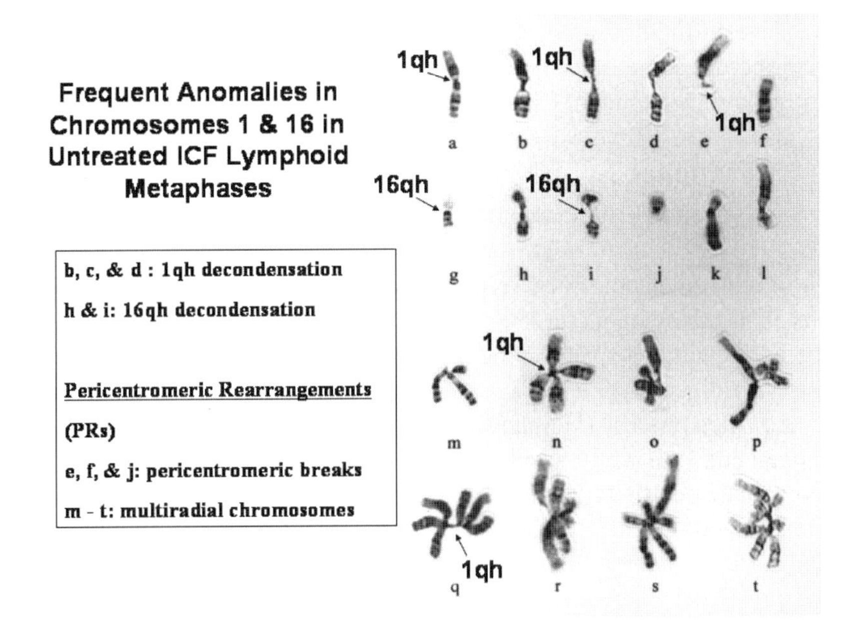

Figure 2. Examples of chromosome 1 and 16 homologues displaying pericentromeric anomalies in metaphases from untreated ICF lymphoblastoid cells. *a*, normal Chr1 and *g*, normal Chr16; *b-d* and *h* and *i*, different extents of decondensation of the qh region of Chr1 or Chr16; *e* and *f*, the two fragments derived from a chrb(1)(qh) in one metaphase with the 1p arm *(e)* and the acentric 1q arm *(f)* (note the decondensed qh below the centromeric constriction of the 1p arm); *j*, del(16)(qh); *k*, der(16)(1;16)(q10;p10); *l*, der(1)t(1;16)(p10;q10). In *m-t* are multiradials from diploid cells, with the exception of *q* which was from a tetraploid LCL B metaphase; *m*, triradial(1;16)(q,q;p); *n*, quadriradial(1)(p,p,q,q); *o*, quadriradial(1;16)(p,q;p,q); *p*, pentaradial(1;16)(p,q;p,q,q); *q*, hexaradial(1)(p,p,q,q,q,q); *r*, hexaradial(1;16)(p,p,q,q;p,q); *s*, hexaradial(1;16)(p,q,q;p,q,q), *t*, octaradial(1;16)(p,p,q,q;p,q,q,q). Note the frequent decondensation of at least one qh region in the multiradials.

ICF is usually linked to mutations in both alleles of *DNMT3B* (Ehrlich, 2003). The biallelic *DNMT3B* mutations in ICF patients usually reside in the catalytically active C-terminal portion of the protein (Gowher and Jeltsch, 2002). Most of these mutations probably give low-level residual enzymatic activity (Gowher and Jeltsch, 2002), and the mutant proteins are still able to engage in normal protein-protein interactions (Geiman et al., 2004). Although DNMT3B has repressor activity that is independent of its DNA methyltransferase activity, repression involves the ATRX-like domain in the central portion of the protein that does not overlap the methyltransferase domain (Bachman et al., 2001).

DNMT3B also forms a complex with DNMT1 and DNMT3B and with small ubiquitin-like modifier 1 but these interactions take place at the N-terminus of DNMT3B (Kang et al., 2001; Kim et al., 2002). In contrast, many ICF patients have a C-terminal missense mutation in one or both of their mutant alleles giving an amino acid substitution in one of ten motifs conserved among all cytosine-C5 methyltransferases (Gowher and Jeltsch, 2002). These findings suggest that it is the loss of DNA methyltransferase activity and not some other function of the protein that is responsible for the syndrome and its attendant Chr1 and Chr16 instability in lymphoid cells.

3.3 Evidence for DNA Hypomethylation Predisposing to Certain Pericentromeric Rearrangements

The involvement of DNA hypomethylation in the phenotype of ICF is supported at the cytogenetic level by our finding that treatment of a normal pro-B LCL with the DNA methylation inhibitors 5-azacytidine or 5-azadeoxycytidine gave ICF-like rearrangements in about 25-50% of the examined metaphases (Hernandez et al., 1997). These chromosomal rearrangements are very similar in their high frequency, spectrum, and chromosomal specificity as those found in ICF lymphocytes stimulated with mitogens or untreated ICF LCLs and were obtained under conditions that yielded no cytotoxicity. About 80% of the induced rearrangements in the treated pro-B cell line of normal origin were in Chr1, and almost 90% of these were in the pericentromeric region. Most of these rearrangements were multiradial chromosomes, whole-arm deletions, or pericentromeric breaks. Standard mutagens and a DNA inhibitor did not cause this targeted Chr1 karyotypic instability in this normal pro-B cell line (Ji et al., 1997). Similarly, treatment of normal lymphocytes with 5-azacytidine caused a high frequency of pericentromeric deletions and translocations, almost exclusively targeted to the pericentromeric regions of chromosomes 1, 9, and 16, with Chr1 pericentromeric rearrangements predominating (Kokalj-Vokac et al., 1993), although in that study multiradial chromosomes were not observed. Interestingly, a much lower frequency of multiradial chromosomes was obtained when we treated a mature B-cell line of normal origin with azacytidine than when we used the normal pro-B cell line (Ji et al., 1997). These findings suggest a cell type-specific response to DNA hypomethylation, which is consistent with ICF studies (Ehrlich, 2003). This may help explain different relationships between Sat2 hypomethylation and chromosome rearrangements in *cis* in different types of tumours as discussed below. There also seems to be a sequence-specific component to the relationship between satellite DNA hypomethylation and chromosome instability. In ICF cells, Sat3 at 9qh is strongly hypomethylated, like Sat2 at 1qh and 16qh. However, only these Sat2 regions frequently display decondensation and rearrangements in ICF lymphoid cultures (Smeets et al., 1994; Tuck-Muller et al., 2000). Similarly, an association between spontaneous Sat2 demethylation in non-ICF LCLs in culture and 1qh and 16qh decondensation was found (Vilain et al., 2000). Both 9qh and Yqh failed to display this decondensation even though their major DNA component, Sat3 became spontaneously hypomethylated in concert with Sat2.

3.4 Hypomethylation of Satellite DNA in Cancer

We demonstrated that there is hypomethylation of centromeric satellite α DNA (Sat α) and juxtacentromeric Sat2 in chromosomes 1 and 16 in the majority of Wilms tumours, breast adenocarcinomas, and ovarian epithelial carcinomas (Ehrlich et al., 2003; Jackson et al., 2004; Narayan et al., 1998; Qu et al., 1999a; Qu et al., 1999b; Widschwendter et al., 2004). This has been confirmed by others for hepatocellular carcinomas (Wong et al., 2001) and extended to Sat3, which is found mostly in 9qh (Saito et al., 2001). In our studies, satellite DNA hypomethylation in cancers was defined as less methylation than in any of the examined postnatal somatic tissues, all of which were highly methylated in this sequence. This was determined by Southern blot analysis with a CpG methylation-sensitive restriction endonuclease. As noted in Fig. 1, complex non-satellite tandem DNA repeats in pericentromeric or interstitial regions are also frequently targeted for hypomethylation in cancers (Itano et al., 2002; Nagai et al., 1999; Thoraval et al., 1996) (R. Nishiyama, K. Jackson, & M. Ehrlich, unpub. results). Some common structure, perhaps constitutive heterochromatin, is probably targeting these dissimilar sequences for hypomethylation in cancer and sperm (Nagai et al., 1999; Narayan et al., 1998). However, it should be noted that ICF syndrome cells also display hypomethylation of many, but not all of these sequences (Kondo et al., 2000). Most notably, usually they are not hypomethylated in Sat α (Miniou et al., 1997; Tuck-Muller et al., 2000).

To test whether cancer-associated satellite DNA demethylation might be an inducer of *de novo* methylation of transcription control regions of tumour suppressor genes or, alternatively, a response to prior *de novo* methylation during tumorigenesis, we looked in Wilms tumours and ovarian epithelial cancers for a positive association between this satellite DNA hypomethylation and hypermethylation of CpG islands at the 5' ends of many genes that are prone to cancer-linked hypermethylation (Ehrlich et al., 2004; Widschwendter et al., 2004) (M. Ehrlich, P. Laird, and M. Yu, unpub. data). There was no general positive association between CpG island hypermethylation and hypomethylation of satellite DNA. Therefore, it is unlikely that cancer-linked satellite DNA hypomethylation acts only as an inducer of or responder to cancer-linked hypermethylation in multiple gene regions, and its prevalence in cancer and the large size of the satellite DNA regions suggest that this hypomethylation facilitates carcinogenesis.

3.5 Satellite DNA Hypomethylation vs. Pericentromeric Chromosomal Rearrangements in Cancer

The model of chromosome instability in lymphoid ICF cells (mitogen-stimulated *in vitro* or lymphoblastoid cell lines) in Sat2 regions (1qh or 16qh) exhibiting DNA hypomethylation and the above-described studies with 5-azadeoxycytidine suggest that Sat2 hypomethylation in juxtacentromeric heterochromatin can favour instability in this region. The relationship of satellite DNA hypomethylation to chromosome instability in cancer has been compared in only a few studies. First, we examined 52 Wilms tumours by quantitative loss-of-heterozygosity analysis and found no significant relationship of 1q gain to Chr1Sat2 hypomethylation (Qu et al., 1999b). However, we did observe a significant association of 1q gain with Chr1 Satα hypomethylation and of Chr16 Sat2 hypomethylation with 16q loss. Clearly, the distribution of satellite hypomethylation among cancers does not parallel chromosome rearrangements in *cis* because there was a very much higher frequency of Chr1 Sat2 and Chr1 Satα hypomethylation than of 1q or 1p imbalances. This was also seen in our subsequent analysis of a different set of 35 karyotyped Wilms tumours (Ehrlich et al., 2003). In both studies, about half of the tumours displayed Chr1 Sat2 hypomethylation and about 90% exhibited Chr1 Satα hypomethylation. In the second study, we determined the methylation status of Satα throughout the centromeres, and not just in Chr1 Satα, by using a Satα probe under low-stringency blot-hybridization conditions on DNA digested with a CpG methylation-sensitive restriction endonuclease. There was a very high degree of concordance between Chr1 Satα hypomethylation and hypomethylation of Saα throughout the centromeres just as we had found for Chr1 Sat2 and Chr16 Sat 2 in cancers (Narayan et al., 1998; Qu et al., 1999a; Qu et al., 1999b). From 35 Wilms tumours in the second study, seven had cytogenetically identified, clonal pericentromeric rearrangements, with five affecting Chr1. These five had extra copies of 1q. Four of the five tumours with pericentromeric Chr1 rearrangements displayed hypomethylation of satellite DNA in the long juxtacentromeric heterochromatin of this chromosome (1qh) and one displayed hypomethylation in only the centromeric satellite DNA of Chr1. To explain the much higher frequencies of satellite DNA hypomethylation than of pericentromeric rearrangements, we proposed that

DNA hypomethylation in the pericentromeric regions predisposes to, but does not suffice for, rearrangements in this region via an indirect mechanism.

In a comparative genomic hybridization analysis of hepatocellular carcinomas, Wong *et al.* (Wong et al., 2001) found a significant association between 1qh gain and Chr1 Sat2 hypomethylation. The highly significant association of Chr1 Sat2 hypomethylation and rearrangements in *cis* in hepatocellular carcinomas (Wong et al., 2001) despite the lack of statistical significance of this association in Wilms tumours (Ehrlich et al., 2003) might reflect cancer-type and/or cell-type specific differences. Even in the ICF syndrome, which involves greatly decreased DNMT3B activity in all cells from the patients due to germline mutations in *DNMT3B*, cell-type specific effects on chromosome stability are observed. ICF-type pericentromeric Chr1 or Chr16 rearrangements have been seen in bone marrow cells from only one of four studied ICF patients (Fasth et al., 1990; Hulten, 1978; Smeets et al., 1994; Turleau et al., 1989), but were absent from four examined ICF fibroblast cultures (Brown et al., 1995; Carpenter et al., 1988; Maraschio et al., 1988; Tiepolo et al., 1979) despite the constitutive hypomethylation of Sat2 DNA in 1qh and 16qh in ICF tissues (Jeanpierre et al., 1993; Miniou et al., 1994; Tuck-Muller et al., 2000). The rearrangements observed in mitogen-stimulated ICF lymphocytes and in untreated ICF LCLs may occur *in vivo,* albeit at a very low rate, as deduced from a study of micronucleus formation in unstimulated bone marrow and lymphocytes from ICF patients (Fasth et al., 1990; Sawyer et al., 1995b; Smeets et al., 1994). Furthermore, there appears to be a special relationship between the pericentromeric rearrangements and *in vitro* mitogen stimulation of lymphocytes that is independent of induction of cell cycling *per se.* A much higher frequency of pericentromeric rearrangements of Chr1 and Chr16 per metaphase is seen 72 or 96 h after mitogen stimulation of ICF lymphocytes than at 48 h, although the frequent abnormal decondensation of 1qh and 16qh can be observed in metaphases at 48 h (Brown et al., 1995; Smeets et al., 1994; Tiepolo et al., 1979). These observations suggest that lymphocytes have a propensity to undergo the ICF-specific rearrangements and that the artificial conditions of mitogen stimulation or short-term *in vitro* culture enhance the formation of these rearrangements.

There are biochemical consequences of satellite DNA hypomethylation that could explain how DNA demethylation could predispose chromatin abnormalities to form in *cis*. As mentioned above, a downstream consequence of DNA hypomethylation can be an increase in histone acetylation and other changes in histone modification. These alterations might predispose to pericentromeric chromatin decondensation and then to rearrangements. This would be consistent with the partial overlap of DNA demethylation and histone acetylation or H3 K9 demethylation pathways for inducing localized decompaction of euchromatin in promoter regions in the human genome (Baylin, 2004; Cervoni et al., 2002; Coombes et al., 2003; Fahrner et al., 2002; Lehnertz et al., 2003; Nguyen et al., 2002; Yan et al.,

2003; Yasui et al., 2003). Given this partial overlap of these two types of epigenetics, during the formation of some tumours, changes in histone acetylation or methylation or changes in the ratios of histone acetylating and histone deacetylating activity associated with carcinogenesis (Cervoni et al., 2002) might favour instability of pericentromeric heterochromatin independently of DNA methylation.

3.6 Satellite DNA Hypomethylation may favour Pericentromeric Instability but is not necessary for Decondensation or Rearrangements in *Cis*: Results from Normal Embryonic Cell Cultures

Surprisingly, we found that untreated cultures from normal chorionic villus (CV) or amniotic fluid-derived (AF) samples displayed dramatic cell passage-dependent increases in ICF-like chromosomal aberrations (Tsien et al., 2002). They showed negligible levels of chromosomal aberrations in primary culture and no other consistent chromosomal abnormality at any passage. By passage 8 or 9, 82α 7% of the CV metaphases from all eight studied samples exhibited 1qh or 16qh decondensation and 25α 16% had rearrangements in these regions with no other consistent chromosomal abnormality at any passage. All six analyzed late-passage AF cultures displayed this regional decondensation and recombination in 54α 16 and 3α 3% of the metaphases, respectively. Late-passage skin fibroblasts did not show these aberrations. There was a high degree of methylation of Sat2 in AF cells at all studied passages, which is attributable to their derivation from embryonic fibroblasts. The frequent 1qh decondensation observed in these cells at high passage, despite their retention of high levels of Chr1 Sat2 methylation indicates that DNA hypomethylation at 1qh is not necessary for 1qh decondensation. Sat2 hypomethylation may nonetheless favour 1qh and 16qh anomalies. In contrast to AF and fibroblast cultures, CV cultures, extraembryonic mesodermal cells, displayed DNA hypomethylation at all passage numbers, as expected, due to their extraembryonic origin. The CV cultures, with their Sat2 hypomethylation, displayed 1qh and 16qh decondensation and rearrangements at significantly lower passage numbers than did AF cultures. Also, in chromosomes 1 and 16, CV cultures had much more ICF-like rearrangement than heterochromatin decondensation. A study of four human LCLs which spontaneously underwent Chr1 Sat2, Chr16 Sat2, Chr9 Sat3, and ChrY Sat3 hypomethylation upon very prolonged culture showed correlations between this hypomethylation and decondensation and rearrangements in Chr1 Sat2 or Chr16 Sat2 (Vilain et al., 2000). However, we have also observed high levels of Chr1 Sat2 decondensation in several normal LCLs with little or no hypomethylation in this satellite DNA (M. Ehrlich, Lixin Qi, Suzana Sogorovic, and Cathy Tuck-Muller, unpub. data). We propose that this can be explained by either

DNA hypomethylation or histone modification changes independent of DNA hypomethylation predisposing to 1qh decondensation.

3.7 Centromeric Satellite DNA Hypomethylation in Cancer is not associated with Aneuploidy

As described above, our study of satellite DNA hypomethylation in 35 karyotyped Wilms tumours, revealed that the vast majority of tumours had hypomethylation of Satα in centromeric heterochromatin throughout the genome (Ehrlich et al., 2003). This hypomethylation was observed in Southern blots of DNA singly digested with any one of three different CpG methylation-sensitive restriction endonucleases. All the normal somatic tissues had similar high levels of methylation at the tested sites. The greater extent of digestion of the cancer DNAs with all three enzymes indicates that hypomethylation affects many centromeric CpG sites in the tumours. For a few of these tumours, hypomethylation of Satα was so extensive that it resembled that of sperm. This centromeric hypomethylation might have affected centromere function in mitosis, but no relationship was apparent between aneuploidy and the methylation status of Satα. However in a study of mouse cells, chromatin epigenetic defects at the protein level were associated with aneuploidy. Fibroblasts from histone H3 K9 methyltransferase-deficient (*Suv39h1*$^{-/-}$, *Suv39h2*$^{-/-}$) mice displayed frequent aneuploidy (Peters et al., 2001).

3.8 DNA Hypomethylation associated with Interstitial Chromosome Rearrangements

DNA methylation changes may affect DNA recombination at interspersed repeats, lymphogenesis-related recombination signals, and at various unique DNA sequences in addition to tandem repeat arrays. With few exceptions, the correlation between methylation changes and chromosome rearrangements is less methylation, more rearrangements. A notable exception is a recent study on an artificially reconstructed transposon (SB) in mouse cells, which provided evidence for increased SB transposition in the germ line and in transfected embryonal stem cells when it was highly methylated (Yusa et al., 2004).

Among the non-tandem DNA repeats that might be sites of rearrangements during carcinogenesis, highly repeated interspersed DNA sequences are good candidates for somatic-cell recombination hotspots by homologous recombination or for insertional mutagenesis by retrotransposition. In human DNA, the most numerous of these repeats are the LINE1 (or L1) repeats and Alu repeats, which constitute ~17% and 10% of the genome, respectively. Retrotransposons or retroviral-derived elements can have their transcription upregulated *in vivo* by DNA demethylation as seen in studies of *Dnmt1* knockout mouse embryos, interspecies mammalian hybrids, and mice with an inherited epigenetically controlled phenotype whose expression is regulated by a genetically linked retrotransposon (IAP) (Morgan et al., 1999; O'Neill et al., 1998; Walsh et al., 1998). Also, there is evidence for frequent activation of expression of full-length transcripts from retrotransposons in certain types of murine cancer (Dupressoir and Heidmann, 1997). However, retrotransposition of endogenous elements is implicated in disease much less frequently for humans than for mice (Kazazian and Moran, 1998).

The human genome's LINE1 repeats are up to 6 kb in length, although usually much shorter. They are retrotransposon-derived sequences, but of the ~4 x 10^5 copies of LINE1 elements in the human genome, only about 30-60 are estimated to be competent for transposition (Sassaman et al., 1997). There have been occasional reports of cancer-associated retrotransposition-like insertions involving LINE1 sequences (Miki et al., 1992; Morse et al., 1988), and they may mobilize cellular RNAs at low frequencies (Wei et al., 2001). LINE1 hypomethylation was observed in chronic lymphocytic leukemia *vs.* normal mononuclear blood cells (Dante et al., 1992), urinary bladder carcinomas compared to normal bladder (Jurgens et al., 1996), hepatocellular carcinomas *vs.* non-tumorous "normal" or cirrhotic tissue (Takai et al., 2000), and prostate carcinomas *vs.* normal prostate and other normal tissues (Santourlidis et al., 1999). Schulz has suggested that frequent hypomethylation of LINE repeats throughout rodent and human genomes in cancer contributes to high levels of recombination at these sequences in solid tumours (Schulz, 1998). Correlative support for this was provided by his group (Schulz et al., 2002) in a study of 54 prostate cancers by comparative genomic hybridization and Southern blot analysis. Two of the most frequent chromosomal alterations seen, loss of 8p and gain of 8q were significantly associated with LINE1 hypomethylation, although no association was seen between this hypomethylation and loss of 13q, the other frequently observed chromosomal anomaly. However, it should be noted that LINE1 hypomethylation might just be associated with hypomethylation globally, in satellite DNAs, and in certain 5' gene regions. Therefore, evidence is needed to link LINE1 hypomethylation causally to increased DNA recombination in cancer. Of even higher copy number than the LINE1 repeats in the human genome are the Alu repeats (size, ~0.3 kb; copy number, ~1.1 x 10^6), which also can be mobilized, thus leading occasionally to cancer-associated gene insertions (Rothberg et al., 1997; Schichman et al., 1994) or other types of Alu-Alu recombination (Schmid, 1996). Alu repeats also show cancer-associated hypomethylation but only in a small percentage of these sequences in Wilms tumours (M. Ehrlich, C. Woods, and L. Qi, unpub.data). At this time no firm evidence links Alu repeat hypomethylation and Alu-Alu recombination.

Retrotransposition might be favoured by cancer-associated hypomethylation of human endogenous retroviruses, especially the HERV-K family. However, there are only about 30-50 full-length HERV-K sequences in the human genome, as compared to an estimated 10,000 solitary long terminal repeats (LTRs) from HERV-K (Leib-Mosch et al., 1993). These repeats are usually highly methylated. In a study of urinary bladder cancers (Florl et al., 1999), cancer-associated hypomethylation of the HERV-K sequences was seen by Southern blotting with an HERV-K *gag* probe. It is possible that such hypomethylation might favour retrotransposition. Although analysis of the presence of HERV-K transcripts in these bladder cancers was not reported, evidence for demethylation of HERV-K *gag* sequences and correlated expression of Gag at the protein level in human testicular tumour samples was provided in a small-scale study involving Southern and western blotting (Gotzinger et al., 1996).

Studies of V(D)J recombination (Ji et al., 2003) suggest that local demethylation favours this type of normally programmed DNA rearrangement (Mostoslavsky et al., 1998; Nakase et al., 2003), although under some circumstances it may not be necessary for it (Sikes et al., 1999). Abnormal V(D)J rearrangements that are found in many lymphoid malignancies (Kirsch et al., 1994) and at low frequencies in the blood of a very large percentage of healthy adults (Ji et al., 1995) might also be promoted by aberrant DNA hypomethylation, although direct evidence for this is lacking. The relationship between DNA methylation changes in *cis* and V(D)J recombination may be partially mediated by changes in histone acetylation and methylation (Ji et al., 2003; Osipovich et al., 2004). DNA methylation might directly inhibit the RAG1/RAG2 recombinase function as well as alter the accessibility of the chromatin configuration to RAG1/RAG2 (Nakase et al., 2003).

There is an interesting confluence of two sources of genetic instability in cancer, namely, DNA hypomethylation and mutation of *TP53*, which is linked to a high rate of gene rearrangement and amplification. It was found that loss of TP53 in cell culture is associated with DNA hypomethylation (Nasr et al., 2003). This inactivation of TP53 function can be seen in Li-Fraumeni syndrome cells carrying a germline mutation of one *TP53* allele and acquiring a mutation of the other allele during growth crisis upon immortalization. In these cells there is a decrease in methylation of the *TROP1* gene (Nasr et al., 2003), which had previously been demonstrated to become prone to amplification (about 4-10 fold) upon treatment *in vivo* with 5-azacytidine prior to stable transfection into mouse L cells (Alberti et al., 1994). Upon stable transfection into L cells, hypomethylated *TROP1* genes were amplified compared to *TROP1* genes that had been more methylated by *in vitro* treatment with *Sss*I methyltransferase or by *in vivo* circumstance, namely, isolation from wild-type cells or from Li-Fraumeni *TP53*-null cells transfected with a wild-type *TP53* gene. It was proposed that TP53 may impact DNA methylation by protein-protein interactions in complexes that contain TP53, DNA methyltransferases, and histone modifying enzymes or by effects of TP53 on gene expression or cell-cycle control (Nasr et al., 2003). Moreover, another key cell-cycle regulatory protein, RB, whose loss is also related to genomic instability (Zheng et al., 2002), similarly interacts with DNMT1. While these proteins may cooperate in maintaining the integrity of the genome, it is unlikely that there is a synergistic interaction of decreased activity of these proteins and of DNA methyltransferases because DNA methyltransferase levels generally do not appear to decrease during oncogenesis, as explained above (Section 2.5.).

Dnmt1 knockout or hypomorphic mutant mice or cell cultures derived from them have provided evidence for the involvement of abnormal DNA hypomethylation (or DNA methyltransferase deficiency) in aberrant recombination. Jaenisch and colleagues showed that homozygous knockout of *Dnmt1*, which caused global DNA hypomethylation, increased abnormal DNA recombination and, thereby, deletion mutagenesis at the *Hprt* locus and at a transgenic viral *tk* locus in murine embryonal stem cells (Chen et al., 1998). The increase in mutation rate (predominantly deletions) associated with the loss of Dnmt1 activity was about 10 fold. However, using different stably transfected murine embryonal stem cells, Chan *et al.* unexpectedly found that homozygous knockout of *Dnmt1*decreased gene loss and point mutagenesis from a chimeric *tk-neo* transgene (Chan et al., 2001). The differences between these studies might be due to chromosome position effects on hypomethylated DNA sequences. In another model system, transgenic mice carrying a hypomorphic *Dnmt1* allele and a null *Dnmt1* allele, most of the *Dnmt1* expression was lost (Gaudet et al., 2003). This resulted in a large extent of global and centromeric DNA hypomethylation in the runted transgenic mice. All of these mice developed T-cell lymphomas. Four of 10 analyzed tumours had a predominant DJ rearrangement, which suggests a monoclonal origin. This indicates that although oncogenic transformation in these mice was frequent at the level of the individual, it was rare at the cellular level. The lack of RNA for a tested endogenous IAP retrovirus and for c-*myc* and the absence of insertional inactivation of the c-*myc* locus suggest that the loss of Dnmt1 activity did not promote oncogenesis by inducing retrotransposition or proto-oncogenes. By comparative genomic hybridization, lymphoma DNA from the Dnmt1-deficient mice and MMLV transgenic mice were compared. There was significantly more gain of Chr 14 and Chr15 in the former mice, which could have resulted from a whole-chromosome gain or an unbalanced translocation in the pericentromeric regions of these acrocentric chromosomes.

Another transgenic mouse model was used to study the oncogenic and chromosome destabilising effects of Dnmt1 deficiency. This involved mice doubly heterozygous for the mutant *Nfl* and *Tp53* genes (Eden et al., 2003). These genes are closely linked on Chr11, and both are often involved together in LOH in murine soft tissue sarcomas. When these mice were also made transgenic for a null allele and a hypomorphic allele of *Dnmt1*, they tended to develop sarcomas at an earlier age. Furthermore, 77% displayed LOH at the *Nfl* and *Tp53* loci in sarcomas of the Dnmt1-deficient mice compared to 45% in isogenic mice that did not have *Dnmt1* mutations. Also, by fluctuation analysis of fibroblasts from the Dnmt1-deficient mice *vs.* the isogenic mice without *Dnmt1* mutations, there was a significant increase, but only 2 fold, in the development of LOH at the *Nf1* and *Tp53* loci. An LOH analysis of five markers along Chr11 suggested either whole chromosome loss or unequal translocation at the acrocentric centromere. Therefore, these murine models indicate that DNA hypomethylation plays a significant, but modest, role in chromosome instability during carcinogenesis.

4. CONCLUSIONS AND IMPLICATIONS

There are diverse lines of evidence linking cancer-associated DNA hypomethylation to increased DNA rearrangements but the effects are not of a large magnitude. The connection between genetic instability and DNA hypermethylation-related inactivation of genes involved in maintaining chromosome stability are much stronger. Nonetheless, it is very likely that DNA hypomethylation, as well as DNA hypermethylation, does make a major contribution to carcinogenesis because of the prevalence of DNA hypomethylation in a wide variety of cancers, its lack of a positive correlation with DNA hypermethylation, and its association with tumour progression and poor prognosis (Ehrlich, 2002; Itano et al., 2002; Widschwendter et al., 2004) (M. Ehrlich, P. Laird, M. Yu, & L. Dubeau, unpub. data).

Other possible roles of DNA hypomethylation in cancer relate to either *cis* or, possibly, *trans* effects on gene expression. Because satellite DNA hypomethylation in ovarian carcinomas, Wilms tumours, and breast adenocarcinomas has been shown to be significantly associated with global DNA hypomethylation (Ehrlich et al., 2002; Qu et al., 1999b) (K. Jackson and M. Ehrlich, unpub. data), there may be waves of DNA hypomethylation that typically include satellite DNA sequences, but also involve gene targets (Ehrlich, 2002) (M. Ehrlich, P. Laird, M. Yu, unpub. data), some of which might impact tumour formation and progression. Satellite DNA hypomethylation could additionally spread to adjacent euchromatin regions. While it does not seem that activation of DNA methylation-repressed retrotransposons plays a major role in cancers (Gaudet et al., 2003), there is growing evidence that some gene targets of cancer-associated demethylation (Cho et al., 2001; Gupta et al., 2003; Sato et al., 2003; Scelfo et al., 2002), but not all of them (Feinberg and Vogelstein, 1983), may get turned on by this hypomethylation and contribute to carcinogenesis. Furthermore, there is a heightened appreciation of the importance of intranuclear localization of chromosomal regions in the regulation of expression of certain genes, and this can involve constitutive heterochromatin (Gasser, 2001), which is frequently the target of cancer-linked DNA hypomethylation. Evidence indicates that centromeric heterochromatin can interact in *trans* with genes dispersed in the genome to help control their expression. This might be mediated by different types of constitutive heterochromatin serving as reservoirs for specific DNA-binding proteins (Sabbattini et al., 2001). Such interactions involving centromeric or juxtacentromeric heterochromatin and distant gene regions could be governed by the state of methylation of the normally highly methylated DNA of these heterochromatic regions.

Whatever the most important biological target of cancer-associated genomic hypomethylation, it should be noted that decreases in DNA methylation induced as part of a therapeutic regimen might contribute to carcinogenesis (Eden et al., 2003; Ehrlich, 2002; Gaudet et al., 2003) or tumor progression (Gaudet et al., 2003). Attempts to decrease DNA methylation in cancers as a therapeutic strategy by using 5-azacytidine or 5-aza-2'-deoxycytidine have been productive in hematologic malignancies but disappointing in solid tumours (Aparicio and Weber, 2002). Moreover, azacytidine has been shown to enhance the formation of lung tumours (Stoner et al., 1973) in mice, testicular and liver cancer in rats (Carr et al., 1984), and to have oncogenic effects on cultured cells (Kerbel et al., 1984). Our finding that an increase in DNA hypomethylation is associated with an increase in aggressiveness of ovarian cancers and with a decrease in patient survival (Widschwendter et al., 2004) and evidence described above for DNA hypomethylation favouring chromosomal instability and possibly having other roles to play in oncogenesis calls for caution in using demethylating agents as anti-cancer drugs.

ACKNOWLEDGEMENT

This study was supported by a grant from the NIH, R01-CA81506.

REFERENCES

Alberti, S., M. Nutini, and L.A. Herzenberg. 1994. DNA methylation prevents the amplification of TROP1, a tumour-associated cell surface antigen gene. *Proc. Natl. Acad. Sci. U. S. A.* 91:5833-5837.

Albor, A., and M. Kulesz-Martin. 2000. *p53* and the Development of Genomic Instability: Possible Role of Topoisomerase I Interaction. *In* DNA Alterations in Cancer: Genetic and Epigenetic Alterations. M. Ehrlich, editor. Eaton Publishing, Natick. 409-422.

Aparicio, A., and J.S. Weber. 2002. Review of the clinical experience with 5-azacytidine and 5-aza-2'-deoxycytidine in solid tumours. *Curr. Opin. Investig. Drugs*. 3:627-633.

Bachman, K.E., M.R. Rountree, and S.B. Baylin. 2001. Dnmt3a and Dnmt3b are transcriptional repressors that exhibit unique localization properties to heterochromatin. *J. Biol. Chem.* 276:32282-32287.

Baylin, S.B. 2004. Reversal of gene silencing as a therapeutic target for cancer--roles for DNA methylation and its interdigitation with chromatin. *Novartis Found Symp.* 259:226-233; discussion 234-237, 285-238.

Baylin, S.B., and J.G. Herman. 2000. Epigenetics and Loss of Gene Function in Cancer. *In* DNA Alterations in Cancer: Genetic and Epigenetic Alterations. M. Ehrlich, editor. Eaton Publishing, Natick. 293-309.

Bhakat, K.K., and S. Mitra. 2003. CpG methylation-dependent repression of the human O6-methylguanine-DNA methyltransferase gene linked to chromatin structure alteration. *Carcinogenesis*. 24:1337-1345.

Bianco, T., G. Chenevix-Trench, D.C. Walsh, J.E. Cooper, and A. Dobrovic. 2000. Tumour-specific distribution of BRCA1 promoter region methylation supports a pathogenetic role in breast and ovarian cancer. *Carcinogenesis*. 21:147-151.

Billard, L.M., F. Magdinier, G.M. Lenoir, L. Frappart, and R. Dante. 2002. MeCP2 and MBD2 expression during normal and pathological growth of the human mammary gland. *Oncogene*. 21:2704-2712.

Bocker, T., J. Ruschoff, and R. Fishel. 1999. Molecular diagnostics of cancer predisposition: hereditary non-polyposis colorectal carcinoma and mismatch repair defects. *Biochim. Biophys. Acta*. 1423:O1-O10.

Brito-Babapulle, V., and N.B. Atkin. 1981. Break points in chromosome #1 abnormalities of 218 human neoplasms. *Cancer Genet. Cytogenet.* 4:215-225.

Brown, D.C., E. Grace, A.T. Summer, A.T. Edmunds, and P.M. Ellis. 1995. ICF syndrome (immunodeficiency, centromeric instability and facial anomalies): investigation of heterochromatin abnormalities and review oaf clinical outcome. *Hum. Genet.* 96:411-416.

Carpenter, N.J., A. Fillpovich, R.M. Blaese, T.L. Carey, and A.I. Berkel. 1988. Variable immunodeficiency with abnormal condensation of the heterochromatin of chromosomes 1, 9, and 16. *J. Ped.* 112:757-760.

Carr, B.I., J.G. Reilly, S.S. Smith, C. Winberg, and A. Riggs. 1984. The tumorigenicity of 5-azacytidine in the male Fischer rat. *Carcinogenesis*. 5:1583-1590.

Cervoni, N., N. Detich, S.B. Seo, D. Chakravarti, and M. Szyf. 2002. The oncoprotein Set/TAF-1beta, an inhibitor of histone acetyltransferase, inhibits active demethylation of DNA, integrating DNA methylation and transcriptional silencing. *J. Biol. Chem.* 277:25026-25031.

Chan, M.F., R. van Amerongen, T. Nijjar, E. Cuppen, P.A. Jones, and P.W. Laird. 2001. Reduced rates of gene loss, gene silencing, and gene mutation in dnmt1- deficient embryonic stem cells. *Mol. Cell. Biol.* 21:7587-7600.

Chen, R.Z., U. Pettersson, C. Beard, L. Jackson-Grusby, and R. Jaenisch. 1998. DNA hypomethylation leads to elevated mutation rates. *Nature*. 395:89-93.

Cho, M., H. Uemura, S.C. Kim, Y. Kawada, K. Yoshida, Y. Hirao, N. Konishi, S. Saga, and K. Yoshikawa. 2001. Hypomethylation of the MN/CA9 promoter and upregulated MN/CA9 expression in human renal cell carcinoma. *Br. J. Cancer*. 85:563-567.

Chuang, L.S., H.I. Ian, T.W. Koh, H.H. Ng, G. Xu, and B.F. Li. 1997. Human DNA-(cytosine-5) methyltransferase-PCNA complex as a target for p21WAF1. *Science*. 277:1996-2000.

Chuang, L.S., E.H. Tan, H.K. Oh, and B.F. Li. 2002. Selective depletion of human DNA-methyltransferase DNMT1 proteins by sulfonate-derived methylating agents. *Cancer Res.* 62:1592-1597.

Coombes, M.M., K.L. Briggs, J.R. Bone, G.L. Clayman, A.K. El-Naggar, and S.Y. Dent. 2003. Resetting the histone code at CDKN2A in HNSCC by inhibition of DNA methylation. *Oncogene*. 22:8902-8911.

Dante, R., J. Dante-Paire, D. Rigal, and G. Roizes. 1992. Methylation patterns of long interspersed repeated DNA and alphoid repetitive DNA from human cell lines and tumours. *Anticancer Res.* 12:559-563.

De Marzo, A.M., V.L. Marchi, E.S. Yang, R. Veeraswamy, X. Lin, and W.G. Nelson. 1999. Abnormal regulation of DNA methyltransferase expression during colorectal carcinogenesis. *Cancer Res.* 59:3855-3860.

De Smet, C., C. Lurquin, B. Lethe, V. Martelange, and T. Boon. 1999. DNA methylation is the primary silencing mechanism for a set of germ line- and tumour-specific genes with a CpG-rich promoter. *Mol. Cell. Biol.* 19:7327-7335.

Dupressoir, A., and T. Heidmann. 1997. Expression of intracisternal A-particle retrotransposons in primary tumours of oncogene-expressing transgenic mice. *Oncogene*. 14:2951-2958.

Dutrillaux, B. 2000. Chromosome and Karyotype Instability in Human Cancers and Cancer-Predisposing Syndromes. *In* DNA Alterations in Cancer: Genetic and Epigenetic Alterations. M. Ehrlich, editor. Eaton Publishing, Natick. 369-382.

Eads, C.A., K.D. Danenberg, K. Kawakami, L.B. Saltz, P.V. Danenberg, and P.W. Laird. 1999. CpG island hypermethylation in human colorectal tumours is not associated with DNA methyltransferase overexpression. *Cancer Res.* 59:2302-2306.

Eads, C.A., R.V. Lord, K. Wickramasinghe, T.I. Long, S.K. Kurumboor, L. Bernstein, J.H. Peters, S.R. DeMeester, T.R. DeMeester, K.A. Skinner, and P.W. Laird. 2001. Epigenetic patterns in the progression of esophageal adenocarcinoma. *Cancer Res.* 61:3410-3418.

Eden, A., F. Gaudet, A. Waghmare, and R. Jaenisch. 2003. Chromosomal instability and tumors promoted by DNA hypomethylation. *Science*. 300:455.

Ehrlich, M. 2002. DNA methylation in cancer: too much, but also too little. *Oncogene*. 21:5400-5413.

Ehrlich, M. 2003. The ICF syndrome, a DNA methyltransferase 3B deficiency and immunodeficiency disease. *Clin. Immunol.* 109:17-28.

Ehrlich, M., L. Dubeau, C. Woods, E. Fiala, B. Youn, T.I. Long, and P. Laird. 2004. Few associations of DNA hypermethylation and hypomethylation in ovarian tumors. *MS in prepn.*

Ehrlich, M., N. Hopkins, G. Jiang, J.S. Dome, M.S. Yu, C.B. Woods, G.E. Tomlinson, M. Chintagumpala, M. Champagne, L. Diller, D.M. Parham, and J. Sawyer. 2003. Satellite hypomethylation in karyotyped Wilms tumors. *Cancer Genet. Cytogenet.* 141:97-105.

Ehrlich, M., G. Jiang, E.S. Fiala, J.S. Dome, M.S. Yu, T.I. Long, B. Youn, O.-S. Sohn, M. Widschwendter, G.E. Tomlinson, M. Chintagumpala, M. Champagne, D.M. Parham, G. Liang, K. Malik, and P.W. Laird. 2002. Hypomethylation and hypermethylation of DNA in Wilms tumors. *Oncogene*. 21:6694-6702.

Esteller, M., P.G. Corn, S.B. Baylin, and J.G. Herman. 2001. A gene hypermethylation profile of human cancer. *Cancer Res.* 61:3225-3229.

Esteller, M., R. Levine, S.B. Baylin, L.H. Ellenson, and J.G. Herman. 1998. MLH1 promoter hypermethylation is associated with the microsatellite instability phenotype in sporadic endometrial carcinomas. *Oncogene*. 17:2413-2417.

Esteller, M., J.M. Silva, G. Dominguez, F. Bonilla, X. Matias-Guiu, E. Lerma, E. Bussaglia, J. Prat, I.C. Harkes, E.A. Repasky, E. Gabrielson, M. Schutte, S.B. Baylin, and J.G. Herman. 2000a. Promoter hypermethylation and BRCA1 inactivation in sporadic breast and ovarian tumors. *J. Natl. Cancer Inst.* 92:564-569.

Esteller, M., M. Toyota, M. Sanchez-Cespedes, G. Capella, M.A. Peinado, D.N. Watkins, J.P. Issa, D. Sidransky, S.B. Baylin, and J.G. Herman. 2000b. Inactivation of the DNA repair gene O6-methylguanine-DNA methyltransferase by promoter hypermethylation is associated with G to A mutations in K-ras in colorectal tumorigenesis. *Cancer Res.* 60:2368-2371.

Fahrner, J.A., S. Eguchi, J.G. Herman, and S.B. Baylin. 2002. Dependence of histone modifications and gene expression on DNA hypermethylation in cancer. *Cancer Res.* 62:7213-7218.

Fasth, A., E. Forestier, E. Holmberg, G. Holmgren, I. Nordenson, T. Soderstrom, and J. Wahlstrom. 1990. Fragility of the centromeric region of chromosome 1 associated with combined immunodeficiency in siblings: a recessively inherited entity? *Acta Paediatr. Scand.* 79:605-612.

Fearon, E.R., D.M. Pardoll, T. Itaya, P. Golumbek, H.I. Levitsky, J.W. Simons, H. Karasuyama, B. Vogelstein, and P. Frost. 1990. Interleukin-2 production by tumor cells bypasses T helper function in the generation of an antitumor response. *Cell*. 60:397-403.

Feinberg, A.P., and B. Vogelstein. 1983. Hypomethylation distinguishes genes of some human cancers from their normal counterparts. *Nature*. 301:89-92.

Ferguson, A.T., E. Evron, C.B. Umbricht, T.K. Pandita, T.A. Chan, H. Hermeking, J.R. Marks, A.R. Lambers, P.A. Futreal, M.R. Stampfer, and S. Sukumar. 2000. High frequency of hypermethylation at the 14-3-3 sigma locus leads to gene silencing in breast cancer. *Proc. Natl. Acad. Sci. U. S. A.* 97:6049-6054.

Filippova, G.N., C.P. Thienes, B.H. Penn, D.H. Cho, Y.J. Hu, J.M. Moore, T.R. Klesert, V.V. Lobanenkov, and S.J. Tapscott. 2001. CTCF-binding sites flank CTG/CAG repeats and form a methylation- sensitive insulator at the DM1 locus. *Nat. Genet.* 28:335-343.

Fleisher, A.S., M. Esteller, S. Wang, G. Tamura, H. Suzuki, J. Yin, T.T. Zou, J.M. Abraham, D. Kong, K.N. Smolinski, Y.Q. Shi, M.G. Rhyu, S.M. Powell, S.P. James, K.T. Wilson, J.G. Herman, and S.J. Meltzer. 1999. Hypermethylation of the hMLH1 gene promoter in human gastric cancers with microsatellite instability. *Cancer Res.* 59:1090-1095.

Florl, A.R., R. Lower, B.J. Schmitz-Drager, and W.A. Schulz. 1999. DNA methylation and expression of LINE-1 and HERV-K provirus sequences in urothelial and renal cell carcinomas. *Br. J. Cancer.* 80:1312-1321.

Fryns, J.P., M. Azou, J. Jacken, E. Eggermont, J.C. Pedersen, and H. Van den Berghe. 1981. Centromeric instability of chromosomes 1, 9, and 16 associated with combined immunodeficiency. *Hum. Genet.* 57:108-110.

Fujita, N., S. Watanabe, T. Ichimura, S. Tsuruzoe, Y. Shinkai, M. Tachibana, T. Chiba, and M. Nakao. 2003. Methyl-CpG binding domain 1 (MBD1) interacts with the Suv39h1-HP1 heterochromatic complex for DNA methylation-based transcriptional repression. *J. Biol. Chem.* 278:24132-24138.

Fuks, F., P.J. Hurd, R. Deplus, and T. Kouzarides. 2003a. The DNA methyltransferases associate with HP1 and the SUV39H1 histone methyltransferase. *Nucleic Acids Res.* 31:2305-2312.

Fuks, F., P.J. Hurd, D. Wolf, X. Nan, A.P. Bird, and T. Kouzarides. 2003b. The methyl-CpG-binding protein MeCP2 links DNA methylation to histone methylation. *J. Biol. Chem.* 278:4035-4040.

Gama-Sosa, M.A., V.A. Slagel, R.W. Trewyn, R. Oxenhandler, K.C. Kuo, C.W. Gehrke, and M. Ehrlich. 1983. The 5-methylcytosine content of DNA from human tumors. *Nucleic Acids Res.* 11:6883-6894.

Gasser, S.M. 2001. Positions of potential: nuclear organization and gene expression. *Cell.* 104:639-642.

Gaudet, F., J.G. Hodgson, A. Eden, L. Jackson-Grusby, J. Dausman, J.W. Gray, H. Leonhardt, and R. Jaenisch. 2003. Induction of tumors in mice by genomic hypomethylation. *Science.* 300:489-492.

Geiman, T.M., U.T. Sankpal, A.K. Robertson, Y. Zhao, and K.D. Robertson. 2004. DNMT3B interacts with hSNF2H chromatin remodeling enzyme, HDACs 1 and 2, and components of the histone methylation system. *Biochem. Biophys. Res. Commun.* 318:544-555.

Girault, I., S. Tozlu, R. Lidereau, and I. Bieche. 2003. Expression analysis of DNA methyltransferases 1, 3A, and 3B in sporadic breast carcinomas. *Clin. Cancer Res.* 9:4415-4422.

Gotzinger, N., M. Sauter, K. Roemer, and N. Mueller-Lantzsch. 1996. Regulation of human endogenous retrovirus-K Gag expression in teratocarcinoma cell lines and human tumours. *J. Gen. Virol.* 77:2983-2990.

Gowher, H., and A. Jeltsch. 2002. Molecular enzymology of the catalytic domains of the Dnmt3a and Dnmt3b DNA methyltransferases. *J. Biol. Chem.* 277:20409-20414.

Guo, G., W. Wang, and A. Bradley. 2004. Mismatch repair genes identified using genetic screens in Blm-deficient embryonic stem cells. *Nature.* 429:891-895.

Gupta, A., A.K. Godwin, L. Vanderveer, A. Lu, and J. Liu. 2003. Hypomethylation of the synuclein gamma gene CpG island promotes its aberrant expression in breast carcinoma and ovarian carcinoma. *Cancer Res.* 63:664-673.

Hansen, R.S., C. Wijmenga, P. Luo, A.M. Stanek, T.K. Canfield, C.M. Weemaes, and S.M. Gartler. 1999. The DNMT3B DNA methyltransferase gene is mutated in the ICF immunodeficiency syndrome. *Proc. Natl. Acad. Sci. USA.* 96:14412-14417.

Heim, S., and M.R. Teixeira. 2000. Clonal Evolution of Neoplastic Cell Populations: Lessons from Solid Tumor Cytogenetics. *In* DNA Alterations in Cancer: Genetic and Epigenetic Alterations. M. Ehrlich, editor. Eaton Publishing, Natick. 383-394.

Hemminki, A., P. Peltomaki, J.P. Mecklin, H. Jarvinen, R. Salovaara, M. Nystrom-Lahti, A. de la Chapelle, and L.A. Aaltonen. 1994. Loss of the wild type MLH1 gene is a feature of hereditary nonpolyposis colorectal cancer. *Nat. Genet.* 8:405-410.

Hernandez, R., A. Frady, X.-Y. Zhang, M. Varela, and M. Ehrlich. 1997. Preferential induction of chromosome 1 multibranched figures and whole-arm deletions in a human pro-B cell line treated with 5-azacytidine or 5-azadeoxycytidine. *Cytogenet. Cell Genet.* 76:196-201.

Hoglund, M., D. Gisselsson, G.B. Hansen, T. Sall, and F. Mitelman. 2003. Ovarian carcinoma develops through multiple modes of chromosomal evolution. *Cancer Res.* 63:3378-3385.

Huang, L.H., R. Wang, M.A. Gama-Sosa, S. Shenoy, and M. Ehrlich. 1984. A protein from human placental nuclei binds preferentially to 5-methylcytosine-rich DNA. *Nature.* 308:293-295.

Hulten, M. 1978. Selective somatic pairing and fragility at 1q12 in a boy with common variable immunodeficiency. *Clin. Genet.* 14:294.

Issa, J.P. 2000. Hypermethylator Phenotypes in Aging and Cancer. *In* DNA alterations in cancer: genetic and epigenetic alterations. M. Ehrlich, editor. Eaton Publishing, Natick. 311-322.

Itano, O., M. Ueda, K. Kikuchi, O. Hashimoto, S. Hayatsu, M. Kawaguchi, H. Seki, K. Aiura, and M. Kitajima. 2002. Correlation of postoperative recurrence in hepatocellular carcinoma with demethylation of repetitive sequences. *Oncogene.* 21:789-797.

Jackson, K., M. Yu, E. Fiala, B. Youn, H.M. Muller, M. Widschwendter, and M. Ehrlich. 2004. DNA hypomethylation is prevalent even in low-grade breast cancers. *submitted.*

Jeanpierre, M., C. Turleau, A. Aurias, M. Prieur, F. Ledeist, A. Fischer, and E. Viegas-Pequignot. 1993. An embryonic-like methylation pattern of classical satellite DNA is observed in ICF syndrome. *Hum. Mol. Genet.* 2:731-735.

Jhanwar-Uniyal, M. 2003. BRCA1 in cancer, cell cycle and genomic stability. *Front. Biosci.* 8:s1107-s1117.

Ji, W., R. Hernandez, X.-Y. Zhang, G. Qu, A. Frady, M. Varela, and M. Ehrlich. 1997. DNA demethylation and pericentromeric rearrangements of chromosome 1. *Mutat. Res.* 379:33-41.

Ji, W., G. Qu, P. Ye, X.-Y. Zhang, S. Halabi, and M. Ehrlich. 1995. Frequent detection of bcl-2/JH translocations in human blood and organ samples by a quantitative polymerase chain reaction assay. *Cancer Res.* 55:2876-2882.

Ji, Y., J. Zhang, A.I. Lee, H. Cedar, and Y. Bergman. 2003. A multistep mechanism for the activation of rearrangement in the immune system. *Proc. Natl. Acad. Sci. U. S. A.* 100:7557-7562.

Jiang, G., C. Sanchez, F. Yang, and M. Ehrlich. 2004. Histone modification in constitutive heterochromatin vs. unexpressed euchromatin in human cells. *J. Cell. Biochem.*:in press.

Jurgens, B., B.J. Schmitz-Drager, and W.A. Schulz. 1996. Hypomethylation of L1 LINE sequences prevailing in human urothelial carcinoma. *Cancer Res.* 56:5698-5703.

Kanai, Y., S. Ushijima, Y. Nakanishi, M. Sakamoto, and S. Hirohashi. 2003. Mutation of the DNA methyltransferase (DNMT) 1 gene in human colorectal cancers. *Cancer Lett.* 192:75-82.

Kanduri, C., G. Fitzpatrick, R. Mukhopadhyay, M. Kanduri, V. Lobanenkov, M. Higgins, and R. Ohlsson. 2002. A differentially methylated imprinting control region within the Kcnq1 locus harbors a methylation-sensitive chromatin insulator. *J Biol. Chem.* 277:18106-18110.

Kane, M.F., M. Loda, G.M. Gaida, J. Lipman, R. Mishra, H. Goldman, J.M. Jessup, and R. Kolodner. 1997. Methylation of the hMLH1 promoter correlates with lack of expression of hMLH1 in sporadic colon tumors and mismatch repair-defective human tumor cell lines. *Cancer Res.* 57:808-811.

Kang, G.H., S. Lee, H.J. Lee, and K.S. Hwang. 2004. Aberrant CpG island hypermethylation of multiple genes in prostate cancer and prostatic intraepithelial neoplasia. *J. Pathol.* 202:233-240.

Kang, G.H., Y.H. Shim, H.Y. Jung, W.H. Kim, J.Y. Ro, and M.G. Rhyu. 2001. CpG island methylation in premalignant stages of gastric carcinoma. *Cancer Res.* 61:2847-2851.

Kazazian, H.H., Jr., and J.V. Moran. 1998. The impact of L1 retrotransposons on the human genome. *Nat. Genet.* 19:19-24.

Kerbel, R.S., P. Frost, R. Liteplo, D.A. Carlow, and B.E. Elliott. 1984. Possible epigenetic mechanisms of tumor progression: induction of high-frequency heritable but phenotypically unstable changes in the tumorigenic and metastatic properties of tumor cell populations by 5-azacytidine treatment. *J. Cell Physiol. Suppl.* 3:87-97.

Kim, G.D., J. Ni, N. Kelesoglu, R.J. Roberts, and S. Pradhan. 2002. Co-operation and communication between the human maintenance and de novo DNA (cytosine-5) methyltransferases. *EMBO J.* 21:4183-4195.

Kimura, F., H.H. Seifert, A.R. Florl, S. Santourlidis, C. Steinhoff, S. Swiatkowski, C. Mahotka, C.D. Gerharz, and W.A. Schulz. 2003. Decrease of DNA methyltransferase 1 expression relative to cell proliferation in transitional cell carcinoma. *Int. J. Cancer.* 104:568-578.

Kinzler, K.W., and B. Vogelstein. 1996. Lessons from hereditary colorectal cancer. *Cell.* 87:159-170.

Kirsch, I.R., J.M. Abdallah, V.L. Bertness, M. Hale, S. Lipkowitz, F. Lista, and D.P. Lombardi. 1994. Lymphocyte-specific genetic instability and cancer. *Cold Spr. Harb. Symp. Quant. Biol.* 59:287-294.

Kokalj-Vokac, N., A. Almeida, E. Viegas-Pequignot, M. Jeanpierre, B. Malfoy, and B. Dutrillaux. 1993. Specific induction of uncoiling and recombination by azacytidine in classical satellite-containing constitutive heterochromatin. *Cytogenet. Cell Genet.* 63:11-15.

Kondo, T., Y. Comenge, M.P. Bobek, R. Kuick, B. Lamb, X. Zhu, A. Narayan, D. Bourc'his, E. Viegas-Pequinot, M. Ehrlich, and S. Hanash. 2000. Whole-genome methylation scan in ICF syndrome: hypomethylation of non-satellite DNA repeats D4Z4 and NBL2. *Hum. Mol. Gen.* 9:597-604.

Kondo, Y., L. Shen, and J.P. Issa. 2003. Critical role of histone methylation in tumor suppressor gene silencing in colorectal cancer. *Mol. Cell Biol.* 23:206-215.

Laird, P.W. 2003. The power and the promise of DNA methylation markers. *Nat. Rev. Cancer.* 3:253-266.

Le Baccon, P., D. Leroux, C. Dascalescu, S. Duley, Marais, E. Esmenjaud, J.J. Sotto, and M. Callanan. 2001. Novel evidence of a role for chromosome 1 pericentric heterochromatin in the pathogenesis of B-cell lymphoma and multiple myeloma. *Genes Chromosomes Cancer.* 32:250-264.

Lee, P.J., L.L. Washer, D.J. Law, C.R. Boland, I.L. Horon, and A.P. Feinberg. 1996. Limited up-regulation of DNA methyltransferase in human colon cancer reflecting increased cell proliferation. *Proc. Natl. Acad. Sci. USA.* 93:10366-10370.

Lehnertz, B., Y. Ueda, A.A. Derijck, U. Braunschweig, L. Perez-Burgos, S. Kubicek, T. Chen, E. Li, T. Jenuwein, and A.H. Peters. 2003. Suv39h-mediated histone H3 lysine 9 methylation directs DNA methylation to major satellite repeats at pericentric heterochromatin. *Curr. Biol.* 13:1192-1200.

Leib-Mosch, C., M. Haltmeier, T. Werner, E.M. Geigl, R. Brack-Werner, U. Francke, V. Erfle, and R. Hehlmann. 1993. Genomic distribution and transcription of solitary HERV-K LTRs. *Genomics.* 18:261-269.

Maraschio, P., O. Zuffardi, T.D. Fior, and L. Tiepolo. 1988. Immunodeficiency, centromeric heterochromatin instability of chromosomes 1, 9, and 16, and facial anomalies: the ICF syndrome. *J. Med. Genet.* 25:173-180.

Markowitz, S., J. Wang, L. Myeroff, R. Parsons, L. Sun, J. Lutterbaugh, R.S. Fan, E. Zborowska, K.W. Kinzler, B. Vogelstein, and et al. 1995. Inactivation of the type II TGF-beta receptor in colon cancer cells with microsatellite instability. *Science*. 268:1336-1338.

Mertens, F., B. Johansson, M. Hoglund, and F. Mitelman. 1997. Chromosomal imbalance maps of malignant solid tumors: a cytogenetic survey of 3185 neoplasms. *Cancer Res.* 57:2765-2780.

Mhawech, P., A. Benz, C. Cerato, V. Greloz, M. Assaly, J.C. Desmond, H.P. Koeffler, D. Lodygin, H. Hermeking, F. Herrmann, and J. Schwaller. 2004. Downregulation of 14-3-3sigma in ovary, prostate and endometrial carcinomas is associated with CpG island methylation. *Mod. Pathol.*

Miki, Y., I. Nishisho, A. Horii, Y. Miyoshi, J. Utsunomiya, K.W. Kinzler, B. Vogelstein, and Y. Nakamura. 1992. Disruption of the APC gene by a retrotransposal insertion of L1 sequence in a colon cancer. *Cancer Res.* 52:643-645.

Miniou, P., M. Jeanpierre, V. Blanquet, V. Sibella, D. Bonneau, C. Herbelin, A. Fischer, A. Niveleau, and E.T. Viegas-Pequignot. 1994. Abnormal methylation pattern in constitutive and facultative (X inactive chromosome) heterochromatin of ICF patients. *Hum. Molec. Genet.* 3:2093-2102.

Miniou, P., M. Jeanpierre, D. Bourc'his, A.C. Coutinho Barbosa, V. Blanquet, and E. Viegas-Pequignot. 1997. Alpha-satellite DNA methylation in normal individuals and in ICF patients: heterogeneous methylation of constitutive heterochromatin in adult and fetal tissues. *Hum. Genet.* 99:738-745.

Mitelman, F., F. Mertens, and B. Johansson. 1997. A breakpoint map of recurrent chromosomal rearrangements in human neoplasia. *Nat. Genet.* 15 Spec No:417-474.

Mizuno, S., T. Chijiwa, T. Okamura, K. Akashi, Y. Fukumaki, Y. Niho, and H. Sasaki. 2001. Expression of DNA methyltransferases DNMT1, 3A, and 3B in normal hematopoiesis and in acute and chronic myelogenous leukemia. *Blood.* 97:1172-1179.

Momparler, R.L. 2003. Cancer epigenetics. *Oncogene*. 22:6479-6483.

Morgan, H.D., H.G. Sutherland, D.I. Martin, and E. Whitelaw. 1999. Epigenetic inheritance at the agouti locus in the mouse. *Nat. Genet.* 23:314-318.

Morse, B., P.G. Rotherg, V.J. South, J.M. Spandorfer, and S.M. Astrin. 1988. Insertional mutagenesis of the myc locus by a LINE-1 sequence in a human breast carcinoma. *Nature*. 333:87-90.

Mostoslavsky, R., N. Singh, A. Kirillov, R. Pelanda, H. Cedar, A. Chess, and Y. Bergman. 1998. Kappa chain monoallelic demethylation and the establishment of allelic exclusion. *Genes Dev.* 12:1801-1811.

Muegge, K., H. Young, F. Ruscetti, and J. Mikovits. 2003. Epigenetic control during lymphoid development and immune responses: aberrant regulation, viruses, and cancer. *Ann. N. Y. Acad. Sci.* 983:55-70.

Nagai, H., Y.S. Kim, T. Yasuda, Y. Ohmachi, H. Yokouchi, M. Monden, M. Emi, N. Konishi, M. Nogami, K. Okumura, and K. Matsubara. 1999. A novel sperm-specific hypomethylation sequence is a demethylation hotspot in human hepatocellular carcinomas. *Gene*. 237:15-20.

Nakagawachi, T., H. Soejima, T. Urano, W. Zhao, K. Higashimoto, Y. Satoh, S. Matsukura, S. Kudo, Y. Kitajima, H. Harada, K. Furukawa, H. Matsuzaki, M. Emi, Y. Nakabeppu, K. Miyazaki, M. Sekiguchi, and T. Mukai. 2003. Silencing effect of CpG island hypermethylation and histone modifications on O6-methylguanine-DNA methyltransferase (MGMT) gene expression in human cancer. *Oncogene*. 22:8835-8844.

Nakase, H., Y. Takahama, and Y. Akamatsu. 2003. Effect of CpG methylation on RAG1/RAG2 reactivity: implications of direct and indirect mechanisms for controlling V(D)J cleavage. *Embo Rep.* 4:774-780.

Narayan, A., W. Ji, X.-Y. Zhang, A. Marrogi, J.R. Graff, S.B. Baylin, and M. Ehrlich. 1998. Hypomethylation of pericentromeric DNA in breast adenocarcinomas. *Int. J. Cancer*. 77:833-838.

Nasr, A.F., M. Nutini, B. Palombo, E. Guerra, and S. Alberti. 2003. Mutations of TP53 induce loss of DNA methylation and amplification of the TROP1 gene. *Oncogene*. 22:1668-1677.

Nguyen, C.T., D.J. Weisenberger, M. Velicescu, F.A. Gonzales, J.C. Lin, G. Liang, and P.A. Jones. 2002. Histone H3-lysine 9 methylation is associated with aberrant gene silencing in cancer cells and is rapidly reversed by 5-aza-2'-deoxycytidine. *Cancer Res*. 62:6456-6461.

O'Neill, R., M. O'Neill, and J.A. Graves. 1998. Undermethylation associated with retroelement activation and chromosome remodelling in an interspecific mammalian hybrid. *Nature*. 393:68-72.

Osipovich, O., R. Milley, A. Meade, M. Tachibana, Y. Shinkai, M.S. Krangel, and E.M. Oltz. 2004. Targeted inhibition of V(D)J recombination by a histone methyltransferase. *Nat. Immunol*. 5:309-316.

Pandis, N., G. Bardi, Y. Jin, C. Dietrich, B. Johansson, J. Andersen, N. Mandahl, F. Mitelman, and S. Heim. 1994. Unbalanced t(1;16) as the sole karyotypic abnormality in a breast carcinoma and its lymph node metastasis. *Cancer Genet. Cytogenet*. 75:151-159.

Peters, A.H., D. O'Carroll, H. Scherthan, K. Mechtler, S. Sauer, C. Schofer, K. Weipoltshammer, M. Pagani, M. Lachner, A. Kohlmaier, S. Opravil, M. Doyle, M. Sibilia, and T. Jenuwein. 2001. Loss of the Suv39h histone methyltransferases impairs mammalian heterochromatin and genome stability. *Cell*. 107:323-337.

Plass, C., and P.D. Soloway. 2002. DNA methylation, imprinting and cancer. *Eur. J. Hum. Genet*. 10:6-16.

Pradhan, S., and G.D. Kim. 2002. The retinoblastoma gene product interacts with maintenance human DNA (cytosine-5) methyltransferase and modulates its activity. *EMBO J*. 21:779-788.

Qian, X.C., and T.P. Brent. 1997. Methylation hot spots in the 5' flanking region denote silencing of the O6-methylguanine-DNA methyltransferase gene. *Cancer Res*. 57:3672-3677.

Qu, G., L. Dubeau, A. Narayan, M. Yu, and M. Ehrlich. 1999a. Satellite DNA hypomethylation vs. overall genomic hypomethylation in ovarian epithelial tumors of different malignant potential. *Mut. Res*. 423:91-101.

Qu, G., P.E. Grundy, A. Narayan, and M. Ehrlich. 1999b. Frequent hypomethylation in Wilms tumors of pericentromeric DNA in chromosomes 1 and 16. *Cancer Genet. Cytogenet*. 109:34-39.

Reese, B.E., K.E. Bachman, S.B. Baylin, and M.R. Rountree. 2003. The methyl-CpG binding protein MBD1 interacts with the p150 subunit of chromatin assembly factor 1. *Mol. Cell Biol*. 23:3226-3236.

Robertson, K.D. 2002. DNA methylation and chromatin - unraveling the tangled web. *Oncogene*. 21:5361-5379.

Roman-Gomez, J., J.A. Castillejo, A. Jimenez, M.G. Gonzalez, F. Moreno, C. Rodriguez Mdel, M. Barrios, J. Maldonado, and A. Torres. 2002. 5' CpG island hypermethylation is associated with transcriptional silencing of the p21(CIP1/WAF1/SDI1) gene and confers poor prognosis in acute lymphoblastic leukemia. *Blood*. 99:2291-2296.

Rothberg, P.G., S. Ponnuru, D. Baker, J.F. Bradley, A.I. Freeman, G.W. Cibis, D.J. Harris, and D.P. Heruth. 1997. A deletion polymorphism due to Alu-Alu recombination in intron 2 of the retinoblastoma gene: association with human gliomas. *Mol. Carcinog*. 19:69-73.

Sabbattini, P., M. Lundgren, A. Georgiou, C. Chow, G. Warnes, and N. Dillon. 2001. Binding of Ikaros to the lambda5 promoter silences transcription through a mechanism that does not require heterochromatin formation. *EMBO J*. 20:2812-2822.

Saito, Y., Y. Kanai, M. Sakamoto, H. Saito, H. Ishii, and S. Hirohashi. 2001. Expression of mRNA for DNA methyltransferases and methyl-CpG-binding proteins and DNA methylation status on CpG islands and pericentromeric satellite regions during human hepatocarcinogenesis. *Hepatology*. 33:561-568.

Saito, Y., Y. Kanai, M. Sakamoto, H. Saito, H. Ishii, and S. Hirohashi. 2002. Overexpression of a splice variant of DNA methyltransferase 3b, DNMT3b4, associated with DNA hypomethylation on pericentromeric satellite regions during human hepatocarcinogenesis. *Proc. Natl. Acad. Sci. USA*. 99:10060-10065.

Salem, C., G. Liang, Y.C. Tsai, J. Coulter, M.A. Knowles, A.C. Feng, S. Groshen, P.W. Nichols, and P.A. Jones. 2000. Progressive increases in de novo methylation of CpG islands in bladder cancer. *Cancer Res*. 60:2473-2476.

Santourlidis, S., A. Florl, R. Ackermann, H.C. Wirtz, and W.A. Schulz. 1999. High frequency of alterations in DNA methylation in adenocarcinoma of the prostate. *Prostate*. 39:166-174.

Sassaman, D.M., B.A. Dombroski, J.V. Moran, M.L. Kimberland, T.P. Naas, R.J. DeBerardinis, A. Gabriel, G.D. Swergold, and H.H. Kazazian, Jr. 1997. Many human L1 elements are capable of retrotransposition. *Nat. Genet*. 16:37-43.

Sato, K., R. Hayami, W. Wu, T. Nishikawa, H. Nishikawa, Y. Okuda, H. Ogata, M. Fukuda, and T. Ohta. 2004. Nucleophosmin/B23 is a candidate substrate for the BRCA1-BARD1 ubiquitin ligase. *J. Biol. Chem*. 279:30919-30922.

Sato, N., A. Maitra, N. Fukushima, N.T. van Heek, H. Matsubayashi, C.A. Iacobuzio-Donahue, C. Rosty, and M. Goggins. 2003. Frequent hypomethylation of multiple genes overexpressed in pancreatic ductal adenocarcinoma. *Cancer Res*. 63:4158-4166.

Sawyer, J.R., C.M. Swanson, B.S. Koller, P.E. North, and S.W. Ross. 1995a. Centromeric instability of chromosome 1 resulting in multibranched chromosomes, telomeric fusions, and "jumping translocations" of 1q in a human immunodeficiency virus-related non-Hodgkin's lymphoma. *Cancer*. 76:1238-1244.

Sawyer, J.R., C.M. Swanson, G. Wheeler, and C. Cunniff. 1995b. Chromosome instability in ICF syndrome: formation of micronuclei from multibranched chromosome 1 demonstrated by fluorescence in situ hybridization. *Am. J. Med. Genet*. 56:203-209.

Scelfo, R.A., C. Schwienbacher, A. Veronese, L. Gramantieri, L. Bolondi, P. Querzoli, I. Nenci, G.A. Calin, A. Angioni, G. Barbanti-Brodano, and M. Negrini. 2002. Loss of methylation at chromosome 11p15.5 is common in human adult tumors. *Oncogene*. 21:2564-2572.

Schichman, S.A., M.A. Caligiuri, M.P. Strout, S.L. Carter, Y. Gu, E. Canaani, C.D. Bloomfield, and C.M. Croce. 1994. ALL-1 tandem duplication in acute myeloid leukemia with a normal karyotype involves homologous recombination between Alu elements. *Cancer Res*. 54:4277-4280.

Schmid, C.W. 1996. Alu: structure, origin, evolution, significance and function of one- tenth of human DNA. *Prog. Nucleic Acid Res. Mol. Biol*. 53:283-319.

Schulz, W.A. 1998. DNA methylation in urological malignancies (review). *Int. J. Oncol*. 13:151-167.

Schulz, W.A., J.P. Elo, A.R. Florl, S. Pennanen, S. Santourlidis, R. Engers, M. Buchardt, H.H. Seifert, and T. Visakorpi. 2002. Genomewide DNA hypomethylation is associated with alterations on chromosome 8 in prostate carcinoma. *Genes Chromosomes Cancer*. 35:58-65.

Sengupta, P.K., M. Ehrlich, and B.D. Smith. 1999. A methylation-responsive MDBP/RFX site is in the first exon of the collagen alpha2(I) promoter. *J. Biol. Chem*. 274:36649-36655.

Sengupta, P.K., J. Fargo, and B.D. Smith. 2002. The RFX family interacts at the collagen (COL1A2) start site and represses transcription. *J. Biol. Chem*. 277:24926-24937.

Sikes, M.L., C.C. Suarez, and E.M. Oltz. 1999. Regulation of V(D)J recombination by transcriptional promoters. *Mol. Cell. Biol*. 19:2773-2781.

Smeets, D.F.C.M., U. Moog, C.M.R. Weemaes, G. Vaes-Peeters, G.F.M. Merkx, J.P. Niehof, and G. Hamers. 1994. ICF syndrome: a new case and review of the literature. *Hum. Genet*. 94:240-246.

Speit, G., and K. Trenz. 2004. Chromosomal mutagen sensitivity associated with mutations in BRCA genes. *Cytogenet. Genome Res*. 104:325-332.

Stoner, G.D., M.B. Shimkin, A.J. Kniazeff, J.H. Weisburger, E.K. Weisburger, and G.B. Gori. 1973. Test for carcinogenicity of food additives and chemotherapeutic agents by the pulmonary tumor response in strain A mice. *Cancer Res.* 33:3069-3085.

Sumner, A.T., A.R. Mitchell, and P.M. Ellis. 1998. A FISH study of chromosome fusion in the ICF syndrome: involvement of paracentric heterochromatin but not of the centromeres themselves. *J. Med. Genet.* 35:833-835.

Tagarro, I., A.M. Fernandez-Peralta, and J.J. Gonzalez-Aguilera. 1994. Chromosomal localization of human satellites 2 and 3 by a FISH method using oligonucleotides as probes. *Hum.Genet.* 93:383-388.

Takai, D., Y. Yagi, N. Habib, T. Sugimura, and T. Ushijima. 2000. Hypomethylation of LINE1 retrotransposon in human hepatocellular carcinomas, but not in surrounding liver cirrhosis. *Jpn. J. Clin. Oncol.* 30:306-309.

Takizawa, T., K. Nakashima, M. Namihira, W. Ochiai, A. Uemura, M. Yanagisawa, N. Fujita, M. Nakao, and T. Taga. 2001. DNA methylation is a critical cell-intrinsic determinant of astrocyte differentiation in the fetal brain. *Dev. Cell.* 1:749-758.

Thoraval, D., J. Asakawa, M. Kodaira, C. Chang, E. Radany, R. Kuick, B. Lamb, B. Richardson, J.V. Neel, T. Glover, and S. Hanash. 1996. A methylated human 9-kb repetitive sequence on acrocentric chromosomes is homologous to a subtelomeric repeat in chimpanzees. *Proc. Natl. Acad. Sci. U. S. A.* 93:4442-4447.

Tiepolo, L., P. Maraschio, G. Gimelli, C. Cuoco, G.F. Gargani, and C. Romano. 1979. Multibranched chromosomes 1, 9, and 16 in a patient with combined IgA and IgE deficiency. *Hum. Genet.* 51:127-137.

Tsien, F., B. Youn, E.S. Fiala, P. Laird, T.I. Long, K. Weissbecker, and M. Ehrlich. 2002. Prolonged culture of chorionic villus cells yields ICF syndrome-like chromatin decondensation and rearrangements. *Cytogen. Genome Res.* 98:13-21.

Tuck-Muller, C.M., A. Narayan, F. Tsien, D. Smeets, J. Sawyer, E.S. Fiala, O. Sohn, and M. Ehrlich. 2000. DNA hypomethylation and unusual chromosome instability in cell lines from ICF syndrome patients. *Cytogenet. Cell Genet.* 89:121-128.

Turleau, C., M.-O. Cabanis, D. Girault, F. Ledeist, R. Mettey, H. Puissant, P. Marguerite, and J. de Grouchy. 1989. Multibranched chromosomes in the ICF syndrome: Immunodeficiency, centromeric instability, and facial anomalies. *Am. J. Med. Genet.* 32:420-424.

Veigl, M.L., L. Kasturi, J. Olechnowicz, A.H. Ma, J.D. Lutterbaugh, S. Periyasamy, G.M. Li, J. Drummond, P.L. Modrich, W.D. Sedwick, and S.D. Markowitz. 1998. Biallelic inactivation of hMLH1 by epigenetic gene silencing, a novel mechanism causing human MSI cancers. *Proc. Natl. Acad. Sci. U. S. A.* 95:8698-8702.

Vilain, A., J. Bernardino, M. Gerbault-Seureau, N. Vogt, A. Niveleau, D. Lefrancois, B. Malfoy, and B. Dutrillaux. 2000. DNA methylation and chromosome instability in lymphoblastoid cell lines. *Cytogenet. Cell Genet.* 90:93-101.

Wade, P.A. 2001. Methyl CpG binding proteins: coupling chromatin architecture to gene regulation. *Oncogene.* 20:3166-3173.

Walsh, C.P., J.R. Chaillet, and T.H. Bestor. 1998. Transcription of IAP endogenous retroviruses is constrained by cytosine methylation. *Nat. Genet.* 20:116-117.

Wang, C., A. Horiuchi, T. Imai, S. Ohira, K. Itoh, T. Nikaido, Y. Katsuyama, and I. Konishi. 2004. Expression of BRCA1 protein in benign, borderline, and malignant epithelial ovarian neoplasms and its relationship to methylation and allelic loss of the BRCA1 gene. *J. Pathol.* 202:215-223.

Wei, W., N. Gilbert, S.L. Ooi, J.F. Lawler, E.M. Ostertag, H.H. Kazazian, J.D. Boeke, and J.V. Moran. 2001. Human L1 retrotransposition: cis preference versus trans complementation. *Mol. Cell. Biol.* 21:1429-1439.

Westermark, U.K., M. Reyngold, A.B. Olshen, R. Baer, M. Jasin, and M.E. Moynahan. 2003. BARD1 participates with BRCA1 in homology-directed repair of chromosome breaks. *Mol. Cell Biol.* 23:7926-7936.

Wheeler, J.M., A. Loukola, L.A. Aaltonen, N.J. Mortensen, and W.F. Bodmer. 2000. The role of hypermethylation of the hMLH1 promoter region in HNPCC versus MSI+ sporadic colorectal cancers. *J. Med. Genet.* 37:588-592.

Widschwendter, M., G. Jiang, C. Woods, H.M. Muller, H. Fiegl, G. Goebel, C. Marth, E.M. Holzner, A.G. Zeimet, P.W. Laird, and M. Ehrlich. 2004. DNA hypomethylation and ovarian cancer biology. *Cancer Res.* 64:4472-4480.

Wong, N., W.C. Lam, P.B. Lai, E. Pang, W.Y. Lau, and P.J. Johnson. 2001. Hypomethylation of chromosome 1 heterochromatin DNA correlates with q-arm copy gain in human hepatocellular carcinoma. *Am. J. Pathol.* 159:465-471.

Xin, Z., M. Tachibana, M. Guggiari, E. Heard, Y. Shinkai, and J. Wagstaff. 2003. Role of histone methyltransferase G9a in CpG methylation of the Prader-Willi syndrome imprinting center. *J. Biol. Chem.* 278:14996-5000.

Yakushiji, T., K. Uzawa, T. Shibahara, H. Noma, and H. Tanzawa. 2003. Over-expression of DNA methyltransferases and CDKN2A gene methylation status in squamous cell carcinoma of the oral cavity. *Int J Oncol*. 22:1201-7.

Yan, P.S., H. Shi, F. Rahmatpanah, T.H. Hsiau, A.H. Hsiau, Y.W. Leu, J.C. Liu, and T.H. Huang. 2003. Differential distribution of DNA methylation within the RASSF1A CpG island in breast cancer. *Cancer Res.* 63:6178-6186.

Yang, F., C. Shao, V. Vedanarayanan, and M. Ehrlich. 2004. Cytogenetic and Immuno-FISH analysis of the 4q subtelomeric region, which is associated with facioscapulohumeral muscular dystrophy. *Chromosoma*. 112:350-359.

Yasui, W., N. Oue, S. Ono, Y. Mitani, R. Ito, and H. Nakayama. 2003. Histone acetylation and gastrointestinal carcinogenesis. *Ann. N. Y. Acad. Sci.* 983:220-231.

Yin, D., D. Xie, W.K. Hofmann, W. Zhang, K. Asotra, R. Wong, K.L. Black, and H.P. Koeffler. 2003. DNA repair gene O6-methylguanine-DNA methyltransferase: promoter hypermethylation associated with decreased expression and G:C to A:T mutations of p53 in brain tumors. *Mol. Carcinog.* 36:23-31.

Yusa, K., J. Takeda, and K. Horie. 2004. Enhancement of Sleeping Beauty transposition by CpG methylation: possible role of heterochromatin formation. *Mol. Cell Biol.* 24:4004-4018.

Zhang, J., H. Willers, Z. Feng, J.C. Ghosh, S. Kim, D.T. Weaver, J.H. Chung, S.N. Powell, and F. Xia. 2004. Chk2 phosphorylation of BRCA1 regulates DNA double-strand break repair. *Mol. Cell Biol.* 24:708-718.

Zhang, X.Y., N. Jabrane-Ferrat, C.K. Asiedu, S. Samac, B.M. Peterlin, and M. Ehrlich. 1993. The major histocompatibility complex class II promoter-binding protein RFX (NF-X) is a methylated DNA-binding protein. *Mol. Cell. Biol.* 13:6810-6818.

Zheng, L., A. Flesken-Nikitin, P.L. Chen, and W.H. Lee. 2002. Deficiency of Retinoblastoma gene in mouse embryonic stem cells leads to genetic instability. Cancer Res. 62:2498-2502.

Chapter 3.7

DEREGULATION OF THE CENTROSOME CYCLE AND THE ORIGIN OF CHROMOSOMAL INSTABILITY IN CANCER

Wilma L. Lingle, Kara Lukasiewicz, and Jeffrey L. Salisbury
Mayo Clinic Cancer Center, Rochester, Minnesota, USA

1. INTRODUCTION

Gross changes in nuclear staining and morphology are hallmarks of cancer, and together with an assessment of tissue differentiation, they are key features used to identify tumour grade. These nuclear changes reflect an underlying *genomic instability* of cancer cells. Genomic instability broadly encompasses diverse mechanisms that give rise to gains and losses of whole chromosomes (*aneuploidy*), to alteration of chromatin structure and nuclear architecture, and to gene-specific changes including gene mutation, translocation, amplification or deletion. Importantly, genomic instability is thought to be responsible for the generation of phenotypic diversity among tumour cells and to be key to tumour progression, the development of chemoresistance, and, ultimately, to poor outcome. Chromosomal instability and consequent aneuploidy has been proposed as a driving mechanism for the generation of large scale genetic alterations thought to be necessary for rapid evolution of tumour cell genotypic and phenotypic diversity. The importance of aneuploidy in cancer development has recently gained strong support with the identification of a cancer predisposition syndrome involving constitutional mosaicism for chromosomal gains and losses (Grimm, 2004; Hanks et al., 2004).

A century ago, Theodor Boveri and W.S. Sutton *independently* recognised that chromosome behaviour provided the mechanistic basis for the transmission of genetic traits (Wilson, 1925). Remarkably, Boveri also recognised a crucial link between centrosome defects, aneuploidy, and the development of cancer (Boveri, 1914). Boveri found that when two sperm, instead of one, fertilised sea urchin eggs, multipolar mitotic spindles formed due to double the number of sperm-derived centrosomes. One consequence

E.A. Nigg (ed.), Genome Instability in Cancer Development, 393-421.

of multipolar mitotic spindle organisation in these cells was unequal chromosome segregation into daughter cells (Figure 1). Aneuploidy in sea urchin embryos resulted in abnormal development or in cell death if the chromosomal imbalance was too severe. Boveri recognised the similarity between the abnormally developing embryos and loss of tissue architecture and development of aneuploidy seen in cancer. Indeed, it is humbling for contemporary cancer researchers to realise that Boveri, armed only with a light microscope and the literature of the day, first proposed that an 'imbalance of chromosomes' might play an important role in tumour development, and, further, for him to recognise that defective centrosome behaviour could drive these changes.

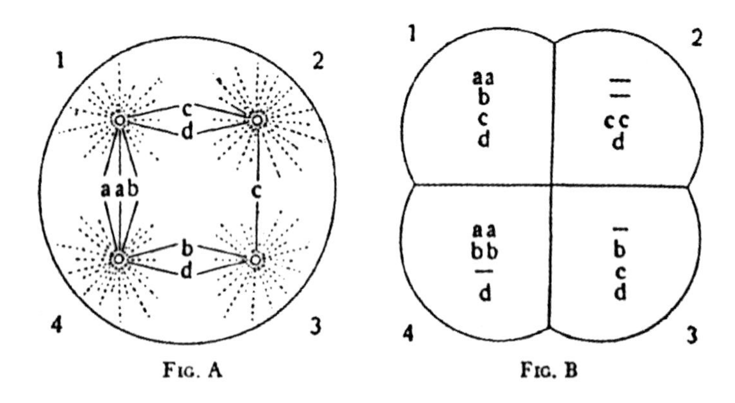

Figure 1. Chromosomal missegregation following multipolar mitosis. Theodor Boveri first proposed that an imbalance of chromosomes may play a causative role in the origin of malignant tumours based on his observations of chromosome missegregation in a sea urchin eggs following dispermic fertilisation (Boveri, 1914)

Despite Boveri's early prediction, the role of abnormal centrosome behaviour in the origin of aneuploidy in cancer has only recently become an area of active investigation. For an interesting aside on why "ten years of neglect" became a century, see (Metcalf, 1925). First described in human breast, neuroectodermal, and prostate tumours (Lingle et al., 1998; Pihan et al., 1998; Weber et al., 1998), and subsequently in other solid tumours and hematopoietic cancers, *centrosome amplification* has been found to be a common event in the development of many cancers. The term "centrosome amplification" designates centrosomes that contain more than four centrioles (i.e., "supernumerary centrioles"), centrosomes that appear significantly larger than normal as defined by the accumulation of structural centrosome components in excess of that seen in the corresponding normal tissue or cell type, and/or when more than two centrosomes are present within a cell (Lingle et al., 1998; Lingle and Salisbury, 1999).

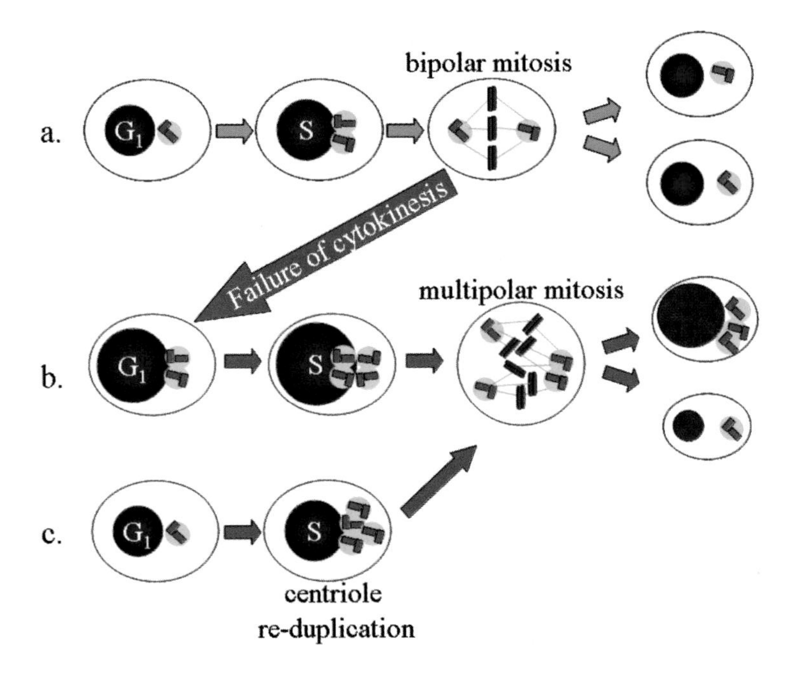

Figure 2. Models for the origin of centrosome defects in cancer. a) Normal centrosome cycle in which centrioles duplicate in S phase and a bipolar mitosis results in equal segregation of chromosomes and centrosomes into each of two daughter cells. b) Failure of cytokinesis followed by centrosome doubling in S phase resulting in a multipolar mitosis with unequal chromosome and centrosome segregation. c) Centriole reduplication in S phase resulting in a multipolar mitosis with unequal chromosome and centrosome segregation.

There are two current models for the origin of centrosome defects in the development of cancer (Figure 2): In the first, centrosome amplification arises through failure of cytokinesis and the consequent failure of equal partition of sister chromatids and spindle poles into daughter cells. In this model, a single 4N daughter cell inherits both spindle poles, instead of just one, to yield two functional centrosomes – this scenario mimics the dispermy experiments of Boveri (Meraldi et al., 2002; Nigg, 2002). The two centrosomes double again in the next cell cycle to yield four functional spindle poles and multipolar mitosis. Centrosome amplification arises in the second model through a deregulation of the centriole duplication cycle leading to centrosomes with supernumerary centrioles. In this model, disruption of key cell and/or centrosome cycle regulators may play a causative role. These models are not mutually exclusive and may operate independently or sequentially in the development of cancer. Here, we present an overview of centrosome behaviour in normal cells and in tumour development, review recent progress in understanding the mechanisms by which centrosome amplification can arise, and discuss the consequences of centrosome amplification in the origin of chromosomal instability in cancer.

2. CENTROSOME STRUCTURE AND FUNCTION

2.1 Centrosome Behaviour in Normal Cells

The centrosome is a cytoplasmic organelle that typically resides in a juxtanuclear position near the cell centre (Wilson, 1925). Intense study over the past century has revealed two defining functional features of centrosomes in normal cells: the ability to nucleate microtubules, and their doubling, once in each cell cycle, to yield two centrosomes that act as spindle poles during mitosis (Nigg 2004). In addition to these fundamental properties, the centrosome also influences the position, orientation and completion of cytokinesis, and it provides an important structural context for coordinating cell cycle regulation (Doxsey, 2001b; Khodjakov and Rieder, 2001; Piel et al., 2001; Salisbury et al., 2002; Sluder and Hinchcliffe, 2000). Understanding the molecular basis for these diverse cellular functions is beginning to emerge through the careful analysis of centrosome genes and proteins, and centrosome formation, structure, and organisation in model systems, early embryo development, and in mammalian somatic cells (Andersen et al., 2003; Avidor-Reiss et al., 2004; Li et al., 2004; Pazour et al., 2000; Rosenbaum and Witman, 2002, Nigg 2004).

Structurally, the centrosome consists of four fundamental components: a core structure consisting of a pair of *centrioles* that serve as a centrosomal organiser (Bobinnec et al., 1998; Preble et al., 2000); a surrounding protein lattice or matrix called *pericentriolar material* (PCM) that serves as a framework to anchor microtubule nucleation sites (Ou et al., 2004); *γ-tubulin complexes* that are responsible for the nucleation of microtubules (Schiebel, 2000); and fibres composed of *Sfi1p and centrin* that act as calcium-sensitive contractile or elastic connections between the various elements of the centrosome that mediate dynamic changes in its overall structure (Salisbury, 2004).

2.2 Coordination of the Centrosome, DNA, and Cell Cycles

Progress in understanding centrosome doubling has recently accelerated. During a normal cell cycle the centrosome doubles once, and only once, to yield two centrosomes. The presence of only two centrosomes in the cell as it enters mitosis ensures the formation of a bipolar spindle and the equal segregation of sister chromatids to each daughter cell. Centrosome doubling is initiated with the semi-conservative duplication of the pair of centrioles resulting in two pairs. Centrosomes increase in size through the recruitment of additional PCM and γ-tubulin in late G_2 of the cell cycle, when a dramatic increase in microtubule nucleating activity also occurs (Khodjakov and Rieder, 1999; Kuriyama and Borisy, 1981; Moudjou et al., 1996; Starita et al., 2004; Stearns et al., 1991). This period coincides with a major redeployment of PCM components that anchor γ-tubulin complexes at the centrosome (Casenghi et al., 2003). During mitotic prophase the two centriole pairs separate from one another, each pair carrying associated PCM, to function in the organisation of two half spindles of the mitotic apparatus (Piel et al., 2000).

2.3 Cell Cycle Regulators Converge on the Centrosome

Centrosome doubling is strictly coordinated with DNA replication, mitosis and cell division (Sluder and Hinchcliffe, 2000). Recent evidence suggests that the centrosome itself may provide a structural platform on which exquisite local control of the cell and centrosome cycles resides. The most compelling observation for this is that ablation of centrosomes by microsurgery or laser treatment result in G_1 cell cycle arrest (Doxsey, 2001a; Doxsey, 2001b; Hinchcliffe et al., 2001; Khodjakov and Rieder, 2001; Sluder and Hinchcliffe, 1998). Similarly, RNAi knockdown of even a single centrosome protein key to centriole duplication, such as the calcium-modulated protein centrin, results in arrest of centriole duplication, mitotic catastrophe, and ultimately cell death (Salisbury et al., 2002). Local activity of many important regulators of cell cycle progression is suggested by their residence at the centrosome, albeit some only transiently. Key cell cycle regulators described at the centrosome include: p53, the cyclin/cdks, protein kinase A (PKA), Aurora-A, Plk1, BRCA1 and BRCA2, and the anaphase promoting complex (APC/cyclosome) (Bailly et al., 1992; DeCamilli et al., 1986; Diviani et al., 2000; Fry et al., 2000; Giannakakou et al., 2000; Hsu and White, 1998; Keryer et al., 2003; Kraft et al., 2003; Matsumoto and Maller, 2004; Matyakhina et al., 2002; Nigg et al., 1986; Pockwinse et al., 1997; Rattner et al., 1990; Tugendreich et al., 1995; Xu et al., 1999). Finally, many PCM proteins harbour one or more structural motifs to anchor cell cycle regulators and thus have the potential to order their activities (Diviani et al., 2000; Diviani and Scott, 2001; Gillingham and Munro, 2000; Keryer et al., 1993). Taken together, these observations suggest a mechanism by which key regulators act locally within the structural context of the centrosome to coordinate or amplify critical steps for cell cycle progression.

2.4 Control of Centrosome Doubling and Deregulation in Cancer

Control of centrosome doubling is tightly coupled to cell cycle progression through multiple interacting regulatory pathways. The first of these operates through activity of the Cdk/cyclin cell cycle regulators, which coordinate the cell, centrosome and DNA cycles (Hinchcliffe et al., 1999; Lacey et al., 1999; Matsumoto et al., 1999; Sluder and Hinchcliffe, 2000).

A second control pathway involves the p53-mediated G_1/S and G_2/M cell cycle checkpoints that monitor DNA integrity and arrest centrosome doubling through induction of $p21^{waf1}$ and consequent inhibition of the cdk/cyclins (D'Assoro et al., 2004; el-Deiry et al., 1993; Meraldi et al., 1999). Finally, centrosome behaviour, particularly at the time of mitosis, operates under the control of several centrosome-associated kinases, including protein kinase A (PKA), Aurora A, Nek2, and the Polo-like kinases Plk1 and Plk2 (Casenghi et al., 2003; Chen et al., 2003; DeCamilli et al., 1986; Fry, 2002; Fry et al., 1998b; Giet et al., 1999; Hamill et al., 2002; Kraft et al., 2003; Logarinho and Sunkel, 1998; Lou et al., 2004; Matyakhina et al., 2002; Meraldi and Nigg, 2001; Nigg et al., 1985; Tsvetkov et al., 2003; Yanai et al., 1997). Importantly, several of these regulatory processes become disrupted during cancer progression, and individually or in combination with other transforming events may lead to the development of amplified centrosomes and consequent chromosomal instability.

2.5 Cdk/cyclins and the Centrosome Cycle

The key stages of cell cycle progression are governed by the subcellular location, and periodic activation and subsequent inactivation of the serine/threonine cyclin-dependent protein kinases (Cdks) (Pines, 1995). Evidence suggesting a direct role for the Cdks in regulating the mitotic activity of centrosomes first came to light in studies on the localisation of cyclin B and Cdk1 (p34cdc2) at the centrosome during G2/M phase, and from experiments using *Xenopus* cell free extracts that implicated cyclins A and B in the control of microtubule dynamics (Bailly et al., 1992; Debec and Montmory, 1992; Verde et al., 1990). A centrosome localisation signal (CLS) has been identified in the cyclin E protein sequence and found to be essential for both centrosomal targeting and DNA synthesis (Matsumoto and Maller, 2004). Evidence for the direct involvement of Cdk2/cyclin activity in regulation of centrosome doubling is also beginning to accumulate. Both centrosome doubling and DNA replication are dependent on Cdk2 activation and are blocked by the Cdk2 inhibitors butyrolactone I and roscovitine (D'Assoro et al., 2004; Keezer and Gilbert, 2002; Matsumoto et al., 1999; Meraldi et al., 1999). Cdk2/cyclin E activity is required for completion of the centrosome cycle, since centrosome doubling can be blocked in *Xenopus* egg extracts by the small protein inhibitors of Cdk2, or by immuno-depletion of Cdk2 or cyclin E, and excess purified cdk2/cyclin E can restore centrosome doubling in the depleted egg extracts (Hinchcliffe et al., 1999; Lacey et al., 1999). Also using *Xenopus* egg extracts, separation of the centriole pair (centriole disjunction) was shown to depend on Cdk2/cyclin E-mediated regulation of centriole pair cohesion (Lacey et al., 1999).

Cdk2/cyclin A has also been implicated in the control of centrosome doubling (D'Assoro et al., 2004; Kronenwett et al., 2003; Lacey et al., 1999; Meraldi et al., 1999). Using the centriole re-duplication assay in mammalian cells (see below) (Balczon et al., 1995), and a series of dominant negative acting mutations of key cell cycle regulatory proteins, Cdk2/cyclin A was shown to be required for centriole duplication (Meraldi et al., 1999). Likewise, cyclin A levels or its overexpression correlate with or drive centrosome doubling in a variety of experimental systems (Balczon, 2001; Kronenwett et al., 2003; Lacey et al., 1999). Importantly, in vivo studies demonstrate that transient elevation of cyclin D1 can result in persistent centrosome amplification, chromosomal instability, and the development of aneuploidy (Nelsen et al., 2004).

It is likely that numerous Cdk/cyclin target substrates play important roles in the control of the centrosome cycle. However, few have been unequivocally identified or the consequences of their phosphorylation determined. Proposed substrates of Cdk2 implicated in the regulation of centrosome duplication include Mps1 kinase, nucleophosmin, and CP110 (Chen et al., 2002b; Fisk et al., 2003; Okuda et al., 2000; Tarapore et al., 2002). Nucleophosmin is a Cdk2/cyclin E substrate that was described to localise to the centrosome and to be released after phosphorylation by Cdk2/cyclin E, coincident with centrosome doubling (Cha et al., 2004; Okuda, 2002; Okuda et al., 2000; Tarapore et al., 2002). CP110 localises to centrosomes and is phosphorylated *in vitro* by Cdk2/cyclin A or E, as well as Cdc2/cyclin B (Chen et al., 2002b). Its expression is strongly induced at the G1-to-S phase transition, coincident with the initiation of centrosome duplication. Long-term disruption of CP110 phosphorylation leads to unscheduled centrosome separation and overt polyploidy.

Finally, a broader role for Cdk/cyclin activation in centrosome doubling operates through the phosphorylation status of retinoblastoma tumour suppressor Rb, which governs the availability of the E2F transcription factor to promote S phase progression (Angus et al., 2002; D'Assoro et al., 2004; Meraldi et al., 1999; Saavedra et al., 2003). Taken together, these observations establish a general mechanism by which DNA replication and the centrosome cycle are coordinated: both DNA replication and centrosome doubling require Cdk/cyclin activation and are controlled by the Rb pathway and downstream transcriptional activity of E2F.

2.6 The Centrosome and DNA Cycles can be Uncoupled

In certain cells, multiple rounds of centriole duplication (called centriole 're-duplication') can occur when DNA synthesis is blocked; thus the centrosome cycle is not strictly dependent on DNA replication per se (Balczon et al., 1995). However, accumulating evidence suggests that the centrosome and DNA cycles can be uncoupled only in cells that are defective in G_1/S or G_2/M checkpoint controls (D'Assoro et al., 2004; Meraldi et al., 2002). Understanding the role of the tumour suppressor protein p53 in this process has begun to clarify a mechanism linking DNA integrity and the centrosome cycle. Centrosome-anchored microtubules transport p53 into the nucleus by a dynein motor driven process in response to DNA damage and provide a mechanism for delivering p53 to the centrosome itself (Giannakakou et al., 2000). Importantly, loss of function of p53 is common in many cancers and can lead to centrosome amplification and aneuploidy in mouse and human cell lines (Carroll et al., 1999; Duensing et al., 2000; Fukasawa et al., 1996; Murphy et al., 2000; Wang et al., 1998). While these studies suggest that p53 is essential for maintenance of centrosome homeostasis, a precise role for p53 in the development of centrosome amplification remains an area of active investigation.

Two distinct p53-related mechanisms can account for centrosome amplification. Interestingly, loss of p53 function alone is not sufficient to drive centrosome amplification. Deregulation of centriole duplication cycle can arise through failure of the p53-mediated G1/S checkpoint, thereby allowing unchecked CDK/cyclin E activity to drive centrosome duplication and subsequent centrosome amplification (D'Assoro et al., 2004; Kawamura et al., 2004; Lacey et al., 1999; Matsumoto et al., 1999). Alternatively, multiple centrosomes may accumulate through failure of cytokinesis and apoptosis in p53 defective cells (Meraldi et al., 2002; Nigg, 2002). In both of these settings, centrosome amplification is not a result of loss of p53 function alone, rather additional cellular stress is necessary to uncouple DNA replication and centriole duplication and this can be exacerbated by deregulation of key cell cycle and/or centrosome kinases (Du and Hannon, 2002; Katayama et al., 2004; Meraldi et al., 2002; Meraldi et al., 2004; Zhou et al., 1998). Taken together, it is evident that alternative pathways operate in parallel and converge onto cell cycle regulators and G_1/S and G_2/M checkpoints to link the centrosome, DNA and cell cycles to ensure their coordinate progression.

Recent genetic studies also suggest that alternative p53-independent pathways leading to centrosome amplification exist. The tumour suppressor gene products BRCA1 and BRCA2 localise to centrosomes and have been implicated in the development of centrosome defects in breast tumours (Deng and Brodie, 2000; Hsu and White, 1998; Tutt et al., 1999; Xu et al., 1999). BRCA1 and BRCA2 both play important roles in transcription and DNA repair, however these proteins have recently been shown to have other apparently unrelated functional properties. BRCA1 and its binding partner BARD1 function together as an E3 ligase to ubiquitinate a variety of target proteins (Hashizume et al., 2001). While the mechanistic relationship between BRCA1 and centrosome function is poorly understood, recent evidence suggests that BRCA1/BARD1 ubiquitinate γ-tubulin at the centrosome and may play a role in regulating microtubule dynamics in normal cells (Starita et al., 2004). Importantly, transient inhibition of BRCA1 function in cell lines derived from mammary tissue caused rapid amplification and fragmentation of centrosomes (Starita et al., 2004). Just how failure of γ-tubulin ubiquitination leads to these centrosome defects is an unresolved issue. Nonetheless, inactivation of BRCA1 is seen in the development of certain familial breast and ovarian cancers, and it is tempting to speculate that centrosome defects resulting from failure of γ-tubulin ubiquitination in these individuals could be related to the etiology of this disease through consequent chromosomal instability (Starita et al., 2004).

2.7 Centrosome Associated Kinases

Centrosome protein phosphorylation increases dramatically at the onset of mitosis and falls precipitously at the metaphase/anaphase transition (Rao et al., 1989; Vandre et al., 2000). While the Cdk/cyclins are paramount in coordinating the centrosome, DNA, and cell cycles, several additional regulators reside at the centrosome and are critical for aspects of centrosome structure and function (Table 1). These centrosome-associated kinases affect one or more fundamental properties of centrosome behaviour, including: *Centriole duplication* – the process leading to procentriole formation around the time of S phase; *Centrosome maturation* – the remodelling of mitotic PCM structure and the recruitment of additional γ-tubulin complexes for microtubule nucleation; and *Centrosome separation* – the partitioning of newly doubled centrosomes into two distinct structures that function as the mitotic spindle poles.

Protein kinase A (PKA) plays an important role in many cellular processes. PKA is localised at the centrosome in vertebrate cells, including HeLa (De Camilli et al., 1986; Nigg et al., 1985) through interaction of its regulatory subunit (RII) and the protein kinase A anchoring coiled-coil domains of AKAP450 and the centrosomal structural protein pericentrin (Diviani et al., 2000; Keryer et al., 1999; Witczak et al., 1999). Tethering of PKA to protein kinase A anchoring proteins is thought to target the enzyme to the proximity of relevant substrates, thereby conveying spatial specificity to cAMP/PKA signalling. Expression of the C-terminus of AKAP450, which contains the centrosome targeting domain of AKAP450 but not its coiled-coil domains or binding sites for signalling molecules, leads to the displacement of the endogenous centrosomal AKAP450 without removing centriolar or PCM components such as centrin, γ-tubulin or pericentrin (Keryer et al., 2003). This expression also impaired cytokinesis, increased ploidy in HeLa cells, and resulted in G1 arrest in normal diploid fibroblasts. Experimental elevation of PKA activity in HeLa resulted in phosphorylation of the centrosome protein centrin and lead to centrosome separation (Lutz et al., 2001). Together, these observations demonstrate that centriole cohesion is sensitive to PKA activity and the association between centrioles and the centrosomal matrix protein AKAP450 is critical for the integrity of the centrosome. Polo kinase was initially identified in *Drosophila* mutants that showed aberrant mitotic spindle organisation (Sunkel and Glover, 1988). In human cells, the polo-like kinase, Plk1 is required for the functional maturation of centrosomes at G2/M and for the establishment of a bipolar spindle (Lane and Nigg, 1996). Subsequently, the ninein-like protein, Nlp,

Table 1. Centrosome Associated Kinases.

Kinase	Symbols	Proposed Centrosomal Function	Substrates	Substrate References	Selected References
Aurora A	AIK ARK1 AURA STK15 STK6 BTAK MGC3453 8	Checkpoint regulation	aurora A BRCA1 CDC25B CENP-A H3 Lats2 ninein p53 TPX2	(Hirota et al., 2003) (Ouchi et al., 2004) (Dutertre et al., 2004) (Kunitoku et al., 2003) (Crosio et al., 2002) (Toji et al., 2004) (Chen et al., 2003) (Katayama et al., 2004) (Kufer et al., 2002)	(Sen et al., 1997) (Dutertre et al., 2002) (Dutertre and Prigent, 2003) (Goepfert and Brinkley, 2000) (Katayama et al., 2003) (Meraldi et al., 2004)
PLK1	PLK STPK13	Centrosome maturation	BRCA2 Cdc16 (subunit of APC) Cdc20 (subunit of APC) Cdc23 (subunit of APC) Cdc25c cyclin B1 MKLP2 Myt1 Nlp (ninein-like protein) Nud C p53 TCTP	(Lee et al., 2004; Lin et al., 2003) (Golan et al., 2002) (Kraft et al., 2003) (Golan et al., 2002) (Toyoshima-Morimoto et al., 2002) (Jackman et al., 2003; Toyoshima-Morimoto et al., 2001) (Neef et al., 2003) (Nakajima et al., 2003) (Casenghi et al., 2003) (Aumais et al., 2003; Zhou et al., 2003) (Ando et al., 2004) (Yarm, 2002)	(Golsteyn et al., 1994) (Golsteyn et al., 1995) (Barr et al., 2004) (Dai et al., 2002) (Dai and Cogswell, 2003)
PLK2	SNK	Not fully defined	None yet identified		(Liby et al., 2001; Warnke et al., 2004)
PKA		Centrosome separation	aurora A centrin E6 ninein	(Walter et al., 2000) (Lutz et al., 2001) (Massimi et al., 2001) (Chen et al., 2003)	
NEK2	NLK1 HsPK21	Centrosome separation	C-Nap1 Hec1 Nek11 Nek2 protein phosphatase 1 (PP1)	(Faragher and Fry, 2003; Fry et al., 1998) (Chen et al., 2002) (Noguchi et al., 2004) (Fry et al., 1999) (Helps et al., 2000)	(Schultz et al., 1994) (Fry, 2002) (O'Connell et al., 2003)
MPS1P	TTK ESK PYT MPS1L1	Not fully defined	None yet identified		(Mills et al., 1992) (Fisk and Winey, 2001) (Stucke et al., 2002) (Winey and Huneycutt, 2002)

was identified as an important centrosomal substrate for Plk1 (Casenghi et al., 2003). In interphase cells, Nlp is a PCM component that anchors γ-tubulin complexes at the centrosome. At the time of the G2/M transition, when Plk1 reaches maximum activity, phosphorylation of Nlp triggers its exchange with an as yet unidentified mitosis-specific γ-tubulin binding protein (Casenghi et al., 2003). This exchange is an important feature of the centrosome maturation process and may represent a more general mechanism that distinguishes interphase and mitotic centrosomes.

Plk2 is another member of the polo-like kinase family important for centrosome function. PLK2 knockout mice are viable, albeit their growth and development are retarded and embryonic fibroblasts show delayed entry into S phase (Ma et al., 2003). Plk2 is activated near the G_1/S transition of the cell cycle, but its activity is not required for centrosome localisation (Warnke et al., 2004). In cultured cells, inactivation of Plk2 interferes with centriole re-duplication, implicating Plk2 kinase activity in regulating this process (Warnke et al., 2004). However, viability of PLK2 knockout mice suggests that essential Plk2 centrosome functions are redundant or can be compensated by other mechanisms.

Mps1 kinase was originally identified as a dual specificity serine/tyrosine kinase required for yeast spindle pole body duplication (Lauze et al., 1995; Winey et al., 1991). There are conflicting reports regarding the centrosomal location and requirement of Mps1 for centrosome doubling in mammalian cells (Fisk et al., 2003; Stucke et al., 2002; 2004). Human Mps1 was found to localise to kinetochores and its maximal activity was seen during M phase (Stucke et al., 2002; 2004). Antibody microinjection and RNAi implicated hMps1 in mitotic checkpoint activation, while centriole re-duplication as well as cell division occurred in the absence of hMps1. These observations suggest that hMps1 is required for the spindle assembly checkpoint but not for centrosome duplication. In contrast, in other studies Mps1 showed centrosomal localisation, and in S phase arrested cells overexpression of GFP-Mps1 supported, while a kinase deficient mutant blocked, centriole re-duplication (Fisk et al., 2003). These later observations suggest a role for Mps1 in centriole duplication during S phase. Disparities in the particular experimental systems employed may contribute to the differing interpretations in these studies.

Nek2 is a NIMA-related kinase whose abundance and activity are tightly regulated in a cell cycle-specific manner, peaking around the time of the G_2/M transition (Fry et al., 1995). Nek2 localisation is concentrated near the proximal end of both centrioles along with protein phosphatase I and C-Nap, which together with Nek2, are substrates for Nek2 kinase activity (Fry et al., 1998a). Functional studies implicate Nek2 in the separation of newly doubled centrosomes to yield two spindle poles at mitotic prophase, possibly through modification of the linkage between the centriole pairs (Fry, 2002). Increased Nek2 expression in cultured cells leads to accumulation of multinucleated cells with supernumerary centrosomes, presumably through failure of cytokinesis (Hayward et al., 2004). Elevated Nek2 may play a role in the origin of chromosomal instability through deregulation of its centrosome functions, or through effects on other mitotic targets, including the kinetochore protein Hec1(Chen et al., 2002a).

The Aurora family of serine/threonine kinases is essential for the control of mitotic progression (Nigg, 2001). Importantly, over-expression of Aurora A (STK15) in NIH 3T3 and breast epithelial cells results in centrosome amplification and aneuploidy (Zhou et al., 1998). Amplification at the Aurora A locus on chromosome 20q13 has been shown in primary breast tumours, as well as in breast, ovarian, colon, prostate, neuroblastoma, and cervical cancer cell lines and its amplification in tumours correlates with poor prognosis (Sen, 2000; Sen et al., 1997; Zhou et al., 1998). These findings suggest that Aurora A is critical for the regulation of the centrosome cycle. While many substrates of Aurora A have been identified that are important in mitotic progression, the recent demonstration (Katayama et al., 2004) that Aurora A phosphorylates p53 leading to its ubiquitination by Mdm2 and proteolysis may largely account for its role in the regulation of centrosome behaviour through down regulation of p53-mediated checkpoint-response pathways.

3. ANEUPLOIDY AND CHROMOSOMAL INSTABILITY

Because centrosome amplification may play a role in the origin of chromosomal instability (CIN) and the development of cancer (see Gisselsson, 2003; Storchova and Pellman, 2004; Wang et al., 2004 for recent reviews), it is important to formally establish a direct causal relationship between the two processes. Flow cytometry, fluorescence in situ hybridisation (FISH), or image cytometry can be used to determine if cells and tissues are aneuploid, polyploid, or diploid. FISH, which allows for enumeration of specific chromosomes in individual interphase cells, is the most sensitive of these methods, and can detect very low frequencies of cells with aberrant chromosome numbers. FISH and image cytometry can be used to calculate the frequency of gains or losses of chromosomes in tumour tissues, irrespective of ploidy. Chromosomal instability, measured using FISH, is calculated as the percent of cells differing from the modal value for a given chromosome (Lengauer et al., 1997). The stemline scatter index (SSI), measured using image cytometry, is approximately equivalent to CIN (Kronenwett et al., 2004). Both CIN and SSI are measures of the *rate* of change in karyotypes and indicate the relative stability of the karyotype in a given population of cells, as opposed to ploidy, which is a *snapshot* of the average DNA content of a population of cells.

3.1 Relationship between Centrosome Amplification and CIN

Centrosome amplification may cause CIN by generating daughter cells with chromosome complements that differ from the parental cell. This can occur via the formation of multipolar mitotic spindles that yield daughter cells, each potentially having a unique set of chromosomes. Both aneuploidy and CIN occur with centrosome amplification in many tumour types and model systems. Aneuploidy has been associated with centrosome amplification in bladder cancer (Jiang et al., 2003), non-Hodgkin's lymphoma (Kramer et al., 2003), testicular germ cell tumours (Mayer et al., 2003), hepatocellular carcinoma (Nakajima et al., 2004), acute myeloid leukemia (Neben et al., 2004), and others (see (D'Assoro et al., 2002; Kramer et al., 2002) for reviews of earlier literature). Studies have also shown an association between CIN and centrosome amplification in osteosarcomas (Al-Romaih et al., 2003), renal cell carcinomas (Gisselsson et al., 2004), bladder cancer (Kawamura et al., 2004; Kawamura et al., 2003), preleukemia (Kearns et al., 2004), breast (Kronenwett et al., 2004; Lingle et al., 2002; Pihan et al., 2003), Kaposi's sarcoma (Pan et al., 2004), pancreas (Sato et al., 2001), and prostate cancers (Pihan et al., 2001). Centrosome amplification affects CIN and ploidy through aberrant spindle pole function. Alternative methods for generating CIN exist, including those that affect kinetochore function via defects in mitotic checkpoint control (Hanks et al., 2004) and telomere erosion that can lead to bridge-fusion-breakage events (Gisselsson et al., 2004). Missegregation of chromosomes during cell division is the feature common to all of these routes to CIN and aneuploidy.

3.2 Centrosome Amplification drives CIN

Several lines of evidence support the concept that centrosome amplification drives CIN. First, the degree of centrosome amplification correlates with CIN levels in breast tumours (Lingle et al., 2002) and bladder cancers (Kawamura et al., 2003). Both of these studies were done using multicolour FISH for simultaneous enumeration of 3 chromosomes to calculate CIN and identify clonal populations of cells. A statistically significant correlation between the level of CIN and the level of centrosome amplification, measured as either a continuous variable (Lingle et al., 2002) or a step-wise grade of amplification (Kawamura et al., 2003), was demonstrated. Kronenwett and co-workers demonstrated that centrosome amplification is associated with high SSI and increased expression of cyclins A and E, suggesting that these cyclins are involved in centrosome amplification in breast cancer (Kronenwett et al., 2004).

Second, centrosome amplification is an early event in breast (Lingle et al., 2002; Pihan et al.), bladder (Jiang et al., 2003), prostate, cervix (Pihan et al., 2003), and testicular germ cell tumours (Mayer et al., 2003).

Centrosome abnormalities have been compared to FISH as a method to screen bladder wash cells from patients with suspected bladder cancer (Jiang et al., 2003). FISH and centrosome amplification both had high specificity, with neither positive in any patients without bladder cancer. Centrosome amplification had greater sensitivity as a marker of bladder cancer than did FISH, especially in low-grade tumours. These results indicate that centrosome amplification is an early event in bladder cancer that precedes ploidy changes. *In situ* carcinomas of the breast (Lingle et al., 2002; Pihan et al., 2003) cervix, prostate (Pihan et al., 2003), and pre-invasive testicular germ cell tumours (Mayer et al., 2003) have centrosome abnormalities, indicating that centrosome amplification is an early event in tumour progression that occurs prior to invasion.

Third, centrosome amplification can precede ploidy changes (Duensing and Munger, 2002b; Huang et al., 2004; Jiang et al., 2003; Ochi et al., 2004). As discussed above, centrosome amplification precedes ploidy changes in bladder cancer as detected by FISH (Jiang et al., 2003). FISH is the most sensitive method available to detect low levels of CIN or aneuploidy in tumour cells, therefore this study is an important milestone that definitively demonstrates that aneuploidy and CIN follow centrosome amplification and that the lack of detected aneuploidy is not due to poor sensitivity of the detection methods. Mouse embryonic fibroblasts deficient for the CCAAT/enhancer binding protein delta (Cepbd) transcription factor develop amplified centrosomes, display genomic instability, become spontaneously immortalised, and have several characteristics indicative of transformation (Huang et al., 2004). Centrosome amplification is present prior to genomic instability in this model. The authors propose that Cepbd is a tumour suppressor gene whose transcriptional activity affects centrosome function. In a Syrian hamster embryo model, cells developed amplified centrosomes within 2 hours of treatment with dimethylarsenide, (Ochi et al., 2004). This rapid timing precludes failure of cytokinesis or tetraploidisation as mechanisms that generate supernumerary centrosomes in this model. Primary cultures of normal human keratinocytes expressing the human papilloma virus type 16 E7 oncoprotein also develop amplified centrosomes prior to nuclear changes associated with changes in ploidy (Duensing and Munger, 2002a).

Taken together, these observations demonstrate that centrosome amplification occurs prior to chromosomal instability in the cancers and model systems discussed above. Other studies show that tetraploidisation and failure of cytokinesis are also routes to centrosome amplification (discussed in (Nigg, 2002)). Cells having amplified centrosomes as a result of any of these processes are vulnerable to further CIN through unequal and multipolar mitoses.

3.3 Success of Multipolar Mitoses and Proliferation of resultant Cells with CIN in Tumours

In living tumour tissues does centrosome amplification lead to CIN and the accumulation of aneuploid cells, or are aneuploid cells removed from the cell population through apoptosis? The latter may be true for well-differentiated liposarcomas (WDLs). WDL tumours have wild-type p53 and tend to be either near-diploid or near-tetraploid. Yet in primary culture, WDL cells have significant frequencies of amplified centrosomes that often result in multipolar mitoses (Perucca-Lostanlen et al., 2004). p53-dependent apoptosis may eliminate resultant aneuploid cells from the tumour population, *in vivo*. WDL cells in culture often clustered their supernumerary centrosomes to form functionally bipolar spindles (see Figure 3A and B for examples of centrosome clustering). Do most WDL tumour cells with multipolar mitoses eventually resolve into bipolar mitoses? *In vivo*, elimination of cells with amplified centrosomes through apoptosis and/or centrosome clustering could aid in maintaining the stable karyotypes characteristic of WDLs and some other tumours, even when

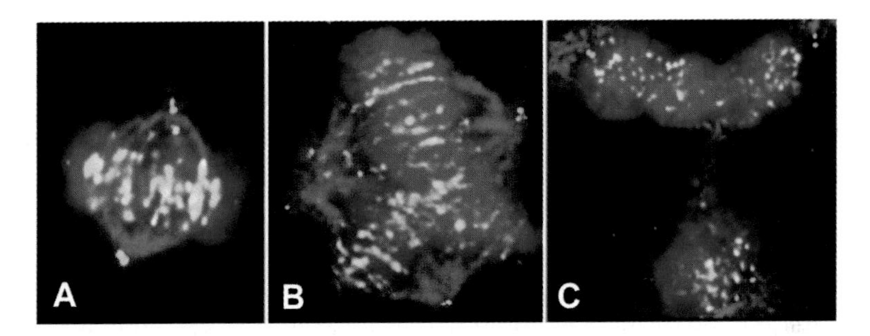

Figure 3A and B. Mitotic defects in mammary epithelial cells. A) This nearly normal bipolar metaphase has a cluster of 3 centrioles at the upper spindle pole. Cells with mitotic spindles such as this are likely to successfully complete mitosis with proper chromosome segregation. B) Three spindles poles are present in this metaphase cell, 2 of which have multiple centrioles. Cells with this configuration could die during mitosis via mitotic catastrophe or could complete mitosis to yield 2 or more daughter cells with attendant aneuploidy through chromosome missegregation. Notice the stretched appearance of the kinetochores, indicating spindle microtubule attachment and tension. C) This telophase cell has 3 spindle poles, each with 2 centrioles. Kinetochores have assumed a globular shape, indicating that spindle microtubule tension has been released. One mass of chromosomes has been separated by karyokinesis from the rest of the chromosomes. A G1 daughter cell resulting from this configuration would have normal centrosomes, but would likely be aneuploid – possibly with significant chromosome loss. The remainder of the chromosomes is in a sausage shaped mass with a spindle pole at each end. This configuration would likely result in a G1 daughter cell with 2 centrosomes and a large, lobed nucleus with extra chromosomes. Cells were immunostained with antibodies against the centrosome protein centrin (green), Eg5, a mitotic kinesin that associates with spindle microtubules (red), and kinetochores (turquoise). DNA is stained with Hoechst 33342.

amplified centrosomes are present. Conversely, aneuploidy has been demonstrated to occur in the absence of centrosome amplification in progesterone-treated p53 null mouse mammary epithelial cells (Goepfert et al., 2000). In this case, other mechanisms such as telomere erosion are likely to be involved in the generation of aneuploidy (Gisselsson et al., 2004).

Live cell microscopy of cultured cells has demonstrated that multipolar mitoses can be completed to yield more than 2 daughter cells (Hut et al., 2003). When Chinese hamster ovary cells were exposed to DNA damaging agents, mitotic cells developed fragmented centrosomes containing only one centriole each. Analysis of time lapse video microscopy of the mitotic process of these cells showed that some cells started off with bipolar spindles that either remained functionally bipolar or fragmented to become multipolar, and other cells were multipolar from the outset of mitosis. Nearly half of the cells with multiple spindle poles finished mitosis to yield more than 2 daughter cells. This study demonstrates that centrosome defects can be manifested during mitosis, and that these defects can produce multiple daughter cells. Other investigators have suggested that centrosome amplification induced by DNA damage is a mechanism to target the cell for death, even if DNA damage or spindle assembly checkpoints are evaded (Dodson et al., 2004).

Genetic lesions may be required to permit cells to enter and complete multipolar mitoses and to escape apoptosis during the next cell cycle (Castedo et al., 2004a; Castedo et al., 2004b; Morrison and Rieder, 2004). In cells with intact cell cycle checkpoints, abnormal mitotic spindle morphology can result in mitotic catastrophe (Castedo et al., 2004a; Castedo et al., 2004b). Mitotic catastrophe is defined as cell death that occurs during mitosis, often due to failure to arrest the cell cycle prior to mitosis. This mitotic cell death involves the p53-independent activation of caspase-2 and/or the release of mitochondrial cell death factors (Castedo et al., 2004a; Castedo et al., 2004b). Suppression of mitotic catastrophe through knockdown of caspase-2 permits completion of mitosis to yield aneuploid daughter cells (Castedo et al., 2004b). Mitotic catastrophe can be evaded by other means as well. For example, knockdown of Aurora-A in HeLa cells leads to mitotic arrest with a tetraploid DNA content and eventual apoptosis (Du and Hannon, 2004). This phenotype is rescued by suppression of p160ROCK, and cells are then able to complete aberrant mitoses resulting in aneuploid daughter cells.

p53 is sometimes implicated in the accumulation of aneuploid or high CIN cells. Defective p53 allows accumulation of cells with more extensive CIN than those with intact p53 in osteosarcomas (Al-Romaih et al., 2003), and hepatocellular carcinomas with mutant p53 had more severe aneuploidy and greater centrosome amplification than did those with wild-type p53 (Nakajima et al., 2004). However, other genetic defects may come into play instead of or in addition to p53. For example, mutations in BRCA1, BRCA2 (Grigorova et al., 2004), and Rb (Hernando et al., 2004) permit proliferation and accumulation of tumour cells with CIN. Centrosome amplification and CIN can occur independently of p53 mutation in some tumours such as breast (Lingle et al., 2002; Pihan et al., 2003), prostate and cervix (Pihan et al., 2003), testicular germ cell tumours (Mayer et al., 2003), and non-Hodgkin's lymphoma (Kramer et al., 2003).

3.4 Moderation of CIN during Tumour Progression

In theory, unbridled CIN would result in self-limiting tumours due to the eventual death of cells deficient in critical genes essential for metabolism or proliferation. This is illustrated by the observation of short-term cultured cells derived from pancreatic carcinomas (Sato et al., 2001). In this system, high frequency of centrosome defects and multipolar mitoses correlated with a low growth rate in culture. A similar phenomenon has been observed in short-term culture of cells derived from six ovarian tumours (Lingle, unpublished). There are several mechanisms by which tumours might achieve re-stabilisation of unstable karyotypes. One mechanism is through the clustering of supernumerary centrosomes to form functionally bipolar spindles (Brinkley, 2001; Lingle and Salisbury, 1999; Perucca-Lostanlen et al., 2004). Another mechanism is a decrease in the expression of gene products known to initiate centrosome amplification. This has been observed for Aurora-A in human breast tumours (Hoque et al., 2003). The highest Aurora-A expression was observed in ductal carcinoma *in situ* lesions when compared to adjacent invasive lesions. This also correlated with higher centrosome amplification in ductal carcinoma *in situ*. The authors suggest that progression from ductal carcinoma *in situ* to invasive ductal carcinoma is associated with concomitant decrease in expression of Aurora-A.

The oncogenic kinase Pim-1 is often over-expressed in prostate tumours and leads to centrosome amplification and chromosomal instability in prostate cancer cells (Roh et al., 2003). High expression of Pim-1 is correlated with a better clinical outcome. The authors propose that early tumours have high chromosomal instability driven by Pim-1 overexpression, and that this instability is moderated during progression to aggressive tumours through down-regulation of Pim-1, thus stabilising acquired abnormalities. In summary, moderation of CIN during tumour progression can occur by clustering supernumerary centrosomes into a single functional unit and/or by reducing the original stimulus that initiated centrosome amplification.

4. SUMMARY

Although we have begun to tap into the mechanisms behind Boveri's initial observation that supernumerary centrosomes cause chromosome missegregation in sea urchin eggs, there is still much left to discover with regard to chromosomal instability in cancer. Many of the molecular players involved in regulation of the centrosome and cell cycles, and the coupling of the two cycles to produce a bipolar mitotic spindle have been identified. One theme that has become apparent is that cross talk and interrelatedness of the pathways serve to provide redundant mechanisms to maintain genomic integrity. In spite of this, cells occasionally fall prey to insults that initiate and maintain the chromosomal instability that results in viable malignant tumours. Deregulation of centrosome structure is an integral aspect of the origin of chromosomal instability in many cancers. There are numerous routes to centrosome amplification including: environmental insults such as ionising radiation and exposure to estrogen (Li et al., 2005); failure of cytokinesis; and activating mutations in key regulators of centrosome structure and function. There are two models for initiation of centrosome amplification (Figure 2). In the first, centrosome duplication and chromosome replication remain coupled and cells enter G2 with 4N chromosomes and duplicated centrosomes. However, these cells may fail to complete mitosis, and thus reenter G1 as tetraploid cells with amplified centrosomes. In the second, the centrosome cycle is uncoupled from chromosome replication and cells go through one or more rounds of centriole/centrosome duplication in the absence of chromosome replication. If these cells then go through chromosome replication accompanied by another round of centrosome duplication, cells complete G2 with 4N chromosomes and more than 2 centrosomes, and therefore are predisposed to generate multipolar mitotic spindles. Fragmentation of centrosomes due to ionising radiation is a variation of the second model.

Once centrosome amplification is present, even in a diploid cell, that cell has the potential to yield viable aneuploid progeny. The telophase cell in Figure 3C illustrates this scenario. In a normal telophase configuration, the total number of chromosomes is 92 (resulting from the segregation of 46 pairs of chromatids), with each daughter nucleus containing 46 individual chromosomes. Based on the number of kinetochore signals present, the lower nucleus in Figure 3C has approximately 28 chromosomes, and the elongate upper nucleus has approximately 60, for a total of 88. Due to superimposition of kinetochores in this maximum projection image, 88 is an underestimate of the actual number of kinetochores and is not significantly different from the expected total of 92. A cell resulting from the lower nucleus with only around 28 chromosomes would probably not be viable, much as Boveri's experiments indicated. However, the upper nucleus with at least 60 chromosomes could be viable. This cell would enter G1 as hypotriploid (69 chromosomes = triploid) with 2 centrosomes. During S and G2, the centrosomes and chromosomes would double, and the following mitosis could be tetrapolar with a 6N chromosome content. When centrosome amplification is accompanied by permissive lapses in cell cycle checkpoints, the potential for malignant growth is present. These lapses could result from specific genetic mutations and amplifications, epigenetic gene silencing, or from massive chromosomal instability caused by the centrosome amplification. Centrosome amplification, therefore, can serve to exacerbate and/or generate genetic instabilities associated with cancers.

REFERENCES

Al-Romaih, K., J. Bayani, J. Vorobyova, J. Karaskova, P.C. Park, M. Zielenska, and J.A. Squire. 2003. Chromosomal instability in osteosarcoma and its association with centrosome abnormalities. *Cancer Genetics and Cytogenetics*. 144:91-99.

Andersen, J.S., C.J. Wilkinson, T. Mayor, P. Mortensen, E.A. Nigg, and M. Mann. 2003. Proteomic characterization of the human centrosome by protein correlation profiling. *Nature*. 426:570-4.

Ando, K., T. Ozaki, H. Yamamoto, K. Furuya, M. Hosoda, S. Hayashi, M. Fukuzawa, and A. Nakagawara. 2004. Polo-like kinase 1 (Plk1) inhibits p53 function by physical interaction and phosphorylation. *J Biol Chem*. 279:25549-61.

Angus, S.P., A.F. Fribourg, M.P. Markey, S.L. Williams, H.F. Horn, J. DeGregori, T.F. Kowalik, K. Fukasawa, and E.S. Knudsen. 2002. Active RB elicits late G1/S inhibition. *Exp Cell Res*. 276:201-13.

Aumais, J.P., S.N. Williams, W. Luo, M. Nishino, K.A. Caldwell, G.A. Caldwell, S.H. Lin, and L.Y. Yu-Lee. 2003. Role for NudC, a dynein-associated nuclear movement protein, in mitosis and cytokinesis. *J Cell Sci*. 116:1991-2003.

Avidor-Reiss, T., A.M. Maer, E. Koundakjian, A. Polyanovsky, T. Keil, S. Subramaniam, and C.S. Zuker. 2004. Decoding cilia function: defining specialized genes required for compartmentalized cilia biogenesis. *Cell*. 117:527-39.

Bailly, E., J. Pines, T. Hunter, and M. Bornens. 1992. Cytoplasmic accumulation of cyclin B1 in human cells: association with a detergent-resistant compartment and with the centrosome. *J Cell Sci*. 101:529-45.

Balczon, R., L. Bao, W.E. Zimmer, K. Brown, R.P. Zinkowski, and B.R. Brinkley. 1995. Dissociation of centrosome replication events from cycles of DNA synthesis and mitotic division in hydroxyurea-arrested Chinese hamster ovary cells. *J Cell Biol*. 130:105-15.

Balczon, R.C. 2001. Overexpression of cyclin A in human HeLa cells induces detachment of kinetochores and spindle pole/centrosome overproduction. *Chromosoma*. 110:381-92.

Barr, F.A., H.H. Sillje, and E.A. Nigg. 2004. Polo-like kinases and the orchestration of cell division. *Nat Rev Mol Cell Biol*. 5:429-40.

Bobinnec, Y., A. Khodjakov, L.M. Mir, C.L. Rieder, B. Edde, and M. Bornens. 1998. Centriole Disassembly In Vivo and Its Effect on Centrosome Structure and Function in Vertebrate Cells. *J. Cell Biol.* 143:1575-1589.

Boveri, T. 1914. Zur Frage der Entstehung Maligner Tumoren. Jena: Fischer Verlag (1929 English translation by M. Boveri reprinted as The Origin of malignant tumours, The Williams and Wilkins Co., Baltimore), Baltimore. 119 pp.

Brinkley, B.R. 2001. Managing the centrosome numbers game: from chaos to stability in cancer cell division. *Trends Cell Biol*. 11:18-21.

Carroll, P.E., M. Okuda, H.F. Horn, P. Biddinger, P.J. Stambrook, L.L. Gleich, Y.Q. Li, P. Tarapore, and K. Fukasawa. 1999. Centrosome hyperamplification in human cancer: chromosome instability induced by p53 mutation and/or Mdm2 overexpression. *Oncogene*. 18:1935-44.

Casenghi, M., P. Meraldi, U. Weinhart, P.I. Duncan, R. Korner, and E.A. Nigg. 2003. Polo-like kinase 1 regulates Nlp, a centrosome protein involved in microtubule nucleation. *Dev Cell*. 5:113-25.

Castedo, M., J.L. Perfettini, T. Roumier, K. Andreau, R. Medema, and G. Kroemer. 2004a. Cell death by mitotic catastrophe: a molecular definition. *Oncogene*. 23:2825-37.

Castedo, M., J.L. Perfettini, T. Roumier, A. Valent, H. Raslova, K. Yakushijin, D. Horne, J. Feunteun, G. Lenoir, R. Medema, W. Vainchenker, and G. Kroemer. 2004b. Mitotic catastrophe constitutes a special case of apoptosis whose suppression entails aneuploidy. *Oncogene*. 23:4362-70.

Cha, H., C. Hancock, S. Dangi, D. Maiguel, F. Carrier, and P. Shapiro. 2004. Phosphorylation regulates nucleophosmin targeting to the centrosome during mitosis as detected by cross-reactive phosphorylation-specific MKK1/MKK2 antibodies. *Biochem J*. 378:857-65.

Chen, C.H., S.L. Howng, T.S. Cheng, M.H. Chou, C.Y. Huang, and Y.R. Hong. 2003. Molecular characterization of human ninein protein: two distinct subdomains required for centrosomal targeting and regulating signals in cell cycle. *Biochem Biophys Res Commun*. 308:975-83.

Chen, Y., D.J. Riley, L. Zheng, P.L. Chen, and W.H. Lee. 2002a. Phosphorylation of the mitotic regulator protein Hec1 by Nek2 kinase is essential for faithful chromosome segregation. *J Biol Chem*. 277:49408-16.

Chen, Z., V.B. Indjeian, M. McManus, L. Wang, and B.D. Dynlacht. 2002b. CP110, a cell cycle-dependent CDK substrate, regulates centrosome duplication in human cells. *Dev Cell*. 3:339-50.

Crosio, C., G.M. Fimia, R. Loury, M. Kimura, Y. Okano, H. Zhou, S. Sen, C.D. Allis, and P. Sassone-Corsi. 2002. Mitotic phosphorylation of histone H3: spatio-temporal regulation by mammalian Aurora kinases. *Mol Cell Biol*. 22:874-85.

D'Assoro, A.B., R. Busby, K. Suino, E. Delva, G.J. Almodovar-Mercado, H. Johnson, C. Folk, D.J. Farrugia, V. Vasile, F. Stivala, and J.L. Salisbury. 2004. Genotoxic stress leads to centrosome amplification in breast cancer cell lines that have an inactive G1/S cell cycle checkpoint. *Oncogene*. 23:4068-4075.

D'Assoro, A.B., W.L. Lingle, and J.L. Salisbury. 2002. Centrosome amplification and the development of cancer. *Oncogene*. 21:6146-53.

Dai, W., and J.P. Cogswell. 2003. Polo-like kinases and the microtubule organization center: targets for cancer therapies. *Prog Cell Cycle Res*. 5:327-34.

Dai, W., Q. Wang, and F. Traganos. 2002. Polo-like kinases and centrosome regulation. *Oncogene*. 21:6195-200.

De Camilli, P., M. Moretti, S.D. Donini, U. Walter, and S.M. Lohmann. 1986. Heterogeneous distribution of the cAMP receptor protein RII in the nervous system: evidence for its intracellular accumulation on microtubules, microtubule-organizing centers, and in the area of the Golgi complex. *J Cell Biol*. 103:189-203.

Debec, A., and C. Montmory. 1992. Cyclin B is associated with centrosomes in Drosophila mitotic cells. *Biol Cell*. 75:121-6.

Deng, C.X., and S.G. Brodie. 2000. Roles of BRCA1 and its interacting proteins. *Bioessays*. 22:728-37.

Diviani, D., L.K. Langeberg, S.J. Doxsey, and J.D. Scott. 2000. Pericentrin anchors protein kinase A at the centrosome through a newly identified RII-binding domain. *Curr Biol*. 10:417-20.

Diviani, D., and J.D. Scott. 2001. AKAP signalling complexes at the cytoskeleton. *J Cell Sci*. 114:1431-7.

Dodson, H., E. Bourke, L.J. Jeffers, P. Vagnarelli, E. Sonoda, S. Takeda, W.C. Earnshaw, A. Merdes, and C. Morrison. 2004. Centrosome amplification induced by DNA damage occurs during a prolonged G2 phase and involves ATM. *EMBO J*.

Doxsey, S. 2001a. Re-evaluating centrosome function. *Nat Rev Mol Cell Biol*. 2:688-98.

Doxsey, S.J. 2001b. Centrosomes as command centres for cellular control. *Nat Cell Biol*. 3:E105-8.

Du, J., and G.J. Hannon. 2002. The centrosomal kinase Aurora-A/STK15 interacts with a putative tumour suppressor NM23-H1. *Nucleic Acids Res*. 30:5465-75.

Du, J., and G.J. Hannon. 2004. Suppression of p160ROCK bypasses cell cycle arrest after Aurora-A/STK15 depletion. *PNAS*. 101:8975-8980.

Duensing, S., L.Y. Lee, A. Duensing, J. Basile, S. Piboonniyom, S. Gonzalez, C.P. Crum, and K. Munger. 2000. The human papillomavirus type 16 E6 and E7 oncoproteins cooperate to induce mitotic defects and genomic instability by uncoupling centrosome duplication from the cell division cycle. *Proc Natl Acad Sci U S A*. 97:10002-7.

Duensing, S., and K. Munger. 2002a. The human papillomavirus type 16 E6 and E7 oncoproteins independently induce numerical and structural chromosome instability. *Cancer Res*. 62:7075-82.

Duensing, S., and K. Munger. 2002b. Human papillomaviruses and centrosome duplication errors: modeling the origins of genomic instability. *Oncogene*. 21:6241-8.

Dutertre, S., M. Cazales, M. Quaranta, C. Froment, V. Trabut, C. Dozier, G. Mirey, J.-P. Bouche, N. Theis-Febvre, E. Schmitt, B. Monsarrat, C. Prigent, and B. Ducommun. 2004. Phosphorylation of CDC25B by Aurora-A at the centrosome contributes to the G2-M transition. *J Cell Sci*. 117:2523-2531.

Dutertre, S., S. Descamps, and C. Prigent. 2002. On the role of aurora-A in centrosome function. *Oncogene*. 21:6175-83.

Dutertre, S., and C. Prigent. 2003. Aurora-A overexpression leads to override of the microtubule-kinetochore attachment checkpoint. *Mol Interv*. 3:127-30.

el-Deiry, W.S., T. Tokino, V.E. Velculescu, D.B. Levy, R. Parsons, J.M. Trent, D. Lin, W.E. Mercer, K.W. Kinzler, and B. Vogelstein. 1993. WAF1, a potential mediator of p53 tumour suppression. *Cell*. 75:817-25.

Faragher, A.J., and A.M. Fry. 2003. Nek2A kinase stimulates centrosome disjunction and is required for formation of bipolar mitotic spindles. *Mol Biol Cell*. 14:2876-89.

Fisk, H.A., C.P. Mattison, and M. Winey. 2003. Human Mps1 protein kinase is required for centrosome duplication and normal mitotic progression. *Proc Natl Acad Sci U S A*. 100:14875-80.

Fisk, H.A., and M. Winey. 2001. The mouse Mps1p-like kinase regulates centrosome duplication. *Cell*. 106:95-104.

Fry, A.M. 2002. The Nek2 protein kinase: a novel regulator of centrosome structure. *Oncogene*. 21:6184-94.

Fry, A.M., L. Arnaud, and E.A. Nigg. 1999. Activity of the human centrosomal kinase, Nek2, depends on an unusual leucine zipper dimerization motif. *J Biol Chem*. 274:16304-10.

Fry, A.M., T. Mayor, P. Meraldi, Y.D. Stierhof, K. Tanaka, and E.A. Nigg. 1998a. C-Nap1, a novel centrosomal coiled-coil protein and candidate substrate of the cell cycle-regulated protein kinase Nek2. *J Cell Biol*. 141:1563-74.

Fry, A.M., T. Mayor, and E.A. Nigg. 2000. Regulating centrosomes by protein phosphorylation. *In* Centrosome in Cell Replication and Early Development. Vol. 49. 291-312.

Fry, A.M., P. Meraldi, and E.A. Nigg. 1998b. A centrosomal function for the human Nek2 protein kinase, a member of the NIMA family of cell cycle regulators. *EMBO J*. 17:470-81.

Fry, A.M., S.J. Schultz, J. Bartek, and E.A. Nigg. 1995. Substrate specificity and cell cycle regulation of the Nek2 protein kinase, a potential human homolog of the mitotic regulator NIMA of Aspergillus nidulans. *J Biol Chem*. 270:12899-905.

Fukasawa, K., T. Choi, R. Kuriyama, S. Rulong, and G.F. Vande Woude. 1996. Abnormal centrosome amplification in the absence of p53. *Science*. 271:1744-7.

Giannakakou, P., D.L. Sackett, Y. Ward, K.R. Webster, M.V. Blagosklonny, and T. Fojo. 2000. p53 is associated with cellular microtubules and is transported to the nucleus by dynein. *Nat Cell Biol*. 2:709-717.

Giet, R., R. Uzbekov, I. Kireev, and C. Prigent. 1999. The Xenopus laevis centrosome aurora/Ipl1-related kinase. *Biol Cell*. 91:461-70.

Gillingham, A.K., and S. Munro. 2000. The PACT domain, a conserved centrosomal targeting motif in the coiled-coil proteins AKAP450 and pericentrin. *Embo Reports*. 1:524-529.

Gisselsson, D. 2003. Chromosome instability in cancer: How, when, and why? *Advances in Cancer Research, Vol 87*. 87:1-29.

Gisselsson, D., L. Gorunova, M. Hoglund, N. Mandahl, and P. Elfving. 2004. Telomere shortening and mitotic dysfunction generate cytogenetic heterogeneity in a subgroup of renal cell carcinomas. *Br J Cancer*. 91:327-32.

Goepfert, T.M., and B.R. Brinkley. 2000. The centrosome-associated Aurora/Ipl-like kinase family. *Curr Top Dev Biol*. 49:331-42.

Goepfert, T.M., M. McCarthy, F.S. Kittrell, C. Stephens, R.L. Ullrich, B.R. Brinkley, and D. Medina. 2000. Progesterone facilitates chromosome instability (aneuploidy) in p53 null normal mammary epithelial cells. *Faseb Journal*. 14:2221-2229.

Golan, A., Y. Yudkovsky, and A. Hershko. 2002. The cyclin-ubiquitin ligase activity of cyclosome/APC is jointly activated by protein kinases Cdk1-cyclin B and Plk. *J Biol Chem*. 277:15552-7.

Golsteyn, R.M., K.E. Mundt, A.M. Fry, and E.A. Nigg. 1995. Cell cycle regulation of the activity and subcellular localization of PLK1, a human protein kinase implicated in mitotic spindle function. *J. Cell Biol.* 129:1617-1628.

Golsteyn, R.M., S.J. Schultz, J. Bartek, A. Ziemiecki, T. Ried, and E.A. Nigg. 1994. Cell cycle analysis and chromosomal localization of human Plk1, a putative homologue of the mitotic kinases Drosophila polo and Saccharomyces cerevisiae Cdc5. *J Cell Sci*. 107 (Pt 6):1509-17.

Grigorova, M., J.M. Staines, H. Ozdag, C. Caldas, and P.A. Edwards. 2004. Possible causes of chromosome instability: comparison of chromosomal abnormalities in cancer cell lines with mutations in BRCA1, BRCA2, CHK2 and BUB1. *Cytogenet Genome Res*. 104:333-40.

Grimm, D. 2004. Genetics. Disease backs cancer origin theory. *Science*. 306:389.

Hamill, D.R., A.F. Severson, J.C. Carter, and B. Bowerman. 2002. Centrosome Maturation and Mitotic Spindle Assembly in C. elegans Require SPD-5, a Protein with Multiple Coiled-Coil Domains. *Dev Cell*. 3:673-84.

Hanks, S., K. Coleman, S. Reid, A. Plaja, H. Firth, D. Fitzpatrick, A. Kidd, K. Mehes, R. Nash, N. Robin, N. Shannon, J. Tolmie, J. Swansbury, A. Irrthum, J. Douglas, and N. Rahman. 2004. Constitutional aneuploidy and cancer predisposition caused by biallelic mutations in BUB1B. *Nat Genet*. 36:1159-61.

Hashizume, R., M. Fukuda, I. Maeda, H. Nishikawa, D. Oyake, Y. Yabuki, H. Ogata, and T. Ohta. 2001. The RING heterodimer BRCA1-BARD1 is a ubiquitin ligase inactivated by a breast cancer-derived mutation. *J Biol Chem*. 276:14537-40.

Hayward, D.G., R.B. Clarke, A.J. Faragher, M.R. Pillai, I.M. Hagan, and A.M. Fry. 2004. The centrosomal kinase Nek2 displays elevated levels of protein expression in human breast cancer. *Cancer Res*. 64:7370-6.

Helps, N.R., X. Luo, H.M. Barker, and P.T. Cohen. 2000. NIMA-related kinase 2 (Nek2), a cell-cycle-regulated protein kinase localized to centrosomes, is complexed to protein phosphatase 1. *Biochem J*. 349:509-18.

Hernando, E., Z. Nahle, G. Juan, E. Diaz-Rodriguez, M. Alaminos, M. Hemann, L. Michel, V. Mittal, W. Gerald, R. Benezra, S.W. Lowe, and C. Cordon-Cardo. 2004. Rb inactivation promotes genomic instability by uncoupling cell cycle progression from mitotic control. *Nature*. 430:797-802.

Hinchcliffe, E.H., C. Li, E.A. Thompson, J.L. Maller, and G. Sluder. 1999. Requirement of Cdk2-cyclin E activity for repeated centrosome reproduction in Xenopus egg extracts [see comments]. *Science*. 283:851-4.

Hinchcliffe, E.H., F.J. Miller, M. Cham, A. Khodjakov, and G. Sluder. 2001. Requirement of a centrosomal activity for cell cycle progression through G1 into S phase. *Science*. 291:1547-50.

Hirota, T., N. Kunitoku, T. Sasayama, T. Marumoto, D. Zhang, M. Nitta, K. Hatakeyama, and H. Saya. 2003. Aurora-A and an interacting activator, the LIM protein Ajuba, are required for mitotic commitment in human cells. *Cell*. 114:585-98.

Hoque, A., J. Carter, W. Xia, M.-C. Hung, A.A. Sahin, S. Sen, and S.M. Lippman. 2003. Loss of Aurora A/STK15/BTAK Overexpression Correlates with Transition of in Situ to

Invasive Ductal Carcinoma of the Breast. *Cancer Epidemiol Biomarkers Prev.* 12:1518-1522.

Hsu, L.C., and R.L. White. 1998. BRCA1 is associated with the centrosome during mitosis. *Proc Natl Acad Sci U S A.* 95:12983-8.

Huang, A.M., C. Montagna, S. Sharan, Y. Ni, T. Ried, and E. Sterneck. 2004. Loss of CCAAT/enhancer binding protein delta promotes chromosomal instability. *Oncogene.* 23:1549-57.

Hut, H.M., W. Lemstra, E.H. Blaauw, G.W. Van Cappellen, H.H. Kampinga, and O.C. Sibon. 2003. Centrosomes Split in the Presence of Impaired DNA Integrity during Mitosis. *Mol Biol Cell.* 14:1993-2004.

Jackman, M., C. Lindon, E.A. Nigg, and J. Pines. 2003. Active cyclin B1-Cdk1 first appears on centrosomes in prophase. *Nat Cell Biol.* 5:143-8.

Jiang, F., N.P. Caraway, A.L. Sabichi, H.Z. Zhang, A. Ruitrok, H.B. Grossman, J. Gu, S.P. Lerner, S. Lippman, and R.L. Katz. 2003. Centrosomal abnormality is common in and a potential biomarker for bladder cancer. *International Journal of Cancer.* 106:661-665.

Katayama, H., W.R. Brinkley, and S. Sen. 2003. The Aurora kinases: role in cell transformation and tumorigenesis. *Cancer Metastasis Rev.* 22:451-64.

Katayama, H., K. Sasai, H. Kawai, Z.M. Yuan, J. Bondaruk, F. Suzuki, S. Fujii, R.B. Arlinghaus, B.A. Czerniak, and S. Sen. 2004. Phosphorylation by aurora kinase A induces Mdm2-mediated destabilization and inhibition of p53. *Nat Genet.* 36:55-62.

Kawamura, K., H. Izumi, Z. Ma, R. Ikeda, M. Moriyama, T. Tanaka, T. Nojima, L.S. Levin, K. Fujikawa-Yamamoto, K. Suzuki, and K. Fukasawa. 2004. Induction of centrosome amplification and chromosome instability in human bladder cancer cells by p53 mutation and cyclin E overexpression. *Cancer Res.* 64:4800-9.

Kawamura, K., M. Moriyama, N. Shiba, M. Ozaki, T. Tanaka, T. Nojima, K. Fujikawa-Yamamoto, R. Ikeda, and K. Suzuki. 2003. Centrosome Hyperamplification and Chromosomal Instability in Bladder Cancer. *European Urology.* 43:505-515.

Kearns, W.G., J.F. Sutton, J.P. Maciejewski, N.S. Young, and J.M. Liu. 2004. Genomic instability in bone marrow failure syndromes. *Am J Hematol.* 76:220-4.

Keezer, S.M., and D.M. Gilbert. 2002. Sensitivity of the origin decision point to specific inhibitors of cellular signalling and metabolism. *Exp Cell Res.* 273:54-64.

Keryer, G., R.M. Rios, B.F. Landmark, B. Skalhegg, S.M. Lohmann, and M. Bornens. 1993. A high-affinity binding protein for the regulatory subunit of cAMP- dependent protein kinase II in the centrosome of human cells. *Exp Cell Res.* 204:230-40.

Keryer, G., B.S. Skalhegg, B.F. Landmark, V. Hansson, T. Jahnsen, and K. Tasken. 1999. Differential localization of protein kinase A type II isozymes in the Golgi-centrosomal area. *Exp Cell Res.* 249:131-46.

Keryer, G., O. Witczak, A. Delouvee, W.A. Kemmner, D. Rouillard, K. Tasken, and M. Bornens. 2003. Dissociating the centrosomal matrix protein AKAP450 from centrioles impairs centriole duplication and cell cycle progression. *Mol. Biol. Cell.* 14:2436-2446.

Khodjakov, A., and C.L. Rieder. 1999. The sudden recruitment of gamma-tubulin to the centrosome at the onset of mitosis and its dynamic exchange throughout the cell cycle, do not require microtubules. *J Cell Biol.* 146:585-96.

Khodjakov, A., and C.L. Rieder. 2001. Centrosomes enhance the fidelity of cytokinesis in vertebrates and are required for cell cycle progression. *J Cell Biol.* 153:237-42.

Kraft, C., F. Herzog, C. Gieffers, K. Mechtler, A. Hagting, J. Pines, and J.M. Peters. 2003. Mitotic regulation of the human anaphase-promoting complex by phosphorylation. *EMBO J.* 22:6598-6609.

Kramer, A., K. Neben, and A.D. Ho. 2002. Centrosome replication, genomic instability and cancer. *Leukemia.* 16:767-75.

Kramer, A., S. Schweizer, K. Neben, C. Giesecke, J. Kalla, T. Katzenberger, A. Benner, H.K. Muller-Hermelink, A.D. Ho, and G. Ott. 2003. Centrosome aberrations as a possible mechanism for chromosomal instability in non-Hodgkin's lymphoma. *Leukemia.* 17:2207-13.

Kronenwett, U., J. Castro, U.J. Roblick, K. Fujioka, C. Ostring, F. Faridmoghaddam, N. Laytragoon-Lewin, B. Tribukait, and G. Auer. 2003. Expression of Cyclins A, E and Topoisomerase II alpha correlates with centrosome amplification and genomic instability and influences the reliability of cytometric S-phase determination. *BMC Cell Biol.* 4:8.

Kronenwett, U., S. Huwendiek, C. Ostring, N. Portwood, U.J. Roblick, Y. Pawitan, A. Alaiya, R. Sennerstam, A. Zetterberg, and G. Auer. 2004. Improved Grading of Breast Adenocarcinomas Based on Genomic Instability. *Cancer Res.* 64:904-909.

Kufer, T.A., H.H.W. Sillje, R. Korner, O.J. Gruss, P. Meraldi, and E.A. Nigg. 2002. Human TPX2 is required for targeting Aurora-A kinase to the spindle. *J. Cell Biol.* 158:617-623.

Kunitoku, N., T. Sasayama, T. Marumoto, D. Zhang, S. Honda, O. Kobayashi, K. Hatakeyama, Y. Ushio, H. Saya, and T. Hirota. 2003. CENP-A phosphorylation by Aurora-A in prophase is required for enrichment of Aurora-B at inner centromeres and for kinetochore function. *Dev Cell.* 5:853-64.

Kuriyama, R., and G.G. Borisy. 1981. Microtubule-nucleating activity of centrosomes in Chinese hamster ovary cells is independent of the centriole cycle but coupled to the mitotic cycle. *J Cell Biol.* 91:822-6.

Lacey, K.R., P.K. Jackson, and T. Stearns. 1999. Cyclin-dependent kinase control of centrosome duplication. *Proc Natl Acad Sci U S A.* 96:2817-2822.

Lane, H.A., and E.A. Nigg. 1996. Antibody microinjection reveals an essential role for human polo-like kinase 1 (Plk1) in the functional maturation of mitotic centrosomes. *J Cell Biol.* 135:1701-13.

Lauze, E., B. Stoelcker, F.C. Luca, E. Weiss, A.R. Schutz, and M. Winey. 1995. Yeast spindle pole body duplication gene MPS1 encodes an essential dual specificity protein kinase. *EMBO J.* 14:1655-63.

Lee, M., M.J. Daniels, and A.R. Venkitaraman. 2004. Phosphorylation of BRCA2 by the Polo-like kinase Plk1 is regulated by DNA damage and mitotic progression. *Oncogene.* 23:865-72.

Lengauer, C., K. Kinzler, and B. Vogelstein. 1997. Genetic instability in colorectal cancers. *Nature.* 386:623-627.

Li, J.B., J.M. Gerdes, C.J. Haycraft, Y. Fan, T.M. Teslovich, H. May-Simera, H. Li, O.E. Blacque, L. Li, C.C. Leitch, R.A. Lewis, J.S. Green, P.S. Parfrey, M.R. Leroux, W.S. Davidson, P.L. Beales, L.M. Guay-Woodford, B.K. Yoder, G.D. Stormo, N. Katsanis, and S.K. Dutcher. 2004. Comparative genomics identifies a flagellar and basal body proteome that includes the BBS5 human disease gene. *Cell.* 117:541-52.

Li, J.J., J.S. Weroha, L.L. Lingle, D. Papa, J.L. Salisbury, and S.A. Li. 2005. Estrogen mediates Aurora-A over-expression, centrosome amplification, chromosomal instability, and breast cancer in female ACI rats. *Proc Natl Acad Sci U S A.* (*in press*).

Liby, K., H. Wu, B. Ouyang, S. Wu, J. Chen, and W. Dai. 2001. Identification of the human homologue of the early-growth response gene Snk, encoding a serum-inducible kinase. *DNA Seq.* 11:527-33.

Lin, H.R., N.S. Ting, J. Qin, and W.H. Lee. 2003. M phase-specific phosphorylation of BRCA2 by Polo-like kinase 1 correlates with the dissociation of the BRCA2-P/CAF complex. *J Biol Chem.* 278:35979-87.

Lingle, W.L., S.L. Barrett, V.C. Negron, A.B. D'Assoro, K. Boeneman, W. Liu, C.M. Whitehead, C. Reynolds, and J.L. Salisbury. 2002. Centrosome amplification drives chromosomal instability in breast tumour development. *Proc Natl Acad Sci U S A.* 99:1978-1983.

Lingle, W.L., W.H. Lutz, J.N. Ingle, N.J. Maihle, and J.L. Salisbury. 1998. Centrosome hypertrophy in human breast tumours: implications for genomic stability and cell polarity. *Proc Natl Acad Sci U S A.* 95:2950-5.

Lingle, W.L., and J.L. Salisbury. 1999. Altered centrosome structure is associated with abnormal mitoses in human breast tumours. *Am J Pathol.* 155:1941-51.

Logarinho, E., and C.E. Sunkel. 1998. The Drosophila POLO kinase localises to multiple compartments of the mitotic apparatus and is required for the phosphorylation of MPM2 reactive epitopes. *J Cell Sci*. 111:2897-909.

Lou, Y., W. Xie, D.F. Zhang, J.H. Yao, Z.F. Luo, Y.Z. Wang, Y.Y. Shi, and X.B. Yao. 2004. Nek2A specifies the centrosomal localization of Erk2. *Biochem Biophys Res Commun*. 321:495-501.

Lutz, W., W.L. Lingle, D. McCormick, T.M. Greenwood, and J.L. Salisbury. 2001. Phosphorylation of Centrin during the Cell Cycle and Its Role in Centriole Separation Preceding Centrosome Duplication. *J. Biol. Chem*. 276:20774-20780.

Ma, S., J. Charron, and R.L. Erikson. 2003. Role of Plk2 (Snk) in mouse development and cell proliferation. *Mol Cell Biol*. 23:6936-43.

Massimi, P., D. Pim, C. Kuhne, and L. Banks. 2001. Regulation of the human papillomavirus oncoproteins by differential phosphorylation. *Mol Cell Biochem*. 227:137-44.

Matsumoto, Y., K. Hayashi, and E. Nishida. 1999. Cyclin-dependent kinase 2 (Cdk2) is required for centrosome duplication in mammalian cells. *Curr Biol*. 9:429-32.

Matsumoto, Y., and J.L. Maller. 2004. A centrosomal localization signal in cyclin E required for Cdk-2independent S phase entry. *Science*. 306:885-888.

Matyakhina, L., S.M. Lenherr, and C.A. Stratakis. 2002. Protein kinase A and chromosomal stability. *Ann N Y Acad Sci*. 968:148-57.

Mayer, F., H. Stoop, S. Sen, C. Bokemeyer, J.W. Oosterhuis, and L.H. Looijenga. 2003. Aneuploidy of human testicular germ cell tumours is associated with amplification of centrosomes. *Oncogene*. 22:3859-66.

Meraldi, P., R. Honda, and E.A. Nigg. 2002. Aurora-A overexpression reveals tetraploidization as a major route to centrosome amplification in p53-/- cells. *EMBO J*. 21:483-492.

Meraldi, P., R. Honda, and E.A. Nigg. 2004. Aurora kinases link chromosome segregation and cell division to cancer susceptibility. *Curr Opin Genet Dev*. 14:29-36.

Meraldi, P., J. Lukas, A. Fry, J. Bartek, and E. Nigg. 1999. Centrosome duplication in mammalian somatic cells requires E2F and Cdk2-cyclin A. *Nature Cell Biol*. 1:88-93.

Meraldi, P., and E.A. Nigg. 2001. Centrosome cohesion is regulated by a balance of kinase and phosphatase activities. *J Cell Sci*. 114:3749-3757.

Metcalf, M.M. 1925. Boveri's work on cancer. *JAMA*. 84:1140.

Mills, G.B., R. Schmandt, M. McGill, A. Amendola, M. Hill, K. Jacobs, C. May, A.M. Rodricks, S. Campbell, and D. Hogg. 1992. Expression of TTK, a novel human protein kinase, is associated with cell proliferation. *J Biol Chem*. 267:16000-6.

Morrison, C., and C.L. Rieder. 2004. Chromosome damage and progression into and through mitosis in vertebrates. *DNA Repair*. 3:1133-1139.

Moudjou, M., N. Bordes, M. Paintrand, and M. Bornens. 1996. gamma-Tubulin in mammalian cells: the centrosomal and the cytosolic forms. *J Cell Sci*. 109:875-87.

Murphy, K.L., A.P. Dennis, and J.M. Rosen. 2000. A gain of function p53 mutant promotes both genomic instability and cell survival in a novel p53-null mammary epithelial cell model. *Faseb J*. 14:2291-302.

Nakajima, H., F. Toyoshima-Morimoto, E. Taniguchi, and E. Nishida. 2003. Identification of a consensus motif for Plk (Polo-like kinase) phosphorylation reveals Myt1 as a Plk1 substrate. *J Biol Chem*. 278:25277-80.

Nakajima, T., M. Moriguchi, Y. Mitsumoto, S. Sekoguchi, T. Nishikawa, H. Takashima, T. Watanabe, T. Katagishi, H. Kimura, T. Okanoue, and K. Kagawa. 2004. Centrosome aberration accompanied with p53 mutation can induce genetic instability in hepatocellular carcinoma. *Mod Pathol*. 17:722-7.

Neben K., B. Tews, G. Wrobel, M. Hahn, F. Kokocinski, C. Giesecke, U. Krause, A.D. Ho, A. Kramer, P. Lichter. 2004. Gene expression patterns in acute myeloid leukemia correlate with centrosome aberrations and numerical chromosome changes. [Article] *Oncogene*. 23:2379-2384.

Neef, R., C. Preisinger, J. Sutcliffe, R. Kopajtich, E.A. Nigg, T.U. Mayer, and F.A. Barr. 2003. Phosphorylation of mitotic kinesin-like protein 2 by polo-like kinase 1 is required for cytokinesis. *J Cell Biol.* 162:863-75.

Nelsen, C.J., R. Kuriyama, B. Hirsch, V.C. Negron, W.L. Lingle, M.M. Goggin, M.W. Stanley, and J.H. Albrecht. 2004. Short-term cyclin D1 overexpression induces centrosome amplification, mitotic spindle abnormalities, and aneuploidy. *J Biol Chem.*

Nigg, E., G. Schäfer, H. Hilz, and H. Eppenberger. 1985. Cyclic-AMP-dependent protein kinase type II is associated with Golgi complex and with centrosomes. *Cell.* 41:1039-1051.

Nigg, E.A. 2001. Mitotic kinases as regulators of cell division and its checkpoints. *Nat Rev Mol Cell Biol.* 2:21-32.

Nigg, E.A. 2002. Centrosome aberrations: cause or consequence of cancer progression? *Nat Rev Cancer.* 2:815-25.

Nigg, E.A. (Editor) 2004. Centrosomes in Development and Disease. Wiley-VCH, Weinheim, Germany.

Noguchi, K., H. Fukazawa, Y. Murakami, and Y. Uehara. 2004. Nucleolar Nek11 is a novel target of Nek2A in G1/S-arrested cells. *J Biol Chem.* 279:32716-27.

O'Connell, M.J., M.J. Krien, and T. Hunter. 2003. Never say never. The NIMA-related protein kinases in mitotic control. *Trends Cell Biol.* 13:221-8.

Ochi, T., T. Suzuki, J.C. Barrett, and T. Tsutsui. 2004. A trivalent dimethylarsenic compound, dimethylarsine iodide, induces cellular transformation, aneuploidy, centrosome abnormality and multipolar spindle formation in Syrian hamster embryo cells. *Toxicology.* 203:155-163.

Okuda, M. 2002. The role of nucleophosmin in centrosome duplication. *Oncogene.* 21:6170-4.

Okuda, M., H.F. Horn, P. Tarapore, Y. Tokuyama, A.G. Smulian, P.K. Chan, E.S. Knudsen, I.A. Hofmann, J.D. Snyder, K.E. Bove, and K. Fukasawa. 2000. Nucleophosmin/B23 is a target of CDK2/cyclin E in centrosome duplication. *Cell.* 103:127-40.

Ou, Y., M. Zhang, and J.B. Rattner. 2004. The centrosome: The centriole-PCM coalition. *Cell Motil Cytoskeleton.* 57:1-7.

Ouchi, M., N. Fujiuchi, K. Sasai, H. Katayama, Y.A. Minamishima, P.P. Ongusaha, C. Deng, S. Sen, S.W. Lee, and T. Ouchi. 2004. BRCA1 phosphorylation by Aurora-A in the regulation of G2 to M transition. *J Biol Chem.* 279:19643-8.

Pan, H., F. Zhou, and S.J. Gao. 2004. Kaposi's sarcoma-associated herpesvirus induction of chromosome instability in primary human endothelial cells. *Cancer Res.* 64:4064-8.

Pazour, G.J., B.L. Dickert, Y. Vucica, E.S. Seeley, J.L. Rosenbaum, G.B. Witman, and D.G. Cole. 2000. Chlamydomonas IFT88 and its mouse homologue, polycystic kidney disease gene tg737, are required for assembly of cilia and flagella. *J Cell Biol.* 151:709-18.

Perucca-Lostanlen, D., P. Rostagno, J. Grosgeorge, S. Marcie, P. Gaudray, and C. Turc-Carel. 2004. Distinct MDM2 and P14ARF expression and centrosome amplification in well-differentiated liposarcomas. *Genes Chromosomes Cancer.* 39:99-109.

Piel, M., P. Meyer, A. Khodjakov, C.L. Rieder, and M. Bornens. 2000. The respective contributions of the mother and daughter centrioles to centrosome activity and behaviour in vertebrate cells. *J Cell Biol.* 149:317-30.

Piel, M., J. Nordberg, U. Euteneuer, and M. Bornens. 2001. Centrosome-dependent exit of cytokinesis in animal cells. *Science.* 291:1550-3.

Pihan, G.A., A. Purohit, J. Wallace, H. Knecht, B. Woda, P. Quesenberry, and S.J. Doxsey. 1998. Centrosome defects and genetic instability in malignant tumours. *Cancer Res.* 58:3974-85.

Pihan, G.A., A. Purohit, J. Wallace, R. Malhotra, L. Liotta, and S.J. Doxsey. 2001. Centrosome Defects Can Account for Cellular and Genetic Changes That Characterize Prostate Cancer Progression. *Cancer Res.* 61:2212-2219.

Pihan, G.A., J. Wallace, Y. Zhou, and S.J. Doxsey. 2003. Centrosome Abnormalities and Chromosome Instability Occur Together in Pre-invasive Carcinomas. *Cancer Res.* 63:1398-404.

Pines, J. 1995. Cyclins, CDKs and cancer. *Seminars in Cancer Biology*. 6:63-72.

Pockwinse, S.M., G. Krockmalnic, S.J. Doxsey, J. Nickerson, J.B. Lian, A.J. van Wijnen, J.L. Stein, G.S. Stein, and S. Penman. 1997. Cell cycle independent interaction of CDC2 with the centrosome, which is associated with the nuclear matrix-intermediate filament scaffold. *Proc Natl Acad Sci U S A*. 94:3022-7.

Preble, A.M., T.M. Giddings, and S.K. Dutcher. 2000. Basal bodies and centrioles: Their function and structure. *In* Centrosome in Cell Replication and Early Development. Vol. 49. 207-233.

Rao, P.N., J.Y. Zhao, R.K. Ganju, and C.L. Ashorn. 1989. Monoclonal antibody against the centrosome. *J Cell Sci*. 93:63-9.

Rattner, J.B., J. Lew, and J.H. Wang. 1990. p34cdc2 kinase is localized to distinct domains within the mitotic apparatus. *Cell Motil Cytoskeleton*. 17:227-35.

Roh, M., B. Gary, C. Song, N. Said-Al-Naief, A. Tousson, A. Kraft, I.-E. Eltoum, and S.A. Abdulkadir. 2003. Overexpression of the Oncogenic Kinase Pim-1 Leads to Genomic Instability. *Cancer Res*. 63:8079-8084.

Rosenbaum, J.L., and G.B. Witman. 2002. Intraflagellar transport. *Nat Rev Mol Cell Biol*. 3:813-25.

Saavedra, H.I., B. Maiti, C. Timmers, R. Altura, Y. Tokuyama, K. Fukasawa, and G. Leone. 2003. Inactivation of E2F3 results in centrosome amplification. *Cancer Cell*. 3:333-46.

Salisbury, J.L. 2004. Centrosomes: sfi1p and centrin unravel a structural riddle. *Curr Biol*. 14:R27-9.

Salisbury, J.L., K.M. Suino, R. Busby, and M. Springett. 2002. Centrin-2 is required for centriole duplication in mammalian cells. *Curr Biol*. 12:1287-92.

Sato, N., K. Mizumoto, M. Nakamura, N. Maehara, Y.A. Minamishima, S. Nishio, E. Nagai, and M. Tanaka. 2001. Correlation between centrosome abnormalities and chromosomal instability in human pancreatic cancer cells. *Cancer Genet Cytogenet*. 126:13-9.

Schiebel, E. 2000. gamma-tubulin complexes: binding to the centrosome, regulation and microtubule nucleation. *Curr Opin Cell Biol*. 12:113-8.

Schultz, S.J., A.M. Fry, C. Sutterlin, T. Ried, and E.A. Nigg. 1994. Cell cycle-dependent expression of Nek2, a novel human protein kinase related to the NIMA mitotic regulator of Aspergillus nidulans. *Cell Growth Differ*. 5:625-35.

Sen, S. 2000. Aneuploidy and cancer. *Curr Opin Oncology*. 12:82-88.

Sen, S., H. Zhou, and R.A. White. 1997. A putative serine/threonine kinase encoding gene BTAK on chromosome 20q13 is amplified and overexpressed in human breast cancer cell lines. *Oncogene*. 14:2195-200.

Sluder, G., and E.H. Hinchcliffe. 1998. The apparent linkage between centriole replication and the S phase of the cell cycle. *Cell Biol Int*. 22:3-5.

Sluder, G., and E.H. Hinchcliffe. 2000. The coordination of centrosome reproduction with nuclear events during the cell cycle. *In* Centrosome in Cell Replication and Early Development. Vol. 49. 267-289.

Starita, L.M., Y. Machida, S. Sankaran, J.E. Elias, K. Griffin, B.P. Schlegel, S.P. Gygi, and J.D. Parvin. 2004. BRCA1-Dependent Ubiquitination of {gamma}-Tubulin Regulates Centrosome Number. *Mol Cell Biol*. 24:8457-8466.

Stearns, T., L. Evans, and M. Kirschner. 1991. g-Tubulin is a highly conserved component of the centrosome. *Cell*. 65:825-836.

Storchova, Z., and D. Pellman. 2004. From polyploidy to aneuploidy, genome instability and cancer. *Nat Rev Mol Cell Biol*. 5:45-54.

Stucke, V.M., H.H. Sillje, L. Arnaud, and E.A. Nigg. 2002. Human Mps1 kinase is required for the spindle assembly checkpoint but not for centrosome duplication. *EMBO J*. 21:1723-32.

Stucke, V.M., C. Baumann, and E.A. Nigg. 2004. Kinetochore localization and microtubule interaction of the human spindle checkpoint kinase Mps1. *Chromosoma*. 113:1-15.

Sunkel, C.E., and D.M. Glover. 1988. polo, a mitotic mutant of Drosophila displaying abnormal spindle poles. *J Cell Sci*. 89:25-38.

Tarapore, P., M. Okuda, and K. Fukasawa. 2002. A mammalian in vitro centriole duplication system: evidence for involvement of CDK2/cyclin E and nucleophosmin/B23 in centrosome duplication. *Cell Cycle*. 1:75-81.

Toji, S., N. Yabuta, T. Hosomi, S. Nishihara, T. Kobayashi, S. Suzuki, K. Tamai, and H. Nojima. 2004. The centrosomal protein Lats2 is a phosphorylation target of Aurora-A kinase. *Genes Cells*. 9:383-97.

Toyoshima-Morimoto, F., E. Taniguchi, and E. Nishida. 2002. Plk1 promotes nuclear translocation of human Cdc25C during prophase. *EMBO Rep*. 3:341-8.

Toyoshima-Morimoto, F., E. Taniguchi, N. Shinya, A. Iwamatsu, and E. Nishida. 2001. Polo-like kinase 1 phosphorylates cyclin B1 and targets it to the nucleus during prophase. *Nature*. 410:215-20.

Tsvetkov, L., X. Xu, J. Li, and D.F. Stern. 2003. Polo-like Kinase 1 and Chk2 Interact and Co-localize to Centrosomes and the Midbody. *J. Biol. Chem.* 278:8468-8475.

Tugendreich, S., J. Tomkiel, W. Earnshaw, and P. Hieter. 1995. CDC27Hs colocalizes with CDC16Hs to the centrosome and mitotic spindle and is essential for the metaphase to anaphase transition. *Cell*. 81:261-8.

Tutt, A., A. Gabriel, D. Bertwistle, F. Connor, H. Paterson, J. Peacock, G. Ross, and A. Ashworth. 1999. Absence of Brca2 causes genome instability by chromosome breakage and loss associated with centrosome amplification. *Curr Biol*. 9:1107-10.

Vandre, D.D., Y. Feng, and M. Ding. 2000. Cell cycle-dependent phosphorylation of centrosomes: Localization of phosphopeptide specific antibodies to the centrosome. *Microscopy Research and Technique*. 49:458-466.

Verde, F., J.C. Labbe, M. Doree, and E. Karsenti. 1990. Regulation of microtubule dynamics by cdc2 protein kinase in cell-free extracts of Xenopus eggs. *Nature*. 343:233-8.

Walter, A.O., W. Seghezzi, W. Korver, J. Sheung, and E. Lees. 2000. The mitotic serine/threonine kinase Aurora2/AIK is regulated by phosphorylation and degradation. *Oncogene*. 19:4906-16.

Wang, Q., Y. Hirohashi, K. Furuuchi, H.W. Zhao, Q.D. Liu, H.T. Zhang, R. Murali, A. Berezov, X.L. Du, B. Li, and M.I. Greene. 2004. The centrosome in normal and transformed cells. *DNA and Cell Biology*. 23:475-489.

Wang, X.J., D.A. Greenhalgh, A. Jiang, D. He, L. Zhong, B.R. Brinkley, and D.R. Roop. 1998. Analysis of centrosome abnormalities and angiogenesis in epidermal- targeted p53172H mutant and p53-knockout mice after chemical carcinogenesis: evidence for a gain of function. *Mol Carcinog*. 23:185-92.

Warnke, S., S. Kemmler, R.S. Hames, H.-L. Tsai, U. Hoffmann-Rohrer, A.M. Fry, and I. Hoffmann. 2004. Polo-like Kinase-2 Is Required for Centriole Duplication in Mammalian Cells. *Current Biology*. 14:1200-1207.

Weber, R.G., J.M. Bridger, A. Benner, D. Weisenberger, V. Ehemann, G. Reifenberger, and P. Lichter. 1998. Centrosome amplification as a possible mechanism for numerical chromosome aberrations in cerebral primitive neuroectodermal tumours with TP53 mutations. *Cytogenet Cell Genet*. 83:266-9.

Wilson, E.B. 1925. *The Cell in Development and Heredity*. Macmillan Company, New York. 1232 pp.

Winey, M., L. Goetsch, P. Baum, and B. Byers. 1991. MPS1 and MPS2: novel yeast genes defining distinct steps of spindle pole body duplication. *J Cell Biol*. 114:745-54.

Winey, M., and B.J. Huneycutt. 2002. Centrosomes and checkpoints: the MPS1 family of kinases. *Oncogene*. 21:6161-9.

Witczak, O., B.S. Skalhegg, G. Keryer, M. Bornens, K. Tasken, T. Jahnsen, and S. Orstavik. 1999. Cloning and characterization of a cDNA encoding an A-kinase anchoring protein located in the centrosome, AKAP450. *EMBO J*. 18:1858-68.

Xu, X., Z. Weaver, S.P. Linke, C. Li, J. Gotay, X.W. Wang, C.C. Harris, T. Ried, and C.X. Deng. 1999. Centrosome amplification and a defective G2-M cell cycle checkpoint induce genetic instability in BRCA1 exon 11 isoform-deficient cells. *Mol Cell.* 3:389-95.

Yanai, A., E. Arama, G. Kilfin, and B. Motro. 1997. ayk1, a novel mammalian gene related to Drosophila aurora centrosome separation kinase, is specifically expressed during meiosis. *Oncogene.* 14:2943-50.

Yarm, F.R. 2002. Plk phosphorylation regulates the microtubule-stabilizing protein TCTP. *Mol Cell Biol.* 22:6209-21.

Zhou, H., J. Kuang, L. Zhong, W.L. Kuo, J.W. Gray, A. Sahin, B.R. Brinkley, and S. Sen. 1998. Tumour amplified kinase STK15/BTAK induces centrosome amplification, aneuploidy and transformation. *Nat Genet.* 20:189-93.

Zhou, T., J.P. Aumais, X. Liu, L.Y. Yu-Lee, and R.L. Erikson. 2003. A role for Plk1 phosphorylation of NudC in cytokinesis. *Dev Cell.* 5:127-38.

Part 4

Genome Integrity Checkpoints

Chapter 4.1

MAMMALIAN DNA DAMAGE RESPONSE PATHWAY

Zhenkun Lou and Junjie Chen
Department of Oncology, Mayo Clinic and Foundation, Rochester, USA

1. INTRODUCTION

The genome of an organism is under constant attack from exogenous and endogenous DNA damaging factors such as radiation, carcinogens and reactive radicals. Cells have developed an elaborate DNA damage response system, which is responsible for sensing DNA damage, halting the ongoing cell cycle, and repairing DNA lesions. Failure to detect and repair DNA damage will lead to genomic instability, which help drive the development of cancer (Hahn and Weinberg, 2002). Many human genetic cancer predisposition syndromes are linked to defective DNA damage responses. For example, mutations in MLH1 and MSH2, proteins involved in DNA mismatch repair, account for about half of hereditary nonpolyposis colorectal cancer (HNPCC) cases. Mutations in the BRCA1 gene account for about 50% of familial breast cancer cases. Therefore, prompt and correct response to DNA damage is crucial for the well being of the organism.

There are many excellent reviews about the DNA damage response pathway (Iliakis et al., 2003; Motoyama and Naka, 2004; Rouse and Jackson, 2002; Sancar et al., 2004; Zhou and Elledge, 2000). This review will discuss recent advance in our understanding of mammalian DNA damage response, with a focus on checkpoint activation. DNA damage repair will be discussed in other chapters of this book.

2. ORGANISATION OF THE DNA DAMAGE RESPONSE PATHWAY

The DNA damage response pathway can be divided into three categories: cell cycle checkpoint, DNA damage repair and adaptation/apoptosis. The

E.A. Nigg (ed.), Genome Instability in Cancer Development, 425-455.

exact sequence of the signalling events following DNA damage is still not clear. A possible scenario is that cells detect DNA damage, initiate cell cycle checkpoints to halt the cell cycle and then initiate DNA damage repair. However, if DNA damage is minor, cells might quickly repair the damage without triggering the checkpoint. On the other hand, if the DNA damage is extensive, mammalian cells will undergo apoptosis, since it is beneficial to eliminate cells with unrepairable DNA rather than allow them to propagate incorrect genetic information. Unicellular organisms, on the other hand, will undergo adaptation, which allows them to resume the cell cycle in the presence of DNA damage. Adaptation has been poorly studied in multicellular organisms, although a recent report demonstrates an adaptation process in *Xenopus laevis* (Yoo et al., 2004).

On the molecular level, the DNA damage response pathway contains several key components: sensors, transducers, mediators, and effectors (Figure 1). Although it is often presented in a linear pathway, the

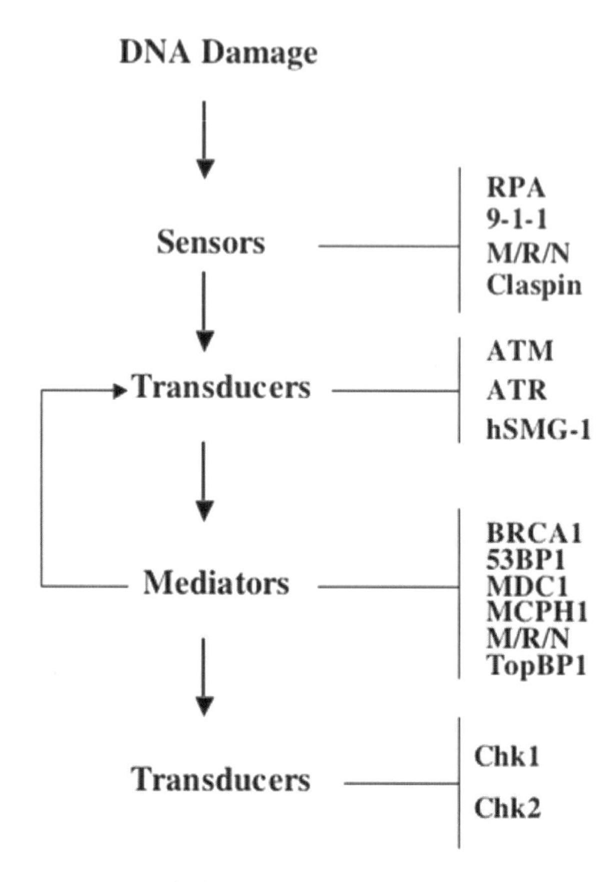

Figure 1. Organisation of DNA Damage Response Pathway.

DNA damage response pathway is more like a network wired with many feedback loops. Often times, proteins play multiple roles in the pathway, therefore the terms upstream and downstream become less clear. For example, NBS1 is proposed to act as a double strand break (DSB) sensor and play a role in regulating ATM autophosphorylation. On the other hand, NBS1 itself is a substrate of ATM and mediates an ATM-dependent intra-S phase checkpoint. Thus, it is hard to assign NBS1 in a linear pathway with one clearly defined role.

3. DNA DAMAGE INDUCED NUCLEAR FOCI

An interesting finding among players in the DNA damage response pathway is that many of them form nuclear foci in response to DNA damage. The number of nuclear foci observed following ionizing radiation (IR) correlates well with the estimated number of DSBs per cell, suggesting that damage induced nuclear foci correspond to the sites of DNA damage. The phosphorylated form of histone H2AX (γH2AX) turns out to be a reliable marker for the presence of DSBs (Fernandez-Capetillo et al., 2003a). Studies using a laser beam to introduce a path of DNA breaks reveal that γH2AX form foci along the path of DNA breaks (Rogakou et al., 1999). γH2AX foci are also observed when cells are treated with many other DNA damaging agents, such as bleomycin, mitomycin C (MMC), hydorxyurea, and topoisomerase II inhibitors. Interestingly, γH2AX foci can be detected in cycling cells, probably reflecting DSBs formed due to the collapse of stalled replication forks. Both ATM and DNA-PK phosphorylate H2AX *in vivo* after IR (Burma et al., 2001; Stiff et al., 2004) and the phosphorylation site has been mapped to Ser139 at the C-terminal tail of human H2AX (Rogakou et al., 1998). Many proteins involved in the DNA damage response pathway form damage-induced nuclear foci that colocalise with γH2AX. These include ATM, BRCA1, MRE11/NBS1/Rad50, 53BP1, MDC1, and phosphorylated Chk2 (Bakkenist and Kastan, 2003; Goldberg et al., 2003; Lou et al., 2003b; Paull et al., 2000; Rappold et al., 2001a; Schultz et al., 2000; Stewart et al., 2003; Ward et al., 2001). Therefore, nuclear protein foci formation following DNA damage is a natural response to DNA damage.

Many studies have shown that H2AX is required for foci formation of BRCA1, 53BP1, MDC1 and NBS1 (Bassing et al., 2002; Celeste et al., 2003b; Stewart et al., 2003). It is proposed that γH2AX regulates foci formation of checkpoint proteins by acting as a docking site. This could be due to a direct interaction between γH2AX and these checkpoint proteins. The FHA and BRCT domains of NBS1 have been shown to interact directly with γH2AX (Kobayashi et al., 2002). Similarly, the region on 53BP1 required for its damage-induced focus formation also binds directly to a phosphorylated H2AX peptide, but not to a control unphosphorylated H2AX (Ward et al., 2003a). Alternatively, it is also proposed that γH2AX may modulate chromatin structure, thus facilitating accumulation of checkpoint proteins (Fernandez-Capetillo et al., 2003b). More recently, kinetic studies using a laser beam revealed that the initial recruitment of BRCA1 and NBS1 is normal in H2AX$^{-/-}$ cells, and only the later accumulation of BRCA1 and NBS1 is defective (Celeste et al., 2003b). These studies suggest that H2AX functions to accumulate, but not recruit proteins involved in the DNA damage response at the sites of DNA damage.

Accumulation of checkpoint and DNA repair proteins at the sites of DNA damage was thought to facilitate signal transduction events by bringing kinases and their substrates together. However, in many cases, there is no obvious connection between the formation of nuclear foci and protein phosphorylation. For example, phosphorylation of 53BP1 is normal in H2AX$^{-/-}$ cells in spite of defective foci formation. Similarly, mutation of ATM phosphorylation sites on 53BP1 did not affect its ability to form foci (Ward et al., 2003a). Thus, it is not clear why these checkpoint proteins need to accumulate at the sites of DNA damage. However, H2AX does play an important role in maintaining genomic stability, since H2AX haploinsufficiency in a p53$^{-/-}$ background shows a dramatic increase in tumour formation (Bassing et al., 2003; Celeste et al., 2003a). Whether or not the role of H2AX in maintaining genomic stability is related to its role in the accumulation of proteins involved in the DNA damage response pathway is still not clear. H2AX$^{-/-}$ cells show only a mild checkpoint defect (Fernandez-Capetillo et al., 2002), and V(D)J recombination remains largely intact in H2AX$^{-/-}$ mice (Bassing et al., 2002; Celeste et al., 2002). Therefore, the functional significance of foci formation of proteins involved in the DNA damage response needs to be further clarified.

4. COMPONENTS OF THE DNA DAMAGE RESPONSE PATHWAY

4.1 Sensors

Despite intensive investigation, the identities of DNA damage sensors remain elusive. It is possible that various types of DNA damage are detected by distinct sensor proteins. Alternatively, DNA damage may first be processed to a common structure by DNA repair proteins and then recognised by common sensors. Here, we will discuss several candidates that have been proposed to be DNA damage sensors.

4.1.1 DNA-PKcs/Ku

DNA-PKcs (DNA-PK catalytic subunit)/Ku have been implicated as DSB sensors (Bradbury and Jackson, 2003). Ku70/80 heterodimers directly bind double strand breaks (DSBs), and in turn recruit DNA-PKcs to the sites of DNA damage. Once it binds DNA ends, DNA-PKcs is activated, and participates in nonhomologous end joining (NHEJ) through a still unknown mechanism. However, even though DNA-PKcs and Ku are essential for NHEJ, they do not seem to play a major role in checkpoint activation. Based on these findings DNA-PKcs/Ku has been proposed as the sensor for DNA damage repair, with another sensor being important for checkpoint activation (Bradbury and Jackson, 2003).

4.1.2 ATM

A recent paper suggest a role of ATM in sensing DNA damage (Bakkenist and Kastan, 2003). In resting cells, ATM exists as an inactive dimer. Upon DNA damage, ATM undergoes autophosphorylation at Ser1981, which causes the dissociation of the ATM dimer and exposes the ATM active site for substrate access. One evidence supporting ATM as a sensor is that ATM responds to minor DNA damage at the earliest detectable time point (Bakkenist and Kastan, 2003). In addition, ATM also responds to changes in chromatin structure, since ATM could be activated by treatment with hypotonic shock or histone deacetylase inhibitor, which does not cause detectable DNA damage (Bakkenist and Kastan, 2003). These findings raise interesting questions as to what DSB sensors really recognise: changes in chromatin structure or exposed DNA ends. This remains an important question that needs to be addressed further.

4.1.3 The M/R/N Complex

The MRE11/Rad50/NBS1 (M/R/N) complex is also proposed as a DSB sensor (Petrini and Stracker, 2003). NBS1 forms a stable complex with MRE11 and Rad50. NBS1 contains FHA and BRCT domains, both of which are required for NBS1 localisation and function (Cerosaletti and Concannon, 2003; Zhao et al., 2002b). Mre11 has 3'-5' exonuclease activity, which could process DSBs and facilitate DNA damage repair (Paull and Gellert, 1998; Trujillo et al., 1998). Rad50 contains Walker A and B motifs that confer ATPase activity and a coil-coil domain that is required for intermolecular interactions. Rad50 and Mre11 form a flexible complex that could tether the broken DNA ends (Costanzo et al., 2004; de Jager et al., 2001; Hopfner et al., 2002). Both NBS1 and MRE11 have been found mutated in human genetic disorders. NBS1 is mutated in patients with Nijmegen Breakage Syndrome (NBS) and MRE11 mutations were found in patients with ataxia-telangiectasia-like disease (ATLD) (Carney et al., 1998; Stewart et al., 1999). NBS and ATLD share many similar phenotypes with ataxia telangictasia (AT), such as radiosensitivity, immunodeficiency and predisposition for cancer, suggesting an important role of the M/R/N complex in the ATM pathway. Consistent with a role of the MRN complex in DNA damage sensing, Mre11- and NBS1-deficient cells show decreased ATM phosphorylation and defective accumulation of phospho-ATM at the sites of DNA damage (Carson et al., 2003; Costanzo et al., 2004; Horejsi et al., 2004; Uziel et al., 2003). In addition, the MRN complex directly stimulates ATM kinase activity toward its substrates (Lee and Paull, 2004). On the other hand, NBS1 has been shown to be a substrate of ATM and regulate the intra-S checkpoint (Gatei et al., 2000; Lim et al., 2000; Wu et al., 2000; Zhao et al., 2000). Thus, it is possible that the M/R/N complex may function in a positive feed-back loop and facilitate the activation of ATM following DNA damage.

4.1.4 RPA

A more recent finding in the field concerns the role of RPA in DNA damage sensing. RPA is a single strand DNA-binding protein that plays an essential role in DNA replication and DNA damage repair. It has been demonstrated in yeast that extensive single strand DNA exists during replication block or following DNA damage. More recently, RPA has been shown to facilitate the recruitment of ATR and its binding partner ATRIP to single stranded DNA (Zou and Elledge, 2003). Down-regulation of RPA results in defective Chk1 phosphorylation and G2/M checkpoint control, suggesting a role of RPA in the activation of the ATR-Chk1 pathway (Zou and Elledge, 2003). Consistent with these findings, depletion of xRPA in *Xenopus laevis* also prevents the loading of xATR and xHus1 onto chromatin (You et al., 2002). Since single-stranded DNA also occurs during the normal replication process, the question that arises from these observations is how cells differentiate normal replication intermediates from damage-induced single-stranded DNA. A possible explanation is that the length of single-stranded DNA and the amount of RPA binding determine a threshold for checkpoint activation (Lupardus et al., 2002; Zou and Elledge, 2003). When replication forks reach the sites of damaged DNA, DNA replication and strand unwinding become uncoupled (Walter and Newport, 2000). DNA polymerases are stalled at the sites of DNA damage, while replication helicases keep moving ahead, thereby leaving long patches of single-stranded DNA that triggers checkpoint activation. While this has been suggested to be mediated by RPA, RPA-independent binding of ATR to chromatin following DNA damage has also been reported (Barr et al., 2003; Bomgarden et al., 2004; Unsal-Kacmaz and Sancar, 2004).

4.1.5 The 9-1-1 Complex

The Rad9/Rad1/Hus1 complex (the 9-1-1 complex) has long been considered a DNA damage sensor. The 9-1-1 complex forms a ring-like clamp, similar to that of a PCNA homotrimer (Bermudez et al., 2003; Griffith et al., 2002; Venclovas and Thelen, 2000). The RFC-like protein Rad17 acts as the clamp loader that is responsible for the loading of the 9-1-1 complex (Bermudez et al., 2003; Lindsey-Boltz et al., 2001; Rauen et al., 2000; Zou et al., 2002). In response to DNA damage, the 9-1-1 complex becomes retained on chromatin (Zou et al., 2002). Interestingly, the loading of the 9-1-1 complex and the ATR-ATRIP complex seem to be independent events (Zou et al., 2002), although both complexes are required for efficient Chk1 activation and the G2/M checkpoint (Bao et al., 2001; Roos-Mattjus et al., 2002; Weiss et al., 2002). It is possible that two independent potential DNA damage sensors might constitute a mechanism to guarantee the specificity of checkpoint activation. One might imagine that since cells constantly encounter DNA intermediates similar to those generated following DNA damage, the existence of multiple sensors will ensure the specificity of checkpoint activation and allow cells to go through normal cell cycle progression without constantly putting on the break.

4.1.6 Claspin

Recent findings also implicate Claspin as a DNA damage sensor. Claspin was initially cloned in *Xenopus* (Kumagai and Dunphy, 2000). In response to replication stress, Claspin relocalises to chromatin in a RPA and ATR-independent manner (Lee et al., 2003). Claspin interacts with ATR, Chk1 and BRCA1, and regulates BRCA1 and Chk1 phosphorylation following DNA damage (Chini and Chen, 2003; Kumagai and Dunphy, 2000; Lin et al., 2004). Depletion of Claspin results in defective S phase and G2/M checkpoints in response to replication blocks or UV radiation (Chini and Chen, 2003; Kumagai and Dunphy, 2000). Interestingly, purified Claspin forms a ring structure *in vitro* (Sar et al., 2004). Therefore, Claspin could function in a similar fashion to the 9-1-1 complex and act as another DNA damage sensor.

Despite these new discoveries, it is still not clear how DNA damage repair and checkpoint activation are coordinated. In contrast to DNA-PKcs/Ku, ATM is required for checkpoint activation but only plays a minor role in DNA repair. It is puzzling as to why double strand breaks would require two kinds of sensors, one for DNA repair, one for checkpoint activation. A plausible explanation is that certain types of DNA damage activate DNA-PKcs/Ku and are immediately repaired without activating a checkpoint. Only DNA damage that is not immediately repaired would then be processed and activate the DNA damage checkpoint pathway.

4.2 Transducers

DNA damage sensors transmit signals to transducers, which amplify and transduce signals to downstream effectors. ATM/ATR and the downstream kinases Chk1/Chk2 are well-studied transducers in the DNA damage response pathway. These are serine/threonine kinases that initiate a cascade of phosphorylation events, eventually resulting in cell cycle checkpoint activation, DNA repair and apoptosis.

4.2.1 ATM/ATR/hSMG-1

ATM and ATR are the centre pieces of the DNA damage checkpoint pathway (Abraham, 2001). ATM and ATR, together with DNA-PK, mTOR and hSMG-1, are members of the phosphoinositide 3-kinase–like kinase (PIKK) family (Abraham, 1996). Although their kinase domains share homology with PI3 kinase, none of them has been shown to possess phosphoinositol kinase activity. Instead, all of the PIKKs possess serine/threonine kinase activity. ATM exists as an inactive dimer in resting cells. Following IR, ATM is activated through autophosphorylation of ATM and disassociation of the ATM dimer (Bakkenist and Kastan, 2003). However, how ATR is activated is still not clear. ATR kinase activity does not increase following DNA damage or upon binding to DNA like ATM or DNA-PK (Abraham, 2001; Unsal-Kacmaz and Sancar, 2004). Therefore, the activation of ATR is not likely to be associated with an increase in kinase activity *per se*, but rather with an altered access of ATR to its substrates. For example, ATRIP, the binding partner of ATR (Cortez et al., 2001), could stimulate ATR kinase activity toward RPA by binding both ATR and RPA (Unsal-Kacmaz and Sancar, 2004).

Both ATM and ATR phosphorylate SQ/TQ motifs, and share almost the same spectrum of substrates. These include BRCA1, NBS1, p53, Chk1, Chk2, SMC1 etc. (Kastan and Lim, 2000). However, ATM and ATR have distinct roles in the DNA damage response pathway. ATM preferentially responds to DSB, while ATR preferentially responds to UV and replication block.

Both ATM and ATR have been linked to human diseases. ATM is mutated in AT patients, which show cerebellar degeneration, genomic instability, higher tumour incidence, immunodeficiency, and radiation hypersensitivity (Shiloh, 2003). At the cellular lever, $ATM^{-/-}$ cells show gross chromosomal rearrangements and checkpoint defects. An ATR splicing mutation has been found in Seckel syndrome patients (O'Driscoll et al., 2003), which show growth retardation, microcephaly, and increased risk of myelodysplasia and acute myeloid leukemia. Cells derived from a Seckel syndrome patient show hypersensitivity to UV and mitomycin C (MMC), and defective phosphorylation of p53 and NBS1 following DNA damage (O'Driscoll et al., 2003). In addition, overexpression of dominant-negative ATR abrogates the G2/M checkpoint in response to IR (Cliby et al., 1998). Animal models with disrupted ATM or ATR genes have been established, confirming the important role of ATM and ATR in checkpoint activation. ATM knockout mice recapitulate many of the AT phenotypes, such as genomic instability, checkpoint defects and increased incidence of lymphoma (Barlow et al., 1996; Elson et al., 1996). ATR knockout mice are embryonic lethal, suggesting an essential role of ATR in embryonic development (Brown and Baltimore, 2000; de Klein et al., 2000). Conditional knockout cells and mice of ATR have been generated (Brown and Baltimore, 2003; Cortez et al., 2001). ATR knockout cells show a defective G2/M checkpoint in response to IR (Brown and Baltimore, 2003; Cortez et al., 2001). In addition, deletion of both ATM and ATR completely abolishes the G2/M checkpoint, suggesting that ATM and ATR cooperate in regulating this checkpoint (Brown and Baltimore, 2003). However, checkpoint delay in response to stalled DNA replication is intact in ATR knockout cells and ATR/ATM double-knockout cells (Brown and Baltimore, 2003), suggesting that there are additional checkpoint kinases involved in DNA damage responses.

More recently, another PIKK family member, hSMG-1, has been shown to be involved in DNA damage responses. HSMG-1 is required for full p53 activation and depletion of hSMG1 results in spontaneous DNA damage and radiosensitivity (Brumbaugh et al., 2004), implying that hSMG1 is another key upstream kinase in a DNA damage response pathway.

4.2.2 Chk1/Chk2

Chk2 is a key transducer in the ATM-dependent pathway (Bartek et al., 2001; McGowan, 2002). Chk2 is the mammalian homolog of budding yeast Rad53 and fission yeast Cds1, which play critical roles in checkpoint activation and adaptation. Chk2 mutations have been identified in a subtype of Li-Fraumeni Syndrome (LFS) patients (Bell et al., 1999). LFS is characterised by multiple tumours at young age and is often linked to p53 mutations. In a subset of Li-Fraumeni patients without p53 mutations, germline mutations of Chk2 have been identified (Bell et al., 1999). In addition, a Chk2 truncation (1100delC) has also been found in hereditary breast cancer patients (Meijers-Heijboer et al., 2002; Vahteristo et al., 2002). These genetic studies suggest a role of Chk2 in tumour suppression. Unexpectedly, Chk2 knockout mice did not show a cancer-prone phenotype (Hirao et al., 2002; Jack et al., 2002; Takai et al., 2002). It is possible that Chk2 function is compensated by other checkpoint kinases such as Chk1. A more recent report using Chk2/BRCA1 double knockout mice does suggest that Chk2 cooperates with BRCA1 in tumour suppression (McPherson et al., 2004).

In response to DNA damage, Chk2 is phosphorylated and activated in an ATM-dependent manner. ATM phosphorylates Chk2 at Thr68, which is critical for Chk2 activation (Ahn et al., 2000; Matsuoka et al., 2000; Melchionna et al., 2000). Phosphorylation at Thr68 also serves to create a binding site for other proteins. Both MDC1 and Chk2 FHA domains have been found to bind the phosphorylated Thr68 site (Ahn et al., 2002; Lou et al., 2003b; Xu et al., 2002b). In addition, Chk2 autophosphorylates itself at Thr383 and Thr387 sites, located within its kinase activation loop. These autophosphorylation sites of Chk2 are important for Chk2 activity (Lee and Chung, 2001). Chk2 also undergoes autophosphorylation at Ser516, which is involved in optimal activation of Chk2 and IR-induced apoptosis (Schwarz et al., 2003; Wu and Chen, 2003). Many functional studies suggest that Chk2 plays an important role in DNA repair, multiple checkpoints and apoptosis. Its substrates include BRCA1, p53, Plk1, E2F1 and Cdc25A (Bartek et al., 2001; McGowan, 2002).

Chk1 is a key transducer of the ATR signalling pathway (Chen and Sanchez, 2004). Similar to those of ATR, Chk1 knockout mice are embryonic lethal (Liu et al., 2000; Takai et al., 2000). Cells from heterozygous Chk1 background show gross genomic instability, suggesting an important role of Chk1 in maintaining genomic stability (Lam et al., 2004). DNA replication block and DNA damage failed to arrest the cell cycle before initiation of mitosis in Chk1$^{-/-}$ ES cells (Liu et al., 2000; Takai et al., 2000), supporting the view that Chk1 is the key kinase involved in cell cycle checkpoint control. Chk1 is phosphorylated in response to DNA damage in an ATM/ATR dependent manner. Two ATM/ATR dependent phosphorylation sites have been identified: Ser317 and Ser345 (Gatei et al., 2003; Liu et al., 2000; Zhao and Piwnica-Worms, 2001). The phosphorylation of Ser345 has been shown to be critical for Chk1 activation and checkpoint function (Liu et al., 2000). Just like ATM and ATR share common substrates, Chk1 and Chk2 share many of the common substrates, and probably function together in checkpoint controls. The substrates of Chk1 include p53, Cdc25A and Cdc25C.

Much progress has been made towards understanding the activation of DNA damage transducers and the identification of their substrates. The generation of various knockout mice also revealed much insight into the molecular mechanism of DNA damage transducers. It also raised some interesting questions that remain to be answered. These include the role of Chk2 in tumourigenesis and the ATM/ATR-independent replication checkpoint control.

4.3 Mediators

Recently, proteins characterised as mediators have been found to play important roles in the DNA damage response. Like adaptor and anchor proteins in protein tyrosine kinase (PTK) signalling pathways, mediators serve to integrate and amplify signalling from upstream kinases. Most of the mediators contain protein-protein interaction domains. For example, NBS1 and MDC1 contain a FHA domain, a protein domain that recognises phosphorylated Serine/Threonine (pS/T) motifs (Durocher and Jackson, 2002). In addition, BRCA1, 53BP1, TopBP1, MCPH1, MDC1 and NBS1 all contain a BRCA1 associated C-terminal (BRCT) domain. Similar to the FHA domain, the BRCT domain also recognises pS/T motif (Manke et al., 2003; Yu et al., 2003). Thus, the current hypothesis is that these mediators facilitate the transduction of DNA damage signals through a series of protein-protein interactions.

4.3.1 BRCA1

The BRCA1 gene was originally mapped and cloned from a cohort of familial breast cancer patients (Miki et al., 1994). BRCA1 mutations were found in about 50% of the familial breast cancer patients, suggesting the important role of BRCA1 as a tumour suppressor. BRCA1 knockout mice are embryonic lethal (Gowen et al., 1996; Hakem et al., 1996; Liu et al., 1996; Ludwig et al., 1997; Shen et al., 1998). Gross chromosomal abnormalities have been observed in BRCA1$^{-/-}$ embryos. Conditional knockout mice or mice carrying a truncation 3' of BRCA1 exon 11 have been generated (Ludwig et al., 2001; Xu et al., 1999), and these animals show increased incidence of mammary tumour development. Disruption of BRCA1 in a p53 heterozygous background increased mammary tumour frequency and latency (Xu et al., 2001b). These genetic studies using animal models confirm the role of BRCA1 in tumour suppression.

BRCA1 contains tandem BRCT domains at its C-terminus and a RING domain at its N-terminus. Mutations within the RING domain of BRCA1 have been found in several breast cancer patients, suggesting that the RING domain of BRCA1 is essential for its tumour suppression function. However, how the RING domain contributes to BRCA1 function has not yet been resolved. The RING domain of BRCA1 interacts with the RING domain of BARD1, forming a heterodimer that has E3 ubiquitin ligase activity (Brzovic et al., 2003; Chen et al., 2002; Hashizume et al., 2001; Kentsis et al., 2002; Lorick et al., 1999; Mallery et al., 2002; Ruffner et al., 2001; Xia et al., 2003). The BRCA1-BARD1 E3 ligase seems to preferentially catalyze monoubiquination at K6 *in vitro* and *in vivo* (Morris and Solomon, 2004; Wu-Baer et al., 2003). It is possible that this catalytic activity of BRCA1 is required for it tumour suppression function. However, the *in vivo* substrates of the BRCA1/BARD1 E3 ligase have not yet been identified. One clue about the potential substrates of BRCA1/BARD1 is that this complex also forms nuclear foci in response to DNA damage (Morris and Solomon, 2004), suggesting that its potential substrates may also be components of the DNA damage response pathway. The C-terminal BRCT domain of BRCA1 is also mutated in patients with early-onset breast and ovarian cancers. The BRCT domain of BRCA1 interacts with BACH1, a protein that contains DNA helicase activity (Cantor et al., 2001). Mutations of the BRCA1 BRCT domain identified in breast cancer patients disrupt the BRCA1-BACH1 interaction. In addition, BACH1 mutations have been found in breast cancer patients (Cantor et al., 2001). These results suggest that the BRCA1-BACH1 interaction contributes to the tumour suppression function of BRCA1. Consistent with its role as a mediator, BRCA1 exists as a complex called BRCA1-associated genome surveillance complex (BASC) (Wang et al., 2000), which contains tumour suppressors and DNA damage repair proteins MSH2, MSH6, MLH1, ATM, BLM, and the M/R/N complex. In addition, the associations of BRCA1 with many cellular proteins including BRCA2, CtIP, FANCD2, ZBRK1 SWI/SNF, Rad51,

RNA polymerase have also been reported (Deng and Brodie, 2000; Kerr and Ashworth, 2001; Scully et al., 2004). As a result, BRCA1 has been implicated in multiple aspects of the DNA damage response, such as homologous recombination (HR) and checkpoint activation following DNA damage (Kerr and Ashworth, 2001; Lou and Chen, 2003; Scully and Livingston, 2000; Scully et al., 2004).

4.3.2 53BP1

Initially identified as a p53 binding protein (Iwabuchi et al., 1994), 53BP1 was thought to be a homologue of budding yeast Rad9. 53BP1 contains tandem BRCT domains at its C- terminus. In response to DNA damage, 53BP1 is phosphorylated and quickly forms nuclear foci (Anderson et al., 2001; Rappold et al., 2001b; Schultz et al., 2000; Xia et al., 2001). Knockdown of 53BP1 using small interference RNA (siRNA) reveals that 53BP1 regulates the intra-S and G2/M checkpoints (DiTullio et al., 2002; Wang et al., 2002). Consistent with these studies, $53BP1^{-/-}$ cells show a defective G2/M checkpoint at low doses of IR (Fernandez-Capetillo et al., 2002). Increased genomic instability and tumour incidence are also observed in $53BP1^{-/-}$ mice, suggesting that 53BP1 acts as a tumour suppressor (Morales et al., 2003; Ward et al., 2003b). In addition, $53BP1^{-/-}$ mice have defective class switching recombination but apparently normal V(D)J recombination (Manis et al., 2004; Ward et al., 2004), suggesting a role of 53BP1 in certain aspects of DSB repair.

4.3.3 MDC1/NFBD1

MDC1/NFBD1 is another protein that contains tandem BRCT domains. In addition to the BRCT domains, MDC1 contains a FHA domain at its N terminus and a middle repeat region with 14 repeats of 41 residues. MDC1 binds to chromatin and is phosphorylated in an ATM-dependent manner (Goldberg et al., 2003; Lou et al., 2003a; Lou et al., 2003b; Shang et al., 2002; Stewart et al., 2003; Xu and Stern, 2003). Following DNA damage, MDC1 rapidly forms nuclear foci and this focus formation requires γH2AX (Goldberg et al., 2003; Lou et al., 2003a; Lou et al., 2003b; Shang et al., 2002; Stewart et al., 2003; Xu and Stern, 2002; Xu and Stern, 2003). Interestingly, γH2AX foci formation and H2AX phosphorylation are defective in cells depleted of MDC1 (Stewart et al., 2003). This could be due to a role of MDC1 in the recruitment of activated ATM to the sites of DNA damage (Mochan et al., 2003), thereby forming a positive feedback loop from MDC1 to γH2AX. MDC1 interacts with many proteins, such as NBS1, Chk2, ATM, SMC1 and BRCA1 (Goldberg et al., 2003; Lou et al., 2003a; Lou et al., 2003b; Peng and Chen, 2003; Stewart et al., 2003). The FHA domain of MDC1 mediates its interactions with NBS1 and Chk2 (Goldberg et al., 2003; Lou et al., 2003b). The binding partners of the MDC1 BRCT domains are currently unknown. Interestingly, the repeat region in the middle of MDC1 is also a protein-protein interaction domain and mediates its interaction with DNA-PKcs/Ku (Lou et al., in press). Similar to other mediator proteins, MDC1 has roles in multiple aspects of the DNA damage response pathway, including the intra-S and G2/M checkpoints, and apoptosis (Goldberg et al., 2003; Lou et al., 2003a; Lou et al., 2003b; Peng and Chen, 2003; Stewart et al., 2003).

4.3.4 TopBP1

TopBP1 was initially identified as a Topoisomerase IIβ interacting protein that contains eight BRCT repeats (Yamane et al., 1997). TopBP1 forms foci following DNA damage (Makiniemi et al., 2001; Yamane et al., 2002), and interacts with Rad9 in a DNA damage dependent manner (Makiniemi et al., 2001). *Xenopus* Cut5 (TopBP1) is required for the recruitment of ATR and activation of Chk1 following DNA damage (Parrilla-Castellar and Karnitz, 2003). Consistent with the findings in *Xenopus*, down-regulation of TopBP1 and BRCA1 in mammalian cells results in decreased Chk1 activation and therefore defective G2/M checkpoint control, suggesting that TopBP1 cooperates with BRCA1 in regulating the G2/M checkpoint (Yamane et al., 2003). Additionally, TopBP1 has also been reported to negatively regulate E2F1 and inhibit apoptosis in response to DNA damage (Liu et al., 2003; Liu et al., 2004).

4.3.5 MCPH1

Microcephalin (MCPH1) is a new member of the family of BRCT-containing proteins involved in the DNA damage response (Jackson et al., 2002). MCPH1 is the first gene identified among six loci that contribute to primary microcephaly, a congenital reduction in brain size. MCPH1 was also identified as a transcriptional suppressor of the human telomerase catalytic subunit and named BRIT1 (Lin and Elledge, 2003). A role of MCPH1 in the regulation of BRCA1 and Chk1 expression and in DNA damage checkpoint activation has been reported recently (Xu et al., 2004).

In summary, the mediator proteins are involved in multiple aspects of DNA damage checkpoints. The molecular mechanisms of how individual mediator protein participate in DNA damage responses are not yet fully understood. Detailed spatial-temporal studies are needed to reveal the exact roles of these mediator proteins in DNA damage response pathways.

5. DNA DAMAGE INDUCED CHECKPOINT ACTIVATION AND APOPTOSIS

DNA damage-induced cell cycle checkpoints are signal transduction pathways that monitor the integrity of the genome in coordination with cell cycle transitions (Abraham, 2001; Zhou and Elledge, 2000). The mammalian DNA damage checkpoints are usually composed of the G1/S checkpoint, the intra-S checkpoint and the G2/M checkpoint (Figure 2 below).

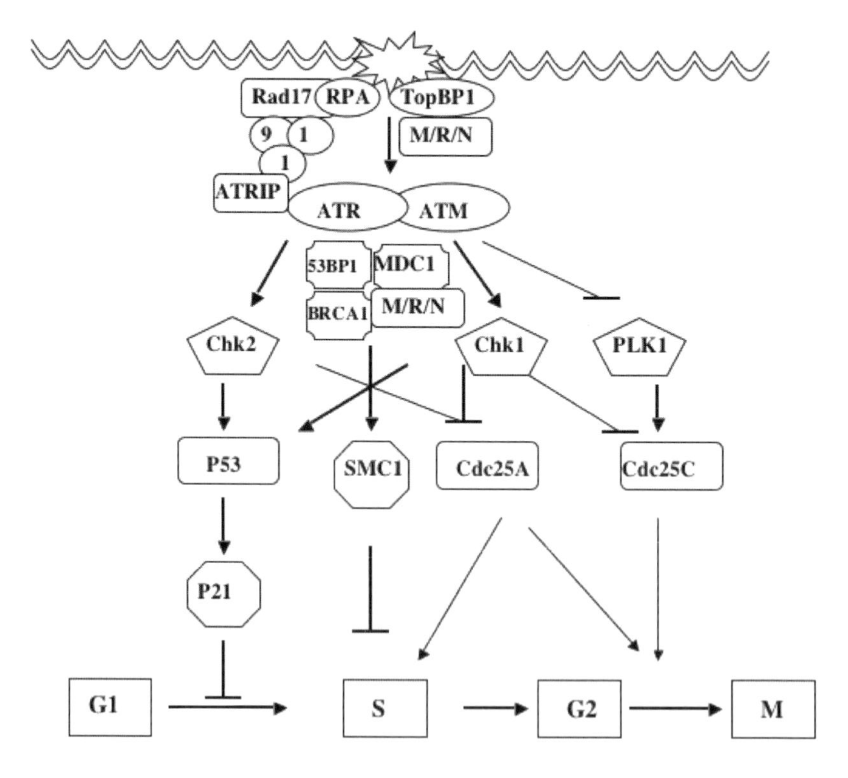

Figure 2. DNA Damage Checkpoint Pathway.

5.1 The G1/S Checkpoint

The G1/S checkpoint helps ensure the integrity of the genome before the genetic information is replicated. p53 is the major effector of the G1/S checkpoint (Fei and El-Deiry, 2003; Giaccia and Kastan, 1998; Ko and Prives, 1996; Vogelstein et al., 2000). p53 acts as a transcription factor that induces the expression of many proteins involved in DNA damage responses. A major target of p53 in the G1/S checkpoint is the CDK inhibitor p21, which binds CDK2/CyclinA/E to block theG1/S transition (el-Deiry et al., 1994; el-Deiry et al., 1993; Harper et al., 1993; Xiong et al., 1993). p53- or p21-deficient cells show complete abolishment of the G1/S checkpoint (Deng et al., 1995; Kuerbitz et al., 1992), suggesting the critical role of the p53-p21 pathway in the G1/S checkpoint. Intriguingly, in contrast to p53 knockout mice, p21 knock out mice did not show increased tumour incidence (Deng et al., 1995), suggesting that inactivation of the G1/S checkpoint is not sufficient for the development of tumours.

p53 activity is mainly regulated by its stability. In normal cycling cells, p53 levels are kept low by MDM2, which is an E3 ubiquitin ligase. MDM2 binds p53, resulting in the ubiquitination and degradation of p53. In addition to MDM2, p53 polyubiquitination and degradation also requires p300 and YY1, which work together with MDM2 to promote polyubiquination of p53 (Grossman et al., 2003; Sui et al., 2004). In response to DNA damage, p53 stability is significantly increased. The stability of p53 can be regulated by several mechanisms. Chk2 phosphorylates p53 at Ser20, causing the disruption of a p53-MDM2 interaction and thus stabilizing p53 (Chehab et al., 2000; Chehab et al., 1999; Hirao et al., 2000). In addition, ATM phosphorylates MDM2 at Ser395, resulting in the stabilization of p53 by an unknown mechanism (Maya et al., 2001). Therefore, a signalling pathway of ATM-Chk2-p53 is critical for p53 stabilization and the G1/S checkpoint. Genetic studies confirmed the critical role of this pathway. Both ATM$^{-/-}$ and Chk2$^{-/-}$ cells show defective p53 stabilization and G1/S checkpoint (Hirao et al., 2002; Takai et al., 2002; Xu and Baltimore, 1996). However, conflicting results have been reported recently. Using Chk2 knockout or knockdown cells, several groups reported intact p53 responses following DNA damage (Ahn et al., 2003; Jack et al., 2002; Jallepalli et al., 2003), suggesting the existence of Chk2-independent regulation of p53 following DNA damage. Further investigation is required to address the exact role of Chk2 in the G1/S checkpoint.

In addition to increased stability, p53 transcriptional activity is also regulated by posttranslational modifications. p53 is phosphorylated by ATM and ATR at Ser15 in response to IR and UV (Banin et al., 1998; Canman et al., 1998; Khanna et al., 1998; Tibbetts et al., 1999). The phosphorylation of Ser15 is not critical for p53 stability, instead it modulates p53 transcriptional activity (Dumaz and Meek, 1999). In addition, Pin1 binds to phosphorylated p53 and modulates the transcription activity of p53 (Zacchi et al., 2002; Zheng et al., 2002). As a result, the G1/S checkpoint in response to UV is defective in Pin1$^{-/-}$ cells (Zheng et al., 2002). Thus, it is apparent that the G1/S checkpoint is controlled by the activation of p53 following DNA damage through several different pathways.

5.2 The Intra-S Phase Checkpoint

The intra-S checkpoint blocks the firing of replication forks following DNA damage. Failure of the intra-S phase checkpoint results in a radioresistant DNA synthesis (RDS) phenotype. One of the characteristic phenotypes of AT cells is RDS (Painter and Young, 1980), suggesting a critical role of ATM in the intra-S phase checkpoint. ATM activates Chk2, which in turn phosphorylates Cdc25A (Falck et al., 2001). Cdc25A is a dual-specificity phosphatase that is required for the dephosphorylation of the inhibitory phospho-T14Y15 residues of Cdk2. The phosphorylation of Cdc25A at Ser123 by Chk2 results in the down-regulation of Cdc25A (Falck et al., 2001), thus blocking the activation of CDK2/Cyclin E and initiation of replication. These studies suggest an important role of ATM-Chk2-Cdc25A in the intra-S checkpoint. However, the role of Chk2 in the intra-S checkpoint has also been challenged. Although Chk2-deficient HCT15 cells show a RDS phenotype (Falck et al., 2001), cells derived from Chk2 knockout mice do not demonstrate a defective intra-S checkpoint (Hirao et al., 2002; Takai et al., 2002). Instead, Chk1 turns out to be a predominant transducer of the intra-S checkpoint, functioning downstream of ATM. In response to IR, Chk1 phosphorylates Cdc25A at Ser123, 178, 278, and 292 (Sorensen et al., 2003; Zhao et al., 2002a). Mutation of these phosphorylation sites stabilises Cdc25A, resulting in a defective intra-S phase checkpoint. These studies suggest that Cdc25A is an effector of Chk1 in intra-S checkpoint control. The mechanism of Cdc25A down-regulation is emerging. It was found that the hyperphosphorylated Cdc25A is recognised by the F-box protein β-TrCP, which targets Cdc25A for degradation by the Skp1/Cul1/F-box (SCF) protein complex (Busino et al., 2004). Thus, the ATM-Chk1-Cdc25A pathway appears to be the major pathway controlling the intra-S phase checkpoint.

Cells from NBS and ADLR patients also show a RDS phenotype similar to those derived from AT patients (Carney et al., 1998; Stewart et al., 1999), suggest that the M/R/N complex is important for the intra-S checkpoint. NBS1 is phosphorylated by ATM in response to DNA damage and the phosphorylation of BRCA1 and NBS1 are important for intra-S phase checkpoint control (Gatei et al., 2000; Lim et al., 2000; Xu et al., 2001a). NBS1 regulates the intra-S phase checkpoint potentially through the regulation of SMC1 phosphorylation (Kim et al., 2002; Kitagawa et al., 2004). SMC1 is a member of cohesin proteins that are involved in holding sister chromatid together following replication, and has emerged as another regulator involved in the intra-S checkpoint. In response to IR, ATM phosphorylates SMC1 at Ser957 and Ser966 in a NBS1-dependent manner (Kim et al., 2002; Yazdi et al., 2002). Overexpression of phosphorylation-mutant of SMC results in defective intra-S checkpoint and increased radiation sensitivity. Additionally, targeted knock-in of SMC phosphorylation-mutant confirms the critical role of SMC phosphorylation in the intra-S phase checkpoint (Kitagawa et al., 2004). However, the molecular mechanism by which SMC regulates the intra-S checkpoint has not yet been elucidated.

Recent studies suggest important roles of mediator proteins in the intra-S phase checkpoint control concerning both the Cdc25A and SMC pathway. The detailed molecular mechanism of how these mediator proteins regulate the intra-S checkpoint is still not clear. Given the putative structural function of the mediator proteins, it is possible that they facilitate interactions between downstream signalling molecules and upstream kinases. BRCA1 has been shown to regulate the intra-S checkpoint (Xu et al., 2002a), probably by affecting SMC1 phosphorylation (Kitagawa et al., 2004). 53BP1 has also been shown to regulate the intra-S checkpoint by regulating BRCA1 and Chk2 phosphorylation and foci formation (Wang et al., 2002). However, cells from 53BP1 knockout mice only show a moderate defect in the intra-S phase checkpoint (Ward et al., 2003b), suggesting a minor role of 53BP1 in this checkpoint. MDC1 interacts with both BRCA1 and NBS1, and regulates the foci formation of NBS1 and BRCA1 in response to IR (Goldberg et al., 2003; Lou et al., 2003a; Stewart et al., 2003). Interestingly, MDC1 also regulates the phosphorylation of SMC1 and BRCA1 following DNA damage, but not that of NBS1. In addition, MDC1 interacts with activated Chk2 through the FHA domain of MDC1 (Lou et al., 2003b). Therefore, MDC1 interacts with the two parallel pathways that regulate the intra-S checkpoint (Falck et al., 2002). It is no surprising that down-regulation of MDC1 results in a defective intra-S checkpoint (Goldberg et al., 2003; Lou et al., 2003b; Stewart et al., 2003). These results confirm the important role of mediator proteins in the activation and phosphorylation of downstream transducers and effectors. A twist to this paradigm is the recent finding that mediator proteins also regulate the localisation and autophosphorylation of ATM. 53BP1 and MDC1 have been shown to regulate ATM autophosphorylation (Mochan et al., 2003). However, it may be that they are not required for ATM activation *per se*. Instead, they may function to recruit activated ATM to the sites of DNA damage and thus amplify ATM signalling through a positive feedback loop.

5.3 The G2/M Checkpoint

The G2/M checkpoint arrests cells at G2 phase, and prevents the propagation of unrepaired DNA to the next generation. Cdc25C was originally believed to be a major target involved in the G2/M checkpoint. Like Cdc25A, Cdc25C is a phosphatase that removes inhibitory phosphorylation from T14Y15 of CDK1, resulting in CDK1 activation. Both Chk2 and Chk1 have been implicated in phosphorylating Cdc25C (Matsuoka et al., 1998; Sanchez et al., 1997; Zhou et al., 2000). The phosphorylated Cdc25C is sequestered to the cytoplasm by the binding of 14-3-3 proteins (Peng et al., 1997). As a result, CDK1 remains inactive. Unexpectedly, though, CDK1 phosphorylation and cellular responses to DNA damage are normal in Cdc25C$^{-/-}$ cells (Chen et al., 2001), suggesting the existence of a redundant pathway involved in the G2/M checkpoint. Recently, Cdc25A has been implicated in G2/M progression, targeted by Chk1 (Zhao et al., 2002a). Therefore, both Cdc25A and Cdc25C could contribute to regulate the G2/M transition. Consistent with a role of Chk1 in Cdc25A/C regulation, Chk1$^{-/-}$ cells show a defective G2/M checkpoint (Liu et al., 2000; Takai et al., 2000), while Chk2$^{-/-}$ cells have intact G2/M checkpoint control (Hirao et al., 2002; Takai et al., 2002). Therefore, similar to of the situation with the intra-S checkpoint, Chk1 probably plays a predominant role in the G2/M checkpoint. Upstream of Chk1, ATM and ATR are also essential for the G2/M checkpoint (Brown and Baltimore, 2003; Liu et al., 2000). Genetic studies using knockout mice suggest that ATR and ATM cooperate to cause an early G2 arrest but ATR plays a major role in the late G2 phase arrest (Brown and Baltimore, 2003). Double deletion of ATR and ATM eliminates all G2 arrests following DNA damage, supporting the hypothesis that both of these kinases are key regulators of the DNA damage response.

ATM/ATR also regulate G2/M entry through controlling the activity of Polo like kinase 1 (PLK1). PLK1 positively regulates mitotic entry by phosphorylating Cdc25C (Kumagai and Dunphy, 1996; Qian et al., 1998). Following DNA damage, Plk1 activity is inhibited in an ATM- and ATR-dependent manner (Smits et al., 2000; van Vugt et al., 2001). Expression of a constitutively active mutant of PLK1 overrides G2 arrest in response to IR (Smits et al., 2000), suggesting that inhibition of PLK1 activity is required for the G2/M checkpoint.

As for all of the checkpoint responses, mediator proteins are important for the activation of the G2/M checkpoint through regulating the phosphorylation and activation of downstream transducers. BRCA1 regulates the G2/M checkpoint through its role in Chk1 activation in response to IR (Yarden et al., 2002). MDC1 regulates BRCA1 foci formation and phosphorylation in response to IR. Knockdown of MDC1 using siRNA results in decreased BRCA1 and Chk1 phosphorylation and a defective G2/M checkpoint (Lou et al., 2003a; Stewart et al., 2003). Cells from $H2AX^{-/-}$ and $53BP1^{-/-}$ mice also show a defective G2/M checkpoint at low doses of IR (Fernandez-Capetillo et al., 2002).

5.4 Apoptosis

Apoptosis is an important aspect of the DNA damage response. Cells containing extensive DNA damage will undergo apoptosis to prevent the propagation of incorrect genetic material. The essential role of p53 in apoptosis has been well established (Burns and El-Deiry, 1999; Ko and Prives, 1996; Vousden and Lu, 2002). Similar to that of the G1-S checkpoint, the Chk2-p53 pathway is critical for DNA damage-induced apoptosis (Figure 3). Chk2 phosphorylates and stabilises p53 in response to DNA damage. $Chk2^{-/-}$ cells show defective p53 stabilisation and defective transcription of p53 target genes (Hirao et al., 2002; Hirao et al., 2000; Jack et al., 2002; Takai et al., 2002). As a result, IR-induced apoptosis is abolished in $Chk2^{-/-}$ cells (Hirao et al., 2002; Hirao et al., 2000; Jack et al., 2002). Numerous p53 target genes involved in apoptosis, including Noxa, Bax, p53AIP1 and PUMA, have been identified (Vousden and Lu, 2002). Among them, PUMA (p53 upregulated modulator of apoptosis) and Noxa are "BH3-only" proteins that have been proposed to play essential role in the initiation of apoptosis. Importantly, disruption of PUMA and to a lesser degree Noxa, abolish the IR-induced apoptosis (Jeffers et al., 2003;

Villunger et al., 2003), establishing the important role of PUMA and Noxa in p53-dependent apoptosis following DNA damage. Similar to $p21^{-/-}$ mice, $PUMA^{-/-}$ mice did not show increased tumour incidence in a time-frame when tumours in $p53^{-/-}$ mice were already evident. These results suggest that disruption of either the G1/S checkpoint or apoptosis alone is not sufficient for tumorigenesis. It will be interesting to know the tumour incidence of p21/PUMA double knockout mice. Besides its ability to upregulate genes involved in apoptosis, p53 has also been shown to translocate to mitochondria and directly induce apoptosis (Dumont et al., 2003; Mihara et al., 2003). In addition to regulating p53, Chk2 also phosphorylates and stabilises E2F1 (Stevens et al., 2003). E2F1 in turn induces apoptosis, probably by up-regulating Arf, which could stabilise p53 (Stevens et al., 2003). A p53-independent Chk2-PML pathway has also been reported to be important for IR-induced apoptosis (Yang et al., 2002).

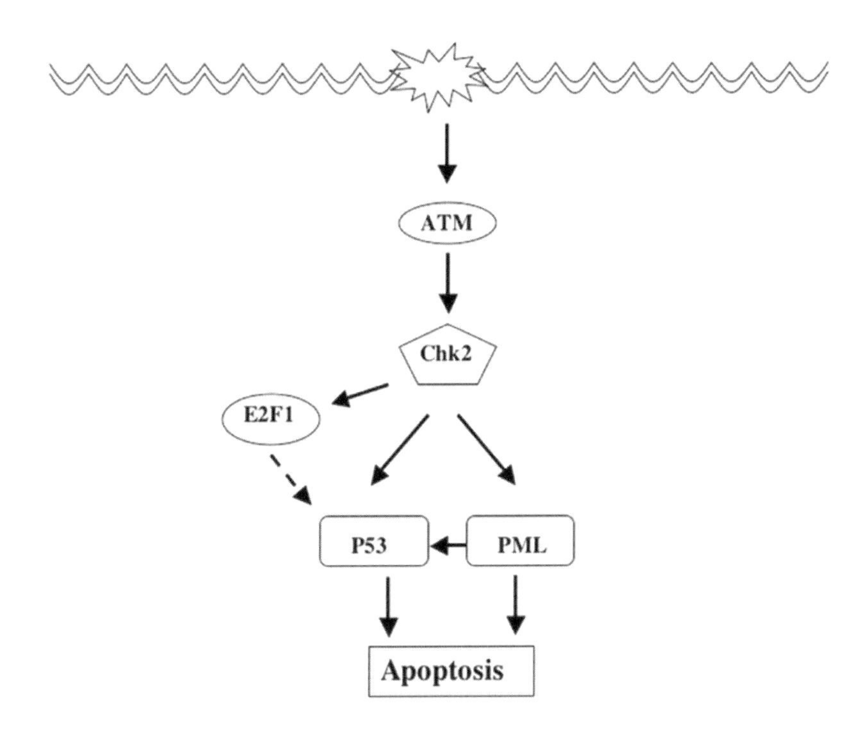

Figure 3. IR Induced Apoptosis.

6. CONCLUSIONS

The DNA damage response pathway is critical for the maintenance of genomic stability. Many components of this pathway are tumour suppressors, such as p53, ATM, BRCA1, NBS1, Mre11 and Chk2, underlining the critical role of the DNA damage response pathway in tumour suppression. The elucidation of the DNA damage response pathway will not only help us understand the mechanism of tumour suppression, but also allow for the discovery of new drug targets for cancer therapy. A recent example is the development of MDM2 inhibitors that could activate the p53 pathway in cancer cells (Vassilev et al., 2004). With the help of proteomics and advances in imaging techniques, we should have a better understanding of the molecular events mediating the DNA damage response in the near future.

REFERENCES

Abraham, R.T. 1996. Phosphatidylinositol 3-kinase related kinases. *Curr Opin Immunol.* 8:412-8.

Abraham, R.T. 2001. Cell cycle checkpoint signalling through the ATM and ATR kinases. *Genes Dev.* 15:2177-96.

Ahn, J., M. Urist, and C. Prives. 2003. Questioning the role of checkpoint kinase 2 in the p53 DNA damage response. *J Biol Chem.* 278:20480-9.

Ahn, J.Y., X. Li, H.L. Davis, and C.E. Canman. 2002. Phosphorylation of threonine 68 promotes oligomerization and autophosphorylation of the Chk2 protein kinase via the forkhead-associated domain. *J Biol Chem.* 277:19389-95.

Ahn, J.Y., J.K. Schwarz, H. Piwnica-Worms, and C.E. Canman. 2000. Threonine 68 phosphorylation by ataxia-telangiectasia mutated is required for efficient activation of Chk2 in response to ionizing radiation. *Cancer Res.* 60:5934-6.

Anderson, L., C. Henderson, and Y. Adachi. 2001. Phosphorylation and rapid relocalization of 53BP1 to nuclear foci upon DNA damage. *Mol Cell Biol.* 21:1719-29.

Bakkenist, C.J., and M.B. Kastan. 2003. DNA damage activates ATM through intermolecular autophosphorylation and dimer dissociation. *Nature.* 421:499-506.

Banin, S., L. Moyal, S. Shieh, Y. Taya, C.W. Anderson, L. Chessa, N.I. Smorodinsky, C. Prives, Y. Reiss, Y. Shiloh, and Y. Ziv. 1998. Enhanced phosphorylation of p53 by ATM in response to DNA damage. *Science.* 281:1674-7.

Bao, S., R.S. Tibbetts, K.M. Brumbaugh, Y. Fang, D.A. Richardson, A. Ali, S.M. Chen, R.T. Abraham, and X.F. Wang. 2001. ATR/ATM-mediated phosphorylation of human Rad17 is required for genotoxic stress responses. *Nature.* 411:969-74.

Barlow, C., S. Hirotsune, R. Paylor, M. Liyanage, M. Eckhaus, F. Collins, Y. Shiloh, J.N. Crawley, T. Ried, D. Tagle, and A. Wynshaw-Boris. 1996. Atm-deficient mice: a paradigm of ataxia-telangiectasia. *Cell.* 86:159-71.

Barr, S.M., C.G. Leung, E.E. Chang, and K.A. Cimprich. 2003. ATR kinase activity regulates the intranuclear translocation of ATR and RPA following ionizing radiation. *Curr Biol.* 13:1047-51.

Bartek, J., J. Falck, and J. Lukas. 2001. CHK2 kinase--a busy messenger. *Nat Rev Mol Cell Biol.* 2:877-86.

Bassing, C.H., K.F. Chua, J. Sekiguchi, H. Suh, S.R. Whitlow, J.C. Fleming, B.C. Monroe, D.N. Ciccone, C. Yan, K. Vlasakova, D.M. Livingston, D.O. Ferguson, R. Scully, and F.W. Alt. 2002. Increased ionizing radiation sensitivity and genomic instability in the absence of histone H2AX. *Proc Natl Acad Sci U S A.* 99:8173-8.

Bassing, C.H., H. Suh, D.O. Ferguson, K.F. Chua, J. Manis, M. Eckersdorff, M. Gleason, R. Bronson, C. Lee, and F.W. Alt. 2003. Histone H2AX: a dosage-dependent suppressor of oncogenic translocations and tumors. *Cell.* 114:359-70.

Bell, D.W., J.M. Varley, T.E. Szydlo, D.H. Kang, D.C. Wahrer, K.E. Shannon, M. Lubratovich, S.J. Verselis, K.J. Isselbacher, J.F. Fraumeni, J.M. Birch, F.P. Li, J.E. Garber, and D.A. Haber. 1999. Heterozygous germ line hCHK2 mutations in Li-Fraumeni syndrome. *Science*. 286:2528-31.

Bermudez, V.P., L.A. Lindsey-Boltz, A.J. Cesare, Y. Maniwa, J.D. Griffith, J. Hurwitz, and A. Sancar. 2003. Loading of the human 9-1-1 checkpoint complex onto DNA by the checkpoint clamp loader hRad17-replication factor C complex in vitro. *Proc Natl Acad Sci U S A*. 100:1633-8.

Bomgarden, R.D., D. Yean, M.C. Yee, and K.A. Cimprich. 2004. A novel protein activity mediates DNA binding of an ATR-ATRIP complex. *J Biol Chem*. 279:13346-53.

Bradbury, J.M., and S.P. Jackson. 2003. The complex matter of DNA double-strand break detection. *Biochem Soc Trans*. 31:40-4.

Brown, E.J., and D. Baltimore. 2000. ATR disruption leads to chromosomal fragmentation and early embryonic lethality. *Genes Dev*. 14:397-402.

Brown, E.J., and D. Baltimore. 2003. Essential and dispensable roles of ATR in cell cycle arrest and genome maintenance. *Genes Dev*. 17:615-28.

Brumbaugh, K.M., D.M. Otterness, C. Geisen, V. Oliveira, J. Brognard, X. Li, F. Lejeune, R.S. Tibbetts, L.E. Maquat, and R.T. Abraham. 2004. The mRNA surveillance protein hSMG-1 functions in genotoxic stress response pathways in mammalian cells. *Mol Cell*. 14:585-98.

Brzovic, P.S., J.R. Keeffe, H. Nishikawa, K. Miyamoto, D. Fox, 3rd, M. Fukuda, T. Ohta, and R. Klevit. 2003. Binding and recognition in the assembly of an active BRCA1/BARD1 ubiquitin-ligase complex. *Proc Natl Acad Sci U S A*. 100:5646-51.

Burma, S., B.P. Chen, M. Murphy, A. Kurimasa, and D.J. Chen. 2001. ATM phosphorylates histone H2AX in response to DNA double-strand breaks. *J Biol Chem*. 276:42462-7.

Burns, T.F., and W.S. El-Deiry. 1999. The p53 pathway and apoptosis. *J Cell Physiol*. 181:231-9.

Busino, L., M. Chiesa, G.F. Draetta, and M. Donzelli. 2004. Cdc25A phosphatase: combinatorial phosphorylation, ubiquitylation and proteolysis. *Oncogene*. 23:2050-6.

Canman, C.E., D.S. Lim, K.A. Cimprich, Y. Taya, K. Tamai, K. Sakaguchi, E. Appella, M.B. Kastan, and J.D. Siliciano. 1998. Activation of the ATM kinase by ionizing radiation and phosphorylation of p53. *Science*. 281:1677-9.

Cantor, S.B., D.W. Bell, S. Ganesan, E.M. Kass, R. Drapkin, S. Grossman, D.C. Wahrer, D.C. Sgroi, W.S. Lane, D.A. Haber, and D.M. Livingston. 2001. BACH1, a novel helicase-like protein, interacts directly with BRCA1 and contributes to its DNA repair function. *Cell*. 105:149-60.

Carney, J.P., R.S. Maser, H. Olivares, E.M. Davis, M. Le Beau, J.R. Yates, 3rd, L. Hays, W.F. Morgan, and J.H. Petrini. 1998. The hMre11/hRad50 protein complex and Nijmegen breakage syndrome: linkage of double-strand break repair to the cellular DNA damage response. *Cell*. 93:477-86.

Carson, C.T., R.A. Schwartz, T.H. Stracker, C.E. Lilley, D.V. Lee, and M.D. Weitzman. 2003. The Mre11 complex is required for ATM activation and the G2/M checkpoint. *EMBO J*. 22:6610-20.

Celeste, A., S. Difilippantonio, M.J. Difilippantonio, O. Fernandez-Capetillo, D.R. Pilch, O.A. Sedelnikova, M. Eckhaus, T. Ried, W.M. Bonner, and A. Nussenzweig. 2003a. H2AX haploinsufficiency modifies genomic stability and tumor susceptibility. *Cell*. 114:371-83.

Celeste, A., O. Fernandez-Capetillo, M.J. Kruhlak, D.R. Pilch, D.W. Staudt, A. Lee, R.F. Bonner, W.M. Bonner, and A. Nussenzweig. 2003b. Histone H2AX phosphorylation is dispensable for the initial recognition of DNA breaks. *Nat Cell Biol*. 5:675-9.

Celeste, A., S. Petersen, P.J. Romanienko, O. Fernandez-Capetillo, H.T. Chen, O.A. Sedelnikova, B. Reina-San-Martin, V. Coppola, E. Meffre, M.J. Difilippantonio, C. Redon, D.R. Pilch, A. Olaru, M. Eckhaus, R.D. Camerini-Otero, L. Tessarollo, F. Livak,

K. Manova, W.M. Bonner, M.C. Nussenzweig, and A. Nussenzweig. 2002. Genomic instability in mice lacking histone H2AX. *Science*. 296:922-7.

Cerosaletti, K.M., and P. Concannon. 2003. Nibrin forkhead-associated domain and breast cancer C-terminal domain are both required for nuclear focus formation and phosphorylation. *J Biol Chem*. 278:21944-51.

Chehab, N.H., A. Malikzay, M. Appel, and T.D. Halazonetis. 2000. Chk2/hCds1 functions as a DNA damage checkpoint in G(1) by stabilizing p53. *Genes Dev*. 14:278-88.

Chehab, N.H., A. Malikzay, E.S. Stavridi, and T.D. Halazonetis. 1999. Phosphorylation of Ser-20 mediates stabilization of human p53 in response to DNA damage. *Proc Natl Acad Sci U S A*. 96:13777-82.

Chen, A., F.E. Kleiman, J.L. Manley, T. Ouchi, and Z.Q. Pan. 2002. Autoubiquitination of the BRCA1*BARD1 RING ubiquitin ligase. *J Biol Chem*. 277:22085-92.

Chen, M.S., J. Hurov, L.S. White, T. Woodford-Thomas, and H. Piwnica-Worms. 2001. Absence of apparent phenotype in mice lacking Cdc25C protein phosphatase. *Mol Cell Biol*. 21:3853-61.

Chen, Y., and Y. Sanchez. 2004. Chk1 in the DNA damage response: conserved roles from yeasts to mammals. *DNA Repair (Amst)*. 3:1025-32.

Chini, C.C., and J. Chen. 2003. Human claspin is required for replication checkpoint control. *J Biol Chem*. 278:30057-62.

Cliby, W.A., C.J. Roberts, K.A. Cimprich, C.M. Stringer, J.R. Lamb, S.L. Schreiber, and S.H. Friend. 1998. Overexpression of a kinase-inactive ATR protein causes sensitivity to DNA-damaging agents and defects in cell cycle checkpoints. *EMBO J*. 17:159-69.

Cortez, D., S. Guntuku, J. Qin, and S.J. Elledge. 2001. ATR and ATRIP: partners in checkpoint signalling. *Science*. 294:1713-6.

Costanzo, V., T. Paull, M. Gottesman, and J. Gautier. 2004. Mre11 assembles linear DNA fragments into DNA damage signalling complexes. *PLoS Biol*. 2:E110.

de Jager, M., J. van Noort, D.C. van Gent, C. Dekker, R. Kanaar, and C. Wyman. 2001. Human Rad50/Mre11 is a flexible complex that can tether DNA ends. *Mol Cell*. 8:1129-35.

de Klein, A., M. Muijtjens, R. van Os, Y. Verhoeven, B. Smit, A.M. Carr, A.R. Lehmann, and J.H. Hoeijmakers. 2000. Targeted disruption of the cell-cycle checkpoint gene ATR leads to early embryonic lethality in mice. *Curr Biol*. 10:479-82.

Deng, C., P. Zhang, J.W. Harper, S.J. Elledge, and P. Leder. 1995. Mice lacking p21CIP1/WAF1 undergo normal development, but are defective in G1 checkpoint control. *Cell*. 82:675-84.

Deng, C.X., and S.G. Brodie. 2000. Roles of BRCA1 and its interacting proteins. *Bioessays*. 22:728-37.

DiTullio, R.A., Jr., T.A. Mochan, M. Venere, J. Bartkova, M. Sehested, J. Bartek, and T.D. Halazonetis. 2002. 53BP1 functions in an ATM-dependent checkpoint pathway that is constitutively activated in human cancer. *Nat Cell Biol*. 4:998-1002.

Dumaz, N., and D.W. Meek. 1999. Serine15 phosphorylation stimulates p53 transactivation but does not directly influence interaction with HDM2. *EMBO J*. 18:7002-10.

Dumont, P., J.I. Leu, A.C. Della Pietra, 3rd, D.L. George, and M. Murphy. 2003. The codon 72 polymorphic variants of p53 have markedly different apoptotic potential. *Nat Genet*. 33:357-65.

Durocher, D., and S.P. Jackson. 2002. The FHA domain. *FEBS Lett*. 513:58-66.

el-Deiry, W.S., J.W. Harper, P.M. O'Connor, V.E. Velculescu, C.E. Canman, J. Jackman, J.A. Pietenpol, M. Burrell, D.E. Hill, Y. Wang, and et al. 1994. WAF1/CIP1 is induced in p53-mediated G1 arrest and apoptosis. *Cancer Res*. 54:1169-74.

el-Deiry, W.S., T. Tokino, V.E. Velculescu, D.B. Levy, R. Parsons, J.M. Trent, D. Lin, W.E. Mercer, K.W. Kinzler, and B. Vogelstein. 1993. WAF1, a potential mediator of p53 tumor suppression. *Cell*. 75:817-25.

Elson, A., Y. Wang, C.J. Daugherty, C.C. Morton, F. Zhou, J. Campos-Torres, and P. Leder. 1996. Pleiotropic defects in ataxia-telangiectasia protein-deficient mice. *Proc Natl Acad Sci U S A*. 93:13084-9.

Falck, J., N. Mailand, R.G. Syljuasen, J. Bartek, and J. Lukas. 2001. The ATM-Chk2-Cdc25A checkpoint pathway guards against radioresistant DNA synthesis. *Nature*. 410:842-7.

Falck, J., J.H. Petrini, B.R. Williams, J. Lukas, and J. Bartek. 2002. The DNA damage-dependent intra-S phase checkpoint is regulated by parallel pathways. *Nat Genet*. 30:290-4.

Fei, P., and W.S. El-Deiry. 2003. P53 and radiation responses. *Oncogene*. 22:5774-83.

Fernandez-Capetillo, O., A. Celeste, and A. Nussenzweig. 2003a. Focusing on foci: H2AX and the recruitment of DNA damage response factors. *Cell Cycle*. 2:426-7.

Fernandez-Capetillo, O., H.T. Chen, A. Celeste, I. Ward, P.J. Romanienko, J.C. Morales, K. Naka, Z. Xia, R.D. Camerini-Otero, N. Motoyama, P.B. Carpenter, W.M. Bonner, J. Chen, and A. Nussenzweig. 2002. DNA damage-induced G2-M checkpoint activation by histone H2AX and 53BP1. *Nat Cell Biol*. 4:993-7.

Fernandez-Capetillo, O., S.K. Mahadevaiah, A. Celeste, P.J. Romanienko, R.D. Camerini-Otero, W.M. Bonner, K. Manova, P. Burgoyne, and A. Nussenzweig. 2003b. H2AX is required for chromatin remodeling and inactivation of sex chromosomes in male mouse meiosis. *Dev Cell*. 4:497-508.

Franchitto, A., P. Pichierri, R. Piergentili, M. Crescenzi, M. Bignami, and F. Palitti. 2003. The mammalian mismatch repair protein MSH2 is required for correct MRE11 and RAD51 relocalization and for efficient cell cycle arrest induced by ionizing radiation in G2 phase. *Oncogene*. 22:2110-20.

Gatei, M., K. Sloper, C. Sorensen, R. Syljuasen, J. Falck, K. Hobson, K. Savage, J. Lukas, B.B. Zhou, J. Bartek, and K.K. Khanna. 2003. Ataxia-telangiectasia-mutated (ATM) and NBS1-dependent phosphorylation of Chk1 on Ser-317 in response to ionizing radiation. *J Biol Chem*. 278:14806-11.

Gatei, M., D. Young, K.M. Cerosaletti, A. Desai-Mehta, K. Spring, S. Kozlov, M.F. Lavin, R.A. Gatti, P. Concannon, and K. Khanna. 2000. ATM-dependent phosphorylation of nibrin in response to radiation exposure. *Nat Genet*. 25:115-9.

Giaccia, A.J., and M.B. Kastan. 1998. The complexity of p53 modulation: emerging patterns from divergent signals. *Genes Dev*. 12:2973-83.

Goldberg, M., M. Stucki, J. Falck, D. D'Amours, D. Rahman, D. Pappin, J. Bartek, and S.P. Jackson. 2003. MDC1 is required for the intra-S-phase DNA damage checkpoint. *Nature*. 421:952-6.

Gowen, L.C., B.L. Johnson, A.M. Latour, K.K. Sulik, and B.H. Koller. 1996. Brca1 deficiency results in early embryonic lethality characterized by neuroepithelial abnormalities. *Nat Genet*. 12:191-4.

Griffith, J.D., L.A. Lindsey-Boltz, and A. Sancar. 2002. Structures of the human Rad17-replication factor C and checkpoint Rad 9-1-1 complexes visualized by glycerol spray/low voltage microscopy. *J Biol Chem*. 277:15233-6.

Grossman, S.R., M.E. Deato, C. Brignone, H.M. Chan, A.L. Kung, H. Tagami, Y. Nakatani, and D.M. Livingston. 2003. Polyubiquitination of p53 by a ubiquitin ligase activity of p300. *Science*. 300:342-4.

Hahn, W.C., and R.A. Weinberg. 2002. Rules for making human tumor cells. *N Engl J Med*. 347:1593-603.

Hakem, R., J.L. de la Pompa, C. Sirard, R. Mo, M. Woo, A. Hakem, A. Wakeham, J. Potter, A. Reitmair, F. Billia, E. Firpo, C.C. Hui, J. Roberts, J. Rossant, and T.W. Mak. 1996. The tumor suppressor gene Brca1 is required for embryonic cellular proliferation in the mouse. *Cell*. 85:1009-23.

Harper, J.W., G.R. Adami, N. Wei, K. Keyomarsi, and S.J. Elledge. 1993. The p21 Cdk-interacting protein Cip1 is a potent inhibitor of G1 cyclin-dependent kinases. *Cell*. 75:805-16.

Hashizume, R., M. Fukuda, I. Maeda, H. Nishikawa, D. Oyake, Y. Yabuki, H. Ogata, and T. Ohta. 2001. The RING heterodimer BRCA1-BARD1 is a ubiquitin ligase inactivated by a breast cancer derived mutation. *J Biol Chem*. 276:14537-40.

Hirao, A., A. Cheung, G. Duncan, P.M. Girard, A.J. Elia, A. Wakeham, H. Okada, T. Sarkissian, J.A. Wong, T. Sakai, E. De Stanchina, R.G. Bristow, T. Suda, S.W. Lowe, P.A. Jeggo, S.J. Elledge, and T.W. Mak. 2002. Chk2 is a tumor suppressor that regulates apoptosis in both an ataxia-telangiectasia mutated (ATM)-dependent and an ATM-independent manner. *Mol Cell Biol*. 22:6521-32.

Hirao, A., Y.Y. Kong, S. Matsuoka, A. Wakeham, J. Ruland, H. Yoshida, D. Liu, S.J. Elledge, and T.W. Mak. 2000. DNA damage-induced activation of p53 by the checkpoint kinase Chk2. *Science*. 287:1824-7.

Hopfner, K.P., L. Craig, G. Moncalian, R.A. Zinkel, T. Usui, B.A. Owen, A. Karcher, B. Henderson, J.L. Bodmer, C.T. McMurray, J.P. Carney, J.H. Petrini, and J.A. Tainer. 2002. The Rad50 zinc-hook is a structure joining Mre11 complexes in DNA recombination and repair. *Nature*. 418:562-6.

Horejsi, Z., J. Falck, C.J. Bakkenist, M.B. Kastan, J. Lukas, and J. Bartek. 2004. Distinct functional domains of Nbs1 modulate the timing and magnitude of ATM activation after low doses of ionizing radiation. *Oncogene*. 23:3122-7.

Iliakis, G., Y. Wang, J. Guan, and H. Wang. 2003. DNA damage checkpoint control in cells exposed to ionizing radiation. *Oncogene*. 22:5834-47.

Iwabuchi, K., P.L. Bartel, B. Li, R. Marraccino, and S. Fields. 1994. Two cellular proteins that bind to wild-type but not mutant p53. *Proc Natl Acad Sci U S A*. 91:6098-102.

Jack, M.T., R.A. Woo, A. Hirao, A. Cheung, T.W. Mak, and P.W. Lee. 2002. Chk2 is dispensable for p53-mediated G1 arrest but is required for a latent p53-mediated apoptotic response. *Proc Natl Acad Sci U S A*. 99:9825-9.

Jackson, A.P., H. Eastwood, S.M. Bell, J. Adu, C. Toomes, I.M. Carr, E. Roberts, D.J. Hampshire, Y.J. Crow, A.J. Mighell, G. Karbani, H. Jafri, Y. Rashid, R.F. Mueller, A.F. Markham, and C.G. Woods. 2002. Identification of microcephalin, a protein implicated in determining the size of the human brain. *Am J Hum Genet*. 71:136-42.

Jallepalli, P.V., C. Lengauer, B. Vogelstein, and F. Bunz. 2003. The Chk2 tumor suppressor is not required for p53 responses in human cancer cells. *J Biol Chem*. 278:20475-9.

Jeffers, J.R., E. Parganas, Y. Lee, C. Yang, J. Wang, J. Brennan, K.H. MacLean, J. Han, T. Chittenden, J.N. Ihle, P.J. McKinnon, J.L. Cleveland, and G.P. Zambetti. 2003. Puma is an essential mediator of p53-dependent and -independent apoptotic pathways. *Cancer Cell*. 4:321-8.

Kastan, M.B., and D.S. Lim. 2000. The many substrates and functions of ATM. *Nat Rev Mol Cell Biol*. 1:179-86.

Kentsis, A., R.E. Gordon, and K.L. Borden. 2002. Control of biochemical reactions through supramolecular RING domain self-assembly. *Proc Natl Acad Sci U S A*. 99:15404-9.

Kerr, P., and A. Ashworth. 2001. New complexities for BRCA1 and BRCA2. *Curr Biol*. 11:R668-76.

Khanna, K.K., K.E. Keating, S. Kozlov, S. Scott, M. Gatei, K. Hobson, Y. Taya, B. Gabrielli, D. Chan, S.P. Lees-Miller, and M.F. Lavin. 1998. ATM associates with and phosphorylates p53: mapping the region of interaction. *Nat Genet*. 20:398-400.

Kim, S.T., B. Xu, and M.B. Kastan. 2002. Involvement of the cohesin protein, Smc1, in Atm-dependent and independent responses to DNA damage. *Genes Dev*. 16:560-70.

Kitagawa, R., C.J. Bakkenist, P.J. McKinnon, and M.B. Kastan. 2004. Phosphorylation of SMC1 is a critical downstream event in the ATM-NBS1-BRCA1 pathway. *Genes Dev*. 18:1423-38.

Ko, L.J., and C. Prives. 1996. p53: puzzle and paradigm. *Genes Dev*. 10:1054-72.

Kobayashi, J., H. Tauchi, S. Sakamoto, A. Nakamura, K. Morishima, S. Matsuura, T. Kobayashi, K. Tamai, K. Tanimoto, and K. Komatsu. 2002. NBS1 localizes to gamma-H2AX foci through interaction with the FHA/BRCT domain. *Curr Biol*. 12:1846-51.

Kuerbitz, S.J., B.S. Plunkett, W.V. Walsh, and M.B. Kastan. 1992. Wild-type p53 is a cell cycle checkpoint determinant following irradiation. *Proc Natl Acad Sci U S A*. 89:7491-5.

Kumagai, A., and W.G. Dunphy. 1996. Purification and molecular cloning of Plx1, a Cdc25-regulatory kinase from Xenopus egg extracts. *Science*. 273:1377-80.

Kumagai, A., and W.G. Dunphy. 2000. Claspin, a novel protein required for the activation of Chk1 during a DNA replication checkpoint response in Xenopus egg extracts. *Mol Cell*. 6:839-49.

Lam, M.H., Q. Liu, S.J. Elledge, and J.M. Rosen. 2004. Chk1 is haploinsufficient for multiple functions critical to tumor suppression. *Cancer Cell*. 6:45-59.

Lee, C.H., and J.H. Chung. 2001. The hCds1 (Chk2)-FHA domain is essential for a chain of phosphorylation events on hCds1 that is induced by ionizing radiation. *J Biol Chem*. 276:30537-41.

Lee, J., A. Kumagai, and W.G. Dunphy. 2003. Claspin, a Chk1-regulatory protein, monitors DNA replication on chromatin independently of RPA, ATR, and Rad17. *Mol Cell*. 11:329-40.

Lee, J.H., and T.T. Paull. 2004. Direct activation of the ATM protein kinase by the Mre11/Rad50/Nbs1 complex. *Science*. 304:93-6.

Lim, D.S., S.T. Kim, B. Xu, R.S. Maser, J. Lin, J.H. Petrini, and M.B. Kastan. 2000. ATM phosphorylates p95/nbs1 in an S-phase checkpoint pathway. *Nature*. 404:613-7.

Lin, S.Y., and S.J. Elledge. 2003. Multiple tumor suppressor pathways negatively regulate telomerase. *Cell*. 113:881-9.

Lin, S.Y., K. Li, G.S. Stewart, and S.J. Elledge. 2004. Human Claspin works with BRCA1 to both positively and negatively regulate cell proliferation. *Proc Natl Acad Sci U S A*. 101:6484-9.

Lindsey-Boltz, L.A., V.P. Bermudez, J. Hurwitz, and A. Sancar. 2001. Purification and characterization of human DNA damage checkpoint Rad complexes. *Proc Natl Acad Sci U S A*. 98:11236-41.

Liu, C.Y., A. Flesken-Nikitin, S. Li, Y. Zeng, and W.H. Lee. 1996. Inactivation of the mouse Brca1 gene leads to failure in the morphogenesis of the egg cylinder in early postimplantation development. *Genes Dev*. 10:1835-43.

Liu, K., F.T. Lin, J.M. Ruppert, and W.C. Lin. 2003. Regulation of E2F1 by BRCT domain-containing protein TopBP1. *Mol Cell Biol*. 23:3287-304.

Liu, K., Y. Luo, F.T. Lin, and W.C. Lin. 2004. TopBP1 recruits Brg1/Brm to repress E2F1-induced apoptosis, a novel pRb-independent and E2F1-specific control for cell survival. *Genes Dev*. 18:673-86.

Liu, Q., S. Guntuku, X.S. Cui, S. Matsuoka, D. Cortez, K. Tamai, G. Luo, S. Carattini-Rivera, F. DeMayo, A. Bradley, L.A. Donehower, and S.J. Elledge. 2000. Chk1 is an essential kinase that is regulated by Atr and required for the G(2)/M DNA damage checkpoint. *Genes Dev*. 14:1448-59.

Lorick, K.L., J.P. Jensen, S. Fang, A.M. Ong, S. Hatakeyama, and A.M. Weissman. 1999. RING fingers mediate ubiquitin-conjugating enzyme (E2)-dependent ubiquitination. *Proc Natl Acad Sci U S A*. 96:11364-9.

Lou, Z., and J. Chen. 2003. BRCA proteins and DNA damage checkpoints. *Front Biosci*. 8:s718-21.

Lou, Z., C.C. Chini, K. Minter-Dykhouse, and J. Chen. 2003a. Mediator of DNA damage checkpoint protein 1 regulates BRCA1 localization and phosphorylation in DNA damage checkpoint control. *J Biol Chem*. 278:13599-602.

Lou, Z., K. Minter-Dykhouse, X. Wu, and J. Chen. 2003b. MDC1 is coupled to activated CHK2 in mammalian DNA damage response pathways. *Nature*. 421:957-61.

Ludwig, T., D.L. Chapman, V.E. Papaioannou, and A. Efstratiadis. 1997. Targeted mutations of breast cancer susceptibility gene homologs in mice: lethal phenotypes of Brca1, Brca2, Brca1/Brca2, Brca1/p53, and Brca2/p53 nullizygous embryos. *Genes Dev*. 11:1226-41.

Ludwig, T., P. Fisher, S. Ganesan, and A. Efstratiadis. 2001. Tumorigenesis in mice carrying a truncating Brca1 mutation. *Genes Dev*. 15:1188-93.

Lupardus, P.J., T. Byun, M.C. Yee, M. Hekmat-Nejad, and K.A. Cimprich. 2002. A requirement for replication in activation of the ATR-dependent DNA damage checkpoint. *Genes Dev.* 16:2327-32.

Makiniemi, M., T. Hillukkala, J. Tuusa, K. Reini, M. Vaara, D. Huang, H. Pospiech, I. Majuri, T. Westerling, T.P. Makela, and J.E. Syvaoja. 2001. BRCT domain-containing protein TopBP1 functions in DNA replication and damage response. *J Biol Chem.* 276:30399-406.

Mallery, D.L., C.J. Vandenberg, and K. Hiom. 2002. Activation of the E3 ligase function of the BRCA1/BARD1 complex by polyubiquitin chains. *EMBO J.* 21:6755-62.

Manis, J.P., J.C. Morales, Z. Xia, J.L. Kutok, F.W. Alt, and P.B. Carpenter. 2004. 53BP1 links DNA damage-response pathways to immunoglobulin heavy chain class-switch recombination. *Nat Immunol.* 5:481-7.

Manke, I.A., D.M. Lowery, A. Nguyen, and M.B. Yaffe. 2003. BRCT repeats as phosphopeptide-binding modules involved in protein targeting. *Science.* 302:636-9.

Matsuoka, S., M. Huang, and S.J. Elledge. 1998. Linkage of ATM to cell cycle regulation by the Chk2 protein kinase. *Science.* 282:1893-7.

Matsuoka, S., G. Rotman, A. Ogawa, Y. Shiloh, K. Tamai, and S.J. Elledge. 2000. Ataxia-telangiectasia-mutated phosphorylates Chk2 in vivo and in vitro. *Proc Natl Acad Sci U S A.* 97:10389-94.

Maya, R., M. Balass, S.T. Kim, D. Shkedy, J.F. Leal, O. Shifman, M. Moas, T. Buschmann, Z. Ronai, Y. Shiloh, M.B. Kastan, E. Katzir, and M. Oren. 2001. ATM-dependent phosphorylation of Mdm2 on serine 395: role in p53 activation by DNA damage. *Genes Dev.* 15:1067-77.

McGowan, C.H. 2002. Checking in on Cds1 (Chk2): A checkpoint kinase and tumor suppressor. *Bioessays.* 24:502-11.

McPherson, J.P., B. Lemmers, A. Hirao, A. Hakem, J. Abraham, E. Migon, E. Matysiak-Zablocki, L. Tamblyn, O. Sanchez-Sweatman, R. Khokha, J. Squire, M.P. Hande, T.W. Mak, and R. Hakem. 2004. Collaboration of Brca1 and Chk2 in tumorigenesis. *Genes Dev.* 18:1144-53.

Meijers-Heijboer, H., A. van den Ouweland, J. Klijn, M. Wasielewski, A. de Snoo, R. Oldenburg, A. Hollestelle, M. Houben, E. Crepin, M. van Veghel-Plandsoen, F. Elstrodt, C. van Duijn, C. Bartels, C. Meijers, M. Schutte, L. McGuffog, D. Thompson, D. Easton, N. Sodha, S. Seal, R. Barfoot, J. Mangion, J. Chang-Claude, D. Eccles, R. Eeles, D.G. Evans, R. Houlston, V. Murday, S. Narod, T. Peretz, J. Peto, C. Phelan, H.X. Zhang, C. Szabo, P. Devilee, D. Goldgar, P.A. Futreal, K.L. Nathanson, B. Weber, N. Rahman, and M.R. Stratton. 2002. Low-penetrance susceptibility to breast cancer due to CHEK2(*)1100delC in noncarriers of BRCA1 or BRCA2 mutations. *Nat Genet.* 31:55-9.

Melchionna, R., X.B. Chen, A. Blasina, and C.H. McGowan. 2000. Threonine 68 is required for radiation-induced phosphorylation and activation of Cds1. *Nat Cell Biol.* 2:762-5.

Mihara, M., S. Erster, A. Zaika, O. Petrenko, T. Chittenden, P. Pancoska, and U.M. Moll. 2003. p53 has a direct apoptogenic role at the mitochondria. *Mol Cell.* 11:577-90.

Miki, Y., J. Swensen, D. Shattuck-Eidens, P.A. Futreal, K. Harshman, S. Tavtigian, Q. Liu, C. Cochran, L.M. Bennett, W. Ding, and et al. 1994. A strong candidate for the breast and ovarian cancer susceptibility gene BRCA1. *Science.* 266:66-71.

Mochan, T.A., M. Venere, R.A. DiTullio, Jr., and T.D. Halazonetis. 2003. 53BP1 and NFBD1/MDC1-Nbs1 function in parallel interacting pathways activating ataxia-telangiectasia mutated (ATM) in response to DNA damage. *Cancer Res.* 63:8586-91.

Morales, J.C., Z. Xia, T. Lu, M.B. Aldrich, B. Wang, C. Rosales, R.E. Kellems, W.N. Hittelman, S.J. Elledge, and P.B. Carpenter. 2003. Role for the BRCA1 C-terminal repeats (BRCT) protein 53BP1 in maintaining genomic stability. *J Biol Chem.* 278:14971-7.

Morris, J.R., and E. Solomon. 2004. BRCA1 : BARD1 induces the formation of conjugated ubiquitin structures, dependent on K6 of ubiquitin, in cells during DNA replication and repair. *Hum Mol Genet.* 13:807-17.

Motoyama, N., and K. Naka. 2004. DNA damage tumor suppressor genes and genomic instability. *Curr Opin Genet Dev*. 14:11-6.

O'Driscoll, M., V.L. Ruiz-Perez, C.G. Woods, P.A. Jeggo, and J.A. Goodship. 2003. A splicing mutation affecting expression of ataxia-telangiectasia and Rad3-related protein (ATR) results in Seckel syndrome. *Nat Genet*. 33:497-501.

Painter, R.B., and B.R. Young. 1980. Radiosensitivity in ataxia-telangiectasia: a new explanation. *Proc Natl Acad Sci U S A*. 77:7315-7.

Parrilla-Castellar, E.R., and L.M. Karnitz. 2003. Cut5 is required for the binding of Atr and DNA polymerase alpha to genotoxin-damaged chromatin. *J Biol Chem*. 278:45507-11.

Paull, T.T., and M. Gellert. 1998. The 3' to 5' exonuclease activity of Mre 11 facilitates repair of DNA double-strand breaks. *Mol Cell*. 1:969-79.

Paull, T.T., E.P. Rogakou, V. Yamazaki, C.U. Kirchgessner, M. Gellert, and W.M. Bonner. 2000. A critical role for histone H2AX in recruitment of repair factors to nuclear foci after DNA damage. *Curr Biol*. 10:886-95.

Peng, A., and P.L. Chen. 2003. NFBD1, like 53BP1, is an early and redundant transducer mediating Chk2 phosphorylation in response to DNA damage. *J Biol Chem*. 278:8873-6.

Peng, C.Y., P.R. Graves, R.S. Thoma, Z. Wu, A.S. Shaw, and H. Piwnica-Worms. 1997. Mitotic and G2 checkpoint control: regulation of 14-3-3 protein binding by phosphorylation of Cdc25C on serine-216. *Science*. 277:1501-5.

Petrini, J.H., and T.H. Stracker. 2003. The cellular response to DNA double-strand breaks: defining the sensors and mediators. *Trends Cell Biol*. 13:458-62.

Qian, Y.W., E. Erikson, C. Li, and J.L. Maller. 1998. Activated polo-like kinase Plx1 is required at multiple points during mitosis in Xenopus laevis. *Mol Cell Biol*. 18:4262-71.

Rappold, I., K. Iwabuchi, T. Date, and J. Chen. 2001a. Tumor suppressor p53 binding protein 1 (53BP1) is involved in DNA damage-signalling pathways. *J Cell Biol*. 153:613-20.

Rappold, I., K. Iwabuchi, T. Date, and J. Chen. 2001b. Tumor suppressor p53 binding protein 1 (53BP1) is involved in DNA damage-signalling pathways. *J Cell Biol*. 153:613-20.

Rauen, M., M.A. Burtelow, V.M. Dufault, and L.M. Karnitz. 2000. The human checkpoint protein hRad17 interacts with the PCNA-like proteins hRad1, hHus1, and hRad9. *J Biol Chem*. 275:29767-71.

Rogakou, E.P., C. Boon, C. Redon, and W.M. Bonner. 1999. Megabase chromatin domains involved in DNA double-strand breaks in vivo. *J Cell Biol*. 146:905-16.

Rogakou, E.P., D.R. Pilch, A.H. Orr, V.S. Ivanova, and W.M. Bonner. 1998. DNA double-stranded breaks induce histone H2AX phosphorylation on serine 139. *J Biol Chem*. 273:5858-68.

Roos-Mattjus, P., B.T. Vroman, M.A. Burtelow, M. Rauen, A.K. Eapen, and L.M. Karnitz. 2002. Genotoxin-induced Rad9-Hus1-Rad1 (9-1-1) chromatin association is an early checkpoint signalling event. *J Biol Chem*. 277:43809-12.

Rouse, J., and S.P. Jackson. 2002. Interfaces between the detection, signalling, and repair of DNA damage. *Science*. 297:547-51.

Ruffner, H., C.A. Joazeiro, D. Hemmati, T. Hunter, and I.M. Verma. 2001. Cancer-predisposing mutations within the RING domain of BRCA1: loss of ubiquitin protein ligase activity and protection from radiation hypersensitivity. *Proc Natl Acad Sci U S A*. 98:5134-9.

Sancar, A., L.A. Lindsey-Boltz, K. Unsal-Kaccmaz, and S. Linn. 2004. Molecular Mechanisms of Mammalian DNA Repair and the DNA Damage Checkpoints. *Annu Rev Biochem*. 73:39-85.

Sanchez, Y., C. Wong, R.S. Thoma, R. Richman, Z. Wu, H. Piwnica-Worms, and S.J. Elledge. 1997. Conservation of the Chk1 checkpoint pathway in mammals: linkage of DNA damage to Cdk regulation through Cdc25. *Science*. 277:1497-501.

Sar, F., L.A. Lindsey-Boltz, D. Subramanian, D.L. Croteau, S.Q. Hutsell, J.D. Griffith, and A. Sancar. 2004. Human claspin is a ring-shaped DNA binding protein with high affinity to branched DNA structures. *J Biol Chem*.

Schultz, L.B., N.H. Chehab, A. Malikzay, and T.D. Halazonetis. 2000. p53 binding protein 1 (53BP1) is an early participant in the cellular response to DNA double-strand breaks. *J Cell Biol*. 151:1381-90.

Schwarz, J.K., C.M. Lovly, and H. Piwnica-Worms. 2003. Regulation of the Chk2 protein kinase by oligomerization-mediated cis- and trans-phosphorylation. *Mol Cancer Res*. 1:598-609.

Scully, R., and D.M. Livingston. 2000. In search of the tumour-suppressor functions of BRCA1 and BRCA2. *Nature*. 408:429-32.

Scully, R., A. Xie, and G. Nagaraju. 2004. Molecular Functions of BRCA1 in the DNA Damage Response. *Cancer Biol Ther*. 3.

Shang, Y., A. Bodero, and P. Chen. 2002. NFBD1, a novel nuclear protein with signature motifs of FHA and BRCT, and an internal 41 amino acid repeat sequence, is an early participant in DNA damage response. *J. Biol. Chem*. 278:6323-6329.

Shen, S.X., Z. Weaver, X. Xu, C. Li, M. Weinstein, L. Chen, X.Y. Guan, T. Ried, and C.X. Deng. 1998. A targeted disruption of the murine Brca1 gene causes gamma-irradiation hypersensitivity and genetic instability. *Oncogene*. 17:3115-24.

Shiloh, Y. 2003. ATM and related protein kinases: safeguarding genome integrity. *Nat Rev Cancer*. 3:155-68.

Smits, V.A., R. Klompmaker, L. Arnaud, G. Rijksen, E.A. Nigg, and R.H. Medema. 2000. Polo-like kinase-1 is a target of the DNA damage checkpoint. *Nat Cell Biol*. 2:672-6.

Sorensen, C.S., R.G. Syljuasen, J. Falck, T. Schroeder, L. Ronnstrand, K.K. Khanna, B.B. Zhou, J. Bartek, and J. Lukas. 2003. Chk1 regulates the S phase checkpoint by coupling the physiological turnover and ionizing radiation-induced accelerated proteolysis of Cdc25A. *Cancer Cell*. 3:247-58.

Stevens, C., L. Smith, and N.B. La Thangue. 2003. Chk2 activates E2F-1 in response to DNA damage. *Nat Cell Biol*. 5:401-9.

Stewart, G.S., R.S. Maser, T. Stankovic, D.A. Bressan, M.I. Kaplan, N.G. Jaspers, A. Raams, P.J. Byrd, J.H. Petrini, and A.M. Taylor. 1999. The DNA double-strand break repair gene hMRE11 is mutated in individuals with an ataxia-telangiectasia-like disorder. *Cell*. 99:577-87.

Stewart, G.S., B. Wang, C.R. Bignell, A.M. Taylor, and S.J. Elledge. 2003. MDC1 is a mediator of the mammalian DNA damage checkpoint. *Nature*. 421:961-6.

Stiff, T., M. O'Driscoll, N. Rief, K. Iwabuchi, M. Lobrich, and P.A. Jeggo. 2004. ATM and DNA-PK function redundantly to phosphorylate H2AX after exposure to ionizing radiation. *Cancer Res*. 64:2390-6.

Sui, G., B. Affar el, Y. Shi, C. Brignone, N.R. Wall, P. Yin, M. Donohoe, M.P. Luke, D. Calvo, and S.R. Grossman. 2004. Yin Yang 1 is a negative regulator of p53. *Cell*. 117:859-72.

Takai, H., K. Naka, Y. Okada, M. Watanabe, N. Harada, S. Saito, C.W. Anderson, E. Appella, M. Nakanishi, H. Suzuki, K. Nagashima, H. Sawa, K. Ikeda, and N. Motoyama. 2002. Chk2-deficient mice exhibit radioresistance and defective p53-mediated transcription. *EMBO J*. 21:5195-205.

Takai, H., K. Tominaga, N. Motoyama, Y.A. Minamishima, H. Nagahama, T. Tsukiyama, K. Ikeda, K. Nakayama, and M. Nakanishi. 2000. Aberrant cell cycle checkpoint function and early embryonic death in Chk1(-/-) mice. *Genes Dev*. 14:1439-47.

Tibbetts, R.S., K.M. Brumbaugh, J.M. Williams, J.N. Sarkaria, W.A. Cliby, S.Y. Shieh, Y. Taya, C. Prives, and R.T. Abraham. 1999. A role for ATR in the DNA damage-induced phosphorylation of p53. *Genes Dev*. 13:152-7.

Trujillo, K.M., S.S. Yuan, E.Y. Lee, and P. Sung. 1998. Nuclease activities in a complex of human recombination and DNA repair factors Rad50, Mre11, and p95. *J Biol Chem*. 273:21447-50.

Unsal-Kacmaz, K., and A. Sancar. 2004. Quaternary structure of ATR and effects of ATRIP and replication protein A on its DNA binding and kinase activities. *Mol Cell Biol*. 24:1292-300.

Uziel, T., Y. Lerenthal, L. Moyal, Y. Andegeko, L. Mittelman, and Y. Shiloh. 2003. Requirement of the MRN complex for ATM activation by DNA damage. *EMBO J.* 22:5612-21.

Vahteristo, P., J. Bartkova, H. Eerola, K. Syrjakoski, S. Ojala, O. Kilpivaara, A. Tamminen, J. Kononen, K. Aittomaki, P. Heikkila, K. Holli, C. Blomqvist, J. Bartek, O.P. Kallioniemi, and H. Nevanlinna. 2002. A CHEK2 genetic variant contributing to a substantial fraction of familial breast cancer. *Am J Hum Genet.* 71:432-8.

van Vugt, M.A., V.A. Smits, R. Klompmaker, and R.H. Medema. 2001. Inhibition of Polo-like kinase-1 by DNA damage occurs in an ATM- or ATR-dependent fashion. *J Biol Chem.* 276:41656-60.

Vassilev, L.T., B.T. Vu, B. Graves, D. Carvajal, F. Podlaski, Z. Filipovic, N. Kong, U. Kammlott, C. Lukacs, C. Klein, N. Fotouhi, and E.A. Liu. 2004. In vivo activation of the p53 pathway by small-molecule antagonists of MDM2. *Science.* 303:844-8.

Venclovas, C., and M.P. Thelen. 2000. Structure-based predictions of Rad1, Rad9, Hus1 and Rad17 participation in sliding clamp and clamp-loading complexes. *Nucleic Acids Res.* 28:2481-93.

Villunger, A., E.M. Michalak, L. Coultas, F. Mullauer, G. Bock, M.J. Ausserlechner, J.M. Adams, and A. Strasser. 2003. p53- and drug-induced apoptotic responses mediated by BH3-only proteins puma and noxa. *Science.* 302:1036-8.

Vogelstein, B., D. Lane, and A.J. Levine. 2000. Surfing the p53 network. *Nature.* 408:307-10.

Vousden, K.H., and X. Lu. 2002. Live or let die: the cell's response to p53. *Nat Rev Cancer.* 2:594-604.

Walter, J., and J. Newport. 2000. Initiation of eukaryotic DNA replication: origin unwinding and sequential chromatin association of Cdc45, RPA, and DNA polymerase alpha. *Mol Cell.* 5:617-27.

Wang, B., S. Matsuoka, P.B. Carpenter, and S.J. Elledge. 2002. 53BP1, a mediator of the DNA damage checkpoint. *Science.* 298:1435-8.

Wang, Y., D. Cortez, P. Yazdi, N. Neff, S.J. Elledge, and J. Qin. 2000. BASC, a super complex of BRCA1-associated proteins involved in the recognition and repair of aberrant DNA structures. *Genes Dev.* 14:927-39.

Ward, I.M., K. Minn, K.G. Jorda, and J. Chen. 2003a. Accumulation of checkpoint protein 53BP1 at DNA breaks involves its binding to phosphorylated histone H2AX. *J Biol Chem.* 278:19579-82.

Ward, I.M., K. Minn, J. van Deursen, and J. Chen. 2003b. p53 Binding protein 53BP1 is required for DNA damage responses and tumor suppression in mice. *Mol Cell Biol.* 23:2556-63.

Ward, I.M., B. Reina-San-Martin, A. Olaru, K. Minn, K. Tamada, J.S. Lau, M. Cascalho, L. Chen, A. Nussenzweig, F. Livak, M.C. Nussenzweig, and J. Chen. 2004. 53BP1 is required for class switch recombination. *J Cell Biol.* 165:459-64.

Ward, I.M., X. Wu, and J. Chen. 2001. Threonine 68 of Chk2 is phosphorylated at sites of DNA strand breaks. *J Biol Chem.* 276:47755-8.

Weiss, R.S., S. Matsuoka, S.J. Elledge, and P. Leder. 2002. Hus1 acts upstream of chk1 in a mammalian DNA damage response pathway. *Curr Biol.* 12:73-7.

Wu, X., and J. Chen. 2003. Autophosphorylation of checkpoint kinase 2 at serine 516 is required for radiation-induced apoptosis. *J Biol Chem.* 278:36163-8.

Wu, X., V. Ranganathan, D.S. Weisman, W.F. Heine, D.N. Ciccone, T.B. O'Neill, K.E. Crick, K.A. Pierce, W.S. Lane, G. Rathbun, D.M. Livingston, and D.T. Weaver. 2000. ATM phosphorylation of Nijmegen breakage syndrome protein is required in a DNA damage response. *Nature.* 405:477-82.

Wu-Baer, F., K. Lagrazon, W. Yuan, and R. Baer. 2003. The BRCA1/BARD1 heterodimer assembles polyubiquitin chains through an unconventional linkage involving lysine residue K6 of ubiquitin. *J Biol Chem.* 278:34743-6.

Xia, Y., G.M. Pao, H.W. Chen, I.M. Verma, and T. Hunter. 2003. Enhancement of BRCA1 E3 ubiquitin ligase activity through direct interaction with the BARD1 protein. *J Biol Chem*. 278:5255-63.

Xia, Z., J.C. Morales, W.G. Dunphy, and P.B. Carpenter. 2001. Negative cell cycle regulation and DNA damage-inducible phosphorylation of the BRCT protein 53BP1. *J Biol Chem*. 276:2708-18.

Xiong, Y., G.J. Hannon, H. Zhang, D. Casso, R. Kobayashi, and D. Beach. 1993. p21 is a universal inhibitor of cyclin kinases. *Nature*. 366:701-4.

Xu, B., S. Kim, and M.B. Kastan. 2001a. Involvement of Brca1 in S-phase and G(2)-phase checkpoints after ionizing irradiation. *Mol Cell Biol*. 21:3445-50.

Xu, B., A.H. O'Donnell, S.T. Kim, and M.B. Kastan. 2002a. Phosphorylation of serine 1387 in Brca1 is specifically required for the Atm-mediated S-phase checkpoint after ionizing irradiation. *Cancer Res*. 62:4588-91.

Xu, X., J. Lee, and D.F. Stern. 2004. Microcephalin is a DNA damage response protein involved in regulation of CHK1 and BRCA1. *J Biol Chem*. 279:34091-4.

Xu, X., W. Qiao, S.P. Linke, L. Cao, W.M. Li, P.A. Furth, C.C. Harris, and C.X. Deng. 2001b. Genetic interactions between tumor suppressors Brca1 and p53 in apoptosis, cell cycle and tumorigenesis. *Nat Genet*. 28:266-71.

Xu, X., and D.F. Stern. 2002. NFBD1/KIAA0170 is a chromatin-associated protein involved in DNA damage signalling pathways. *J. Biol. Chem*. 278:8795-8803.

Xu, X., and D.F. Stern. 2003. NFBD1/MDC1 regulates ionizing radiation-induced focus formation by DNA checkpoint signalling and repair factors. *Faseb J*. 17:1842-8.

Xu, X., L.M. Tsvetkov, and D.F. Stern. 2002b. Chk2 activation and phosphorylation-dependent oligomerization. *Mol Cell Biol*. 22:4419-32.

Xu, X., K.U. Wagner, D. Larson, Z. Weaver, C. Li, T. Ried, L. Hennighausen, A. Wynshaw-Boris, and C.X. Deng. 1999. Conditional mutation of Brca1 in mammary epithelial cells results in blunted ductal morphogenesis and tumour formation. *Nat Genet*. 22:37-43.

Xu, Y., and D. Baltimore. 1996. Dual roles of ATM in the cellular response to radiation and in cell growth control. *Genes Dev*. 10:2401-10.

Yamane, K., J. Chen, and T.J. Kinsella. 2003. Both DNA topoisomerase II-binding protein 1 and BRCA1 regulate the G2-M cell cycle checkpoint. *Cancer Res*. 63:3049-53.

Yamane, K., M. Kawabata, and T. Tsuruo. 1997. A DNA-topoisomerase-II-binding protein with eight repeating regions similar to DNA-repair enzymes and to a cell-cycle regulator. *Eur J Biochem*. 250:794-9.

Yamane, K., X. Wu, and J. Chen. 2002. A DNA damage-regulated BRCT-containing protein, TopBP1, is required for cell survival. *Mol Cell Biol*. 22:555-66.

Yang, S., C. Kuo, J.E. Bisi, and M.K. Kim. 2002. PML-dependent apoptosis after DNA damage is regulated by the checkpoint kinase hCds1/Chk2. *Nat Cell Biol*. 4:865-70.

Yarden, R.I., S. Pardo-Reoyo, M. Sgagias, K.H. Cowan, and L.C. Brody. 2002. BRCA1 regulates the G2/M checkpoint by activating Chk1 kinase upon DNA damage. *Nat Genet*. 30:285-9.

Yazdi, P.T., Y. Wang, S. Zhao, N. Patel, E.Y. Lee, and J. Qin. 2002. SMC1 is a downstream effector in the ATM/NBS1 branch of the human S-phase checkpoint. *Genes Dev*. 16:571-82.

Yoo, H.Y., A. Kumagai, A. Shevchenko, and W.G. Dunphy. 2004. Adaptation of a DNA replication checkpoint response depends upon inactivation of Claspin by the Polo-like kinase. *Cell*. 117: 575-88.

You, Z., L. Kong, and J. Newport. 2002. The role of single-stranded DNA and polymerase alpha in establishing the ATR, Hus1 DNA replication checkpoint. *J Biol Chem*. 277:27088-93.

Yu, X., C.C. Chini, M. He, G. Mer, and J. Chen. 2003. The BRCT domain is a phospho-protein binding domain. *Science*. 302:639-42.

Zacchi, P., M. Gostissa, T. Uchida, C. Salvagno, F. Avolio, S. Volinia, Z. Ronai, G. Blandino, C. Schneider, and G. Del Sal. 2002. The prolyl isomerase Pin1 reveals a mechanism to control p53 functions after genotoxic insults. *Nature*. 419:853-7.

Zhao, H., and H. Piwnica-Worms. 2001. ATR-mediated checkpoint pathways regulate phosphorylation and activation of human Chk1. *Mol Cell Biol*. 21:4129-39.

Zhao, H., J.L. Watkins, and H. Piwnica-Worms. 2002a. Disruption of the checkpoint kinase 1/cell division cycle 25A pathway abrogates ionizing radiation-induced S and G2 checkpoints. *Proc Natl Acad Sci U S A*. 99:14795-800.

Zhao, S., W. Renthal, and E.Y. Lee. 2002b. Functional analysis of FHA and BRCT domains of NBS1 in chromatin association and DNA damage responses. *Nucleic Acids Res*. 30:4815-22.

Zhao, S., Y.C. Weng, S.S. Yuan, Y.T. Lin, H.C. Hsu, S.C. Lin, E. Gerbino, M.H. Song, M.Z. Zdzienicka, R.A. Gatti, J.W. Shay, Y. Ziv, Y. Shiloh, and E.Y. Lee. 2000. Functional link between ataxia-telangiectasia and Nijmegen breakage syndrome gene products. *Nature*. 405:473-7.

Zheng, H., H. You, X.Z. Zhou, S.A. Murray, T. Uchida, G. Wulf, L. Gu, X. Tang, K.P. Lu, and Z.X. Xiao. 2002. The prolyl isomerase Pin1 is a regulator of p53 in genotoxic response. *Nature*. 419:849-53.

Zhou, B.B., P. Chaturvedi, K. Spring, S.P. Scott, R.A. Johanson, R. Mishra, M.R. Mattern, J.D. Winkler, and K.K. Khanna. 2000. Caffeine abolishes the mammalian G(2)/M DNA damage checkpoint by inhibiting ataxia-telangiectasia-mutated kinase activity. *J Biol Chem*. 275:10342-8.

Zhou, B.B., and S.J. Elledge. 2000. The DNA damage response: putting checkpoints in perspective. *Nature*. 408:433-9.

Zou, L., D. Cortez, and S.J. Elledge. 2002. Regulation of ATR substrate selection by Rad17-dependent loading of Rad9 complexes onto chromatin. *Genes Dev*. 16:198-208.

Zou, L., and S.J. Elledge. 2003. Sensing DNA damage through ATRIP recognition of RPA-ssDNA complexes. *Science*. 300:1542-8.

Chapter 4.2

ATM AND CELLULAR RESPONSE TO DNA DAMAGE

Martin F. Lavin[1,2], Sergei Kozlov[1], Nuri Gueven[1], Cheng Peng[1], Geoff Birrell[1], Phillip Chen[1] and Shaun Scott[1]
[1.] The Queensland Institute of Medical Research, PO Royal Brisbane Hospital, Herston, Queensland, Australia, [2.] Central Clinical Division, University of Queensland, PO Royal Brisbane Hospital, Herston, Queensland, Australia

1. INTRODUCTION

ATM is the gene mutated in the human genetic disorder ataxia – telangiectasia (A-T) which is characterized by genome instability and predisposition to develop tumours (Boder, 1985, Savitsky et al 1995). The gene product is a serine/threonine kinase, a member of the phosphoinositide 3-kinase-like kinases (PIKK) that phosphorylates a variety of proteins involved in the damage response to DNA double strand breaks (Shiloh, 2003). Recognition of the double strand break in DNA by ATM is achieved at least in part by the sensor complex Mrell/Rad50/Nbsl (Uziel et al., 2003). This leads to activation of ATM by autophosphorylation which enables it to phosphorylate in turn a spectrum of substrates involved in cell cycle checkpoint activation and DNA repair (Bakkenist and Kastan, 2003, Kozlov et al., 2003; Goodarzi et al., 2003). The end result is the maintenance of genome integrity and the minimization of risk of malignancy. This is very evident in patients with A-T where ATM is absent or mutated. Lymphocytes from these individuals are characterized by a high frequency of chromosome abnormalities including translocations and inversions, primarily involving sites of T-cell receptor and immunoglobulin heavy chain genes (Carbonari et al., 1990). This instability is even more pronounced in response to ionizing radiation exposure (Higurashi and Conen, 1973). The genetic instability observed in A-T lymphocytes correlates well with the high incidence of lymphoid malignancies observed in these patients (Spector et al., 1982; Hecht and Hecht, 1990). While the majority of such tumours are lymphoid in origin as many as 25% of the cancers seen in A-T patients

E.A. Nigg (ed.), Genome Instability in Cancer Development, 457-476.

are solid tumours originating in a variety of different tissues (Morrell et al., 1986) Even though A-T is an autosomal recessive syndrome there is evidence for some penetration of the phenotype in A-T carriers including cancer predisposition (Swift et al., 1991).

It is evident that ATM in a key player in the cellular response to DNA double strand breaks. It is rapidly activated to orchestrate radiation – induced signalling through multiple pathways designed to maintain genome integrity. In this chapter we review the central role of ATM in protecting cells against DNA damage and reducing the risk of malignant development.

2. ATAXIA – TELANGIECTASIA

Ataxia – telangiectasia (A-T) is one of an increasing number of chromosomal breakage syndromes characterized by inherent genome instability or instability exaggerated by exposure to one of a number of DNA damaging agents (Lavin and Lederman 2004). A sub-group of these syndromes with increased sensitivity to ionizing radiation include A-T; A-T like disorder (deficient in Mre 11), Nijmegen breakage syndrome (*Nbs1*); Bloom syndrome (*BLM*) and Fanconis anemia (*FA* genes) (Table 1).

Table 1. Chromosomal Breakage Syndromes.

Syndrome	Radiosensitivity	Cancer Predisposition
Ataxia-telangiectasia	++++	+
Ataxia-telangiectasia-like disorder	++	?
Bloom syndrome	+	+
Nijmegen breakage syndrome	++++	+
Fanconi's anemia	+	+
DNA ligase IV deficiency	+	?

A-T was described as a distinct clinical entity by Boder and Sedgwick (1957) and Biemond (1957), with the aid of autopsies they described organ developmental abnormalities, neurological manifestations and a third major characteristic of the disease, recurrent sinopulmonary infection. The main clinical features of the disease are outlined in Table 2 (Boder, 1985).

Ataxia, generally the presenting symptom in this syndrome, becomes evident when a child begins to walk at the end of the first year of life, manifesting ataxic gait and truncal movements. As with others major characteristics of A-T, ataxia is progressive, spreading to affect the extremities and then speech. The underlying pathology is primarily progressive cerebellar cortical degeneration. Cortical cerebellar degeneration involves primarily Purkinje and granular cells. While degenerative changes in the brain are seen predominantly in the cerebellum, it is clear from an increasing number of autopsies that changes to the CNS in A-T are more widespread (Boder, 1985).

Table 2. Clinical features in ataxia-telangiectasia.

Clinical Feature	Cases With Feature (No.)	Cases With Data Available (No.)	Cases With Feature (%)
I. Neurological Abnormalities			
Cerebellar ataxia, infantile or childhood onset	101	101	100
Diminished or absent deep reflexes	54	61	89
Flexor or equivocal plantar response	60	61	98
Negative Romberg sign	28	36	78
Intact deep and superficial sensation	51	52	98
Choreoathetosis	61	67	91
Oculomotor			
Apraxia of eye movements	47	56	84
Fixation of gaze nystagmus	48	58	83
Strabismus	7	15	47
Dysarthric speech	70	70	100
Drooling	43	49	88
Characteristic facies and postural attitudes	60	61	98
II. Telangiectasia, Oculocutaneous	101	101	100
III. Frequent Sinopulmonary Infection	60	72	83
IV. Familial Occurrence	43	96	45
V. Reported Mental Deficiency	22	66	33
VI. Equable Disposition	34	34	100
VII.Retardation of Somatic Growth	42	50	72
VIII. Progeric Changes of Hair and Skin	46	52	88

Taken from Boder (1985)

A second major clinical manifestation of the disease is telangiectasia. It usually has a later onset than ataxia, between 2 and 8 years (McFarlin et al., 1972; Boder, 1985).

Another feature of A-T is abnormal susceptibility to infections. Recurrent infections have been described in up to 80% of patients in some studies (Waldmann, 1982). McFarlin et al. (1972) demonstrated a correlation between the severity of the respiratory infections and reduced immune responses. The observation of Boder and Sedgwick (1957) that the thymus is absent or poorly developed in A-T, together with reports of hypogammaglobulinemia suggested a basis for the predisposition to infection. It is now obvious that there is a more generalized defect in multiple aspects of the immune response in A-T patients (Waldmann, 1982).

3. GENOME INSTABILITY AND RADIOSENSITIVITY

Lymphocytes from A-T patients are characterized by chromosomal fragility (Hecht et al., 1966; Miller, 1967). Chromosomal breakage products include dicentric and other forms of rearrangement such as acentrics, rings, inversions and translocations. This instability is manifested *in vivo* in direct bone marrow preparations and has been shown to be clonal. Instability is most obvious in T lymphocytes but is also observed in B cells and fibroblasts (Hecht et al., 1973). Adverse response to radiation was also demonstrated in A-T cells *in vitro*. Higurashi and Conen (1973) reported a higher level of radiation-induced chromosomal changes in lymphocytes from A-T patients than in controls, and a greater increase in chromosomal aberrations after G2 irradiation was also demonstrated (Rary et al., 1975). Following G0 irradiation of A-T cells a marked increase in chromatid type damage was observed which was not seen in normal cells (Taylor et al., 1976). A 7-15-fold increase in chromatid gaps and breaks occurred and a 20-fold increase in chromatid interchanges. A similar pattern of increase was observed when lymphocytes were irradiated in G2 phase.

Radiosensitivity was firmly established for A-T cells when Taylor et al., (1975) demonstrated that A-T fibroblasts were three to four times more sensitive to ionizing radiation than controls. These observations were confirmed using fibroblasts and other cell types after radiation exposure (Paterson and Smith, 1979; Edwards and Taylor, 1981; Lehmann et al., 1982), and after treatment with bleomycin (Cohen and Simpson, 1982; Morris et al., 1983) or neocarzinostatin (Cohen and Simpson, 1982; Shiloh et al., 1983). The response of A-T cells to UV radiation is normal or near normal (de Wit et al., 1981). Nevertheless, while there is no evidence of a gross defect in repair of DNA double strand breaks in A-T cells some reports of a more subtle defect in DNA strand break repair in A-T cells have been described (Cornforth and Bedford, 1985; Pandita and Hittleman, 1992; Foray et al., 1997). This position was enforced by cytogenetic studies which revealed both basal chromosome instability and higher levels of radiation-induced chromosome breakage in A-T cells (Hecht et al., 1966; Higurashi and Conen, 1973; Taylor, 1982).

4. CANCER PREDISPOSITION

A major hallmark in A-T is the propensity to develop a range of lymphoid malignancies (Boder and Sedgwick, 1963). The association between a defective thymus, immunodeficiency, and the high frequency of lymphoid malignancy became evident early (Peterson et al., 1964; Miller, 1967). An explanation for the increased incidence of malignancy was suggested to be chromosomal instability when it was observed that leukemia cells from an A-T patient had a translocation involving chromosomes 12 and 14 (Hecht et al., 1966). Regular monitoring of this patient provided evidence for a progressive increase in these abnormal lymphocytes to 78% of the total lymphocytes prior to the patient succumbing to infection (Hecht et al., 1973). Lymphoid malignancies in A-T are of both B cell and T cell origin and include non-Hodgkin's lymphoma, Hodgkin's lymphoma, and different forms of leukemia (Spector et al., 1982; Hecht and Hecht 1990). Non-Hodgkin's lymphoma accounts for approximately 40% of neoplasms detected, leukemias about 20%, and Hodgkin's lymphomas 10%. Hecht and Hecht (1990), in an analysis of 108 A-T patients with 119 neoplasms, reported that 31 of these (26%) were solid tumours varying in type and location. Determination of subsequent risk in A-T patients diagnosed with one type of neoplasm revealed that approximately 25% of patients with solid tumours subsequently developed non-Hodgkin's lymphoma or leukemia. A low risk of subsequent neoplasms existed when the first tumour was lymphoid in origin. In a retrospective United States study, mortality from all causes in A-T was 50-fold and 147-fold higher for white and black A-T patients, respectively, than expected based on overall U.S. mortality rates (Morrell et al., 1986). Neoplasia is the second most frequent cause of death in A-T after pulmonary disease. Of 62 complete autopsy reports (Sedgwick and Boder, 1991), 29 deaths (47%) were caused by pulmonary complications, 14 (22%) by malignancy, and 16 (26%) by a combination of both. The lifetime cancer risk among A-T patients has been estimated to be between 10% and 38% (Morrell et al., 1990; Spector et al., 1982).

5. CANCER IN A-T HETEROZYGOTES

The risk for cancer among heterozygotes is controversial. Swift and colleagues (1991; Athma et al., 1996) reported a 3.8-fold increased risk for breast cancer among A-T heterozygotes, particularly women older than age 60, compared with controls. Several other epidemiologic studies also support an increased risk in A-T heterozygotes (Janin et al., 1999; Olsen et al., 2001). Fitzgerald et al. (1997) and other reports, determining truncating ATM mutations in breast tumours occurring among the general population, failed to find an increased incidence of mutations compared to that in controls. However, accumulating evidence implicates missense ATM mutations in cancer. These mutations account for approximately 15% of those seen in A-T patients (http://www.benaroyaresearch.org/bri_investigators/atm.html). In families with multiple cases of breast cancer, two ATM missense mutations segregated with the disease with an estimated average penetrance of 60% to the age of 70 years, equivalent to an increase in relative cancer risk of approximately 16-fold (Chenevix-Trench et al., 2002). A previous study found increased loss of heterozygosity (LOH) of the wildtype allele associated with one missense mutation and six rare ATM variants in matched breast/tumour tissue, implying possible involvement of these mutations in the development of the tumours (Izatt et al., 1999). Another ATM missense mutation, R1054P, was also shown to be associated with breast cancer with an overall odds ratio of 4.5, which increased to 6.9 in the presence of a rare *HRAS1* allele (Larson et al., 1998). A substantially greater frequency of ATM mutations was found in individuals with breast cancer who had early onset, a family history of breast cancer or both. All of these mutations were missense mutations (Teraoka et al., 2001). Finally, a significant excess of missense substitutions has been reported in unselected individuals with breast cancer (Dork et al., 2001). Two families with a missense mutation 7271T→G that allows expression of full-length ATM protein at a level comparable with that in unaffected individuals have been identified (Stewart et al., 2001). Interference with normal ATM could well account for the increased risk of breast cancer in these families. Mutations in the kinase domain have been demonstrated to confer dominant-negative activity upon ATM (Lim et al., 2000). In addition, other data demonstrate that missense mutations outside the kinase domain also interfere in a dominant-negative fashion with ATM kinase activity, probably through a mechanism involving ATM-ATM interaction (Scott et al., 2002).

6. ATM AND RESPONSE TO DNA DOUBLE STRAND BREAKS

6.1 The Mre11 Complex as a Sensor of DNA Double Strand Breaks

DNA double strand breaks arise as part of the normal process of T cell receptor and immunoglobulin gene rearrangement, during disruption of replication forks or after exposure of cells to damaging agents such as ionizing radiation (Pfeiffer et al., 2004). DNA damage induced breaks are repaired by non-homologous end-joining (NHEJ) or by homologous recombination between sister chromatids (Jeggo et al., 1999; Johnson and Jasin, 2000). Recognition of double strand breaks in DNA involves the co-ordinated action of several proteins and complexes including DNA-PK$_{CS}$, the catalytic subunit of DNA-PK, that is recruited to DNA free ends by the Ku heterodimer as part of the NHEJ mechanism (Meek et al., 2004). One of these complexes, BRCA1 – associated genome surveillance complex (BASC) has been shown to contain ATM, the Mre11 complex and a variety of other proteins involved in DNA damage recognition and repair (Wang et al., 2000). Another complex, Mre11/Rad50/Xrs2, also participates in NHEJ in *Saccharoymces cerevisiae* (Moore and Haber, 1996). Homologs of Mre11 and Rad50 are present in mammalian cells and the functional counterpart of Xrs2 (non-conserved) is Nbs1, mutated in Nijmegen Breakage Syndrome (Carney et al., 1998). This complex is localized to small granular foci at sites of DNA damage within 10 min post-irradiation (Mirzoeva and Petrini, 2001). The Mre11 complex is also involved in homologous recombination, meiotic recombination and telomere maintenance (Stracker et al., 2004). Once localized to double strand breaks it is likely that a number of *in vitro* activities described for this complex, exonuclease, single strand endonuclease and DNA unwinding, contribute to end processing of the break as part of the repair mechanism (Trujillo et al., 1998; Paull and Gelbert, 1998). This association with DNA damaged sites and its involvement in cell cycle checkpoint activation provides strong support for a role for the Mre11 complex as a sensor of DNA double strand breaks. A number of recent reports further substantiate this and highlight the importance of the complex for the efficient activation of ATM kinase (Uziel et al., 2003; Carson et al 2003., Lee and Paull 2004; Costanzo et al., 2004). Previous data here shown that the Mre11 complex associates with damaged DNA independent of ATM (Mirzoeva and Petrini, 2001). In response to DNA strand breaks induced by the radiomimetic agent neocarcinostatin (NCS), activation of ATM, as determined by autophosphorylation, was defective in both NBS and A-TLD cells (Uziel et al., 2003). This defect was most apparent in A-TLD cells homozygous for a truncating mutation in the Mre11 gene (A-TLD 1/2). Nuclear retention of ATM was reduced in these mutant lines and the extent of NCS-induced phosphorylation was decreased. Transfection of these cells with full-length Nbs1 or Mre11 cDNA restored a normal damage response. These data point to an important role for the Mre11 complex in initiating the ATM-mediated response to double strand breaks (Figure 1).

Infection of cells with adenovirus leads to degradation of the Mre11 complex but when this virus is incapacitated by deletion of the E4 gene its genome is joined into concatamers and there is evidence for activation of ATM signalling pathways (Carson et al., 2003). This represents additional evidence for a critical role for the Mre11 complex in ATM activation. ATM kinase can also be activated *in vitro* by incubation with Mre11 complex proteins expressed in a baculovirus system (Lee and Paull, 2004). When the A-TLD1/2 mutant form of Mre11 replaced the wild-type protein only partial activation of ATM kinase was observed. Addition of fragmented DNA to *Xenopus* egg extracts reconstitutes an ATM-dependent functional cell cycle checkpoint (Costanzo et al., 2000). Extracts depleted of Mre11 showed abrogation of DNA double strand break – dependent phosphorylation of H2AX (Costanzo et al., 2004). This provided additional evidence that the complex was required for ATM activation since ATM was shown to be the major contributor of DNA-end induced H2AX phosphorylation.

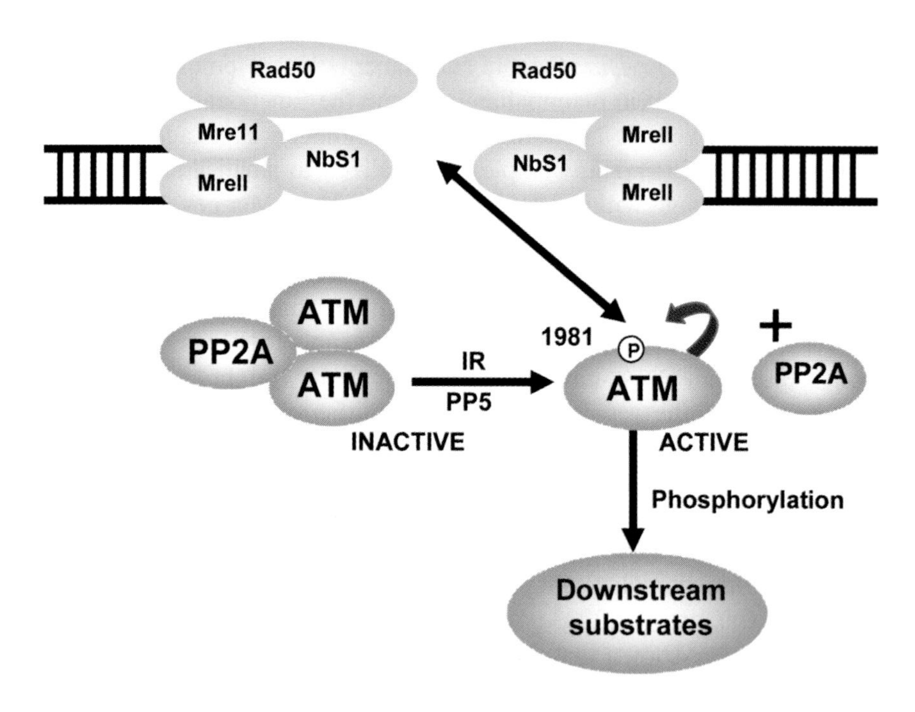

Figure 1. Activation of ATM kinase. ATM is rapidy recruited to sites of DNA damage via the Mre11 sensor complex. Activation involves transphorylation of an ATM dimer to release active monomers. This activated form of ATM is then free to phosphorylate downstream substrates.

6.2 ATM Activation

As outlined above the Mre11 complex plays an important role in sensing double strand breaks in DNA and passing this information to ATM to initiate the activation of multiple pathways responsible for cell cycle checkpoint activation (Shiloh, 2003). A two-way functional interaction exists between these molecules since the Mre11 complex transduces the signal for ATM activation but once activated ATM phosphorylates members of this complex to intiate downstream signalling (Lavin, 2004, Figure 1). It is still not clear what the stimulus is for the activation of ATM, the presence of a double strand break in DNA or the relaxation of chromatin structure which eventually leads to the recruitment of ATM to the Mre11 complex. It should be stressed that this recruitment is not essential for ATM activation since mutation in Mre11 complex proteins only affects the efficiency of ATM activation (Uziel et al., 2003). This is also evident from a recent report showing that while the Mre11/Rad50 complex is necessary for ATM activation after low doses of radiation at short times post-irradiation, activation is independent of the complex at later times and after exposure to higher radiation doses (Cerosaletti and Concannon, 2004).

ATM is activated as a pre-existing protein by DNA double strand breaks (Shiloh, 2003). However there is also evidence that other stimuli can also lead to the activation of ATM (Lavin et al., 2003). It remains unclear as to how the process is initiated but activation involves autophosphorylation (Bakkenist and Kastan, 2003; Kozlov et al., 2003) and this autophosphorylation is regulated by phosphatase activities (Ali et al., 2004; Goodarzi et al., 2004). It appears that ATM is present as an inactive dimer in unirradiated cells and after DNA damage rapid intermolecular autophosphorylation on serine 1981 (Ser1981) leads to dimer dissociation and activation of ATM so that it is capable of phosphorylating downstream targets (Bakkenist and Kastan, 2003). However, Goodarzi and Lees-Miller (2004) failed to observe an alteration in apparent molecular weight of isolated ATM protein, regardless of the phosphorylation state of Ser1981. At a minimal radiation dose used to activate ATM (0.5Gy) it can be estimated that 15-20 double strand breaks arise in a cell (Cedervall et al., 1995). Under these conditions >50% of the ATM was autosphosphorylated. Thus it is evident that most of the ATM is activated remote from the site of a double strand break. This might be achieved by ATM responding to alterations in superhelicity of chromatin or having access to specific chromatin proteins. Bakkenist and Kastan, (2003) were able to demonstrate a weak autophosphorylation of ATM in response to hypotonicity, chloroquine treatment and in the presence of histone deacetylase inhibitors, all of which are capable of altering chromatin structure. Under these conditions Ser1981 phosphorylated ATM was diffusely distributed throughout the nucleus. This diffuse pattern was also seen immediately after γ-irradiation but some of this form of ATM localized to foci at later times, consistent with sites of DNA damage. It is unlikely that the agents that altered chromatin structure caused breaks in DNA since no γH2AX foci were observed (Bakkenist and Kastan, 2003).

More recently two other reports revealed that phosphatase activity plays a role in the activation of ATM (Ali et al., 2004; Goodarzi et al., 2004). Incubation of cells with the phosphatase inhibitor okadaic acid (OA), at concentrations that inhibit protein phosphatase PP2A, induced Ser1981 ATM autophosphorylation (Goodarzi et al., 2004). DNA double strand breaks were not induced with OA since there was no appearance of γH2AX foci. PP2A was active and constitutively bound to ATM and dissociated from ATM with loss of ATM-associated phosphatase activity in response to radiation exposure. These data suggest that PP2A plays an important role in regulating the autophosphorylation of ATM. A second phosphatase, protein serine-threonine phosphatase 5 (PP5), interacts with ATM in response to DNA damage (Ali et al., 2004). When this protein was downregulated DNA damage induced activation of ATM was reduced and a mutant form of PP5 interfered with Ser1981 autophosphorylation of ATM as well as phosphorylation of downstream substrates. Transfection of cells with this mutant PP5 also abrogated the S phase checkpoint. In short autophosphorylation is a key initial step in the activation of ATM and this is regulated by both PP2A and PP5 phosphatases. It is likely that this activation is more complex involving other changes in phosphorylation status of the protein since it has been shown by phosphopeptide mapping that several sites in ATM are phosphorylated in response to radiation (Kozlov et al., 2003). In addition, no Ser1981 autophosphorylation was observed in ATM during the first meiotic prophase even though ATM was active as illustrated by its localization to the meiotic chromosome (Hamer et al., 2004). Finally, while E2F1 overexpression induced ATM-dependent phosphorylation of Chk2, no Ser1981 phosphorylation of ATM was detected (Powers et al., 2004).

6.3 Cell Cycle Checkpoint Activation

ATM recognises double strand breaks in DNA and acts primarily as a cell cycle checkpoint activator (Lavin and Shiloh, 1997). Once activated it phosphorylates multiple substrates or indirectly regulates their phosphorylation to delay the progress of cells through the cycle, allowing time for DNA repair and ensuring chromosomal integrity (Figure 2).

6.3.1 G1/S Checkpoint

When A-T cells are exposed to radiation they fail to show the normal delay in progression from G1 into S phase (Nagasawa and Little, 1983; Beamish and Lavin, 1994). The nature of the defect involved was subsequently elucidated when it was shown that the radiation signal transduction pathway operating through p53 was defective in A-T cells (Kastan et al., 1992; Khanna and Lavin, 1993). Stabilization and activation of p53 leads to the induction of a number of effector proteins including p21/WAF1 which binds to cyclin E-Cdk2 kinase inhibiting its activity for downstream substrates, delaying the passage of cells from G1 into S phase (El-Deiry et al., 1993; Harper et al., 1993). All levels of this signalling pathway were shown to be defective in A-T cells (Canman et al., 1994; Khanna et al., 1995). While the rate of stabilization of p53 is defective in all A-T cells, it is nevertheless variable and in all cases stabilization is evident at later times post-irradiation. It is not clear what protein kinase takes over this role of ATM but both ataxia-telangiectasia and rad3-related protein (ATR) and SMG-1 have been suggested as candidates (Abraham, 2001; Abraham, 2004). ATM exerts its control on the G1/S checkpoint through several discrete modifications (Figure 2). After radiation exposure it directly phosphorylates p53 on Ser15 (Banin et al., 1998; Canman et al., 1998; Khanna et al., 1998) and Mdm2 on Ser395 (Khosravi et al., 1999). ATM- dependent phosphorylation of a second site on p53, Ser46, occurs as well as Thr68 phosphorylation of Chk2 (Saito et al., 2002; Hirao et al., 2000).

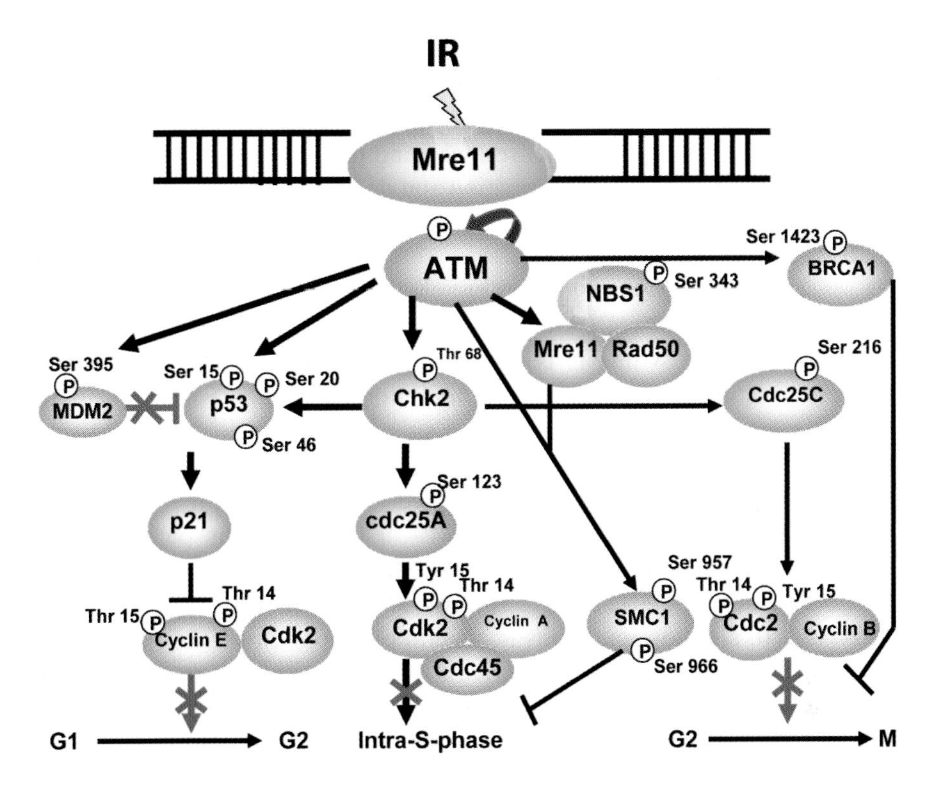

Figure 2. Cell cycle checkpoint activation in response to double strand breaks in DNA. ATM is activated by autophosphorylation to phosphorylate a number of downstream substrates involved in cell cycle checkpoint activation. Phosphorylation of Chk2 plays a role in activation of the G1/S, intra-S phase and G2/M checkpoints. Stabilization and transcriptional activation of p53 is central to the G1/S checkpoint. This is achieved by direct phosphorylation of p53 on Ser 15 and Ser 46 by ATM, and ATM-dependent phosphorylation at Ser 20 by Chk2. These modifications are part of a complex series of changes to p53 as part of its stabilization/activation. Stabilization is also facilitated by ATM-dependent phosphorylation of Mdm2 which reduces its affinity for p53. For the intra-S phase checkpoint Chk2 induced phosphorylation of CdC25A destabilizes this phosphatase and in turn inhibits the activation of the cyclinA-cdk2 complex to drive cells through S phase. A second parallel pathway involves ATM- and Nbs1-dependent phosphorylation of SMC1. Another member of the Cdc25 group (Cdc25c) is phosphorylated by Chk2 for G2/M checkpoint activation. The regulation of this checkpoint is also complex involving ATM-dependent phosphorylation of BRCA1.

These modifications to p53 are only a small proportion of the total changes that occur in the molecule in response to DNA damage and none of the ATM-induced changes *per se* appear to alter the stability of p53 (Blattner et al., 1997; Thompson et al., 2004). Nevertheless, early radiation-induced phosphorylation of p53 on Ser15 is a reliable indicator of ATM activation and together with the host of other modifications to p53 must have some influence on the transcriptional activity of p53, perhaps in the recruitment of co-activators (Zacchi et al., 2002). In addition to modifying p53 to contribute to its stabilization/activation, ATM also phosphorylates Mdm2 in (Ser395) a negative regulator of p53 (Khosravi et al., 1999).

This phosphorylation alters reactivity with an antibody directed against this region of the molecule pointing to a conformational change in Mdm2 (Maya et al., 2001). It is suggested that this may alter the affinity of Mdm2 for p53 making it less capable of re-localising p53 to the cytoplasm and less effective in promoting its ubiquitination and degradation.

6.3.2 S-Phase Checkpoint

When mammalian cells are exposed to radiation a biphasic pattern of inhibition of DNA synthesis is observed with the most marked effect on replicon initiation. A much reduced inhibition of DNA synthesis occurs in A-T cells in response to radiation and this has been called radioresistant DNA synthesis (RDS) (Houldsworth and Lavin, 1980). This phenomenon is still not elucidiated but can be explained by defects in more than one signalling pathway controlled by ATM. The two major pathways that control the S-phase checkpoint are ATM/Chk2/Cdc25A/cyclin A-Cdk2/ Cdc45 (Falck et al., 2001) and ATM/Nbs1/SMC1 (Kim et al., 2002; Yazdi et al., 2002). In addition mutations in several other proteins (BRCA1, 53BP1, FANCD2) also give rise to a defective S-phase checkpoint (Xu et al., 2001; Wang et al., 2002; Nakanashi et al., 2002). All of these undergo ATM-dependent phosphorylation post-irradiation and may participate in the two established pathways or be involved in some other pathway of S-phase control.

Falck et al., (2001) have established a functional link between ATM, the checkpoint signalling kinase, Chk2, and Cdc25A phosphatase and its downstream target for activation, Cdk2, in S phase (Fig.2). Radiation-induced degradation of Cdc25A requires both ATM and Chk2-mediated phosphorylation of Cdc25A on Ser123 and this prevents dephosphorylation of Cdk1 (also known as Cdc2), leading to a transient block in DNA replication. Exposure of A-T cells to radiation failed to cause an increase in Cdk2 Tyr 15 dephosphorylation or inhibition of cyclin E-Cdk2 kinase activity, consistent with the radioresistant DNA synthesis phenotype. Furthermore, Chk2 alleles defective in catalytic activity or ability to interact with Cdc25A had a dominant interfering effect and abrogated the S-phase checkpoint.

The second arm of the S-phase checkpoint is dependent on both ATM and Nbs1 (Kim et al., 2002; Yazdi et al., 2002). As discussed above there is a requirement for the Mrell/Nbs1/Rad50 complex for the efficient activation of ATM and ATM in turn phosphorylates Nbs1 for activation of the S-phase and G2/M checkpoints (Fig.2). Yazdi et al. (2002) showed that SMC1 phosphorylation was necessary for the S-phase checkpoint. They established that ATM phosphorylates SMC1 at two sites, Ser957 and Ser966, after irradiation and ATM – dependent phosphorylation of Nbs1 was required for this phosphorylation establishing a role for NbS1 as an adapter in this pathway. While this pathway is defective in NBS cells, the other S-phase checkpoint pathway (ATM/Chk2/Cdc25A) is intact. This may explain why it is that RDS is not as marked in NBS cells as it is in A-T cells. This is the case also for A-TLD patients with mutations in the Mre11 gene (Stewart et al., 1999).

6.3.3 G2/M Checkpoint

Suppression of the mitotic index is observed in response to radiation damage (Zampetti-Bosseler and Scott, 1981). This is due to arrest in G2 phase and abrogation of this checkpoint sensitizes cells to radiation (Bache et al., 2001). This checkpoint is defective in A-T cells as is evident by a reduced delay of cells in progressing into mitosis after irradiation in G2 phase (Scott and Zampetti-Bosseler, 1982). G2 phase delay is detected by irradiating cells, adding colchicine to block mitosis and by determining rate of accumulation of mitotic indices with time post-irradiation (Beamish and Lavin, 1994). Alternatively, it is possible to label cells in S phase with BrdU, block their passage into mitosis with nocodazole prior to irradiation (in G2 phase), and score for their ability to enter the next G1 phase (Beamish and Lavin, 1994). A third method involves distinguishing G2 phase cells from mitotic cells using histone H3 phosphospecific antibody (Xu et al., 2002). As with S and G1 phase checkpoints, it appears that ATM influences G2 arrest by more than one pathway (Fig.2). Chk2 is again a key player as for the two other checkpoints. Once activated by ATM, Chk2 phosphorylates Cdc25C on Ser216 which allows binding to 14-3-3 protein and sequestration to the cytoplasm (Peng et al., 1997; Dalal et al., 1999). This involves overlap with another pathway of ATM signalling, transcriptional activation of p53 which is responsible for 14-3-3 induction. Other effectors of p53, GADD45 and p21 also contribute to the eventual targeting of cyclin B-Cdk1 so that it remains phosphorylated, inactive and incapable of phosphorylating key substrates to drive cells from G2 into mitosis. Control of this checkpoint is even more complex since it also involves inactivation of Cdc25A by degradation and possibly Cdc25B by interaction with 14-3-3 (Bulavin et al., 2001).

A G_2/M checkpoint abnormality, similar to that seen in A-T cells, has been reported for the BRCA1-null cell line HCC1937 and a normal checkpoint was restored to these cells with BRCA1 transfection (Xu et al., 2001). Transfection of HCC1937 cells with a mutant form of a major ATM phosphorylation site in BRCA1 (S1423A) failed to complement the defective G_2/M checkpoint. Thus, it appears likely that radiation-induced phosphorylation of BRCA1 on Ser1423 by ATM is important in the regulation of the G_2/M checkpoint (Fig.2).

7. FUTURE DIRECTION

The processes by which a cell protects its genome against instability are complex and involves multiple pathways. ATM is a central player in responding to DNA double strand breaks. While great advances have been made in elucidating its mechanism of action questions remain as to how it is activated and the context in which it recognises individual substrates. While the Mrell complex is required for the efficient activation of ATM it is not an essential requirement since ATM becomes active by autophosphorylation on Ser1981 even in the absence of functional Mre11 or Nbs1. This raises the issue as to how and where ATM is so rapidly activated. The extent and distribution of ATM activation suggests that it is mostly activated remote from double strand breaks. What remains to be determined is whether this is achieved by interaction of ATM with altered chromatin structure and, if so, whether this involves interaction with a specific chromatin protein(s). To date only one modification, Ser1981 phosphorylation, has been implicated in the activation of ATM. However, there is evidence that multiple phosphorylations occur on ATM in response to exposure of cells to radiation (Bakkenist and Kastan, 2003; Kozlov et al., 2003). It will be of interest to identify other phosphorylation sites in ATM and determine whether these arose as a consequence of autophosphorylation or through an associated protein kinase. The functional significance of these modifications for ATM signalling will shed further light on signalling to DNA repair and cell cycle checkpoint activation.

A multitude of substrates have been described for ATM and it has been possible to associate the phosphorylation events involved with specific functions. No doubt there are additional substrates waiting in the wings to be included in existing signalling pathways or for which additional pathways will emerge. Of the existing substrates for ATM some are reliant on Nbs1 as an adapter. It will be important to establish why it is that only some substrates have this requirement and how it relates to the activation of ATM by the Mrell complex. In relation to cell cycle control, one emerging theme is the degree of complexity and the existence of parallel pathways to ensure that damaged cells delay in transiting the various cell cycle stages. It is evident that further intermediates/interactors will be added to the various pathways involved. The relative importance of these pathways in response to specific DNA lesions awaits further elucidation. It is not unreasonable to assume that defects in all these cell cycle checkpoints contribute to the greater propensity for chromosomal instability and in turn the higher risk of cancer in A-T patients.

ACKNOWLEDGEMENTS

The author is supported by grants from the Australian NHMRC, the Australian Research Council, the A-T Children's Foundation, the Queensland Cancer Fund and the NIH, Washington. Thanks to Tracey Laing for typing the manuscript.

REFERENCES

Abraham, R.T. 2001. Cell cycle checkpoint signalling through the ATM and ATR kinases. *Genes Dev.* 15: 2177-2196.

Abraham, R.T. 2004. The ATM-related kinase, hSMG-1, bridges genome and RNA surveillance pathways. *DNA Repair.* 3: 919-925.

Ali, A., J. Zhang, S. Bao, L. Liu, D. Otterness, N.M. Dean, R.T. Abraham, and X.F. Wang. 2004. Requirement of protein phosphatase 5 in DNA damage-induced ATM activation. *Genes Dev.* 18: 249-254.

Athma, P., R. Rappaport, and M. Swift. 1996. Molecular genotyping shows that ataxia-telangiectasia heterozygotes are predisposed to breast cancer. *Genet Cytogenet.* 92: 130-134.

Bache, M., S. Pigorsch, J. Dunst, P. Wurl, A. Meye, F. Bartel, H. Schmidt, F.W. Rath, and H. Taubert. 2001. Loss of G2/M arrest correlates with radiosensitization in two human sarcoma cell lines with mutant p53. *Int. J. Cancer.* 96: 110-117.

Bakkenist, C.J., M.B. Kastan. 2003. DNA damage activates ATM through intermolecular autophosphorylation and dimer dissociation. *Nature.* 421: 499-506.

Banin, S., L. Moyal, S. Shieh, Y. Taya, C.W. Anderson, L. Chessa, N.I. Smorodinsky, C. Prives, Y. Reiss, Y. Shiloh, and Y. Ziv. 1998. Enhanced phosphorylation of p53 by ATM in response to DNA damage. *Science.* 281: 1674-1677.

Beamish, H., M.F. Lavin. 1994. Radiosensitivity in ataxia-telangiectasia: anomalies in radiation-induced cycle delay. *Int J. Radiat Biol.* 65: 175-184.

Biemond, A. 1957. A Palaeocerebellar atrophy with extra-pyramidal manifestations in association with bronchiectasis and telangiectasis of the conjunctivabulbi as a familial syndrome. In van Bogaert L, Radermacker J, eds. Proc 1st Inter, Congr of Neurological Sciences, Brussels. London. *Pergamon Press.* 4: 206.

Blattner, F.R., G. Plunkett 3rd, C.A. Bloch, N.T. Perna, V. Burland, M. Riley, J. Collado-Vides, J.D. Glasner, C.K. Rode, G.F. Mayhew, J. Gregor, N.W. Davis, H.A. Kirkpatrick, M.A. Goeden, D.J. Rose, B. Mau, and Y. Shao. 1997. The complete genome sequence of Escherichia coli K-12. *Science.* 277: 1453-1474.

Boder, E. 1985. Ataxia-telangiectasia: an overview. In Gatti, R.A., Swift M, eds. Ataxia-Telangiectasia. New York: Alan R Liss, Kroc Foundation Series: 19: 1-63.

Boder, E., R.P. Sedgwick. 1957. Ataxia-telangiectasia. A familial syndrome of progressive cerebellar ataxia, oculocutaneous telangiectasia and frequent pulmonary infection. A preliminary report on 7 children, an autopsy, and a case history. *Univ S Calif Med Bull.* 9: 15-28.

Boder, E., R.P. Sedgwick. 1963. Ataxia-telangiectasia. A review of 101 cases. In: Walsh G, ed. *Little Club Clinics in Developmental Medicine no.8. London: Heinemann Medical Books.* 110-118.

Bulavin, D.V., Y. Higashimoto, I.J. Popoff, W.A. Gaarde, V. Basrur, O Potapova, E. Appella, and A.J. Fornace Jr. 2001. Initiation of a G2/M checkpoint after ultraviolet radiation requires p38 kinase. *Nature.* 411: 102-107.

Canman, C.E., A.C. Wolff, C.Y. Chen, A.J.Jr. Fornace, and M.B. Kastan. 1994. The p53-dependent G1 cell cycle checkpoint pathway and ataxia-telangiectasia. *Cancer Res.* 54: 5054-5058.

Canman, C.E., D.S. Lim, K.A. Cimprich, Y. Taya, K. Tamai, K. Sakaguchi, E. Appella, M.B. Kastan, and J.D. Siliciano. 1998. Activation of the ATM kinase by ionizing radiation and phosphorylation of p53. *Science.* 281: 1677-1679.

Carbonari, M., M. Cherchi, R. Paganelli, G. Giannini, E. Galli, C Gaetan, C. Papetti, and M. Fiorilli. 1990. Relative increase of T cells expressing the gamma/delta rather than the alpha/beta receptor in ataxia-telangiectasia. *N Engl J Med.* 322: 73-76.

Carney, J.P., R.S. Maser, H. Olivares, E.M. Davis, M. Le Beau, Yates, Jr., 3rd, L. Hays, W.F. Morgan, and J.H. Petrini. 1998. The hMRE11 protein complex and Nijmegen breakage syndrome: linkage of double-strand break repair to the cellular DNA damage response. *Cell.* 93: 477-486.

Carson, C.T., R.A. Schwartz, T.H. Stracker, C.E. Lilley, D.V. Lee, and M.D. Weitzman. 2003. The Mre11 complex is required for ATM activation and the G2/M checkpoint. *EMBO J.* 22: 6610-6620.

Cedervall, B., P. Kallman, and W.C. Dewey. 1995. Repair of DNA double-strand breaks: errors encountered in the determination of half-life times in pulsed-field gel electrophoresis and neutral filter elution. *Radiat Res.* 142: 23-28.

Cerosaletti, K., P. Concannon. 2004. Independent roles for nibrin and Mre11-Rad50 in the activation and function of Atm.. *J. Biol Chem.* 279: 38813-38819.

Chenevix-Trench, G., A.B. Spurdle, M. Gatei, H. Kelly, A. Marsh, X. Chen, K. Donn, M. Cummings, D. Nyholt, M.A. Jenkins, C. Scott, G.M. Pupo, T. Dork, R. Bendix, J. Kirk, K. Tucker, M.R. McCredie, J.L. Hopper, J. Sambrook., G.J. Mann, and K.K. Khanna. 2002. Dominant negative ATM mutations in breast cancer families. J. Natl. *Cancer Inst.* 94: 205-215.

Cohen, M.M., S.J. Simpson. 1982. The Effect of bleomycin on DNA synthesis in ataxia-telangiectasia lymphoid cells. Envi*ron Mutagen.* 4: 27-36.

Cornforth, M.W., J.S. Bedford. 1985. On the nature of a defect in cells from individuals with ataxia-telangiectasia. *Science.* 227: 1589-1591.

Costanzo, V., K. Robertson, C.Y. Ying, E. Kim, E. Avvedimento, M. Gottesman, D. Grieco, and J. Gautier. 2000. Reconstitution of an ATM-dependent checkpoint that inhibits chromosome DNA replication following DNA damage. *Mol Cell.* 6: 649-659.

Costanzo, V., K. Robertson, and J. Gautier. 2004. Xenopus cell-free extracts to study the DNA damage response. *Methods Mol Biol.* 280: 213-227.

Dalal, S.N., C.M. Schweitzer, J. Gan, and J.A. DeCaprio. 1999. Cytoplasmic localization of human cdc25C during interphase requires an intact 14-3-3 binding site. *Mol Cell Biol.* 19: 4465-4479.

De Wit, J., N.G.L. Jaspers, and D. Bootsma. 1981. The rate of DNA synthesis in normal and ataxia-telangiectasia cells after exposure to X-irradiation. *Mutat Res.* 80: 221-226.

Dork, T., R. Bendix, M. Bremer, D. Rades, K. Klopper, M. Nicke, B. Skawran, A. Hector, P. Yamini, D. Steinmann, S. Weise, M. Stuhrmann, and J.H. Karstens. 2001. Sprectrum of ATM gene mutations in a hospital-based series of unselected breast cancer patients. *Cancer Res.* 61: 7608-7615.

Edwards, M.J., A.M.R. Taylor. 1981. Unusual levels of (ADP-ribose) and DNA synthesis in ataxia-telangiectasia cells following γ-ray irradiation. *Nature.* 287: 745-747.

El-Deiry, W.S., T. Tokino, V.E. Velculescu, D.B. Levy, R. Parsons, J.M. Trent, D. Lin, W.E. Mercer, K.W. Kinzler, and B. Vogelstein. 1993. WAF1, a potential mediator of p53 tumour suppression. *Cell.* 75: 817-825.

Falck, J., N. Mailand, R.G. Syljuasen, J. Bartek, and J. Lukas. 2001. The ATM-Chk2-Cdc25A checkpoint pathway guards against radioresistant DNA synthesis. *Nature.* 410: 842-847.

Fitzgerald, M.G., J.M. Bean, S.R. Hedge, H. Unsal, D.J. MacDonald, D.P. Harkin, D.M. Finkelstein, K.J. Isselbacher, and D.A. Haber. 1997. Heterozygous ATM mutations do not contribute to early onset breast cancer. *Nature.* 15: 307-310.

Foray, N., A. Priestly, G. Alsbeith, C. Badie, E.P. Capulas, C.F. Arlett, and E.P. Malaise. 1997. Hypersensitivity of ataxia-telangiectasia fibroblasts to ionizing radiation is associated with a repair deficiency of DNA double-strand breaks. *Int J Radiat Biol.* 72: 271-283.

Goodarzi, A.A., W.D. Block, and S.P. Lees-Miller. 2003. The role of ATM and ATR in DNA damage-induced cell cycle control. *Prog Cell Cycle Res.* 5: 393-411.

Goodarzi, A.A., J.C. Jonnalagadda, P. Douglas, D. Young, R. Ye, G.B. Moorhead, S.P. Lees-Miller, and K.K. Khanna. 2004. Autophosphorylation of ataxia-telangiectasia mutated is regulated by protein phosphatase 2A. *EMBO J.* 23: 4451-4461.

Goodarzi, A.A., S.P. Lees-Miller. 2004. Biochemical characterization of the ataxia-telangiectasia mutated (ATM) protein from human cells. *DNA Repair*. 3: 753-767.

Hamer, G., H.B. Kal, C. H. Westphal, T. Ashley, and D. G. de Rooij. 2004. Ataxia-telangiectasia mutated expression and activation in the Testis. *Biology of Reproduction*. 70: 1206-1212.

Harper, J.W., G.R. Adami, N. Wei, K. Keyomarsi, and S.J. Elledge. 1993. The p21 Cdk-interacting protein Cip1 is a potent inhibitor of G1 cyclin-dependent kinases. *Cell*. 75: 805-816.

Hecht, F., R.D. Koler, D.A. Rigas, G.S. Dahnke, M.P. Case, V. Tisdale, and R.W. Miller. 1966. Leukaemia and lymphocytes in ataxia-telangiectasia. *Lancet* 2: 1193.

Hecht, F., B.K. Hecht. 1990. Cancer in ataxia-telangiectasis patients. *Cancer Genet Cytogenet*. 46: 9-19.

Hecht, F., R.D. Koler, D.A. Rigas, G.S. Dahnke, M.P. Case, V. Tisdale, and R.W. Miller. 1966. Leukaemia and lymphocytes in ataxia-telangiectasis. *Lancet*. 2: 1193.

Hecht, F., B.K. McCaw, and R. Koler. 1973. Ataxia-telangiectasia-clonal growth of translocation lymphocytes. *N Engl J Med*. 289: 286-291.

Higurashi, M., P.E. Conen. 1973. *In vitro* chromosomal radiosensitivity in 'chromosomal breakage syndromes'. *Cancer*. 32: 380-383.

Hirao, A., Y.Y. Kong, S. Matsuoka, A. Wakeham, J. Ruland, H. Yoshida, D. Liu, S.J. Elledge, and T.W. Mak. 2000. DNA damage-induced activation of p53 by the checkpoint kinase Chk2. *Science*. 287: 1824-1827.

Houldsworth, J., M.F. Lavin. 1980. Effect of ionizing radiation on DNA synthesis in ataxia-telangiectasia cells. *Nucleic Acids Res*. 8: 3709-3720.

Izatt, L., J. Greenman, S. Hodgson, D. Ellis, S. Watt, G. Scott, C. Jacobs, R. Liebmann, M.J. Zvelebil, C. Mathew, and E. Solomon. 1999. Identification of germline missense mutations and rare allelic variants in ATM gene in early-onset breast cancer. *Genes Chromosomes Cancer*. 26: 286-294.

Janin, N., N. Andrieu., K. Ossian, A. Lauge, M.F. Croquette, C. Griscelli, M. Debre, B. Bressac-de-Paillerets, A. Aurias, and D. Stoppa-Lyonett. 1999. Breast cancer risk in ataxia-telangiectasia (AT) heterozygotes: haplotype study in French AT families. *Br. J. Cancer*. 80: 1042-1045.

Jeggo, P., B. Singleton, H. Beamish, and A. Priestley. 1999. Double strand break rejoining by the Ku-dependent mechanism of non-homologous end-joining. *CR ACAD Sci III*. 322: 109-112.

Johnson, R.D., M. Jasin. 2000. Sister chromatid gene conversion is a prominent double-strand break repair pathway in mammalian cells. *EMBO J*. 19: 3398-3407.

Kastan, M.B., O. Zhan, W.S. EL-Deiry, F Carrier, T. Jacks, W.V. Walsh, B.S. Plunkett, B. Vogelstein, and A.J. Fornace, 1992. A mammalian cell cycle checkpoint pathway utilizing p53 and GADD45 is defective in ataxia-telangiectasia. *Cell*. 71: 587-597.

Khanna, K.K. M.F. Lavin. 1993. Ionizing radiation and UV induction of p53 protein by different pathways in ataxia-telangiectasia cells. *Oncogene*. 8: 3307-3312.

Khanna, K.K., H. Beamish, Y. Yan, K. Hobson, R. Williams, I. Dunn, and M.F. Lavin. 1995. Nature of G1/S cell cycle checkpoint defect in ataxia-telangiectasia. *Oncogene* 17: 609-618.

Khanna, K.K., K.E. Keating, S. Kozlov, S. Scott, M. Gatei, K. Hobson, Y. Taya, B. Gabrielli, D. Chan, S.P. Lees-Miller, and M.F. Lavin. 1998. ATM associates with and phosphorylates p53: mapping the region of interaction. *Nat Genet*. 20: 398-400.

Khosravi, R., R. Maya, T. Gottlieb, M. Oren, Y. Shiloh, and D. Shkedy. 1999. Rapid ATM-dependent phosphorylation of MDM2 precedes p53 accumulation in response to DNA damage. *Proc Natl Acad Sci USA* 96: 14973-14977.

Kim, J.S., T.B. Krasieva, V. LaMorte, A.M. Taylor, and K. Yokomori. 2002. Specific recruitment of human cohesin to laser-induced DNA damage. *J. Biol. Chem*. 277: 45149-45153.

Kim, S.T., B. Xu, and M.B. Kastan. 2002. Involvement of the cohesin protein, Smc1, in Atm-dependent independent responses to DNA damage. *Genes Dev.* 16: 560-570.
Kozlov, S., N. Gueven, K. Keating, J. Ramsay, and M.F. Lavin. 2003. ATP activates ataxia-telangiectasia mutated (ATM) in vitro. Importance of autophosphorylation. *J. Biol Chem.* 278: 9309-9317.
Larson, G. 1998. An allelic variant at the ATM locus is implicated in breast cancer susceptibility. *Genetic Testing.* 1: 165-170.
Lavin, M.F., S. Scott, P. Chen, S. Kozlov, N. Gueven, and G. Birrell. 2003. In Handbook of Cell Signaling vol 3. (Eds. R.A. Bradshaw and E.A. Dennis) Elsevier Science, pp 225-236.
Lavin, M.F. 2004. The Mre11 complex and ATM: a two-way functional interaction in recognising and signalling DNA double strand breaks. *DNA Repair.* 3: 1515-1520.
Lavin, M.F., H. Lederman. 2004. Chromosomal Breakage Syndromes associated with immunodeficiency. In: Stiehm E.R. ed. *Immunologic Disorders in Infants & Children.* 580-605.
Lavin, M.F., Y. Shiloh. 1997. The genetic defect in ataxia-telangiectasia. *Annu Rev Immunol.* 15: 177-202.
Lee, J.H. T.T. Paull. 2004. Direct activation of the ATM protein kinase by the Mre11/Rad50/Nbs1 complex. *Science.* 304: 93-96.
Lehmann, A.R., M.R. James, and S. Stevens. 1982. Miscellaneous observations on DNA repair in ataxia-telangiectasia. In Bridges B A, Harnden D G, eds. A Cellular and Molecular Link with Cancer. *New York: John Wiley and Sons.* 347-353.
Lim, D.S., S.T. Kim, B. Xu, R.S. Maser, J. Lin, J.H. Petrini, and M.B. Kastan. 2000. ATM phosphorylates p95/nbs1 in an S-phase checkpoint pathway. *Nature.* 404: 613-617.
Maya, R., M. Balass, S.T. Kim, D. Shkedy, J.F. Leal, O. Shifman, M. Moas, T. Buschmann, Z. Ronai, Y. Shiloh, M.B. Kastan, E. Katzir, and M. Oren. 2001. ATM-dependent phosphorylation of Mdm2 on serine 395: role in p53 activation by DNA damage. *Genes Dev.* 15: 1067-1077.
McFarlin, D.E., W. Strober, and T.A. Waldmann. 1972. Ataxia-telangiectasia. *Medicine.* 51: 281-314.
Meek, K., S. Gupta, D.A. Ramsden, and S.P. Lees-Miller. 2004. The DNA-dependant protein kinase: the director at the end. *Immunol Rev.* 200: 132-141.
Miller, R.W. 1967. Persons with exceptionally high risk of leukaemia. *Cancer Res.* 27: 2420-2423.
Mirzoeva, O.K. J.H. Petrini. 2001. DNA damage-dependent nuclear dynamics of the Mre11 complex. *Mol Cell Biol.* 21: 281-288.
Moore, J.K. J.E. Haber. 1996. Cell cycle and genetic requirements of two pathways of nonhomologous joining repair of double-strand breaks in Saccharomyces cerevisiae. *Mol Cell Biol.* 16: 2164-2173.
Morrell, D., C.L. Chase, L.L. Kupper, and M. Swift. 1986. Diabetes mellitus in ataxia-telangiectasia, Fanconis anemia, xeroderma pigmentosum, common variable immune deficiency, and severe combined immune deficiency families. *Diabetes.* 35: 143-147.
Morrell, D., C.L. Chase, and M. Swift. 1990. Cancers in 44 families with ataxia-telangiectasia. *Cancer Genet Cytogenet.* 50: 119-123.
Morris, C., R. Mohamed, and M.F. Lavin. 1983. DNA replication and repair in atxia-telangiectasia cells exposed to bleomycin. *Mutat Res.* 112: 67-74.
Nagasawa, H., J.B. Little. 1983. Comparison of kinetics of X-ray-induced cell killing in normal, ataxia-telangiectasia and hereditary retinoblastoma fibroblasts. *Mutat Res.* 109: 297-308.
Nakanishi, K., T. Taniguchi, V. Ranganathan, H.V New, L.A. Moreau, M. Stotsky, C.G. Mathew, M.B. Kastan, D.T. Waever, and A.D. D'Andrea. 2002. Interaction of FANCD2 and NBS1 in the DNA damage response. *Nat, Cell Biol.* 4: 913-920.
Olsen, J.H., J.M. Hahnemann, A.L. Borresen-Dale, K. Brondum-Nielsen, L. Hammarstrom, R. Kleinerman, H. Kaariainen, T. Lonnqvist, R. Sankila, N. Seersholm, S. Tretli, J. Yuen,

J.D. Boice Jr, and M. Tucker. 2001. Cancer in patients with ataxia-telangiectasia and in their relatives in the nordic countries. *J. Natl. cancer Inst.* 93: 121-127.

Pandita, T.K., W.N. Hittleman. 1992. Initial chromosome damage but not DNA damage is greater in ataxia-telangiectasia cells. *Radiat Res.* 130: 94-103.

Paterson, M.C., P.J. Smith. 1979. Ataxia-telangiectasia, an inherited human disorder involving hypersensitivity to ionizing radiation and related DNA-damaging chemicals. *Annu Rev Genet.* 13: 291-318.

Paull, T.T., M. Gellert. 1998. The 3' to 5' exonuclease activity of Mre11 facilitates repair of DNA double strand breaks. *Mol Cell.* 1: 969-979.

Peng, C.Y., P.R. Graves, R.S. Thoma, Z. Wu, A.S. Shaw, and H. Piwnica-Worms. 1997. Mitotic and G2 checkpoint control: regulation of 14-3-3 protein binding by phosphorylation of Cdc25C on Ser-216. *Science.* 277: 1501-1505.

Peterson, R.D., W.D. Kelly, and R.A. Good. 1964. Ataxia-telangiectasia: its association with a defective thymus, immunological-defective thymus, immunological-deficiency disease, and malignancy. *Lancet.* 1: 1189-1193.

Pfeiffer, P., W. Goedecke, S. Kuhfittig-Kulle, and G. Obe. 2004. Pathways of DNA double-strand break repair and their impact on the prevention and formation of chromosomal aberrations. *Cytogenet Genome Res.* 104: 7 -13.

Powers, J.T., S. Hong, C.N. Mayhew, P.M. Rogers, E.S. Knudsen, and D.G. Johnson. 2004. E2F1 uses the ATM signaling pathway to induce p53 and Chk2 phosphorylation and apoptosis. *Mol Cancer Res.* 2: 203-214.

Rary, J.M., M.A. Bender, and T.E. Kelly. 1975. A 14/14 marker chromosome lymphocyte clone in ataxia-telangiectasia. *J Hered.* 66: 33-35.

Saito, H., Y. Minamiya, S. Saito, and J. Ogawa. 2002. Endothelial Rho and Rho kinase regulate neutrophil migration via endothelial myosin light chain phosphorylation. *J. Leuko Biol.* 72: 829-836.

Savitsky, K., A. Bar-Shira, G. Gilad, G. Rotman, Y. Ziv, L. Vanagaite, D.A Tagle, S. Smith, T. Uziel, S. Sfez, M. Ashkenazi, I. Pecker, M. Fryman, R. Harnik, S.R. Patanjali, A. Simmons, G.A. Clines, A. Sartiel, R.A. Gatti, L. Chessa, O. Sanal, M.F. Lavin, N.G.J. Jaspers, A.M.R. Taylor, C.F. Arlett, T. Miki, S.M. Weissman, M. Lovett, F.S. Collins, and Y. Shiloh (1995) A single ataxia-telangiectasia gene with a product similar to PI-3 kinase. *Science.* 268: 1749-1753.

Scott, S.P., R. Bendix, P. Chen, R. Clark, T. Dork, and M.F. Lavin. 2002. Missense mutations but not allelic variants alter the function of ATM by dominant interference in patients with breast cancer. *Proc Natl Acad Sci USA.* 99: 925-930.

Scott, D. and F. Zampetti-Bosseler. 1982. Cell cycle dependence of mitotic delay in X-irradiated normal and ataxia-telangiectasia fibroblasts. *Int. J. Radiat. Biol. Relat. Stud. Phys. Chem. Med.* 42: 679-683.

Sedgwick, R.P. and E. Boder (1991) Ataxia-telangiectasia. In Vianney De Jong J.M.B., ed. Hereditary Neuropathies and Spinocerebellar Atrophies. Amsterdam: Elsevier Science, pp 347-423. 1991.

Shiloh, Y., E. Tabor, and Y. Becker. 1983. Abnormal response of ataxia-telangiectasia cells to agents that break the deoxyribose moiety of DNA via a targeted free radical mechanism. *Carcinogenesis.* 4: 1317-1322.

Shiloh, Y. 2003. ATM and related protein kinases: safeguarding genome integrity. *Nature Rev Cancer.* 3: 155-168.

Spector, B.D., A.H. Filipovich, G.S.III. Perry, and J.H. Kersey. 1982. Epidemiology of cancer in ataxia-telangiectasia. In Bridges B A, Harnden D G, eds. Ataxia-Telangiectasia. *Chichester: Wiley.* 103-138.

Stewart, G.S., R.S. Maser, T. Stankovic, D.A. Bressan, M.I. Kaplan, N.G. Jaspers, A. Raams, P.J. Byrd, J.H. Petrini, and A.M. Taylor. 1999. The DNA double-strand break repair gene hMRE11 is mutated in individuals with an ataxia-telangiectasia-like disorder. *Cell* 99: 577-587.

Stewart, G.S., J.I. Last, T. Stankovic, N. Haites, A.M. Kidd, P.J. Bryd, and A.M.Taylor. 2001. Residual ataxia-telangiectasia mutated protein function in cells from ataxia-telangiectasia patients, with 5762ins 137 and 7271T >G mutations showing a less severe phenotype. *J. Biol Chem.* 276: 30133-30141.

Stracker, T.H., J.W. Theunissen, M. Morales, and J.H. Petrini. 2004. The Mre11 complex and the metabolism of chromosome breaks: the importance of communicating and holding things together. *DNA Repair (Amst).* 3: 845-854.

Swift, M., D. Morrell, R.B. Massey, and C.L. Chase. 1991. Incidence of cancer in 161 families affected by ataxia-telangiectasia. *N Eng J Med.* 325: 1831-1836.

Taylor, A.M.R. 1982. Cytogenetics of ataxia-telangiectasia. In bridges B A, Harnden D G, eds. Ataxia-Telangiectasia- A Cellular and Molecular Link Between Cancer, Neuropathology and Immune Deficiency. *New York: Wiley.* 53-83.

Taylor, A.M., D.G. Harnden, C.F. Arlett, S.A. Harcourt, A.R. Lehmann, S. Stevens, and B.A. Bridges. 1975. Ataxia-Telangiectasia: a human mutation with abnormal radiation sensitivity. *Nature.* 4: 427-429.

Taylor A.M., J.R. Matcalfe, J.M. Oxford, and D.G. Harnden. 1976. Is chromatid-type damage in ataxia-telangiectasia after irradiation at G0 a consequence of defective repair? *Nature.* 260: 441-443.

Teraoka, S.N., K.E. Malone, D.R. Doody, N.M. Suter, E.A. Ostrander, J.R. Daling, and P. Concannon 2001. Increased frequency of ATM mutations in breast carcinoma patients with early onset disease and positive family history. *Cancer.* 92: 479-487.

Thompson, T., C. Tovar, H. Yang, D. Carvajal, B.T. Vu, Q. Xu, G.M. Wahl, D.C. Heimbrook, and L.T. Vassilev. 2004. Phosphorylation of p53 on key serines is dispensable for transcriptional activation and apoptosis. *J. Biol Chem.* (October 6, published ahead of print).

Trujillo, K.M., S.S. Yuan, E.Y. Lee, and P. Sung. 1998. Nuclease activities in a complex of human recombination and DNA repair factors Rad50, Mre11 and p95. *J. Biol Chem.* 273: 21447-21450.

Uziel, T., Y. Lerenthal, L. Moyal, Y. Andegeko, L. Mittelman, and L. Shiloh. 2003. Requirement of the MRN complex for ATM activation by DNA damage. *EMBO J.* 22: 5612-5621.

Waldmann T.A. Immunological abnormalities in ataxia-telangiectasia. 1982. In Bridges B A, Harden D G, eds. Ataxia-Telaniectasia: A Cellular and Molecular Link Between Cancer Neuropathology and Immune Deficiency. *New York: Wiley.* 37-51.

Wang, Y., D. Cortez, P. Yazdi, N. Neff, S.J. Elledge, and J. Qin. 2000. BASC, a super complex of BRCA1-associated proteins involved in the recognition and repair of aberrant DNA structures. *Genes Dev.* 14: 927-939.

Wang, B., S. Matsuoka, P.B. Carpenter, and S.J. Elledge. 2002. 53BP1, a mediator of the DNA damage checkpoint. *Science.* 298: 1435-1438.

Xu, B., S.T. Kim, and M.B. Kastan. 2001. Involvement of Brca1 in S-phase and G(2)-phase checkpoints after ionizing irradiation. *Mol Cell Biol.* 21: 3445-3450.

Xu, B. S. T. Kim, D.S. Lim, and M.B. Kastan. 2002. Two molecularly distinct g(2)/m checkpoints are induced by ionizing irradiation. *Mol. Cell Biol.* 22: 1049-1059.

Zacchi, P., M. Gostissa, T. Uchida, C. Salvagno, F. Avolio, S. Volinia, Z. Ronai, G. Blandino, C. Schneider, and G. Del Sal. 2002. The prolyl isomerase Pin1 reveals a mechanism to control p5 functions after genotoxic insults. *Nature.* 419: 853-857.

Yazdi, P.T., Y. Wang, S. Zhao, N. Patel, E.Y. Lee, and J. Qin. 2002. SMC1 is a downstream effector in the ATM/NBS1 branch of the human phase checkpoint. *Genes Dev.* 16: 571-582.

Zampetti-Besseler, F. and D. Scott. 1981. Cell death, chromosome damage and mitotic delay in normal human, ataxia-telangiectasia and retinoblastoma fibroblasts after x-irradiation. *Int. J. Radiat. Biol. Relat. Stud. Phys. Chem. Med.* 39: 547-558.

Chapter 4.3

MITOTIC CHECKPOINT, ANEUPLOIDY AND CANCER

Tim J. Yen[1] and Gary D. Kao[2]
[1] *Fox Chase Cancer Center, Philadelphia, USA;* [2] *Dept of Radiation Oncology, University of Pennsylvania, Philadelphia, USA*

1. INTRODUCTION

The mitotic checkpoint is a failsafe mechanism that prevents cells with unaligned chromosomes from prematurely exiting mitosis. As chromosome instability (CIN) and aneuploidy are features common amongst many cancers, the mitotic checkpoint may play a pivotal role in promoting tumorigenesis. The discovery of an evolutionarily conserved set of mitotic checkpoint genes has stimulated efforts to examine their importance in the origin of cancer and their use in potential new therapies for cancer.

The transformation of a normal cell into a malignant tumour is a multi-step process that cannot be accounted for by spontaneous mutations that arise from the inherent error rates that are associated with DNA replication and repair. Indeed, it has been estimated that the natural mutation rate is so low that it cannot generate enough mutations for cancer to develop within a human lifespan (Loeb, 1991; Orr-Weaver and Weinberg, 1998). That a majority of human cancers exhibit gross chromosome abnormalities and gene mutations strongly suggests that carcinogenesis is driven by mechanisms that actively destabilise the genome. Microsatellite instability (MIN) is one such mechanism whereby increased mutation rate at the nucleotide level is attributed to defects in DNA mismatch repair genes (Lengauer et al., 1998). Chromosome instability (CIN) is a second mechanism that promotes genome instability through the loss or gain of chromosomes. Indeed, many types of tumours are aneuploid and *in vitro* studies of colorectal cancer cell lines have shown a defect in maintaining a stable karyotype (Cahill et al., 1998; Pihan and Doxsey, 1999; Takahashi et al., 1999). Only recently have there been mechanistic advances in understanding the molecular basis for CIN. These insights came to light

E.A. Nigg (ed.), Genome Instability in Cancer Development, 477-499.

largely from a convergence of genetic and biochemical studies of the mitotic checkpoint, a mechanism that prevents aneuploidy by ensuring that cells with even a single unaligned chromosome cannot exit mitosis. This chapter will review our current knowledge of the mitotic checkpoint and examine its relationship with tumorigenesis.

2. CHROMOSOME SEGREGATION

2.1 Kinetochore Functions

The kinetochore is a macromolecular complex that is localised at the centromeres of chromosomes where it plays an essential role in mediating attachment of the chromosome to the spindle. Kinetochores interact with microtubules differently than other organelles (i.e. vesicles) in that they bind to the highly dynamic plus ends of microtubules, rather than along the side or the lattice of the microtubule (Rieder and Salmon, 1994). Thus, unlike organelles that rely on motors to translocate them along a static microtubule surface, the motility of chromosomes is specified by the kinetochore's ability to remain attached to the end of a microtubule that is rapidly switching between elongating and shortening states. How this is achieved is not clear but is likely to be specified by the combined and coordinated activities of a plethora of microtubule binding proteins and molecular motors that localise to kinetochores (Biggins and Walczak, 2003). Despite the large numbers of microtubule binding proteins at kinetochores, microtubule connections are established by chance encounters that depend on the location of the chromosome relative to the spindle. Chromosomes situated near the centre of the spindle rapidly establish bipolar connections as both kinetochores encounter microtubules at a high frequency. Chromosomes situated near a pole will rapidly establish a monopolar attachment but attachment to the opposite pole requires significantly more time as the frequency at which they encounter microtubules from the opposite pole is low.

As with many situations that rely on chance, there is the potential for mistakes. Kinetochores are not an exception as they can establish non-productive interactions as in cases when both kinetochores are connected to the same pole or one kinetochore is connected to microtubules from both poles and combinations of the two. These aberrant connections, if unresolved, can lead to chromosome fragmentation or nondisjunction (Cimini et al., 2001).

2.2 The Mitotic Checkpoint

The stochastic and error-prone nature by which chromosomes establish connections to the spindle explains why chromosomes cannot achieve metaphase alignment synchronously (Nicklas, 1997; Nicklas and Ward, 1994). The cell is therefore confronted with the problem of knowing when all of its chromosomes are aligned before it decides to proceed into anaphase. This problem is solved by the mitotic checkpoint which is a failsafe mechanism that monitors kinetochore microtubule attachments so that a single defective kinetochore will delay the onset of anaphase. In order to satisfy this checkpoint, kinetochores must be fully saturated with microtubules (~25 microtubules per kinetochore for mammals) and sufficient tension develops between sister kinetochores as a result of opposing poleward forces that attempt to pull apart the sister chromatids. If either of these parameters is not fulfilled, the checkpoint must then execute a program that inhibits mitotic exit. This program can be envisioned as a signalling cascade whereby a localised defect at a single kinetochore alters the global biochemical status of the cell. The nature by which the kinetochore generates the "wait anaphase" inhibitory signal is not entirely clear. However, the target of the "wait anaphase" signal is the Anaphase Promoting Complex/Cyclosome (APC/C) (Skibbens and Hieter, 1998), an E3 ubiquitin ligase that promotes the degradation of key proteins to irreversibly drive cells from metaphase to anaphase (King et al., 1995; Sudakin et al., 1995). Thus, the mitotic checkpoint is a highly complex program that consists of multiple modules, defects in any of its components will result in chromosome instability that promotes tumorigenesis.

2.2.1 Mitotic Checkpoint Proteins Monitor Kinetochore Attachments

The molecular components of the mitotic checkpoint are specified by a collection of evolutionarily conserved genes that include Mad1, Mad2, Mad3 (BubR1), Bub1, Bub3 and Mps1 (Musacchio and Hardwick, 2002). With the exception of Mps1, Bub1 and BubR1, which are protein kinases, the biochemical functions of the remaining proteins are not well understood. Cytological studies have shown that all of these proteins bind to kinetochores and are thus likely to be involved in monitoring attachments or generating the "wait anaphase" signal. In the case of Mad1 and Mad2, their preferential localisation at unattached kinetochores (Chen et al., 1996; Li and Benezra, 1996; Waters et al., 1998) suggests that they may be part of a counting mechanism that monitors microtubule occupancy and contributes to the generation of the "wait anaphase" signal. As kinetochores become saturated with microtubules, these proteins are released and the checkpoint signalling is silenced.

The existence of a tension-sensitive checkpoint was first demonstrated in insect spermatocytes where application of an external force to the unattached kinetochore of a monopolar chromosome would relieve the checkpoint induced delay in meiosis (Li and Nicklas, 1995). The spermatocytes proceeded into anaphase because the external force exerted tension that would normally be applied by microtubule attachments. Although this result was inconsistent with the idea that checkpoint is silenced only when both kinetochores are saturated with microtubules and develop tension, this discrepancy may be attributed to the difference between mitotic versus meiotic systems. Evidence supporting a tension-sensitive checkpoint in somatic cells has come from experiments that examined the mechanism by which the anti-cancer drug, Taxol arrests mammalian cells in mitosis (Waters et al., 1998). Cells treated with Taxol are able to establish a full complement of kinetochore microtubule attachments. However, tension did not develop because the drug suppressed microtubule dynamics that normally contributed to the generation of poleward force. As the vast majority of the bipolar attached kinetochores lacked detectable Mad2, it suggested that Mad2 does not respond to loss of tension and that other components were maintaining the checkpoint arrest. Bub1 and BubR1 (Bub1-related) kinases are candidates as they are present at kinetochores with reduced tension (Skoufias et al., 2001). However, it is important to point out that unlike Mad2, neither of these proteins completely dissociate from kinetochores of aligned chromosomes (Hoffman et al., 2001). Thus, their presence alone cannot be used as an indicator of whether the checkpoint is active.

2.2.2 Tension Sensing helps Resolve Aberrant Kinetochore Attachments

The importance of a tension-sensitive checkpoint is not to merely allow cells to arrest in response to microtubule poisons such as taxol. The need arises because kinetochores can form aberrant attachments where both sister kinetochores are attached to the same pole (syntelic) or the same kinetochore is attached to both poles (merotelic). In both instances, kinetochores are fully saturated with microtubules but lack tension (Cimini et al., 2001). Clearly, if microtubule occupancy was the only criterion that is used by the checkpoint, these aberrant connections would go unchecked and thus lead to non-disjunction or broken chromosomes. Recent studies in both yeast and human cells indicate that the aurora B/Ipl1 kinase is responsible for monitoring tension in order to resolve merotelic and syntelic attachments (Ditchfield et al., 2003; Hauf et al., 2003; Tanaka et al., 2002). How aurora B responds to lack of tension is unclear but it is thought to stimulate the release of microtubules by regulating the microtubule depolymerase activity of the kinesin-like MCAK protein (Andrews et al., 2004; Lan et al., 2004). Despite these intriguing findings, there is uncertainty as to whether aurora B indeed defines the tension-sensitive arm of the mitotic checkpoint as it may be part of an elaborate self-correcting mechanism (Andrews et al., 2003). That aurora B is essential for taxol-induced mitotic arrest may result from an indirect action on the checkpoint (Ditchfield et al., 2003; Hauf et al., 2003). The mitotic checkpoint may be responding not directly to loss of tension but rather to the presence of detached kinetochores that are induced by aurora B. This possibility is consistent with the observation that there are always a few kinetochores that retain Mad2 in Taxol arrested cells (Waters et al., 1998).

2.2.3 The Anaphase Promoting Complex is the Target of the Mitotic Checkpoint

There are two models proposed to explain how defective kinetochores can inhibit APC/C activity (Chan and Yen, 2003). In the "Sequestration Model" a dynamic pool of Mad2 cycles through unattached kinetochores where it is proposed to undergo a conformational change that increases its affinity for Cdc20, a WD repeat protein, that normally recruits substrates to the APC/C. In vitro, Mad2 can bind Cdc20 and prevent it from activating the APC/C (Fang et al., 1998). Structural studies showed that Mad2 undergoes a major conformational change when bound to Cdc20 or to its other partner, Mad1 (Luo et al., 2000; Luo et al., 2004). *In vivo*, Mad2 has been shown to cycle rapidly through kinetochores at approximately 2000 molecules per minute (Howell et al., 2000). How Mad2 exchanges its partner between Mad1 at the kinetochores and Cdc20 in the cytosol remains a challenging problem. Clarification of this issue may address the more important question as to what the fate of Mad2 is after release its release from kinetochores.

The alternative model posits that there are two distinct steps to the inhibition of the APC/C. This model came about through the biochemical purification of a factor from Hela cells that inhibited mitotic APC/C (Sudakin et al., 2001). The Mitotic Checkpoint Complex (MCC) consisting of BubR1, Bub3, Cdc20 and Mad2 forms independently of kinetochores and is thus not the "wait anaphase" signal. MCC exists in near stochiometry with APC/C in Hela cells and is the only known inhibitor of APC/C that is purified from mitotic cells. It is proposed that a preformed pool of MCC in interphase provides cells with a rapid way to inhibit the APC/C when it is activated upon entry into mitosis. The affinity of these two complexes for each other cannot be very high as inhibition of the APC/C must be readily reversible to allow cells to exit mitosis. The duration of the MCC/APC/C interaction may be extended if there are unattached kinetochores. In this case, the "wait anaphase" signal is postulated to directly act on the APC/C to sensitise it to prolonged inhibition by the MCC.

The ability to directly test the two-step model *in vivo* is challenging as the same proteins (BubR1, Bub3, Cdc20, Mad2) are involved in both steps. However, two recent reports provided *in vivo* evidence that Mad2 may have kinetochore dependent and independent roles in the mitotic checkpoint. Hela cells depleted of the kinetochores proteins HEC1/Ndc80 (Martin-Lluesma et al., 2002) or CENP-I, accumulate unaligned chromosomes and delay in mitosis (Liu et al., 2003) et al.,). Surprisingly, this delay occurred despite the loss of Mps1, Mad1 and Mad2 from kinetochores. This finding was inconsistent with studies where direct inhibition of Mad2 not only abrogated the mitotic checkpoint but accelerated cells out of mitosis before chromosomes could align properly (Gorbsky et al., 1998; Meraldi et al., 2004; Shannon et al., 2002). This discrepancy could be resolved if one argued that cells lacking Mad2 at kinetochores were able to delay in mitosis through the kinetochore-independent mechanism. Indeed, both studies demonstrated that the delay was still dependent on Mad2 even though it had been depleted from kinetochores. It is therefore likely, that this Mad2-dependent delay reflected the action of the MCC. The caveat of both studies is the degree to which Mad2 was depleted from kinetochores. In CENP-I depleted cells, there was a twenty-fold reduction while a subsequent study of HEC1 depleted cells showed about five-fold reduction (DeLuca et al., 2003). As there can be a hundred-fold difference in the level of Mad2 between unattached and fully attached kinetochores (Hoffman et al., 2001), it is possible that the reduction achieved by HEC1 and CENP-I depletion was insufficient to completely silence the production of the "wait anaphase" signal. However, the reduction of Mad2 at kinetochores likely reduced the output of the "wait anaphase" signal as a prolonged mitotic arrest could not be attained when there were only a few unattached kinetochores (Liu et al., 2003). Only when the number of unattached kinetochores was increased was a prolonged delay achieved. As other checkpoint proteins such as BubR1 and Bub1 kinases remained at kinetochores in these cells, it is likely that they were producing the "wait anaphase" signal when Mad2 levels were reduced. However, a threshold level of "wait anaphase" signal was not achieved to sustain a prolonged inhibition of the APC/C.

2.2.4 Spatial and Temporal Regulation of APC/C

The mitotic checkpoint models presented here do not explain two important observations. The degradation of cyclin A, like cyclin B, depends on Cdc20 and APC/C (Geley et al., 2001). Yet, cyclin A is degraded early in prometaphase and is not inhibited when cells are delayed in mitosis (den Elzen and Pines, 2001). The simplest explanation is that the mitotic checkpoint consists of additional layers that may directly act on APC/C substrates. Our current understanding of the mitotic checkpoint also cannot account for how spatial control of APC/C is achieved. Using GFP-cyclin B as a real-time reporter for APC/C activity (Clute and Pines, 1999), it was clear that APC/C does not appear to be activated throughout the cell upon achieving metaphase. Interestingly, GFP-cyclin B that was localised over the spindle was preferentially lost from the pole towards the chromosomes. While this observation supports the idea that checkpoint inhibition of the APC/C may be spatially confined to the spindle, an alternative explanation is that the sensitivity of APC/C substrates may be spatially regulated. This possibility may also account for why cyclin A is insensitive to checkpoint inhibition.

2.2.5 Factors that Influence the Mitotic Checkpoint

Genetic studies in mice illustrate an important concept about the mitotic checkpoint. Mice that are haplodeficient for a variety of mitotic checkpoint genes (see below) are viable and grow to adulthood with minimal problems. However, mouse embryo fibroblasts (MEF's) derived from these animals exhibit increased rates of chromosome loss that would seem to be incompatible with normal development. This paradox could be resolved if one considers that the mitotic checkpoint is not an essential process. Cells only need the checkpoint when their chromosomes encounter problems with attaching to the spindle. One could imagine that if the spindle in mouse embryos was highly efficient at capturing chromosomes, they may well tolerate a reduction (not elimination) in their capacity to delay mitosis in face of attachments defects. Indeed, the early embryonic cell cycle of *Drosophila* and *Xenopus* are not subject to mitotic checkpoint control. The situation may be different in mice as homozygous checkpoint mutants are lethal. The variable nature by which different cell types rely on the mitotic checkpoint is ultimately due to the complex interplay amongst components of the spindle, checkpoint proteins, the APC/C and its substrates. The biochemical activities of each of these processes must balance each other to achieve coordination between chromosome alignment and mitotic exit. For example, a moderate overexpression of Cdc20 in budding yeast can drive cells prematurely out of mitosis (Pan and Chen, 2004). The focus on profiling just the mitotic checkpoint proteins in cancer cells may be inadequate to address the origin of their aneuploidy.

If we were to consider the kinetochore, underexpression of a component that is important for microtubule capture would impose demands on the mitotic checkpoint that must be capable of inhibiting all APC/C in the cell. CENP-E is a kinetochore associated kinesin-like protein (McEwen et al., 2001; Schaar et al., 1997; Yao et al., 2000) and its loss causes cells to accumulate monopolar chromosomes and thus delay mitotic exit. Although CENP-E is an essential gene in mice, haplodeficient mice are viable (Putkey et al., 2002; Weaver et al., 2003). However, MEF's derived from these mutant mice showed that they cannot sustain a prolonged mitotic arrest but exited mitosis after a transient delay (Weaver et al., 2003). While one interpretation is that CENP-E is a component of the checkpoint, the explanation lies in its interactions with other kinetochores proteins. Comparison between haplodeficient versus wild type MEF's showed that the levels of BubR1, Mad1 and Mad2 at kinetochores of CENP-E depleted MEF's were reduced by up to 50%. This outcome is similar to that reported when Hela cells were depleted of the kinetochore proteins Hec1 and CENP-I. Their loss led to reduction in the amounts of Mps1, Mad1 and Mad2 at kinetochores. Consequently, these kinetochores cannot generate sufficient amounts of the "wait anaphase" signal to sustain prolonged inhibition of the APC/C. The theme that emerges from the three studies is that unattached kinetochores can vary their production of "wait anaphase" signal depending on the amount or activities of the checkpoint proteins present there.

Adenomatous polyposis coli (APC) is a tumour suppressor gene that is frequently mutated in colorectal carcinomas. While APC is well recognised for its role in the Wnt signalling pathway, recent studies have shown that it is important for kinetochore microtubule attachments (Fodde et al., 2001; Kaplan et al., 2001). APC was found at the tips of microtubules that are attached to kinetochores. Furthermore, it was found to form a complex with checkpoint proteins Bub1 and Bub3 (Kaplan et al., 2001). Significantly, expression of mutant APC diminished the mitotic checkpoint response of once checkpoint proficient cells (Tighe et al., 2001). Thus, the combination of defective kinetochore attachments and a reduced mitotic checkpoint in cells expressing mutant APC is likely to result in aneuploidy in colorectal cancers.

These situations underscore two important points. First, disruption of the mitotic checkpoint does not occur exclusively by mutating the bona fide checkpoint genes. Mutations that alter any component that feeds into the pathway can influence the mitotic checkpoint. Second, the mitotic checkpoint is not simply an on/off switch as its capacity to delay mitosis varies as a function of complex interactions with the kinetochore and the APC/C. That cells can exhibit variable lengths of delay suggests that the ratio between the "wait anaphase" signal and its target, the APC/C, is an important parameter in dictating how long a cell can delay mitosis in response to unaligned chromosomes.

3. MITOTIC CHECKPOINT, ANEUPLOIDY AND CANCER

3.1 Genetic Evidence in Human Cancers

The identification of mitotic checkpoint genes has contributed significantly towards a molecular understanding of aneuploidy, and mechanisms that might be associated with increased carcinogenesis. Interest in such mechanisms is underscored by the recent identification of mutations in *Bub1B*, the gene encoding the BubR1 protein, in families with mosaic variegated aneuploidy (MVA) (Hanks et al, 2004). MVA is a rare autosomal recessive disorder marked by a high predisposition to mitotic non-disjunction, the probable cause of the high levels of aneuploidy of multiple different chromosomes and tissues in each affected individual. The phenotype of this condition has been quite consistent in cases reported to date, and which has included severe microcephaly, growth deficiency, mild physical anomalies, eye anomalies and mental retardation. The risk of malignancy seems to be elevated, and have included rhabdomyosarcoma,

Wilms tumour, and leukemia. Hanks et al. assessed eight pedigrees with MVA, and identified biallelic mutations in *Bub1B* in five families. In all cases, mutations in one allele results in the inactivation of the gene while missense mutations were found in the second allele. Four of the five missense mutations occurred in the catalytic domain and thus suggest a dysfunctional BubR1 kinase. The fifth missense mutation was found in a region of the protein with no ascribed function. Nevertheless, this missense mutation along with one found in the kinase domain were associated with two cases of embryonal rhabdomyosarcoma. Why only two cases of cancer were identified amongst the five MVA families is unclear but suggests additional factors are likely to be involved in the cancer development.

The MVA study was predated by one of the first studies to draw attention to a potential role for perturbed mitotic checkpoint function in the etiology of human cancer (Cahill et al., 1998). Heterozygous mutations in Bub1 and BubR1 kinases were found in a few of the nineteen aneuploid colorectal cancer cell lines that were examined. Although the BubR1 mutations were not pursued, the effects of the Bub1 mutants were examined in more detail. In V400 cells, a mutation at an intronic splice donor site created a frameshift mutation that led to a premature termination codon. In V429 cells, a missense mutation that converted a serine to a tyrosine was identified. When the Bub1 mutants were transfected into a checkpoint proficient colorectal cancer cells (HCT116), the transfectants were no longer able to block mitosis when challenged with spindle poisons. It was therefore concluded that these were dominant mutants that were responsible for the defective checkpoint response of the V400 and V429 cells. The caveat is that this result was obtained by overexpressing the mutant proteins to non-physiological levels. It is therefore unclear if the amount of mutant proteins expressed in the V400 and V429 cells can effectively compete against the wild type protein.

Since that report, there have surprisingly few reports of mutated mitotic checkpoint genes in human cancers cell lines, or cancers freshly biopsied or resected from patients in the clinic. For example, no mutations in either Bub3, BubR1 or Bub1 were found in a large number of glioblastomas and lung cancers derived from patients or cell lines (Reis et al., 2001; Sato et al., 2000). One sample from a series of surgically resected colorectal, hepatocellular, and renal tumours was found to contain a missense mutation in Bub1 (Shichiri et al., 2002). Whether this mutation disrupted checkpoint function of the protein is unknown. However, quantitation of transcript levels by real-time polymerase chain reaction identified a subset of tumours with depressed levels of mRNA that was postulated to be due to epigenetic silencing of the genes. This subset of tumours was associated with a significantly higher recurrence rate, suggesting that low levels of expression of BubR1 or Bub1 might confer a growth advantage to the tumours of the subset. Examples of heterozygous mutations in the Bub1 gene have been reported in T lymphoblastic leukemia cell lines and in patients with acute lymphoblastic leukemia (ALL) and Hodgkin's lymphoma (Ru et al., 2002).

Of five patient samples examined, three were found to harbour deletion mutations in Bub1. A similar study of adult T-cell leukemia (ATLL) that exhibited aneuploidy showed that in 4 out of 10 cases, mutations in either Bub1 and BubR1 were found (Ohshima et al., 2000). All except one mutation, which resulted in a truncated BubR1, were mutations that resulted in an amino acid substitution. It remains to be seen if the biochemical functions of these mutant proteins were affected.

The search for Mad1 and Mad2 mutations also showed that they did not occur at high frequency. A screen for Mad1 mutations involving a large panel of 44 cancer cell lines and 133 primary tumours consisting of lymphomas, bladder, breast and gliomas identified only eight mutations that potentially disrupted Mad1 function (Tsukasaki et al., 2001). Two of the eight mutations resulted in premature termination while the other six mutations led to amino acid substitutions. A study of Mad2 in a group of 96 human primary tumours comprised of 44 transitional-cell carcinomas of the bladder, 42 adult soft-tissue sarcomas and 10 hepatocellular carcinomas identified one missense mutation in a bladder tumour where an isoleucine was mutated to a valine (Hernando et al., 2001). This alteration did not appear to alter protein function as transfection of the mutant Mad2 cDNA into cells did not result in a phenotype different from wild type cDNA. Similar screens have shown that Mad2 mutations are rare in cancer cells obtained from breast, lung (Gemma et al., 2001; Percy et al., 2000; Takahashi et al., 1999), and digestive tract (Imai et al., 1999). However, reduced expression of Mad2 protein and mRNA has been reported in breast (Li and Benezra, 1996; Percy et al., 2000), nasopharyngeal and ovarian cancer cell lines (Wang et al., 2002; Wang et al., 2000) that exhibited a defect in their mitotic checkpoint response to spindle poisons. The molecular explanation for why some cancer cells do not express sufficient levels of Mad2 remains unknown but illustrates the earlier point that inactivation of the mitotic checkpoint can be achieved in many different ways.

3.1.1 Mutations in Mitotic Checkpoint Genes are Rare

It may be premature to conclude from these studies whether disruption of the mitotic checkpoint genes promotes tumorigenesis. As many of the studies did not exhaustively screen all of the known checkpoint genes, it is possible that mutations may still be found. However, one study that examined all of the known checkpoint genes in nineteen aneuploid cell lines did not uncover any mutations (Cahill et al., 1999). In these cases, the mitotic checkpoint defects may be due to mutations that affect other components (spindle, kinetochore, APC/C and its substrates) that influence the mitotic checkpoint. Of the mutations that have been identified in the mitotic checkpoint genes, the majority are missense mutations whose effects on the stability or biochemical activity of the protein is not known. Western blots to determine the expression levels of various checkpoint proteins would be very informative. Equally informative are immunocytological assays that monitor the localisation at kinetochores of key mitotic checkpoint proteins. In the case of Bub1, we now know that it is important for assembling other proteins to the kinetochore (Johnson et al., 2004). Those proteins may be informative biomarkers that indirectly monitor Bub1 activity. Similarly, the localisation of Mad2 at kinetochores may be an informative biomarker for mitotic checkpoint status. These immunocytological assays are not only a simple way to functionally assess mutant checkpoint genes, it may be quite effective in screening for the molecular defects in cells that are phenotypically defective for the mitotic checkpoint. The immunocytological data may help to define the pathways that are affected in these cells.

3.2 Mouse Models that Test the Link between Aneuploidy and Cancer

Efforts to directly test the link between aneuploidy and tumorigenesis have been to conduct targeted disruption of mitotic checkpoint genes in mice. A strikingly consistent finding has been that mice with targeted knockouts of these genes are nonviable, dying early in embryogenesis. To circumvent the early lethality of the total knockouts, partial knockouts have been generated, in which the mice are haplodeficient for genes encoding the mitotic checkpoint proteins (Table 1).

Table 1. Characteristics of Mice Deficient for Mitotic Checkpoint Genes

Gene Target	BubR1	Bub1	Mad2	Bub3/Rae1
Genotype	+/-	h/h[1]	+/-	+/-
Expression	Reduced	Much reduced	Reduced	Reduced
Checkpoint	Loss	Loss	Loss	Loss
Aneuploidy in MEF's	Yes	Yes	Yes	Yes
Spontaneous tumours in young animals	No	No	No	No
Tumour formation during lifespan	Carcinogen induced lung and colonic tumours	5% of moribund or deceased animals have solitary tumour	Lung tumours in minority of animals after long latency	Carcinogen induced lung tumours only
Other phenotypes	NA[2]	Early aging, Shortened lifespan.	NA	NA

[1] hypomorphic
[2] Not Assessed

A number of other common themes have emerged between the efforts targeting different proteins:

- Haplodeficient MEFs show reductions of 25-75% in the expressed levels of the targeted proteins, indicating that the remaining allele does not fully compensate for the missing allele.
- The reduction in expressed protein is sufficient to inactivate the mitotic checkpoint in response to microtubule disruption (e.g. by nocodazole).
- Haplodeficient MEFs show increased aneuploidy at the earliest stages post-coitus when they were harvested.
- Haplodeficient mice are phenotypically normal in utero, survive birth, and grow well into adulthood, during which their body mass is similar to wildtype mice.
- No spontaneous tumour formation (except for papillary lung cancers late in life of Mad2-haplodeficient mice).

3.2.1 Mad2

Mad2 null mice die 6.5 days after coitus (Dobles et al., 2000), while Mad2 haplodeficient mice are viable well into adulthood. Mouse embryo fibroblasts (MEFs) from the Mad2 haplodeficient mice were found on metaphase spread analysis to have a high frequency of cells with premature sister chromatid separation (30-57% of cells), and a likewise high proportion of cells that were aneuploid (33-60%), suggesting a link between the chromosomal missegregation and development of aneuploidy. Despite the high proportion of aneuploidy in embryonic cells, animals were not found to have early spontaneous tumour development, and live comparatively long lifespans. Upon sacrifice at 18-19 months, 14/57 (27%) of Mad2-deficient animals were found to have papillary lung tumours, a tumour that is extremely rare in wildtype animals. No increased incidence of lymphomas or other tumours were noted in the haplodeficient animals.

3.2.2 Bub3 and Rae1

Rae1 is a member of the superfamily of WD proteins that is most similar to Bub3. Findings that included interactions between Rae1 and Bub1 in cultured cells suggested it may be involved in the mitotic checkpoint (Wang et al., 2001). Targeted deletion of either Bub3 or Rae1 in mice showed that they were essential genes as homozygous Bub3 and Rae 1 null mutants died by days E8.5 and E5.5, respectively (Babu et al., 2003). Rae1 haplodeficient mice were viable and survived to adulthood. MEF's derived from Rae1 +/- mice exhibited increased aneuploidy relative to wildtype MEFs. Nevertheless, spontaneous tumours were not detected in the animals. Similar to the Rae1-haplodeficient MEFs, the Bub3-haplodeficient MEFs were deficient for the nocodazole-induced mitotic arrest and showed increased aneuploidy. Retroviral-mediated gene-transfer of full-length Rae1 cDNA into both Rae1- and Bub3-haplodeficient MEFs restored the nocodazole-induced mitotic checkpoint. To further assess the relationship between Rae1 and Bub3, mice haplodeficient for both genes were generated. MEFs derived from embryos that were haplodeficient for both Rae1 and Bub3 showed considerably greater aneuploidy than either targeted haplodeficient MEF alone. Despite the increased rates of aneuploidy of the compound heterozygote MEF's, the mice were viable and showed no reduced body mass or spontaneous tumour formation. Lung tumours were obtained only after the animals were exposed to a potent carcinogen from an early age. The frequency of tumours was: WT: 50%, Bub3 +/-: 72%, Rae1 +/-: 80%, Bub3 +/- Rae1 +/-: 90%. Thus, partial abrogation of Rae1 and Bub3, either separately or together, did not result in increased spontaneous tumour formation. A notable increase in the proportion of animals with tumours was only seen after exposure to a potent carcinogen.

3.2.3 BubR1

Targeted deletion of BubR1 in mice have shown that this is also an essential gene for embryogenesis (Baker et al., 2004; Dai et al., 2004). Analysis of haplodeficient BubR1 MEF's isolated at day E14.5 showed that they expressed only about 25%, not the expected 50%, of the level of BubR1 protein compared to wildtype BubR1 +/+ MEFs. The BubR1 heterozygous MEF's are haploinsufficient as they failed to arrest in mitosis in response to nocodazole. Despite the loss of the mitotic checkpoint, no spontaneous tumours were identified in any of the heterozygous mice. Application of a potent colonic carcinogen resulted in a higher average number of colonic microadenomas, adenomas, and adenocarcinomas in the BubR1 haplodeficient mice compared to wildtype mice (the proportion of haplodeficient animals which became afflicted with these tumours was not stated). Other major organs were also searched for tumour formation, and only the lung and liver showed tumours (the incidence and average number of these tumours per mouse was not stated).

A fascinating aspect of these animal studies is that the level of expression of the mitotic checkpoint proteins appears to dictate in a dramatic fashion the resulting cellular phenotype and embryonic development. Complete absence of expression of these proteins as in the homozygous knockouts is incompatible with embryonic development. Partial expression (levels of protein that are 25-50% of wildtype cells) of protein in the haplodeficient animals is insufficient to mediate the mitotic checkpoint in response to microtubule-disrupting drugs, yet it enables normal development, birth, and growth, including well into adulthood. Even a reduction in the dosage of a checkpoint protein can be tolerated as long as the cells can delay for sufficient amounts of time for chromosomes to align properly.

The importance of dosage was demonstrated dramatically in a study of mice that were engineered to allow for graded expression of BubR1 (Baker et al., 2004). This was accomplished by crossing mice with alleles for *Bub1b* (encoding for BubR1 protein) that were knockout (*Bub1b*$^{-}$), hypomorphic (*Bub1b* ^{H}), or wildtype (*Bub1b* $^{+}$). MEFs were generated that expressed no BubR1 protein (*Bub1b* $^{-/-}$), extremely low levels (4% of wildtype) (*Bub1b* $^{-/H}$), or very low levels (11% of wildtype) (*Bub1b* $^{H/H}$), and finally low levels of protein. *Bub1b* $^{H/+}$ and *Bub1b* $^{-/+}$ MEFs showed protein levels respectively 29% and 42% of wildtype MEFs.

Interestingly, *Bub1b* $^{-/H}$ (4% of wild type) mice developed unimpeded, but died within several hours of birth of respiratory failure, suggesting deficient development of some tissue that is critical to maintain respiratory fitness. This observation indicates that specific organ systems might have different thresholds of BubR1 protein expression to ensure sufficient development and growth. In contrast, *Bub1b* $^{H/H}$ mice showed slow postnatal growth, but survived to adulthood. Finally, the *Bub1b* $^{H/+}$ and *Bub1b* $^{-/+}$ mice both showed no discernable abnormal phenotype. The level of BubR1 protein expression that was sufficient to ensure development and growth to adulthood could therefore be established as between 4 and 11% of wildtype BubR1 levels. The *Bub1b* $^{H/H}$, *Bub1b* $^{H/+}$, *Bub1b* $^{-/+}$, and *Bub1b*$^{+/+}$ MEFs were assessed for nocodazole-induced mitotic arrest, persistence of cdc2/cyclin B1 kinase activity, and presence of lagging chromosomes (also known as premature sister chromatid separation) as an indicator of chromosomal missegregation. Of these, only *Bub1b* $^{H/H}$ MEFs showed deficient nocodazole-induced arrest, decreased cdc2/cyclin B1 kinase activity, and increased lagging chromosomes. Interestingly, in contrast to an earlier study, the *Bub1b* $^{-/+}$, and *Bub1b*$^{+/+}$ MEFs and animals appeared similar in every regard, and no increased aneuploidy or deficient mitotic checkpoint response was noted in the *Bub1b* $^{-/+}$ MEFs.

The hypomorphic mice showed further surprises. The *Bub1b* $^{H/H}$ mice generated were followed until 15-16 months of age. Six of the moribund or deceased *Bub1b* $^{H/H}$ mice were found to have solitary tumours, one of which was life-threatening. Even more striking was the shortened lifespan and the appearance of accelerated aging in mice starting at 2-3 months of age. The mice showed progressive development of cataracts, thinning of dermis and subcutaneous fat, cachexia, and lordokyphosis, all hallmarks of aging. None of these were observed in *Bub1b* $^{H/+}$, *Bub1b* $^{-/+}$, or *Bub1b* $^{+/+}$ mice. The physical appearance of the *Bub1b* $^{H/H}$ mice strikingly suggested early aging. But how was this related to a deficient mitotic checkpoint? Was the checkpoint deficiency somehow leading to increased cell death? Mice were therefore assessed for senescence and apoptosis. Beta-galactosidase activity is increased in senescent cells and so can be usefully employed as a visual assay when tissue sections are exposed to a substrate that turns blue to indicate activity of the enzyme. The kidney sections of five month old *Bub1b* $^{H/H}$ mice showed abundant beta-galactosidase activity, which was barely detectable in the tissues of wildtype and other BubR1-deficient backgrounds. MEFs from *Bub1b* $^{H/H}$ mice were also investigated for the response to hypoxia and were found to readily undergo apoptosis when oxygen concentration was lowered from the normal 20 to 3%. This suggested the lack of BubR1 resulted in heightened apoptosis under conditions of oxidative stress. To further establish the link between lack of BubR1 and aging, BubR1 protein expression levels in the testis of wildtype mice were assessed via immunoblotting. Together these results suggested that deficient mitotic checkpoint control due to lack of BubR1 led to increased apoptosis and early senescence.

One additional aspect the BubR1 hypomorphic mice bears special mention. Karyotypic analyses of passage five MEFs from *Bub1b* $^{H/H}$ mice showed that 36% were aneuploid, almost four times the incidence in wildtype mice. Karyotypic analyses of adult splenocytes showed 33% were aneuploid by twelve months of age, with an increased incidence detectable even as early as two months. The degree of aneuploidy is far greater than the eventual incidence of tumour formation in these animals.

4. DOES ANEUPLOIDY DIRECTLY PROMOTE CANCER?

The mouse models clearly showed that aneuploidy by itself often does not lead to tumour formation during an average lifespan. Similarly, the low frequency of mutations in mitotic checkpoint genes in human cancers suggests that they may not be the primary event that triggered chromosome instability. Consistent with the relative inefficiency of aneuploidy as a direct cause of cancer, in the report linking BubR1 mutations to mosaic variegated aneuploidy discussed earlier, malignancies were not identified in six of the eight families studied. The combined data therefore suggest that aneuploidy resulting from impairment of the mitotic checkpoint appears to be at best an inefficient mechanism in promoting tumorigenesis. Perhaps the events giving rise to the development of a cancer phenotype concurrently gave rise to aneuploidy. It has been proposed that defects in the mitotic checkpoint may confer a growth advantage to cancer cells, by enabling cells to tolerate chromosomal anomalies that normally would invoke a cell cycle arrest (see below).

If aneuploidy in many or most cancer cell lines does not arise from deficient mitotic checkpoint control, then where does it stem from? In recent years, it has become apparent that aneuploidy and carcinogenesis can arise from defects in DNA replication or recombination control. In contrast to the surprising lack of early transformation in cell lines deficient in the mitotic checkpoint proteins or early tumorigenesis in animals haplodeficient for these genes, a number of syndromes involving defects in DNA replication or recombination control have been identified in which chromosomal instability, aneuploidy, and increased tumorigenesis are prominent hallmarks. These include ataxia-telangiectasia, xeroderma pigmentosum, Nijmegen breakage syndromes, Bloom's syndrome, and Werner's syndrome, (Modesti and Kanaar, 2001; Thompson and Schild, 2002). Defects in mitotic checkpoint control have not been described for these syndromes, nor do they seem required for the chromosomal instability characteristic of these syndromes. Rather, aneuploidy appears to stem from aberrant chromosomal duplication and deletions during homologous recombination occurring prior to the onset of mitosis. The net effect of such "unnatural acts" repeated over many cell cycles could be the deletion or duplication of large parts of or entire chromosomes, and potentially the deletion of genes encoding components of the mitotic checkpoint. Thus, the development of aneuploidy may antedate mitosis itself.

5. THE MITOTIC CHECKPOINT AS A TARGET FOR CANCER TREATMENT: WALKING A TIGHTROPE

Microtubule inhibitors are widely used in the clinic to treat a variety of cancers. Given our current understanding of the mitotic checkpoint, it would seem that this may be an important factor that dictates sensitivity of tumours to these drugs. Indeed, there appears to be some correlation between expression of mitotic checkpoint proteins and sensitivity to anti-microtubule agents (Masuda et al., 2003). Breast and ovarian cancer cell lines that showed little or no expression of one or more mitotic checkpoint proteins were sensitive to rapid killing by nocodazole and paclitaxel (Lee et al., 2004). In contrast, cells derived from cervical, colorectal, and renal cancers that showed stronger expression of these proteins and intact mitotic checkpoint control, were found to be relatively more resistant to killing by the same drugs. Importantly, abrogation of the mitotic checkpoint by RNA interference (RNAi) efficiently reversed drug resistance. Similarly, abrogation of mitotic checkpoint via the stable expression of RNAi targeting Mad2 and BubR1 led to massive chromosomal loss and cell death within six cell divisions (Kops et al., 2004). Antisense and ribozyme-mediated inhibition of Bub1 in normal human fibroblasts resulted in chromosome instability and massive nuclear fragmentation in many cells (Musio et al., 2003). Cells developed anchorage independence in soft agar and did not form tumours when injected into nude mice. Additional evidence supporting the idea that highly aneuploid cells are often non-viable came from the analysis of mSds3, an essential component of the mSin3/HDAC corepressor complex (David et al., 2003). mSds3 is essential in mice and MEF's lacking mSds3 exhibited defects in pericentric heterochromatin formation that interfered with centromere function. These cells were massively aneuploid and were largely inviable.

That disruption of the mitotic checkpoint leads to cell death may be an oversimplification as there is ample evidence supporting the idea that loss of the checkpoint promotes cell proliferation. The most striking example comes from the studies of the breast cancer susceptibility gene BRCA2. BRCA2 is an essential gene in mice and MEF's isolated from functionally null BRCA2 mutant embryos exhibit poor growth in vitro as a result of elevated p53 and p21 that were induced by DNA damage (Lee et al., 1999). This growth defect was overcome by expressing either dominant negative p53 or Bub1 mutants. More interestingly, lymphomas isolated from the rare BRCA2 deficient mice that survived to adulthood, were found to be mutated in p53, Bub1 and BubR1. Thus, disruption of p53 and the mitotic checkpoint must cooperate with the BRCA2 mutation to promote cell transformation and uncontrolled proliferation. The intriguing relationship between BRCA2 and mitotic checkpoint control also highlights the notion that aneuploidy by itself does not promote cellular transformation efficiently. Thus, disruption of the mitotic checkpoint may be a secondary event that provides added growth advantage to cells that have undergone a transforming event.

How can we reconcile the difference as to whether cells proliferate or die when the mitotic checkpoint is disrupted? One reasonable explanation might be the degree to which the mitotic checkpoint is inhibited. If the checkpoint is completely eliminated, chromosomes have little to no time to align before cells exit mitosis. Consequently, cells undergo massive chromosome loss (or gain) that is incompatible with life. This would be consistent with the fact that mitotic checkpoint genes are essential in mice (and probably all mammals). On the other hand, heterozygous mutations, such as those identified in the BRCA2 mutant mice, or those reported in some human cancers, may retain sufficient checkpoint activity to allow cells to proceed through mitosis normally most of the time. The chromosome loss rate per generation may be sufficiently low that a large proportion of the population continues to proliferate. Along this line, it is also noteworthy that there is also selective pressure for mutations that cripple but do not obliterate the mitotic checkpoint. In other words, the mutations found in checkpoint genes of tumours may have been selected for so as to allow cells to proliferate in face of chromosomal defects.

6. CONCLUDING REMARKS

The recent efforts to understand the mechanism of action of the mitotic checkpoint mechanism can provide some insights into the mechanism of aneuploidy and its relationship to cancer. The current information indicates that mutations in mitotic checkpoint genes are rare in cancers. Despite this, many cancer cell lines still exhibit defects in the mitotic checkpoint (cannot arrest in response to microtubule inhibitors). This would suggest that additional genes are involved in the mitotic checkpoint. On the other hand, an important consideration is that the activity of the mitotic checkpoint is dosage dependent. There are examples where inactivation of the checkpoint was attributed to the reduced expression of one of the mitotic checkpoint genes. There is also experimental evidence that mutations in genes that are responsible for recruiting checkpoint proteins to the kinetochore could effectively reduce their capacity to generate the "wait anaphase" signal. The mitotic checkpoint potential of a cell is therefore not merely dictated by the level of its checkpoint proteins but must include all the components of the system that include the APC/C, the amount of APC/C substrates, and the efficiency by which kinetochores establish productive interactions with the spindle.

The difference in the mitotic checkpoint status of cancer cells may be correlated with their sensitivity to drugs that interfere with spindle functions. The taxanes have shown modest or low efficacy in controlling sarcoma, colorectal, (squamous cell) cervical and renal cancers, (Edmonson et al., 1996; Hartmann and Bokemeyer, 1999; McGuire et al., 1996; Patel et al., 1996), cancers that appear to retain the mitotic checkpoint response when grown as cell lines in the laboratory. In contrast, the taxanes have been effective for and have become or will become standard chemotherapeutic treatment for patients afflicted with lung, breast, prostate, and ovarian cancers (Crown et al., 2004; Nowak et al., 2004; Patel et al., 1996; Petrylak and de Wit, 2002; Piccart et al., 2003; Picus and Schultz, 1999; Rigas, 2004; Shepherd, 2004). Consequently, effective clinical screens should be developed to profile the mitotic checkpoint status of tumours. This information would be of value in predicting outcome to treatment with current drugs such as the taxanes. Cells that have intact checkpoints may require longer regimens of drug infusion to ensure that the cancer cells do not simply rely on their checkpoint to overcome the drug treatment. Thus longer periods of drug exposure may improve tumour response. As increased exposure to drugs increases undesirable side effects, pharmacological inhibitors of the mitotic checkpoint should significantly enhance sensitivity of cancer cells to existing microtubule inhibitors. For example, inhibitors of the aurora kinases, seemed to selectively prevent cells treated with Taxol from arresting in mitosis (Ditchfield et al., 2003; Hauf et al., 2003). As Taxol treatment reduces kinetochore tension that is normally monitored by aurora B kinase, inhibitors of aurora B would be expected to sensitise cells to Taxol treatment. As inhibition of aurora B results in the loss of other kinetochore proteins that contribute towards the mitotic checkpoint, it remains to be seen if inhibition of those proteins might also sensitize cells to Taxol treatment. Regardless, it is clear that significant advances in cancer treatment will be achieved through continued efforts to elucidate the molecular and biochemical mechanisms that ensure accurate chromosome segregation.

ACKNOWLEDGEMENTS

The authors would like to thank the respective lab members for advice and discussion in preparing this chapter. Research in TJY's lab is supported by the NIH (GM44762, CA099423, Core Grants CA75138, CA06927) U.S. Department of Defense. Research in GDK's lab is supported by Office of

Research and Development Medical Research Services, Department of Veterans Affairs (ARCD-024-02F), the NIH (CA-R01CA107956), the University of Pennsylvania Cancer Center Foundation. Both labs are also supported by an appropriation from the Commonwealth of Pennsylvania.

REFERENCES

Andrews, P.D., E. Knatko, W.J. Moore, and J.R. Swedlow. 2003. Mitotic mechanics: the auroras come into view. *Curr Opin Cell Biol.* 15:672-83.

Andrews, P.D., Y. Ovechkina, N. Morrice, M. Wagenbach, K. Duncan, L. Wordeman, and J.R. Swedlow. 2004. Aurora B regulates MCAK at the mitotic centromere. *Dev Cell.* 6:253-68.

Babu, J.R., K.B. Jeganathan, D.J. Baker, X. Wu, N. Kang-Decker, and J.M. van Deursen. 2003. Rae1 is an essential mitotic checkpoint regulator that cooperates with Bub3 to prevent chromosome missegregation. *J Cell Biol.* 160:341-53.

Baker, D.J., K.B. Jeganathan, J.D. Cameron, M. Thompson, S. Juneja, A. Kopecka, R. Kumar, R.B. Jenkins, P.C. de Groen, P. Roche, and J.M. van Deursen. 2004. BubR1 insufficiency causes early onset of aging-associated phenotypes and infertility in mice. *Nat Genet.* 36:744-9.

Biggins, S., and C.E. Walczak. 2003. Captivating capture: how microtubules attach to kinetochores. *Curr Biol.* 13:R449-60.

Cahill, D.P., L.T. da Costa, E.B. Carson-Walter, K.W. Kinzler, B. Vogelstein, and C. Lengauer. 1999. Characterization of MAD2B and other mitotic spindle checkpoint genes. *Genomics.* 58:181-7.

Cahill, D.P., C. Lengauer, J. Yu, G.J. Riggins, J.K. Willson, S.D. Markowitz, K.W. Kinzler, and B. Vogelstein. 1998. Mutations of mitotic checkpoint genes in human cancers. *Nature.* 392:300-3.

Chan, G.K., and T.J. Yen. 2003. The mitotic checkpoint: a signaling pathway that allows a single unattached kinetochore to inhibit mitotic exit. *Prog Cell Cycle Res.* 5:431-9.

Chen, R.H., J.C. Waters, E.D. Salmon, and A.W. Murray. 1996. Association of spindle assembly checkpoint component XMAD2 with unattached kinetochores. *Science.* 274:242-6.

Cimini, D., B. Howell, P. Maddox, A. Khodjakov, F. Degrassi, and E.D. Salmon. 2001. Merotelic kinetochore orientation is a major mechanism of aneuploidy in mitotic mammalian tissue cells. *J Cell Biol.* 153:517-27.

Clute, P., and J. Pines. 1999. Temporal and spatial control of cyclin B1 destruction in metaphase. *Nat Cell Biol.* 1:82-7.

Crown, J., M. O'Leary, and W.S. Ooi. 2004. Docetaxel and paclitaxel in the treatment of breast cancer: a review of clinical experience. *Oncologist.* 9 Suppl 2:24-32.

Dai, W., Q. Wang, T. Liu, M. Swamy, Y. Fang, S. Xie, R. Mahmood, Y.M. Yang, M. Xu, and C.V. Rao. 2004. Slippage of mitotic arrest and enhanced tumor development in mice with BubR1 haploinsufficiency. *Cancer Res.* 64:440-5.

David, G., G.M. Turner, Y. Yao, A. Protopopov, and R.A. DePinho. 2003. mSin3-associated protein, mSds3, is essential for pericentric heterochromatin formation and chromosome segregation in mammalian cells. *Genes Dev.* 17:2396-405.

DeLuca, J.G., B.J. Howell, J.C. Canman, J.M. Hickey, G. Fang, and E.D. Salmon. 2003. Nuf2 and Hec1 are required for retention of the checkpoint proteins Mad1 and Mad2 to kinetochores. *Curr Biol.* 13:2103-9.

den Elzen, N., and J. Pines. 2001. Cyclin A is destroyed in prometaphase and can delay chromosome alignment and anaphase. *J Cell Biol.* 153:121-36.

Ditchfield, C., V.L. Johnson, A. Tighe, R. Ellston, C. Haworth, T. Johnson, A. Mortlock, N. Keen, and S.S. Taylor. 2003. Aurora B couples chromosome alignment with anaphase by targeting BubR1, Mad2, and Cenp-E to kinetochores. *J Cell Biol*. 161:267-80.

Dobles, M., V. Liberal, M.L. Scott, R. Benezra, and P.K. Sorger. 2000. Chromosome missegregation and apoptosis in mice lacking the mitotic checkpoint protein Mad2. *Cell*. 101:635-45.

Edmonson, J.H., L.P. Ebbert, A.G. Nascimento, S.H. Jung, H. McGaw, and J.B. Gerstner. 1996. Phase II study of docetaxel in advanced soft tissue sarcomas. *Am J Clin Oncol*. 19:574-6.

Fang, G., H. Yu, and M.W. Kirschner. 1998. The checkpoint protein MAD2 and the mitotic regulator CDC20 form a ternary complex with the anaphase-promoting complex to control anaphase initiation. *Genes Dev*. 12:1871-83.

Fodde, R., J. Kuipers, C. Rosenberg, R. Smits, M. Kielman, C. Gaspar, J.H. van Es, C. Breukel, J. Wiegant, R.H. Giles, and H. Clevers. 2001. Mutations in the APC tumour suppressor gene cause chromosomal instability. *Nat Cell Biol*. 3:433-8.

Geley, S., E. Kramer, C. Gieffers, J. Gannon, J.M. Peters, and T. Hunt. 2001. Anaphase-promoting complex/cyclosome-dependent proteolysis of human cyclin A starts at the beginning of mitosis and is not subject to the spindle assembly checkpoint. *J Cell Biol*. 153:137-48.

Gemma, A., Y. Hosoya, M. Seike, K. Uematsu, F. Kurimoto, S. Hibino, A. Yoshimura, M. Shibuya, S. Kudoh, and M. Emi. 2001. Genomic structure of the human MAD2 gene and mutation analysis in human lung and breast cancers. *Lung Cancer*. 32:289-95.

Gorbsky, G.J., R.H. Chen, and A.W. Murray. 1998. Microinjection of antibody to Mad2 protein into mammalian cells in mitosis induces premature anaphase. *J Cell Biol*. 141:1193-205.

Hanks S, Coleman K, Reid S, Plaja A, Firth H, Fitzpatrick D, Kidd A, Mehes K, Nash R, Robin N, Shannon N, Tolmie J, Swansbury J, Irrthum A, Douglas J,Rahman N. 2004. Constitutional aneuploidy and cancer predisposition caused by biallelic mutations in BUB1B. *Nat Genet*. 36 (11):1159-61.

Hartmann, J.T., and C. Bokemeyer. 1999. Chemotherapy for renal cell carcinoma. *Anticancer Res*. 19:1541-3.

Hauf, S., R.W. Cole, S. LaTerra, C. Zimmer, G. Schnapp, R. Walter, A. Heckel, J. van Meel, C.L. Rieder, and J.M. Peters. 2003. The small molecule Hesperadin reveals a role for Aurora B in correcting kinetochore-microtubule attachment and in maintaining the spindle assembly checkpoint. *J Cell Biol*. 161:281-94.

Hernando, E., I. Orlow, V. Liberal, G. Nohales, R. Benezra, and C. Cordon-Cardo. 2001. Molecular analyses of the mitotic checkpoint components hsMAD2, hBUB1 and hBUB3 in human cancer. *Int J Cancer*. 95:223-7.

Hoffman, D.B., C.G. Pearson, T.J. Yen, B.J. Howell, and E.D. Salmon. 2001. Microtubule-dependent changes in assembly of microtubule motor proteins and mitotic spindle checkpoint proteins at PtK1 kinetochores. *Mol Biol Cell*. 12:1995-2009.

Howell, B.J., D.B. Hoffman, G. Fang, A.W. Murray, and E.D. Salmon. 2000. Visualization of Mad2 dynamics at kinetochores, along spindle fibers, and at spindle poles in living cells. *J Cell Biol*. 150:1233-50.

Imai, Y., Y. Shiratori, N. Kato, T. Inoue, and M. Omata. 1999. Mutational inactivation of mitotic checkpoint genes, hsMAD2 and hBUB1, is rare in sporadic digestive tract cancers. *Jpn J Cancer Res*. 90:837-40.

Johnson, V.L., M.I. Scott, S.V. Holt, D. Hussein, and S.S. Taylor. 2004. Bub1 is required for kinetochore localization of BubR1, Cenp-E, Cenp-F and Mad2, and chromosome congression. *J Cell Sci*. 117:1577-89.

Kaplan, K.B., A.A. Burds, J.R. Swedlow, S.S. Bekir, P.K. Sorger, and I.S. Nathke. 2001. A role for the Adenomatous Polyposis Coli protein in chromosome segregation. *Nat Cell Biol*. 3:429-32.

King, R.W., J.M. Peters, S. Tugendreich, M. Rolfe, P. Hieter, and M.W. Kirschner. 1995. A 20S complex containing CDC27 and CDC16 catalyzes the mitosis-specific conjugation of ubiquitin to cyclin B. *Cell*. 81:279-88.

Kops, G.J., D.R. Foltz, and D.W. Cleveland. 2004. Lethality to human cancer cells through massive chromosome loss by inhibition of the mitotic checkpoint. *Proc Natl Acad Sci U S A*. 101:8699-704.

Lan, W., X. Zhang, S.L. Kline-Smith, S.E. Rosasco, G.A. Barrett-Wilt, J. Shabanowitz, D.F. Hunt, C.E. Walczak, and P.T. Stukenberg. 2004. Aurora B phosphorylates centromeric MCAK and regulates its localization and microtubule depolymerization activity. *Curr Biol*. 14:273-86.

Lee, E.A., M.K. Keutmann, M.L. Dowling, E. Harris, G. Chan, and G.D. Kao. 2004. Inactivation of the mitotic checkpoint as a determinant of the efficacy of microtubule-targeted drugs in killing human cancer cells. *Mol Cancer Ther*. 3:661-9.

Lee, H., A.H. Trainer, L.S. Friedman, F.C. Thistlethwaite, M.J. Evans, B.A. Ponder, and A.R. Venkitaraman. 1999. Mitotic checkpoint inactivation fosters transformation in cells lacking the breast cancer susceptibility gene, Brca2. *Mol Cell*. 4:1-10.

Lengauer, C., K.W. Kinzler, and B. Vogelstein. 1998. Genetic instabilities in human cancers. *Nature*. 396:643-9.

Li, X., and R.B. Nicklas. 1995. Mitotic forces control a cell-cycle checkpoint. *Nature*. 373:630-2.

Li, Y., and R. Benezra. 1996. Identification of a human mitotic checkpoint gene: hsMAD2. *Science*. 274:246-8.

Liu, S.T., J.C. Hittle, S.A. Jablonski, M.S. Campbell, K. Yoda, and T.J. Yen. 2003. Human CENP-I specifies localization of CENP-F, MAD1 and MAD2 to kinetochores and is essential for mitosis. *Nat Cell Biol*. 5:341-5.

Loeb, L.A. 1991. Mutator phenotype may be required for multistage carcinogenesis. *Cancer Res*. 51:3075-9.

Luo, X., G. Fang, M. Coldiron, Y. Lin, H. Yu, M.W. Kirschner, and G. Wagner. 2000. Structure of the Mad2 spindle assembly checkpoint protein and its interaction with Cdc20. *Nat Struct Biol*. 7:224-9.

Luo, X., Z. Tang, G. Xia, K. Wassmann, T. Matsumoto, J. Rizo, and H. Yu. 2004. The Mad2 spindle checkpoint protein has two distinct natively folded states. *Nat Struct Mol Biol*. 11:338-45.

Martin-Lluesma, S., V.M. Stucke, and E.A. Nigg. 2002. Role of hec1 in spindle checkpoint signaling and kinetochore recruitment of mad1/mad2. *Science*. 297:2267-70.

Masuda, A., K. Maeno, T. Nakagawa, H. Saito, and T. Takahashi. 2003. Association between mitotic spindle checkpoint impairment and susceptibility to the induction of apoptosis by anti-microtubule agents in human lung cancers. *Am J Pathol*. 163:1109-16.

McEwen, B.F., G.K. Chan, B. Zubrowski, M.S. Savoian, M.T. Sauer, and T.J. Yen. 2001. CENP-E is essential for reliable bioriented spindle attachment, but chromosome alignment can be achieved via redundant mechanisms in mammalian cells. *Mol Biol Cell*. 12:2776-89.

McGuire, W.P., J.A. Blessing, D. Moore, S.S. Lentz, and G. Photopulos. 1996. Paclitaxel has moderate activity in squamous cervix cancer. A Gynecologic Oncology Group study. *J Clin Oncol*. 14:792-5.

Meraldi, P., V.M. Draviam, and P.K. Sorger. 2004. Timing and checkpoints in the regulation of mitotic progression. *Dev Cell*. 7:45-60.

Modesti, M., and R. Kanaar. 2001. Homologous recombination: from model organisms to human disease. *Genome Biol*. 2:REVIEWS1014.

Musio, A., C. Montagna, D. Zambroni, E. Indino, O. Barbieri, L. Citti, A. Villa, T. Ried, and P. Vezzoni. 2003. Inhibition of BUB1 Results in Genomic Instability and Anchorage-independent Growth of Normal Human Fibroblasts. *Cancer Res*. 63:2855-63.

Nicklas, R.B. 1997. How cells get the right chromosomes. *Science*. 275:632-7.

Nicklas, R.B., and S.C. Ward. 1994. Elements of error correction in mitosis: microtubule capture, release, and tension. *J Cell Biol*. 126:1241-53.

Nowak, A.K., N.R. Wilcken, M.R. Stockler, A. Hamilton, and D. Ghersi. 2004. Systematic review of taxane-containing versus non-taxane-containing regimens for adjuvant and neoadjuvant treatment of early breast cancer. *Lancet Oncol*. 5:372-80.

Ohshima, K., S. Haraoka, S. Yoshioka, M. Hamasaki, T. Fujiki, J. Suzumiya, C. Kawasaki, M. Kanda, and M. Kikuchi. 2000. Mutation analysis of mitotic checkpoint genes (hBUB1 and hBUBR1) and microsatellite instability in adult T-cell leukemia/lymphoma. *Cancer Lett*. 158:141-50.

Orr-Weaver, T.L., and R.A. Weinberg. 1998. A checkpoint on the road to cancer. *Nature*. 392:223-4.

Pan, J., and R.H. Chen. 2004. Spindle checkpoint regulates Cdc20p stability in Saccharomyces cerevisiae. *Genes Dev*. 18:1439-51.

Patel, S.R., N.E. Papadopoulos, C. Plager, K.A. Linke, S.H. Moseley, C.H. Spirindonidis, and R. Benjamin. 1996. Phase II study of paclitaxel in patients with previously treated osteosarcoma and its variants. *Cancer*. 78:741-4.

Percy, M.J., K.A. Myrie, C.K. Neeley, J.N. Azim, S.P. Ethier, and E.M. Petty. 2000. Expression and mutational analyses of the human MAD2L1 gene in breast cancer cells. *Genes Chromosomes Cancer*. 29:356-62.

Petrylak, D.P., and R. de Wit. 2002. Editorial: the coming revolution in the treatment of prostate cancer patients. *Semin Urol Oncol*. 20:1-3.

Piccart, M.J., K. Bertelsen, G. Stuart, J. Cassidy, C. Mangioni, E. Simonsen, K. James, S. Kaye, I. Vergote, R. Blom, R. Grimshaw, R. Atkinson, K. Swenerton, C. Trope, M. Nardi, J. Kaern, S. Tumolo, P. Timmers, J.A. Roy, F. Lhoas, B. Lidvall, M. Bacon, A. Birt, J. Andersen, B. Zee, J. Paul, S. Pecorelli, B. Baron, and W. McGuire. 2003. Long-term follow-up confirms a survival advantage of the paclitaxel-cisplatin regimen over the cyclophosphamide-cisplatin combination in advanced ovarian cancer. *Int J Gynecol Cancer*. 13 Suppl 2:144-8.

Picus, J., and M. Schultz. 1999. Docetaxel (Taxotere) as monotherapy in the treatment of hormone-refractory prostate cancer: preliminary results. *Semin Oncol*. 26:14-8.

Pihan, G.A., and S.J. Doxsey. 1999. The mitotic machinery as a source of genetic instability in cancer. *Semin Cancer Biol*. 9:289-302.

Putkey, F.R., T. Cramer, M.K. Morphew, A.D. Silk, R.S. Johnson, J.R. McIntosh, and D.W. Cleveland. 2002. Unstable kinetochore-microtubule capture and chromosomal instability following deletion of CENP-E. *Dev Cell*. 3:351-65.

Reis, R.M., M. Nakamura, J. Masuoka, T. Watanabe, S. Colella, Y. Yonekawa, P. Kleihues, and H. Ohgaki. 2001. Mutation analysis of hBUB1, hBUBR1 and hBUB3 genes in glioblastomas. *Acta Neuropathol (Berl)*. 101:297-304.

Rieder, C.L., and E.D. Salmon. 1994. Motile kinetochores and polar ejection forces dictate chromosome position on the vertebrate mitotic spindle. *J Cell Biol*. 124:223-33.

Rigas, J.R. 2004. Taxane-platinum combinations in advanced non-small cell lung cancer: a review. *Oncologist*. 9 Suppl 2:16-23.

Ru, H.Y., R.L. Chen, W.C. Lu, and J.H. Chen. 2002. hBUB1 defects in leukemia and lymphoma cells. *Oncogene*. 21:4673-9.

Sato, M., Y. Sekido, Y. Horio, M. Takahashi, H. Saito, J.D. Minna, K. Shimokata, and Y. Hasegawa. 2000. Infrequent mutation of the hBUB1 and hBUBR1 genes in human lung cancer. *Jpn J Cancer Res*. 91:504-9.

Schaar, B.T., G.K.T. Chan, P. Maddox, E.D. Salmon, and T.J. Yen. 1997. CENP-E function at kinetochores is essential for chromosome alignment. *J. Cell Biol.* 139:1373-1382.

Shannon, K.B., J.C. Canman, and E.D. Salmon. 2002. Mad2 and BubR1 function in a single checkpoint pathway that responds to a loss of tension. *Mol Biol Cell*. 13:3706-19.

Shepherd, F.A. 2004. Current paradigms in first-line treatment of non-small-cell lung cancer. *Oncology (Huntingt)*. 18:13-20.

Shichiri, M., K. Yoshinaga, H. Hisatomi, K. Sugihara, and Y. Hirata. 2002. Genetic and epigenetic inactivation of mitotic checkpoint genes hBUB1 and hBUBR1 and their relationship to survival. *Cancer Res*. 62:13-7.

Skibbens, R.V., and P. Hieter. 1998. Kinetochores and the checkpoint mechanism that monitors for defects in the chromosome segregation machinery. *Annu Rev Genet*. 32:307-37.

Skoufias, D.A., P.R. Andreassen, F.B. Lacroix, L. Wilson, and R.L. Margolis. 2001. Mammalian mad2 and bub1/bubR1 recognize distinct spindle-attachment and kinetochore-tension checkpoints. *Proc Natl Acad Sci U S A*. 98:4492-7.

Sudakin, V., G.K. Chan, and T.J. Yen. 2001. Checkpoint inhibition of the APC/C in HeLa cells is mediated by a complex of BUBR1, BUB3, CDC20, and MAD2. *J Cell Biol*. 154:925-36.

Sudakin, V., D. Ganoth, A. Dahan, H. Heller, J. Hershko, F.C. Luca, J.V. Ruderman, and A. Hershko. 1995. The cyclosome, a large complex containing cyclin-selective ubiquitin ligase activity, targets cyclins for destruction at the end of mitosis. *Mol Biol Cell*. 6:185-97.

Takahashi, T., N. Haruki, S. Nomoto, A. Masuda, S. Saji, and H. Osada. 1999. Identification of frequent impairment of the mitotic checkpoint and molecular analysis of the mitotic checkpoint genes, hsMAD2 and p55CDC, in human lung cancers. *Oncogene*. 18:4295-300.

Tanaka, T.U., N. Rachidi, C. Janke, G. Pereira, M. Galova, E. Schiebel, M.J. Stark, and K. Nasmyth. 2002. Evidence that the Ipl1-Sli15 (Aurora kinase-INCENP) complex promotes chromosome bi-orientation by altering kinetochore-spindle pole connections. *Cell*. 108:317-29.

Thompson, L.H., and D. Schild. 2002. Recombinational DNA repair and human disease. *Mutat Res*. 509:49-78.

Tighe, A., V.L. Johnson, M. Albertella, and S.S. Taylor. 2001. Aneuploid colon cancer cells have a robust spindle checkpoint. *Embo Rep*. 2:609-14.

Tsukasaki, K., C.W. Miller, E. Greenspun, S. Eshaghian, H. Kawabata, T. Fujimoto, M. Tomonaga, C. Sawyers, J.W. Said, and H.P. Koeffler. 2001. Mutations in the mitotic check point gene, MAD1L1, in human cancers. *Oncogene*. 20:3301-5.

Wang, X., J.R. Babu, J.M. Harden, S.A. Jablonski, M.H. Gazi, W.L. Lingle, P.C. de Groen, T.J. Yen, and J.M. van Deursen. 2001. The mitotic checkpoint protein hBUB3 and the mRNA export factor hRAE1 interact with GLE2p-binding sequence (GLEBS)-containing proteins. *J Biol Chem*. 276:26559-67.

Wang, X., D.Y. Jin, R.W. Ng, H. Feng, Y.C. Wong, A.L. Cheung, and S.W. Tsao. 2002. Significance of MAD2 expression to mitotic checkpoint control in ovarian cancer cells. *Cancer Res*. 62:1662-8.

Wang, X., D.Y. Jin, Y.C. Wong, A.L. Cheung, A.C. Chun, A.K. Lo, Y. Liu, and S.W. Tsao. 2000. Correlation of defective mitotic checkpoint with aberrantly reduced expression of MAD2 protein in nasopharyngeal carcinoma cells. *Carcinogenesis*. 21:2293-7.

Waters, J.C., R.H. Chen, A.W. Murray, and E.D. Salmon. 1998. Localization of Mad2 to kinetochores depends on microtubule attachment, not tension. *J Cell Biol*. 141:1181-91.

Weaver, B.A., Z.Q. Bonday, F.R. Putkey, G.J. Kops, A.D. Silk, and D.W. Cleveland. 2003. Centromere-associated protein-E is essential for the mammalian mitotic checkpoint to prevent aneuploidy due to single chromosome loss. *J Cell Biol*. 162:551-63.

Yao, X., A. Abrieu, Y. Zheng, K.F. Sullivan, and D.W. Cleveland. 2000. CENP-E forms a link between attachment of spindle microtubules to kinetochores and the mitotic checkpoint. *Nat Cell Biol*. 2:484-91.

INDEX